PULP AND PAPER

Third Edition
Volume I

PULP AND PAPER
Chemistry and Chemical Technology
Third Edition-Revised and Enlarged

Volumes in preparation—

PULP AND PAPER
Chemistry and Chemical Technology
Third Edition, Volume I

JAMES P. CASEY
Editor

A WILEY-INTERSCIENCE
PUBLICATION

JOHN WILEY & SONS
New York Chichester
Brisbane Toronto

Library of Congress Cataloging in Publication Data:

Main entry under title:

Pulp and paper.

   First-2d ed. written by J. P. Casey.
   "A Wiley-Interscience publication."
   Includes bibliographies and index.
   1. Paper making and trade. 2. Wood-pulp.
I. Casey, James P., 1915–
TS1105.C29 1979   676   79–13435
ISBN 0–471–03175–5 (v. 1)

Printed in the United States of America

10 9 8 7 6 5 4 3 2 1

CONTRIBUTORS

Walter J. Bublitz, Ph.D.

Professor, Pulp and Paper—Department of Forest Products
Oregon State University—Corvallis, Oregon

J. R. G. Bryce, Ph.D.

Manager, Pulp, Paper and Packaging Research
Domtar, Inc.—Senneville, Quebec, Canada

Warren F. Daniell

Pulp and Paper Consultant
Hanover, New Hampshire

H. P. Dixson, Ph.D.

Technical Advisor to President
Fox River Paper Company—Appleton, Wisconsin

A. J. Felton

Director, Systems Technology—Shartle and Pandia Divisions
The Black Clawson Company—Middletown, Ohio

Wolfgang G. Glasser, Ph.D.

Associate Professor, Wood Chemistry—Department of
  Forest Products
Virginia Polytechnic Institute and State University—Blacksburg, Virginia

J. A. Kurdin

Technical Director, Pulp, Paper and Board Division
Sprout-Waldron, Koppers Company, Inc.—Muncy, Pennsylvania

V. Lorås

Research Director
The Norwegian Pulp and Paper Research Institute—Oslo, Norway

W. W. Marteny, Ph.D.

Senior Research Scientist—Forest Products Division
Owens-Illinois, Inc.—Toledo, Ohio

G. D. McGinnis, Ph.D.

Professor, Forest Products Laboratory—Department of Chemistry
Mississippi State University—Mississippi State, Mississippi

John N. McGovern, Ph.D.

Emeritus Professor, Department of Forestry
The University of Wisconsin—Madison, Wisconsin

D. K. Misra, Ph.D.

Director General
Thessalian Pulp and Paper Industries Ltd.—Athens, Greece

F. Shafizadeh, Ph.D.

Professor and Director of Wood Chemistry Laboratory—
  Department of Chemistry
University of Montana—Missoula, Montana

To my wife, Ruth

in appreciation, belatedly,
for typing the first and second editions
and, on a more timely basis,
for assistance in the preparation of this edition.

PREFACE

PULP AND PAPER when first published in 1951 was an attempt to bring together in book form the known science and technology of pulp and papermaking and to focus attention at the interface of the scientific and technological aspects of this industry.

This edition, the third one to be published, is an attempt to update the knowledge, using the same approach as that followed in the first two editions. There are, however, some differences. The first two editions were written alone; this edition has been written by multiple authors who are experts in their field. The number of volumes has expanded. This edition has four volumes, a change that was made necessary by the vast growth in the body of technical knowledge on pulp and paper manufacture.

Another difference is somewhat more subtle; it involves a shift in emphasis, a change brought about by problems that have ballooned in importance and are in urgent need of solution.

These problems are well known: to reduce the capital intensity; to reduce the use of chemicals, energy, and water; to reduce environmental impact; to reduce operating costs; to increase automation; to use virgin fibers more efficiently; to make greater use of secondary fibers; and to employ wood for multiple use as a source of energy and chemicals, as well as for fibers in papermaking.

The basic science is just as important as ever. More basic knowledge is still needed for improving the fiberization process, for understanding the fiber–water relationship (dry-forming of paper has not yet arrived), and for finding new ways to improve the functional properties of paper. Basic understanding, creative thinking, and innovative solutions are needed now more than ever. The problems are deeper and broader and they call for integrative solutions derived from a combination of the disciplines of science, engineering, and economics.

JAMES P. CASEY

P.O. Box 1598
Tryon, N.C.
November 1979

PREFACE TO THE SECOND EDITION

Any technical book is likely to be out of date ten years after publication. This is certainly true of the first edition of *Pulp and Paper.*

Pulp and Paper is, as stated before, an attempt to present the technical side of papermaking from a fundamental viewpoint. One might think this would lend permanence, but it appears that the first edition of *Pulp and Paper* is more out of date than if it had been merely a presentation of the technology of the industry. Evidently ideas are changing faster than practice in the paper industry. This is a healthy sign because it indicates not only stability of operation, but a progressive interest in change. It is not so much that the old facts are wrong, but that so many new facts have been uncovered that a book acceptable ten years ago is too far behind the times today.

The greatest development over the past ten years has taken place in the areas of pulping, sheet formation, and coating. The great amount of study given these major operations denotes a driving determination on the part of the technical man to uncover the fundamentals underlying the important steps in papermaking.

In sheet formation, particularly, new knowledge is being brought to light so rapidly that it is hard to grasp what effect this will have on future paper machines. In the foreseeable future, paper will continue to be made by separation of fibers from water on a moving wire, but the techniques for doing this, the speeds that will be obtained, and the properties that will be built into the paper of the future is beyond our grasp at the present time. No doubt, the fundamentals are now at hand for the changes that will take place over the next ten years, just as all the fundamentals were there ten years ago for the creation of stretchable paper.

This revised edition of *Pulp and Paper* follows the original fundamental approach, and it presents the present state of knowledge insofar as it is possible for the author to put this together. This fundamental knowledge can help solve everyday problems, and it can, in the hands of the ingenious, lead to innovations that will change the operations of converting fibers into useful paper products.

J.P.C.

Columbus, Indiana
June 1960

PREFACE
TO THE FIRST EDITION

It is possible to divide the principles of papermaking into those which are chemical and those which belong in the field of engineering. This book is concerned primarily with the subjects relating to chemistry.

Papermaking is essentially a chemical process, although the remarkable engineering achievements of the industry have often obscured this fact. The maze of intricate mechanical operations necessary in the modern paper mill makes it difficult, for the casual observer at least, to realize that the final object of all this effort is the modification and rearrangement of the papermaking fibers. It should be realized, however, that these intricate machines are merely tools for carrying out the various chemical operations necessary to the formation of a sheet of paper. This method of thought will become more widespread since, as seems likely, the important developments of the future will be in the field of applied chemistry.

Unlike many of our modern chemical industries, the paper industry had its origin in antiquity. The knowledge and methods of papermaking were handed down to a select few in an attempt to keep them as secret as possible, and this practice was carried into modern times. As a result, our knowledge has increased slowly, and many of the processes used today show but little change over those used by the medieval papermaker. In the last thirty years, however, a marked change has taken place. Secrecy has given way to cooperation in the sharing of ideas. The change has been due to the entrance of scientifically trained personnel from such well-known institutions as the College of Forestry of the State University of New York, the Institute of Paper Chemistry, the University of Maine, Massachusetts Institute of Technology, and many others. Another important factor has been the growth of the Technical Association of the Pulp and Paper Industry, which has promoted research and interchange of ideas through its various publications and other activities. Much has been done to make available technical facts, data, and standards that have been highly useful to the paper mill chemist. There are several good books available on the technological aspects of making and using paper, and much has been published in current literature.

However, in the author's opinion, there has been too little emphasis placed on the fundamental chemistry of paper, its manufacture, and use. This book is an attempt to present such fundamental information on paper. It approaches the study of the papermaking raw materials and processes from the standpoint of the colloid and physical chemist. Although there is a general awakening to the fact that all forces of science are the same, it is still convenient and useful to study the sciences separately, and further to divide the science of chemistry into several branches. As long as this division is desirable, it is logical to study papermaking with the special concepts and tools of colloid and physical chemistry. Colloid chemistry is that branch of chemistry that deals with systems in which surface is the most important factor, and the process of papermaking is therefore a colloidal phenomenon, involving as it does the control and interaction of surfaces in the beating and felting of fibers. Dyeing, sizing, filling and coating are colloid chemical processes utilizing materials that are almost without exception classified as colloids. It is the concepts of colloid chemistry which are of the greatest importance to an understanding of the papermaking processes and which, at the same time, are least understood by the average paper mill chemist.

An attempt has been made to present the material in fundamental terms, and yet at the same time to bridge the gap between strict theory on one hand and extreme empiricism on the other. This middle ground is the fertile field for the growth of new ideas which may eventually become worth while, and an effort has been made to keep this book within that scope. The new concepts of cellulose and lignin, the host of new sizing agents, coatings, adhesives, fillers, etc., warrant considerable study by paper mill chemists. A thoroughgoing understanding of the chemistry of these new materials, as well as the older, more familiar ones, will open up new possibilities in the paper field. It is with this idea in mind that the book has been written.

The author believes that the day is past when it is necessary to write a book on papermaking which must, of necessity, be understandable to the paper mill machine tender and superintendent. The chemist has taken his rightful place in the paper mill. He is specially trained for, and should have the time to devote his energies to, process development, improvement, and control, whereas the superintendent is often not equipped by training, and is too busy with current production, to do so. That the chemists' efforts have been fruitful is attested to by the recent developments of the paper industry, such as the broadening in the number of wood species used, improved pulping and bleaching processes, the development of wet-strength papers, waterproof papers, paper-base plastics, coated papers, and many others, all of which have been the results of the cooperative efforts of the chemists of the industry. This book has been written for the chemist who has more than a casual interest in the paper industry. It is hoped that it will be of value both to advanced students (that is, seniors and graduate students) and to chemists who are already working in the paper and allied industries. It has been written for use as a textbook and a reference book.

In conclusion, the author wishes to point out that, although he wrote the original manuscript himself, the final work is the result of many helpful comments by a number of experts in the paper industry. In attempting to write the book by himself, the author realized the hazards of one man's trying to cover such an extensive industry. On the other hand, there were the disadvantages of group writing to be considered, particularly in the lack of continuity which usually results. An attempt was made to find at least one qualified expert to review each section of the book. To all these men, the author owes a debt of gratitude.

J.P.C.

Syracuse, New York
June 1951

CONTENTS

PULP AND PAPER

Third Edition
Volume I

1 CELLULOSE AND HEMICELLULOSE

GARY D. McGINNIS, Ph.D.

Forest Products Laboratory
Department of Chemistry
Mississippi State University
Mississippi State, Mississippi

F. SHAFIZADEH, Ph.D.

Wood Chemistry Laboratory
Department of Chemistry
University of Montana
Missoula, Montana

One of the most momentous events in the development of mankind must have been the discovery of a method for generating fire by ignition of cellulosic materials.[1] The use of camp fire for protection and supplementary heat by early man led to the arts of cooking, baking, pottery, and eventually production of charcoal and naval stores by partial combustion and pyrolysis of wood. It is believed that the destructive distillation of wood has a long industrial history dating back to the ancient Chinese and Egyptians who used the tarry products for embalming. The use of cellulose materials for paper making came much later in history.

Historical Development

Sheets of papyrus made by pressing the pith tissue of a sedge, *Cyperus papyrus,* were used for writing as early as 3000 B.C. in Egypt. In China strips of bamboo or wood were used for drawing and writing until the discovery of paper, which is attributed to Ts'ai Lun in A.D. 105[2-4] The original paper was made in China from rags, bark fiber, and bamboo. Pieces of bamboo were soaked for more than 100 days and boiled in milk of lime for approximately 8 days and nights to release the fibers. The art of papermaking finally reached Persia by A.D. 751 and from there spread to the Mediterranean countries from whence the Moors took the industry to Europe in the twelfth century.

Acceptance of paper and the spread of the paper industry in Europe created a chronic shortage of rags for use as a raw material. The shortage of paper eventually was relieved by production of pulp from wood. The credit for starting this

1

immense industry is obviously shared by many scientists and inventors, but it is interesting to note that the seeds for the idea of extracting the cellulosic fiber from woody tissues not only can be found in the Chinese practice of boiling bamboo in milk of lime, but also in a treatise submitted to the French Royal Academy by René Antoine Ferchault de Réaumur in 1719. In this treatise Réaumur, a noted physicist and naturalist, observed that: The American wasps form very fine paper, like ours. They extract the fibers of common wood of the countries where they live; they teach us that paper can be made from the fibers of plants without the use of rags and linen, and seem to invite us to try whether we cannot make fine and good paper from the use of certain woods.

Another 120 years elapsed before the French chemist, Anselme Payen, demonstrated that a fibrous substance, which he called cellulose (in 1839), could be isolated by treatment of wood with nitric acid. Isolation of this substance opened the door for production of wood pulp by commercial methods of delignification including soda processes patented by Watt and Burgers (1853), the sulfite process invented by Tilghman (1866), the kraft process developed by Eaton (1870) and Dahl (1879), and various bleaching methods. Numerous refinements of these processes in the twentieth century have led to the rapid growth and adaptation of paper not only for writing and printing, but also for wrapping, packaging, and a variety of disposable products.

Isolation of cellulose from plant tissues by laboratory and industrial processes naturally focused scientific interest on defining the structure, composition, properties, and the biogenesis of this material. It is interesting to note that after more than a century of scientific investigation, numerous findings, controversies, and debates, the term cellulose (in pure form) still means different things to different groups. To organic chemists, it means β-D-(1 $\rightarrow$ 4)-linked glucopyran. To the technologists, it means an asymptotic entity, often called α-cellulose,. which represents the alkali insoluble portion of wood pulp. To the biologists, it means the fine microfibrils of plant cell walls that reach a high degree of purity and perfection in a group of green algae, including *Valonia, Cladophora,* and *Chaetomorpha.* These groups have been concerned not only with the chemical structure of cellulose and its reactions, but also with its interrelated physical, morphological, and biological properties. Through various types of inquiries, they have tried to find out: how cellulose is crystallized and packed in the fibrils to give it fibrous and other physical properties; what the fibrils and microfibrils are and how they are organized within the layers and lamella of the cell wall; and whether the microfibrils are produced and organized by inanimate physical forces or by the biological influence of the living cell.

The isolation and chemical investigation of cellulose led to the production of derivatives, including cellulose nitrate, cellulose acetate, rayon, and cellophane, which were the forerunners of the modern plastic and polymer industry. Cellulose nitrate is the oldest known derivative of cellulose. The efforts of John Wesley Hyatt to produce synthetic ivory for billiard balls led to the development of the first synthetic plastic known as celluloid. In 1870 this material was produced from a mixture of partially nitrated cellulose (called pyroxylin) with camphor. Pyroxylin was also used in the manufacture of lacquers, films, and adhesives. Highly nitrated cellulose has been developed as a modern explosive and propellant. Because of the high flammability of cellulose nitrate, cellulose acetate was developed as a safer substitute, first for coating wing fabrics of World War I airplanes and subsequently for photographic and motion-picture films. Cellulose acetate was one of the first materials to be used for manufacturing plastic objects by the injection molding process. In 1892 Charles Frederick Cross and E. J. Bevan discovered cellulose xanthate. This forms a viscous solution from which cellulose can be regenerated as a continuous fiber (rayon) or film (cellophane). Development of industrial processes for production of rayon for textiles and cellophane film for packaging led to the development of the dissolving pulp industry in 1911. Until then, the pulp industry was concerned only with the manufacture of fibers for the paper industry; chemical derivatives of cellulose were produced from purified rags or cotton linters. Development of dissolving pulp from wood made available a relatively cheap and pure source of cellulose as a raw material for the expanding chemical industry.

The above industrial developments involve the isolation, modification, and application of cellulose in the form of a fibrous or structural polymer. However the fact that cellulosic material can be converted to its sugar component and used as a source of food, alcohol, and industrial chemicals has created another kind of industrial opportunity. This has been repeatedly explored, but practiced only within the controlled economy of the USSR and Japan during the two World Wars. The increasing demand for food, chemicals, and fuel as well as the limited supply of petroleum has made the conservation and efficient utilization of cellulosic materials a more urgent problem. The investigation of the conversion of cellulosic material to a wide variety of chemicals is gaining momentum. The efforts are directed toward the development of pyrolytic or enzymatic methods as a substitute for the simple acid hydrolysis that has been practiced in the past. Since only a fraction of the annual harvest of cellulosic material produced by forests and agricultural lands is being utilized, conversion of this material to food and chemicals may some day provide a fundamental solution to the problems of supply for these products.

Isolation of Cellulose and Hemicelluloses

Plant cell walls in wood are composed of cellulose and hemicelluloses bound together by lignin, a highly oxygenated aromatic polymer with a repeating phenylpropane skeleton. On this matrix is deposited a mixture of low-molecular-weight compounds called extractives. In the laboratory the extractives are removed with neutral organic solvents. Owing to the diversity of extractive substances that occur in various species, no single solvent is effective for all woods. Solvents that have been used to remove the extractives include petroleum ether, ethyl ether, methanol, ethanol, acetone, benzene, ethanol-benzene (1:2), and water. The extraction is generally performed using a soxhlet-type extractor.

A minor portion of the hemicelluloses can be removed before delignification.

However, in most cases it is necessary to remove most of the lignin before major portions of the hemicellulose or the cellulose can be isolated. Laboratory processes of delignification are mostly based on the reaction of moist wood with chlorine gas or on digestion with an acidified solution of sodium chlorite. The classical method of Cross and Bevan[5] for isolation of cellulose preparations comprises alternate chlorination and extraction with a hot, aqueous, sodium sulfite solution. The procedure leads to removal of a considerable part of the hemicelluloses along with the lignin. Treatments that remove essentially all the lignin, while leaving the carbohydrate materials unattacked, lead to the preparation of holocellulose. Holocellulose can be prepared in the laboratory by alternate chlorination and extraction with a hot alcoholic solution of mono-ethanolamine.[6] Another method used for preparation of "chlorite holocellulose" depends on digestion of wood meal with an acidified solution of sodium chlorite.[7] The temperature of treatment is varied from about 70 to 90°C. The extent of degradation of cellulose and holocellulose caused by these different methods of isolation has been compared and reviewed by several workers.[8,9]

The hemicelluloses are separated from cellulose by extraction, preferably of a holocellulose preparation, with aqueous alkaline solutions at approximately room temperature. Effective extraction of the hemicelluloses and partial fractionation with extracting solvents are accomplished with solvents, such as 5% sodium hydroxide, 16 to 18% sodium hydroxide, or 24% potassium hydroxide.[10,11] Addition of boric acid or metaborates to the alkaline solutions enhances the extraction of mannan-containing hemicelluloses.

For a quantitative isolation of wood cellulose, the lignin is first removed with chlorous acid. The resulting holocellulose is exhaustively extracted with 24% aqueous sodium hydroxide containing 4% boric acid to remove the hemicelluloses. The cellulose, which is obtained in almost quantitative yield, normally contains only traces of mannose and xylose residues and on hydrolysis gives D-glucose. Its X-ray diagram is that of Cellulose II. The cellulose is, however, severely degraded, the original number-average degree of polymerization having been lowered from approximately 5000 to values in the range of 200 to 700.[12] For a quantitative isolation of an undegraded cellulose, the wood must first be directly nitrated with a nondegrading acid mixture so that the cellulose obtained is in the form of a nitrate. This material can be used for determing the molecular weight of cellulose.[13]

Molecular Structure of Cellulose

Cellulose was first isolated and recognized as a distinct chemical substance in the 1830s by the noted French agricultural chemist, Anselme Payen.[14] Payen concluded, more or less correctly, that cellulose and starch were isomeric substances because both had the same carbon and hydrogen analysis and, subjected to hydrolysis, both yielded D-glucose. Yet the passage of nearly three-quarters of a century was required before the precise empirical formula of cellulose was established as $(C_6H_{10}O_5)x$.[15] Results from earlier studies on acetylation[16] and nitration[17,18] had indicated that cellulose had three free hydroxyl groups per $(C_6H_{10}O_5)$ unit.

The next major question in unraveling the overall structure of cellulose was to determine if cellulose contained only D-glucose units or if it consisted mainly of D-glucose units with trace amounts of other sugars. When cellulose was hydrolyzed directly with acids, over a 90.7% yield of D-glucose was obtained.[19] This was good evidence that the only repeating units in the cellulose polymer were anhydro-D-glucose. Even better evidence was obtained when the cellulose was initially converted into cellulose acetate and then hydrolyzed to a mixture of methyl glycosides. With this procedure, over 95.5% yield of a crystalline mixture of methyl α- and β-D-glucosides was obtained.[20,21] The reaction is shown below:

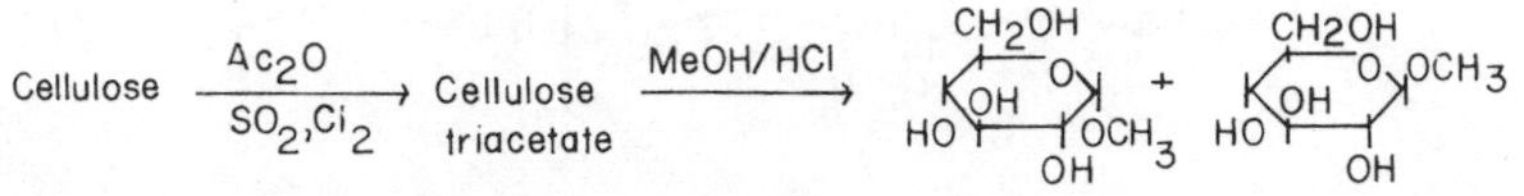

The next step in determining the overall structure of cellulose involved determining how the anhydro-D-glucose units were linked together. To determine this, samples of cellulose were methylated and then hydrolyzed to the individual monomeric units.[22,23] The position of the methyl groups in the hydrolyzed units corresponds to the position of the free hydroxyl group within the cellulose molecule. The major product obtained under these conditions was 2,3, 6-tri-O-methyl-D-glucose, the formula for which is:

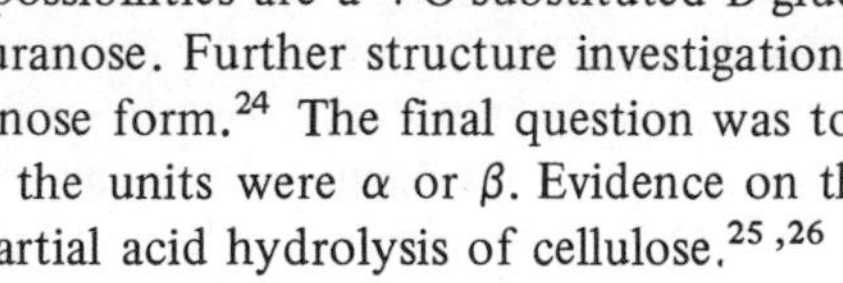

This formula indicated that the free hydroxyl groups in cellulose were located at the 2, 3, and 6 positions and that the 1, 4, and 5 positions were linked by chemical bonds. The two possibilities are a 4-O-substituted D-glucopyranose or a 5-O-substituted D-glucofuranose. Further structure investigation settled the problem in favor of the pyranose form.[24] The final question was to determine whether the linkage between the units were α or β. Evidence on this point came from experiments using partial acid hydrolysis of cellulose.[25,26] This treatment converted cellulose into a series of cellulose oligomers, which include cellobiose and cellotriose. Their formulas are:

Cellobiose

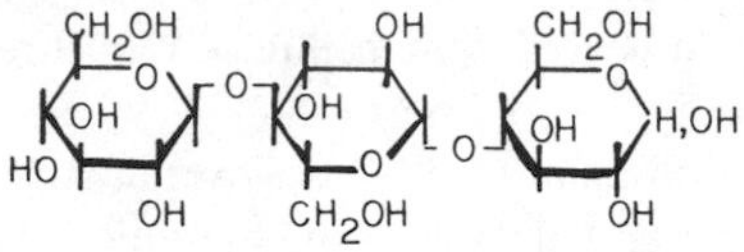

Cellotriose

Structural studies of these two compounds indicated that the linkage connecting the individual units was β. Therefore, cellulose is a linear polysaccharide consisting of anhydro-D-glucopyranose units linked between the 1 and 4 position of adjacent sugar units by a β linkage, as shown below:

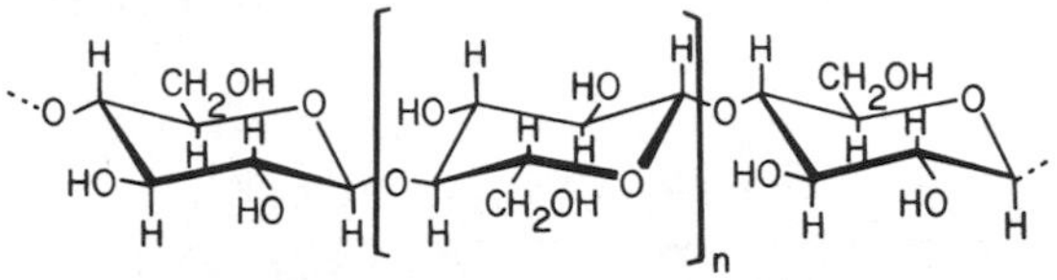

In the determination of the average molecular weight of cellulose, the usual methods for polymers have been used including osmometry, light-scattering measurements, ultracentrifugation, gel permeation, and viscometric determinations.[27-33] The degree of polymerization of cellulose, which is polydispersed, seems to vary with its source and method of isolation. Of the various methods tried, the direct nitration of the wood has been the most successful.[34] In contrast, holocellulose preparations contain cellulose that is partially degraded. Nitration is carried out in a mixture of nitric acid and phosphorus pentoxide or in a mixture of nitric acid and acetic anhydride[35] in which practically no degradation occurs. Studies involving nitrate derivatives give an average degree of polymerization (DP) from viscosimetry of about 4000 to 5000 for both softwoods and hardwoods.[34-36] Even higher values are obtained on the nitrate derivatives using light-scattering measurements.[37]

Cellular Structure of Cellulose

It would be impossible to understand the chemical reactions occurring in the pulping process, account for the physical properties of wood, or even fully appreciate the complex process of biogenesis without some understanding of the arrangement of cellulose, hemicelluloses, and lignin in the cell wall. There are several excellent monographs and articles[38-42 b] in which the subject is presented in great detail. The present discussion will briefly cover some features relevant to the pulping of wood.

Plant cells are distinguished from animal cells by the presence of true cell walls containing polysaccharides as the major structural material. Cellulose is the major structural component of plant cell walls. It exists in the cell wall as long, thread-like fibers (microfibrils). The cellulose microfibrils in mature wood cells are embedded in a matrix composed mainly of hemicelluloses and lignin.

Electron-microscope studies of mature wood cells show that they consist of several layers of cell wall surrounded by an amorphous, intercellular substance (I). A simplified drawing of the organization of a typical softwood tracheid or hardwood fiber is seen in Figure 1-1[42 b]

Between the cells is a region, called the compound middle lamella, that contains mainly lignin and pectic substances. The primary wall (P), which is only 0.1 to 0.2 μm in thickness, contains a randomly and loosely organized network of cellulose microfibrils embedded in a matrix that for many years was considered to consist of amorphous pectins and hemicelluloses lacking structural orientation. However, later studies have shown that the hemicelluloses are partially oriented.[43] Immediately below the primary wall is the secondary wall comprising, in fact, nearly all of the cell wall. The secondary wall is divided into three layers called S_1, S_2, and S_3. The outer layer of the secondary wall (S_1) is 0.1- to 0.3-μm thick and has a cross-hatch pattern of microfibrils. The S_2 layer of the secondary wall is 1- to 5-μm thick and accounts for the major part of the cell-wall volume. The microfibrils in this portion of the secondary wall are oriented almost parallel to the fiber axis. In the thin S_3 layer (0.1 μm) the microfibrils form a flat helix in the transverse direction. The innermost portion of the cell wall consists of the so called warty layer, most likely formed from protoplasmic debris.

The distribution of lignin, cellulose, hemicelluloses, and pectin over the middle lamella and the cell wall in wood fibers has been found to be quite heterogenous. Early studies by Lange[44] indicated that approximately 70% of the lignin in softwoods and 90% in hardwoods is located in the compound middle lamella. The remaining lignin is distributed within the secondary wall with higher concentrations generally being found in the S_3 layer of the cell wall.[45] Meier,[46-48] using an indirect approach, has determined the polysaccharide distribution across "average" cell walls of tracheids and fibers from three different species of wood. The results are shown in Table 1-1. As one would expect, the compound middle lamella contains large amounts of lignin and pectic materials.

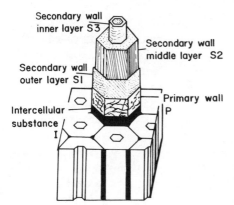

Figure 1-1. Simplified structure of the cell wall of a softwood tracheid or a hardwood fiber.

TABLE 1-1 RELATIVE PERCENTAGE OF POLYSACCHARIDES IN DIFFERENT LAYERS OF THE CELL WALL[48]

Polysaccharide	M + P[a]	S_1	S_2 Outer	$S_2 + S_3$ Inner
Pinus sylvestris				
Galactan	20.1%	5.2%	1.6%	3.2%
Cellulose	35.5	61.5	66.5	47.5
Glucomannan	7.7	16.9	24.6	27.2
Arabinan	29.4	0.6	Nil	2.4
Arabino-glucurono-xylan	7.3	15.7	7.4	19.4
Picea abies				
Galactan	16.4	8.0	Nil	Nil
Cellulose	33.4	55.2	64.3	63.6
Glucomannan	7.9	18.1	24.4	23.7
Arabinan	29.3	1.1	0.8	Nil
Arabino-glucurono-xylan	13.0	17.6	10.7	12.7
Betula verrucosa				
Galactan	16.9	1.2	0.7	0.0
Cellulose	41.4	49.8	48.0	60.0
Glucomannan	3.1	2.8	2.1	5.1
Arabinan	13.4	1.9	1.5	0.0
Arabino-glucurono-xylan	25.2	44.1	47.7	35.1

[a] Also contains a high percentage of pectic acid.

Arabinan and galactans are generally associated with pectic materials. The concentration of cellulose is highest in the inner portion of the secondary wall in birch, but in the outer portion of the S_2 layer in pine and spruce. The hemicellulose is distributed throughout the secondary wall, the glucomannans generally having higher concentrations in the inner part of the S_2 and S_3 layers.

Microfibrils

In the cell wall cellulose chains aggregate to form long thin threads called microfibrils. The microfibrils, in combination with the other matrix materials, provide the necessary rigidity and stress resistance in the plant. According to Freudenberg[49] the microfibrils in the cell wall act in the same way as reinforcing rods in prestressed concrete. A better mathematical model, suggested by Mark,[50] is furnished by filament-wound, reinforced plastic structures, such as those used in the construction of pressure vessels.

In wood the microfibrils are embedded in a matrix of polysaccharides and amorphous lignin. The noncellulosic materials can be removed by a variety of chemical treatments and the microfibrils observed by using the electron microscope. An electron micrograph of wood microfibrils is shown in Figure 1-2[51] This micrograph, from a tracheid of *Picea jezoensis*, illustrates the differences in

Figure 1-2. A replica of an earlywood tracheid showing the microfibrillar orientation in the S_2 and S_3 layers of *Picea jezoensis*. (Courtesy of Syracuse University Press.)

orientation of the microfibrils in the S_2 and S_3 layers of the plant cell. In various species of green algae, the lamella structure of the cell wall and the orderly arrangement of the microfibrils reach a high degree of perfection. This is illustrated in the electron micrograph of *Chaetomorpha melangonium* shown in Figure 1-3 by Frei and Preston.[52] The cell wall consists of alternating lamellae. These contain parallel-oriented microfibrils that cross each other approximately at right angles. A long series of chemical, physical, and microscopic investigations have provided an understanding of the size and shape of the microfibril, the arrangement of the cellulose molecules within the microfibril, and the biological processes involved in the biogenesis of the microfibrils within the plant cell.

Dimensions of Microfibrils. The dimensions of microfibrils are not uniform,

Figure 1-3. A replica showing the high degree of orientation of microfibrils in the cell wall of the algae *Chaetomorpha melangonium.*

but vary depending on the source of the microfibrils and their position within the cell wall.[53] In algae, Preston[54] measured diameters between 8.3 and 38 nm. For wood cell walls, Hodge and Wardrop[55] reported widths between 5.0 and 10.0 nm. Vogel,[56] measuring ramie microfibrils, found values between 17 and 20 nm. Each of these studies demonstrates the lack of uniformity in the width of microfibrils. Since the initial studies, there have been improvements in the resolving power of the electron microscope as well as better and more precise ways of sample preparation for cellulosic materials. Many of the later electron-microscope studies indicated that the original microfbrils observed were made up of aggregates of smaller units, called elementary fibrils or protofibrils. Studies by X-ray diffraction[57,58] and by electron-microscope observation[53,59−65] indicate that the microfibril has a width of approximately 3.5 nm in higher plants. It has been proposed by Mühlethaler[53] and others[66] that these elementary fibrils are the true structural units of the higher plant cell wall and the variation in the observed sizes of the microfibrils from various plant sources are due to differences in amounts of aggregation of a common 3.5 nm elementary fibril.

The idea that the 3.5 nm fibril is the ultimate cellulose fibril in higher plants is an area of much controversy.[67] Franke and Ermen,[68] Fengel,[69] and Hanna and Côté[70] have observed fibrils that are much smaller than 3.5 nm. Hanna and Coté have suggested that the 1.0-nm fibril, which they called the subelementary fibril, is at an intermediate stage in the development of the final 3.5-nm elementary fibril. Other workers believe that the ultimate size of the fibril depends on the degree of lignification. Marton,[71] Wardrop,[72] and others[70,71] suggested that the lignin matrix restricts the lateral movement of the cellulose fibril and thus prevents crystalline growth from proceeding unrestricted.

Structure of Microfibrils. Nägeli[73] in 1858 proposed that cellulose exists in plants as crystalline micelles. Many years elapsed, however, before the crystalline structure of cellulose was confirmed by X-ray diffraction methods.[74-77] The ensuing investigations showed that, with the possible exception of cellulose in the *Halicystis* plant, cellulosic materials from all sources, including bacteria and animals, have the same crystalline structure, which is called Cellulose I. When the crystalline structure is destroyed by solution or chemical treatment, the regenerated cellulose exhibits a different X-ray pattern. Based on the original X-ray diffraction pattern, Meyer and Misch[77] proposed a unit cell for native cellulose. In their proposed unit cell the adjacent cellulose molecules were antiparallel; in other words, the adjacent cellulose molecules in the microfibril were aligned in opposite directions. More and more evidence has accumulated that makes this conclusion somewhat questionable.

Studies by Honjo and Watanabe[78] using a low temperature electron diffraction of *Volania* cellulose indicated that the unit cell, as originally conceived by Meyer and Misch, is probably not correct and that the actual unit-cell dimensions are somewhat larger. These conclusions were confirmed by X-ray diffraction studies by Gardner and Blackwell[79] with *Valonia* cellulose, electron-diffraction studies by Dobb[80] using cotton cellulose, and neutron-diffraction studies by Beg and coworkers[81] using cotton cellulose. It has also been suggested, based on electron-diffraction studies,[82] that cellulose may contain two different types of crystalline unit cells, that is ramie and cotton cellulose may contain the unit cell similar to the one proposed by Meyer and Misch, while algal and bacterial cellulose may contain the larger unit cell suggested by Honjo and Watanabe. The antiparallel order of the adjacent cellulose molecules, as proposed by Meyer and Misch, is also being questioned. Evidence accumulated by Gardner and Blackwell[79] and others[83-86] using X-ray, electron, and neutron diffraction indicates that the cellulose molecules in the microfibril may actually be parallel. If the chains are parallel, this would be support for an extended-chain concept for biogenesis.

Crystalline and Amorphous Cellulose

Studies based on a variety of physical and chemical methods have indicated that the microfibrils are not completely crystalline, but instead contain two distinctly different regions. One region consists of highly ordered cellulose molecules; this is called the crystalline area. Another portion of the microfibril consists of less highly ordered cellulose molecules; this is called the amorphous or para-crystalline region.

Originally, it was proposed that cellulose had a brick-like structure in which crystalline micelles (partially ordered regions) were surrounded or fringed by amorphous materials.[73] The micellar theory, which was subsequently modified to the fringed micellar theory, provided a ready explanation for: the diffuse X-ray diffraction pattern; absorption of moisture without modification of the crystalline structure; partial hydrolysis of the microfibrils to rod-like micelles or crystallites of 5.0 to 10 nm in diameter and 50 to 60 nm in length, commonly known as microcrystalline cellulose; and numerous other observations that indicated a partial accessibility or crystallinity of cellulose. According to the fringed micellar theory, the microfibrils are composed of statistically distributed crystalline and amorphous regions formed by the transition of the cellulose chain from an orderly arrangement in the microfibrils in the crystalline regions to a less orderly orientation in the amorphous area. This is shown schematically in Figure 1-4.[87]

As more information on the microfibril has accumulated, it has become obvious that the fringed micellar model can not account for all of the observed properties of the microfibril. Consequently, other models have been proposed that are more consistent with the observed properties of the microfibril. One of the most unusual models was originally proposed by Manley[66] and later by Asunmaa,[88] Marx-Figini and Schulz,[89] and by Bittiger and coworkers.[90] This model proposes that the cellulose microfibrils consist of a folded-chain structure. According to Manley, the molecular chain forms a ribbon 3.5-nm wide by folding in concertina fashion. With the ribbon wound as a tight helix, the

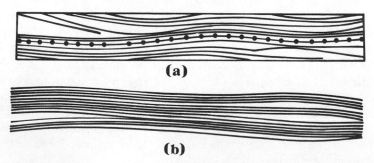

(a)

(b)

Figure 1-4. Schematic arrangement of cellulose molecules according to (a) the continuous theory and (b) the fringed micellar theory.

straight segments of the molecular chain become parallel to the helical, fibril axis, as shown in Figure 1-5, according to Manley.[66] One of the notable features of the folded-chain conformation is that it readily explains the antiparallel arrangement of the cellulose molecules within the crystalline structure of the microfibrils, as proposed by Meyer and Misch.[77] There is a considerable amount of evidence against the folded-chain concept of the microfibril. The extremely small dimensions of the subelementary fibril suggests the greater likelihood of an extended-chain conformation.[70] Other studies by Muggli[91,92] on the molecular-weight distribution before and after sectioning as well as the increasing evidence of parallel orientation of the cellulose molecules within the microfibrils strongly support an extended-chain concept for the microfibril.

A model has been proposed by Mühlethaler,[93,94] based on electron-microscopic studies using a negative-staining technique. In this method, the substrate is treated with phosphotungstic acid that penetrates the capillary spaces and loose textures and gives them a dark contrast on electron-microscope examination. This is a much better method than metal shadowing for studying the internal structure of microfibrils. The results indicate that microfibrils are composed of isodiametric, elementary fibrils having an average width of 3.5 nm. The elementary fibrils seem to be crystalline along their entire length as there is no sign of penetration of the phosphotungstic acid. If crystalline and paracrystalline sections were present, alternating dark and light segments would be expected. Based on his study Mühlethaler has proposed a model for the elementary fibril as shown in Figure 1-6.[93,94] The cross section of the elementary fibril, according to Mühlethaler, is 3.5 nm × 3.5 nm and contains about 36 cellulose chains. According to this model there is no true amorphous area. The observed diffuse X-ray diffraction pattern is due to chain-end dislocations.

Rowland and coworkers have proposed still another model for the elementary fibrils of cellulose.[95] This is similar to Mühlethaler's model in that there are no true amorphous areas, but only areas of slight disorder. The evidence for this model is based upon a series of chemical accessibility studies.[95-98] The authors conclude that the microfibril consists of a completely crystalline structure in which the irregularities are accounted for by various types of surface imperfections, such as distorted surfaces and twisted or strained regions of the crystalline elementary fibrils as shown in Figure 1-7, according to Rowland.[95]

No one model can completely account for all of the observed properties and characteristics of the microfibrils. There is still a considerable amount of controversy and no clear agreement on the structure of the microfibrils. It has even been suggested that the microfibril is an artifact produced by electron micros-

Figure 1-5. Schematic folded-chain structure of protofibrils as proposed by Manley.

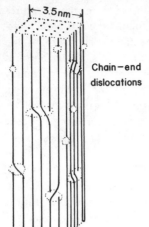

Figure 1-6. Schematic structural model of elementary fibrils as proposed by Mühlethaler showing chain-end dislocations.

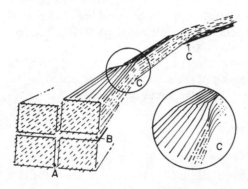

Figure 1-7. Schematic representation of the elementary fibril as proposed by Rowland showing (A) coalesced surfaces of high order, (B) readily accessible slightly disordered surfaces, and (C) readily accessible surfaces of strain-distorted tilt and twist regions.

copy and that its ultimate diameter is the width of the molecule, as has been shown to be true for extended-chain, polyethylene crystals.

Biogenesis of the Cell-Wall Polysaccharides

The process of cell-wall biogenesis includes formation of the precursors of cell-wall polymers, biosynthesis of the polymers, and assembly of the polymers in the cell wall. From the previous discussion about the structure of the cell wall, it is apparent that this process is very complex. A proposed mechanism for biogenesis must account not only for the precise arrangement of the cellulose molecules in the microfibril, but also for the orientation of the microfibril within the cell. The large number of events that must be coordinated during synthesis and assembly implies a complex control system. Unfortunately, relatively little is known about how such processes are controlled at the molecular level. Another factor, which should be noted, is that there are probably two different types of cellulose synthesized within the plant. Studies by Marx-Figini[89,99−101] and others[102] have indicated that cellulose from the primary cell wall consists of a fairly low-molecular-weight cellulose with broad molecular-weight distribution. For example, cellulose from the primary wall of cotton has a degree of polymerization (DP_w) ranging from 2000 to 6000, whereas the secondary wall contained cellulose with a much higher molecular weight of 13,000 DP_w and with a very narrow molecular-weight distribution. In the next paragraphs biogenesis of the cell wall will be discussed briefly. More detailed descriptions of this extremely important process can be found in reviews by Hassid and Nikaido,[103,104] Karr,[105] Shafizadeh and McGinnis,[76] and Villemez.[106]

Sugar nucleotides are the carbohydrate precursors that form the cell-wall polysaccharides. The nucleotides are formed by the combination of purine or pyrimidine bases linked to sugars, which in turn are esterified with phosphoric acid. The two nucleotides that have been found to be involved in cellulose synthesis are uridine, 5′-(α-D-glucopyranosyl pyrophosphate) (UDP-D-glucose), and guanosine, 5′-(α-D-glucopyranosyl pyrophosphate) (GDP-D-glucose).

The first reports of the synthesis of cellulose-like molecules by a cell-free system was by Glaser.[107] He found that a cell-free, particulate enzyme preparation from *Acetobacter xylinum* catalyzed incorporation of the D-glucopyranosyl group from UDP-D-glucose into cellulose. Since this initial work, there have been numerous reports of cell-free, particulate enzymes from mung beans,[103,105,108−110] maturing cotton bolls,[111,112] lupin,[113,114] oat coleoptiles,[115] peas,[109,110] string beans,[109] corn,[109] and squash,[109] all of which use GDP-D-glucose and/or UDP-D-glucose to form cellulose-like materials.

The precursor for the hemicellulose is also nucelotides. It has been found that xylans are formed from UDP-D-xylose[103,116,117] and that the precursors of arabino-xylan are UDP-D-xylose and UDP-L-arabinose,[103,117] while the synthesis of a glucomannan has been shown to involve the precursor GDP-D-mannose and GDP-D-glucose.[118−120]

There is general agreement that the precursors for all plant cell-wall polysaccharides involve sugar nucleotides. Hassid has postulated that cellulose is formed in higher plants by the sequence of reactions:

$$\text{GTP} + \alpha\text{-D-glucopyranosyl phosphate} \xrightarrow{\text{pyrophosphorylase}}$$

$$\text{GDP-D-glucose} + \text{PPi}$$

$$n(\text{GDP-D-glucose}) + \text{acceptor} \xrightarrow{\text{glucotransferase}}$$

$$\text{acceptor} - [(1 \rightarrow 4) - \beta\text{-D-glucosyl}]_n + n(\text{GDP}) \text{ cellulose}$$

It is not clear if the sugar nucleotides are the direct glycosyl donors during the formation of cell-wall polysaccharides or if the glycosyl unit is transferred to another intermediate, such as a glycolipid. It has been shown that the synthesis of the bacterial cell-wall polysaccharides[121] and a mannan from *Micrococcus lysodeikticus*[122] involves the formation of glycolipid intermediates. A glycolipid intermediate has also been postulated by Colvin[123] and others.[103-105,124] It has been suggested that the glycosyl unit is transferred from the nucleoside D-glucose pyrophosphate to the glycolipid at or on the cytoplasmic membrane, and that the D-glucosyl unit is then translocated outside the cell by the lipid moiety. That is, the glycolipid acts as a carrier from within the cell to the external medium where the D-glucosyl unit is polymerized to cellulose. Villemez[125] reported discovery of two types of compounds, a glycolipid and a glycoprotein, that could conceivably act as intermediates in the formation of a number of plant polysaccharides. The importance of these compounds in the biosynthesis of cell-wall polysaccharides remains to be shown.

A considerable amount of study has been done in order to locate the site of polysaccharide synthesis in the cell. Most of the evidence, based on autoradiographic studies[106,126,127] and chemical studies, indicates that the hemicelluloses and pectic materials are synthesized within the Golgi bodies in contrast to cellulose synthesis. Most evidence indicates that cellulose synthesis occurs outside the cytoplasm at the plasma-membrane and cell-wall interface, that is, at the site of microfibril deposition. The most convincing evidence for this is derived from autoradiographic studies *in vivo*,[126,128-132] which indicate synthesis of cellulose only outside the protoplast. A diagrammatic representation of the synthesis and transportation of plant polysaccharides as proposed by Northcote[133] is shown in Figure 1-8.

It is not known how the cell controls the molecular size and the orientation of the microfibrils. Some of the earlier theories proposed that the high degree of orientation of the microfibrils was due to the stress and strain produced by the turgor pressure of the growing cell or by the streaming of the flowing cytoplasm over the cell wall. Both of these theories have largely been discarded as more information has become available about the function and structure of the plant cell.

Because of the precise molecular size of the cellulose located in the secondary wall and because of the orientation of both the cellulose molecules and the microfibrils, it has been proposed by Preston[134] and others[135] that plant cells contain some type of template in which the microfibrils are synthesized. After plasmolysis, the inner face of the wall carries granular aggregates. It was suggested by Preston that these granular aggregates may represent parts of the microfibril-synthesizing system in which the granules constitute an enzyme complex that transfers D-glycosyl groups to the growing end of the microfibrils. In this way, the microfibrils would be propagated from one granule to the next by the process of end-wise synthesis, as shown in Figure 1-9, according to Mühlethaler.[93] A template system would explain both the orientation of the molecules and the precise molecular size of the molecule. Other researchers[135] have suggested that the template for microfibril orientation may be the microtubules within the plant cell. This is supported by microscopic observa-

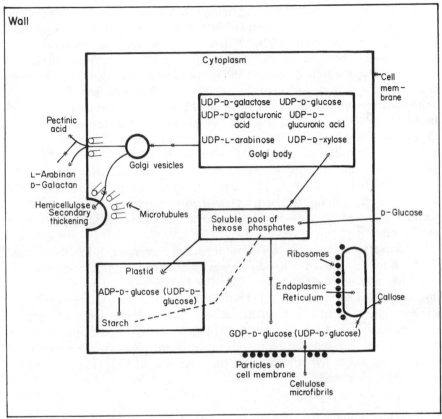

Figure 1-8. Diagrammatic representation of the synthesis and transportation of various polysaccharides in a growing plant cell.

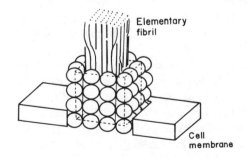

Figure 1-9. A hypothetical model proposed by Mühlethaler for the biogenesis of cellulosic elementary fibrils.

17

tions that the direction of the microtubules is the same as that of microfibrils in the adjacent cell wall. However, microtubules have been observed where their orientation does not coincide with that of the cellulose microfibrils. Studies[136] with the drug colchicine have provided some information about the relationship between microfibril orientation and the microtubules because colchicine reacts with and destroys the microtubules. In studies with the green algae *Oocystis solitaria,* the presence of colchicine causes the cells to lose their ability to change the direction of microfibril orientation. Normally, the microfibrils in the secondary wall contain 90° alternating lamellae of microfibril, similar to *Chaetomorpha* (see Figure 1-3). However, after addition of colchicine, the lamallae are all parallel, with no alternating lamallae. This suggests that the microtubules do not completely control the orientation of the microfibrils, but that they do play a part in determining the direction of the microfibrils. In addition to the template theory, there have been several other theories proposed to account for microfibril orientation, including glycolipid involvement.[137]

In summary, there is no precise information concerning either the control mechanisms that govern cell-wall biogenesis or the interactions between cell-wall biogenesis processes and general cellular metabolism. The number of steps involved in the formation of a polysaccharide from a sugar nucleotide is not known. It is not clear how cellular control is extended beyond the plasma membrane or how the cell wall is formed from the component polymers. Indeed, the field of cell-wall biogenesis provides more questions than answers. Most of the major hypotheses about the operation of the biogenesis process have not yet been made.

Sorption, Swelling, and Solution of Cellulose

The free hydroxyl groups of cellulose have a strong affinity for polar solvents and solutes that can reach them. An example of this type of interaction is the swelling of cellulose with water. When cellulose is dry, it absorbs moisture from the air reaching an equilibrium moisture content that increases with the increasing ambient relative humidity. The absorbed moisture swells the cellulose, but does not change the crystalline structure, indicating that it enters the accessible regions rather than the crystalline regions of the microfibril. If absorption is carried to saturation and the relative humidity is then progressively decreased, the proportion of sorbed moisture decreases, but the new values at any given relative humidity are a little higher than those obtained for the absorption curve. The explanation for the difference between the absorption and desorption curve (hysteresis loop) is based on the interconversion of cellulose-water and cellulose-cellulose hydrogen bonds. During swelling, the hydrogen bonds between cellulose molecules are broken and replaced by hydrogen bonds between the cellulose molecule and water. Conversely, desorption and shrinking reform the broken cellulose-cellulose hydrogen bonds, but with a lag that gives a higher equilibrium moisture content when the cellulose is being dried. This phenomenon accounts for the characteristic hysteresis loop observed on absorption and desorption of moisture and the corresponding changes in the physical properties of cellulosic materials at different relative humidities. A similar phenomenon is observed with other polar liquids, for example, methanol, ethanol, aniline, and benzaldehyde. In general, the higher the polarity of the liquid, the greater the amount of swelling produced, but in all the above examples, swelling is less than that with water. More detailed discussions about the swelling of cellulose with liquids and gases can be found in articles by Browning[138] and Skaar.[139]

The swelling with water is an example of intercrystalline swelling or swelling that involves only the accessible portion of the cellulose microfibrils. The most convincing evidence for this is the fact that the X-ray pattern of native cellulose does not change on wetting. There are, on the other hand, many liquids, mainly aqueous solutions of strong acids, strong bases, and a few salts that do change the X-ray pattern. This type of swelling is called intracrystalline swelling and it involves the penetration and swelling of both the accessible and the crystalline regions of the microfibril. The most common example of a liquid that can lead to intracrystalline swelling is 15 to 20% aqueous sodium hydroxide. The interaction of cellulose with highly polar solutions of acids, alkali, salts, or complex-forming reagents may cause intramolecular swelling and, under the right conditions, may lead to a solution. The solution, however, is not a true cellulose solution, but rather it is an addition product formed by the cellulose and the components present in the liquid.

Reagents that form an addition product and that dissolve cellulose include amino and ammonium complexes of copper, cadmium, and other multivalent ions, such as cupriethylenediamine $Cu(H_2N-C_2H_4-NH_2)_2$ $(OH)_2$ and cuprammonium hydroxide $CU(NH_3)_4$ $(OH)_2$. The latter reagent, discovered by Schweizer, is used for the preparation of cuprammonium rayon and for the characterization of cellulose by viscosity measurement.

Cellulose dissolves in 44% hydrochloric acid, 72% sulfuric acid, and 85% phosphoric acid, all of which produce homogeneous hydrolysis of cellulose. Concentrated nitric acid (66%) does not dissolve cellulose but, like sodium hydroxide, forms an addition compound known as Knecht compound, which is an intermediate in the nitration of cellulose. Cellulose also dissolves in some concentrated aqueous salt solutions, but this may require heating, which leads to partial degradation of the cellulose.

Degradation Reactions of Cellulose

The fact that cellulose degrades under a variety of circumstances is important from several points of view. For the manufacturer of cellulosic products, degradation is both helpful and undesirable. For chemical uses a certain amount of degradation, such as that which occurs during the aging of alkali cellulose, may be desirable since it provides a means of controlling the properties of the final product. However, for the pulp and paper producer the degradation of the cellulose and hemicelluloses must be kept at a minimum in order to obtain high yields and to retain many of the physical and mechanical properties of the fiber.

There are several different types of degradation:

1. Hydrolytic
2. Oxidative
3. Alkaline
4. Thermal
5. Microbiological
6. Mechanical

Of these, only the first five are important and these will be briefly discussed. More detailed discussion of cellulosic degradation can be found in reviews by Bolker,[39] Rydholm,[140] and Ward.[141]

Hydrolytic Degradation. This type of degradation refers to cleavage at the glycosidic linkage between the ^1C carbon and oxygen by an acid. If the process is carried to completion, the final product will be the monosaccharide, glucose. Acid hydrolysis may proceed as a homogeneous process with strong acids, which dissolve the substrate, or as a heterogeneous process with weaker acids.

In the heterogeneous process, cellulose maintains its fibrous structure and the reaction occurs in two distinct phases, a rapid initial reaction followed by a much slower reaction. It is believed that the easily accessible regions are hydrolyzed first and then the less accessible regions are hydrolyzed, but more slowly. The hydrolysis in the initial stage removes 10 to 12% of the substrate and rapidly reduces the DP to a leveling off or limiting valve. Depending on the history of the sample, the leveling off usually occurs at approximately 200 to 300 D-glucose units/chain. Commercially, the process of heterogeneous hydrolysis is used to produce hydrocellulose and gel-forming microcrystalline cellulose. In several places in the papermaking process, prevailing conditions are sufficiently acidic to cause degradation of the fibers. Sulfite pulping is the most notable example. The loss of strength of paper upon aging, particularly when the pH of the paper is low, is another example of acid degradation.

The homogeneous process is a much simpler reaction and it proceeds at a uniform rate. Several processes have been developed for industrial saccharification using concentrated hydrochloric or sulfuric acid. However, these processes have only been used during wartime or in countries with a controlled economy, such as the Soviet Union.[142] The main product from the reaction is D-glucose, which can be converted into a wide range of other products by fermentation and hydrogenation.[143] The homegeneous process is also used in the laboratory for the analysis of wood pulp. The procedure consists of an initial brief treatment with strong sulfuric acid followed by treatment with hot dilute acid to complete the hydrolysis.[144]

Oxidative Degradation. Cellulose is highly susceptible to oxidizing agents. The extent of the degradation depends on the nature of the reagent and on the conditions under which oxidation occurs. The hydroxyl groups and the terminal reducing ends are the sites most susceptible to attack. Most oxidations are random processes and lead to the introduction of carbonyl and carboxyl groups at various positions in the anhydro-D-glucose units of cellulose. There are also a variety of secondary reactions that can occur including chain scission. Oxycellulose is the term used for the product obtained from cellulose by any oxidizing procedure that introduces carbonyl and/or carboxyl groups anywhere in the molecule.

Certain oxidizing agents are fairly specific. For example, periodic acid, sodium metaperiodate, and acidic lead tetraacetate cleave the anhydroglucose units between the ^2C and ^3C positions and form aldehyde groups at these two positions to produce a product called dialdehyde cellulose. Nitrogen dioxide — or, more properly, dinitrogen tetroxide — converts the hydroxy groups at ^6C with a fair degree of specificity into carboxyl groups, although there is some oxidation at the ^2C and ^3C to form ketone groups.[145] Hypoiodite, chlorine dioxide, and acidified chlorite solutions are mild oxidizing agents that have little or no effect on the hydroxyl groups, but that oxidize the terminal reducing group (or any internal aldehyde group) to carboxyl groups.[140,146] These oxidations are specific enough to have been used for quantitative measurement of such reducing groups.[146]

Most oxidants are less specific than those mentioned and produce a variety of different combinations of carbonyl and carboxyl groups in cellulose. A good example is the chlorine and hypochlorite systems that are widely used in bleaching wood pulp. Depending on the pH, these bleaching agents can cause oxidation of the hydroxyl groups to aldehyde, ketone, and carboxyl groups, as well as cause depolymerization of the polysaccharide.[140]

Oxygen causes oxidation of cellulose in alkaline solutions. This process is used industrially in the manufacture of viscose rayon. In this process, cellulose sheets are mixed with a concentrated solution of sodium hydroxide, shredded, and allowed to age. Aging reduces the molecular weight of the cellulose and results in the formation of carboxyl groups. Ozone, hydrogen peroxide, permanganate, and many other reagents cause a nonspecific oxidation of cellulose.

Alkaline Degradation. There are three major types of base degradation of cellulose. One type is an oxidative process that occurs when alkaline solutions of cellulose come in contact with air. This process occurs only at relatively high temperatures and leads to chain scission and the introduction of carboxyl groups. The other two types of degradation are nonoxidative processes.

Probably the most important type of base degradation is the so-called peeling reaction. This reaction involves the gradual shortening of cellulose chains by a β-elimination mechanism. This reaction normally proceeds from the reducing end of the cellulose molecule. However, if the cellulose has been oxidized at any point in the molecule and a carbonyl or carboxyl group is present, alkaline degradation will proceed at this point with chain scission. The major steps in the peeling reaction are shown in Figure 1-10 (Reaction I). The initial step, occurring at the reducing end, involves the extraction of a proton from ^2C and the formation of the ketose, which is in equilibrium with the corresponding enediols (Lobry de Bruyn-Alberda van Ekenstein enolization).[147-149] The next step involves the elimination of the ^4C substituent. This leads to a new reducing end and the formation of a dicarbonyl compound, shown in step (5), which in turn undergoes a rearrangement to form isosaccharinic acid, shown in step (6). The cellulose chain, which is eliminated, contains a new reducing end. Therefore, this process will be repeated with progressive shortening of the cellulose chain. The peeling reaction will remove about 50 to 60 glucose units[150] from each starting point until finally another reaction mechanism, called the stopping reaction,[148] provides the end unit with a configuration stable to alkali, as shown in Figure 1-10 (Reaction II). The stopping reaction leads to the formation of a substituted metasaccharinic acid, shown in step (11). It follows a very similar mechanism as the peeling reaction, except that elimination occurs at the ^3C position, and thus stabilizes the cellulose chain to further eliminations.

The hemicelluloses also undergo the peeling reaction. In kraft pulping, the reaction rate of noncellulosic carbohydrates, such as xylan, glucomannan, and galactoglucomannan, varies with the accessibility, branching, type of sugar, and glycosodic bonds involved.[140,148,151] Hamilton and Thompson[151] followed the degradation rate of arabinoxylan in kraft liquor and compared it with that of xylan and glucuronoxylan. They found that the arabinoxylan was much more stable toward alkaline degradation than the other hemicelluloses. Apparently, the presence of the arabinofuranose unit attached to ^3C of the anhydroxylose unit caused the end-group peeling reaction to be terminated at that point by the formation of a stable metasaccharinic acid during alkaline digestion. Such an explanation has been postulated[150] and confirmed by experimental studies using 3-O-methyl-D-xylose as a model compound.[152] The glucomannan has been found to be more susceptible to alkaline degradation than the xylose-containing wood polysaccharides,[153] and only a small fraction of the original glucomannan survives the kraft digestion. In contrast the galactoglucomannan is relatively stable to the alkaline solution used in the kraft process.[153]

Another important carbohydrate reaction with alkali is the alkaline hydrolysis of glycosidic bonds. This reaction is responsible for a shortening of the polysaccharide chains, which is likely to have some bearing on the strength properties of the final product. This type of alkaline degradation occurs at relatively high temperatures of around 160-180°C.[140,154] This reaction is independent of the presence of small amounts of oxygen and is, therefore, entirely an alkaline hydrolysis of glycosidic bonds. These same conclusions have been reached with studies of model glycosides.[155]

Reaction I

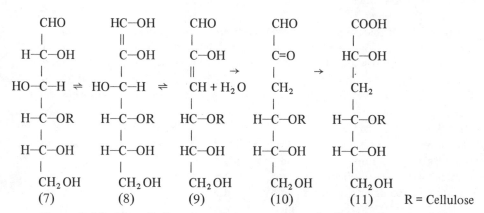

R = Cellulose

Reaction II

Figure 1-10. The alkaline degradation reactions of cellulose. Reaction I shows peeling reaction. Reaction II shows stopping reaction.

Thermal Degradation. Thermal degradation of cellulose proceeds through two types of reactions: (1) a gradual degradation, decomposition, and charring on heating at low temperature and (2) a rapid volatilization accompanied by the formation of levoglucosan and a variety of other organic products on heating at high temperatures.

At the lower temperatures, ambient to 200°C, it is difficult to draw a line of demarcation between thermal degradation and the normal aging of cellulose, which is accelerated by heating. According to Richter,[156] when rag paper is heated at 38°C for about six months, the accelerated aging results in a 19%

reduction of the folding strength. Farquhar and coworkers[157] heated raw cotton for 4 to 24 hr at temperatures of 75 to 220°C in air and nitrogen. This and other investigations[158-160] show that oxygen and water have a profound effect in enhancing the deterioration and the thermal degradation of cellulosic materials. The thermal degradation in air or oxygen apparently is caused primarily by oxidation reactions. These reactions provide a type of oxycellulose, or pyrocellulose, which on further heating decomposes with evolution of H_2O, CO_2, and CO. Low-temperature heating will also cause a large change in the degree of polymerization. Under oxygen the degree of polymerization rapidly decreases and then levels off at a value of 200, which corresponds to the size of the cellulose microcrystallites. In a nitrogen atmosphere, the degree of polymerization decreases, but at a much slower rate. Carbonyl and carboxyl groups are also formed in the cellulose chain during heating under oxygen.

Rapid pyrolysis of cellulose takes place at temperatures above 250°C and gives levoglucosan and its thermolysis products, including various quantities of char, tar, and volatile products.[160] This reaction is particularly important because it is the basic reaction that provides the gaseous fuel for the burning of cellulosic materials. Controlling the pyrolytic reactions provides a basis for the flameproofing of cellulose material by reducing the production of levoglucosan and other volatile combustible products.[161]

Enzymatic Degradation. There are many enzymes in fungi, bacteria, plants, and animals that can cause hydrolytic decomposition of cellulose and hemicellulose. These enzymes are classified according to the substrate with which they react. Cellulose-hydrolyzing enzymes are called cellulases and hemicellulose-hydrolyzing enzymes are called hemicellulases. Cellulases and hemicellulases cause degradation of wood, cotton, and paper and are responsible for losses of millions of dollars each year.[162,163] However, both types of enzymes also play a very necessary role in nature. Cellulases in organisms that catabolize cellulose play an important role in balancing the carbon cycle. Cellulases and hemicellulases are probably involved, along with related enzymes, in the germination of seeds and in the ripening as well as the rotting of fruits and vegetables. Cellulase, excreted by microflora present in the digestive tracts of herbivores, enable these animals to utilize cellulose as a prime source of energy.

Because of their economic impact and potential, the cellulases of microorganisms have been the most widely investigated. When a fungal spore germinates on wood, there is general dissolution of the primary wall caused by a thread-like structure called hyphae that secretes enzymes as it penetrates the cell wall by way of bore holes or through pits present in the wood. The cellulases and hemicellulases initially attack at the lumen wall and gradually work outward as the polysaccharides are removed from each successive wall layer. Bacteria also produce cellulase enzymes that are capable of degrading cellulose. In contrast to the bore holes and inner-fiber damage of fungal growth, the bacterial attack appears to involve only the surfaces of fibers. Much of the research on cellulases has concentrated on developing an economic enzymatic process for converting waste cellulose into D-glucose. The most common sources of cellulase are from *Trichoderma viride* mutants,[163,164] which have their greatest activity at 50°C. A high-temperature cellulase that works at 60 to 70°C isolated from *Thielatia terrestris* has been reported.[165]

The susceptibility of cellulose to enzymatic hydrolysis is determined largely by its accessiblity. It is necessary during enzymatic hydrolysis that there be direct physical contact between the cellulose and the enzyme. Since cellulose is an insoluble and structurally complex substrate, this contact can be achieved only by diffusion of the enzyme into the complex structural matrix of the cellulose. Any structural feature that limits the accessibility of cellulose to enzymes will diminish its susceptibility to hydrolysis. Probably the two most important factors that limit the accessibility of wood to enzymatic degradation are the lignin and the crystallinity of the cellulose. Other factors that are important are the moisture content, degree of polymerization, and the presence of extraneous constituents in wood. More information about enzymatic degradation can be found in reviews by Reese and others.[163,164,166,167]

Cellulose Derivatives

Various modifications and derivatives of cellulose are produced in large quantities and used for industrial production of fiber, film, plastics, explosives, coatings, and thickeners. Comprehensive discussion of these products can be found in reviews by Spurlin,[30] Bolker,[39] Harris,[168] and Ward.[141] Only a brief description of the raw material and the reactions involved in making the major commercial cellulose derivatives will be given.

Raw Materials for Derivatives. Production of cellulose derivatives generally requires a raw material having a high content of pure cellulose, technically referred to as α-cellulose. The α-cellulose content of a given raw material is measured by the amount of cellulose that remains undissolved in 18% sodium hydroxide solution that is subsequently diluted by water to a lower concentration.

Originally, bleached rags were used for production of cellulose nitrate, the first derivative produced commercially. As the industry developed, the lack of availability and uniformity of this material led to the use of cellulose obtained from cotton linters, which are the short fibers (about 2 to 3 mm) left on the seeds after removal of the long cotton fibers. These short fibers are removed from the seeds at different stages in the processing and are cleaned and purified to give chemical cotton. The latter process involves digestion of the fibers with dilute sodium hydroxide, bleaching with chlorine, neutralization, and drying. During World War I, increased demand for cellulose nitrate for explosives led to the use of wood pulp as a raw material, but in the United States chemical cotton remained the preferred raw material for production of cellulose derivatives until the dissolving-pulp industry was developed during World War II.

Dissolving pulp can be prepared by removing the remaining lignin, hemicelluloses, and resin from hardwoods pulped by the sulfite process. The acidic condition used in sulfite pulping is responsible for removing most of the hemi-celluloses. Subsequent purification involves extensive bleaching and extraction with a dilute solution of 1 to 2% sodium hydroxide at temperatures of 90 to 170°C or extraction with a concentrated alkali solution (10 to 15%) at room temperature. The kraft pulping process can also be used to prepare dissolving pulp from both hardwoods and softwoods, providing a prehydrolysis step is used. The purified cellulose obtained by these processes is used for the manu-facturing of a variety of commercial products.

Reactions in Derivatization.   A series of products can be obtained when cellu-lose is converted into an ester or ether. The chemical and physical properties of the final product are dependent on:

1. The type of substituting group, that is, methyl, ethyl, or acetyl.
2. The degree of substitution, that is, the relative number of substituted and free hydroxyl groups.
3. The uniformity of substitution.
4. The uniformity and the length of the cellulose molecules.

In practice it is very difficult to completely convert all the hydroxyl groups in cellulose into ester or ether groups. This is due in part to a steric factor. When one or two of the hydroxyl groups are substituted, steric crowding may inhibit complete reaction. Another factor is accessibility. Since the reactions are conducted under heterogeneous conditions, not all the hydroxyl groups may be accessible to reaction, particularly those in the crystalline regions. Finally, not all the accessible hydroxyl groups are equally reactive. The three hydroxyl groups of the glucose units differ in their reactivity and the relative rates are not necessarily the same for other reagents. As a general rule, most commercial cellulose derivatives are the products of only partial reaction, that is, they contain a certain proportion of unchanged hydroxyl groups. Therefore, the expression methyl cellulose or cellulose acetate by no means characterizes a product.

The physical and chemical properties of cellulose derivatives are determined largely by the degree of substitution (DS) and the degree of polymerization (DP). The term degree of substitution is used to denote the extent of the reaction, and it is defined as the average number of hydroxyl groups substi-tuted of the three available in the anhydro-glucose units. Properties that are most strongly affected by changing the degree of substitution are the solu-bility, swelling, and plasticity. Derivatives of a low degree of substitution are often more sensitive to water than the original cellulose and may even be dis-persible in water. In the case of derivatives having a high degree of substitution with nonpolar substituents, the water solubility is decreased, the sorption of water is decreased, and the solubility in organic solvents is increased. On the other hand the plasticity is increased by the substitution of nonpolar groups. The greater the chain length of the substituent group, the higher the plasticity is because the individual chains are forced further apart. The physical and chemical properties of cellulose derivatives are also strongly affected by the average chain length or DP. The DP can be controlled in two ways: (1) by selecting a dissolving pulp of the correct DP and (2) by carefully controlling the reaction conditions during formation of the cellulose derivative. During commercial esterification or etherification of cellulose, two reactions occur: one a substitution reaction at the hydroxyl groups and the other a glycosidic-cleavage reaction. For derivatives to be used in spinning fibers, casting films, molding plastics, or spraying lacquers, a low-viscosity (a low DP) derivative is required. On the other hand, if the mechanical properties are important in the final product, a balance must be struck between low DP, which will confer ease of fabrication, and the somewhat higher DP necessary for acceptable mechanical properties.

Cellulose Esters. Over 100 esters of cellulose have been prepared and described. It is beyond the scope of this chapter to describe in detail the chemical and physical properties of each of them. A discussion of the cellulose ethers and esters used in the paper industry is covered in other chapters. Details on the esters used in other industries can be found in specialized reviews.[30]

The three most important commercial esters are cellulose xanthate, cellulose nitrate, and cellulose acetate. Cellulose xanthate is not a final product; it is formed as an intermediate product in the process for making viscose rayon and cellophane. The manufacturing of these two products involves steeping of pulp in a 17.5% solution of sodium hydroxide to obtain activated alkali cellulose, removing the excess sodium hydroxide, and aging to reduce the molecular weight. After aging, the activated cellulose is treated with carbon disulfide and converted to a partially xanthated derivative that dissolves in alkali to give the viscose solution. The viscose solution is then forced through an orifice (or a slit) into an acid bath where the xanthate groups are decomposed and regenerated cellulose is obtained as a filament or sheet. The chemical reactions involved are:

$$\text{Cellulose} - \text{OH} + \text{NaOH} \longrightarrow \text{Cellulose} - \text{OH} - \text{OH}^-\text{Na}^+$$

$$\text{Cellulose} - \text{OH} - \text{OH}^-\text{Na}^+ + \text{CS}_2 \rightarrow \text{Cellulose} - \text{OCS}_2^-\text{Na}^+ + \text{H}_2\text{O}$$

$$\text{Cellulose} - \text{OCS}_2^-\text{Na}^+ + \text{H}^+ \longrightarrow \text{Cellulose} - \text{OH} + \text{CS}_2$$

Highly nitrated cellulose (DS 2.4 to 2.8) is used as an explosive and propellant; less highly nitrated products have been used for the production of films, adhesives, lacquers, and plastics. Commercially the nitrate esters are prepared by reacting cellulose with a mixture of nitric acid, sulfuric acid, and water. The DS and DP may be controlled by addition of water and sulfuric acid. The reaction is:

$$\text{Cellulose} - \text{OH} + \text{HNO}_3 \xrightarrow{\text{H}_2\text{SO}_4 + \text{H}_2\text{O}} \text{Cellulose} - \text{O}-\text{NO}_2 + \text{H}_2\text{O}$$

If undegraded nitrates and high nitrogen contents are desired, for example, in determination of molecular weight, nitric acid and phosphorus pentoxide or nitric acid and acetic anhydride may be used.[169,170]

Cellulose acetate is widely used for the manufacture of films, plastics, fibers and coatings. The most common industrial process for making cellulose acetate involves a two-stage process. In the first stage, the cellulose is activated by swelling in a mixture of sulfuric and acetic acids. In the second stage, excess acetic acid is removed and acetic anhydride is added to begin the reaction:

$$\text{Cellulose} - OH + (CH_3CO)_2O \xrightarrow{H_2SO_4} Cell\!-\!O\!-\!COCH_3 + H_2O + CH_3COOH$$

Cellulose acetate

The cellulose acetate can be precipitated directly from solution and redissolved in ethylene dichloride. This material can be forced through an orifice to make cellulose triacetate fibers. If the product is not precipitated and water is added to the solution, hydrolysis occurs and the DS is reduced to approximately 2.0. This product is sometimes called secondary acetate; it is used to make fiber or photographic films.

Cellulose Ethers. Ether derivatives of cellulose are mainly used as thickeners, dispersants, adhesives, extenders, and films because of their water solubility and gel-forming properties. Introduction of a few ether groups disrupts the regularity and the hydrogen bonding of the crystalline structure without seriously affecting the polarity of the molecules, and the product becomes alkali and water soluble. The polarity can, in fact, be increased by introduction of ionizing groups, such as a carboxymethyl group. However, as the degree of substitution increases, the material becomes more soluble in organic solvents.

Ether derivatives are generally produced by the reaction of alkali cellulose with the corresponding alkyl halides. Methyl chloride and carboxymethyl chloride give methyl cellulose and carboxymethyl cellulose (CMC) respectively, two of the best-known cellulose ether derivatives. The highly reactive and accessible alkali cellulose can also react with ethylene oxide and acrylonitrile to give hydroxyethyl cellulose and cyanoethyl cellulose, which introduce new functions to the polysaccharide structure. The reactions are:

$$\text{Cellulose} - OH\!-\!\overset{-}{O}H\,\overset{+}{Na} + RCl \longrightarrow \text{Cellulose} - OR + NaCl + H_2O$$

$$\text{Cellulose} - OH\!-\!\overset{-}{O}H\,\overset{+}{Na} + CH_2\!-\!CH_2 \longrightarrow \text{Cellulose} - O\!-\!CH_2\!-\!CH_2\!-\!OH$$

$\underset{O}{\diagdown\diagup}$ + NaOH

$$\text{Cellulose} - OH\!-\!\overset{-}{O}H\,\overset{+}{Na} + CH_2 = CH\!-\!CN \longrightarrow \text{Cellulose} - O\!-\!CH_2\!-\!CH_2\!-\!CN$$

+ NaOH

Graft and Cross-Linked Derivatives. In addition to the ester and ether derivatives, the derivatives made by crosslinking and graft polymerization of cellulose are also of considerable industrial and academic interest. For crosslinking, the substrate is treated with bi- or poly-functional reagents generally derived from formaldehyde in order to bridge the neighboring molecules and reduce their hygroexpansivity. These treatments not only improve the dimensional stability, but also favorably affect the related properties of the fabrics, such as drying and wrinkle resistance. In graft polymerization the cellulose molecule is subjected to radiation or specific oxidants to create free radicals that can initiate free radical polymerization of vinyl monomers. The combination with polymers substantially modifies the properties of cellulose.

Hemicelluloses

The cellulose and lignin of plant cell walls are closely interpenetrated by a mixture of polysaccharides called hemicelluloses. The name hemicelluloses was originally proposed by Schulze[171] in 1891 to designate those polysaccharides extractable from plants by aqueous alkali. Today most workers limit the term hemicelluloses to designate cell-wall polysaccharides of land plants, excluding the cellulose and pectin components. The hemicelluloses are generally water-insoluble, alkali-soluble substances that are more readily hydrolyzed by acid than is cellulose. Structurally, the hemicelluloses differ from cellulose in that they are branched and have much lower molecular weights. The hemicelluloses that are found in the stalk or supporting tissue of woody plants are primarily modified xylans, galactoglucomannans, glucomannans, and arabinogalactans. The arabinogalactans are soluble in water and are commonly classed among the extractives. They are discussed in this chapter because of their widespread distribution in conifers and for comparison of their structure and composition to other nonglucose polysaccharides. All of these polysaccharides are built up from a relatively limited number of sugar residues; the principal ones are: D-xylose, D-mannose, D-glucose, D-galactose, L-arabinose, and 4-0-methyl-D-glucuronic acid.

Although related, the hemicelluloses in wood from gymnosperms and angiosperms are not the same. In angiosperms (hardwoods) the predominant hemicellulose is a partially acetylated (4-0-methylglucurono) xylan with minor amounts of a glucomannan. In gymnosperms (softwoods) the hemicelluloses consist of partially acetylated galactoglucomannans with smaller, but substantial, amounts of an arabino-(4-0-methylglucurono) xylan. The larches occupy a unique position among the gymnosperms in that they contain arabinogalactan as a major constituent. The difference between angiosperms and gymnosperms can best be illustrated by showing the chemical composition of some typical hardwoods and softwoods. This is shown in Table 1–2 for hardwoods and Table 1–3 for softwoods, taken from Timell.[38]

Numerous reviews of the chemistry and biochemistry of the hemicelluloses are available.[10,11,38,172,173] Several detailed reviews on their biogenesis,[105,126] X-ray analysis,[86] and morphology[38] have been published.

TABLE 1–2 CHEMICAL COMPOSITION OF WOOD FROM FIVE HARDWOODS (ANGIOSPERMS)[a]

Component	Acer rubrum	Betula papyrifera	Fagus grandifolia	Populus tremuloides	Ulmus americana
Cellulose	45	42	45	48	51
Lignin	24	19	22	21	24
O-Acetyl (4-O-methylglucurono) xylan	25	35	26	24	19
Glucomannan	4	3	3	3	4
Pectin, starch, ash, etc.	2	1	4	4	2

[a] All values in percent of extractive-free wood.

TABLE 1–3 CHEMICAL COMPOSITION OF WOOD FROM FIVE SOFTWOODS (GYMNOSPERMS)[a]

Component	Abies balsamea	Pinus glauca	Pinus strobus	Tsuga canadensis	Thuja occidentalis
Cellulose	42	41	41	41	41
Lignin	29	27	29	33	31
Arabino-(4-O-methylglucurono) xylan	9	13	9	7	14
O-Acteyl-galactoglucomannan	18	18	18	16	12
Pectin, starch, ash, etc.	2	1	3	3	2

[a] All values in percent of extractive-free wood.

Structure and Properties of Hemicelluloses

Many of the chemical and physical properties of the individual wood hemi-celluloses are summarized in Table 1-4.[38] These values were obtained by sum-marizing the properties of the polysaccharides from a large number of different species.

Xylans. The basic skeleton of the xylans found in the stalk or supporting tissue of all land plants contains a linear (or singly branched) backbone of 1 → 4-linked anhydro-β-D-xylopyranose units. There are only a few examples of xylans, such as the esparto grass xylan, [10,11] that are made up completely of anhydro-xylose units. Most xylans found in land plants contain short side chains consisting of one or two sugar units linked to the main backbone of anhydro-xylose units. The most common carbohydrate in the side chain is 4-0-methyl-α-D-glucopyranosyluronic acid.

Xylan consisting of xylose and 4-0-methyl-α-D-glucopyranosyluronic acid is the major hemicellulose component of hardwoods. The dominant structure of these polymers is a linear or singly branched β-D-xylopyranose backbone linked together by (1 → 4) glycosidic bonds. Branch points consisting of the pyranose forms of 4-0-methyl-D-glucuronic acid are attached by an alpha linkage mainly to the C-2 position of the xylose unit. Timell[10,11] has shown that there is also a small proportion of C-3 linkage. In hardwoods the xylose-to-uronic acid ratio varies from 3:1 to 20:1; however, the most common ratio is from 7:1 to 12:1. Besides the uronic acid groups, the hardwood xylans also contain acetyl groups. The amount of acetyl groups varies: silver birch xylan contains 11.4 to 16.9% acetyl groups depending on extraction conditions; yellow birch contains 16.5%. Most evidence indicates that the acetyl groups are randomly attached to the oxygen at both the ^{2}C and ^{3}C position of the xylose units. The degree of polymerization reported for hardwood xylans has varied from 50 to 300. However, studies [38] in which special precautions have been taken to minimize contamination and degradation have indicated a number-average degree of polymerization in the range of 150 to 200.

Softwood xylans are an arabino-(4-0-methylglucurono) xylan. They make up 7 to 15% of softwoods. The first pure softwood xylan was isolated in 1956. Structurally it consists of an anhydro-xylose backbone with branch points containing both α-linked 4-0-methyl-D-glucuronic acid groups and α-linked L-arabinofuranose units. The ratio of xylose to uronic acid varies from 4:1 to 9:1, while the most common xylose-to-arabinose ratio ranges from 5:1 to 7:1. The arabinose units are extremely acid labile and can be removed under mild acid conditions in which the main xylose chain remains intact. The best evidence available indicates that the uronic acid groups are linked mainly through the 2-position of the anhydro-xylose chain, while the arabinose units are linked predominantly at the 3-position of the main xylan chain. The softwood xylan does not contain acetyl groups and is more highly branched and more acidic than the corresponding hardwood xylan.

TABLE 1-4 CHEMICAL AND PHYSICAL PROPERTIES OF WOOD HEMICELLULOSES

Polysaccharide	Occurrence	Percentage of Wood	Composition	Parts	Linkages	Linear or Branched	DP_n	DP_w
O-Acetyl-(4-O-methylglucurono)xylan	Hardwoods	10–35	β-D-Xylp 4-O-Me-α-D-GlupA O-Acetyl	10 1 7	1 → 4 1 → 2 –	Slightly branched	200	180–250
Glucomannan	Hardwoods	3–5	β-D-Manp β-D-Glup	1–2 1	1 → 4 1 → 4	Undecided	>70	>120
Arabino-(4-O-methylglucurono)xylan	Softwoods	10–15	β-D-Xylp 4-O-Me-α-D-GlupA L-Araf	10 2 1.3	1 → 4 1 → 2 1 → 3	Undecided	>120	
Galacto-glucomannan (water soluble)	Softwoods	5–10	β-D-Manp β-D-Glup α-D-Galp O-Acetyl	3 1 1 0.24	1 → 4 1 → 4 1 → 6 –	Probably branched	>100	>150
Galactoglucomannan (alkali soluble)	Softwoods	10–15	β-D-Manp β-D-Glup α-D-Galp O-Acetyl	3 1 0.1 0.24	1 → 4 1 → 4 1 → 6 –	Probably branched	>100	>150
Arabinogalactan	Larchwood	10–20	β-D-Galp L-Araf β-L-Arap β-D-GlupA	6 2/3 1/3 Few	1 → 3, 1 → 6 1 → 6 1 → 3 1 → 6	Heavily branched	200	100 and 600

Glucomannans (Hardwoods). Hardwoods usually contain from 3 to 5% of glucomannan. This polysaccharide consists of both D-glucose and D-mannose residues generally in a ratio of 1:2, with the exception of the genus *Betula* for which the ratio is usually 1:1. This molecule contains a backbone of $(1 \rightarrow 4)$-linked β-D-glucopyranose and β-D-mannopyranose units with no D-galactose being present.

Galactoglucomannans (Softwoods). These polysaccharides are the major hemicellulose component of softwoods. Structurally, they consist of a backbone of $(1 \rightarrow 4)$-linked β-D-mannopyranose units and β-D-glucopyranose units. Branch points occur at the ^6C position of the glucose and mannose units where α-D-galactopyranose units are attached. In many wood species two distinctly different galactoglucomannans occur. Generally the major galactoglucomannan is a water-soluble polysaccharide containing galactose, glucose, and mannose residues in a ratio of 1:1:3. The other galactoglucomannan fraction found in many softwoods is insoluble in water and consists of residues of galactose, glucose, and mannose in a ratio of 0.1:1:3. Evidently, softwoods contain a whole family of related galactoglucomannans that differ mainly in their galactose content.

Arabinogalactans. The arabinogalactans are water-soluble polysaccharides present in the wood of conifers. With the exception of the genus *Larix*, which may contain up to 25%, most gymnosperms contain only small proportions of this polysaccharide. Arabinogalactans are mainly extracellular and found largely in the fiber lumens. They are the only major wood polysaccharide that can be isolated in good yield by extraction of wood with water before delignification. Because of their ease of extraction and their useful properties as gums, these materials have been used commercially under the trade name Stractan. The arabinogalactans found in larch consist of a basic backbone of $(1 \rightarrow 3)$-linked β-D-galactopyranose residues, each of which contains various side chains linked through the ^6C positions of the galactose units. The exact nature of all the side chains is not known, but the majority of these side chains are composed of $(1 \rightarrow 6)$-linked β-D-galactopyranose residues with two such units present per average chain. Another common side chain consists of a 3–0-β-L-arabinopyranosyl-L-arabinofuranose unit.

The arabinogalactans from other conifer species have been found to have a similar structure. More detailed information on the structure of arabinogalactans can be found in reviews by Timell[10,38] and others. [172,173]

REFERENCES

1. O. C. Stewart in "Man's Role in Changing the Face of the Earth," W. L. Thomas, Jr., Ed., U. of Chicago Press, Chicago, Ill., 1956, p 115.
2. Li Ch'iao-p'ing, "The Chemical Arts of Old China," Mack Printing Co., Easton, Pa., 1948.

3. D. Hunter, "Papermaking through Eighteen Centuries," William Edwin Rudge, New York, N.Y., 1930.
4. "Fiber of Enlightenment," *MD*, 222, (1968).
5. C. F. Cross and E. J. Bevan, *J. Chem. Soc.*, **38**, 666 (1880).
6. W. G. Van Beckum and G. J. Ritter, *Paper Trade J.*, **105** (18), 127 (1937).
7. L. E. Wise, M. Murphy, and A. A. D'Addieco, *Paper Trade J.*, **122** (2), 35 (1946).
8. J. W. Green, *Methods Carbohyd. Chem.*, **3**, 9 (1963).
9. B. L. Browning and L. O. Bublitz, *Tappi*, **36**, 452 (1953).
10. T. E. Timell, *Advan. Carbohyd. Chem.*, **19**, 247 (1964).
11. T. E. Timell, *Advan. Carbohyd. Chem.*, **20**, 410 (1965).
12. T. E. Timell, *Methods Carbohyd. Chem.*, **5**, 100 (1965).
13. H. A. Swenson, *Methods Carbohyd. Chem.*, **3**, 84 (1963).
14. A. Payen, *Compt. Rend.*, **7**, 1052 (1838).
15. R. Willstätter and L. Zechmeister, *Chem. Ber.*, **46**, 2401 (1913).
16. H. Ost, *Z. Angew Chem.*, **19**, 993 (1906).
17. W. Crum, *Phil. Mag.*, **30**, 426 (1847).
18. W. Crum, *Justus Liebigs Ann. Chem.*, **62**, 233 (1847).
19. G. W. Monier-Williams, *J. Chem. Soc.*, **119**, 803 (1921).
20. J. C. Irvine and C. W. Soutar, *J. Chem. Soc.*, **117**, 1489 (1920).
21. J. C. Irvine and E. L. Hirst, *J. Chem. Soc.*, **121**, 1585 (1922).
22. J. C. Irvine and E. L. Hirst, *J. Chem. Soc.*, **123**, 518 (1923).
23. K. Freudenberg, E. Plankenhorn, and H. Boppel, *Chem. Ber.*, **71**, 2435 (1938).
24. W. N. Haworth, *Nature*, **116**, 430 (1925).
25. H. Friese and K. Hess, *Justus Liebigs Ann. Chem.*, **456**, 38 (1927).
26. C. C. Spencer, *Cellulosechemie*, **10**, 61 (1921).
27. I. Danishefsky, R. L. Whistler, and F. A. Bettelheim in "The Carbohydrate," W. Pigman and D. Horton, Eds., Academic, New York, N.Y., 1970, p 375.
28. J. Dyer in "Cellulose Technology Research," A. F. Turbak, Ed., ACS Symp. Series No. 10, American Chemical Society, Washington, D.C., 1975, p 181.
29. E. H. Immergut in "The Chemistry of Wood," B. L. Browning, Ed., Interscience, New York, N.Y., 1963, 4, p 104.
30. P. M. Doty and H. M. Spurlin in "Cellulose and Cellulose Derivatives," Part III, E. Ott, H. M. Spurlin, and M. W. Grafflin, Eds., Interscience, New York, N.Y., 1955, p 1133.
31. W. J. Alexander and T. E. Mullen, *J. Polym. Sci. Part C*, **30**, 87 (1971).
32. K. H. Altgelt and L. Segal, "Gel Permeation Chromatography," Marcel Dekker, New York, N.Y., 1971, p 429
33. S. H. Churms, *Advan. Carbohyd. Chem. and Biochem.*, **25**, 13 (1970).
34. T. E. Timell and J. L. Snyder, *Svensk Paperstidn.*, **58**, 851, (1955); **59**, 1 (1956); *Tappi*, **40**, 25 (1957); *Pulp Paper Mag. Can.*, **56** (7), 104 (1955).
35. W. G. Harland, *Shirley Inst. Mem.*, **28**, 167 (1955); *J. Textile Inst.*, **45**, T–678 (1954).
36. T. E. Timell, and J. L. Snyder, *Textile Res. J.*, **25**, 870 (1955); *Tappi*, **40**, 749 (1957); *Ind. Eng. Chem.*, **47**, 2166 (1955).
37. D. A. I. Goring and T. E. Timell, *Tappi*, **45**, 454 (1962).
38. T. E. Timell, *Wood Science Technol.*, **1**, 45 (1967).
39. H. I. Bolker, "Natural and Synthetic Polymers," Marcel Dekker, New York, N.Y., 1974, p 37.
40. W. A. Côté, Jr., Ed., "Cellular Ultrastructure of Woody Plants," Syracuse U. Press, Syracuse, N.Y., 1965.

41. H. F. J. Wenzl, "The Chemical Technology of Wood," Academic, New York, N.Y., 1970, p 32.
42a. E. B. Cowling in "Cellulose as a Chemical and Energy Resource," Interscience, New York, N.Y., 1975, p 163.
42b. A. B. Wardrop and D. E. Bland in "Biochemistry of Wood," K. Kratzl and G. Billek, Eds., Pergamon, London, England, 1959, p 92.
43. C. Y. Liang, K. H. Bassett, E. A. McGinnes, and R. H. Marchessault, *Tappi,* **43,** 1017 (1960).
44. P. W. Lange, *Svensk Paperstidn.,* **57,** 525 (1954).
45. I. B. Sachs, I. T. Clark, and J. C. Pew, *J. Polym. Sci., Part C,* **2,** 203 (1963).
46. H. Meier, *Pure Appl. Chem.,* **5,** 37 (1962).
47. H. Meier and K. C. B. Wilkie, *Holzforschung,* **13,** 177 (1959).
48. H. Meier, *J. Polym. Sci.,* **51,** 11 (1961).
49. K. Freudenberg, *J. Chem. Educ.,* **9,** 1171 (1932).
50. R. Mark, in ref. 40, p 493.
51. H. Harada, "Ultrastructure and Organization of Gymnosperm Cell Walls" in "Cellular Ultrastructure of Woody Plants," W. A. Côté, Jr., Ed., Syracuse U. Press, Syracuse, N.Y., 1965, p 221. By permission of the publisher.
52. E. Frei and R. D. Preston, *Proc. Roy. Soc., Ser. B,* **155,** 55 (1961).
53. K. Mühlethaler, in ref. 40, p 191.
54. R. D. Preston, *Faraday Soc. Disc.,* **11,** 165 (1951).
55. A. J. Hodge and A. B. Wardrop, *Nature,* **165,** 272 (1950).
56. A. Vogel, *Makromol. Chem.,* **11,** 111 (1953).
57. A. N. J. Heyn, *J. Appl. Phys.,* **26,** 519 (1955).
58. G. Tsoumis, "Wood as Raw Material," Pergamon, London, England, 1968, p 60.
59. I. Ohad, D. Danon, and S. Hestrin, *J. Cell Biol.,* **17,** 321 (1963).
60. I. Ohad and D. Mejzler, *J. Polym. Sci., Part A,* **3,** 339 (1965).
61. D. Fengel, *J. Polym. Sci., Part. C,* **36,** 383 (1971).
62. A. Frey-Wyssling and K. Mühlethaler, "Ultrastructural Plant Cytology," Elsevier, Amsterdam, The Netherlands, 1965.
63. A. N. J. Heyn, *J. Ultrastructure Res.,* **26,** 52 (1969).
64. D. J. Johnson and D. Crawford, *J. Microscopy,* **98,** 313 (1973).
65. I. Ohad and D. Danon, *J. Cell Biol.,* **22,** 302 (1964).
66. R. St. J. Manley, *Nature,* **204,** 1155 (1964).
67. R. Preston, *J. Microscopy,* **93,** 7 (1971).
68. W. W. Franke and B. Ermen, *Z. Naturforsch.,* **24b,** 918 (1969).
69. D. Fengel, *Naturwissenschaften,* **61,** 31 (1974).
70. R. B. Hanna and W. A. Côté, Jr., *Cytobiologie,* **10** (1), 102 (1974).
71. R. Marton, P. Rushton, J. Sacco, and K. Sumiya, *Tappi,* **55,** 1499 (1972).
72. A. B. Wardrop, in ref. 40, p 61.
73. C. Nägeli, "Die Stärkekörner," F. Schulthess, Zürich, Switzerland, 1858.
74. R. O. Herzog and W. Jancke, *Z. Phys.,* **3,** 196 (1920).
75. H. Mark, *Chem. Rev.,* **26,** 169 (1940).
76. F. Shafizadeh and G. D. McGinnis, *Advan. Carbohyd. Chem. and Biochem.,* **26,** 297 (1971).
77. K. H. Meyer and L. Misch, *Helv. Chim. Acta.,* **20,** 232 (1937).
78. G. Honjo and M. Watanabe, *Nature,* **181,** 326 (1958).
79. K. H. Gardner and J. Blackwell, *Biopolymers,* **13,** 1975 (1974).
80. M. G. Dobb, L. D. Fernando, and J. Sikorski, *Proc. 8th Intl. Cong. Electron Microscopy,* Canberra, Australia, **1,** 364 (1974).
81. M. M. Beg, S. Rolandson, and A. U. Ahmed, *J. Polym. Sci.,* **12,** 311 (1974). [Polym. letters ed.]
82. J. J. Hebert and L. L. Muller, *J. Appl. Polym. Sci.,* **18,** 3373 (1974).

83. K. H. Gardner and J. Blackwell, *Biochim. Biophys. Acta,* **343,** 232 (1974).
84. A. Sarko and R. Muggli, *Macromol.,* **7,** 486 (1974).
85. W. Claffey and J. Blackwell, *Biopolymers,* **15,** 1903 (1976).
86. R. D. Preston, *Phys. Rep.,* **21,** (4), 183 (1975).
87. J. A. Howsman and W. A. Sisson in "Cellulose and Cellulose Derivatives," Part I, E. Ott, H. M. Spurlin, and M. W. Grafflin, Eds., Interscience, New York, N.Y., 1954, p 231.
88. S. K. Asunmaa, *Tappi,* **49,** 319 (1966).
89. M. Marx-Figini and G. V. Schulz, *Biochim. Biophys. Acta,* **112,** 81 (1966).
90. H. Bittiger, E. Husemann, and A. Kuppel. *J. Polym. Sci., Part C,* **28,** 45 (1969).
91. R. Muggli, *J. Cellulose Chem. Technol.,* **2,** 549 (1969).
92. R. Muggli, H. G. Elias, and K. Mühlethaler, *Makromol. Chem.,* **121,** 290 (1969).
93. K. Mühlethaler, *J. Polym. Sci., Part C,* **28,** 305 (1969).
94. K. Mühlethaler and Z. Schweiz, *Forstv.,* **30,** 53 (1960).
95. S. P. Rowland and E. J. Roberts, *J. Polym. Sci., Part A-1,* **10,** 2447 (1972).
96. J. L. Bose, E. J. Roberts, and S. P. Rowland, *J. Appl. Polym. Sci.,* **15,** 2999 (1971).
97. S. P. Rowland and E. J. Roberts, *J. Polym. Sci., Part A-1,* **10,** 2447 (1972).
98. S. P. Rowland, E. J. Roberts, J. L. Bose, and C. P. Wade, *J. Polym. Sci., Part A-1,* **9,** 1623 (1971).
99. M. Marx-Figini, *J. Polym. Sci., Part C,* **28,** 57 (1969).
100. M. Marx-Figini, *Biochim. Biophys. Acta,* **177,** 27 (1969).
101. M. Marx-Figini, *Nature,* **210,** 754 (1966).
102. F. S. Spencer and G. A. MacLachlan, *Plant Physiol.,* **49,** 58 (1972).
103. W. Z. Hassid in "The Carbohydrate," Vol. II A, W. Pigman and D. Horton, Eds., Academic, New York, N.Y., 1970, p 301.
104. H. Nikaido and W. Z. Hassid, *Advan. Carbohyd. Chem. and Biochem.,* **26,** 351 (1971).
105. A. L. Karr in "Plant Biochemistry," 3rd ed., J. Bonner and J. E. Varner, Eds., Academic, New York, N.Y., 1976, p 405.
106. C. L. Villemez, *Ann. Proc., Phytochem. Soc.,* **12,** 183 (1974).
107. L. Glaser, *J. Biol. Chem.,* **232,** 627 (1958).
108. A. D. Elbein, G. A. Barber, and W. Z. Hassid, *J. Amer. Chem. Soc.,* **86,** 309 (1964); *J. Biol. Chem.,* **239,** 4056 (1964).
109. T. Liu and W. Z. Hassid, *J. Biol. Chem.,* **245,** 1922 (1970).
110. G. A. Barber and W. Z. Hassid, *Biochim. Biophys. Acta,* **86,** 397 (1964).
111. G. A. Barber and W. Z. Hassid, *Nature,* **207,** 295 (1965).
112. D. P. Delmer, C. A. Beasley, and L. Ordin, *Plant Physiol.,* **53,** 149 (1974).
113. D. O. Brummond and A. P. Gibbons, *Biochem. Biophys. Res. Commun.,* **17,** 156 (1964).
114. D. O. Brummond and A. P. Gibbons, *Biochem. Z.,* **342,** 308 (1965).
115. L. Ordin and M. A. Hall, *Plant Physiol.,* **42,** 205 (1967); **43,** 473 (1968).
116. W. G. Slater and H. Beevers, *Plant Physiol.,* **33,** 146 (1958).
117. J. B. Pridham and W. Z. Hassid, *Biochem. J.,* **100,** 21P (1966).
118. A. D. Elbein, *J. Biol. Chem.,* **244,** 1608 (1969).
119. J. S. Heller and C. L. Villemez, *Biochem. J.,* **129,** 645 (1972).
120. C. L. Villemez, *Arch. of Biochem. Biophys.,* **165,** 407 (1974).
121. J. S. Anderson, M. Matsuhashi, M. A. Haskin, and J. L. Strominger, *Proc. Nat. Acad. Sci. U.S.,* **53,** 881 (1965).
122. M. Scher and W. J. Lennarz, *J. Biol. Chem.,* **224,** 2777 (1969).

123. J. Kjosbakken and J. R. Colvin, *Can. J. Microbiol.,* **21,** (2), 111 (1975).
124. W. T. Forsee and A. D. Elbein, *J. Biol. Chem.,* **248,** 2858 (1973).
125. C. L. Villemez, *Biochem. Biophys. Res. Commun.,* **40,** 636 (1970).
126. D. H. Northcote, *Ann. Proc. Phytochem. Soc.,* **10,** 165 (1974).
127. H. Kauss, *Ann. Proc. Phytochem. Soc.,* **13,** 191 (1974).
128. F. B. P. Wooding, *J. Cellulose Sci.,* **3,** 71 (1968).
129. G. Shore and G. A. MacLachlan, *J. Cell Biol.,* **64,** 557 (1975).
130. D. T. Dennis and J. R. Colvin, in ref. 40, p 199.
131. D. H. Northcote, *Proc. Roy. Soc.,* **173,** 21 (1969).
132. D. J. Bowles and D. H. Northcote, *Biochem. J.,* **130,** 1133 (1972).
133. D. H. Northcote in "Plant Cell Organelles," J. B. Pridham, Ed., Academic, New York, N.Y., 1968, p 179.
134. R. D. Preston, *Endeavour,* **23,** 153 (1964).
135. M. Marx-Figini and G. V. Schulz, *Makromol. Chem.,* **62,** 49 (1963).
136. I. Grimm, H. Sachs, and D. G. Robinson, *Cytobiologie,* **14,** 61 (1976).
137. W. G. Haigh, H. J. Förster, K. Biemann, N. H. Tattrie, and J. R. Colvin, *Biochem. J.,* **135,** 145 (1973).
138. B. L. Browning, in ref 29, p 406.
139. C. Skaar, "Water in Wood," Syracuse U. Press, Syracuse, N.Y., 1972.
140. S. A. Rydholm, "Pulping Processes," Interscience, New York, N.Y., 1965.
141. K. Ward, Jr., and P. A. Seib in "The Carbohydrate," Academic, New York, N.Y., 1970, p 413.
142. W. Sandermann, "Chemische Holzverwertung," BLV Verlag, München, West Germany, 1963.
143. C. R. Wilke, "Cellulose as a Chemical and Energy Resource," Interscience, New York, N.Y., 1975.
144. J. F. Saeman, W. E. Moore, R. L. Mitchell, and M. A. Millett, *Tappi,* **37,** 336 (1954).
145. P. A. McGee, W. F. Fowler, Jr., E. W. Taylor, C. C. Unruh, and W. O. Kenyon, *J. Amer. Chem. Soc.,* **69,** 355 (1947).
146. B. Alfredsson, W. Czerwinsky, and O. Samuelson, *Svensk Papperstidn.,* **64,** 812 (1961).
147. R. L. Whistler and W. M. Corbett, *J. Amer. Chem. Soc.,* **78,** 1003 (2956).
148. R. L. Whistler and J. N. BeMiller, *Advan. Carbohyd. Chem.* **13,** 289 (1958).
149. G. Machell and G. N. Richards, *J. Chem. Soc.,* 4500 (1957); 1199 (1958); 1924 (1960).
150. O. Samuelson and A. Wennerblom, *Svensk Papperstidn.,* **57,** 827 (1954).
151. J. K. Hamilton and N. S. Thompson, *Pulp Paper Mag. Can.,* **61** (4), T–263 (1960).
152. W. M. Corbett, G. N. Richards, and R. L. Whistler, *J. Chem. Soc.,* **11** (1957).
153. H. Meier, *Svensk Papperstidn.,* **65,** 589 (1962).
154. R. E. Brandon, L. R. Schroeder, and D. C. Johnson in "Cellulose Technology Research," A. F. Turbak, Ed., ACS Symp. Series No. 10, American Chemical Society, Washington, D.C., 1975, p 125.
155. B. Lindberg, *Svensk Papperstidn.,* **59,** 531 (1956).
156. G. A. Richter, *Ind. Eng. Chem.,* **26,** 1154 (1934).
157. R. L. W. Farquhar, D. Pesant, and B. A. McLaren, *Can. Textile J.,* **73,** 51 (1956).
158. W. D. Major, *Tappi,* **41,** 530 (1958).
159. L. F. McBurney, in ref. 87, p 99.
160. F. Shafizadeh, *Advan. Carbohyd. Chem.,* **23,** 419 (1968).
161. F. Shafizadeh in "Thermal Uses and Properties of Carbohydrates and

Lignins," F. Shafizadeh, K. V. Sarkanen, and D. A. Tillman, Eds., Academic, New York, N.Y., 1976; F. Shafizadeh and P. P. S. Chin in "Wood Technology: Chemical Aspects," I. S. Goldstein, Ed., ACS Symp. Series No. 43, American Chemical Society, Washington, D.C., 1977.

162. P. Koch, "Utilization of the Southern Pines," USDA Forest Service Agriculture Handbook, No. 420, 1, U.S. Government Printing Office, Washington, D.C., 1972, p 643.

163. E. T. Reese and M. Mandels, "Enzymatic Degradation" in "Cellulose and Cellulose Derivatives," Part V, Wiley-Interscience, New York, N.Y., 1971, p 1088.

164. D. F. Durso and A. Villarreal, in ref. 154. p 106.

165. *Chem. Eng. News,* **56** (32), 22 (1978).

166. R. F. H. Dekker and G. N. Richards, *Advan. Carbohyd. Chem. and Biochem.,* **32,** 277 (1976).

167. M. Mandels and J. Weber in "Cellulases and Their Applications," Advances in Chemistry Series No. 95, American Chemical Society, Washington, D.C., 1969, p 391.

168. J. F. Harris, J. F. Saeman, and E. G. Locke, in ref. 29, p 536.

169. R. L. Mitchell, *Ind. Eng. Chem.,* **45,** 2526 (1953).

170. C. F. Bennett and T. E. Timell, *Svensk Papperstidn.* **58,** 281 (1955).

171. E. Schulze, *Chem. Ber.,* **24,** 2277 (1891).

172. C. Schuerch, in ref. 29, p 191.

173. R. L. Whistler and E. L. Richards, in ref. 141, p 447.

2 LIGNIN

WOLFGANG G. GLASSER ,Ph.D.

Associate Professor of Wood Chemistry
Department of Forest Products
Virginia Polytechnic Institute and State University
Blacksburg, Virginia

To the pulp and papermaker, lignin is the unwanted ingredient of wood that creates most problems encountered during pulping. If it were not for lignin, it would not be necessary to apply strong alkaline or acidic reagents in the chemical delignification of wood in order to obtain pulp and paper products. Inasmuch as the primary objective of pulping is the liberation of cellulosic fibers, pulping must be understood as the art of lignin removal from woody plant tissues. Delignification is the foremost goal of pulping. Furthermore, lignin undergoes more severe chemical changes during pulping than cellulose. Thus it is apparent that a thorough understanding of the structure of lignin, its purposes, its functions in the tree, its physical and chemical properties, and its reaction behavior are of the utmost importance for an understanding and a rational control of pulping processes. This chapter attempts to provide such an understanding in four principal sections: (1) the formation of lignin in trees, (2) the separation of lignin from woody tissues, (3) the analysis of lignin, and (4) the utilization of lignin. It will not be possible to cover all important aspects and contributions to the subject of lignin; the reader, therefore, is referred to the monographs by Freudenberg and Neish[1] and by Sarkanen and Ludwig.[2]

Lignification: Formation of Lignin in Wood

In this section an evaluation will be made of the current status of our understanding of lignin formation in woody tissues. The subject of plant lignification will be discussed in terms of biological and biochemical aspects, focusing on the emergence of lignin in the plant kingdom, its distribution throughout cell systems, and its presently conceived mechanism of deposition. This discussion is followed by a review of the chemical aspects of lignin formation, and the chemical structure of lignin is considered.

Biological and Biochemical Aspects of Lignin Formation

The current status of plant lignification in terms of its biological and biochemical aspects has been reviewed by Freudenberg and Neish,[1] by Wardrop,[3,4] and

39

by Higuchi,[5] among others. In contrast to cellulose, which is formed by all plants, the formation of lignin is essentially unique to vascular plants, which develop tissues that specialize in such functions as transportation of aqueous solutions and mechanical support. Primitive plants without differentiated cell tissues, such as fungi, algae, and mosses, do not contain lignin,[6] apparently because their cell agglomeration does not require the protective and supportive characteristics provided by lignin. Beyond protecting vascular elements in wood against cell collapse by mechanically reinforcing its cell walls, lignin assumes the role of an important mechanical reinforcement agent for the entire tree. Lignin is the structural component that provides wood with its unique elastic and strength properties. Lignification thus becomes of consequence not only to the development of a water-conducting system, but also to the tree's need to support a crown many feet above ground level. This need is met by reinforcing cellulosic, tensile-strong fibers with a material that absorbs compressive forces—lignin. In addition to lignin's acknowledged roles as sealant and structural-reinforcing agent, lignification has also been described as a disposal mechanism for metabolic wastes. These ideas have been advanced by Neish[7] as well as Wardrop.[3]

Studies concerning the distribution of lignin by quantity and type throughout cell walls and intercellular spaces have been advanced by Goring and coworkers on the basis of ultraviolet (UV) microscope techniques. Goring, and others,[8-11] have quantitatively determined the distribution of lignin in birch and Douglas fir cell tissues. According to these studies, earlywood secondary walls (Douglas fir) comprise 86% of the total tracheid-tissue volume and contain 71% of the total lignin. Values are higher for latewood secondary walls. By contrast, only 7 to 14% of the total tracheid-tissue volume consists of intercellular middle lamella with lignin concentrations of 56 to 88% (W/W). Similar peak lignin concentrations were observed in secondary tracheid walls of compression wood. Figure 2-1 shows the distribution of lignin across two tracheid cell walls and one middle lamella.

The UV microscopy of ultra-thin wood cross sections proved to be a versatile tool when it revealed that hardwood lignins differ structurally with their location in the tracheid matrix. By comparing UV absorbance values with OCH_3/C_9 values for total isolated lignin, Musha and Goring[12] found that hardwood secondary walls contain lignin with a high syringyl content, and that their syringyl content increases with increasing syringyl:guaiacyl ratio of the total lignin. In contrast, lignin of cell corners and middle lamellae seem to be characterized by a high guaiacyl content, and the syringyl content appears only to rise slowly with an increasing syringyl:guaiacyl ratio of their total lignin. Lignin is concentrated in tangential layers concentric with the cell axis, as pointed out by Stone, and others,[13] following measurements of the dry cell wall dimensions of black spruce tracheids made by using electron microscopy, gas adsorption, and mercury porosity.

In spite of an extensive literature on the biochemical aspects of lignification, the understanding of lignin deposition remains incomplete. For many years it

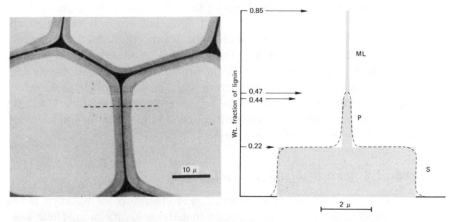

Figure 2-1. Lignin distribution in earlywood tracheids of black spruce. Left: cross section photographed in UV light of wavelength 240 nm (from Fergus, and others[8]) right: corrected distribution diagram (from Sarkanen and Hergert[21]).

was Freudenberg's lignification hypothesis, based on chemical and cytological evidence, that satisfactorily explained all experimental observations. This hypothesis[1–4] locates the initial formation of lignin precursors of the phenylpropane glucoside type (coinferin and analogs) in the region of the cambium. These water-soluble, nonphenolic lignin precursors diffuse centripetally across several cell layers into the differentiating xylem. Derivatization by glucosidation aids both their solubility in aqueous mediae as well as their resistance to premature oxidation and polymerization. Release of the phenylpropanoid aglycone from its glucoside is reportedly accomplished by hydrolytic enzymes (β-glucosidases) located in the wall of the lignifying cell. Following aglycone liberation, phenol oxidation proceeds enzymatically with peroxidases located in the cell walls of differentiating xylem. The exclusive participation of peroxidases in this process has unequivocally been demonstrated by Harkin and Obst.[14] Lignification is completed when the oxidized phenolic lignin precursors have polymerized.

Later it was found that the lignification hypothesis of Freundenberg was unable to explain some anatomical observations that are summarized by Wardrop:[4] (1) the selective lignification of particular wall structures, of layers of the secondary wall in the reaction xylem of angiosperms, and in lignified collenchyma; and (2) the predominance of syringyl-type lignin in the secondary xylem of angiosperms and in the xylem fibers of birch in contrast to the predominance of guaiacyl-type lignin in the primary xylem of angiosperms as well as in xylem vessels of birchwood. In order to explain those experimental observations, Wardrop hypothesizes[4] that lignification, starting with the biosynthesis of lignin precursors, is inherently an intracellular process in which the lignin precursors arise most probably within the lignifying cell and not in adjacent meristems, such as the cambium. Wardrop furthermore finds indications that β-glucosidase is located mainly in the cytoplasm and that oxidation and polymerization of lignin precursors take place within the cell wall. Wardrop's disclaimer of Freudenberg's intercellular lignification hypothesis in favor of an intracellular process enables one to explain differences in lignin structure depending on its original location. It remains a matter of conjecture whether Wardrop's hypothesis of confining the lignification process to the cell as limited by the primary wall is in conflict with the observation that lignification is initiated in the intercellular layers and cell corners of differentiating xylem.[15,16]

Chemical Aspects of Lignin Formation

Once we understand the purpose, function, and emergence of lignin, it is of interest to learn about the chemical aspects underlying lignification. These will be dealt with in three sections: the biosynthesis of precursors, dehydrogenative polymerization, and nondehydrogenative polymerization.

Biosynthesis of Precursors. The biosynthesis of lignin precursors and their dehydrogenative polymerization have been comprehensively covered in several excellent reviews. Among them are those by Freudenberg and Neish,[1] Harkin,[18] Sarkanen,[17] Higuchi,[5] and Adler.[19] A brief summary of the metabolic pathways leading to lignin precursors is given in Figure 2-2. Like most, although not all, aromatic plant constituents, lignin precursors are formed by way of the shikimic acid pathway. Shikimic acid, formed by fusion of phosphoenolpyruvic acid and erythrose-4-phosphate, becomes the major stepping stone of the biosynthesis of the amino acids L-tyrosine and L-phenylalanine, which are formed by reductive amination. Deaminating enzymes (deaminases) subsequently convert the two amino acids into their respective cinnamic acid counterparts. Stepwise hydroxylation by hydroxylases and eventual methylation by O-methyl-transferases transform the cinnamic acids into the three parahydroxy cinnamic acids known to exist as lignin precursors. The extent of hydroxylation and methylation has a significant impact on the structure of the lignin, as it determines whether the lignin will be a guaiacyl lignin or a guaiacyl-syringyl lignin.[20,21] Higuchi, and others,[22-24] carried out extensive studies on the extent of substitution by methoxyl groups of lignin precursors and found that this is based on differences in substrate specificities of O-methyl-transferases as well as on differences in activity levels of ferulic acid hydroxylases, which are lower in gymnosperms than in angiosperms. It thus appears that enzyme specificity and activity control whether a plant is lignified with softwood (guaiacyl) or hardwood (guaiacyl-syringyl) lignin. In general, it can be stated that softwood lignins consist of 85 to 90% guaiacyl aromatic units, whereas hardwood lignins share evenly in guaiacyl and syringyl units. Most biosynthetic pathways have been established by tracer techniques that employ ^{14}C-labeled intermediate substances administered to the tree in a variety of ways. These studies have been reviewed in detail.[1,5,17,25]

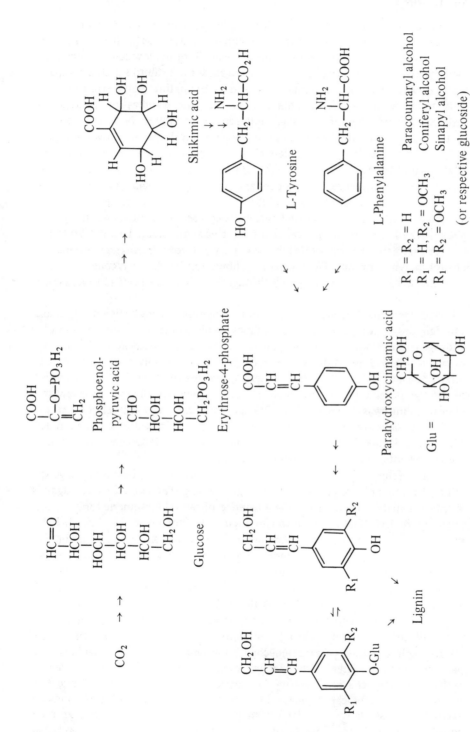

Figure 2-2. Simplified scheme of the metabolic pathway of lignin formation.

43

Parahydroxy cinnamyl alcohols differ greatly from saccharidic precursors with respect to elemental composition, solubility in aqueous mediae, and chemical reactivity towards oxidizing enzyme systems. Both the low solubility and the high reactivity towards oxidants make it crucial for the cell to stabilize the parahydroxy cinnamyl alcohol monomers against premature polymerization, and thus termination of life-sustaining activities. This stabilization is achieved through the formation of glycosides between the phenolic monomers and sugar units. The first structural elucidation of the glycoside coniferin, shown in Figure 2-3, is credited to Tiemann and Mendelsohn;[26] the first detection of a structural relationship between coniferyl alcohol and lignin is credited to Klason.[27] These studies greatly influenced the understanding of lignin chemistry. Coniferin is characterized by an improved solubility in aqueous media. It also has an increased stability against oxidizing systems because the most labile group, the phenolic hydroxy group, is protected by a glycosidic bond. This glycoside can be freely stored and transported by the plant without endangering premature phenol polymerization. The hydrolytic liberation of the aglycone from the glycoside is accomplished by the cell through enzymatic means (β-glucosidase).

Dehydrogenative Polymerization. Contact between phenolic lignin precursors and dehydrogenating enzymes (peroxidases) leads to the initial abstraction of a hydrogen atom from the phenolic precursor. This enzymatic H-atom abstraction, or phenol dehydrogenation, sets in motion the entire polymerization process. First Erdtman[28,29] and then Freudenberg[30] directed the scientific understanding of the lignification process towards the dehydrogenation theory. However, this understanding was dramatically influenced by the ingenious studies at the Heidelberg laboratories under Freudenberg's direction without which much present-day lignin chemistry would be without foundation. These studies have been excellently reviewed by Freudenberg and others.[1,2,7,19,31-34]

The involvement of enzymes during lignification seems to be limited to generating phenoxy radicals at the locale of lignification. The complexity of lignin's chemical structure results from the existence of several mesomeric forms of one phenoxy radical. These are shown in Figure 2-3 for coniferyl alcohol. Polymerization occurs by random coupling of phenoxy radicals in several mesomeric forms. If the existence of five mesomeric forms is assumed for each phenoxy radical, reflecting five sites of greatest electron spin density, the theoretical number of dimeric structures or interunit linkages between monomers is 25. This is illustrated in Table 2-1 using the nomenclature of coupling modes suggested by Sarkanen and Ludwig.[35] The relative frequency with which individual sites will be involved in a phenolic coupling reaction depends on their relative electron-spin densities. Quantum mechanical calculations of several lignin-related phenoxy radicals agree with experience on parahydroxy cinnamyl alcohol radical coupling behavior.[36] Results of such calculations are given in Table 2-2. Several observations can be made from these data: (1) The phenoxy radicals have (in all calculated models) the highest π-electron-spin densities at their phenolic oxygen atoms. This results in a greater probability of forming arylether

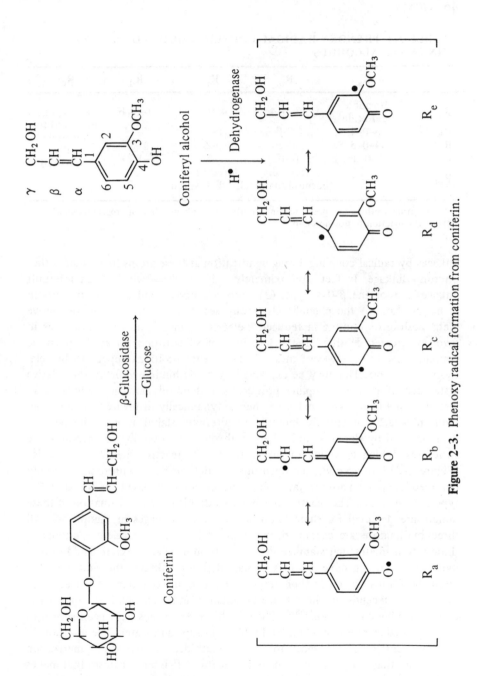

Figure 2-3. Phenoxy radical formation from coniferin.

TABLE 2-1 PHENOXY RADICAL COUPLING MODES OF para-OH CINNAMYL ALCOHOLS

	R_a	R_b	R_c	R_d	R_e
R_a	unstable peroxide	β–0–4	4–0–5	1–0–4[a]	sterically hindered or thermodynamically disfavored
R_b	β–0–4	β–β	β–5	β–1[a]	
R_c	4–0–5	β–5	5–5	1–5[a]	
R_d	1–0–4[a]	β–1[a]	1–5[a]	1–1(?)[a]	
R_e		sterically hindered or thermodynamically disfavored			

[a]Some other options are presented in Figure 2–6; a phenol dienone rearrangement may divert the reaction to position C–6.

linkages by radical coupling during lignification at these atoms than at any other interunit linkage. In fact, approximately half of all softwood lignin interunit linkages are of the β–0–4 type. (2) Alkoxy-, aroxy-, and aryl-substituents in ortho position to the phenolic OH group seem to deactivate their respective meta positions, resulting in reduced π-electron-spin densities at those sites in favor of greater densities at the locations of substitution. In actual phenolic coupling reactions, however, these electron-dense positions appear to be relatively unreactive. This may be explained by steric hindrance or by the relative instability of primary coupling products on thermodynamical grounds, as illustrated in Figure 2–4. That such thermodynamically disfavored dimeric configurations are not entirely immune to alternate stabilization pathways was demonstrated by Connors, and others,[37] through the isolation of catechol-type structures; taking crowding or instability at the reactive sites of R_d and R_e (Figure 2–3) into account, the quantum calculations agree with experimentally observed coupling patterns favoring interunit linkages based on R_b- and R_c-type radical forms. The reaction mechanisms that lead to the formation of these dimers are depicted for three important interunit linkages in Figure 2–5. All three mechanisms are characterized by the existence of transient quinonemethide intermediates that stabilize through intramolecular or external addition of hydroxy-containing moieties. In the case of β–0–4 linkages, this addition may be in the form of water or aliphatic or aromatic hydroxy groups. The frequency of addition depends on the relative abundance and nucleophilicity of the hydroxy carrying compound.[1,38,39] Phenolic hydroxy groups have been shown to add at a higher rate than aliphatic hydroxy groups and water. The addition of hydroxy-containing substances to quinonemethides involves a polymerization mechanism that is not based on dehydrogenation. It is a mechanism that makes possible the incorporation of nonphenolic compounds that have not undergone radicalization. The intimate contact between quinonoid, dimeric, and oligomeric intermediates with hemicellulosic and cellulosic substances in the cell wall make it likely that a certain number of carboxy and primary and secondary hydroxy groups from carbohydrates become covalently bound to the lignin component. These linkages may be of an ether or ester type, as indicated in Figure 2–5.

TABLE 2-2 DISTRIBUTION OF π-ELECTRON-SPIN DENSITIES OF LIGNIN-RELATED PHENOXY RADICALS[a]

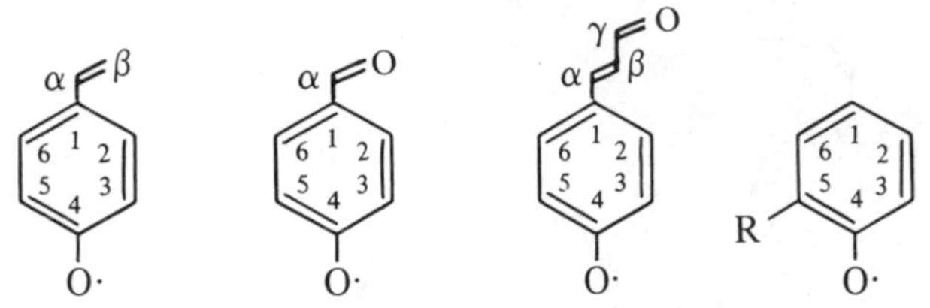

I   II   III   IV R = OCH₃
V R = Aryl
VI R = O—Aryl

Position	I	II	III	IV	V	VI
Phenolic oxygen	.23	.25	.22	.30	.21	.23
1	.20	.22	.19	.19	.17	.18
2	.03	.03	.02	.02	.05	.05
3	.14	.16	.14	.14	.11	.11
4	.05	.04	.04	.07	.05	.06
5	.15	.16	.14	.19	.15	.15
6	.02	.02	.01	.00	.01	.02
α	.00	.02	−.02	—	—	—
β	.18	—	.16	—	—	—
γ	—	—	.01	—	—	—
Carbonyl oxygen	—	.11	.08	—	—	—
Aroxy oxygen	—	—	—	—	—	.04
Methoxy oxygen	—	—	—	.05	—	—
Methoxy carbon	—	—	—	.00	—	—

[a]In the closed shell model with coulomb repulsion integrals according to Mataga-Nishimoto, except for compound IV, which is according to Ohno. (According to ref.[36].)

In contrast to β-0-4 combinations, which lead mostly to guaiacyl-glycerol-β-aryl ether linkages, the combination β-5 leads almost entirely to internal, intramolecular addition of the phenolic hydroxy group of the reaction partner to the qunonemethide. This coupling results in a coumaran-type structure depicted in Figure 2-5. A similar intramolecular stabilization of the quinoneme-thide intermediates is experienced in the cases of β-β combinations leading to pinoresinols, also shown in Figure 2-5.

A certain complication results from high electron-spin densities (Table 2-2) in position 1 of aromatic rings of saturated, nonvinylic phenylpropane structures. Following coupling in position 1, the quinoid intermediate may stabilize in one of two alternate routes that are depicted in Figure 2-6.[40-42] The outcome

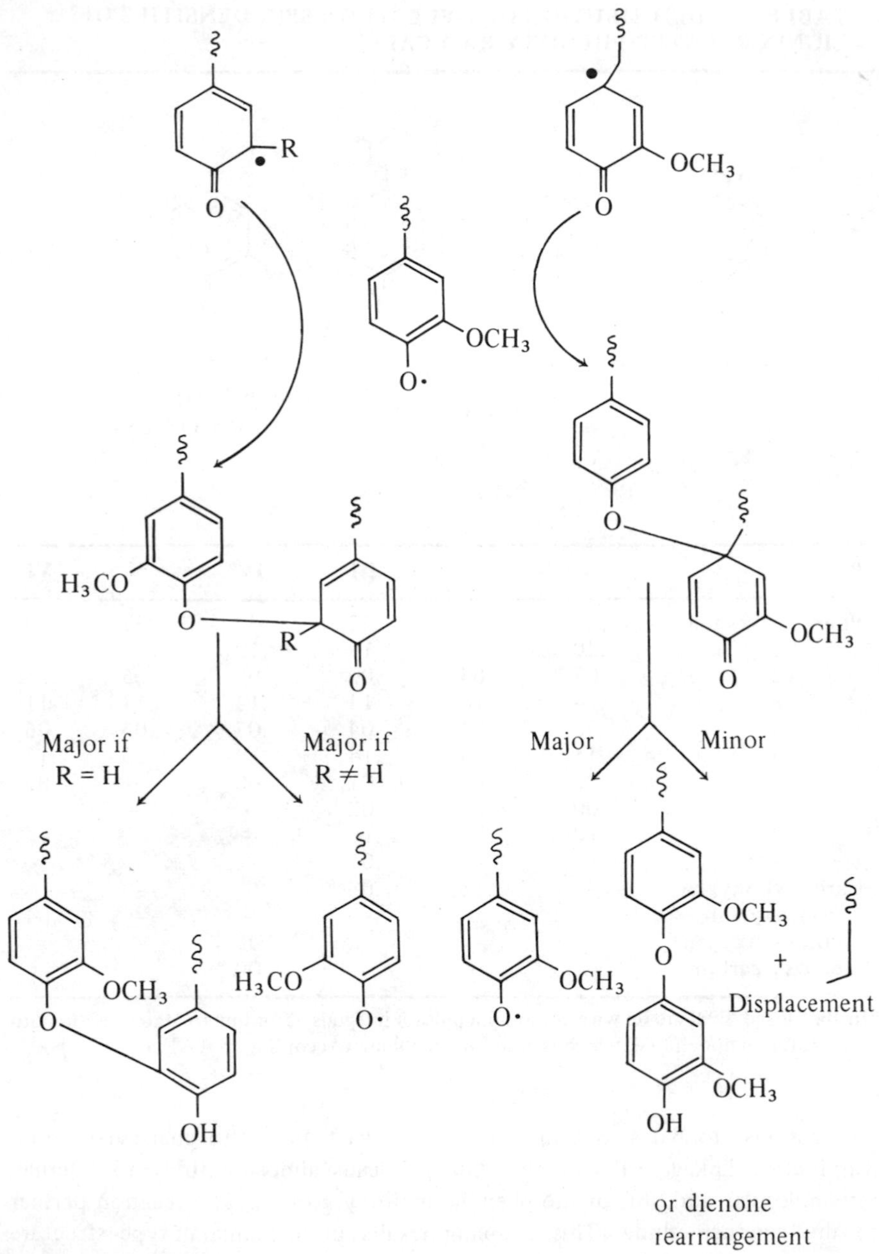

Figure 2–4. Hypothetical dependence of coupling behavior of lignin-related phenoxy radicals on equilibrium constants of primary interunit linkages.

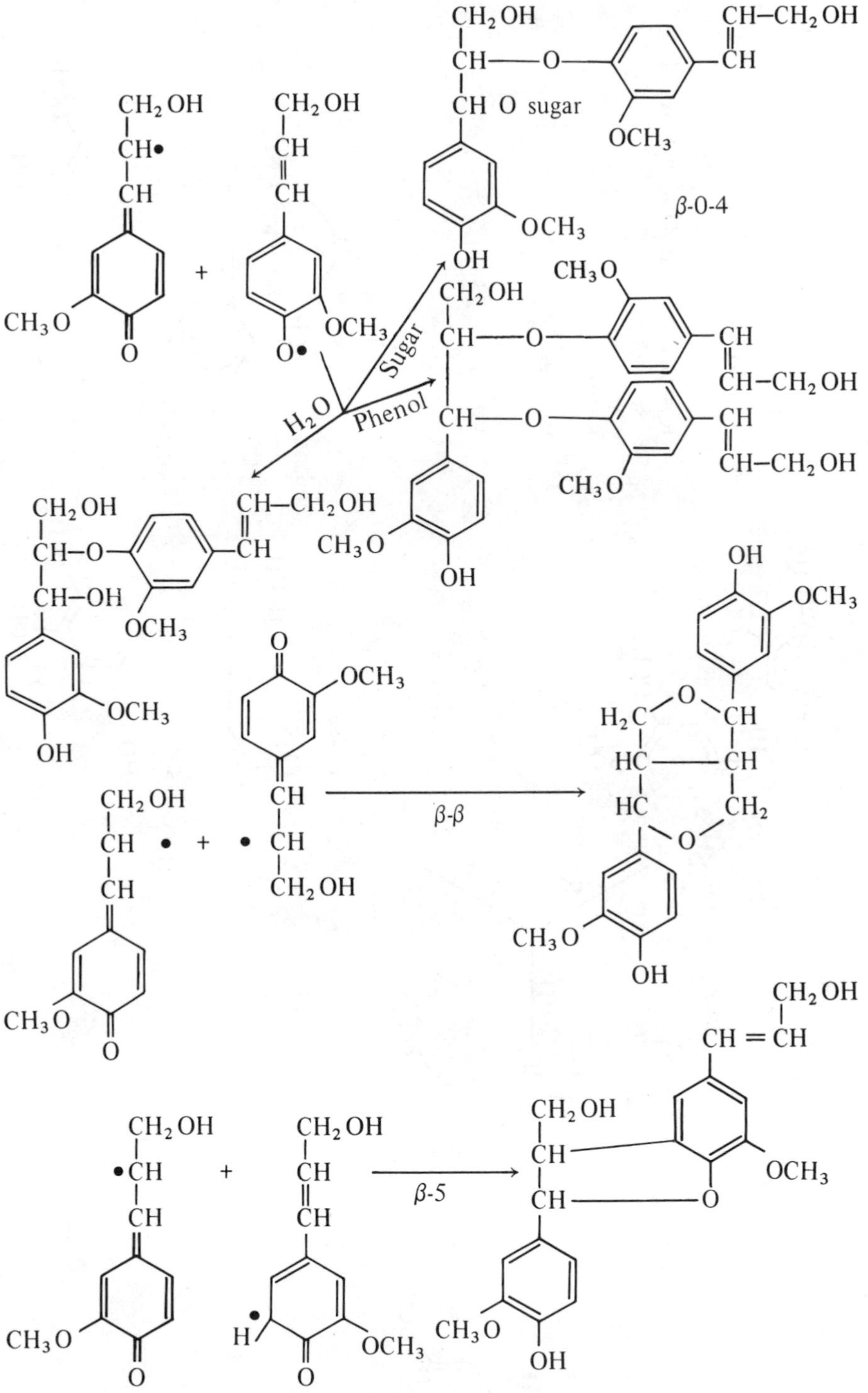

Figure 2-5. Coupling of phenoxy radicals leading to prominent interunit linkages in lignin.

49

50

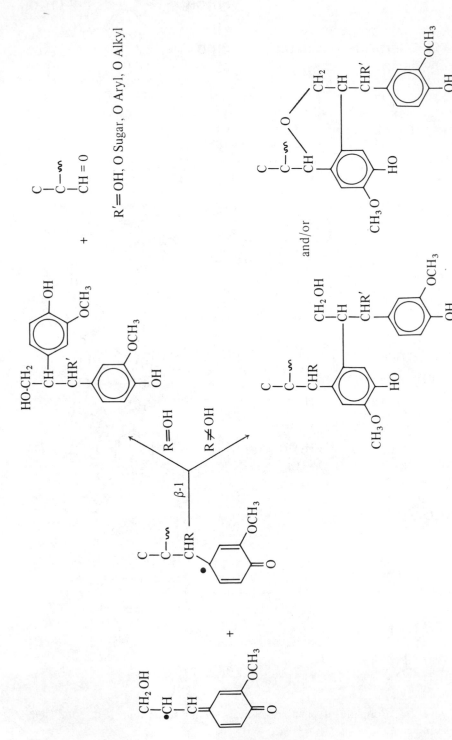

Figure 2-6. Example of a side-chain displacement and phenol dienone rearrangement reaction.

of this stabilization depends on the substituent in a α position of the compound reacting in position 1. If this substituent is a hydroxy group, the side chain will be displaced. If the substituent is something other than a hydroxy group, a phenol-dienone rearrangement will occur avoiding fracture of the C_9 molecule; the result is a β-6 linkage.[43,44] Side-chain displacement will most often lead to a β-1 structure shown in Figure 2-6. With regard to enzymatic dehydrogenation, it is of interest to note that dehydrogenation does not in this case lead to an increase in degree of polymerization, but rather leads to polymer fragmentation. Even in the case of fragmentation, however, it is unlikely that any of the molecule parts are lost from the polymeric system; even displaced side chains remain attached to other molecules. Thus, in summary, dehydrogenation in lignin formation leads to either increased molecular weight or polymer fragmentation; and polymerization exists that is not based on dehydrogenation, but rather on addition to quinonemethide intermediates.[45-47] That alcohol addition to quinonemethides may lead not only to benzylaryl ethers, but also to benzylalkyl ethers, has been concluded from computer simulation studies of lignin formation.[47] Mild hydrogenolysis studies confirmed this hypothesis through the isolation of an α-0-γ linked dimer.[44]

Nondehydrogenative Polymerization Hypotheses. Coniferyl alcohol may form long-lived transient quinonemethides of the type shown in Figure 2-7 under such conditions as flash photolysis.[48,49] The formation of similar extended quinonemethides during lignification may conceivably give rise to a series of yet unknown dialkyl- and alkyl-arylethers with structures indicated in Figure 2-7. The hydroxy-addition behavior of these quinoid substances has been studied extensively by Leary, Thomas, and Hemmingson.[39,50-52]

In addition to the dehydrogenative and additive polymerization mechanisms, it has been proposed that lignification may involve a nondehydrogenative polymerization mechanism based on cationic vinyl polymerization.[53] This type of polymerization and its three possible end products are shown in Figure 2-8.

Structure and Properties of Lignin

The polymerization mechanisms described above refer to reactions with monomeric lignin precursors. The difference in coupling with oligomeric lignin precursors is that there are a smaller number of sites within the phenylpropane molecule available for formation of interunit linkages, as well as a lower solubility and diffusibility of the larger molecules. However, polymer buildup proceeds with the same principal mechanisms as described for monomers. The basic understanding of the coupling mechanisms involved during lignification is credited to Karl Freudenberg, who has done extensive work studying the enzymatic dehydrogenation of coniferyl alcohol *in vitro*.

Chemical Structure and Physical Properties. Continued polymerization of lignin precursors, according to mechanisms outlined in the previous section,

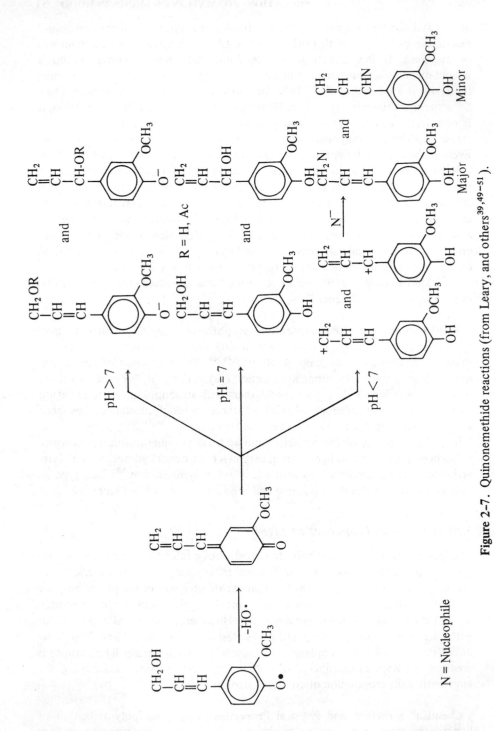

Figure 2-7. Quinonemethide reactions (from Leary, and others[39,49-51]).

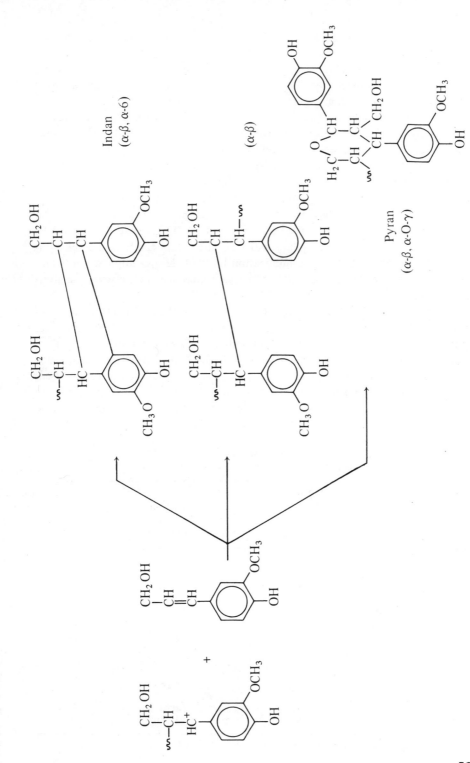

Figure 2–8. Hypothetical vinyl polymerization (from Glasser, Glasser and Nimz[53]).

53

leads to lignin, a polyphenolic, three-dimensional branched network polymer. In designing a configurational structure for lignin, one has to keep in mind that, in the absence of a highly regular, crystalline configuration, a model must be constructed that is based on: (1) proven and/or accepted polymerization mechanisms and (2) all primary experimental observations made on lignin during isolation, analysis, and reactions. Since no structural model available at present completely satisfies all of these constraints, the quality of a structural lignin model has to be judged on how closely it converges on the ideal model. A substantial number of lignin structures have been proposed that are based on the dehydrogenation theory of parahydroxy cinnamyl alcohols.[1,2,19,54-57] These models have been reviewed, modified, and revised many times, yet their concept remains unchanged. The largest structure, a model based on 81 phenylpropane units with a total molecular weight of about 15,000, is shown in Figure 2-9. This model was developed in line with the above modeling criteria by computer simulation.[43,57] Since its publication in 1974, several improvements and revisions have been suggested[47,53,58] to explain more accurately several analytical phenomena.

Common methods of appraising the merits of a lignin model include: (1) listing the types and frequencies of interunit linkages connecting individual phenylpropane units and (2) using Rydholm-type diagrams summarizing qualitatively and quantitatively the substituents in each of the ten critical positions of the phenylpropane building unit. Many other appraisal methods have been used; an intriguing approach using graphs was taken by Faix.[59,60] Table 2-3 shows three appraisals by frequency of interunit linkages for: (1) the lignin structure

TABLE 2-3 FREQUENCY OF INTERUNIT LINKAGES IN LIGNINS CONTAINING 100 PHENYLPROPANE UNITS

	Fig. 2-9 Softwood Model by Glasser-Glasser (ref. 57)	Spruce Lignin by Miksche, et al., (ref. 61)	Beech Lignin by Nimz, (ref. 62)
β-0-4	41	49-51	} 65
α-0-4	5	6-8	
β-5	14	9-15	6
β-1	11	2	15
β-β—all ex. THF[a]	11	2	5.5
—Dibenzyl-THF[a]	—	—	2
5-5	16	9.5	2.3
4-0-5	12	3.5	1.5
α-β	—	—	2.5
β-6	4	4.5-5	—
1-5, 6-5, 1-0-4	5		—
Miscellaneous	—	traces	—
Total	119	85.5-96	99.8

[a]THF stands for tetrahydrofuran structures.

shown in Figure 2-9, (2) similar data proposed by Miksche[61] based on oxidative and acetolytic degradations, and (3) lignin of European beech proposed by Nimz based on depolymerization studies with thioacetic acid.[62] It is obvious that differences exist among the three data sets; however, an evaluation of the significance of these differences is difficult without additional data showing how well each model represents the various analytical and reaction behaviors of lignin. It should be pointed out that the ratio of total number of linkages per total number of units must always be greater than one by the number of interunit linkages giving rise to fracture of the C_9 skeleton plus the number of macrocyclic configurations in the model structure. Thus, the average number of linkages per C_9 should exceed 1.0 in spite of the fact that the average degree of dehydrogenation per C_9 unit falls just short of 2.0.[1,30-34] The interrelationship between structural and analytical information has been demonstrated by computer simulation techniques.[47,53,58] The structural evaluation of lignin is shown in Table 2-4 using the Rydholm diagram to appraise comprehensively the structure in terms of the types and frequencies of interunit linkages as well as the concentration of the functional groups. Some vital statistics about the structure can be quickly detected, such as the number of phenolic hydroxy groups, end groups, carbonyl groups, total ethers, and condensed versus uncondensed units. Other important structural features are, however, totally ignored by this type of structural evaluation, including the concentration of free radicals,[63] such quinoid systems as shown by Harkin[56] and Miksche,[61] and the number of aryl glycerol units.[64]

In sharp contrast to the relative abundance of studies concerning the biosynthesis and chemical structure of lignin, studies concerning the physical properties of lignin have remained few, although an excellent review has been prepared by Goring.[65] One of the critical factors influencing the physical properties is molecular weight. Although much work has been done concerning molecular-weight distribution, lignin may, in fact, exist as one single molecule in its native environment. Since lignin is formed by enzymatic dehydrogenation, followed presumably by random oxidative coupling of monomeric and oligomeric phenols, and since lignin is never completely deprived of dehydrogenatable phenolic hydroxyl groups, or even of free phenoxy radicals, the lignin structure may never cease to grow. In fact, wood may be described as a multicomponent, cross-linked polymer system in which carbohydrates and lignin are intimately associated through hydrogen and ether and ester covalent bonds.[66] The actual molecular size of lignin may be irrelevant to pulping processes as lignin will in any event have to be degraded and modified in order to be dissolved. No matter what one does to lignin, whether it involves photoirradiation or mechanical, chemical, or biological action, the intimate mixture of polymeric wood components will suffer alteration. These changes affect physical appearance, mechanical properties, and chemical reactivities alike. Molecular weights of isolated lignins, such as Bjorkman lignin, kraft lignin, or lignin sulfonates, have been determined; they range anywhere between 2000 to more than 1 million.[65]

A mathematical relationship between the hydrophilic and solution properties

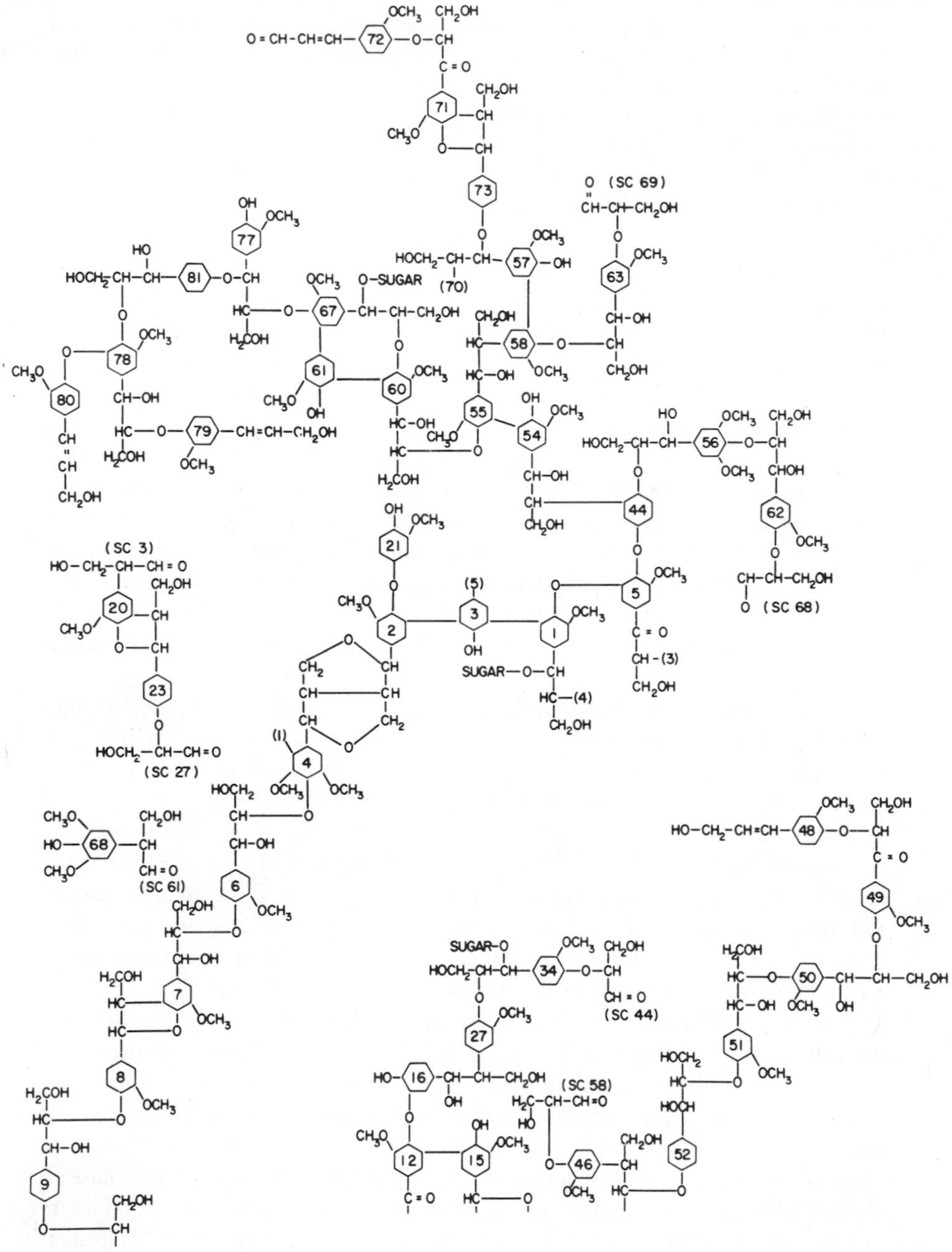

Figure 2-9. Softwood lignin model structure (from Glasser and Glasser[57]).

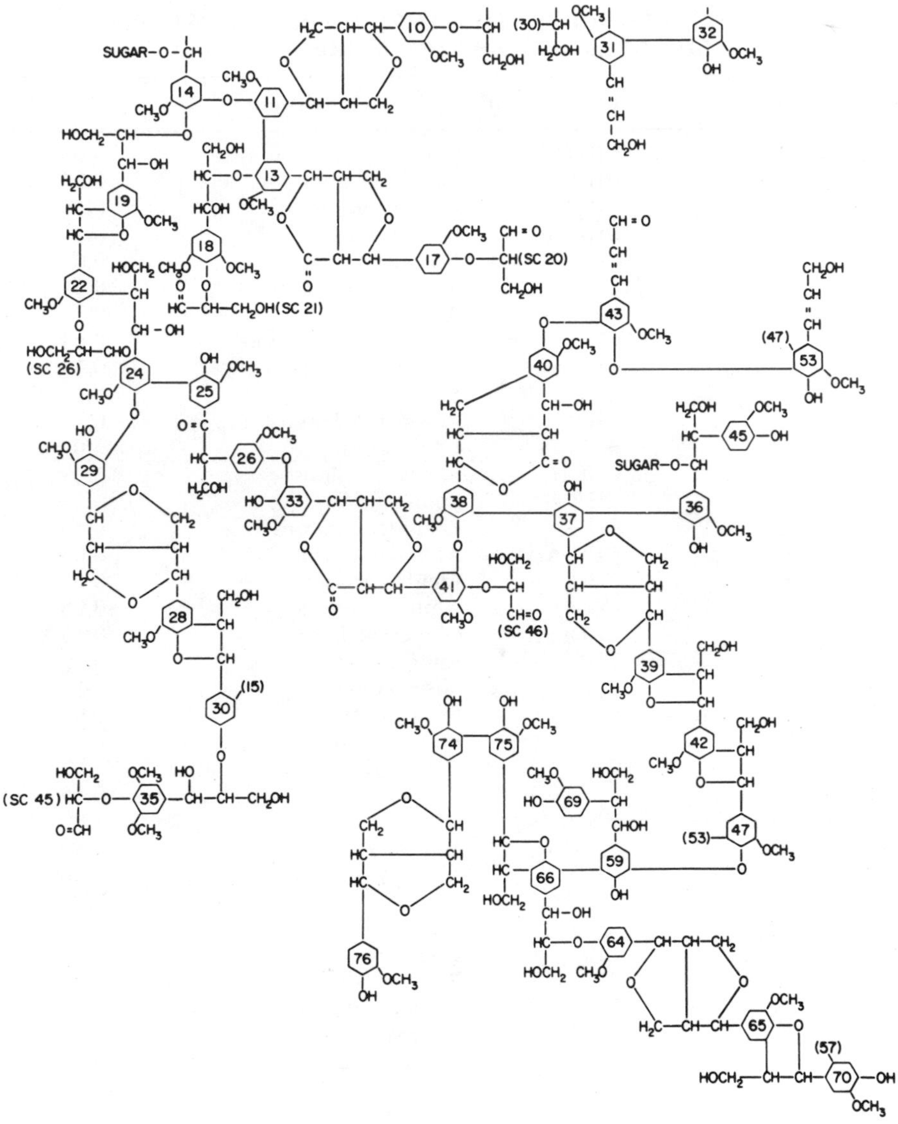

Figure 2-9. Continued

57

TABLE 2–4 LIGNIN-MODEL-STRUCTURE APPRAISAL IN FORM OF A RYDHOLM DIAGRAM (FIG 2–9 STRUCTURE)

Position	Configuration	Description	Content
C-Gamma	CH_2OH	Coniferyl alcohol	0.06
	CHO	Coniferyl aldehyde	0.02
	CH_2OH	Primary hydroxyls	0.69
	CH_2OR	Pinoresinols	0.17
	CO	Lactones	0.04
	CH_2	Gamma-6 condensation	0.01
C-Beta	CH	Beta ethers	0.41
	CH	Beta-6 condensation	0.04
	CH	Phenyl coumarans	0.14
	CH	Pinoresinols	0.22
	CH	Beta-1 condensation	0.11
	CHD	End groups	0.09
C-Alpha	CHOH	Benzyl alcohols	0.29
	CH	Pinoresinols	0.21
	CH	Phenylcoumarans	0.10
	CHO-Aryl	Alpha-aryl ether	0.05
	CHO-Sugar	Lignin-carbohydrate bond	0.06
	CO	Alpha-carbonyl	0.06
	CHO	Glyceraldehydes	0.15
	CHD	End groups	0.09
		Miscellaneous	—
1 Aromatic ring	C	5–1 condensation	0.01
	C	Beta-1 condensation	0.11
	C	Biphenylether condensation	0.02
	C	Side chain	0.85
2 Aromatic ring	CHD		1.00
3 Aromatic ring	$C\text{-}OCH_3$	Methoxy groups	0.85
	CHD		0.10
	C		0.05
4 Aromatic ring	C		1.00
5 Aromatic ring	CHD	Uncondensed units	0.38
	C	Biphenyls	0.28
	C	Biphenylethers	0.12
	C	Coumarans	0.14
	C	5–1 condensation	0.01
	C	5–6 condensation	0.01
	$C\text{-}OCH_3$	Syringyl units	0.06
6 Aromatic ring	CHD	Uncondensed units	0.95
	C	6–5 condensation	0.01
	C	Beta-6 condensation	0.04
phenolic hydroxyl	O	Biphenylether condensation	0.02
	O	Beta-aryl ethers	0.41
	CHO-Aryl	Alpha-aryl ether	0.05
	O	Biphenyl ethers	0.12
	O	Phenylcoumaran	0.10
	OH	Phenolic hydroxyls	0.30

of lignin and its molecular weight, its degree of condensation, and its content of hydrophilic functional groups was established by Kudo and Kondo.[67] The wetting and interfacial properties of wood, however were shown to depend more on surface topology than on the physical and chemical properties of its three major components: lignin, cellulose, and hemicelluloses.[68]

The advent of thermomechanical pulping resulted in an increased interest in the thermal behavior of lignin.[69-72] Hatakeyama and coworkers[70] studied the glass transition temperatures of softwood lignins and found that methoxyl groups caused a depression in the softening temperature, whereas cross-links as well as hydroxyl groups had the opposite effect. An increase in the molecular weight of kraft lignins from softwood and hardwood was found to be responsible for the increase in glass transition temperatures from 109 to 124°C. An almost linear relationship was found between the T_g and the molecular weight of the respective lignin fractions.[71-72] The primary relaxation of conifer lignin at above 160°C was attributed to glass transition temperature in a study of the mechanical properties of softwood lignin by tortional braid analysis.[73] The mechanical properties of lignin (Young's and shear moduli) were found to increase linearly as the moisture content of lignin decreased from 12 to 3.6%.[74,75]

Lignin-Carbohydrate Bonds. The lignin and carbohydrate components exist in intimate association in wood. This closeness is an inherent feature that results from biosynthesis and can be illustrated by microscopy.[76] In view of this close association between the aromatic and saccharidic wood components, it is not surprising that bonds exist between the lignin and carbohydrate components. In line with lignin's function as a mechanical reinforcement agent, the observation that lignin covalently crosslinks polysaccharide chains[77] may be interpreted in terms of an ingenious structural design in which cross-links result in a ladder-type structure with unique physical properties (e.g., reduced slippage). Even though the precise nature, by type and amount, of lignin-carbohydrate linkages is far from being completely understood, experimental observations strongly suggest that there exist at least two major forces, one based on absorption and chemosorption and the other on covalent bonds.

Physical absorption between lignin and carbohydrates of the type provided by hydrogen bonds and van der Waals forces has been reported for wood[78] and alkaline pulps.[79] The existence of chemosorption, formed by the development of donor-acceptor complexes between lignin and carbohydrates, has been reported for alkaline pulps.[80]

A review of lignin-to-carbohydrate bonds by Grushnikov and Shorygina[81] concluded that ester-, ether-, phenyl-glycoside, and acetal bonds are all possible covalent linkages. The existence of three types of lignin-to-carbohydrate bridges is indicated in studies by Efendilva and Shorygina,[82] Koshijima, and others,[83] and Eriksson and Lindgren.[84] Almost identical fractionation studies of Bjorkman lignin-carbohydrate complexes resulted in the separation of a galacto- or acetyl-glucomannan fraction in 50 to 70% yield, an acidic xylan component in 23 to

26% yield, and a lignin-rich arabino-(4-0-methylglucorono) xylan fraction in 2 to 4.5% yield.[82,83] Most of the lignin-to-carbohydrate linkages were of glycosidic nature.[83] In overall agreement with these results, Eriksson and Lindgren[84] found that lignin is linked to arabino side chains of xylans, to galacto side chains of galactoglucomannans, and that there is an additional linkage between lignin and xylan. That lignin is linked to carbohydrates by other bonds in addition to glycosidic bonds has been reported.[85-89] The existence of such additional linkages seems to be indicated from model compound studies by Freudenberg, and others,[90,91] in which it was demonstrated that quinonemethide intermediates of lignin formation may add hydroxy groups of carbohydrates. It was also pointed out that the rate of addition of carboxylic hydroxy groups is faster than aliphatic ones.[1] These findings seem to suggest the formation of ether and ester bonds between lignin and carbohydrates during lignification.

The nature of the bonds between lignin and carbohydrates is highly complex. The stability of such bonds and their resistance to acid hydrolysis not only depends on the type of linkage involved, but also on the types of sugars associated with the linkage,[92] and the chemical structure of the lignin unit attached to the sugar.[93-95]

Lignin Heterogeneity. The use of the generic term, lignin, represents a simplification that does not reflect the many variations that exist in lignin structure. The following is a brief summary of the differences between lignins.

Lignin samples isolated from the sapwood as well as heartwood portions of softwood[96] and tropical hardwood species[97] have shown: (1) that heartwood lignin has a higher molecular weight, (2) that it is more heavily condensed,[96] and (3) that angiosperm heartwood contains a greater percentage of syringyl units than its sapwood counterpart. Compression wood lignin contains more unmethoxylated parahydroxy phenylpropane structures, has fewer conjugated carbonyl groups, and is richer in condensed aromatic structures.[98,99] Lignin from the roots of a hardwood species (*Fagus sylvatica*)[100] was found to have more guaiacyl units than its counterpart from the xylem. The lignin of bark[101,102] has a greater frequency of unmethoxylated parahydroxy phenylpropane units. The polyphenolic constituents of leaves, needles, and some conifer fruit cones also satisfy the criteria for lignin.[103] Lignin structure was also found to vary with regard to the location in the stem;[104] lignin in the older parts of the stem of the reed *Arundo donax L.* was richer in syringyl units than the total lignin. The lignin from poplar callus tissue had a lower ratio of syringyl-to-guaiacyl units and a lower frequency of biphenyl linkages than normal lignin.[105-107]

Temperature during formation was found not to influence lignin properties.[108] However, other external, stress-inducing conditions may exert a significant impact on lignification. Thus it was found[109,110] that attempted fungal penetration of reed canary grass resulted in increased activities of the enzymes involved in lignin biosynthesis as well as in higher lignin content within 18 hr of inoculation with the fungus. A relationship was thus established between externally induced, abnormal levels of peroxidase activity and, at least, degree of lignification. That variations in peroxidase activities may also lead to differences in lignin structure was demonstrated with peroxidases isolated from different parts of the plant.[111] The influence of the various peroxidases differed markedly in their capacity to catalyze bond formation between carbohydrates and ferulic acid and its condensation products.[111]

The lignins associated with milled-wood lignin [MWL], the lignin-carbohydrate complex, and the wood residue were found to differ in having a heterogeneous distribution of condensed-type aromatic structures.[112,113] Angiosperm wood has guaiacyl lignin concentrated in the vessels and guaiacyl-syringyl lignin in the fibers and ray cells.[114] Species variations between softwood lignins are relatively negligible[115,116] in contrast to findings with hardwoods.[100,117] In the latter, major differences relate to the ratio of syringyl to guaiacyl units, a ratio that was found to have a marked and directly proportional influence on the rate of kraft delignification.[118]

No description of lignin structure can be complete without mentioning that the concept of random coupling of lignin precursors leading to a disordered polymer structure is not shared by all lignin chemists. Some scientists believe lignin may have a very ordered nature, similar to cellulose and proteins. The following constitutes a brief discussion of three such unconventional concepts of lignin structure.

Forss and others,[119-123] using results obtained with the sulfonation of spruce wood and the fractionation of lignin sulfonates over ion-exchange resins and dextrans (Sephadex), developed the concept of lignin comprising approximately 80% of hololignin with the remainder being termed hemilignin. The lignin portion was found to represent an agglomerate of repeating units, each having a degree of polymerization of 18. Lignin heterogeneity in terms of two distinct parts was confirmed by Meshitsuka and Nakano on the basis of its behavior towards hydrolysis.[124] Wayman and Obiaga,[125] using kinetic studies on the polymerization of coniferyl alcohol with horseradish peroxidase, produced products that had a distinct bimodal molecular-weight distribution. When a similar bimodal molecular-weight distribution was observed for two softwood lignins subjected to alkaline degradation, it was concluded that lignin is an assembly of subassemblies (modules) with a degree of polymerization (DP) of approximately 20. Bolker and Brenner,[126] based on the molecular weights of successive lignin fractions obtained from the acetolytic and sulfitolytic wood delignification and on calculations according to the Flory-Stockmayer theory, concluded that the primary lignin chains have a DP of 18 and an average of five cross-links per primary chain.

In view of experimental results that seem to leave little doubt about lignin heterogeneity, one must question the need to postulate structural regularity. Even though it is conceivable that enzymes exert more control over the process of lignification than simply generating phenoxy radicals by dehydrogenation, available experimental data and conceptual considerations do not make it necessary to postulate structural orderedness.

Delignification: Separation from Woody Tissues and Fibers

Inasmuch as a plant deposits lignin with the object of reinforcing tensile-strong fibers with a rigidity-building binder, lignin removal, or delignification, can be regarded as fiber-delamination. Nature has designed a superb laminant in lignin, which penetrates fibers in such a fashion that delamination is very difficult. Delamination is made difficult by the gigantic molecular size of lignin and by the existence of covalent bonds between the binder and the carbohydrate components of the fiber. On the other hand, delignification is facilitated by the magnitude of differences in chemical properties between the lignin and the other fiber components or, in chemical terms, between the phenolic and the saccharidic wood component. Whereas cellulose is a linear molecule of certain crystallinity and hydrophilicity, lignin is a polyaromatic three-dimensionally branched network polymer. However, if the composition of the average phenylpropane unit in a lignin macromolecule is examined, one finds that about 1/3 of the basic lignin unit resembles a carbohydrate; the C_3 side chain of lignin's phenylpropane unit comes close to being a triose, as is depicted in Figure 2-10.

In viewing delignification in terms of separating two structural components on the basis of chemical differences, a compromise is needed. Complete separation of the polysaccharidic from the polyphenolic component without simultaneous alteration of the chemical structure of lignin and/or degradation of carbohydrates is not possible. In fact there may be a direct relationship between the yield in which lignin is recovered and the degree to which the chemical structure of lignin has been changed and/or the carbohydrates have been degraded.

To design an optimal delignification process for wood in aqueous media for completeness of separating lignin from carbohydrates, one would attempt to increase the saccharidic and/or hydrophilic characteristics of lignin, while simultaneously degrading its molecular dimensions. Present industrial delignification processes achieve this goal only partly as illustrated in Figure 2-10. In fact, commercial pulping processes appear less than optimal by these standards; however, chemical recovery systems, side reactions detrimental to pulp yields and properties, and overall economic considerations seem to favor alkaline pulping processes.

Considering the problems associated with the separation of lignin, it is not surprising that compromises exist in lignin isolation procedures, which range all the way from recovering only minimal amounts of lignin in a so-called native state to removing lignin from woody tissues in large quantities, but changed to a much greater degree. In commercial pulping processes the concern is not so much with the extent to which lignin structure is altered, but rather with the yield in which cellulosic fibers can be recovered.

The following section deals with the various choices for delignification. First, mild preparative and analytical isolation procedures will be described, followed by the commercial pulping processes.

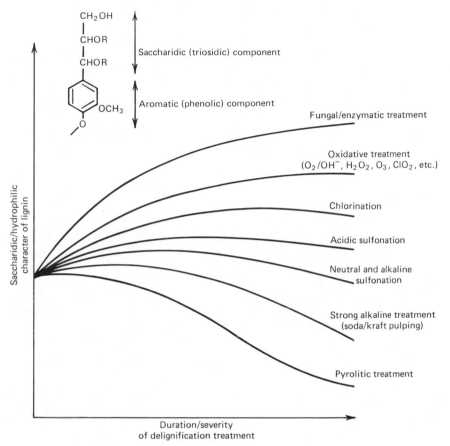

Figure 2-10. The conceptual influence of delignification treatments on the saccharidic/hydrophilic character of lignin.

Laboratory Separations

The isolation of lignin in the laboratory can be accomplished in two principal ways, and these have been the subject of an extensive review.[127] They are: (1) dissolution of carbohydrates with lignin recovered as residue and (2) dissolution of lignin, leaving carbohydrates as residue. In the latter case lignin is recovered by solvent evaporation or by fractional precipitation. The first method, the solubilization of carbohydrates by solvolysis or destructive hydrolysis, employs strong mineral acids in most instances.

The most commonly used lignin-isolation method, based on carbohydrate hydrolysis, employs 72% sulfuric acid in its principal step. This method (named after the Swedish wood chemist, Klason) is the basis of most lignin-determination procedures.[128] Its usefulness is limited to quantitative determinations of lignin content because structural integrity of the lignin is destroyed. When carbohydrate hydrolysis of wood is accomplished by hydrochloric instead of sulfuric acid, a lignin preparation is obtained that is known as Willstätter lignin. The procedure involves the use of supersaturated HC1 at low temperatures for several hours. Some hydrochloric acid lignin preparations are said to be relatively unchanged and fairly representative of the original lignin in wood.[127] Another method, in common use years ago, is the Cuoxam lignin-isolation procedure developed by Freudenberg, and others.[129] This utilizes the solubility of cellulose in cuprammonium hydroxide (Schweitzer's reagent) following cleavage of weak lignin-to-carbohydrate bonds and partial carbohydrate hydrolysis under mild acidic conditions (alternating treatments with 1% sulfuric acid). This method has fallen into disfavor.

The method of lignin isolation involving the dissolution of lignin, leaving carbohydrates and some residual, insoluble lignin fractions as solid residue, has the advantage that the lignin has escaped drastic structural changes during isolation. However, the yields of lignin are generally low, in the range of a few percent to about 1/4 or 1/3 of the entire lignin content. When it was taken for granted that lignin is a homogeneous substance formed by random oxidative coupling reactions of phenols, the yield of lignin recovered was considered to be irrelevant. It was assumed that the isolated lignin would be of essentially the same structural composition as the lignin that could not be solubilized; it only happened to be less accessible to solvents. However, this assumption has become shaky in the light of studies indicating differences between lignins in different locations. Nevertheless, most analytical work on the structure and chemical reactivity of lignin continues to be carried out with samples that have been separated from wood in considerably less than quantitative yields by mild methods involving solvent extraction following minimal pretreatments. Although there are many such isolation methods only a few are in common use.

A problem common to all procedures based on lignin solvolysis is that a compromise must be found between isolating lignin completely, yet structurally altered, and isolating more or less representative fractions of unmodified lignin. All methods of this type focus on cleaving the covalent bonds between lignin and carbohydrates by mild hydrolytic means and dissolving lignin in an appropriate solvent.

One of the mildest procedures was developed by Bjorkman, who separated lignin from wood meal by solvent extraction with aqueous dioxane after ball milling in an inert toluene suspension.[130,131] When this procedure is followed by an extraction with dimethyl formamide, dimethyl sulfoxide, or aqueous acetic acid, an additional lignin-carbohydrate complex (LCC) can be separated from the milled wood.[132] Modifications of the Bjorkman milling procedure have been proposed[133,134] and reviewed.[127] The success of the milling procedure is based on homolytic cleavage of lignin-to-carbohydrate bonds by mechanical force. Free radicals formed during this operation[135] may recombine with other lignin radicals and thereby increase the molecular weight or degree of cross-linking of the lignin; however they have also been found to combine with nitrous oxides[136] or with toluene.[137] Extensive studies have revealed the influence of milling conditions as well as extraction and purification conditions on the resulting MWL structure and properties.[127,138] Thus it was shown with eucalyptus wood that extended milling increases the resulting MWL solubility in alkali, apparently due to a liberation of free phenols.[139] The resulting MWL preparations are reported to contain upwards of 5% carbohydrates, which may be reduced to as little as 0.05% by solvent extraction and/or precipitation.[140] The mechanical action of ball milling may be replaced by sonication with ultrasound. However, some demethylation, especially of syringyl groups, may be encountered in such cases.[141,142]

The solvolytic liberation of lignin from ball-milled wood may be improved by incubating the substrate with cellulolytic enzymes prior to solvent extraction.[143-146] The improved accessibility of lignin to organic solvents is most likely the primary reason for the higher yields obtained. The chemical structure of these lignins is similar, but not identical, to MWL preparations.[145]

Ball milling becomes unnecessary if the lignin-to-carbohydrate bonds are cleaved by mild acid hydrolysis in the presence of a lignin-solubilizing solvent, such as dioxane or methanol.[147] The resulting lignin preparations, known as dioxane lignin, acidolysis lignin, and Pepper lignin,[148] are recovered in yields of 40 to 100% of the Klason-lignin content[148,149] depending on the wood species and hydrolysis time. These lignin preparations appear to be structurally modified as compared to MWL,[147,148] even though they seem to constitute a reasonably good representation of intact protolignin.[150,151]

Structural changes introduced in dioxane lignin during the isolation with anhydrous dioxane can apparently be minimized if the extraction is carried out with a mixture of benzyl ethyl ether and anhydrous dioxane in the presence of trace amounts of HC1.[152-155] However, lower yields (25% of Klason lignin) and a certain amount of transetherification are encountered with this procedure.

The isolation of nearly complete lignin portions from wood and pulp has been achieved in the form of a thioglycolic acid derivative.[156] The derivatization is based on the reaction of the benzyl alcohol groups with thioglycolic acid under acidic conditions.[157] The deprivation of benzyl alcohol functions in pulp lignins is overcome by regenerating benzyl alcohol groups by hydroxymethylation with formaldehyde in the presence of alkali. Thus, lignin is isolated in excellent yields following this selective and controlled derivatization under mild conditions.

It should be reiterated that all lignin-isolation procedures require a compromise between yields and structural alteration.

Commercial Separations

The commercial separation of lignin and carbohydrates from woody tissues to produce pulp fibers for papermaking is discussed in detail in the chapters on pulping and bleaching. In this chapter emphasis will be given to those considerations that relate to the role of lignin in pulping.

Large-scale delignification cannot easily be carried out in organic solvents even though they may be well suited to draw advantage from the structural differences between carbohydrates and lignin. The recovery of millions of gallons of organic solvents would be an almost insurmountable obstacle. By limiting delignification in commercial processes to aqueous systems, conditions must be created that induce lignin to become soluble in water. Since only minute quantities of lignin can be solubilized in an aqueous environment at pH 7, commercial delignification processes mostly involve the use of either acids or alkalis. In the following discussion, a description is given of delignification methods using acidic, alkaline, neutral and nonaqueous, and oxidative systems. Finally, some specialty lignin-isolation procedures will be mentioned.

Before entering into a discussion on acidic and alkaline commercial pulping processes, it is worthwhile to look at the pros and cons of both systems and to compare them. In acidic aqueous systems the acid degrades the carbohydrates through hydrolytic cleavage of glycosidic bonds, resulting in a decreased DP and a decrease in pulp strength. Alkaline systems also degrade the carbohydrates, but hydrolytic degradation seems to be overshadowed by an end-wise depolymerization reaction known as peeling. As a consequence, alkaline pulps generally have higher DP's and are characterized by higher pulp strength than acid (sulfite) pulps. Complete delignification by acids or alkaline solutions would result in poor pulp properties and very low yields. The reason for this is twofold: (1) the increasing resistance of the residual lignins to dissolution in aqueous media and (2) the increased accessibility and vulnerability of the carbohydrate component. Therefore, technical delignification is always interrupted with a few percent of residual lignin remaining in the fibers. This is, if need be, removed by bleaching. In contrast to acid (sulfite) pulps, residual lignin of alkaline pulps is dark and somewhat more difficult to remove by bleaching. Overall, commercial acidic as well as alkaline pulping procedures produce pulps in approximately 45 to 55% yield with the alkaline pulps having a strength advantage.

To make lignin soluble in aqueous solutions, one of the following approaches can be taken: (1) its hydrophilic properties can be increased by raising the number of aliphatic and/or aromatic hydroxy or carboxy groups; (2) its molecular size can be degraded so that the smaller fragments become soluble in water; or (3) solubilizing hydrophilic substituents can be attached to the lignin macromolecule causing it to dissolve in water as a derivative.

Acidic Systems. Strong mineral acids, such as hydrochloric or sulfuric acid, do not cause extensive lignin solubilization. However, they do affect the benzyl-carbon position of the basic phenylpropane unit, which is frequently occupied by either hydroxy or ether groups. Either substituent is acid sensitive and may, under appropriate conditions, form a carbonium ion, as illustrated in Figure 2-11 and discussed in detail later in this chapter. This benzyl-carbonium ion has an affinity for any nucleophile. Nucleophiles may be present in the

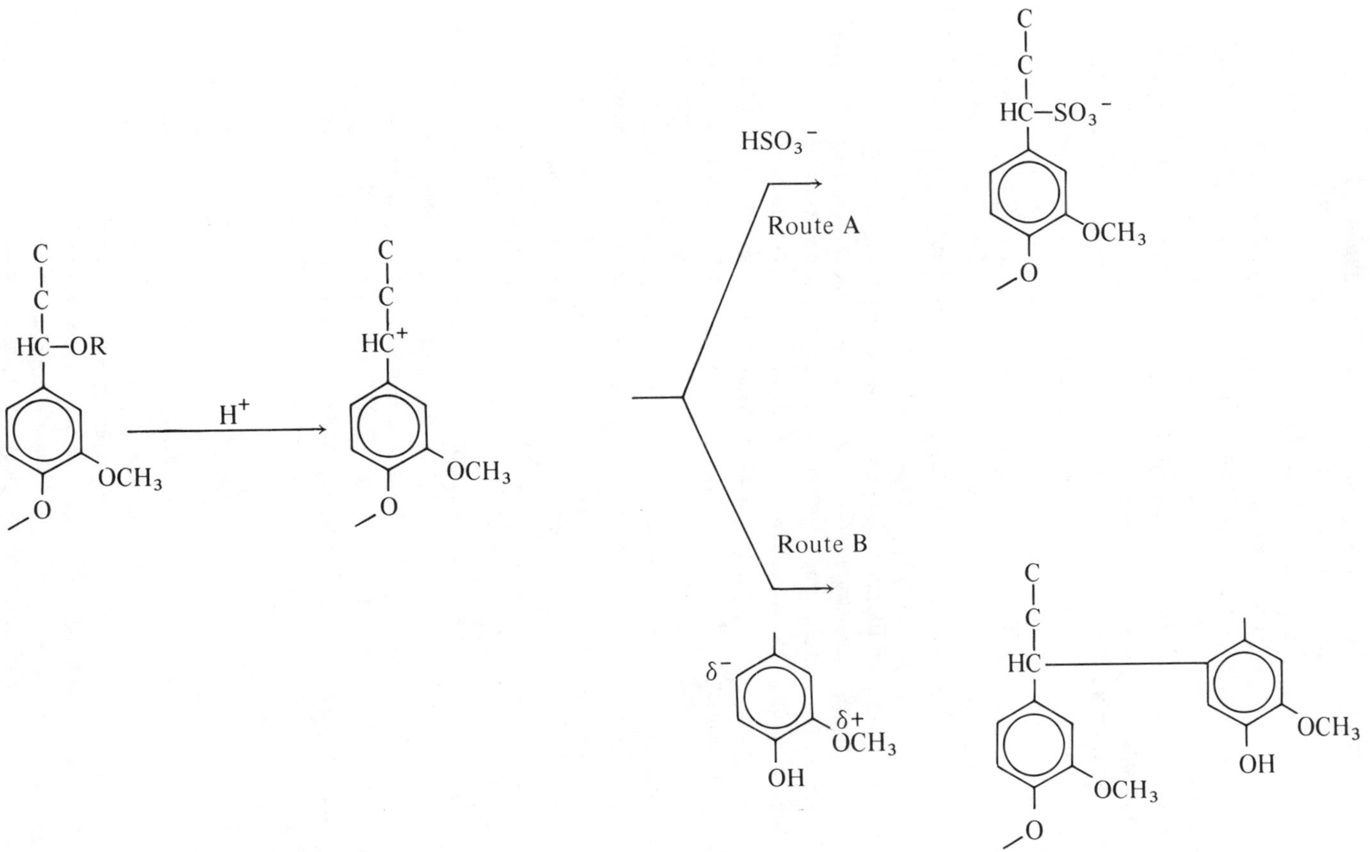

Figure 2-11. Acid-catalyzed reactions of benzyl alcohols with nucleophiles.

aqueous system in the form of phenolic guaiacyl or similar substituents from the wood itself (Route B, Figure 2-11), or in the form of externally supplied nucleophiles such as bisulfite ions (Route A, Figure 2-11). If the first type of nucleophile reacts with a benzylium ion, the solubility of the resulting lignin in water is actually reduced. This results in an increase in molecular weight, the creation of a new carbon-to-carbon bond, an increased C/O ratio, and the reduction of hydrophilicity through the loss of a hydroxy group. However, in the second case, in which a bisulfite ion is attached to the benzylium carbon atom, lignin solubility in aqueous systems greatly increases. The increase in hydrophilicity is a consequence of sulfonate formation and may also be attributed to a certain reduction in molecular size. The contribution to lignin hygroscopicity was found to be ten times more effective with sulfonic acid than with hydroxy groups.[67] Thus it can be concluded that acidic delignification systems operate on the basis of attaching nucleophilic solubilizing substitutents to lignin's intermediary carbonium ions.

Since combinations of sulfur dioxide and water generate nucleophiles, which usually compete successfully with the aromatic constitutents of the lignin molecule, most acidic pulping systems use the nucleophilic action of sulfur dioxide in water. However, other nucleophilic substances may also increase the hydrophilicity of lignin, as was demonstrated when wood was pulped with thioglycolic acid.[158,159] Delignification of wood with various sulfite and bisulfite-containing aqueous solutions has resulted in the development of a substantial number of industrially accepted sulfite pulping processes. The subject of lignin sulfonation has been reviewed by Glennie,[160] Gellerstedt and Gierer,[161] and Gellerstedt.[162] Lignin solubilization by sulfonation at levels less than pH 7 seems to be influenced by degree of sulfonation and condensation, molecular size, and such topochemical factors as pore size of wood cell walls, among others.[160] A correlation between the pH of the cook and the degree of sulfonation and condensation of the resulting lignin sulfonate has been established.[163] Since lignin sulfonic acids constitute a highly polydispersed macromolecular mixture, it has been suspected that the large molecules may have difficulty diffusing through the cell-wall matrix of the fibers,[164] thereby limiting delignification. However, this appears to be in conflict with results obtained by Goring, and others,[165] who found that diffusion does not control the rate of delignification.

Studies concerning the structure and properties of lignin sulfonates have occupied lignin scientists for a great number of years. These studies have contributed significantly to an understanding of lignin and of chemical pulping processes. In addition, these studies have helped lignin sulfonates to become important raw materials for commercial products, ranging from vanillin to adhesives and dispersants. A special mention must be made in this context to the work of J. L. McCarthy. See the sections on "Lignin Sulfonation and Dissolution" and "Recovery of By-Products from the Sulfite Process" in Chapter 4.

Alkaline Systems. The worldwide adoption of alkaline delignification for the manufacture of pulp has spurred much interest among scientists in elucidating the reaction mechanisms underlying this technology. This field has been covered in comprehensive reviews.[166-171]

Alkaline delignification is somewhat more complicated than acid sulfite pulping. For this reason, the following discussion will present a simplified scheme. In contrast to sulfite pulping, where the primary attack is directed at the benzyl position of the C_9 unit, alkaline pulping initially attacks the phenolic (acidic) hydroxy groups. Following dissociation of the phenolic hydroxy groups to phenoxide ions, the phenylpropane units appear to form either quinonemethide intermediates or anions of the type shown in Figure 2-12.[172-175] Quinonemethide intermediates are formed from β-0-4, β-5, β-β, and β-1 linkages alike.[176-178] Subsequent stabilization of β-0-4 type quinonemethides proceeds to the formation of enolethers, which may involve retro-aldol condensation resulting in the elimination of formaldehyde from the terminal C-γ position. The enolethers are relatively alkali stable, and release their β-arylether substituents slowly.[175] The alkoxide ion, by contrast, displaces the aroxy substituent in neighboring β-position by intramolecular nucleophilic substitution; the epoxide intermediate is easily hydrolyzed in alkaline solution. Other interunit linkages behave similarly under alkaline conditions and often lead to highly conjugated, stilbene-type configurations.[176-178] Except for the formation of minor quantities of aryl-glycerols and phenols, the reaction, if terminated at this point, would have an overall negative effect on delignification. Aside from a few new phenolic and aliphatic hydroxy groups from some hydrolyzed linkages, the elimination of large quantities of primary and secondary hydroxy groups would leave lignin less hydrophilic and without a significant reduction in molecular weight. However, alkaline delignification systems become effective primarily through a combination of molecular-weight degradation and phenolic hydroxy group formation. The acidity of the latter, with alkali, causes generation of phenoxides and imparts water solubility, particularly if the molecular weight of the aromatic component is sufficiently small. The most effective increase in phenolic hydroxy group concentration is achieved through the cleavage of the abundant β-0-4 linkages, a reaction that occurs only to a minor extent if straight sodium hydroxide or caustic soda solutions are used (soda pulping). In fact Gierer, and others, observed in model-compound studies that only about 40% of the β-0-4 linkages were cleaved by 2N NaOH at 170°C during 3 hr (soda pulping) as compared to almost 100% under the same conditions with the white liquor of the kraft pulping process.[174] The difference rests with the efficient addition of SH$^-$ ions to quinonemethides, and a greater nucleophilicity of the resulting mercaptide ions as compared to the alkoxide ions. The ions lead the attack on the neighboring aroxy substituent and displace it. This difference in β-0-4 ether cleavage between alkaline solutions with or without SH$^-$ ions is the basis of differences in delignification efficiencies between soda and kraft pulping.

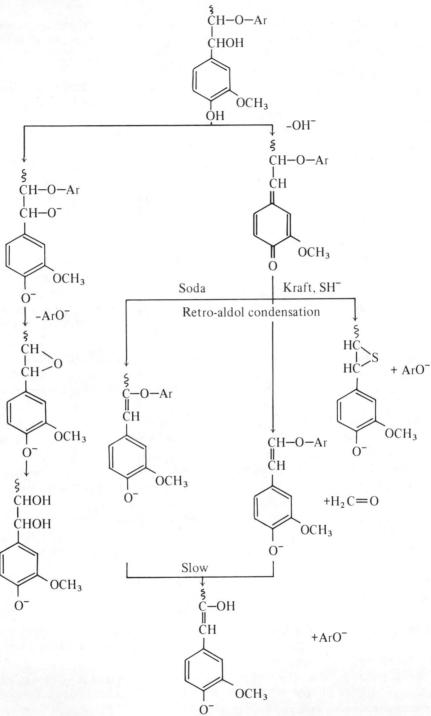

Figure 2-12. Delignification reactions in alkaline solutions.

The side effects of eliminating hydrophilic hydroxy groups from α and γ positions are detrimental to the solubility of the lignin molecule in neutral aqueous solutions. However, the disadvantage of reduced hydrophilicity is overcome by molecular-weight reduction and phenolic hydroxy increase, causing lignin to dissolve in alkali. Depolymerization, however, is counteracted by some competing reactions of secondarily formed styrene derivatives.[168,169] These vinylic phenols are capable of forming nucleophilic carbanions that compete with sulfide ions for the quinonemethide intermediates. The basic mechanism of these carbon-to-carbon bond forming, alkali-catalyzed conden-sation reactions is depicted in Figure 2–13. Figure 2–14 illustrates two other side reactions of alkaline pulping. One concerns the formation of hydroxy-methylated phenols through alkaline formaldehyde condensation; such methy-lols can subsequently condense and form phenol-formaldehyde resin-type polymers.[168,179,180] The second undesirable side reaction illustrated in Figure 2–14 involves the formation of volatile reduced organic sulfur compounds, such as methylmercaptan, dimethylsulfide, or dimethyldisulfide from demethyla-tion of methoxy groups. These highly malodorous gaseous substances are difficult to contain and their release to the environment creates the annoying odor from kraft mills.[181]

Alkaline degradation studies with isolated MWL and subsequent polymer anal-ysis demonstrated general agreement with the mechanisms outlined above.[182,183] Somewhat diverging ideas about alkaline delignification of wood, however, have been developed by Kleinert, who believes that bulk delignification is based on free radical reactions triggered by thermal homolysis of lignin.[184] A sharp increase in free radical concentration during the initial stages of alkaline de-lignification is indicative of such a mechanism. This controversy seems to stress the limitations that underlie the application of results from model-compound studies to the complex multicomponent-system wood.

It can be concluded that alkaline pulping operates on the basis of hydrolytic (or possibly homolytic) depolymerization of phenylalkyl ethers, thereby re-ducing the size of the lignin molecule and simultaneously generating new solubilizing phenoxide ions. The addition of sodium sulfide to the white liquor (kraft process) results in an undesired weakening of the alkalinity of the system. However, the sulfide addition reaches an optimum where nucleophile concen-tration is optimal with regard to quinonemethide and β-ether concentration. If the sulfidity is raised beyond that point, delignification becomes more diffi-cult due to lessened alkalinity. It thus appears likely that each wood species may have its own optimum sulfidity level of the white liquor, which is deter-mined by the concentration of β-arylethers and quinonemethides formed. To raise the sulfidity level beyond that point will be detrimental to delignification in the kraft process.

Unconventional Alkaline Systems. The conventional alkaline delignification processes, soda and kraft, have been practiced with only minor modifications for several generations. Pressures to make industrial pulping processes less

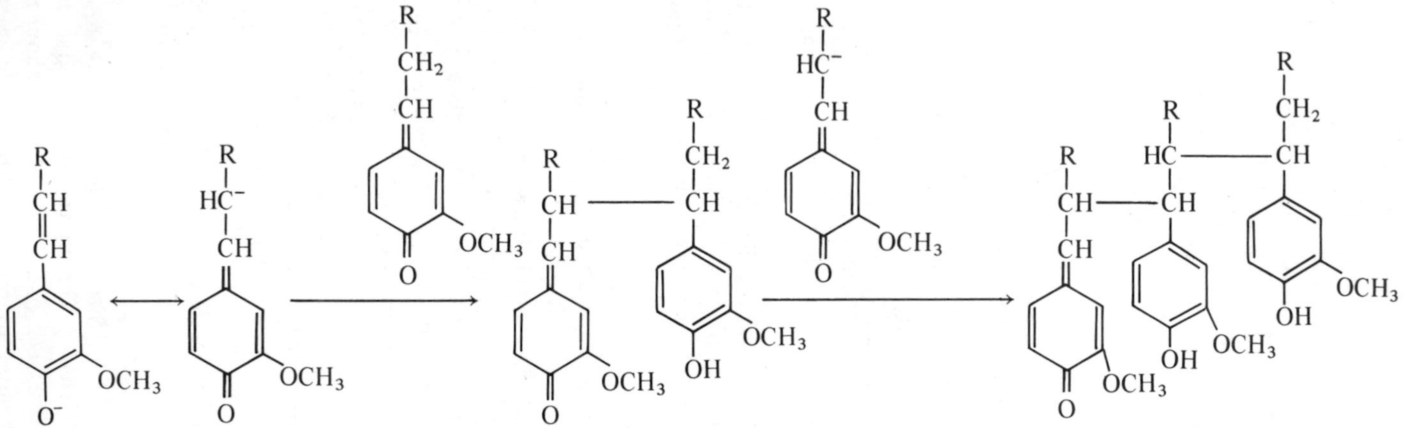

Figure 2-13. Condensation reactions during alkaline delignification (from Gierer, and others[168]).

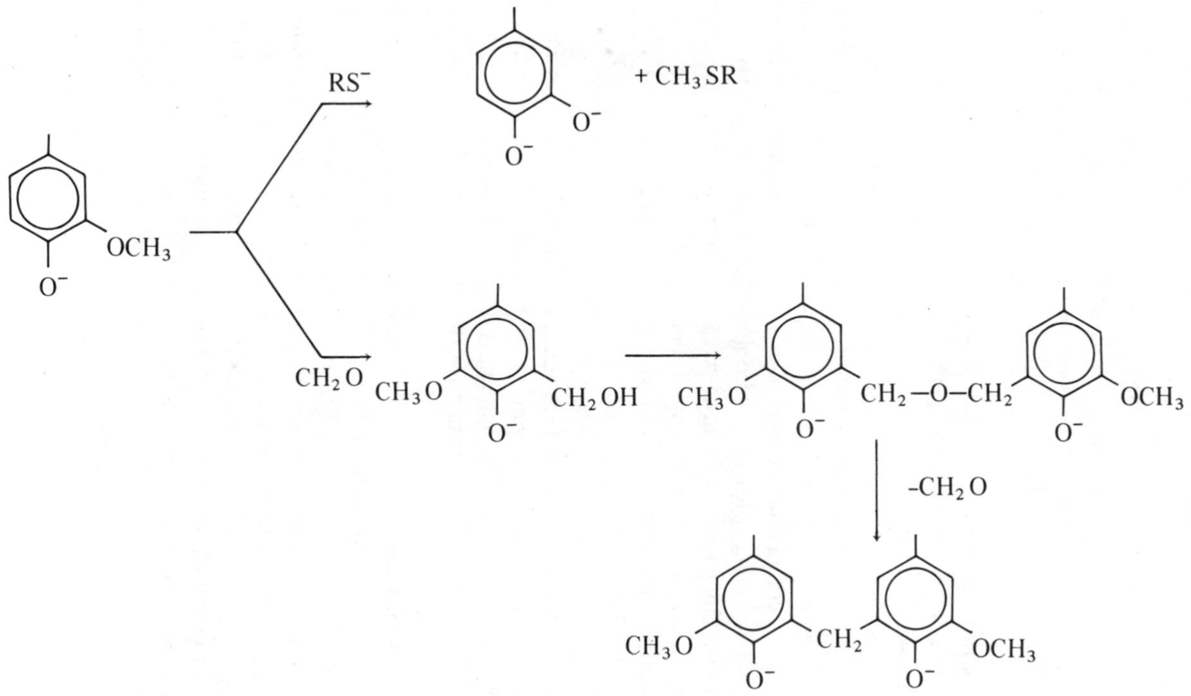

Figure 2-14. Base-catalyzed side reactions in kraft pulping.

detrimental to the environment, along with sharply rising capital and raw-material costs, have spurred scientific and technological explorations into novel delignification methods, which might ultimately be adopted by existing alkaline pulp mills. The following summarizes some of the more promising unconventional alkaline delignification systems. Coverage of this subject matter has been given by Shelfer and Hartley.[185]

Alkaline Sulfite. Attempts to develop a process that combines the universal applicability, the high paper strength, and the ease of chemical recovery of the kraft process with the good features of the sulfite process (namely absence of unpleasant odors and high brightness as well as easily bleachable pulps) has led to the development of an alkaline-sulfite-process technology.[186] From the standpoint of lignin chemistry, alkaline sulfite pulping seems to offer efficient solubilization by combining the phenol-dissolving and depolymerizing characteristics of kraft pulping with the hydrophilicity-creating properties of sulfite processes. Even though relatively little work has been done on the chemistry of alkaline sulfite pulping, it may be anticipated that condensation reactions are prevented by the presence of the strong nucleophilic hydroxyl and sulfite ions.[187]

Soda-Oxygen. With a choice of cleaning up the kraft process through the addition of antipollution equipment or the closing of water cycles versus developing a new and sulfur-free pulping system, it is the soda-oxygen process that appears to be the most promising for the future. A vast number of publications exist in this area.[188]

Investigations into the oxidation of phenols are exceedingly complex due to the many competing reaction pathways. These may be grouped into three categories: (1) alkaline hydrolysis, (2) ionic (heterolytic) oxidation processes, and (3) free radical (homolytic) oxidation processes.[189] Figure 2-15 summarizes the three major competing pathways observed with β-0-4-type dimeric model compounds.[190] The initial fragmentation entails either alkaline hydrolysis-type degradation leading to phenolic aldehydes, or the formation of ortho- and para-quinones by way of hydroperoxide intermediates. Subsequent degradation reactions involve side-chain elimination, ring splitting, and demethoxylation reactions leading primarily to low-molecular-weight aliphatic carboxylic acids. Other studies with model compounds are in general accord with this outline.[191,192]

Degradation studies with isolated lignins demonstrated that: (1) not much change can be discerned during the initial stages of the reaction in terms of aromaticity, functional groups, and molecular weight, with the exception of a decline in solubility and an increase in free radical concentration[193] and (2) more extensive degradation produces the expected low-molecular-weight and aliphatic carboxylic acids, along with some phenolic intermediates in addition to some high-molecular-weight, unreactive, polymeric lignin material.[194]

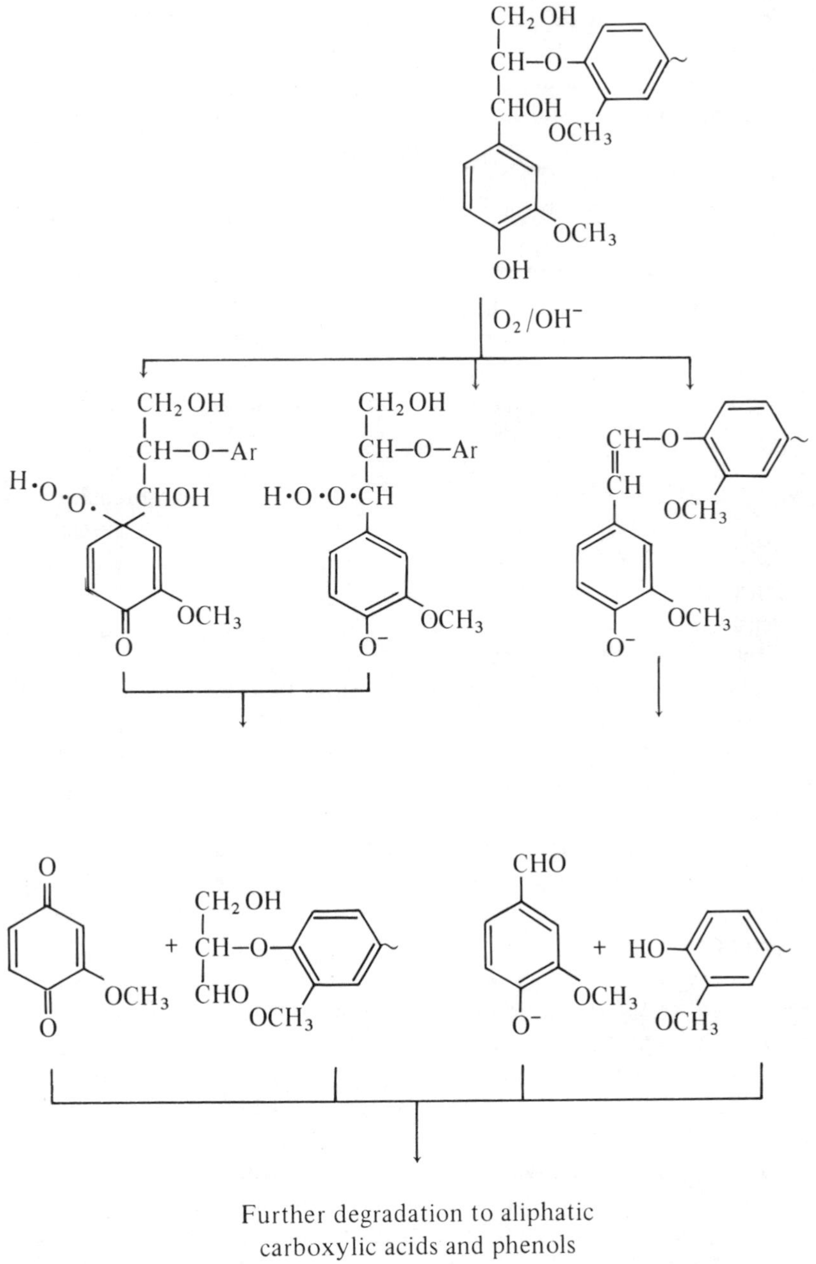

Further degradation to aliphatic
carboxylic acids and phenols

Figure 2-15. Initial fragmentation pathways of β-0-4-type dimers in soda-oxygen treatments (from Gierer, and others[190]).

Amines. Another method for overcoming some of the deficiencies of the conventional kraft process, which results from its use of reduced inorganic sulfur compounds, exists in alkaline pulping in the presence of amines. Mono-ethanolamine and ethylene diamine, both good lignin solvents in themselves, have a beneficial effect on wood delignification when they are added to soda liquors in concentrations of 10 to 80% based on wood dry substance. Improvements occur in terms of operating conditions, product quality, and pulp bleachability.[195] The effectiveness of the amines seems to result from an increase in the lignin-solvation capacity of the liquor. Also, there is an increased stability of carbohydrates towards alkaline degradation. According to work by Nikitin, and others, cited by Kubes and Bolker,[195] the amines apparently prevent secondary condensation reactions in lignin as well as the formation of functional groups giving rise to absorption bands in the range of 1680 to 1710 cm^{-1} in the infrared (IR) spectrum.

Quinones. Alkaline delignification in the presence of anthraquinone and anthraquinone monosulfonate has been reported to result in significant gains in pulp yield, delignification rates, and pulp quality.[196,197] Levels of quinone addition were minimal, ranging between 0.05 and 0.2% on wood dry substance. Apparently, the effectiveness of quinone is based on its redox properties combined with its stability towards alkali. The redox potential seems to facilitate lignin solubilization, resulting in higher delignification rates and apparent hemicellulose conservation.

Neutral Aqueous and Solvent Systems. There has been considerable interest in delignification processes operating in the pH range from 4 to 10. Neutral pulping seems to offer fiber separation with minimal dissolution of wood substance. Whereas acid and alkaline pulping liquors indiscriminately degrade hemicelluose and lignin, neutral solutes offer the advantage of greater lignin specificity.

Neutral Sulfite Semichemical (NSSC) Pulping. Lignin dissolution mechanisms in NSSC pulping have been studied extensively by Gierer and Gellerstedt.[187,198] The particular success of the NSSC process can be attributed to the fact that it combines the beneficial, molecular-weight-degrading effect of an alkaline process with the attachment of hydrophilic functional groups to the molecular fragments. The principal reactions responsible for delignification are illustrated in Figure 2-16 for a β-0-4-type dimer. It is obvious that lignin reactivity is restricted to phenolic units that are capable of forming quinonemethides. The latter add nucleophilic sulfite ions conducive to sulfitolytic cleavage of β-aryl ethers. This important depolymerization step is greatly facilitated by pH values greater than 7, whereas it seems to proceed with great reluctance at pH values below 7. Overall, it appears that delignification with sulfite solutions is extremely pH-sensitive. There exist numerous side reactions, some of which limit lignin solubility.

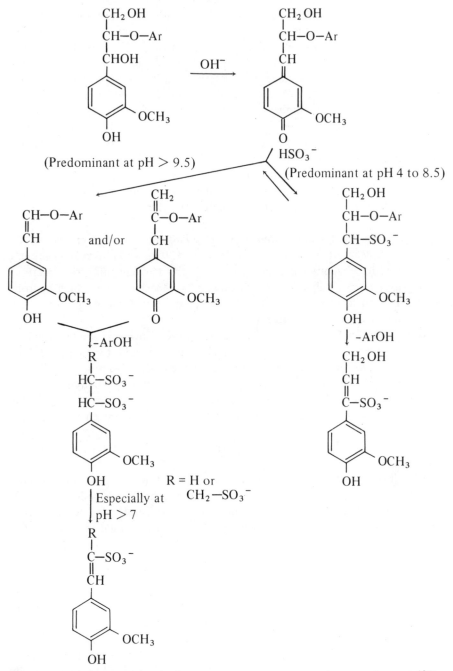

Figure 2-16. Major dissolution reactions in NSSC pulping (from Gellerstedt[187]).

Studies of isolated NSSC pulp lignins agree with the results obtained from model compounds insofar as the pulp lignins were found to have suffered hydrolytic depolymerization and an increase in phenolic-hydroxy-group content.[199] In apparent contrast to residual kraft pulp lignins,[200] NSSC pulp lignins were found to undergo thorough modifications during pulping. However, it was found that the degree of sulfonation had little influence on lignin solubility.[199]

NSSC pulping offers a number of attractive features. The high yield is desirable from the standpoint of raw-material utilization. However, it should be recognized that these pulps are produced in high yield because they contain a considerable amount of lignin that impairs pulp properties, particularly interfiber bonding and color. It is primarily the latter that retards a more widespread use of high-yield pulps. Considerations pertaining to the origin of color and color reversion are discussed in the section on lignin utilization.

Hydrotropic, Phenol, and Maleic Anhydride Pulping. Lignin is effectively dissolved by concentrated aqueous solutions (*circa* 30 to 40%) of xylenesulfonate with or without additional acid catalysis.[201-206] This delignification method, named hydrotropic pulping, solubilizes lignin by a mechanism in which hydrolysis is suspected to be involved [204] and in which phenolic hydroxy groups appear to play important functional roles.

Another sulfur-free approach to delignification involves condensation of lignin with phenols. This method, which was restudied by Schweers, and others,[207,208] is based on acid-catalyzed condensation reactions of phenols with lignin.[209] The phenol-lignin condensate, its lignin-to-carbohydrate bonds cleaved, becomes soluble in methanol. The combined mixture of phenol and phenol-lignin is recovered, and the pulping chemicals are regenerated by solvent evaporation and distillation. Phenol recovery is sufficient to make the process self-supporting and eventually produce a surplus of simple by-product phenols.[210]

Delignification can be achieved by alkaline extraction of a lignin-maleic anhydride copolymer from a sawdust-maleic anhydride condensate.[211,212] The condensation is carried out in an autoclave (digester) at 170°C with no solvent present. The resulting lignin-polycarboxylate can be recovered from the spent pulping liquor and has demonstrated usefulness as raw material for the manufacture of polyurethane products.[213-215]

Solvent Pulping. Aqueous systems are somewhat less than ideal for separation of lignin from woody materials. They usually succeed only after significant chemical alteration in the lignin structure and after sizable carbohydrate losses have occurred. Delignification systems using organic solvent-rich liquors have concentrated on alcohol-water mixtures, mostly 40% methanol or ethanol solutions.[216-220] Lignin-to-carbohydrate bonds are cleaved either homolytically during the high-temperature cook (185°C) or hydrolytically with additions of hydrochloric acid or alkali. The pH of a neutral liquor has been reported to decline to 5.4 to 5.7 during the course of the cook, thus contributing to an accelerated delignification rate. Solvent pulping generates pulp in higher yields than chemical processes and with properties similar to those of bisulfite pulps. The higher yields are primarily due to carbohydrate conservation. Solvent pulping is attractive from the standpoint of complete wood utilization. Uses for the easily isolatable lignin byproducts must be developed, however, to make this process commercially feasible.

Oxidative Systems.   Inasmuch as phenols are sensitive to oxygen, it comes as no surprise that there exist a wide variety of delignification systems based on phenol oxidation. These include reactions of lignin with aqueous halogen solutions at any pH, reactions with chlorine dioxide, peracetic acid, hydrogen peroxide, ozone, photoinduced oxidation, and even microbiological oxidation processes. Many of these topics have been covered in comprehensive reviews by Dence[221] and Chang and Allan.[222]

Chlorination in Aqueous Medium.   Delignification in commercial chemical pulping is interrupted when the lignin content of fibers has dropped to a few residual percent and the carbohydrate portion of the pulp has become more vulnerable to continued alkaline treatment than the remaining lignin. The remaining, darkly colored lignin, has to be removed from the fibers by delignification methods that are more selective in their attack on lignin than are pulping liquors. This specificity is generally achieved with mild phenol-oxidation systems. One of the most common and most widely used systems involves treatment of unbleached pulp fibers with aqueous solutions of chlorine gas. Delignification is achieved by:

1. Aromatic substitution and electrophilic side-chain displacement with formation of chlorophenols.
2. Oxidative cleavage of aryl-alkyl ether linkages (α-0-4, β-0-4 and methoxy groups) with formation of quinones.
3. Oxidation of aliphatic fragments from displaced and hydrolyzed sidechains.
4. Aromatic ring cleavage by conversion of orthoquinones to muconic acid derivatives.
5. Subsequent alkaline hydrolysis of chlorophenols to hydroxy phenols.[223-226]

The major reactions with lignin prevailing during chlorination and alkaline hydrolysis are illustrated in Figure 2-17. Delignification is thus achieved by oxidative depolymerization amounting to molecule fragmentation along with chlorination and hydroxylation resulting in increased hydrophilicity.

Hypochlorite and Chlorine Dioxide.   At pH 11 or higher the chlorine molecule exists as hypochlorite ion.[221] Hypochlorite solutions are used in commercial bleaching operations due to their relative selectiveness for lignin oxidation.[227] However, alkaline hypochlorite bleaching has been gradually replaced by milder

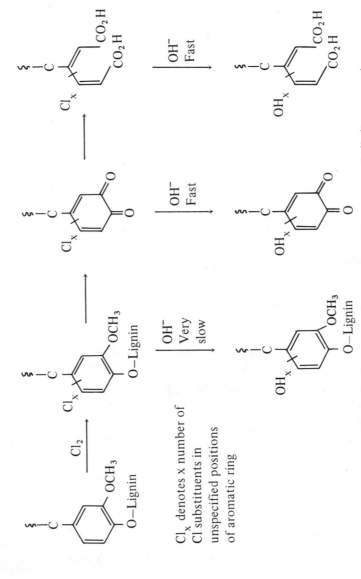

Figure 2-17. Major dissolution reactions during chlorination and alkaline extraction.

Cl_x denotes x number of Cl substituents in unspecified positions of aromatic ring

80

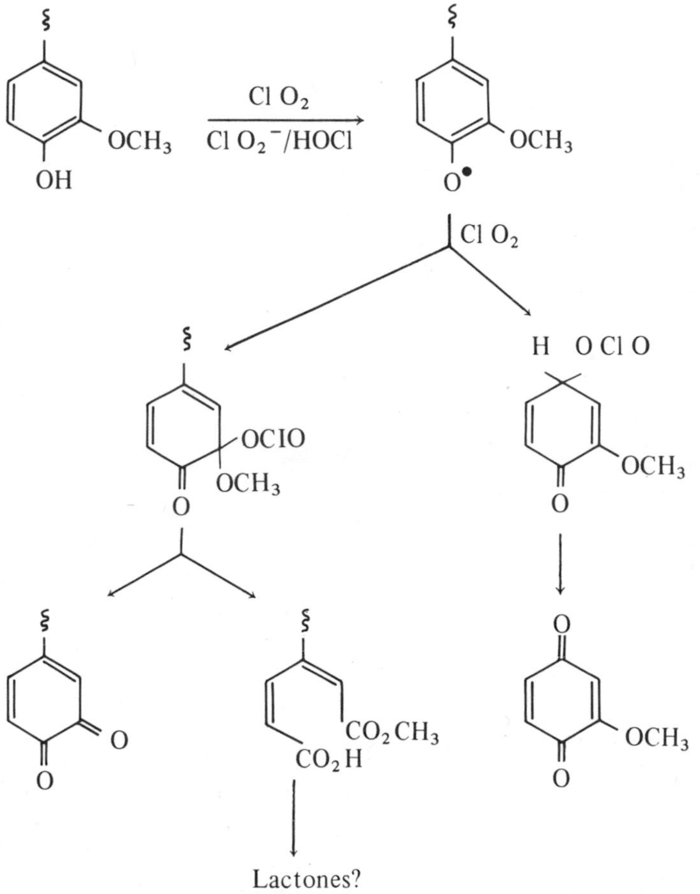

Figure 2-18. Oxidative degradation reactions with chlorine dioxide.

bleaching techniques based on chlorine dioxide, often used in combination with chlorine.[228] Delignification by oxidative lignin degradation with chlorine dioxide is depicted in Figure 2-18.[229] Selective lignin solubilization is achieved through molecular fragmentation and the introduction of carboxy groups. Both actions contribute to an increased solubility in aqueous and alkaline solutes. The end effects of treatment with hypochlorite and chlorine dioxide are apparently quite similar.[230] Chlorine-dioxide treatment of groundwood has been shown to result in drastic lignin modifications, of which dearomatization to the extent of 20% seems to be the most important.[231]

Peracetic Acid and Hydrogen Peroxide. Lignin oxidation by peracetic acid and hydrogen peroxide is of interest from the standpoint of high-yield pulp bleaching. Similar to other oxidation methods, peroxide oxidation proceeds in

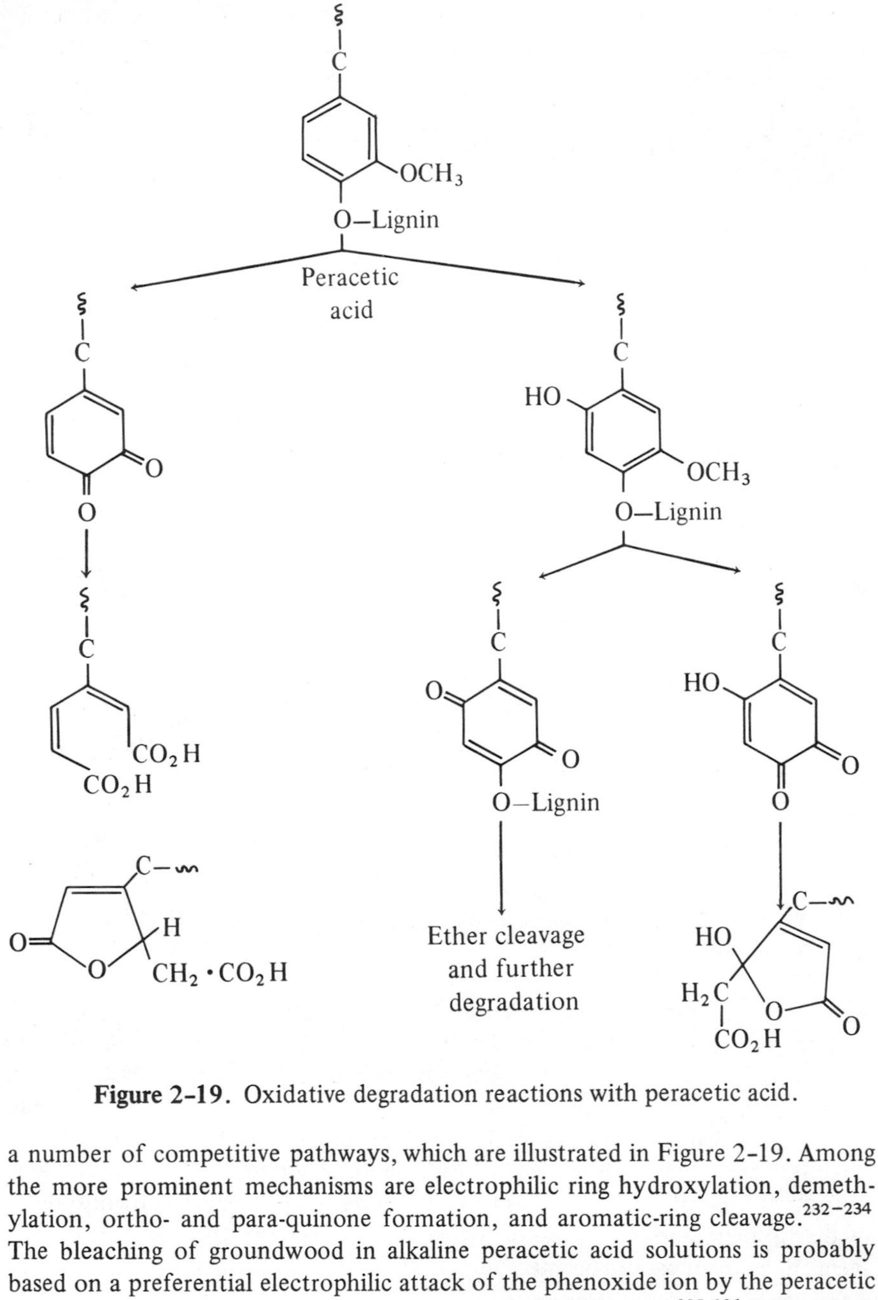

Figure 2-19. Oxidative degradation reactions with peracetic acid.

a number of competitive pathways, which are illustrated in Figure 2-19. Among the more prominent mechanisms are electrophilic ring hydroxylation, demethylation, ortho- and para-quinone formation, and aromatic-ring cleavage.[232-234] The bleaching of groundwood in alkaline peracetic acid solutions is probably based on a preferential electrophilic attack of the phenoxide ion by the peracetic acid. Major reaction pathways seem to involve phenolic sites;[235,236] initial treatment has been reported to result in increased crosslinking, reflected in decreased solubility.[193] These initial reactions are followed by side-chain oxidations[237] and finally by aromatic-ring oxidation and dearomatization.[238]

Hydrogen peroxide appears to be a highly versatile and effective phenol oxidant since it behaves as an electrophile in acidic medium and as nucleophile in alkaline media.[222] As an electrophile, its reaction patterns resemble those of peracetic acids, whereas its nucleophilic attack seems to parallel the reactions depicted in Figure 2-15.[239] In both instances, however, the formation of quinones is followed by aromatic-ring cleavage to low-molecular-weight dicarboxylic acids.

Photoinduced Oxidation. As a substance that strongly absorbs ultraviolet light, lignin is affected by photoirradiation. Its response to photoenergy absorption include darkening as well as brightening reactions.[222] Photoinduced oxidation reactions have been evaluated[240,241] and can be summarzied as illustrated in Figure 2-20. According to this scheme, the reaction is based on the excitation of ring-conjugated, double-bond-containing configurations of the phenone and styrene type, which enter an excited state. These photosensitive, excited groups subsequently give rise to the formation of singlet oxygen and paramagnetic centers (phenoxy radicals). Phenoxy radicals may then stabilize by phenolic coupling reactions or oxidation; the latter will result in the formation of ortho- and para-quinones.

Microbiological Degradation. The literature on lignin reflects an interest in the microbiological degradation of lignin. This activity is interesting, but it is of only minor importance to pulp and papermaking even though the term biological pulping has been introduced in the scientific literature.[242] In view of its marginal importance, this topic will be dealt with briefly. More extensive reviews have been made by Kirk[243,244] and Higuchi.[5]

With fungi, the degradation of lignin appears to start with an enzymatically induced phenol oxidation by such phenol oxidases as laccase or peroxidase. This attack results in polymerization and depolymerization, in demethylation,[245-247] and in quinone formation.[248-250] However, phenol oxidases by themselves seem to be unable to degrade aromatic compounds.[251,252] Further degradation thus proceeds in concert with another enzyme, cellobiase:quinone oxidoreductase,[242,252] which requires cellobiose as a cosubstrate.[253] In contrast, bacterial degradation has been reported to proceed in the absence of phenol oxidase activity.[254] It involves cleavage of β-aryl ethers regardless of whether these configurations are of phenolic or nonphenolic character.[255,256] In addition, bacteria seem to be able to demethoxylate aromatic rings,[257,258] whereas fungi cause demethylation. Following demethoxylation, bacteria are capable of cleaving aromatic rings.[254,259]

Analysis of Lignin

The analysis of lignin deals on the one hand with the qualitative and quantitative analysis of lignin in a given environment and on the other hand with the analysis of the overall structural composition of lignin. Both types of analyses

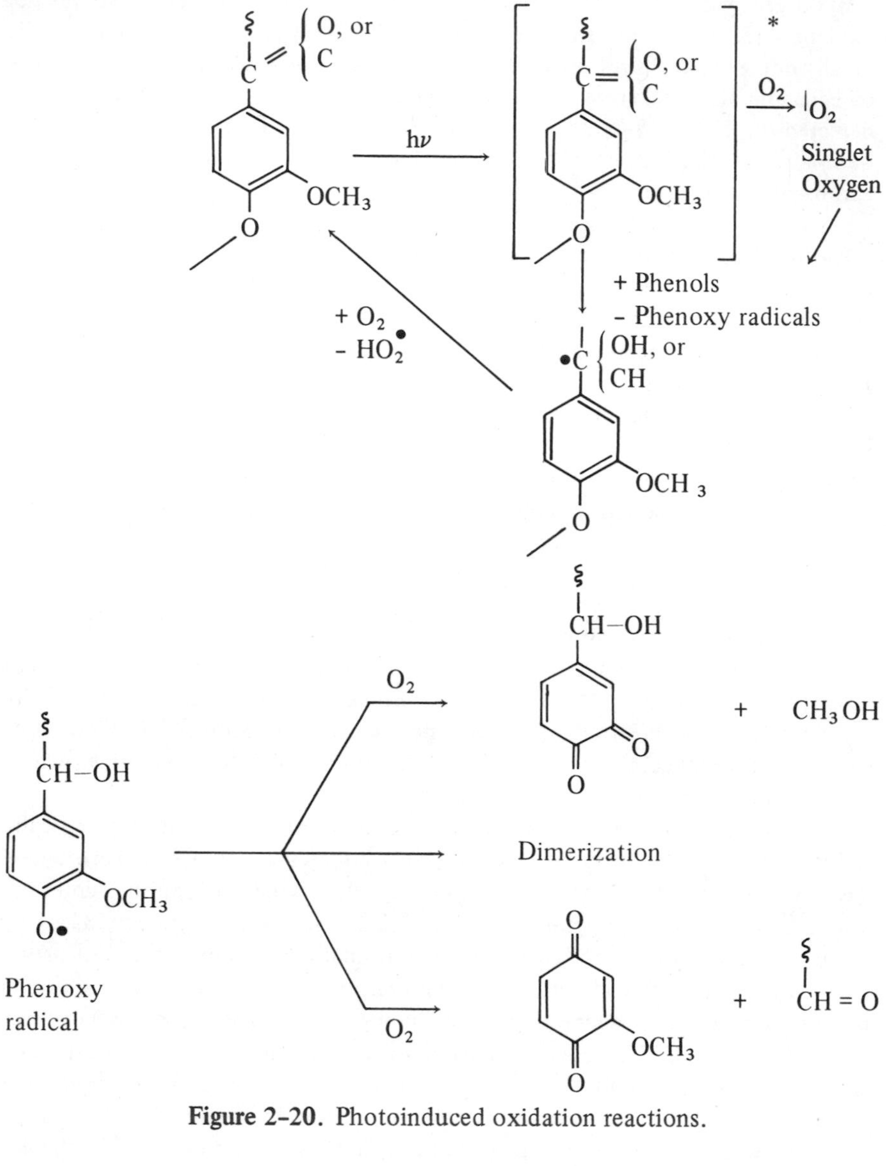

Figure 2-20. Photoinduced oxidation reactions.

have their own particular problems. *In situ* analysis of lignin has to address the question of potential interference from extractable wood components; analysis of isolated lignins has to establish the purity and representativeness of the sample.

The boundaries between lignin and other wood components, such as carbo-hydrates and extractives, are somewhat arbitrary because they represent three chemically inhomogeneous component groups. This is illustrated by the example of a triose attached to an aromatic ring by way of a carbon-to-carbon bond, clearly classifying it as lignin, as opposed to a pentose or hexose sugar linked to an aromatic unit by way of a glycosidic linkage, thereby distinguishing it as a saccharide. The definition of extractives as substances that can be removed from woody tissues with neutral solvents is misleading insofar as there exist substances that are not extractable, but neither do they belong to any other group of wood components. These are simply phytochemicals of high molecular weight, too high for extraction with organic solvents. In the event that these chemicals have phenolic character, they may well be looked on by analytical criteria as lignin, and may even be covalently linked to lignin.

These examples illustrate the need for lignin analysts to consider and respect the sometimes arbitrary divisions that exist among the various wood components. In interpreting results from lignin analyses, it must be recognized that the results will vary from sample to sample as well as from method to method. It thus becomes of major importance to specify what procedures have been employed in an analysis and to adhere closely to the conditions described in standard procedures. Otherwise the reproducibility and reliability of the results will suffer greatly.

The following section is not intended to provide a complete outline of the methods used in wood analysis. If further detail is required, the reader is referred to the monographs by Browning[260] and other literature sources.

Qualitative Analysis

Lignin is qualitatively determined by a number of well-known reactions, which have been reviewed by Sarkanen and Ludwig.[35] These techniques can be divided into color (staining) reactions, spectroscopic methods, and chemical isolation and/or degradation techniques.

Color Reactions. The best-known test for the presence of lignin in a solid carbohydrate-containing environment involves treating the substrate with a mixture of phloroglucinol and hydrochloric acid (Wiesner reaction); a positive reaction is positive only with those lignin preparations that contain coniferyl aldehyde end groups; this constraint seriously limits the application of this test with chemical and high-yield chemical pulps.

Successive treatments of the lignin-containing substrate with aqueous permanganate, hydrochloric acid, and ammonia (Mäule reaction) will result in a deep red color if the lignin contains significant numbers of syringyl groups.

Spectroscopic Methods. Ultraviolet light at 280 nm wavelength is effectively absorbed by lignin; no interference from carbohydrates is detected in this range. Infrared light is absorbed at 1500 and 1600 cm^{-1} by lignin, whereas carbohydrates absorb at 1200 and 1680 cm^{-1}.

Chemical Methods. The presence of lignin in a solid substrate may also be established by isolating a solid residue following acid hydrolysis of carbohydrates. However, some of these reactions are not specific enough to differentiate lignin from some other polyphenolic plant component. An affirmative answer to the latter question requires the identification of phenolic degradation products following treatment of the lignin-containing substrate with acids (to phenylketones); by oxidative treatment with nitrobenzene (to vanillin and related aldehydes and acids); or by hydrogenation or pyrolysis (see the discussion of thermofractometry by Stahl and his coworkers.[261,262]).

Quantitative Analysis

Quantitative lignin analysis techniques are based on the insolubility of lignin in strong mineral acids; on its UV-absorbing properties; on its selective response to halogenation or oxidation; or even on its higher carbon-to-oxygen ratio than is found in other cell-wall polymers. The following is a summary of some of these methods. A more detailed treatment of this subject can be found in the literature.[127]

The classical method for the quantitative determination of lignin is based on Klason's technique involving hydrolysis with 72% sulfuric acid. In this procedure lignin is left as an insoluble residue and is recovered by filtration and gravimetrically determined. While this method works satisfactorily with most softwoods, it requires modification when applied to hardwoods, which contain sizable amounts of acid-soluble lignin fractions that will escape determination unless separately analyzed by UV at 280 nm or 205 nm,[263,264] The latter wavelength is recommended in order to avoid interference from furfural formed by acid hydrolysis and dehydration of pentosans. In addition to the possibility of introducing negative errors due to acid-soluble lignins, positive errors may be encountered through condensation of carbohydrate degradation products, particularly pentosans[265] and proteins,[127] onto the insoluble lignin residue.

In order to avoid some of the problems of the Klason method, Kurschner promoted an acid hydrolysis technique, which is preceded by a mercerization treatment with 25% KOH solution. This technique was found to result in drastically reduced values, such as 21.2% for softwood versus 27.4% by Klason, and 8.5% for a hardwood (beech) versus 19.9% by Klason.[265] These differences were attributed to less carbohydrate condensation and more acid-soluble lignin fractions.[266,267] The surprising and consistent low values for Kurschner lignin kept this procedure from becoming a widely accepted one. The acid-catalyzed hydrolysis of carbohydrates may be performed with other acids as well; hydrofluoric acid has been proposed,[268] as well as mixtures of sulfuric and phosphoric acids (Comecom method).[269]

Quantitative lignin determination by UV absorption involves solubilization of wood in acetylbromide-acetic acid and absorbance measurement at 280 nm.[270] This technique is mainly limited by its solubility requirement and the quality of the standard used. However, it is very attractive from the standpoint of precision, speed, and simplicity.[271]

Chlorination and oxidation are the basis for the common chlorine-number and kappa-number determinations for pulps, respectively.[260] In these procedures lignin is quantitatively oxidized in the presence of carbohydrates by either sodium chlorite or permanganate.

Since the proportions of carbon and hydrogen are different in carbohydrates and lignin, differences in the C/H ratio can be used for quantitative determinations of lignin contents. This method seems to be applicable to wood as well as pulp samples.[272]

The determination of dissolved lignin sulfonates can be accomplished by reacting the sulfonates with nitrite in acidic solutions (Pearl-Benson test). The resulting nitroso derivatives form highly colored quinone-oxime derivatives in alkaline solution by tautomerization; the latter are analyzed colorimetrically.[273] Limitations of the Pearl-Benson test involve interference from several other polyphenolic sources, such as humic acids. These may be overcome by using spectrofluorimetric methods[274] or by oxidizing the dissolved organics to vanillin and related aldehydes and quantitatively analyzing the latter by gas chromatography.[275]

Structural Analysis

Structural analysis begins with the selection of the type of lignin substrate. No matter what lignin preparation is selected representativeness is sacrificed either in terms of yield (quantity) or authenticity.

Model Systems.   Since lignin is perceived by most scientists as a polyphenylpropane molecule of more or less complex statistical constitution, selection of the proper model system is difficult. In lignin chemistry the selection process has followed an established scientific principle, to proceed from simple to more complex model systems.

The simplest approach, but the one most distant from the reality of *in situ* lignin, involves model compounds that are individual chemical substances representing one or another structural feature of lignin. They are often aromatic, phenolic compounds with alkoxy groups, and primary or secondary aliphatic hydroxy groups; sometimes they are dimers reflecting prominent interunit linkages present in lignin. Virtually everything known in lignin chemistry is, directly or indirectly, a result of model-compound studies. This applies in particular to knowledge of lignin reactivity in chemical pulping. Primary criticism of extrapolating results obtained with model compounds concerns the physical state of the reaction, that is, with model compounds one usually works in homogeneous phase with all reactants in solution, thereby totally ignoring problems associated with heterogeneous phases, surface properties, mass transfer, diffusion rates, and the like. Nevertheless, lignin chemistry would be unthinkable without the use of model compounds.

Synthetically prepared dehydrogenation polymers (DHP) of one or all three para-OH cinnamyl alcohols may be visualized as representatives of a model system one step closer to reality. The *in vitro* polymerization method was first introduced by Freudenberg and has become the foundation of the Heidelberg group's great contribution to the field of lignin chemistry.[1,2] This method involves enzymatic or nonenzymatic[276] dehydrogenation (oxidation) of lignin precursors, resulting in polymers or polymer fragments that resemble the lignin in wood in terms of overall composition. Advantages of the DHP model system include: the polymeric character; the simplicity of the polymerization system (enzyme, water, phenylpropanes); the purity of the product in terms of freedom from isolation-induced alteration/degradation and carbohydrate contamination; and the known proportions of guaiacyl, syringyl, and parahydroxy units involved. Ready availability of lignin precursors has increased the scientific interest in this model system. [277-283] The main criticisms are: it ignores the role of carbohydrates in the lignification mechanism;[276,284] the polymer varies greatly with polymerization conditions;[17,278,279,285] the polymers deviate structurally somewhat from isolated lignins.[282,283,286] In view of these shortcomings, it is no surprise that DHP's have much competition from mildly isolated milled-wood or Bjorkman lignins.

The development and refinement of systems analysis techniques was greatly accelerated by the advent of computers. Computer simulation and modeling have become important tools of scientific investigations and it was inevitable that these techniques would be applied to the field of lignin chemistry. The systems approach to lignin chemistry offered the opportunity for the development of a model system of a higher order of complexity by eliminating some of the shortcomings of the simpler model systems. The first computer-aided system analysis approach to the chemistry of lignin formation and reactivity was published in 1974[43,57] and other attempts have been published since.[287-289]

The SIMREL system, the simulation of reactions with lignin by computer, is based on the understanding of biosynthesis mechanisms that lead to the formation of lignin and on the analytical and reaction behavior of lignin. This system employs random coupling reactions between mono- and oligomeric lignin precursors to develop a lignin formula of between 70 and 100 phenylpropane units. The composition of the simulated model polymer is continually monitored and compared to data that describe the overall composition of actual lignin. Deviations from these data are used for constraining the simulated lignification process.[43,57] Both literature data and primary analytical observations may serve as input into the simulation program. By simulating the expected analytical behavior of the lignin polymers, predictions are made that can be correlated with experimentally obtained analytical data.[47,58] Thus the SIMREL system becomes a tool for predicting and estimating chemical behavior and properties of lignin-like molecules. By correlating various types of analytical results with structural details, this method correctly predicted the existence of an α-0-γ interunit linkage that was subsequently isolated.[44,47,53,58] However, it is unlikely that computer-aided systems analysis techniques will remain limited to simulating chemical behavior of lignin; it appears likely that this method will also be adopted for correlating chemical structure with physical properties.

It should be stated that each model system has its distinct advantages and limitations and that the various model systems should be regarded as complementary rather than as competitive.

Analytical Methods. The analytical description of lignin concerns both its chemical and its polymeric structure. Both are important for a complete understanding of its behavior and properties, but lignin analysis has concentrated primarily on its chemical composition. The determination of molecular weights and sizes, of molecular weight distributions and polydispersities, and the qualitative and quantitative identification of macrocyclic configurations in lignin have received relatively little attention by lignin chemists as compared to the analysis of its chemical structure. What is known about lignin polymeric characteristics has been dealt with in reviews by Goring[65] and by Bolker.[290]

Analytical methods that describe chemical structure can be divided into two principal types: nondestructive lignin analysis and analysis following some sort of degradative depolymerization of lignin. Nondestructive methods employ primarily spectrophotometric techniques. Most destructive methods involve the action of aggressive chemicals or concentrated energy forms.

The goal of all structural analyses is qualitatively and quantitatively to describe as completely as possible the changes that the initial mixture of lignin precursors has undergone in the process of lignification. This description usually includes expressions for the elemental composition of the average phenylpropane, C_9 unit, as well as for the overall distribution and frequency of interunit linkages. Since the average elemental composition of the parahydroxy cinnamyl alcohol precursor in the pool of softwood lignin precursors is $C_9H_{9.1}O_2(OCH_3)_{0.9}$, and since lignification appears to involve principally the abstraction of hydrogen and the addition of water, the elemental composition of isolated lignin indicates the formula of $C_9H_{7.1}O_2(H_2O)_{0.4}(OCH_3)_{0.9}$.[1] Thus the average phenylpropane unit will lose approximately two atoms of hydrogen and add 0.4 molecules of water. Additional structural information, particularly that pertaining to interunit linkages, requires more elaborate analyses, which are discussed in the following section. This information is customarily expressed on C_9 basis.

Nondestructive Analysis. The optical properties of lignin in the ultraviolet and infrared region of the electromagnetic spectrum have been reviewed by Goldschmid and Hergert, respectively.[291,292] In addition to the use of UV and IR spectroscopy for the qualitative and quantitative determination of lignin, these methods are also useful for the detection of certain optically active functional groups, such as phenolic hydroxyls and carbonyls. Both methods have greatly assisted in the identification of structural changes induced by chemical pulping and modification reactions.[293]

Nuclear magnetic resonance spectrosocopy (NMR) has developed into a versatile and routine analytical tool in lignin chemistry.[63,294] In its most common form, proton NMR, it provides reliable information on the hydrogen distribution of lignin and lignin sulfonate preparations.[295,296] Improved resolution of hydroxyl groups in proton NMR spectroscopy has been achieved with trimethyl silyl derivatives of lignin.[297,298] The natural abundance of the ^{13}C isotope in organic materials is sufficient for the preparation of well-resolved NMR spectra of lignins and lignin derivatives.[299] All major peaks of ^{13}C NMR spectra of lignins have been assigned to specific structural features of the polymer. This procedure holds the potential of becoming the most powerful spectrophotometric technique applied to lignin.

Electron spin resonance (ESR) spectroscopy, an analytical technique that identifies the existence of paramagnetic centers, has been applied in lignin chemistry to the analysis of free phenoxy radicals, their formation, behavior, and stability. The formation of free radicals during chemical pulping and the existence of free radicals in pulps has been firmly established.[300-302] In addition, ESR spectroscopy has helped clarify reaction mechanisms involving free radical transformations,[303-305] and it has been employed in spin-labeling studies designed to evaluate the physical structure of the lignin polymer and its accessibility to solvents.[304]

Mass spectroscopy has been applied to the analysis and identification of lignin degradation products. Due to the volatility requirements of this technique, only relatively small molecules (molecular weights not exceeding approximately 1000) have been employed. Systematic studies on fragmentation patterns of lignin and lignin-sulfonate-like model compounds during electron bombardment in vacuum have been published.[295,306,307]

Destructive Analysis. In the true meaning of the term destructive analysis, many of the classical chemical analysis techniques should be included in this category, such as elemental analysis and functional group analyses for methoxy, hydroxy, carbonyl, and carboxy groups and for vinylic double bonds. However the most important destructive analysis techniques involve degradative or depolymerization processes. Depolymerization is not a process that regenerates lignin precursors, but rather is one that generates mixtures of identifiable low-molecular-weight polymer fragments that are capable of providing information on the total polymer. The most important lignin depolymerization and degradation reactions are summarized for softwood in Figure 2-21. Only the most prominent degradation products are shown. It should be remembered that the value of each individual analytical procedure depends greatly on the variety of allied degradation products. The following is a brief discussion of these degradation techniques.

Acidolysis, Alcoholysis, and Related Degradations. Lignin degradation and depolymerization by hydrolytic action is one of the longest known and best understood reactions in lignin chemistry. Reviews on this topic are given by Adler,[19] Lundquist,[308] Sakakibara,[44] and Wallis.[309] Hydrolytic degradation usually involves the use of combinations of lignin solvents and trace (catalytic) quantities of mineral acids. If alcohols, such as methanol or ethanol, are used

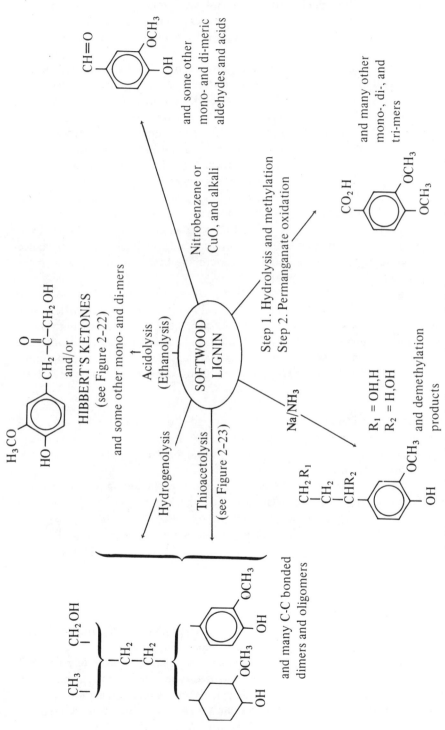

Figure 2–21. Major analytical degradation reactions and their most prominent products.

the procedure is commonly termed alcoholysis, methanolysis, or ethanolysis. During the treatment of wood or lignin with alcoholytic reagents, approximately 10% of the lignin is converted into a distillable oil known as Hibbert's ketones.[310]

If dioxane:water 9:1 is used in the presence of 0.2M HCl, the reaction is known as acidolysis, and ω-hydroxyguaiacylacetone is the primary degradation product recovered after 4 hr of reflux. Hibbert's ketone-analogs will be secondary conversion products of this primary degradation ketol. Acidolysis has proved to be an invaluable analytical tool that has contributed significantly to the understanding of the chemical structure of lignin through the isolation of a vast number of dimeric degradation products.[19,308] The overall reaction mechanism leading to the formation of Hibbert's ketones is depicted in Figure 2-22. Mild aqueous hydrolysis of lignin, occasionally aided by the presence of 2% acetic acid, helped Nimz to isolate a number of structurally important monomers, dimers, trimers, and tetramers following several weeks of reflux. Even though the yield of these degradation products was minimal, their importance cannot be underestimated because they were the first compounds with more or less intact side-chain configurations.[62]

Another type of hydrolytic degradation involves the action of dioxane : water mixtures in the ratio of 1 : 1 at 180°C for 20 min. Under these drastic conditions, the macromolecule is homolytically fragmented, giving rise to a multitude of mono- and oligo-mers by fragmentation and recombination reactions. The isolation and identification of these degradation products has been important in the elucidation of chemical structure. This work has been reviewed by Sakakibara.[44]

Oxidative Degradation.   Lignin can be degraded oxidatively under conditions that retain aromatic rings or under conditions that degrade the aromatic and phenolic moieties in lignin. The latter are not important for structural analytical work in lignin chemistry.

The oxidative degradation reactions that preserve the aromatic ring of lignin employ metal oxides, nitrobenzene, or potassium permanganate as oxidants in combination with alkali. Products from the oxidative degradation of lignin are generally among the best suited for instrumental separation and identification. The development of standard methods for their identification and quantification have given the oxidative degradation of lignin the edge over other depolymerization methods. The vast amount of literature in this field has been reviewed by Chang and Allan.[222] The preferred metal oxide for oxidative degradation of lignin for analytical purposes is cupric oxide in alkali. Oxidation with CuO in strong alkali at temperatures of about 170°C produces a mixture of ether-soluble aldehydes and acids in yields of approximately 35 to 40%.[311] The undesirable formation of carboxy acids in this reaction, which may give rise to an artifactual formation of uncondensed-type degradation products by decarboxylation,[312] seems to be catalyzed by metals of the reaction vessel.[313] Low-molecular-weight degradation products may be separated directly,[314,315]

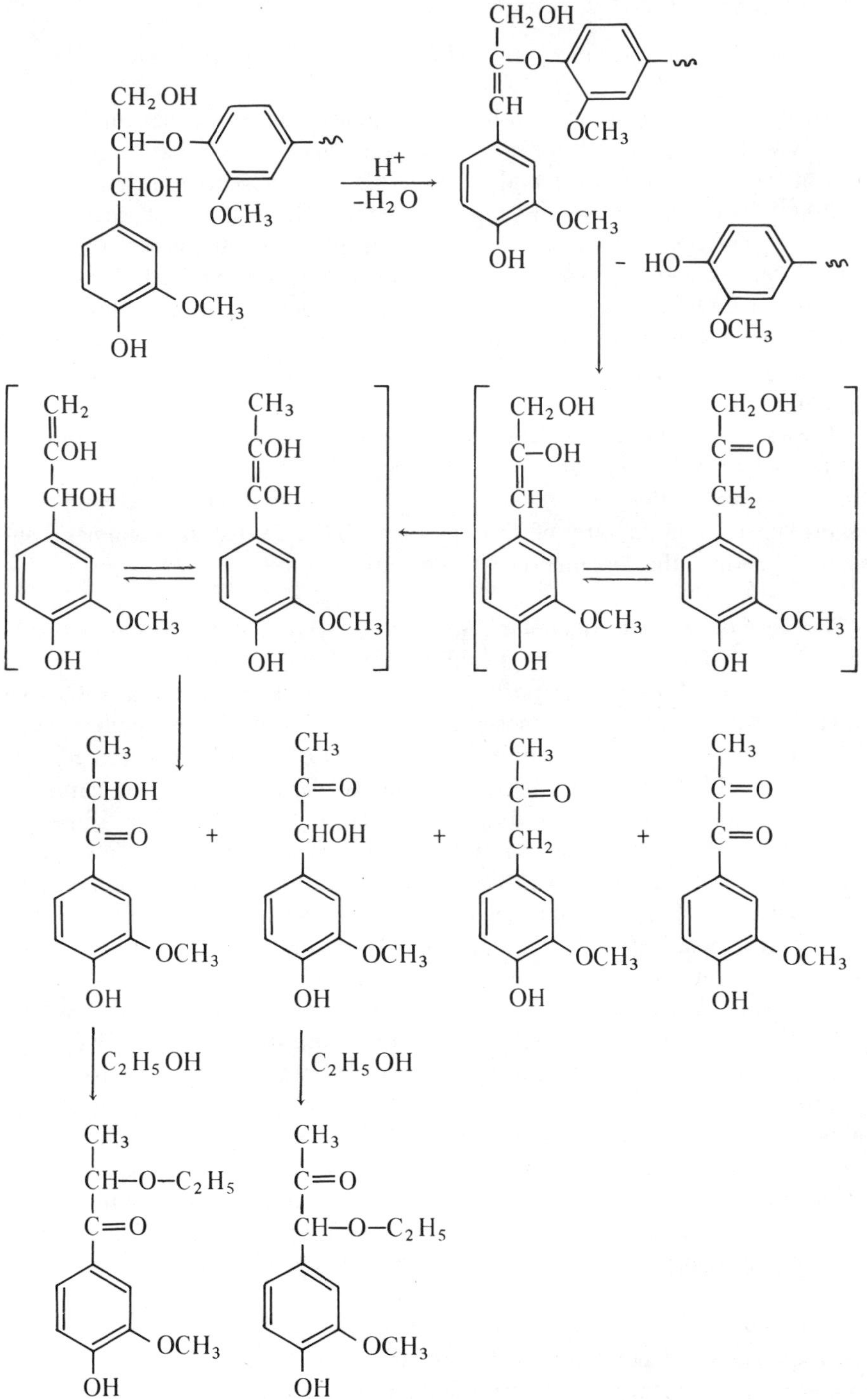

Figure 2-22. Acidolytic and alcoholytic degradation reactions of lignin (from Wallis[309]).

or, following derivatization as methyl ethers and methyl esters, by gas chromatography.[316,317]

Nitrobenzene is an effective two-electron transfer agent that has enjoyed popularity with lignin chemists for some time. It is typically viewed as a replacement for metal oxides since it offers the advantages of homogeneous reaction conditions.[222] Its effects on lignin fragmentation are similar to those of metal oxides.

The degradation of lignin by alkaline permanganate oxidation was first introduced by Freudenberg, et al.,[318] and later was revised and adapted for gas chromatographic separation, identification, and quantification techniques.[319] Since permanganate oxidation converts only phenolic moieties in lignin into identifiable degradation products, oxidation is usually preceded by a hydrolysis step.

The rate of side-chain oxidation seems to depend somewhat on its degree of oxygenation; best results are obtained when the hydrolysis step involves simultaneous preoxidation with cupric oxide in alkali. Under those conditions, approximately 35 to 40% of the lignin can be recovered as monomeric and dimeric methyl ether and methyl ester derivatives.

Hydrogenolysis and Reductive Degradation. Hydrogenolysis and reductive degradation of lignin is of interest from both analytical and industrial viewpoints owing to the relative mildness of the procedure, the high yields and good separatability of degradation products, and the potential use for delignification[320] and by-product phenol production.[321] Drawbacks of this type of lignin degradation concern side reactions, such as demethoxylation and hydrogenation to neutral compounds.[322] In addition, degradation appears to be somewhat solvent and catalyst limited. In spite of these shortcomings, hydrogenolysis has been successfully applied to the characterization and identification of novel types of interunit linkages in lignin not observed with other degradation techniques. The work by Sakakibara,[44] and Hoffmann and Schweers[323] deserve special mention.

The reductive degradation of lignin with sodium in liquid ammonia has been carried out with or without ethanol at temperatures of -33°C for eight days. Yields of 17% ether and chloroform-soluble degradation products have been reported.[324-326] The predominant reaction seems to involve splitting of aryl-alkyl ether linkages, demethylation, and demethoxylation.[327] This technique has been studied extensively by Shorygina, and others,[324-326] and Yamaguchi.[328] Potassium and lithium amalgams have been demonstrated to have similar effects on lignin degradation.[329] Potassium in hexamethyl phosphoramide[330] seems to cleave cyclic benzyl aryl ethers of the β-5 type linkage with particular ease.

Thioacetolysis. A novel, effective, highly selective lignin depolymerization reaction has been developed by Nimz, and others.[331-334] It is illustrated in Figure 2-23. Mild derivatization with thioacetic acid in the presence of boron trifluoride, followed by alkaline hydrolysis, seems to constitute the principal mechanism of this fragmentation reaction. Subsequent desulfurization is accom-

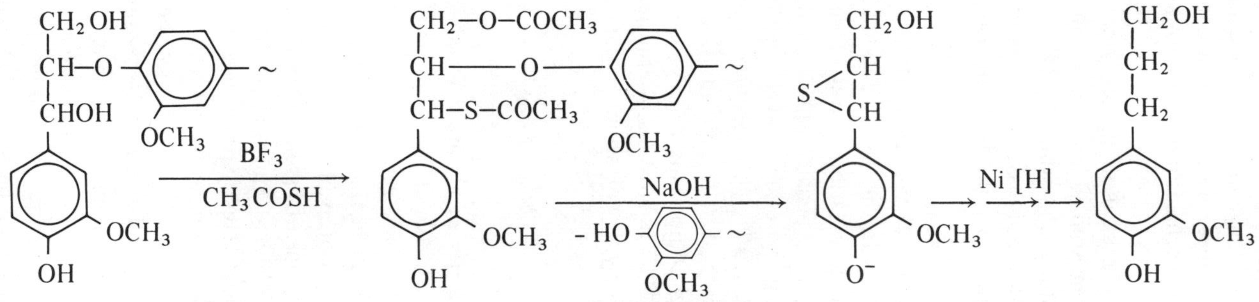

Figure 2–23. Degradation reactions of lignin with thioacetic acid (from Nimz[331]).

plished by hydrogenation over nickel, which has been shown to convert as much as 91% of beechwood lignin to a range of mono- to tetra-meric products. The separation and identification of these degradation products reinforced earlier findings as to the polymerization mechanism of lignin formation.

Thermolysis. Structural fragmentation of lignin by homolytic bond cleavage in high-energy systems has been studied extensively in Russia; some of this work has been reviewed by Domburg, and others,[335,336] and Allan and Mattila.[337] Thermal degradation studies of lignin have both analytical importance[261,262] and technical significance as they pertain to lignin behavior under pyrolytic conditions.[338]

Miscellaneous Depolymerization Techniques. Hydrazinolysis involves the repeated reaction of lignin with hydrazine in dry butanol 1 : 6 under conditions of 115°C for five hours.[339] Such degradation converts lignin into simple phenols, aldehydes, as well as aromatic and long-chain aliphatic acids.

Dienolysis has been termed a β-aryl ether cleaving reaction between lignin and such dienes as tetracyanoethylene.[340a] This reaction is based on a Diels-Alder type of reaction of intermediary styrene configurations. Similar reactions of lignin-related monomeric model compounds have yielded a variety of tetralin (hydronapthalene) derivatives.[340b] Effective β-aryl ether cleavage in model compounds has also been observed with potassium tertiary-butoxide in dimethyl sulfoxide (DMSO).[341]

Utilization of Lignin

Even though lignin is at the center of chemical wood technology, it has not gained great importance as a raw material for commercial products. Nature has invested about 40% of its biologically generated and stored energy, as determined by its enthalpy content, in lignin.[342] The neglect in utilizing lignin is therefore by no means a question of availability. It has been estimated that in 1971 approximately 40 million tons of lignin was generated worldwide in spent pulping liquors[343] and that the United States alone disposed of approximately 16 million tons of kraft lignin.[342] This material was not entirely wasted since most of it was burned in furnaces to recover both chemicals and steam. But at best the lignin was underutilized. This use of lignin for fuel in recovery processes of pulp mills is the primary reason that wood is statistically registered in the United States as raw material for a significant amount of energy.[344] Any new concept for the use of lignin has to recognize that replacing lignin as a fuel in pulp-mill recovery processes involves the use of other fuels or the alteration of recovery operations. However it should be noted: (1) that lignin, like cellulose, is isolated, separated, and purified during chemical pulping and (2) that the corecovery of both polymers for material use seems a logical development in the light of increasing material shortages. One possibility is that low-grade chips from forest residues, such as tree tops and thinnings, might be used to replace lignin as fuel in pulp-mill operations. This appears especially attractive in the light of an increasing trend towards high-yield, lignin-containing pulps that appear to have more stringent requirements for the quality and uniformity of the fibrous raw material.

Conceptually, improved lignin utilization can be accomplished by taking any one of the following approaches:

1. Increased manufacture and use of high-yield, lignin-containing pulps.
2. Breakdown of by-product lignin to chemicals of low molecular weight (feedstock chemicals).
3. Conversion of by-product lignin to special polymeric products without appreciable breakdown or fractionation.

Factors pertaining to each use category are discussed next.

Lignin-Containing Fibers

Papermakers are accustomed to the use of lignin-containing fibers such as stone groundwood, unbleached chemical, and semichemical pulps. In general, lignin-containing fibers find use in products in which inferior optical and sometimes inferior strength characteristics are acceptable. Since the color of lignin and its color stability are the primary reason for the limited use of lignin-containing fibers, the chromophores and their origin, formation, and potential prevention in lignin will be reviewed in the chapter on bleaching.

Considering the magnitude of the problem, it is not surprising that the color of lignin has received much attention.[345] Falkehag, and others, reported in 1966[346] that lignin-chromophores may include a combination of:

1. Vinylic ring-conjugated double bonds
2. Quinonoid systems
3. Chalcone structures
4. Free radicals
5. Metal complexes with catechols

Falkehag later attributed 35 to 60% of MWL light absorption at 457 nm to the presence of 0.7% orthoquinones.[347] By contrast, hydroxylated stilbene structures were reported to account for only a maximum of 5% of color in this region,[348] and free radicals were suspected to account for only a maximum of 7% of absorption at 420 nm.[346] Metal complexes with phenolic hydroxy groups and catechols were held responsible for 65% of the color of lignin sulfonates and for 20% in kraft lignin.[349] In softwood MWL, however, most of the color seems to originate from quinonoid structures.[350] Methylene quinones and phenanthrene derivatives, among others, are suspected of constituting chromophores in lignins from alkaline pulping processes.[351] Their formation is envisaged as stemming from stilbenes (known to be secondary reaction products during alkaline pulping) by oxidation or oxidative photocyclization.[351] The significance of hydroxy stilbenes as precursors of potential chromophores was also pointed out by Gierer, and others.[352]

The mechanism of color formation in lignin is closely related to photo-oxidation discussed earlier. According to this scheme, ring-conjugated carbonyl groups are excited by the absorption of UV light[353,354] and are subsequently photochemically reduced by a free radical mechanism that generates phenoxy and benzyl alcohol radicals. This reaction is likely to involve singlet oxygen as outlined in Figure 2-20 (see also refs. 355 and 356). In accord with the mechanism depicted in Figure 2-20, phenoxy radicals will give rise to chromophoric ortho- and paraquinones.

Photochemical stabilization of lignin in pulp fibers is possible by either one of the following two approaches: (1) by supplying the α-carbonyl-containing irradiated system with a hydrogen-donor, which may be an easily formed resonance-stabilized phenoxy radical capable of forming a photostabilized product with the benzyl alcohol radical by coupling;[355] or (2) by supplying the system with compounds capable of preventing phenoxy radical formation, such as antioxidants, singlet-oxygen quenchers and UV-light absorbers.[241]

Low-Molecular-Weight Chemicals from Lignin

The interest that lignin scientists have shown in lignin degradation to low-molecular-weight chemicals has faded somewhat,[357] possibly because of the fact that fragmentation of the lignin macromolecule by some energy source and the rebuilding it to a polymeric material seems inefficient from an entropic energy standpoint.[342] The concepts of lignin utilization by way of low-molecular-weight chemicals can be summarized in the following schemes:

1. Hydrogenation and hydrocracking: The application of coal liquefaction technology by hydrocracking to lignin has been studied by many investigators and has been reviewed.[358] In the Noguchi process mixtures of simple phenols of the cresol type are obtained from kraft lignins in yields of about 25%. However, yields as high as 50% of distillable products from lignin have been reported[359] under similar conditions, and careful selection of process conditions seems to be vital.[360,361]

2. Thermal decomposition: Pyrolytic degradation of lignin by destructive dry distillation has been reviewed.[337,358] At temperatures of 350 to 400°C, lignin yields mixtures of simple phenols, acetic acid, methanol, wood creosote, thermolysis tars, coke, and charcoal. Lower temperatures of about 300°C and the presence of water have resulted in the formation of phenols by high-temperature hydrolysis.[362]

3. Alkali fusion: The heating of black-liquor organics in the presence of approximately 30% alkali (on organic matter content) to temperatures of about 250 to 300°C leads to the formation of pyrocatechol and its homologs as well as some ether-soluble, water-insoluble diphenolic compounds totally deprived of methoxy groups. This reaction may be carried out in

the presence of water under pressure or under atmospheric or reduced pressure conditions.[337,363]

4. Oxidative degradation: The best-known example of a low-molecular-weight degradation product from lignin is vanillin, obtained by oxidative degradation of lignin sulfonates in alkali. Efficient oxidation and recovery processes result in vanillin yields in excess of 10% on lignin. This has been reviewed by Goheen.[321] Economic opportunities for other low-molecular-weight chemicals from wood have been surveyed by Goldstein.[358]

5. Dimethyl sulfide and DMSO: Nucleophilic cleavage of methoxy groups by mercaptide or sulfide ions has been shown to result in the formation of dimethyl sulfide or methyl mercaptan, respectively. The commercial process employed by the Crown Zellerbach Corp. involves the addition of elemental sulfur to the concentrated black liquor and heating the mixture in a reactor at 200 to 250°C. The DMS formed is flashed from the reactor, condensed, and purified. Most of it is subsequently oxidized to DMSO by the action of NO_2. A detailed review has been published by Goheen.[321]

Polymeric Products from Lignin

The literature is full of examples for converting lignin into polymeric products. Relevant reviews have been prepared by Allan,[364] Hoyt and Goheen[365] and Falkehag.[342] With improved understanding of the structure, properties, and reactivities of lignin, the purely empirical, degradation-oriented approach to utilization has shifted to an approach that is primarily concerned with revitalizing the function that lignin performed in wood. Among the many proposed applications for lignin in polymeric form, only three will be discussed. Each deals with an application that is a direct reflection of an inherent property of lignin:

1. The characteristics that make lignin respond to mechanical stresses in the wood can be adapted to adhesives and binders.

2. The characteristics designed to respond to biochemical (physiological) and chemical stress in trees originating from enzymatic and other attack mechanisms involving redox reactions (e.g., enzymatic and photodegradative mechanisms) have equipped lignin to perform well as a polyelectrolytic antioxidant and dispersant.

3. The characteristics that cause lignin to respond to environmental stresses in natural decay seem to have prepared lignin as an ion exchanger, complexing agent, and controlled release (desorption) agent.

The following summarizes the highlights of these three uses of polymeric products from lignin.

Adhesives and Binders. The use of polyphenolic lignin as an extender of phenol-formaldehyde resins finds constraints in terms of color, resin viscosity, cure rate, and mechanical properties. Some of these problems seem to be surmountable by way of modest fractionation efforts of by-product lignins prior to blending. High-molecular-weight lignin sulfonate and kraft lignin fractions were found to be useful directly as phenol-formaldehyde resins or as extenders in commercial preparations. The results with laboratory and commercial mill runs proved satisfactory.[366,367]

The use of spent sulfite liquor solids as the sole binder in wafer boards was found possible after a brief pretreatment with sulfuric acid.[368] The resulting product satisfied specifications for exterior grade boards in home construction.

In addition to phenol-formaldehyde-type resins, kraft lignin, and lignin sulfonates have been converted into polyols for polyurethane manufacture.[215] Increases in the aliphatic polyether-polyester character of the by-product lignin were achieved following carboxylation (by copolymerization with maleic anhydride) and oxyalkylation. The resulting polyols appeared to be versatile and economically competitive[369] coreagents for the manufacture of polyurethane foams[213] and adhesives and coatings.[214] Other synthetic polymers based on lignin have been reviewed by Lindberg.[370]

Dispersants. The polydispersity of lignin, coupled with the solubility of sulfonated lignins in aqueous systems, makes it a useful dispersant in such applications as drilling mud additives, polydispersed dyes,[371] and rubber additives.[342,372,373] Its polydisperse character significantly influences its flocculation properties. Low-molecular-weight lignin sulfonates performed as dispersants in all concentrations, whereas high-molecular-weight lignin sulfonates were flocculants at low concentrations and dispersants at high concentrations.[374,375] A vast number of dispersant formulations and applications are covered in the patent literature.

Soil Application. Lignin's slow biodegradability, its capacity to exchange ions and form complexes with certain metal ions, and its properties as a controlled release agent have made it an attractive target for research on soil amendments. Gradual release of plant nutrients is accomplished with oxidatively ammoniated lignin made by treatment with ammonia and oxygen. The 18 to 20% of fixed nitrogen is 1/3 ionically and 2/3 organically bound to the macromolecular matrix.[376] The organically bound nitrogen occurs in the form of amides (easily hydrolyzed), amines (hard to hydrolyze), and heterocycles (unhydrolyzable). The result is a fertilizer that gradually releases its nitrogen-containing nutrients into the soil, and whose most noticeable action in terms of nutrient mobility is of a time-release type.[377-379]

The controlled-release concept is not limited to the application of plant nutrients, but has also been applied to pesticides. Lignin as carrier of controlled-release pesticides provides highly desirable long-term protection.[380,381]

REFERENCES

1. K. Freudenberg and A. C. Neish, "Constitution and Biosynthesis of Lignin, Springler-Verlag, New York, N.Y., 1968.

2. K. V. Sarkanen and C. H. Ludwig, Eds. "Lignins—Occurrence, Formation, Structure and Reactions," Wiley-Interscience, New York, N.Y., 1971.
3. A. B. Wardrop, "Occurrence and Formation in Plants" in ref. 2, p. 19.
4. A. B. Wardrop, *J. Appl. Polym. Sci.* (Appl. Polym. Symp.), **28**, 1041 (1976).
5. T. Higuchi, *Advan. Enzymol.*, **34**, 207 (1971).
6. M. Erickson and G. E. Miksche, *Phytochemistry*, **13**, 2295 (1974).
7. A. C. Neish in "Formation of Wood in Forest Trees," M. H. Zimmermann Ed., Academic, New York, N.Y., 1964, p. 219.
8. B. J. Fergus, A. R. Procter, J. A. N. Scott, D. A. I. Goring, *Wood Sci. Technol.*, **3**, 117 (1969).
9. B. J. Fergus and D. A. I. Goring, *Holzforschung*, **24**(4), 118 (1970).
10. J. R. Wood and D. A. I. Goring, *Pulp Paper Mag. Can.*, **72** (3), 61 (1971).
11. J. A. N. Scott and D. A. I. Goring, *Wood Sci. Technol.*, **4** (4), 237 (1970).
12. Y. Musha and D. A. I. Goring, *Wood Sci. Technol.*, **9** (1), 45 (1975).
13. J. E. Stone, A. M. Scallan, and P. A. V. Ahlgren, *Tappi*, **54** (9), 1527 (1971).
14. J. M. Harkin and J. R. Obst, *Science*, **180** (4083), 296 (1973).
15. N. P. Kutscha and J. M. Schwarzmann, *Holzforschung*, **29** (3), 79 (1975).
16. H. Imagawa, K. Fukazawa, and S. Ishida, *Res. Bull. Coll. Exp. For. Hokkaido U.*, **33**(1), 127 (1976) through *Abstr. Bull. Inst. Paper Chem.*, **47** (5), 4753 (1976).
17. K. V. Sarkanen, "Precursors and Their Polymerization" in ref. 2, p. 95.
18. J. M. Harkin in "Oxidative Coupling of Phenols," W. I. Taylor and A. R. Battersby, Eds., Marcel Dekker, New York, N.Y., 1967, p. 243.
19. E. Adler, *Wood Sci. Technol.* **11**, 169 (1977).
20. R. D. Gibbs in "the Physiology of Forest Trees," K. V. Thimann, Ed., Ronald, New York, N.Y., 1958, p. 269.
21. K. V. Sarkanen and H. L. Hergert, "Classification and Distribution" in ref. 2, p. 43.
22. M. Shimada, H. Kuroda, and T. Higuchi, *Phytochemistry*, **12** (12), 2873 (1973).
23. H. Kuroda, M. Shimada, and T. Higuchi, *Phytochemistry*, **14** (8), 1759 (1975).
24. M. Shimada, H. Fushiki, and T. Higuchi, *J. Jap. Wood Res. Soc.* (Mokuzai Gakkaishi), **19** (1), 13 (1973).
25. K. Kratzl, *Holz Roh- u. Werkstoff*, **19** (7), 219 (1961).
26. F. Tiemann and B. Mendelsohn, *Ber. Deutsch. Chem. Ges.*, **8**, 1139 (1875).
27. P. Klason, *Svensk Kem. Tidskr.*, **9**, 133 (1897).
28. H. Erdtman, *Biochem. Z*, **258**, 172 (1933).
29. H. Erdtman, *Justus Liebigs Ann. Chem.*, **503**, 283 (1933).
30. K. Freudenberg, *Fortschr. der Chem. Org. Naturstoffe*, **2**, 1 (1939).
31. K. Freudenberg, *J. Polym. Sci.*, **48**, 371 (1960).
32. K. Freudenberg, *Pure Appl. Chem.* (London), **5**, 9 (1962).
33. K. Freudenberg, "Beiträge zur Biochemie und Physiologie von Naturstoffen: Festschrift Kurt Mothes zum 65. Geburtstag", 1965, p 167.
34. K. Freudenberg, *Science*, **148**, 595 (1965).
35. K. V. Sarkanen and C. H. Ludwig, "Definition and Nomenclature" in ref. 2, p. 1.
36. O. Martensson and G. Karlsson, *Ark. Kemi*, **31**, (2), 5 (1969).
37. W. J. Connors, J. S. Ayers, K. V. Sarkanen, and J. S. Gratzl, *Tappi*, **54** (8), 1284 (1971).
38. F. Nakatsubo, K. Sato, and T. Higuchi, *J. Jap. Wood Res. Soc.* (Mokuzai Gakkaishi), **22** (1), 29 (1976).
39. G. Leary and W. Thomas, *Tetrahedron Lett.* 3631 (1975).

40. K. Lundquist, G. E. Miksche, L. Ericsson, and L. Berndtson, *Tetrahedron Lett.*, 4587 (1967).
41. H. Nimz, *Chem. Ber.*, **98**, 3160 (1965).
42. H. Nimz, *Chem. Ber.*, **99**, 469, 2638 (1966).
43. W. G. Glasser and H. R. Glasser, *Macromol.* **7** (1), 17 (1974).
44. A. Sakakibara, *Rec. Adv. Phytochemistry*, **11**, 117 (1977).
45. K. Freudenberg, J. M. Harkin, and N. K. Werner, *Naturwissenschaften*, **50** (13), 476 (1963).
46. E. Adler, H.-D. Becker, T. Ishihara, and A. Stamvik, *Holzforschung*, **20**, 3 (1966).
47. W. G. Glasser and H. R. Glasser, *Cellulose Chem. Technol.*, **10** (1), 23 (1976).
48. I. J. Miller and G. J. Smith, *Aust. J. Chem.*, **28** (4), 825 (1975).
49. G. Leary, *Chem. Commun.*, **13**, 688 (1971).
50. G. Leary, *JCS Perkin II*, **5**, 640 (1972).
51. J. A. Hemmingson and G. Leary, *JCS Perkin II*, **14**, 1584 (1975).
52. J. A. Hemmingson, *JCS Perkin II*, **5**, 616 (1977).
53. W. G. Glasser, H. R. Glasser, and H. H. Nimz, *Macromol.*, **9** (5), 866 (1976).
54. K. Freudenberg, *Holzforschung*, **18** (1, 2), 3 (1964).
55. K. Freudenberg, *Holzforschung*, **18** (6), 166 (1964).
56. J. M. Harkin, *Rec. Adv. Phytochemistry*, **2**, 35 (1969).
57. W. G. Glasser and H. R. Glasser, *Holzforschung*, **28** (1), 5 (1974).
58. W. G. Glasser and H. R. Glasser, *Cellulose Chem. Technol.*, **10** (1), 39 (1976).
59. O. Faix, *Holzforschung*, **28** (6), 222 (1974).
60. O. Faix, *Papier* (Darmstadt), **30** (10A), V–1 (1976).
61. M. Erickson, S. Larsson, and G. E. Miksche, *Acta Chem. Scand.*, **27** (3), 903 (1973).
62. H. Nimz, *Angew. Chem.* (Intnl. Ed.,) **13**, 313 (1974).
63. C. H. Ludwig, "Magnetic Resonance Sprectra" in ref. 2, p. 299.
64. T. Higuchi, F. Nakatsubo, and Y. Ikeda, *Holzforschung*, **28** (6), 189 (1974).
65. D. A. I. Goring, "Polymer Properties of Lignin and Lignin Derivatives" in ref. 2, p. 695.
66. P. P. Erin'sh, V. A. Tsinite, M. K. Yakobson, and Y. A. Gravitis, *J. Appl. Polym. Sci.* (Appl. Polym. Symp.), **28**, 1117 (1976).
67. M. Kudo and T. Kondo, *Tappi*, **54** (12), 2046 (1971).
68. S. B. Lee and P. Luner, *Tappi*, **55** (1), 116 (1972).
69. H. Hatakeyama and K. Kubota, *Cellulose Chem. Technol.*, **6** (5), 521 (1972).
70. H. Hatakeyama and T. Hatakeyama, *Rep. Progr. Polym. Phys. Jap.*, **17**, 713 (1974).
71. H. Hatakeyama, K. Iwashita, and J. Nakano, *Rep. Progr. Polym. Phys. Jap.*, **17**, 715 (1974).
72. H. Hatakeyama, K. Iwashita, G. Meshitsuka, and J. Nakano, *J. Jap. Wood Res. Soc.* (Mokuzai Gakkaishi), **21** (11), 618 (1975).
73. M. Kimura, H. Hatakeyama, and J. Nakano, *J. Jap. Wood Res. Soc.* (Mokuzai Gakkaishi), **21** (11), 624 (1975).
74. W. J. Cousins, R. W. Armstrong, and W. H. Robinson, *J. Matls. Sci.*, **10** (10), 1655 (1975).
75. W. J. Cousins, *Wood Sci. Technol.*, **10** (1), 9 (1976).
76. D. Fengel, *Holzforschung*, **29** (5), 160 (1975).
77. N. F. Efendieva, N. N. Shorygina, and N. P. Krasovskaya, *Izv. Adv. Nauk*

SSSR, Ser. Khim., **12**, 2812 (1973), through *Abstr. Bull. Inst. Paper Chem.*, **45** (7), 6912 (1973).

78. I. F. Kaimin', M. Ioelovich Ya., Z. N. Kreitsberg, and B. M. Aronchik, *Khim. Drev.* (Riga), **4**, 118 (1975); through *Abstr. Bull. Inst. Paper Chem.*, **46** (8), 7945 (1976).

79. N. G. Moskovtsev, E. I. Chupka, and V. M. Nikitin, *Khim. Drev.* (Riga), **2**, 44 (1976), through *Abstr. Bull. Inst. Paper Chem.*, **47** (6), 5856 (1976).

80. N. G. Moskovtsev, E. I. Chupka, and V. M. Nikitin, *Khim. Drev.* (Riga), **1**, 70 (1976), through *Abstr. Bull. Inst. Paper Chem.*, **47** (5), 4759 (1976).

81. O. P. Grushnikov and N. N. Shorygina, *Usp. Khim.*, **39** (8), 1459, 1970, through *Abstr. Bull. Inst. Paper Chem.* **42** (11), 11139 (1972).

82. N. F. Efendieva and N. N. Shorygina, *Izv. Akad. Nauk SSSR, Ser. Khim.*, **8**, 1860 (1974), through *Abstr. Bull. Inst. Paper Chem.*, **45** (11), 11719 (1975).

83. T. Koshijima, F. Yaku, and R. Tanaka, *J. Appl. Polym. Sci.* (Appl. Polym. Symp.), **28**, 1025 (1976).

84. O. Eriksson and B. O. Lindgren, *Svensk Papperstidn.* **80** (2), 59 (1977).

85. F. Yaku and T. Koshijima, *J. Jap. Wood Res. Soc.* (Mokuzai Gakkaishi), **17** (6), 267 (1971).

86. F. Yaku and T. Koshijima, *J. Jap. Wood Res. Soc.* (Mokuzai Gakkaishi), **18** (10), 519 (1972).

87. T. Koshijima, F. Yaku, and K. Fukube, *J. Jap. Wood Res. Soc.* (Mokuzai Gakkaishi), **17** (6), 238 (1974).

88. B. Kosikova, D. Joniak, and J. Skamla, *Cellulose Chem. Technol.*, **6** (5), 579 (1972).

89. B. Kosikova, J. Polcin, and D. Joniak, *Cellulose Chem. Technol.*, **7**, 605 (1973).

90. K. Freudenberg and G. Grion, *Chem. Ber.*, **92**, 1355 (1959).

91. K. Freudenberg and J. M. Harkin, *Chem. Ber.*, **93**, 2814 (1959).

92. G. V. Davydova, A. V. Lozanova, and N. N. Shorygina, *Khim. Ispol'z. Lignina*, 39, (1974), through *Abstr. Bull. Inst. Paper Chem.*, **46** (11), 11157 (1976).

93. G. V. Davydova, A. V. Lozanova, and N. N. Shorygina, *Izv. Akad. Nauk SSSR, Ser. Khim.*, **8**, 1837 (1972), through *Abstr. Bull. Inst. Paper Chem.*, **44** (9), 9420 (1974).

94. D. Joniak and B. Kosikova, *Chem. Zvesti*, **28** (1), 110 (1974).

95. D. Joniak and B. Kosikova, *Cellulose Chem. Technol.*, **10** (6), 691 (1976).

96. V. V. Elkin, N. N. Shorygina, A. D. Alekseev, and V. M. Reznikov, *Khim. Drev.* (Riga), **14**, 85 (1973), through *Abstr. Bull. Inst. Paper Chem.*, **45** (6), 5741 (1974).

97. N. Parameswaran, O. Faix, and W. Schweers, *Holzforschung*, **29** (1), 1 (1975).

98. M. Erickson, S. Larsson, and G. E. Miksche, *Acta Chem. Scand.*, **27** (5), 1673 (1973).

99. S. Yasuda and A. Sakakibara, *J. Jap. Wood Res. Soc.* (Mokuzai Gakkaishi), **21** (6), 363 (1975).

100. M. Erickson, G. E. Miksche, and I. Somfai, *Holzforschung*, **27** (4), 113 (1973).

101. A. Andersson, M. Erickson, H. Fridh, and G. E. Miksche, *Holzforschung*, **27** (6), 189 (1973).

102. E. D. Swan, *Pulp Paper Mag. Can.*, **67** (1), T–456 (1966).

103. G. E. Miksche and S. Yasuda, *Holzforschung*, **31** (2), 57 (1977).

104. J. P. Joseleau, G. E. Miksche, and S. Yasuda, *Holzforschung*, **31** (1), 19 (1976).

105. T. Fukuda, M. Oota, N. Terashima, and T. Kanda, *J. Jap. Wood Res Soc.* (Mokuzai Gakkaishi), **21** (3), 157 (1975).
106. T. Fukuda and T. Kanda, *J. Jap. Wood Res. Soc.* (Mokuzai Gakkaishi), **22** (2), 112 (1976).
107. T. Fukuda, *J. Jap. Wood Res. Soc.* (Mokuzai Gakkaishi), **22** (11), 638 (1976).
108. D. E. Bland and M. Menshun, *Holzforschung,* **25** (6), 174 (1971).
109. C. P. Vance and R. T. Sherwood, *Plant Physiol.* **57** (6), 915 (1976).
110. C. P. Vance, J. O. Anderson, and R. T. Sherwood, *Plant Physiol.* **57** (6), 920 (1976).
111. F. W. Whitmore, *Phytochemistry,* **15** (3), 375 (1976).
112. N. Morohoshi and A. Sakakibara, *J. Jap. Wood Res. Soc.* (Mokuzai Gakkaishi), **17** (8), 347 (1971).
113. N. Morohoshi and A. Sakakibara, *J. Jap. Wood Res. Soc.* (Mokuzai Gakkaishi), **17** (8), 354 (1971).
114. K. E. Wolter, J. M. Harkin, and T. K. Kirk, *Physiol. Plant.,* **31**, 140 (1974).
115. K. V. Sarkanen, H. -M. Chang, and G. G. Allan, *Tappi,* **50** (12), 583 (1967).
116. M. Erickson and G. E. Miksche, *Holzforschung,* **28** (4), 135 (1974).
117. R. D. Bleam and J. M. Harkin, *Wood Sci. Technol.* **8** (2), 122 (1975).
118. H. -M. Chang and K. V. Sarkanen, *Tappi,* **56** (3), 132 (1973).
119. K. Forss and K. -E. Fremer, *Paperi Puu,* **47** (8), 443 (1965).
120. K. Forss, K. -E. Fremer, and B. Stenlund, *Paperi Puu,* **48** (9), 565 (1966).
121. K. Forss, K. -E. Fremer, and B. Stenlund, *Paperi Puu,* **49** (9), 669 (1966).
122. K. Forss and B. Stenlund, *Kem. Teollisuus,* **28** (10), 757 (1971).
123. K. Forss and K. -E. Fremer, *SITRA Symp. Enzymatic Hydrolysis Cellulose,* 41 (1975).
124. G. Meshitsuka and J. Nakano, *J. Jap. Wood Res. Soc.* (Mokuzai Gakkaishi), **18** (10), 503 (1972).
125. M. Wayman and T. I. Obiaga, *Can. J. Chem.,* **52** (11), 2102 (1974).
126. H. I. Bolker and H. S. Brenner, *Science,* **170**, 173 (1970).
127. Y. Z. Lai and K. V. Sarkanen, "Isolation and Structure" in ref. 2, p. 165.
128. *Tappi Standard Method T-13 m,* Technical Association of the Pulp and Paper Industry, Atlanta, Ga.
129. K. Freudenberg, H. Zocher, and W. Dürr, *Chem. Ber.,* **62**, 1814 (1929).
130. A. Bjorkman, *Svensk Papperstidn.,* **59** (13), 477 (1956).
131. A. Bjorkman and B. Person, *Svensk Papperstidn.,* **60** (15), 158 (1951).
132. A. Bjorkman, *Svensk Papperstid.,* **60** (7), 243 (1957).
133. H. H. Brownell and K. L. West, *Pulp Paper Mag. Can.,* **62**, T-374 (1961).
134. H. H. Brownell, *Tappi,* **48** (9), 513 (1965).
135. C. Simionescu, M. Grigoras, C. Ciubotaru, and G. Rozmarin, *Celuloza Hirtie,* **21** (1), 7 (1972).
136. C. Simionescu, M. Grigoras, A. Stoleriu, and G. Rozmarin, *Celuloza Hirtie,* **21** (2), 91 (1972).
137. S. Larsson and G. E. Miksche, *Acta Chem. Scand.,* **23** (10), 3337 (1969).
138. B. Bogomolov, N. D. Babikova, V. A. Pivovarova, and L. P. Stepovaya, *Khim. Ispol'z. Lignina,* 102, (1974), through *Abstr. Bull. Inst. Paper Chem.,* **46** (11), 11152 (1976).
139. D. E. Bland and M. Menshun, *Holzforschung,* **27** (2), 33 (1973).
140. K. Lundquist and R. Simonson, *Svensk Papperstidn.,* **78** (11), 390 (1975).
141. S. A. Saidalimov, L. S. Smirnova, and Kh. A. Abduazimov, *Khim. Drev.* (Riga), **2**, 75 (1976), through *Abstr. Bull. Inst. Paper Chem.,* **47** (6), 5858 (1976).
142. S. A. Saidalimov, L. S. Smirnova, and Kh. A. Abduazimov, *Khim. Prir. Soedin.,* 5, 643 (1976), through *Abstr. Bull. Inst. Paper Chem.,* **48** (2), 1254 (1977).

143. J. C. Pew, *Tappi,* **40** (7), 553 (1957).
144. J. C. Pew and P. Weyna, *Tappi,* **45** (3), 247 (1962).
145. H. Chang, E. B. Cowling, W. Brown, E. Adler, and G. Miksche, *Holzforschung,* **29** (5), 153 (1975).
146. M. Wayman and T. I. Obiaga, *Tappi,* **57** (4), 123 (1974).
147. B. Kosikova and J. Polcin, *Wood Sci. Technol.,* **7** (4), 308 (1973).
148. J. M. Pepper, P. E. T. Baylis, and E. Adler, *Can. J. Chem,* **37**, 1241 (1959).
149. R. Draganova, *Cellulose Chem. Technol.,* **5** (5), 463 (1971).
150. J. Desmet, Ph.D. Thesis, Grenoble U., Grenoble, France, 1971.
151. A. D. Alekseev, L. G. Matusevich, and V. M. Reznikov, *Khim. Drev.* (Riga), **9**, 57 (1971), through *Abstr. Bull. Inst. Paper Chem,* **43** (10), 10418 (1973).
152. N. Terashima and S. Kachi, *J. Jap. Wood Res. Soc.* (Mokuzai Gakkaishi), **20** (1), 31 (1974).
153. S. Kachi and N. Terashima, *J. Jap. Wood Res. Soc.* (Mokuzai Gakkaishi), **20** (9), 453 (1974).
154. S. Kachi, H. Araki, and N. Terashima, *J. Jap. Wood Res. Soc.* (Mokuzai Gakkaishi), **21** (6), 376 (1975).
155. S. Kachi, H. Araki, N. Terashima, and T. Kanda, *J. Jap. Wood Res. Soc.* (Mokuzai Gakkaishi), 21 (12), 669 (1975).
156. I. Mogharab and W. G. Glasser, *Tappi,* **59** (10), 110 (1976).
157. K. Freudenberg, K. Seib, and K. Dall, *Chem. Ber.,* **92**, 807 (1959).
158. W. E. Fisher, B. L. Lenz, and S. A. Hider, *Tappi,* **55** (2), 234 (1972).
159. W. E. Fisher, B. L. Lenz, and S. A. Hider, *Tappi* **55** (2), 240 (1972).
160. D. W. Glennie, "Reactions in Sulfite Pulping." in ref. 2, p. 597.
161. G. Gellerstedt and J. Gierer, *Svensk Papperstidn.,* **74** (5), 117 (1971).
162. G. Gellerstedt, *Svensk Papperstidn.,* **79** (16), 537 (1976).
163. W. Schweers and W. Vorher, *Holzforschung,* **29** (4) 135 (1975).
164. G. Meshitsuka, K. Kawakamji, and J. Nakano, *J. Jap. Tappi,* **24** (10), 523 (1970).
165. A. R. Procter, W. Q. Yean, and D. A. I. Goring, *Pulp Paper Mag. Can.* **68**, T–445 (1967).
166. J. Marton, "Reactions in Alkaline Pulping." in ref. 2, p. 639.
167. J. Gierer, I. Pettersson, L. -A. Smedman, and I. Wennberg, *Acta Chem. Scand.,* **27** (6), 2083 (1973).
168. J. Gierer, F. Imsgard, and I. Pettersson, *J. Appl. Polym. Sci.* (Appl. Polym. Symp.), **28**, 1195 (1976).
169. J. Gierer and I. Pettersson, *Can. J. Chem.,* **55** (4), 593 (1977).
170. G. E. Miksche, *Acta Chem. Scand.,* **27** (4), 1355 (1973).
171. H. Aminoff, G. Brunow, K. Falck, and G. E. Miksche, *Acta Chem. Scand.,* **28B** (3), 373 (1974).
172. J. Gierer and I. Noren, *Acta Chem. Scand.,* **16**, 1976 (1962).
173. J. Gierer and I. Noren, *Acta Chem. Scand.,* **16**, 1713 (1962).
174. J. Gierer, B. Lenz, and N. H. Wallin, *Acta Chem. Scand.,* **18**, 1469 (1964).
175. G. E. Miksche, *Acta Chem. Scand.,* **26** (8), 3275 (1972).
176. J. Gierer and L. -A. Smedman, *Acta Chem. Scand.,* **25** (4), 1461 (1971).
177. G. E. Miksche, *Acta Chem. Scand.,* **26** (8), 3269 (1972).
178. A. F. A. Wallis, *Austl. J. Chem.,* **25** (7), 1529 (1972).
179. M. Tamao and N. Terashima, *J. Jap. Wood Res. Soc.* (Mokuzai Gakkaishi) **16** (6), 284 (1970).
180. M. Tamao and N. Terashima, *J. Jap. Wood Res. Soc.* (Mokuzai Gakkaishi) **17** (1), 10 (1971).
181. K. Kringstad, F. de Sousa, and U. Westermark, *Svensk Papperstidn.,* **79** (18), 604 (1976).
182. N. Morohoshi and A. Sakakibara, *J. Jap. Wood Res. Soc.* (Mokuzai Gakkaishi), **18** (1), 27 (1972).

183. K. Lundquist, *Svensk Papperstidn.*, **76** (18), 704 (1973).
184. T. N. Kleinert, *Tappi,* **49** (3), 126 (1966).
185. J. W. Shelfer and E. M. Hartley, *Proc. 1977 TAPPI Alkaline Pulping Conf.*, 67 (1977).
186. O. Ingruber, *Papier* (Darmstadt), **24** (10A), 711 (1970).
187. G. Gellerstedt, *Svensk Papperstidn.*, **79** (16), 537 (1976).
188. *Proc. Intl Symp. Delignification with Oxygen, Ozone, and Peroxides,* Raleigh, N.C., May 1975.
189. K. Kratzl, P. Claus, W. Lonsky, and J. S. Gratzl, *Wood Sci. Technol.*, **8** (1), 35 (1974).
190. J. Gierer, F. Imsgard, and I. Noren, *Acta Chem. Scand.*, **31** (7), 561 (1977).
191. R. C. Eckert, H. -M. Chang, and W. P. Tucker, *Tappi,* **56** (6), 134 (1973).
192. T. Oki, K. Okubo, and H. Ishikawa, *J. Jap. Wood Res. Soc.* (Mokuzai Gakkaishi), **22** (2), 98 (1976).
193. S. Katuscak, K. Horsky, and M. Mahdalik, *Paperi Puu,* **53** (4), 197 (1971).
194. R. A. Young and J. Gierer, *J. Appl. Polym. Sci.* (Appl. Polym. Symp.), **28**, 1213 (1976).
195. G. J. Kubes and H. I. Bolker, *Proc. 1977 TAPPI Alkaline Pulping Conf.*, 297 (1977).
196. H. H. Holten and F. L. Chapman, *Proc. 1977 TAPPI Alkaline Pulping Conf.*, 311 (1977).
197. K. L. Ghosh, V. Venkatesh, W. J. Chin, and J. S. Gratzl, *Proc. 1977 TAPPI Alkaline Pulping Conf.*, 211 (1977).
198. G. Gellerstedt and J. Gierer, *Acta Chem. Scand.*, **29B** (5), 561 (1975).
199. R. Draganova and Ts. Khristov, *God. Vissh. Khim-Tekhnol. Inst.* (Sofia), **15** (3), 253, 1968–1970 (publ. 1972), through *Abstr. Bull. Inst. Paper Chem.*, **44** (9), 9423 (1974).
200. Y. –T. Wu, Ph.D. Thesis, U. of California, Berkeley, 1970.
201. E. L. Springer, J. F. Harris, and W. K. Neill, *Tappi,* **46** (9), 551 (1963).
202. E. L. Springer and L. L. Zoch, *Svensk Papperstidn.*, **69** (16), 513 (1966).
203. E. L. Springer and L. L. Zoch, *Tappi,* **54** (12), 2059 (1971).
204. H. Ishikawa, K. Okubo, and T. Oki, *J. Jap. Wood Res. Soc.* (Mokuzai Gakkaishi), **16** (3), 145 (1970).
205. Yu. A. Zoldners, V. S. Gromov, and Ya. A. Surna, *Khim. Drev.* (Riga), **3**, 129 (1969), through *Abstr. Bull. Inst. Paper Chem.*, **43** (1), 821 (1972).
206. Th. N. Kleinert, *Papier* (Darmstadt), **21** (10A), 653 (1967).
207. W. Schweers and M. Rechy, Papier (Darmstadt), 26(10A), 585 (1972).
208. W. Schweers, Chem. Technol., 132, 490 (1974).
209. H. Nimz, *Holzforschung,* **23** (3), 84 (1969).
210. W. Vorher and W. H. M. Schweers, *J. Appl. Polym. Sci.* (Appl. Polym. Symp.), **28**, 277 (1975).
211. W. Sandermann, *Svensk Papperstidn.*, **52** (15), 365 (1949).
212. W. Sandermann, C. Schlink, W. Czirnich, and W. Weissmann, *Papier* (Darmstadt), **24** (4), 187 (1970).
213. O. H. -H. Hsu and W. G. Glasser, *J. Appl. Polym. Sci.* (Appl. Polym. Symp.), **28**, 297 (1975).
214. O. H. -H. Hsu and W. G. Glasser, *Wood Sci. Technol.*, **9** (2), 97 (1976).
215. W. G. Glasser, O. H. –H. Hsu, Research Corp., U. S. Patent 4017474 (1977).
216. T. N. Kleinert, *Papier* (Darmstadt), **30** (10A), V–18 (1976).
217. T. N. Kleinert, *Tappi,* **57** (8), 99 (1974).
218. T. N. Kleinert, *Tappi,* **58** (8), 170 (1975).
219. T. N. Kleinert, *Holzforschung,* **29** (3), 107 (1975).

220. J. Nakano, C. Takatsuka, and H. Daima, *Jap. Tappi*, **30** (12), 650 (1976).
221. C. W. Dence, "Halogenation and Nitration" in ref. 2, p. 373.
222. H. -M. Chang and G. G. Allan, "Oxidation" in ref. 2, p. 433.
223. J. B. Van Buren and C. W. Dence, *Tappi*, **50** (11), 553 (1967).
224. J. Gierer and L. Sundholm, *Svensk Papperstidn.*, **74** (11), 345 (1971).
225. S. A. Braddon and C. W. Dence, *Tappi*, **51** (6), 249 (1968).
226. J. M. Gess and C. W. Dence, *Tappi*, **54** (7), 1114 (1971).
227. P. A. Ahlgren and D. A. I. Goring, *Can. J. Chem.*, **49** (8), 1272 (1971).
228. B. O. Lindgren, *Papier* (Darmstadt), **28** (10A), V–29 (1974).
229. B. O. Lindgren, *Svensk Papperstidn.*, **74** (3) 57 (1971).
230. S. Hosoya and J. Nakano, *J. Jap. Wood Res. Soc.* (Mokuzai Gakkaishi), **18** (9), 457 (1972).
231. N. G. Vander Linden, Ph.D. Thesis, Institute of Paper Chemistry, Appleton, Wis., 1974.
232. J. C. Farrand and D. C. Johnson, *J. Org. Chem.*, **36** (23), 3606 (1971).
233. G. Strumila and W. H. Rapson, *Pulp Paper Mag. Can.*, **76** (9), 74 (1975).
234. T. Oki, K. Okubo, and H. Ishikawa, *J. Jap. Wood Res. Soc.* (Mokuzai (Gakkaishi), **20** (11), 549 (1974).
235. T. J. McDonough and W. H. Rapson, *CPPA Trans. Tech. Sect.*, **1** (1), 12 (1975).
236. T. J. McDonough and W. H. Rapson, *Extended Abstr., Can. Wood Chem. Symp.*, Quebec, Canada, **4**, 43 1973, Chem. Inst. of Canada/CPPA, Montreal, Canada.
237. J. A. Fleck, Ph.D. Thesis, Institute of Paper Chemistry, Appleton, Wis. 1974.
238. J. S. Albrecht and G. A. Nicholls, *Paperi Puu*, **56** (11), 927 (1974).
239. M. Lukacova, J. Klanduch, and S. Kovac, *Holzforschung*, **31** (1), 13 (1977).
240. G. Gellerstedt and E. L. Pettersson, *Acta Chem. Scand.*, **B-29** (10), 1005 (1975).
241. G. Gellerstedt and E. L. Pettersson, *Svensk Papperstidn.*, **80** (1), 15 (1977).
242. P. Ander and K. -E. Eriksson, *Svensk Papperstidn.*, **78** (18), 643 (1975).
243. T. K. Kirk, *Ann. Rev. Phytopathology*, **9**, 185 (1971).
244. T. K. Kirk, "Biological Transformation of Wood by Microorganisms" in *Proc. 2nd Intl. Cong. Plant Pathology* (Wood Products Pathology Sessions), W. Liese, Ed., Minneapolis, Minn., September 1973, 153 (1975).
245. T. K. Kirk and H. Chang, *Holzforschung*, **28** (6), 217 (1974).
246. T. K. Kirk and H. Chang, *Holzforschung*, **29** (2), 56 (1975).
247. T. K. Kirk, *Holzforschung*, **29** (3), 99 (1975).
248. D. Ciaramitaro and C. Steelink, *Phytochemistry*, **14** (2), 543 (1975).
249. T. Ishihara and M. Miyazaki, *J. Jap. Wood Res. Soc.* (Mokuzai Gakkaishi), **18** (8), 415 (1972).
250. T. Ishihara and M. Ishihara, *J. Jap. Wood Res. Soc.* (Mokuzai Gakkaishi), **22** (6), 371 (1976).
251. J. Gierer and A. E. Opara, *Acta Chem. Scand*, **27** (8), 2909 (1973).
252. P. Ander and K. -E. Eriksson, *Arch. Microbiol.*, **109**, 1 (1976).
253. T. K. Kirk, W. J. Connors, and J. G. Zeikus, *Appl. Environ. Microbiol.*, **32** (1), 192 (1976).
254. H. Kawakami, *J. Jap. Wood Res. Soc.* (Mokuzai Gakkaishi), **21** (2), 92 (1975).
255. J. Trojanowski, M. Wojtas-Wasilewska, and B. Junosza-Wolska, *Acta Microbiol. Polon.*, **2B** (1), 13 (1970).
256. H. Kawakami, *J. Jap. Wood Res. Soc.* (Mokuzai Gakkaishi), **21** (11), 629 (1975).

257. P. Porter and A. G. Singleton, *Brit. J. Nutr.*, **25** (1), 3 (1971).
258. H. Kawakami, *J. Jap. Wood Res. Soc.* (Mokuzai Gakkaishi), **22** (4), 252 (1976).
259. H. Kawakami, *J. Jap. Wood Res. Soc.* (Mokuzai Gakkaishi), **22** (4), 246 (1976).
260. B. L. Browning, "Methods of Wood Analysis," 2 vols, Interscience, New York, N.Y., 1967.
261. E. Stahl, F. Karig, U. Brogmann, H. Nimz, and H. Becker, *Holzforschung*, **27** (3), 89 (1973).
262. F. Karig and E. Stahl, *Holzforschung*, **28** (6), 201 (1974).
263. D. E. Bland and M. Menshun, *Appita*, **25** (2), 110 (1971).
264. Y. Musha and D. A. I. Goring, *Wood Sci. Technol.* **7** (2), 133 (1974).
265. H. Augustin, O. Faix, and W. Schweers, *Holzforschung*, **24** (4), 125 (1970).
266. K. Kurschner, *Holzforschung*, **25** (3), 65 (1971).
267. O. Faix and W. Schweers, *Holzforschung*, **26** (1), 9 (1972).
268. I. T. Clark, *Tappi*, **45** (4), 310 (1962).
269. C. Giurcanu and I. Anton, *Celuloza Hirtie*, **22** (6), 233 (1973).
270. D. B. Johnson, W. E. Moore, and L. C. Zauk, *Tappi*, **44** (11), 793 (1961).
271. M. O. Bagby, R. L. Cunningham, and R. L. Maloney, *Tappi*, **56** (4), 162 (1973).
272. C. R. Pottenger and Potlatch Forests Inc., U. S. Patent 3674434 (1972).
273. V. F. Felicetta and J. L. McCarthy, *Tappi*, **46** (6), 337 (1963).
274. T. Almgren and B. Josefsson, *Svensk Papperstidn.*, **76** (1), 19 (1973).
275. B. F. Hrutfiord, P. Y. Jone, and J. L. McCarthy, *Tappi*, **53** (9), 1746 (1970).
276. H. Evliya and A. Olcay, *Holzforschung*, **28** (4), 130 (1974).
277. W. Schweers and O. Faix, *Holzforschung*, **27** (6), 208 (1973).
278. Y. -Z. Lai and K. V. Sarkanen, *Cellulose Chem. Technol.*, **9** (3), 239 (1975).
279. M. Erickson and G. E. Miksche, *Acta Chem. Scand.*, **26** (8), 3085 (1972).
280. T. Yamasaki, K. Hata, and T. Higuchi, *J. Jap. Wood Res. Soc.* (Mokuzai Gakkaishi), **18** (7), 361 (1972).
281. T. Yamasaki, K. Hata, and T. Higuchi, *J. Jap. Wood Res. Soc.* (Mokuzai (Gakkaishi), **19** (6), 299 (1973).
282. H. Nimz, I. Mogharab, and H. -D. Ludemann, *Makromol. Chem.*, **175** (9), 2563 (1974).
283. D. Gagnare, D. Robert, M. Vignon, and P. Vottero, *Eur. Polym. J.*, **7** (7), 965 (1971).
284. T. Higuchi, K. Ogino, and M. Tanahashi, *Wood Res.* (Kyoto), **51**, 1 (1971).
285. O. Faix, *Mitt. Bundesforsch.* (Reinbek/Hamburg), **92**, 205 pp. (1973).
286. O. Faix and W. Schweers, *Holzforschung*, **29** (2), 48 (1975).
287. K. Forss and K. -E. Fremer, *Extended Abstr., Can. Wood Chem. Symp.*, Quebec, Canada, 87, 1976, Chem. Inst. of Canada/CPPA, Montreal, Canada.
288. K. Forss and K. -E. Fremer, Proceedings of *SITRA-Symp. Enzymatic Hydrolysis Cellulose*, p. 41 (1975).
289. O. Faix, *Papier* (Darmstadt), **30** (10A), V–1 (1976).
290. H. I. Bolker, "Natural and Synthetic Polymers—An Introduction," Marcel Dekker, New York, N.Y., 1974.
291. Otto Goldschmid, "Ultraviolet Spectra" in ref. 2, p. 241.
292. H. L. Hergert, "Infrared Spectra" in ref. 2, p. 267.
293. K. Nilsson-Idner and H. Norrstrom, *Svensk Papperstidn.*, **77** (3), 99 (1974).
294. B. L. Lenz, *Tappi*, **51** (11), 511 (1968).
295. W. G. Glasser, J. S. Gratzl, J. J. Collins, K. Forss, and J. L. McCarthy, *Macromol.*, **6** (1), 114 (1973), and ibid., 8 (5), 565 (1975).

296. C. H. Ludwig, B. J. Nist, and J. L. McCarthy, *J. Am. Chem. Soc.,* **86,** 1196 (1964).
297. A. Sato, T. Kitamura, and T. Higuchi, *J. Jap. Wood Res. Soc.* (Mokuzai Gakkaishi), **18** (5), 253 (1972).
298. A. Sato, T. Kitamura, and T. Higuchi, *Wood Res.* (Kyoto), 59 (1976).
299. H. -D. Ludemann and H. Nimz, *Makromol. Chem.,* **175** (8), 2409 (1974).
300. T. N. Kleinert and J. R. Morton, *Nature,* **196** (4852), 334 (1962).
301. T. N. Kleinert, *Tappi,* **50** (3), 120 (1967).
302. B. Ranby, K. Kringstad, E. B. Cowling, and S. Y. Lin, *Acta Chem. Scand.,* **23** (9), 3257 (1969).
303. S. I. Clare and C. Steelink, *Tappi,* **56** (5), 119 (1973).
304. P. Tormala, J. J. Lindberg, and S. Lehtinen, *Paperi Puu,* **57** (9), 601 (1975).
305. W. G. Glasser and W. Sandermann, *Holzforschung,* **24** (3), 73 (1970).
306. V. Kovacik, J. Skamla, D. Joniak, and B. Kosikova, *Chem. Ber.,* **102,** 1513 (1969).
307. V. Kovacik and J. Skamla, *Chem. Ber.,* **102,** 3623 (1969).
308. K. Lundquist, *J. Appl. Polym. Sci.* (Appl. Polym. Symp.), **28,** 1393 (1976).
309. A. F. A. Wallis, "Solvolysis by Acids and Bases" in ref. 2, p. 345.
310. A. B. Cramer, J. M. Hunter, and H. Hibbert, *J. Amer. Chem. Soc.,* **61,** 509 (1939).
311. J. M. Pepper and J. C. Karapally, *Pulp Paper Mag. Can.,* **73** (9), 88 (1972).
312. K. Kratzl, I. Wagner, and O. Ettingshausen, *Holzforschung,* **24** (2), 32 (1972).
313. Ya. G. Mileshkevich, M. V. Latosk, and V. M. Reznikov, *Khim. Drev.* (Riga), **12,** 21 (1972), through *Abstr. Bull. Inst. Paper Chem.,* **44** (7), 7089 (1974).
314. R. D. Hartley, *J. Chromatogr.,* **54** (3), 335 (1971).
315. S. Kagawa, *J. Jap. Tappi,* **26** (11), 550 (1972).
316. D. L. Brink, Y. T. Wu, H. P. Naveau, J. G. Bicho, and M. M. Merriman, *Tappi,* **55** (5), 719 (1972).
317. H. P. Naveau, Y. T. Wu, D. L. Brink, M. M. Merriman, and J. G. Bicho, *Tappi,* **55** (9), 1356 (1972).
318. K. Freudenberg, K. Engler, E. Flickinger, A. Sobek, and F. Klink, *Chem. Ber.,* **71,** 1810 (1938).
319. M. Erickson, S. Larsson, and G. E. Miksche, *Acta Chem. Scand.,* **27** (1), 127 (1973).
320. H. I. Bolker and C. M. Li, *Extended Abstr. Can. Wood Chem. Symp.,* Quebec, Canada, 39, 1973, Chem. Inst. of Canada/CPPA, Montreal, Canada.
321. D. W. Goheen, "Low-Molecular-Weight Chemicals" in ref. 2, p. 797.
322. P. Hoffmann and W. Schweers, *Holzforschung,* **29** (1), 18 (1975).
323. P. Hoffmann and W. Schweers, *Paperi Puu,* **58** (4a), 227 (1976).
324. N. P. Mikhailov, N. N. Shorygina, and N. K. Vorob'eva, *Khim Ispol'z. Lignina* 321 (1974), through *Abstr. Bull. Inst. Paper Chem.,* **46** (11), 11185 (1976).
325. A. F. Semechkina and N. N. Shorygina, *Izv. Akad. Nauk SSSR, Otd. Khim. Nauk,* **4,** 715 (1963) through *Abstr. Bull. Inst. Paper Chem.,* **43** (10), 10077 (1972).
326. V. V. Elkin, A. F. Semechkina, and N. N. Shorygina, *Izv. Akad. Nauk SSSR, Ser. Khim.,* **1,** 187 (1971), through *Abstr. Bull. Inst. Paper Chem.,* **43** (10), 10062 (1972).
327. P.-A. Pernemalm and C. W. Dence, *Acta Chem. Scand,* **28B** (4), 453 (1974).
328. A. Yamaguchi, *J. Jap. Wood Res. Soc.* (Mokuzai Gakkaishi), **21** (7), 422 (1975).

329. E. I. Chupka, G. I. Stromskaya, and L. V. Bronov, *Khim. Drev.* (Riga) **6**, 46 (1976), through *Abstr. Bull. Inst. Paper Chem.*, **48** (3), 2321 (1977).
330. J. Rouger and A. Robert, *Bull. Soc. Chim. Fr.*, **10**, 4039 (1972).
331. H. Nimz, *Chem. Ber.*, **102**, 799 (1969).
332. H. Nimz, *Chem. Ber.*, **102**, 3803 (1969).
333. H. Nimz, K. Das, and N. Minemura, *Chem. Ber.*, **104** (6), 1871 (1971).
334. H. Nimz and K. Das, *Chem. Ber.*, **104** (8), 2359 (1971).
335. G. E. Domburg, V. N. Sergeeva, and G. A. Zheibe, *J. Thermal Anal.* **2** (4), 419 (1970).
336. G. E. Domburg, V. N. Sergeeva, and A. I. Kalnish, *Proc. 3rd Intl. Cong. on Thermal Anal.* Davos, Switzerland, 1971, **3**, 327 (1972).
337. G. G. Allan and T. Mattila, "High Energy Degradation" in ref. 2, p. 575.
338. R. A. Ripley and D. P. C. Fung, *Wood Sci. Technol.* **4** (1), 25 (1971).
339. A. P. Lapan, V. B. Chekhovskaya, and A. I. Gritsa, *Khim Drev.* (Riga), **6**, 42 (1976), through *Abstr. Bull. Inst. Paper Chem.*, **48** (3), 2328 (1977).
340a. W. G. Glasser, *Polymer*, **14** (8), 395 (1973).
340b. W. G. Glasser and W. Sandermann, *Svensk Papperstidn.*, **73** (8), 246 (1970).
341. T. J. Fullerton, *Svensk Papperstidn.*, **78** (6), 224 (1975).
342. S. I. Falkehag, *J. Appl. Polym. Sci.* (Appl. Polym. Symp.), **28**, 247 (1975).
343. W. Schweers, *Mitt. Bundesforsch.* (Reinbek/Hamburg), **82**, 61 (1971).
344. K. V. Sarkanen in "Materials: Renewable and Nonrenewable Resources," P. H. Abelson and A. L. Hammond, Eds., American Association for the Advancement of Science, Washington, D.C., 179, 1976 p. 179.
345. J. M. Harkin, *Tappi*, **55** (7), 991 (1972).
346. S. I. Falkehag, J. Marton, and E. Adler, *Advan. Chem. Ser.*, **59**, 75 (1966).
347. F. Imsgard, S. I. Falkehag, and K. P. Kringstad, *Tappi*, **54** (10), 1680 (1971).
348. L. Lonsky, W. Lonsky, K. Kratzl, and I. Falkehag, *Monatsh. Chem.*, **107** (3), 685 (1976).
349. G. Meshitsuka and J. Nakano, *Tappi*, **56** (7), 105 (1973).
350. K. Iiyama and J. Nakano, *J. Jap. Tappi*, **27** (11), 530 (1973).
351. H. Nimz, *Tappi*, **56** (5), 124 (1973).
352. J. Gierer, I. Pettersson, and I. Szabo-Lin, *Acta Chem. Scand.*, **28B** (10), 1129 (1974).
353. S. Y. Lin and K. P. Kringstad, *Norsk Skogind*, **25** (9), 252 (1971).
354. G. Brunow and B. Eriksson, *Acta Chem. Scand.*, **25** (7), 2779 (1971).
355. J. Gierer and S. Y. Lin, *Svensk Papperstidn.*, **75** (7), 233 (1972).
356. G. Brunow and M. Sivonen, *Paperi Puu*, **57** (4a), 215 (1975).
357. W. G. Glasser, O. H. -H. Hsu, and J. Nakano, *Svensk Papperstidn.*, **79** (16), 527 (1976).
358. I. S. Goldstein, in "Materials: Renewable and Nonrenewable Resources," P. H. Abelson and A. L. Hammond, Eds., American Association for the Advancement of Science, Washington, D.C., 179, 1976 p 179.
359. I. S. Goldstein, *J. Polymer Sci.* (Appl. Polym. Symp.), **28**, 259 (1975).
360. J. D. Benigni and I. S. Goldstein, *J. Polym. Sci. C.*, **36**, 477 (1971).
361. K. Hastbacka and J. B-son. Bredenberg, *Paperi Puu*, **55** (3), 129 (1973).
362. J. D. Benigni and I. S. Goldstein, *J. Polym. Sci.* (Chem. Polym. Symp.), **36**, 467 (1971).
363. T. Enkvist, *J. Appl Polym. Sci.* (Appl. Polym. Symp.), **28**, 285 (1975).
364. G. G. Allan, "Modification Reactions" in ref. 2, p. 511.
365. C. H. Hoyt and D. W. Goheen, "Polymeric Products" in ref. 2, p. 833.
366. H. Kilpelainen, K. Forss, and A. Fuhrmann, *Proc. IUFRO Conf. Wood Gluing,* Madison, Wis., 79 (1975).

367. K. Forss and A. Fuhrmann, *Paperi Puu,* **58** (11), 817 (1976).

368. K. C. Shen, *For. Prod. J.,* **24** (2), 38 (1974).

369. J. Peter Clark, *Chem. Technol.,* **6,** 235 (1976).

370. J. J. Lindberg, V. A. Era, and T. P. Jauhiainen, *J. Appl. Polym. Sci.* (Appl. Polym. Symp.), **28,** 269 (1975).

371. G. Prazak, *Amer. Dyestuff Rep.,* **59** (11), 44 (1970).

372. A. F. Sirianni and I. E. Puddington, *Pulp Paper Mag. Can.,* **73** (4), 58 (1972).

373. A. F. Sirianni, C. M. Barker, G. R. Barker, I. E. Puddington, *Pulp Paper Mag. Can.,* **73** (11), 61 (1972).

374. S. L. H. Chan, C. G. J. Baker, and J. M. Beeckmans, *Powder Technol.,* **13** (2), 223 (1976).

375. G. G. Allan and D. D. Halabisky, *Pulp Paper Mag. Can.,* **71** (2), T–50 (1970).

376. P. H. Wiesner, *Wochbl. Papierfabr.,* **99** (18), 740 (1971).

377. A. M. Kazarnovskii, A. V. Antipova, and N. A. Ivanova, *Khim. Drev.* (Riga), **6,** 49 (1976), through *Abstr. Bull. Inst. Paper Chem.,* **48** (3), 2327 (1977).

378. A. M. Kazarnovskii and M. I. Chudakov, *Gidroliz. Lesokhim. Prom.,* **7,** 8 (1973), through *Abstr. Bull. Inst. Paper Chem.,* **44** (10), 10613 (1974).

379. W. Flaig, *Chem. Ind.,* (London), **16,** 553 (1973).

380. G. G. Allan, C. S. Chopra, A. Neogi, and R. M. Wilkins, *Intl. Pest Control,* (1971).

381. G. G. Allan, C. S. Chopra, A. Neogi, and R. M. Wilkins, *Tappi,* **54** (8) 1293 (1971).

3 PULPWOOD

WALTER J. BUBLITZ, Ph.D.

Professor
Oregon State University
Corvallis, Oregon

Wood is the preferred source of fibers for the paper industry, even though paper historically has been made from nonwoody plants, such as canes and straws. The development of successful methods for preparing pulp from wood, both mechanically and chemically, occurred late in the nineteenth century and paved the way for the rapid growth and diversification of the paper industry in the twentieth century. Among the reasons for using wood have been relative availability, low cost, convenience in handling and storage, high-quality pulp content, and the versatility of the fiber properties from different wood species.

Introduction

Many of the advantages of wood are diminishing as the easily available wood supply is being consumed. The world-wide picture is complex, but those countries that traditionally have been large producers and consumers of paper (the United States, Canada, Western Europe, and Scandinavia) are pushing the limits of their readily available forest resources.[1-3] The undeveloped countries have varying forest resources; Russia and Siberia have vast forests that are relatively untouched, but that are beginning to be exploited. Other countries, notably China and those in the Middle East, have negligible forests to harvest. Nonetheless, the appetite for paper continues to grow in both the developed as well as undeveloped countries; by the year 2000 the world may be consuming four to five times as much paper as in the late 1970s.[4-8] The pressures on world forest resources will thus increase tremendously, because concurrent with the demand for pulpwood will come competition for wood as a source of lumber, plywood, and pressed fiberboard. Similarly, the demand for wood as a source of energy, both as a direct fuel (the largest single world-wide use of wood today) and for conversion into combustible oils and gases, is certain to increase as the world's oil supply dwindles and prices escalate.

All these factors indicate the need for serious study of the source of fibers for tomorrow's paper industry. It is imperative that every effort be made to utilize present supplies of wood as efficiently as possible and to seek other fiber sources. First the general subject of forestry, silviculture, and wood utilization

113

will be briefly discussed. Then, other fiber sources that involve the use of re-cycled paper and nonwoody plants will be covered.

Many changes have occurred in fiber-procurement methods for the paper industry in the past 25 years; these will be discussed in some detail. Due to the wide variation in wood quality, geography, and climate, fiber-procurement methods vary greatly around the world. Discussion of these methods will be concentrated primarily on United States practices with mention of deviations from these practices in other areas. Production of pulp from nonwoody plants and some estimate of potential production that could become available will be examined.

Raw-material Flow. There are many more diverse methods of wood procure-ment today than existed 25 years ago when nearly all pulp mills received their raw material in log form and performed their own debarking and chipping. The introduction of residual chips, that is, chips generated by solid wood manu-facturers from their waste material, presented a new method of chip procure-ment. Whole-tree logging and chipping, which covers a variety of processing methods and techniques, was introduced in the early 1970s and is a third method of chip procurement. A flow diagram showing the various routes from the forest to the pulp mill and illustrating the relationship of the three chip-procurement methods is given in Figure 3-1. This is a generalized picture of the major patterns of movement and does not cover all conceivable situations that may exist. It primarily relates to American and Canadian practices. There is no attempt to cover world-wide practices.

As shown, there are at least three different types of chips that are produced: residual, roundwood, and whole-tree chips. Few mills are utilizing all three types simultaneously; most United States pulp mills, depending on their locale, operate exclusively on either roundwood or residual chips in normal times. Whole-tree chips have been used as an auxiliary source of fiber during periods of chip shortage, but these chips are generally considered to be inferior in quality to the other two types and thus are utilized only when necessary. This will be discussed later.

The division of logging into conventional and whole-tree is quite arbitrary. In the first category are included such practices as clear cutting and selective cutting, without making a distinction as to the source of the trees (from pri-vate, public, or farmer-owned forests). The second category encompasses the more recent developments in logging with emphasis on utilization of thinnings, logging wastes, cull logs, slash, or parts of trees not normally utilized for either solid wood conversion or pulping. There is some overlapping in our use of the term selective logging; in one case, it may involve cutting and chipping of thinnings, in another case it may involve harvesting of mature logs for lumber (with the resultant generation of residual chips).

Logging and Chip Procurement

Until recent times, American logging practices have been quite wasteful; as a result, little virgin timber remains in the United States. The major portion of

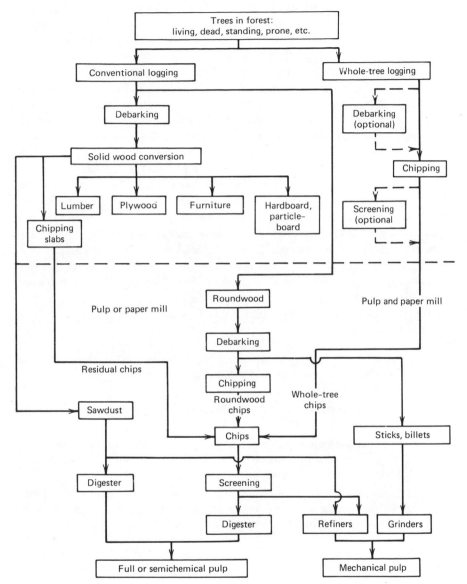

Figure 3-1. Flow sheet of fiber procurement for pulp mills.

wood for pulping comes from young-growth trees, two or more generations away from virgin timber. Forestry practices are undergoing radical changes to accommodate the increasing demand for wood from all sides, and thus such practices as genetic tree selection, fertilization, thinning, and rapid replanting are becoming routine. Trees are being cultivated as crops to be harvested on some finite time scale rather than as expendable and nonrenewable resources.[9] This trend must be accelerated if the future demand for fiber is to be met.

A revolutionary change in chip-procurement practices has occurred in American industry. Prior to 1950, nearly 100% of the chips were prepared from round-wood, that is, from the entire trunk or stem of a tree. Pulp mills would bring in solid logs and store them whole until ready for debarking and chipping. The entire stem would be chipped regardless of basic suitability for other purposes.

The practice of generating chips for pulping as a residue or by-product of some other form of solid wood manufacturing originated in the Pacific Northwest of the United States, primarily because the trees there were so large (1.2 to 2.4 m or 4 to 8 ft diameter at breast height (DBH) is not uncommon in virgin timber). The trees were harvested primarily for plywood and dimension lumber.[10] In the latter instance the log was trued into a cant by sawing off slabs on four sides to produce a square or rectangular piece of wood that could be resawed into the final dimension lumber. The slabs were considered waste material (too small for further conversion into lumber) and were either burned or dumped. To a great extent, the slabs are some of the soundest wood in the tree since they are mainly sapwood and in most cases when rot is present, it is concentrated in the heartwood. Thus a tremendous amount of sound wood, free from bark, was available for conversion to pulp. Therefore, sawmills started to debark logs prior to conversion into lumber. The slabs and waste wood produced were then run through a chipper, and the chips were transported to nearby pulp mils for conversion into pulp. This practice was successful for a number of reasons:

1. Conservation of raw material: sound wood, previously burned, was now being utilized for an economically valuable product, thus conserving timber resources.

2. Value to the sawmill: waste material was now bringing an economic return to the sawmills.

3. Value to the pulp mills: this practice relieved the mills from the responsibility of procuring their own wood supply and from the job of converting solid wood into chips on their premises.

4. Environmental cleanup: wigwam burners became obsolete in the 1960s and 1970s because of air-pollution problems; this provided an additional incentive for converting wood wastes to chips.

As might be expected, such practices are not without their problems. The reliance of the pulp mills on sawmills has resulted in some severe cycles of over-supply and undersupply of chips, depending on the dimension-lumber market and the health of the paper industry. The growth of the chip market for export has had a major effect on the supply and price of chips.[11] As is true in any business deal, there are continuing disagreements between buyer and seller regarding quality, price, and delivery of the items in question. The advantages outweigh the disadvantages, however, and this is rapidly being adopted as a worldwide practice, wherever the timber available has value as a source of solid lumber or plywood. The concept of maximum utilization of the tree is thus approaching reality and this practice is not only valid from the standpoint of best utilization, but also generally results in the greatest profit to the industry.

Logging Practices. The basic logging methods vary considerably worldwide, depending on such factors as climate, terrain, and size of timber. In North America, climate strongly influences the season of harvesting. Harvesting is done in the winter in eastern and central Canada because the swampy terrain is frozen solid in the winter. In the Northwest, logging is done in the summer because in the mountainous terrain there are heavy rains and deep snows that make travel hazardous or impossible in the winter. In the Southeast, logging can proceed year-round because of the mild climate and flat terrain. Only those practices will be mentioned that affect pulpwood procurement; detailed information on logging can be found in other treatises devoted to the subject.

Mechanization of Harvesting. In the early days of logging, trees were manually felled and delimbed, trimmed, and cut into convenient lengths while still in the woods by using one- or two-man saws, axes, and similar equipment. The hauling was done by animal power either to a railhead or body of water where it was transported to a mill. Mechanization of this process has been progressing gradually for the past 100 years, starting with the introduction of steam engines in the woods to facilitate hauling and loading of logs. The introduction of caterpillar-tread tractors for moving logs and of large-capacity trucks for hauling sizable loads further streamlined the operation. Gasoline-powered chain saws were developed in the 1950s and proved to be a great asset to loggers. Cutting time was greatly reduced and delimbing and trimming operations much simplified through their use. Productivity increased proportionately. A man could cut five times as many trees in a shift as previously done with handsaws and axes. Clear cutting became easier and this practice spread rapidly following the introduction of the chain saw.[12] The 1960s saw the advent of rubber-tired or caterpillar-tread mechanical harvestors that could cut, delimb, and stack trees in one continuous operation with only one man at the controls. These machines were the "progeny" of the tractors and bulldozers previously used for clearing brush, cultivating, and moving logs in the forests. They were modified to include a variety of attachments, such as a front-end device for cutting through a tree at the base. Some of these devices resemble giant shears or scissors that can snip through the tree trunk in a single motion; others are saws, either chain or circular, for cutting through the trunk.[13]

Before the tree is cut at the base, it must be secured vertically by clamps that are mounted on movable arms. After cutting, the tree is rotated to a horizontal position and lowered to the ground or stacked on a truck. Some machines incorporate devices that encircle the trunk and strip off limbs and branches by moving rapidly along the trunk. When a tree is lowered to the ground it can be quickly topped, cut into desired lengths, and loaded on a truck or piled on a stack for later pickup. Machines of this type caused a marked jump in productivity. One man can produce several times more output with these machines than he could previously. The machines are not without their drawbacks, however, since they are mechanically complex, subject to various breakdowns, and costly to purchase and operate. They are best suited to flat terrain and relatively small trees (about 0.3 m or 12 in. DBH maximum). Thus these machines seem to operate best where these conditions prevail: in the southeastern United States, parts of New England and the Great-Lake States, and in central and eastern Canada. They are least suited to logging in the Northwest where larger trees and mountainous terrain restrict their utility. Ideally one can picture a plantation of trees on flat land with uniformly spaced trees where such harvestors would be the perfect machines for logging. The final choice as to what is used is based on economics; these new machines are profitable mainly where the conditions described prevail. Other situations are best served with chain saws. The very antithesis of mechanical harvestors, the use of horses for logging, has been suggusted as a profitable method for one- or two-man logging operations on small tree farms.[14] Horses have the ability to skid logs through a forest with minimum damage to the standing trees and this can be a practical consideration during a thinning operation in a dense grove of young trees.

Time of Harvesting. The time of year when the tree is cut is of secondary importance to the pulp mill. Most pulpwood is stored for indefinite periods after harvesting and the time lapse between cutting and pulping is not of primary importance. However, certain species will debark more readily in the spring and early summer, rather than at other times of the year.[15] Cutting and barking these species is most efficient at these critical periods. Logs cut in winter may be frozen or coated with ice and snow; this can affect the chipping operation. Conversely, logs can be cleaner in winter since they mainly gather snow and ice from the ground as compared to the gathering of sand and dirt in the summer. Dirt can cause severe mechanical erosion of chipping and chip-handling equipment as well as cause contamination of pulp and paper.

Clear Cutting. The practice of cutting all trees in a specific area, as opposed to selective cutting, is known as clear cutting. It has been most widely practiced in the western United States where relatively pure stands of coniferous trees are found.[16] All merchantable logs are then shipped to a central area where they are sorted by species or quality. The best quality logs go to sawmills and plywood mills; the off-species, or cull logs, are shunted to the pulp mills. Clear cutting is attacked on an environmental basis; it is defended on the basis of harvesting economics and as a silvicultural benefit. Douglas fir, for example, needs sunlight for best growth of seedlings; they grow better in an area after clear cutting rather than under cover of an existing forest.[17] Not all areas can be successfully reforested by natural means; it is frequently necessary to replant seedlings by hand, wholly or in part.

The practice of clear cutting may or may not affect the quality of wood for pulping. Those mills using residual chips are least affected since they are one or more steps removed from the wood's operation and depend on the sawmills to furnish them with adequate quantities of chips of specified quality and species. If the pulp mill receives roundwood for chipping, it may be more directly affected by the results of clear-cutting procedures.

Selective Cutting. This subject covers a wide range of practices, from cutting giant trees in the virgin forest to modern-day silvicultural practices, such as thinning. In the past, logging was selective because it was not practical to clear cut with manual saws and primitive hauling equipment. Many trees were damaged in the felling and removal of selected trees, but the damage was incidental to harvesting the selected giant trees.

Clear cutting became more economical in recent years with the advent of the chain saw and better equipment for mass movement of logs out of the forest. The use of such equipment may have reached its peak and be declining now because of environmental pressure and economic factors. In a well-managed forest, there are, broadly speaking, two options: (1) treat the forest on a short-term basis (5 to 20 yr) and harvest all trees at the end of the period or (2) treat the forest on a long-term basis (20 to 80 yr) and permit one or more selective harvests of the poor-quality trees (called thinnings) in this period. In an area of good natural reforestation, seedlings may take root literally inches apart, resulting in tremendous competition for nutrients, water, sunlight, as well as causing a high mortality rate. The survivors are stunted in growth. However, if the smaller and deformed trees are harvested at different points in life of the stand (that is, at 15, 30, and 45 yr), and there is a final harvest at from 60 to 70 yr, a much greater total harvest can be expected. The survivors of the thinnings experience periods of rapid growth due to the removal of competition for food and water. Since only selected trees remain standing, the best of the trees are then available for the final harvest.[18]

The disposal of the thinnings is a multifaceted problem. When the thinnings are left in the forest, they ultimately decay, return nutrients and humus to the soil, and help perpetuate the forest. They generally are too small for lumber, but they may be suited for fence posts and fuel. Debarking these smaller and frequently malformed trees is more difficult and costly per unit of wood than debarking larger, relatively well-formed trees. Nevertheless, thinnings can represent an important fiber source for the paper industry under certain conditions. The wood is generally sound and free from rot. Although a larger proportion of juvenile fibers is present in thinnings than is found in mature wood, this can be an advantage in forming paper with good bonding strength (tensile, burst, and fold).[19] Whether thinnings should be used by a pulp mill is primarily a question of economics. It has been shown that thinnings can produce quality chips and good paper. The costs of cutting, debarking, chipping, and transportation must be established. These costs will vary according to the nature of the forest and the pulp mill. Although there is little question that forests of the future will be more intensively managed, (thus more thinnings will be available), it is not certain that the best or most economical use for these trees will be conversion to chips in all cases.

Trees as a Crop. The increase in the intense use of forest management means that trees will be regarded more and more as crops to be harvested on some scheduled basis. In the past, the rotation period was normally considered in terms of decades, varying from a low of 10 to 15 yr for southern pines to 60 to 80 yr for Douglas fir in the Northwest.[20] Certain species of hardwoods, however, such as sycamore and cottonwood, can reproduce vegetatively by cloning, that is, new trees will sprout from roots or suckers rather than from seeds.[21,22] Under the right conditions, in the southeastern United States, these trees grow rapidly, adding as much as 4540 to 9080 kg/hectare (2 to 4 tn/per acre) of wood annually, and these trees are presently being examined for harvesting on a 1 to 3 yr basis. The advantage of this mode of management is that with small-diameter trees planted in rows, mechanized harvesting is not only possible, but also economically feasible. In addition the cropped area does not need to be reseeded since new trees will be generated from the existing root system.[23] The quality of pulp made from such wood is not equal to that of pulp from normal softwoods, but it is adequate for use as filler pulp in grades of paper where bulk and smoothness are important and where high strength is not paramount. Once again, economics is a major consideration, but more research is needed to establish the value of this logging method relative to other traditional methods.

Tissue Culture Trees. Whereas trees normally reproduce from seeds or in some species vegetatively, geneticists have been able to reproduce trees in the laboratory by growing tissue from living trees in various microbiological cultures. The main value of this technique is that one can produce an exact match of the parent tree by replication of the chromosomes, since a new tree may be produced from a single cell. Normal production by way of seeds introduces the complexities of genetics with the inherent variation of the progeny due to mixing of the chromosomes of the parent trees. Using tissue-culture methods, there is no limit to the number of perfectly formed replicates that can be made from one selected tree. It is possible to visualize whole forests of identical trees, all derived from one parent tree and all matching it in every respect, thus contributing to immediate short-term upgrading of the forest and to future cost savings.

Research on tissue culture of trees began in the 1960s. Since then, it has been intensively investigated and is rapidly progressing to the nursery and field-planting stage.[24,25] Present techniques are too costly for commercial forestry operations, but costs will obviously decline with increasing experience.

Whole-Tree Chipping. The term whole-tree chipping is somewhat misleading, because this category of chip preparation encompasses a wide variety of practices, from literal chipping and pulping of every part of the tree (roots, stem, branches, and leaves) to chipping of only selected parts of the tree.[26] The common denominator in most of the methods is that the tree is converted to chips on or near the logging site. By contrast, residual chips are produced from waste material at sawmills and roundwood chips at pulp mills. The concept of utilizing the whole tree has been expounded for many years. However, it was never put into practice commercially until the early 1970s, when the paper industry was operating at full capacity but the lumber business was in a depressed state. Residual chips were thus in short supply and the paper industry looked for new wood sources. In the Northwest, for example, large quantities of wood have been wasted because of inefficient logging. Rotted wood, windfalls, split logs, crooked wood, and trimmings have been generally left in the woods during clear cutting, since the economic return from converting them to lumber or plywood did not justify the cost of collection and transportation.[27] Some of this wasted material contained sound wood adequate for pulping and the paper industry has begun to utilize it.

Logistically it is easier to convert these cull materials to chips at or near the point of pickup rather than converting them at mill sites. Portable chippers powered by diesel engines have been developed; these are capable of being transported into the woods to any place that a chip truck can travel.[28] A wide variety of wood can be put into the chippers in the woods, ranging from the entire tree, to rotted logs, and to sound wood from thinning. From a papermaker's point of view the resultant chip quality varies from excellent to unacceptable. There are chips from mixed species, with and without bark, with large quantities of rot, and much oversized or undersized material. There are also contaminants; these include leaves, twigs, roots, and, most critically, sand and dirt. The effects of rot, bark, and dirt (more thoroughly discussed elsewhere in this chapter) have negative connotations with respect to pulping operations and pulp quality. Rot and bark consume the cooking chemical without a concomitant return in yield of pulp, and the pulp may be dirtier, darker, and weaker because of them. Dirt and sand can play havoc with chip-handling-and-feeding equipment as well as with paper quality and kraft liquor recovery.[29,30]

Whole-tree chipping is a misnomer since most operations involve chipping the stems and/or trunks, but not the branches and roots. Pulping studies have been made on these tree components individually; they all conclude that pulp yields and quality are generally lower when tree components other than the trunk are chipped and pulped.[31,32] Puckerbrush, which are small, bushy hardwoods, frequently take over clear cut areas and have been studied as a source for pulpwood.[33,34] Substantial quantities of puckerbrush are available in many regions of the United States, but different harvesting techniques are required and the pulp yield and quality are lower than from regular stem wood. The problems are similar to those encountered in pulping branches and roots from full-size trees. Although there is a substantial amount of good wood substance available in this form, the cost of utilizing it is high. Also, since most of the mineral nutrients in trees are present in the leaves and needles, removal of the latter tends to deplete the mineral balance of the forest and requires more frequent artifical fertilization, an increasingly costly operation.[35]

Whole-tree chips, because of the variability of their sources, are thus extremely variable and usually lower in quality than residual or roundwood chips. This has two major effects:

1. Whole-tree chips are more easily utilized in the coarser grades of paper, such as corrugating medium and linerboard, where dirt and lower-quality pulp is less critical than in fine grades of paper.

2. Whole-tree chips can be utilized for fine grades of paper, such as bleached paper, but extra measures in the pulp and paper mill are required to produce acceptable products.

The additional measures required in pulping whole-tree chips include more frequent maintenance of equipment, extra pulping chemicals, extra bleaching chemicals, and more screening equipment, all of which add to manufacturing costs. In some cases, whole-tree chips are as costly to produce as normal residual chips.[36] Most mills will not use whole-tree chips if it can be avoided. However, as the demand for wood grows, many mills will return to their usage in order to maintain full production, regardless of their cost or the inherent operating problems connected with them. Several measures can be taken to alleviate the problems that arise with the use of whole-tree chips. Lifting logs during harvesting, rather than dragging them along the ground, can minimize the pickup of dirt and sand. The logs can be washed prior to debarking or the chips can be washed and screened prior to pulping in order to hold down the amount of dirt that enters the pulp mill. Segregation by species and by wood quality (rot, shape, and dirt) will help assure that chips are utilized in the most efficient manner, both technically and economically.

Whole-tree chips would thus seem to be a part of the pulpwood supply picture of the future, but due to the special inherent problems, more development work is needed to make them a viable and economic source of pulpwood. New developments in machinery and processing should help assure the pulp mills of the future a steady and uniform flow of raw material from this source.

Transportation

Movement of pulpwood, either in log or chip form, has undergone significant changes in the past several decades. As pulpwood has evolved into a crop, special techniques have been developed to provide the most efficient methods of transportation to the pulp mill. Detailed information on log hauling is available in many books on forestry; only the high spots will be covered here.

Rail Hauling. This method of log transportation has declined with the advent of better trucks and highways. Logging railroads were popular in the late nineteenth century when there was no competitive hauling methods. Many spur lines were built into the forests to haul the logs to the sawmills and pulp mills. Construction and maintenance of such lines was and is expensive; the only lines still in use today were constructed years ago.

Rail hauling is primarily confined to the main and branch lines of railroad companies rather than to private company lines.[37] Normally, the distance of hauling is 200 to 300 mi maximum, although pulpwood has been shipped from western Montana to Minnesota and central Wisconsin, a distance of over 1000 mi. Much pulpwood is being shipped by rail in the southeastern states, where railroads are trying to win back lost business by constructing special cars to facilitate the loading and unloading of the logs.

Chips are frequently shipped to the pulp mill by rail in the northwest of the United States, although there has been a decline in the railroad's share of this market.[38] Originally gondolas designed for hauling bulk cargo, such as sand, rock, and coal, were used, but their sides were extended to accomodate a larger quantity of chips. Now there are chip cars that have been specially designed for hauling chips. Capacities range from about 4000 ft^3 (113 m^3) to about 7500 ft^3 (212 m^3) in the latest cars. Net loads would vary from 90,000 to 180,000 lb of green chips (40 to 80 MT) or 40,000 to 80,000 lb of dry wood (18 to 36 MT).

Chips are loaded into the cars at the originating sawmill from an overhead hopper. The chips are continuously blown or conveyed to this hopper from the chipper, but the car may be loaded on an intermittent basis. To obtain greatest packing efficiency, the chips are blown into the car as it shuttles back and forth underneath the delivery spout. Mechanical vibrators mounted on the sides of the car promote more dense and even packing of the chips throughout the body of the car. Usually the car is overfilled volumetrically since the chips will settle during movement to the pulp mill. The amount of such compaction is determined by chip size, wood density, distance of haul, speed of travel, and roughness of the roadbed. Open-mesh plastic nets are placed over the top of the car to prevent chips blowing off during transit.

Chip cars may be unloaded by a variety of means, depending on their construction. Hinged bottoms, sides, or ends may be opened to allow the chips to fall by gravity into open pits from which they are conveyed to storage piles. Mechanized scrapers or scoops can be used to push the chips out of the open sides or ends. Some mills empty chip cars by turning them upside down, using a cradle to clamp and rotate one car at a time. Capital cost of this type of unloader is high, but is operates rapidly and avoids both the problem of the bridging of chips (an occasional problem with bottom unloaders) as well as the problem of the freezing of chips in the winter.

Pulpwood is normally shipped in flatcars or gondolas.[39] In the eastern half of the United States and Canada, where pulp trees rarely are more than 0.3 m (12 in.) in diameter, the stems are traditionally cut into 2.4-m (8-ft) lengths and loaded transversely to the axis of the railcar. Flatcars may have side and end stakes to secure the load; modern flatcars have built-up bulkheads at each end of the car to prevent shifting of the load. Loading and unloading of pulpwood cars are frequently done with cranes, although the car may be tilted or inverted in order to expedite unloading. Logs in the Pacific Northwest are generally much larger in diameter and are usually cut to 4.9-m to (16-ft) lengths or longer. Loading of these logs on flatcars is done longitudinally and they are typically unloaded by rolling them off one side of the railcar.

Truck Hauling. This mode of transportation has grown rapidly as trucks have grown in size, power, and efficiency and more and better public highways have increased ease of access to forest lands.[40] In addition, it is cheaper and quicker to construct logging roads in the forest than to construct rail lines. Trucking is less fuel-efficient than railroading, which may mitigate against trucking in the future, but it is an important part of the pulpwood transportation picture today. Trucking of logs and chips normally does not exceed 100 to 150 mi.

Trucks can be loaded close to the logging site in the forests, either longitudinally with 6-m (20-ft) or longer logs or transversely with 2.4-m (8-ft) logs. In the Pacific Northwest, where 6- to 12-m (20- to 40-ft) logs are hauled, the truck consists of a tractor and a trailer. When empty, the trailer is secured in piggyback fashion to the rear of the tractor, shortening the length of the rig by 6 to 9 m (20 to 30 ft) and providing for more rapid movement. At the loading site, a crane picks up the trailer, positions it onto the tractor, and the combination is then loaded with logs. Cables or chains are fastened laterally across the load to prevent log slippage during transit. These trucks are generally unloaded from the side by loosening the restraining bindings and side stakes; then the logs roll off due to gravity, usually into a pond. Trucks with 2.4-m (8-ft) pulpwood are frequently unloaded by overhead cranes, or the load may be pushed off the side mechanically into ponds or jack ladders.

Chip trucks are specially designed vans to hold 15 to 20 units of chips (about 85 to 115 m^3 or 3000 to 4000 ft^3). The most common sites of loading are the sawmills and plywood mills where the residual chips and sawdust are being generated. Loading of chip vans is similar to the loading of rail-chip cars, but almost all vans are unloaded by tilting and dumping from the back. The van or entire truck is driven onto a platform that tilts the van backwards to a 45° angle; the chips then tumble out into a pit for further transportation within the pulp-mill site. A load of chips can be dumped in five minutes with this system; samples are taken for testing and analysis at this time.

Chips that are produced in the woods can be blown directly into a waiting chip van and sent directly to a pulp mill.[41] Although this system saves manpower by minimizing wood-handling operations, it has the limitation that chip vans are less mobile and flexible than log trucks, thus restricting their forest access to the better logging roads and less punishing terrain.

Water Transportation. This subject can be considered in two parts: (1) flotation of logs in the water and (2) movement of wood by barge or ship.

Movement of pulpwood by flotation is a dying practice.[42] Logging in the northern United States and adjacent areas of Canada in the past was accomplished by cutting the trees in winter, skidding the logs to the nearest river bank, and floating the logs down river to the pulp mill during the spring thaw when the river was at its peak. Many "sinkers" were lost by submersion and logjams frequently occurred and had to be broken manually or with dynamite. This practice is nearly extinct in North America in 1977 due to depletion of the forests and restrictions on the use of rivers due to water-pollution problems.[43] Water flotation is still practiced on large bodies of water, usually in the form of rafts towed behind barges. Pulpwood from Canada is shipped across Lake Superior in the summer to pulp mills in northern Michigan and Wisconsin; there is also considerable movement of logs by rafting along the sheltered waterways of Washington, British Columbia, and Alaska.[44] Logs that are rafted and/or stored in seawater will absorb salt from the water. Kraft mills accumulate this salt from the wood in the pulping-liquor system. Although not directly injurious to the process, salt can be corrosive to equipment and interfere with chemical recovery if the salt concentration exceeds certain limits.[45]

Movement by barge or ship is also restricted to large bodies of water, such as the Great Lakes or the coastal areas. Barging of chips on inland waterways (the Ohio or Mississippi rivers, for example) has not become prominent because few pulp mills are located directly on these bodies of water. Barging of chips and sawdust is quite popular for many of the Canadian pulp mills located on salt water in British Columbia and on Vancouver Island; many of these mills are isolated, having neither highway or rail access to the mainland.[46] Loading and unloading of these barges can be done with conveyors or pneumatic devices. Logs are conveyed in special ships that can be tilted, dumped, or partially submerged in order to unload their cargo rapidly and with minimum handling.[47]

Miscellaneous Methods. Belt conveyors can be used for moving chips and small logs short distances within mill yards. Pneumatic tubes are another method of chip movement. Moving chips long distances (up to 100 mi) by pipeline has been discussed, but never implemented.[48] The chips would be moved in a water slurry, necessitating pumping equipment and screening facilities at the receiving terminal. There would be unavoidable losses of water-soluble materials from the wood and the contaminated water would have to be treated or utilized by the pulp mill.[49] In the woods, limited aerial movement of logs is done with balloons and helicopters, but it is expensive and thus restricted to short hauls of several miles.[50,51] It is used only where normal terrestrial movement is impractical or uneconomical, such as on steep terrain, or where selective logging is practiced and logs must be removed with minimum damage to the environment.

Log and Chip Storage

One of the advantages of using wood for pulping is the relative stability of stored wood, at least in solid form. Logs may be stored for several years under average conditions, with minimal losses in quality or quantity. Although it was originally assumed that chip storage would be equally trouble-free, this has not been true.

Storage of Solid Wood. Pulpwood has traditionally been stored in log form and converted to chips just prior to pulping. Storage and handling methods for logs are largely dependent on the size and length of the logs available. Pulpwood from small-diameter trees 0.3 m (12 in.) or less are typically cut in 1.2- or 2.4-m (4- or 8-ft) lengths. Many mills store these logs in carefully constructed rows, all logs lying parallel to each other, forming stacks as high as 15 m (50 ft) and as long as 30 to 150 m (100 to 500 ft). With proper spacing, such storage permits easy mechanized retrieval of logs when needed. In addition, air can circulate among the logs in such stacks, a vital process in the storage of certain coniferous species destined for acid sulfite pulp mills. Such mills located in eastern Canada and the Great Lakes area of the United States must season the wood for periods of 6 to 12 months. During this period, oxidation of certain oleoresins occurs in the spruce and fir wood that is used. If the wood is pulped unaged, severe pitch problems occur in the pulp and paper mills. Aging of the wood prior to pulping minimizes or eliminates the pitch problem.[52] Another advantage to spacing log stacks is the reduction of fire hazards. If one stack should catch fire, equipment can be moved close to combat the blaze and, with luck, isolate it to only one stack. By contrast, some mills store pulpwood in jackstraw piles, dumping the logs into one or more large mounds with no effort at stacking.[53] This practice is most prevalent in the southeastern United States where southern pines are used mainly for kraft pulping and aging of the wood is neither needed nor desirable. This is a cheaper storage method than stacking.

Storage of logs in water is not as popular as it once was, particularly in inland freshwater bodies. It can retard decay, but it can cause contamination of the water. If the water is ultimately returned to a lake or river, federal and state laws require treatment before discharge.[54] Water-soluble materials, such as tannins and low-molecular-weight carbohydrates, can be leached out of the bark and the wood. Since some of these materials are biodegradable, they will consume oxygen in the receiving body of water. Other extracted materials may be toxic to aquatic life.[55]

Storage in water is still in use at coastal mills located on salt water. Logs are most frequently delivered to these mills by water, either by raft or by ship, and it is most convenient to store them in the water adjacent to the mill site. Contamination of the water may be of little consequence, due to tidal action that periodically sweeps the contaminants away to the ocean. Two problems should be reiterated: (1) Some logs become water saturated and sink to the bottom; although recovery is possible, it is an additional expense. (2) All logs will soak up a certain amount of water. If it is seawater, the wood becomes permeated with salt. This salt accumulates in the liquor system of kraft mills, causing corrosion problems, and eventually overloading the chemical recovery operations.

If there is a large pickup of salt by the logs, special measures have to be taken to remove the salt from the system continuously so that the salt content of the white liquor does not exceed about 25 g/l.[56] One of the simplest methods of doing this is by bleeding part of the liquor from the system and replacing it with fresh liquor. While somewhat wasteful of chemicals, the salt content can be held to an acceptable level by this method. Another method involves concentration of the white liquor to the point of saturation of the NaCl so that the salt precipitates and can be filtered off.[57]

Chip Storage. Large-scale and lengthy chip storage came into vogue in the 1950s.[58-60] Before that time, pulpwood was generally stored as roundwood and the logs were chipped just prior to pulping. With the advent of sawmill residue chips as a major fiber source, pulp mills evolved from roundwood-storage sites to chip-storage sites. Many new mills were designed to operate wholly on purchased chips. In such mills chips are received by truck, rail, or ship (barge); unloaded, using appropriate equipment; and then conveyed to large storage piles, normally outdoors. Chip piles vary widely in size, but may cover more than a hectare (several acres) at the base and be 15- to 30-m (50- to 100-ft) high. Chip inventories may be as large as one year's supply for the pulp mill and may run into millions of dollars.

Initially, little thought was paid to storage problems because it was assumed that chips would store as well as roundwood in terms of physical and chemical deterioration.[61,62] One of the first clues that all was not well was combustion in the chip piles.[63] Even if direct flames were not visible, digging into the piles revealed blackened, charred, smoking chips. Laboratory tests proved that such chips were unsuited for pulping, producing low-yield, high-lignin content pulps that were dark-colored and refractory to bleaching, as well as being low in paper strength. Wide scale investigation into the cause and ultimate prevention of such occurrences has disclosed:[64-67]

1. Chip and sawdust piles begin to generate heat from their inception, regardless of wood species or wood quality.

2. Heat generation is biological and/or biochemical in origin and is caused by: (a) biological respiration of the living cells in the chips;[68] (b) biodegradation of the wood by fungi and bacteria;[69] or (c) exothermal oxidation of certain materials in the wood, mainly extractives.[70]

3. Temperatures within the pile rise for a period of from several weeks to several months. Normally the temperatures do not exceed about 60 to 70°C and then they slowly cool down to ambient levels. On occasion, however, the temperatures do not stabilize but continue to rise to a level where wood is destroyed by pyrolytic action and the pile is "on fire."

4. Degradation of wood substance accompanies the heating in the piles. Some wood is literally consumed in the pile and the remaining wood is normally less suited for pulping than fresh wood. As a result, gravimetric pulp yield is lower and the pulp is normally of lower quality. The loss in quality is related to the amount of heating in the pile and the subsequent degree of wood degradation.[71]

5. Efforts to minimize or halt the heating and resultant degradation have been minimally successful.[72,73] As the pile is constructed, various chemicals have been applied to the chips to minimize heating, but this has only met with varying degrees of success and with variable costs. Electron bombardment of chips going into the pile in order to sterilize the wood was tried on a commercial scale, but this has been abandoned because of inconclusive proof of benefits.[74]

In lieu of some type of preventive treatment, a loss of about 1% of the wood per month seems inevitable during storage of wood chips. There is no unanimous agreement on the relative importance of the causes for heating and degradation.

Chips from freshly cut trees might contribute strongly to mechanism 2a above, but chips from old, rotted wood mixed with fresh chips might contribute strongly to mechanism 2b. Most mills exercise little control over the quality of incoming chips, so there may be no dependable method of predicting the extent of heating in a pile and thus appropriate steps to control it cannot be taken.

The reason that chips in piles deteriorate so much more rapidly than logs is the large specific surface of chips relative to logs. Heating is most rapid when oxygen and/or microorganisms have easy access to the chips. Oxygen consumption within a typical chip pile is quite rapid. As the temperature of the pile increases, air and oxygen are drawn in from the sides, thus continuing to fuel degradative actions. The pile acts as a chimney and the heat rises to the top where the greatest degradation takes place.[75] Should the temperature of the pile exceed 60°C, the possibility exists that the pile will become biologically sterile.[76] Further temperature rise must then be due to autocatalytic reactions involving oxidation of chemical substances in the wood. The environment within a chip pile is probably aerobic. On the other hand, sawdust piles may be anaerobic due to the close packing of the particles and the resultant exclusion of air. Anaerobic organisms can operate within this environment and sawdust can degrade quite rapidly in a matter of a few months.[77] The more rapid degradation is attributed to the larger specific surface of sawdust compared to chips.

At present, complete prevention of heating in chip and sawdust piles is not possible, but proper control of piles can minimize the heat buildup and the resultant wood degradation and economic loss that inevitably occurs. Among the suggestions for proper control are:

1. Use chemical treatment of chips during buildup of the pile. This can reduce the heating, but the economic value is questionable.[78]

2. Avoid production of fines. Heavy motorized equipment continuously running over the top of a pile breaks up chips into smaller pieces. Building a pile by blowing chips from a pipe has a tendency to classify chips, the smaller ones falling directly beneath the delivery spout and the larger ones landing further away. Both practices tend to concentrate smaller particles in specific areas and these may turn into hot spots because of the greater surface area of the fines.

3. Follow the practice of first chips in, first chips out. In this way all chips will be in storage for about an equal length of time, suffer about the same degree of degradation, and no chip will be strongly degraded because of excessive storage time. This practice is not always feasible, since chip inventories are rarely constant in size over extended periods of time. Sales of lumber and paper are typically cyclical and the two cycles are frequently out of phase so that the flow of incoming chips does not match that of the outgoing chips.

4. Minimize the height of the pile because the temperature is highest at the top of the pile.[79]

5. Store chips in several smaller piles rather than in one large pile. In this way, it is relatively easy to utilize the oldest pile before continuing to the next oldest pile.

Pile management is an inexact science and some mills prefer to absorb the economic loss stemming from little or no management rather than spend capital and operating money to put any of the above practices in effect. In order to arrive at a rational decision, the additional operating costs for good management of piles must be computed and compared against the potential economic savings due to lower losses in wood substance and quality. The savings are not always clearly identifiable and the rationale for better pile management may be rather intangible.

Debarking

Bark, an unwanted contaminant in a pulp mill, causes lower digester efficiency, greater pulping-chemical consumption, lower gravimetric and volumetric pulp yields, greater bleach-chemical consumption, and ultimately dirtier and weaker pulp.[80,81] The amount of bark that can be tolerated depends on the pulp-mill operations and equipment as well as on the intended use of the pulp.[82] Most pulp mills prefer to have 1% or less bark by weight with their chips, although frequently they do exceed this limit.

The goal of barking is to remove the maximum amount of the bark, leaving the good wood intact and undamaged. Obviously no barker is 100% efficient, and the choice of barker depends on the size of the wood being processed as well as the barker efficiency. The three principal types of barkers are drum, ring, and hydraulic.

Drum barkers. These barkers are high volume units usually restricted to handling short 1.2- to 2.4-m (4- to 8-ft) long and small diameter logs.[83] The logs enter the high end of a slightly inclined rotating drum that is 3.7 to 4.6 m (12 to 15 ft) in diameter and 15 to 21 m (50 to 70 ft) in length. As the drum rotates, the logs tumble and rub against each other, removing the bark by abrasion. Water showers spaced along the drum flush the loose pieces of bark through the perforated sides of the drum into a disposal area. After exiting the drum the logs are inspected and those only partially debarked are returned to the drum for another passage through. The accepted logs normally go to the chipper. Drum barkers work best with small short logs, preferably straight, since bark removal is accomplished only by friction against other logs. Crooked sticks are poorly debarked as are areas around knots. A certain amount of water is needed for efficient operation. In contrast to past practice, this water cannot be dumped without primary and/or secondary treatment. Most mills filter off the solid particles and recycle the filtered water to the drum. Ultimately this water will end up in the mill waste and will have to be treated. (See chapter Environmental Control in Volume II.)

Ring Barkers. Ring barkers normally operate with little or no water, thus avoiding the contamination problem of the other two types of barker.[84] Logs are debarked in these units individually, being fed horizontally into the eye of a vertically mounted ring. A series of scrapers are mounted around the inner circumference of the ring and move inward until they touch the log. Either the log is rotated or the ring and scrapers rotate and thus the bark is removed by mechanical abrasion. The log is slowly fed through the ring and all sides are exposed to the abrasive action of the scrapers. Because they operate on a single log at a time, ring barkers are normally used only for large-diameter logs. They are popular in the northwestern United States at sawmills and plywood mills that generate residual chips for pulp mills. Length of log is no problem, nor are crooked logs, since the scrapers accommodate themselves to changes in diameter and form. Logs can be returned for another pass if not satisfactorily debarked in one pass. The biggest objection to the ring barker is the damage done to the surface of the log; considerable gouging and scaping of the cambial and inner bark layer occurs and the adjacent fibers suffer physical damage. The relative importance of this wood damage depends on many factors, such as the percentage of damaged fibers in the final pulp. Ring barkers are mechanically simple and reasonably portable. Thus they are one of the first choices for debarking in the woods and can be mounted in tandem ahead of a chipper.

Hydraulic Barkers. These units became quite popular in the period 1940–1960 but have not kept pace with other types of barkers.[85] They are best suited for large diameter logs 6- to 12-m (20- to 40-ft) long; this type of log is disappearing. In one design of hydraulic barker, the log enters an enclosed chamber and is rigidly held while an overhead nozzle passes along the length of the log. A jet of water issues from the nozzle at pressures of about 7000 to 10,000 kPa (1000 to 1500 psi), blasting the bark away from the log. At the end of the pass, the log is mechanically rotated a few degrees and the nozzle makes another pass. An operator separated by a safety window controls the procedure and may cause the nozzle to make two passes over a given strip if one pass does not adequately debark the strip. When the log has been debarked over the entire 360° of its circumference and over its entire length, it is ejected and a new log is introduced. Hydraulic barkers are relatively costly to install and maintain because the chamber is large and they require sophisticated auxiliary equipment. Pumps capable of generating the high pressures needed are costly and require frequent maintenance. Erosion of the nozzle caused by the high-pressure water jet necessitates frequent replacement. The water jet is extremely powerful, dangerous, and can chew its way through a thick log in seconds if the travelling carriage should stall before the water can be turned off. Large quantities of water are required for operation, and the water must be filtered before reuse or treated before dumping. Nonetheless, given the proper conditions, hydraulic barkers make good economic sense. They are very efficient in removing the last traces of bark with minimum damage to the log surface and are least affected by crooked stems and wood defects. They are definitely not suited for in-the-woods operations, but for sawmills or pulp mills processing large-diameter logs, they do an excellent job.

Chipping

Pulp quality can be drastically affected by the chipping operation, yet very little attention is paid to this phase of production. Brooming of chip ends and crushing of fibers cause damaged fibers, which weaken pulp and paper, and it behooves chipper operators to maintain their equipment in the best possible condition.

Chipper Design. Most chippers operate on the principle of a flat revolving disk with knives set radially on the surface and projecting slightly from the plane of the disk.[86] The log or slab is fed into this rotating disk, normally at a 45° angle to the plane of the disk. The knives slice off chips that then pass through holes in the disk on the back side of the knives and are next conveyed to screens or to storage. There are modifications in chippers, depending on the feed material (whole logs versus slabs), but the fundamental principles are basically the same.

Commercial log chippers are designed with disks from 230 to 380 cm (90 to 150 in.) in diameter, whereas chippers for waste wood (slabs and edgings from sawmills) are somewhat smaller. Power requirements for the drive motors are established by the type of material being chipped and vary from a low of 135 kw (180 hp) for 15-cm (6-in.) diameter logs to a high of 1870 kw (2500 hp) for 50-cm (20-in.) diameter logs.[87] The sharpness and cutting quality of the knives is vital to the production of quality chips; thus chippers are designed for easy access to the knives so that they can be easily replaced on a periodic basis.[88] This occurs about once per 8-hr shift, although damaged knives should be immediately replaced to minimize the output of defective chips. Damage to the cutting surfaces can occur at any time if foreign materials, such as nails, spikes, rocks, or sand, are allowed to pass through the chipper with the logs.

Integrated Chipping and Sawing. Machinery was introduced in the 1960s to convert logs into dimension lumber and into chips in a unit operation.[89] The debarked log is converted into a square or rectangular cant by means of rotary planer knives that remove chips from each of the four sides of the log in a way that maximizes lumber production. The cant then goes immediately into a multisaw headrig where it is converted in one pass into a set of dimension lumber, which can be further trimmed and edged in conventional machinery. This type of log converter has many potential advantages, the prime ones being the saving of space and the integration of two operations into one. Unfortunately, the quality of chips from this type of operation is substantially different from, and generally lower than, the quality of chips produced from conventional disk chippers. The difference lies in the chipping action; conventional disk chippers with radially mounted knives slice across the grain of the log, whereas the planer knife heads, which are perpendicular to the axis of the log, cut chips parallel to the grain. Chipping with the planer knife seems to cause damage to the fibers, which results in weaker pulp and paper.[90]

Integration of chipping and sawing has much to recommend it. As mature timber is depleted and the lumber industry turns to the utilization of smaller young-growth trees, machines of this type will undoubtedly become more prevalent. Manufacturers are working on their development. Therefore, improvements in chip quality can be anticipated in the future.

Chip Screening. Size distribution is the most important specification for chip quality and for some mills the only one.[91] Material produced by chippers ranges from small particles to long slivers and the mass must be screened to segregate the chips of the proper dimensions. (The question of chip specifications is covered more thoroughly in the section on chip quality.)

Screening of chips can be done at any of several points, depending on the source and flow of raw materials. Residual chips are frequently screened by the sawmill prior to delivery to the pulp mill and the chips may be further screened at the pulp mill. Chips produced within a pulp mill are normally screened only once. Screening is relatively simple: a stream of unclassified chips is passed through a series of screens that have openings of decreasing size; the sizes are selected at the discretion of the operator. Flatbed screens are most common and the openings may be round, square, or slots. The entire screen moves in a reciprocal or rotating motion to facilitate movement of the chips. Two screens may suffice for most commercial operations: (1) a large opening screen—generally about 25 mm (1 in.) in diameter—to remove "overs" and (2) a small screen—about 6 to 9 mm (1/4 to 3/8 in.)—to remove the "unders" or fines. Specifications for such chips would read, "minus 1 inch, plus 1/4 to 3/8 inch," meaning that the accepted fraction would pass through a 1-in. screen, but be retained on a 1/4- or 3/8-inch screen. Chip size is mostly established by the pulping process.[92] Chip-size specifications vary among mills; no industry-wide standards have been adopted.

Disposal of the overs and unders is a matter of individual preference. Overs are frequently returned for rechipping or fed to a hammermill or hog mill, which accomplishes the same purpose. Unders or fines can be blended with sawdust for pulping or can be fed into other wood production lines, such as hardboard. The increasing cost of energy has caused some mills to burn the fines, and occasionally the overs, in hog-fuel furnaces to recover energy. In any case, the fractions of the chips rejected for pulping are not wasted in a well-managed operation.

Removal of Bark from Chips. Traditionally, bark has been removed from logs prior to chipping, whether the log was chipped in its entirety (roundwood) or whether the chips were generated from sawmill residues. The amount of bark remaining with the chips was usually small, normally about 1% or less by weight. With the advent of whole-tree utilization in its many diverse aspects, bark removal before chipping became more difficult, if not impossible. If literally the whole tree was being chipped, the main stem or trunk could be debarked, but the smaller branches and limbs could not be similarly processed simply because machinery was not available to do the job. Debarking of small and/or crooked logs becomes less efficient and less profitable as the diameter decreases, Thus, there has been considerable pressure on the pulp mills to accept chips produced from unbarked wood.

Since chip quality suffers when appreciable quantities of bark are mixed in, several methods have been proposed to remove and segregate the bark from the chips.[93] There are two aspects to the problem: (1) breaking the bond between the bark and the wood and (2) removing the loose bark from the chips. The first part of the problem is complicated by the fact that there are wide variations in the tenacity of the bond between bark and wood. Some species have weak bonds and bark can be easily detached; other species have tough bonds. Within species the tenacity of the bond can vary with moisture content of the wood, individual trees, age of tree, and season of the year—to name a few variables.[94] Hence, any proposed method for bark removal must be tested over a significant period of time with many different samples of wood in order to evaluate its utility.

Various proposals have been made for breaking the bark-wood bond. Most of these involve a rolling or crushing of the chips. The softer bark pops off the chip during this stage and may be significantly reduced in particle size[95] so that it can be removed by mechanical screening. Other methods involve flotation in water frequently combined with a cycle of alternating vacuum and pressure in order to saturate the bark with water and cause it to sink preferentially.[96,97] As is true with many physical treatments of wood, the results are quite variable. The percentage of bark can be significantly reduced, but only at the price of losing a certain amount of good wood along with the bark. Two- and three-stage processing produce cleaner chips than one-stage processing, but at the expense of greater wood loss. No method has been developed that will reduce the bark content, without catastrophic losses of wood, to the 1% or less level that is typical of chips produced from debarked logs. The amount of bark that can be tolerated with chips is a variable for each pulp mill and is influenced by the pulping and papermaking process equipment as well as the end use of the product.

The acceptance of high-bark-content chips by the industry has been slow, primarily because of the inefficiency of the segregation process. If the separation can be substantially improved, the idea may take hold because a strong economic case can be made for not removing bark from logs prior to chipping. As with most new concepts, the acceptance or rejection will depend on economic considerations; if the benefits of chipping logs with bark exceed the costs of removing bark from the chips, then this method will be accepted, at least by certain segments of the industry.

Wood and Chip Mensuration

Measurement of the absolute amount of wood present in trees, logs, lumber, chips, or sawdust is a complex and inexact task. Starting with its microstructure and up to the final form of the tree, wood exhibits wide variation in form and structure. When wood is comminuted into particulate form, such as chips or sawdust, other modifications of size and shape are imposed on the inherent variability of the wood that further complicate the picture.

Volumetric Measurement. Traditionally wood has been assessed on a volumetric basis. Since a major use for logs is conversion to lumber, timber is measured in terms of board feet (bd ft); one bf equals a solid piece of lumber 12-in. square and 1-in. thick. Trees and logs are scaled in terms of the estimated amount of lumber that can be derived from them by an average sawmill, *not* on the basis of the amount of solid wood actually present in the logs. Such a scaling system necessarily involves assumptions about and allowances for log diameter, defects, and other variables, which prevent the system from being a rational method of measuring true wood content.[98] Another measure, derived primarily from the sale of wood for fuel and fence posts, is the cord. The standard cord is the amount of wood that can be stacked in a space 4-ft wide by 4-ft high by 8-ft long (128 ft^3); the sticks of wood measured are typically 4-ft long. The true volume of wood present in a cord is determined by the average diameter of the sticks and the degree of straightness of the sticks. The true volume typically ranges from 2.5 to 3.0 m^3 (90 to 105 ft^3) of solid wood per cord. Nonstandard cords are sometimes used commercially; these have gross volumes of up to 144 ft^3. If a nonstandard cord is intended, it must be specified; otherwise, 128 ft^3 is considered the standard.

These mensuration methods serve their purpose reasonably well, but there is a big gap between measurement and performance when wood is converted to chips and pulp. Pulp and paper are produced, measured, and sold primarily on a weight basis. Pulp yields, for example, are nearly always expressed on a weight basis. The volume of solid wood or chips introduced into a digester to produce a ton of pulp is normally of secondary importance. What is more important to the pulp producer is the dry weight of wood converted to chips. Accurate information for this is difficult to obtain, whether one uses the board-foot measure of logs or the cord.

In the simplest manner, the weight of pulpwood can be calculated as follows: the true volume of wood to be converted to chips is multiplied by the wood density to obtain an estimated dry weight of wood. This estimated dry weight of wood is divided by the solids content of the sample $[100 - (MC_1)]/100$ to estimate the wet or green weight of the wood, where $MC - 1$ = (weight of water)/weight of wet wood $\times$ 100.

Three problems are inherent in this calculation: (1) the variability of specific gravity within a species, (2) the variation of moisture content, and (3) mechanical losses during chipping. For best accuracy, all three should be established for any one sample or batch of wood, but this can be time consuming. Hence many mills have developed arbitrary tables of conversion from solid wood to chip form based on their experience. It should be recognized that these tables are based on average values for wood density, moisture content, and conversion efficiency; gross variations from these averages are frequently encountered in actual practice.

If roundwood is to be chipped in its entirety (after debarking), then the conversion of wood measurement to chip measurement is relatively simple: the volume of debarked logs equals the gross volume of wood in chip form. The major source of error is due to the variable quantity of solid wood within the cord. Some mills make no correction for this variation; others take a more sophisticated approach, such as photographing each load of logs received from the side.[99] An instant print of the photograph is made. By using an overlay grid on the photograph, a rapid and reasonably good estimation of the amount of solid wood present can be made. Some mills buy pulpwood by green weight and use average values for density and moisture content (obtained by experience) to calculate the dry-wood weight.

Estimation of the yield of residual chips from a sawlog is less precise. Tables are available that give estimates of the nominal amount of lumber that can be produced from a log scaled by any of several different methods, such as the Scribner system in the Northwest.[98] These tables also give an estimate of the true volume of wood that can be recovered in solid lumber form as well as the true volume of wood originally available. By calculating the difference, an estimate is obtained of the volume of wood available as residue material. Part of this, however, will be in the form of sawdust generated by the lumber conversion, and the amount of sawdust is determined by the type of equipment used and its operation. The remaining solid wood fraction theoretically is capable of being chipped, but it should be realized that not all chips are acceptable since chipping efficiency is variable with the operational conditions. Hence, it is impossible to estimate accurately the yield of pulpwood chips from a sawmill operation and recourse must be made to experience.[100]

Each mill has developed methods of estimation for converting solid wood to chips, and tables of conversion have been published that give average values and ranges. Examples are given below to illustrate typical values for volumes and weights of unbarked logs per cord, using common North American units:

Species	Diameter of Bolt (in.)	Solid Wood (ft^3)	Board Feet (Scribner)	Green Weight (lb)
Douglas	8	81	365	4350
fir	20	91	650	5000
Western	8	81	370	4850
hemlock	26	89	595	5350

One cord of slabs and edgings from hemlock contains 67 to 75 ft^3 of solid wood. Average green weight per 1000 bd ft, Scribner scale of logs:

Douglas fir	8,200 lb or 3,722 kg
Western hemlock	10,200 lb or 4,630 kg
Western red cedar	8,410 lb or 3,818 kg

One cubic foot of solid wood is equivalent to 5 to 6 bd ft of logs, Scribner decimal C scale. One cubic meter equals $0.0283 \times$ ft^3. In making conversions it should be stated whether the wood is solid, stacked, or straw-piled. The above data are illustrative only and by no means are complete.

Chip Mensuration. Chips, sawdust, and hog fuel can be measured volumetrically as well as gravimetrically, and the same problems of interconversion exist as in the case of solid wood. In the early years of residual-chip development, it was tempting to use volumetric measures, and in some instances this is still used. The chip supplier is concerned about the volume aspect because of limited storage space at his mill and because trucks and railcars have finite capacity and the space occupied by a given volume (or weight) of chips or sawdust is of prime concern in transportation.

Wood chips occupy more space than the original solid wood. As a rule of thumb, chips will occupy 2 to 2.5 times as much space as solid wood; Douglas fir with a specific gravity of 0.48, for example, will weigh in log form about 480 kg/m^3 (29 lb/ft^3) dry basis, whereas a cubic meter of chips will weigh about 190 to 208 kg (12 to 13 lb/ft^3). The bulk density of chips is determined by many variables, such as the original wood density, average moisture content, and average size and size distribution of the chips.[101] It can also be influenced by mechanical factors, such as the depth of the chip bed, the method of forming the chip bed, and settling which may occur during transportation. For this reason, the bulk density of Douglas fir chips can vary as much as twofold, from 128 to 256 kg/m^3 (8 to 16 lb/ft^3) dry basis. Thus, measuring chips volumetrically is subject to the same kind of errors as measuring logs by volume. Chip volume, while important for transportation reasons, is of little concern to the pulp mill in terms of storage, since most mills usually have adequate chip-storage capacity. It does become important during pulping since low-bulk-density chips occupy more digester space per unit of weight of pulp produced than high-density chips.

Most Pacific Northwest mills have turned to gravimetric measurement as the basis for purchasing and accounting for chips.[102] Delivery carriers are weighed before and after dumping to establish the green weight of chips delivered. A representative sample of chips is taken during or just after dumping and a moisture determination is immediately made from which the mill can then estimate the dry weight of wood delivered. The seller is paid and pulp-mill balances made on this basis.

The word unit, used for the measurement of chips in the northwest, is spreading to other parts of the United States. Originally, it was a volumetric term and was defined as 200 ft³ (5.66 m³) of comminuted material, such as chips. This is still used today for other bulky material such as sawdust or hog fuel. More recently, the term has become gravimetric and is defined as 2400 lb of bone-dry wood material, abbreviated as BDU (bone-dry unit). This value was derived from experience that showed that 12 lb/ft³, or 2400 lb/200 ft³, was an *average* value for the bulk density of Douglas fir chips, the major commodity chip item in this market. Such a basis for commerce is quite arbitrary, however, and the use of 2000-lb ton or a metric ton (MT) of 2204 lb is equally sound and fundamentally correct. Use of the gravimetric definition of unit enables the pulp mill to begin processing wood correctly on a weight basis.

Some southeastern United States pulp mills purchase chips on a green or wet basis and make no effort to measure the moisture content of the chips as received.[103] To the extent that the moisture content is uniform over long periods of time, this procedure yields meaningful information and is fair both to buyer and seller. The greater the variation in moisture, the less fair it becomes to one party or the other, and the more misleading it becomes in pulp-mill operation.

Some typical conversion factors from solid wood to chips are:

1000 bd ft of lumber converts directly to 200 ft³ of chips.

1000 bd ft of logs, Scribner Decimal C scale, converts to 400 to 470 ft³ of chips (11 to 13 m³).

A volumetric unit of sawdust contains 80 ft³ of solid wood (2.26 m³).

A volumetric unit of chips contains 72 ft³ of solid wood (2 m³).

In summary, measuring wood for pulping is fundamentally most correct and probably fairer to both buyer and seller when figured on a dry-weight basis.[104] Accurate determination of the true dry weight at times can be very difficult or impossible. Then it may be more expedient to mensurate volumetrically, whether the wood is in tree, log, lumber, chip, or sawdust form, and to make the appropriate conversions to arrive at a dry-basis estimate. One must be aware of the inherent variations in wood quality that can be misleading if rules of thumb and overall averages are followed; estimates of wood can be off by a factor of two or even more if several factors were to compound each other. Reasonable limits for mensuration should be kept in mind at all times, and if a spot check reveals values that approach or exceed the limits, further investigation is warranted. As wood becomes scarcer and more costly in the future, accurate mensuration methods assume even greater importance to the paper industry.

Chip Quality

The quality of the chips used for pulping receives relatively little attention during normal mill operating conditions, yet this is an extremely important factor in the operation of the pulp mill and in final pulp quality. The degree of control and testing of chips varies considerably within the industry. Some mills recognize the importance of chip quality and others are content to utilize whatever comes their way. Following are some of the more important parameters of chip quality. The subject is discussed further in the chapter on pulping.

Chip-Size Distribution. This is the one parameter of chip quality that has been most carefully studied and commercially controlled. The fundamental purpose of chipping is to reduce the average size of the wood particles so that penetration of the pulping liquor may be relatively rapid and uniform. Carried to the extreme, wood ideally would be reduced to a fine flour or meal before pulping, except that in the process of mechanical attrition so many fibers would be physically damaged that the resultant pulp would make very weak paper. Sawdust pulp, for example, has lower strength properties than chip pulp, all other things being equal.

Each mill's experience over the years has led to the establishment of standards for chip size, which is a compromise between the goals of rapid pulping and best pulp strength. It has been assumed that the penetration of liquors into the wood was primarily in the longitudinal direction, parallel to the fiber axes. The lumen of a fiber would become filled with pulping liquor and the liquor would then pass to the lumen of an adjacent fiber through the pits connecting the fibers. Thus, the primary criterion of chip size has generally been chip length; chip width and thickness were considered to be of secondary importance or of no concern.[105] Work by Hatton and Keays[106] has suggested that the above hypothesis may be true for chips destined for acid sulfite pulping, but that chip length is not the most critical parameter for chips to be pulped by the kraft process. The strong alkaline kraft liquor apparently can penetrate the chip about equally well from all directions, due to the swelling action on the wood caused by the alkali, and chip thickness has thus been shown to be the most critical parameter for kraft chips. Uniformity of chip thickness will do more to assure uniform liquor penetration and consequent uniform delignification than uniformity of chip length. Because it has been shown that conventional chip screens with round or square holes segregate chips more on the basis of length or width than on chip thickness, special slotted screens have been developed that will classify chips by their thickness.[107] The concept of screening chips by thickness may grow rapidly in commercial practice in the future.

Absolute standards for chip size vary for different mills. Such factors as digester equipment, pulp-washing equipment, bleach requirements, and product specifications enter into the establishment of chip-size standards for a given mill. Continuous Kamyr digesters are not tolerant of small chips and fines since these have a tendency to plug the liquor-circulating screens within the digester.[108] Other continuous digesters are not so sensitive to the presence of fines. Batch digesters are also less sensitive, particularly if direct steaming is practiced with no forced circulation of liquor. Fines generally produce weaker pulp than large chips and should be avoided where strong paper is required.

Typical chip sizes, specified in terms of the screen sizes, range from minus 25 mm (1 in.) to plus 6 mm (1/4 in.). It should be emphasized, however, that the size of the hole in the screen has little relation to actual chip length. A 150-mm (6-in.) long chip could easily pass a 25-mm (1-in.) screen if the thickness and width were each less than 25 mm (1 in.). In a sulfite mill, such a chip could end up as a long sliver in the rejects. Mills typically specify chip size in terms of distribution: no more than a certain percentage above a given limit for different-size screens. This procedure has many drawbacks, but works reasonably well if rigorously enforced.

Shredded Chips. Various schemes for reducing particle size of chips have been proposed in a number of studies. In one such proposal, the standard-size chips are passed between the revolving disks of a refiner and reduced to matchstick size.[109] The goal is to reduce the particle size and thus to improve liquor penetration without the accompanying physical damage to the wood and fibers that would occur if this were done by more severe chipping. The concept is apparently a sound one for kraft pulping, and increases in screened yields and reduced cooking times have been reported when using shredded chips as compared to whole chips.[110] Strength losses are minimal until the plates are set too close to each other, at which point severe damage occurs to the wood and the resultant pulp. A similar study showed comparable benefits in the screened yields and cooking times for bisulfite pulping, but significant pulp-strength losses were noticed even from chips refined at the widest plate-gap setting.[111] The concept seems economically feasible for kraft linerboard pulps, but to date the industry has not adopted it.[112]

Shredded chips may be a necessity in oxygen-pulping operations.[113] Oxygen pulping with alkali requires small and uniformly sized wood particles to assure uniform penetration of the oxygen into the interior of the particles. This process has turned out to be more difficult with gases than with liquids. Thus, many researchers are recommending that normal chips be shredded in disk refiners before treatment with oxygen and alkali, and this may become one of the future methods of chip preparation.

Chip Damage. To summarize some of the points already mentioned, chips should be as free from mechanical damage as possible. The less the physical damage to the fibers, the higher the quality of pulp. Since nearly all phases of pulpwood procurement result in some type of wood damage, the question is not one of zero damage, but one of minimizing the damage to the wood.

Wood itself is full of unavoidable physical imperfections that show up eventually in poor-quality pulp. Furthermore, every wood-handling operation—from felling, skidding, and debarking to chipping—introduces additional wood and fiber damage. Chipping is probably the worst offender since it is here that the log or slab is "opened up" to create a large specific surface and the greatest numbers of fibers are exposed to mechanical damage. The importance of sharp knives and proper adjustment of the feed to the chipper cannot be overstressed if high-quality pulp is desired. Drum barking is a degrader of wood because the tumbling action of tons of logs has a tendency to broom the ends and crush fibers on the surfaces of the logs.

For a given amount of chip damage, the kraft process is the most forgiving and the acid sulfite process the least forgiving. It has been shown that alkali pulping does not badly degrade damaged chips, whereas the acid cooking liquor of the sulfite process attacks those areas already damaged and accentuates the degree of damage.[114] Thus, it is encumbent on chip suppliers to sulfite mills to deliver the highest-quality (that is least-damaged) chips possible to maintain good-strength pulp and paper, whereas it is less critical for the suppliers of kraft chips. Unfortunately, there are no generally accepted standards for evaluating physical damage to chips. Damage must be evaluated visually, and the estimation of the degree of damage is quite subjective. As previously mentioned, the effect of damaged chips on pulp and paper quality cannot be readily quantified because of the many variables in pulping and papermaking operations. There is no question that damaged chips are an economic drain on the paper industry and that good management involves a careful watch of incoming chip quality.

Chip Moisture. Knowledge of the moisture content of chips is important for two reasons: (1) pulp mills cannot accurately estimate true wood input to digesters without knowing the moisture content of the chips and (2) best pulping control can be achieved only if the liquor-to-wood ratios and the chemical concentrations in the liquors are constant for all cooks. Variable moisture content of the incoming chips can cause significant fluctuations in these two pulping parameters.[115]

If the mill is operating on residual chips, moisture content is normally determined at the time of delivery of the chips. Sample collection techniques vary quite widely, however, and there is little uniformity either in sampling methods or in moisture-determination methods. Some mills continuously collect samples as the truck or railcar is being dumped, while others collect grab samples from the load. Usually a 1-to-2-kg moisture-free sample is weighed and dried in an oven at 100 to 105°C to establish the moisture content, although there is little agreement between mills on the standard technique for drying.[116] If the mill is paying on the basis of dry wood purchased, a technique of drying that results in an erroneously high chip-moisture content would favor the buyer, but an erroneously low chip-moisture content would favor the seller. No method has been developed that has been accepted by all parties concerned, even within local regions.

Moisture content of wood chips is normally expressed in the paper industry as a percentage of the wet (total) weight of the chips:

$$\text{Moisture content} = \frac{\text{Weight of water}}{\text{Weight of wet wood}} \times 100 \ (MC_1)$$

$$\text{Solids content} = 100 - \text{moisture content}$$

Another expression is used by the logging and lumber industries:

$$\text{Moisture content} = \frac{\text{Weight of water}}{\text{Weight of dry wood}} \times 100 \ (MC_2)$$

The two expressions are related as follows:

$$MC_1 = \frac{MC_2}{100 + MC_2} \times 100 \text{ and } MC_2 = \frac{MC_1}{100 - MC_1} \times 100$$

MC_1 for freshly cut chips rarely exceeds 60% and MC_2 rarely exceeds 150%.

Some mills, primarily in the southeastern United States, purchase chips on the wet basis and never measure the moisture content. Such a policy is justified on the basis that the moisture content of incoming chips is relatively uniform, and that the cost of monitoring moisture content offsets any savings that could be realized by this procedure. Mills that chip their own roundwood measure moisture either continuously or intermittently, frequently on a once-per-workshift basis.

Although few mills meter the moisture content of the chips as they are fed to the digesters, most mills weigh chips as they are conveyed by belt to the digester, producing an integrated reading that is an estimate of the total wet-chip charge to the digester. When the moisture content of the chips is known, it is then simple to calculate the total dry charge. Lacking real-time knowledge, mills frequently use average chip-moisture values collected over a period of time. On the other hand, there are commercial devices that continuously measure the moisture content of chips as they are fed to the digester.[117] One such device operates on an electrical conductivity principle. This device may be used to meter the chip moisture content and to send a continuous signal to a computer, which then regulates the amount of black liquor used in sulfate pulping to dilute the concentrated white liquor, thus obtaining uniform liquor-to-wood and chemical-to-wood ratios in the digester.

Contamination of Chips. This covers a broad area and involves the definition of the term contaminant. Dirt and metal obviously qualify as contaminants, but mixtures of different wood species may or may not be thus classified. There are few generally accepted standards for chip contamination; most mills have developed their own specifications.

Identification of wood species can be a tedious and contentious task when in chip form. Most mills utilizing coniferous species object to mixing hardwood chips with softwood chips. Measuring hardwood content must then be done manually by an operator trained in the details of wood identification. Similarly, certain coniferous species are undesirable; for example, cedar chips mixed with Douglas fir or western hemlock chips are unwanted because of the low density of cedar and possible digester corrosion from it.

Tappi Standard T-14 os-74[118] lists procedures for quantifying the dirt content of chips. This method integrates the effect of different contaminants, such as bark, knots, brown and white rot, various stains, insect holes, and resin pockets, into a single value known as the dirt index. Pictures are given in the *Standard* for different kinds of stain, but in actual practice the stains might not be easily detected and identified. Most mills prefer to establish their own standards for bark and rot. Typical standards would be about 1% maximum for both rot and bark individually. In any case, problems arise from interpretation of definitions and methodology.

Some mills wash chips prior to pulping, a practice particularly recommended for continuous digesters. Sand and grit should be washed out to prevent excessive wear and erosion of chip-feeding mechanisms. Tramp metal is troublesome; ferrous metals may be removed by passing chip streams over magnets, but nonferrous metals may pass on through with the chips and cause damage.

Bulk Density. This is the weight of wood chips (or sawdust) contained in a certain volume and is most commonly expressed in terms of pounds per cubic foot. It can be expressed on either a wet- or dry-wood basis, and it is helpful in estimating the amount of wood within a given container, such as a bin, truck, and, *most important,* a digester. Bulk density is quite variable and is influenced by wood density, moisture content, chip size, chip-size distribution, and the degree of mechanical compaction, and so on.[119] Ranges of bulk density are 8 to 16 lb/ft^3 dry basis to 15 to 40 lb/ft^3 wet basis. Wood density is considerably higher, ranging from 20 to 60 lb dry and 40 to 100 lb wet. Comparable metric values are 128 to 256 kg/m^3 dry and 240 to 640 kg/m^3 wet for chip-bulk density, and 320 to 960 kg/m^3 dry and 640 to 1600 kg/m^3 wet for solid-wood density.

In summary, there is abundant evidence that poor chip quality can seriously affect pulp mill operations. Ultimately, pulp mills must do better in monitoring chip quality in order to keep out unwanted contaminants, or in quantifying the degree of contamination in order to handle it most efficiently and with the least effect on pulp quality.

Relationship between Wood and Paper Quality

The paper industry has known for many years that wood quality affects the quality of pulp and ultimately the paper made from the pulp.[120] A classic example is the preference for softwoods for pulping. They have been preferred over hardwoods because they have a preponderance of tracheids, averaging 3 to 5 mm in length, that reputedly produce stronger paper than hardwood fibers. This picture is somewhat oversimplified. More careful analysis reveals cases where hardwood pulps are capable of certain strength values equal to or even greater than those of softwood pulps.

The advent of multiple correlation statistical techniques, together with high-speed, high-capacity computers, has enabled wood and paper technologists to study more carefully the relationship between wood and pulp (paper) properties. Variations in wood and fiber morphology can be statistically correlated with variations in paper properties. This has great potential value for the industry; for example those wood qualities most important for pulp and paper quality can be selected for future genetic breeding.[121,122] This subject is vast, with many published studies. Only a few of the major conclusions will be discussed here.

Wood Density. High-density wood is preferred for construction and furniture uses and it has generally been assumed to be preferable for pulping.[123] Volumetrically this is true, since more tonnage can be introduced into a digester when heavy chips are used compared to light chips. The gravimetric pulp yield is not necessarily different for woods of different densities from the same species; fibers from dense wood have certain drawbacks for papermaking. Dense wood has more wood substance and less void space per unit volume than does less dense wood. Most of the void space is associated with the open lumens of the tracheids in the case of softwoods. Thus, thin-walled fibers have a large lumen and form low-density wood, whereas the opposite is true for thick-walled fibers. In most cases, earlywood (springwood) fibers are thin walled and latewood (summerwood) fibers are thick walled. The effect of cell-wall thickness will be discussed in the following sections, along with other fiber properties. The subject is also discussed in more detail in the Chapter on Fiber Preparation and Approach Flow in Volume II.

Fiber Length. Long fibers have traditionally, but sometimes erroneously, been associated with strong paper. Modern correlation techniques have shown strong positive correlations between long fibers and tearing strength, but with other strength properties the picture is less clear.[124] Tensile strength, the property most commonly associated with strong paper, is more a function of fiber bonding than of fiber length *per se;* unbeaten softwood fibers produce paper very low in this property, even though they have long fibers. Fiber length may actually be a detriment to good strength properties, as in the case of certain nonwoody fibers, such as cotton, linen, flax, and hemp. In such cases, fibers longer than 5 mm are difficult to handle with normal papermaking machinery because of their tendency to flocculate and form lumps in the paper. These fibers must actually be cut in order to fabricate paper with adequate formation and proper strength properties. There is little question, though, that tough paper requires relatively long fibers, meaning softwood fibers. For products such as linerboard, wrapping paper, and sack paper, short-fiber hardwood pulps are not adequate for the end uses of the products, even though at times they can meet product specifications. The basic problem is that the common laboratory paper tests, such as tensile, stretch, burst, and fold, are frequently poor predictors of performance. For example, bursting strength is the single criterion for linerboard strength. This criterion can be readily met with certain hardwood pulps refined to the proper degree. Nevertheless, linerboard today is made predominantly from softwood pulps, although some mills blend a small quantity of hardwood pulp in the top liner to improve surface smoothness and printing quality.[125] The industry has not developed a test or series of tests that un-equivocably predict the performance of paper in many of these areas that re-quire physical strength and toughness as the prime criterion, and until this is more thoroughly understood, long-fiber pulps remain first choice.

Water drains from long-fiber pulps more rapidly, and this is a point in their favor. The freeness of unbeaten softwood pulps is in the range of 700 to 740 ml Canadian Standard, whereas that of hardwood pulps is somewhat lower, 650 to 700 ml. Both types of pulp need refining in order to develop their maximum strengths, but hardwood pulps reach their optimum strength at freenesses 50 to 100 ml lower than softwood pulps. This means slower production speeds on the paper machines and higher costs. Pulps from straws, such as rice, grass, and wheat, have even lower initial freeness, occasionally as low as 300 ml, pre-senting problems not only in papermaking, but also in the pulping, washing, and bleaching operations where water removal is involved.[126]

Fiber Coarseness. One of the lesser appreciated fiber properties is the coarse-ness of the fiber. This is a broad term and encompasses: (1) length-to-width ratio and (2) cross-sectional morphology. For the same length some fibers are wider than others. The greater the length/width ratio, L/W, the greater the fiber flexibility, and the better the chance of forming well-bonded papers.[127] Cross-sectional morphology can be broken into radial and tangential cell-wall thickness and lumen diameter. Each of these parameters must be considered relative to the other and as one approach to the problem, the Runkel ratio has been proposed:[128]

$$\text{Runkel ratio} = \frac{2 \times \text{(radial cell-wall thickness)}}{\text{lumen diameter}}$$

Latewood (summerwood) fibers have thicker cell walls and smaller lumen diame-ters than earlywood (springwood) fibers; thus the Runkel ratios for the former are greater than for the latter.

Although correlations can be made between strength properties (such as ten-sile, tear, and burst) and each of the cross-sectional fiber parameters separately (earlywood radial cell-wall thickness, tangential lumen diameter, and so on), the multiplicity of the correlations tends to become confusing and lose statistical significance.[129] In general, high Runkel ratio fibers (that is, fibers with relatively thick walls, such as those in summerwood) are stiffer, less flexible, and form bulkier paper of lower-bonded area than low Runkel ratio fibers.[130] This effect is related to the degree of fiber collapse during paper drying, a phenomenon affected by the cell-wall thickness and the degree of refining that the fibers undergo prior to papermaking. If the two types of fibers (or wood) could be economically separated, each type could be used most effectively in paper-making.[131]

Individual Fiber Strength; Fibril Angle. Sophisticated equipment has been developed that permits measurement of the strength properties of individual fibers.[132,133] The most common property tested is the stress-strain behavior of single fibers with expression of derived values, such as Young's modulus of elasticity, ultimate stress and strain, and work to rupture. One of the most important characteristics of the microstructure that affects individual fiber behavior is the fibril angle in the secondary wall. In a high-modulus fiber, the fibrils are inclined at very small angles to the fiber axis; low-modulus fibers have fibrils inclined at steep angles to the axis. High-modulus fibers generally have low strain at rupture, but develop high stress; the opposite is true for low-modulus fibers.

The importance of individual fiber strength has not been conclusively established. It is reasonable to assume, however, that individual fiber strength plays an important role in well-bonded paper made from well-refined pulp. Studies have shown that failure of paper in the various strength tests can be broadly classified into two categories: fiber separation and fiber rupture. Fiber separation occurs when adjacent fibers slide past one another and fiber-to-fiber bonds are broken without major fiber damage. This type of failure occurs in lightly refined and poorly bonded paper, such as tissues. If the refining of the pulp is increased, more and stronger fiber-to-fiber bonds are formed within the sheet. After a certain point of refining, rupture of the paper takes place as a result of fiber rupture, not fiber separation.[134] At this point, stress-strain characteristics of single fibers play their greatest role in determining paper strength characteristics, assuming that the fiber structure remains reasonably intact during the refining. (See the Chapter on Fiber Bonding in Volume II.).

Juvenile and Mature Wood. Juvenile wood is that wood formed during the early stages of growth of the tree and is generally considered to be restricted to the first 10 to 12 annual rings. Mature wood is formed afterwards and is the wood in the outer rings. There is no sharp line of demarcation between juvenile and mature wood in most species, like the one for example that may exist between heartwood and sapwood. The differentiation can be made only by microscopic examination of the fibers. Juvenile softwood fibers normally are somewhat shorter and thinner walled than mature fibers. From this, one would predict that paper from predominantly juvenile fibers would be weaker in tearing strength but equal or possibly stronger in fiber-bonding strengths (such as tensile, burst, and fold) than paper from predominantly mature wood.[135] The relative amount of juvenile wood included in a chip furnish is dependent on the age of the tree at the time of harvest and the method of chipping. If roundwood is chipped, the entire juvenile- and mature-wood fractions would be included; if the chips are residues from sawmills, very little of the chips would be juvenile in origin and nearly all would be composed of mature wood. Thinnings would logically have a high juvenile-wood content, since, if used for pulping, they usually consist of young and small trees that are chipped in their entirety.[136] Overall quality of juvenile wood for pulping is satisfactory, unless rot has set in, and the paper industry should accept it with few reservations.

In summary, although the paper industry has at its disposal considerable information relating wood and single-fiber properties to pulp and paper properties, this knowledge is utilized mainly in research efforts. It is rarely used by production personnel because most mills have little or no control over wood quality *per se.*[137] In the research field, geneticists may utilize this information for tree-breeding programs by selecting those trees that grow fibers with the most desirable papermaking properties. This requires a long-term project with uncertain payoffs, and it is complicated by the lack of agreement in the paper industry as to what constitutes desirable fibers. Some paper products are best made from thick cell-wall or coarse fibers, whereas other products are best made from thin cell-wall fibers that collapse and form strong bonds during manufacture of paper. No one pulp is best for all purposes in the widely diversified paper industry.

Anatomy of Wood

Wood is a cellular, porous material, the cells of which show a great diversity in size and shape. A knowledge of the cellular structure of wood is of value to the pulp chemist since the anatomical structure of wood has an important influence on the penetration of liquids in pulping and affects the course of the pulping reactions. Wood remains intact during the pulping process, and consequently the pulping chemicals and the by-products of pulping must pass through the pores of the wood.

In softwoods, the principal cell or fiber elements are the tracheids, often mistakenly referred to as fibers. Tracheids are highly elongated, lignified cells with tapering ends that have an average length of about 3 to 5 mm and an average diameter of 0.03 mm. Hardwoods are more complex than softwoods and contain two principal cell types: wood fibers and vessel elements. The wood fibers, which are the predominating cell type, are elongated, thick-walled fibers having an average length slightly over 1 mm and an average diameter of about 0.02 mm. The vessels are relatively short in length, but have a large diameter. In some hardwoods, the vessels are very large and exist as a definite band in the springwood (ring-porous woods); in other hardwoods, the vessels are nearly uniform in size and are uniformly distributed within the annual ring (diffuse-porous woods). The wood fibers, vessels, and tracheids are aligned longitudinally in the wood, but in all species there are: (1) fine lines of cells extending radially in the wood known as the wood rays and (2) small, thin-walled parenchyma cells.

Pit membrane structure has a pronounced effect on the permeability of wood.[138] In hardwoods, although penetration is easy in the lengthwise direction because the vessel segments are joined end-to-end and have perforated ends, lateral movement is limited to diffusion through the pit membranes. In softwoods, direct lengthwise penetration cannot take place at all because the tracheid ends are closed, and they do not join end-to-end anyway. Penetration must therefore take place in a devious path by lateral movement through the pits, which is facilitated by openings in the bordered pit membranes.

The principal cell types are described in Table 3–1. These are tabulated under

TABLE 3-1 PRINCIPAL CELL TYPES FOUND IN WOOD

Softwoods

Tracheids

Tracheids (sometimes mistakenly called fibers) constitute over 90% of the volume of most softwoods and are the principal papermaking cells of softwoods. Tracheids average between 3 to 5 mm in length. Their length-to-width ratio often exceeds 100:1, which is the reason they make such good papermaking fibers. All the tracheids in a given tree are not the same length, there frequently being a range of length from 1 or 2 mm to 4 to 5 mm. The central cavity of tracheids is usually several times wider than the cell-wall thickness, which is different from wood fibers that have very small central cavities. As a rule, the springwood tracheids have thinner walls and larger diameter than the summerwood tracheids. In Douglas fir, spiral thickening occurs on the inside surface of the tracheid walls. Pits are present in the walls of the fibers and may be simple or bordered (see Figure 3-2.) Pits generally exist in pairs between two adjacent cells in the wood where the pits facilitate the interchange of soluble materials between the cells. The type of pits depends on whether the pit leads from one tracheid to another or from a tracheid to a longitudinal ray cell. The total number of pits in a given fiber varies, depending on the type of fiber, and may run up to 150 to 200 for some softwood tracheids. The function of the pits in the tracheid is to facilitate the movement of water in wood. In hardwoods, the pits are usually closed and are nonfunctional from the standpoint of water transportation.

Ray cells

Softwoods contain a small amount (usually less than 10% by weight) of ray cells. These may be ray parenchyma or ray tracheids. All ray cells are narrow and compared to the tracheids are short, being less than 0.2 mm in length. The ray tracheids have bordered pits, whereas the ray parenchyma contain only simple pits. As a rule, the ray cells are screened out of chemical pulps, but are present in mechanical pulps.

Epithelial cells

In pine, spruce, Douglas fir, and larch, epithelial cells are present surrounding intercellular spaces known as resin canals. These canals may be either transverse or longitudinal, the latter being most important. They have a size up to 0.3 mm in diameter and contain large quantities of resin. As a rule, the resin canals disintegrate during pulping. They constitute about 1% of the wood volume in pine. They are objectionable since the pitch in the cells causes trouble in certain pulping processes. The cells themselves are usually disintegrated during pulping.

Hardwoods

Hardwood fibers

There are three types of hardwood fibers: tracheids, fiber-tracheids, and libriform fibers, the latter being the main

(continued)

147

TABLE 3-1 CONTINUED

Hardwood fibers (cont.)	fiber. A typical fiber is a long, relatively narrow, empty cell that has thick to thin walls. In the typically thick-walled fibers, the lumen may occupy less than one-half the diameter of the cell. The tracheids may have bordered pits, but the other fibers may have very fine, elongated pits. Wood fibers vary from about 0.5 to 3.0 mm, and average about 1 mm or slightly more in length. They normally account for 25 to 35% of the volumetric composition of hardwoods and a much greater proportion of the fiber number. Fibers and fiber-tracheids together may account for more than 50% of the volume of certain hardwoods.
Vessel elements	These are large, empty cells that are found in porous hardwoods. The size and number of vessel elements depend on growth conditions. They range from 0.1 to 1.0 mm in length and vary considerably in shape. They are much wider in diameter, have a very large central cavity, and usually have thinner walls than the wood fibers. The ends of the cells are perforated and, in some cases, contain bars in the form of a grating across the opening. Spiral thickening occurs in the vessels of certain species. In some species, vessels may account for up to 50 to 60% of the volumetric composition of the wood, but generally less than 10% by weight. The vessel elements are aligned or united vertically in the wood to form "vessels," which function as water-conducting tubes. Water can rise vertically in these tubes, in contrast to the zigzag course necessary in the tracheid system of softwood.
Ray cells	These are parenchymatous cells (thin-walled and rectangular in shape) that grow at right angles to the wood fibers. These may occupy from 5 to 35% of the volume of the original wood. Red oak contains about 20% ray cells by volume, while poplar contains about 10%

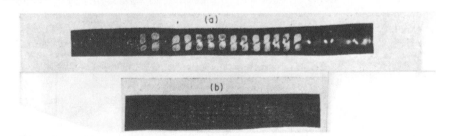

Figure 3-2. Diagrammatic sketch of fiber wall pits: (A) bordered pit and (B) simple pit.

either softwoods or hardwoods, depending on which contains the predominant number of the particular cell types. Some cell types exist in both softwoods and hardwoods, although in greatly different proportions. Thus, tracheids, which are the predominant cell in softwoods, may also be found in hardwoods. Likewise, fibers that make up most of the cells in hardwoods may also be found in

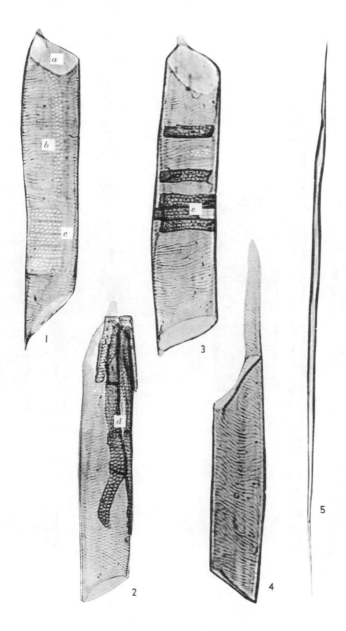

Figure 3-3. Typical vessel elements from red maple (*Acer rubrum*) × 180:
1) vessel element showing a) simple perforation, b) area of contact with longitudinal parenchyma, and c) ray contact area; 2) vessel element showing cells of
longitudinal parenchyma in contact, as in wood; 3) vessel element showing cells
of ray parenchyma in contact as in wood; 4) vessel element; and 5) fiber. Taken
from "Papermaking Fibers." (Courtesy State University of New York, College
of Environmental Science and Forestry.)

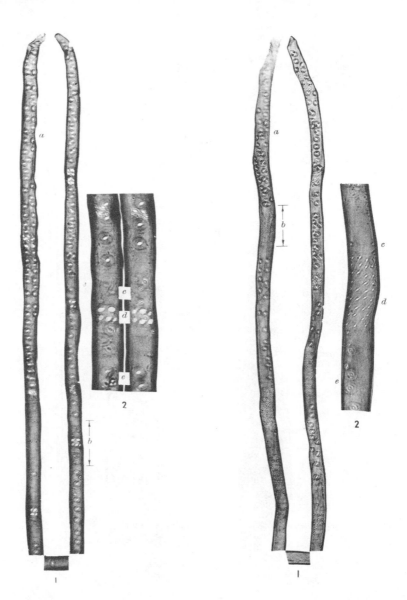

Figure 3-4. Portions of tracheids showing typical ray crossings for softwood species. Shown are shortleaf pine (*Pinus echinata*), white spruce (*Picea glauca*), and Douglas fir (*Pseudotsuga menziesii*). 1) springwood longitudinal tracheid × 90 showing a) intertracheid pits and b) ray contact area, and 2) portions of longitudinal tracheids × 180 showing c) ray-tracheid pits, d) ray-parenchyma pits, and e) intertracheid pits. Spiral thickening of Douglas fir is shown as 2f. Taken from "Papermaking Fibers." (Courtesy State University of New York, College of Environmental Science and Forestry.)

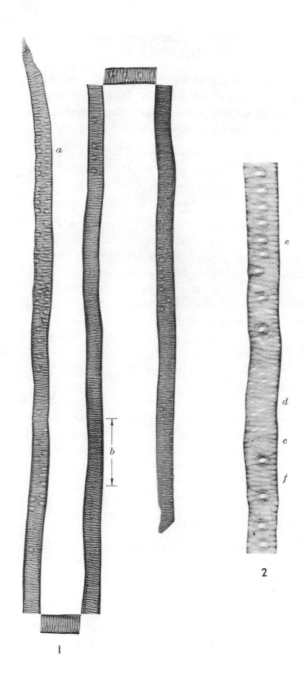

Figure 3–4. Continued.

some softwoods in small numbers. Fiber-tracheids occur in both softwoods and hardwoods in significant numbers. Not all the cell types present in wood are listed, for example longitudinal parenchyma cells, which are often present in both hardwoods and softwoods. These cells rarely constitute more than a few percent of the wood volume. In softwoods, they may contain resin and are then called resin cells. Most of these cells are destroyed and/or mechanically lost during pulping, bleaching, and papermaking and so are rarely found in the final product.

It is frequently desirable to identify the species of wood from which the fibers in a given sample of pulp were obtained. This is a part of pulp and paper testing, rather than wood evaluation, but it is mentioned here, since identification is based on minute cell characteristics visible under a microscope. Detailed instructions for fiber analysis are given in *Tappi Standard T-401 os-74*.[139] It is relatively easy to identify the pulping process employed (kraft, sulfite, and so on), but frequently less easy to identify the wood species or origin of the fibers. In the case of hardwoods identification is based mostly on the size and shape of the vessels, some typical examples of which are shown in Figure 3-3.[140] In the case of softwoods, identification is based mostly on the differences in size, shape, and number of the pits leading from a tracheid to an adjacent ray cell. Portions of different tracheids depicting typical ray crossings for a number of softwood species are shown in Figure 3-4.[140] It is evident, for example, that pine and spruce can be readily distinguished from one another since the pits, in the case of pine, are large and rectangular, whereas those in spruce are smaller and slit-like. Striations in the cell wall, which are due to cell-wall thickening, help in the identification of Douglas fir. In some cases, extreme beating or chemical degradation of the fibers may obliterate the identifying features, but in most samples of pulp or paper, the identifying characteristics are readily apparent. The use of ink as a fiber stain sometimes aids in identification.

Fiber from Nonwoody Plants

Historically, paper was originally made from nonwoody plant materials, such as cotton, linen, and grasses or reeds. The development of chemical and mechanical pulping technology for wood shifted the emphasis of pulping from nonwoody plants to woody materials in the late nineteenth century. Most of the pulp and paper worldwide is made today from wood and this is most true where wood is plentiful and thus relatively cheap. In many areas of the world, timber resources are limited, but there may be substantial quantities of annual plants, some of which are excellent fiber sources for the paper industry. In these areas, pulp is being commercially produced from such plant materials, and as the timber resources of the world decline in the future, even those nations that are now completely dedicated to wood for pulp and paper must reconsider the role that these plants can play in supplying fiber to the industry.[141] In this section, the word "straw" will be used to describe any nonwoody plant whose main stem or stalk can be utilized for pulp.

Advantages of Straw Pulps. Many straws are available as by-products of agriculture and thus can potentially be cheaper than wood.[142] Plants of this type can frequently be grown in areas that will not support trees, often with very limited rainfall and low-quality soil. They generally produce an annual crop with straw tonnages varying from a low of 1 to 2 ton/a (2.2 to 4.4 MT/ha) to a maximum of 8 to 10 ton/a (18 to 22 MT/ha). The upper range surpasses the annual growth of any species of tree, with the possible exception of very-short-rotation trees grown in a coppice, such as sycamore or cottonwood.

Nonwoody plants can generally be harvested within a year or two after planting, compared to the 10 to 20 years needed for trees to become large enough for commercial harvesting. Straw fields can be easily harvested; specialized machinery is usually not required for collection and transportation. If the straw is a by-product, the total cost of production is shared with the main crop and the cost of straw to the pulp mill should not be excessive. Most straws are readily pulped and bleached since they are not heavily lignified, and short processing times and limited chemical demand in these processes add to their attractiveness. Most straw pulps need little refining before papermaking and are well suited for papers where high strength is not needed. Bleached straw pulp is generally excellent as a filler pulp in fine papers to improve smoothness, formation, and printing or writing characteristics, whereas semichemical straw pulps make high-quality corrugating medium.

Drawbacks of Straw Pulps. Even though harvesting is done yearly, it must be accomplished within a short span of time, usually limited to a few weeks in the summer or autumn of the year.[143] This means intensive investment in harvesting machinery, much of which may sit idle for the rest of the year. Straws tend to be bulky, complicating transportation and storage problems, and straw-handling costs are generally higher than wood-handling costs. Densification is possible, but only at substantial capital and operating costs.[144] Storage space must be provided for an annual supply of straw for a pulp mill, whereas wood may be cut continually through the year and brought to the mill as needed in many areas. Straws are quite susceptible to microbiological degradation, particularly when wet, and this degradation can destroy the pulping potential of straw in a matter of a few days, as opposed to the weeks and months needed to degrade wood to a similar degree. For this reason, straw is best stored under shelter from the elements, adding to the cost and complexity of storage.

The fibers of straw pulps are generally much shorter than softwood fibers and tend to drain water slowly. This decreases the production rates during washing and dewatering operations in pulp and bleach mills and can slow down paper machines.[145] Many straws have substantial silica contents; if pulped by strongly alkaline liquors (soda or kraft processes), the silica will dissolve in the pulping liquor and reprecipitate on process equipment in the liquor-recovery operations.[146] Evaporator tubes can become plugged and lime mud settling rates seriously disturbed by this silica. For these reasons, mills now using wood for pulping are quite reluctant to experiment with straw as an alternative fiber source. Certain crops are grown in such quantities that the straw residues have received serious attention as fiber sources. A brief discussion of some of these specific straw crops follows.

Bagasse. This is the residue from the production of cane sugar, the crushed stalk after the juice has been extracted. Fifty million tons of this material is available worldwide annually, much of which is burned for fuel, but some is used for fiber, predominantly in tropical countries.[141] The major obstacle to bagasse pulping is the high pith content of the stalk, which represents about 30% by weight of the stalk. Pith cells are worthless for papermaking and consume chemicals during pulping.[147] Pith cells are best removed mechanically before pulping begins, and this process can be expedited by controlled partial fermentation of the bagasse in storage prior to the depithing operation. Once depithed, the bagasse can be readily pulped by alkaline processes and bleached to a good-quality pulp suitable for fine papers. The average fiber length is 1 to 2 mm and the pulp is generally comparable to hardwood pulps. There is one United States mill producing bagasse pulp and over 100 mills worldwide.

Esparto. Esparto is a grass that grows in parts of southern Europe and northern Africa. The plant may be grown specifically for its utilization in papermaking.[148] It has a longer fiber than most plants of its type, and the pulp is prized for specialty applications where good formation and strength is desired.

Bamboo. This is an extremely rapidly growing plant and can be cultivated as a crop for pulpwood.[141] Substantial quantities of bamboo pulp are made in India and other Asian countries. Pulp quality is average compared to other straw pulps. The high silica content of bamboo complicates alkaline-pulping-liquor recovery.

Kenaf. This is a plant similar in appearance to corn and can be cultivated in semitropical climates with high yields. It has undergone intensive study in the United States as a source of fiber for pulping and may develop into a profitable crop for farmers in the southeast.[149-151] Fibers are typically short, and the pulp is generally competitive with hardwood pulps in papermaking.

Wheat, Barley, Rice, Grasses. This category includes the grass-like plants normally grown for their seeds, but whose stems are generally plowed under, burned, or used for cattle food. Fiber quality varies considerably, as does silica content (rice straw is notoriously high in silica). Quantities of straw available for pulping vary greatly according to local conditions and at present these straws are not utilized to a great degree for pulping. Grasses grown for seed can be successfully pulped and represent an untapped fiber source for the future.[145]

Jute, Ramie, Hemp. These plants are grown in subtropical countries, such as the Philippines, and produce some of the longest fibers used in papermaking. They have outstanding strength and are used for specialty applications, where their properties cannot be matched by conventional wood pulps.[152] Filter papers, saturating papers, condenser tissue, and tea bags are examples of such uses. Used rope has been one of the sources of these fibers.

Cotton Fibers. Strictly speaking, cotton fibers are not in the class of straw pulps since the cotton fibers come from the seed pods rather than from the stalk or stem. Regular cotton fibers are too long and too expensive for conventional papermaking, although cotton rags are a source of prime fiber for rag-content fine papers. Cotton linters, those fibers remaining on the seed pod after the long fibers have been removed for textile use, are used by the paper industry both for conversion to cellulose derivatives, such as cellulose acetate and nitrate, and for rag-content fine paper.[153] The only preparation normally needed for cotton linter fibers is removal of waxes and oils by alkali treatment followed by the proper refining for its end use. Fiber length is about 4 to 5 mm. Linters are capable of making very high quality paper, albeit expensive.

Flax Paper. Flax fibers are capable of making very thin papers with good formation, strength, and opacity. Typical examples of such products are cigarette paper and Bible paper.[154] One mill in the southeastern United States produces these papers using flax shipped to the mill from Minnesota and the Dakotas where it is grown and processed. Flax pulp is expensive, but it can impart certain properties to paper that cannot be obtained from wood pulps.

In summary, nonwoody plants offer promising opportunities for the future of papermaking. Pulp quality runs the gamut from relatively low-quality filler pulps, similar to hardwood pulps, to some of the highest quality pulps that outperform any known wood pulp in certain end uses. Economics has been an obstacle to their acceptance by the paper industry, and where wood is plentiful, these pulps can compete only in very limited instances. The growing scarcity of wood demands that the industry carefully consider all possible replacements for wood as the primary fiber source; the more intense utilization of straw pulps is certainly one viable choice.[155] Further information on nonwood fibers is given in the chapter on pulping under the section on pulping and bleaching of nonwood fibers.

REFERENCES

1. J. L. Keays, "World Developments in Increased Forest Resources for the Pulp and Paper Industry," Report VP-X-64, Forest Products Laboratory, Canadian Forestry Service, Vancouver, Canada, Aug 1970.
2. R. W. Hagemeyer, *Tappi,* **59** (4), 46 (1976).
3. "Development and Forest Resources in the Asia and Far East Region," M-73, FAO, Food and Agriculture Organization of the United Nations, Rome, Italy, 1976.
4. R. Sopko, *Pulp and Paper,* **47** (11), 86 (1973).
5. *Paper Trade J.* **155** (47), 81 (1971).

6. "Statistics of Paper and Paperboard," American Paper Institute, New York, N.Y., 1976.
7. "Wood Fiber Resources and Pulpwood Requirements," FO:PAP/74/6, FAO, Rome, Italy, 1974.
8. R. J. Auchter, *Tappi,* **59** (4), 50 (1976).
9. J. H. Ribe, *Pulp and Paper,* **48** (13), 72 (1974).
10. G. C. Meyer, *Pulp and Paper,* **48** (1), 76 (1974).
11. L. H. Blackerby, *Pulp and Paper,* **45** (8), 84 (1971).
12. C. R. Silversides, *Pulp Paper Mag. Can.,* **67** (11), WR–519 (1966).
13. A. W. J. Dyck, *Amer. Paper Ind.* **48** (6), 22 (1966).
14. A. B. Berg, "A Commercial Thinning in Douglas Fir with a Horse," Research Paper 6, Forest Research Laboratory, School of Forestry, Oregon State U., Corvallis, Oreg., July 1966.
15. D. J. Miller, *For. Prod. J.,* **25** (11), 49 (1975).
16. C. W. Barney and R. E. Dils, "Bibliography of Clearcutting in Western Forests," College of Forestry and Natural Resources, Colorado State U., Ft. Collins, Colo., April 1972.
17. J. F. Franklin and D. S. DeBell, "Effects of Various Harvesting Methods on Forest Regeneration" in "Even-Age Management," Symposium, August 1972, Paper 848, School of Forestry, Oregon State U., Corvallis, Oreg., 1973.
18. A. B. Berg, "Precommercial Thinning in Douglas Fir" in "Management of Young-Growth Douglas Fir and Western Hemlock," Symposium, December, 1970, School of Forestry, Oregon State U., Corvallis, Oreg., 1970.
19. W. J. Bublitz, *Tappi,* **54** (6), 928 (1971).
20. B. J. Zobel and R. C. Kellison, *Tappi,* **55** (8), 1205 (1972).
21. K. Steinbeck and E. N. Gleaton, *Pulp and Paper,* **48** (13), 96 (1974).
22. D. W. Einspahr, *Tappi,* **59** (10), 53 (1976).
23. H. E. Young, *Pulp and Paper,* **48** (13), 46 (1974).
24. B. J. Zobel, et al. *Tappi,* **60** (6), 62 (1977).
25. C. L. Brown and H. E. Sommer, *Tappi,* **60** (6), 72 (1977).
26. K. E. Lowe, *Pulp and Paper,* **47** (13), 42 (1973).
27. J. D. Dell and F. R. Ward, "Logging Residues on Douglas Fir Clearcuts; Weights and Volumes," USDA Forest Service Research Paper PNW–115, Pacific Northwest Forest and Range Station, 1971.
28. J. Morey, *Tappi,* **58** (5), 94 (1975).
29. J. S. Moran and F. L. Schmidt, et al. *Paper Trade J.,* **159** (10), 44 (1975).
30. W. J. Bublitz, *Tappi,* **59** (5), 68 (1976).
31. J. L. Keays, *Pulp Paper Mag. Can.,* **75** (9), 43 (1974).
32. J. V. Hatton and J. L. Keays, *Tappi,* **55** (10), 1505 (1972).
33. A. J. Chase, et al. "Puckerbrush Pulping Studies," Tech. Bull. 49, Life Sciences and Agriculture Experiment Station, U. of Maine, Orono, Maine, September 1971.
34. A. J. Chase and H. E. Young, "Corrugating Pulp from Puckerbrush," *Research in the Life Sciences,* **23** (10), September 1976, Life Sciences and Agriculture Experiment Station, U. of Maine, Orono, Maine, 1976.
35. W. L. Pritchett, *Tappi,* **60** (6), 74 (1977).
36. G. A. Graves, et al. *Tappi,* **60** (4), 94 (1977).
37. J. F. Briody and W. L. Bush, *Tappi,* **59** (4), 60 (1976).
38. P. E. Hyde and S. E. Corder, *For. Prod. J.,* **21** (10), 17 (1971).
39. "Transport and Handling in the Pulp and Paper Industry," Vol. 1., *Proc. 1st Intl. Symp. on Transport and Handling in the Pulp and Paper Industry,* Rotterdam, The Netherlands, April 1974, Miller Freeman, San Francisco, Calif., 1975.

40. A. W. Nelson, Jr., and J. T. Dotts, "Wood Harvesting: Changing Concepts in Fiber Conservation Utilization," *Proc. Pulp and Paper Seminar,* Chicago, Ill., May 1974, Miller Freeman, San Francisco, Calif., 1975, p 163.
41. W. E. Davis, *Pulp and Paper,* **45** (5), 121 (1971).
42. *Can. Pulp and Paper Ind.,* **30** (1), 70 (1976).
43. W. P. Kilfoil, *Can. Pulp Paper Ind.,* **15** (11), 24 (1962).
44. W. J. Bublitz, Report on visit to Canadian pulp and paper mills, June 8–20, 1975, School of Forestry, Oregon State U., Corvallis, Oreg. [Private communication.]
45. P. O. Karjalainen, et al. *Pulp Paper Mag. Can.,* **73** (12), 95 (1972).
46. J. P. Williams, *Pulp Paper Mag. Can.,* **72** (7), 84 (1971).
47. M. Overend, *Pulp Paper Mag. Can.,* **76** (6), 88 (1975).
48. W. A. Hunt, *For. Prod. J.,* **17** (9), 68 (1967).
49. *Paper Age,* **2** (17), 5 (1966).
50. R. D. French, *For. Ind.,* **102** (5), 50 (1975).
51. G. A. MacKenzie, *Pulp Paper Mag. Can.,* **76** (10), 69 (1975).
52. A. Assarson and G. Akerlund, *Svensk Papperstidn.,* **70** (6), 205 (1967); *Abstr. Bull. Inst. Paper Chem.,* **38,** 2855 (1967).
53. R. Grant, *Tappi,* **51** (10), 54A (1968).
54. C. C. Walden and T. E. Howard, *Tappi,* **60** (1), 122 (1977).
55. J. M. Leach and A. N. Thakore, *Tappi,* **59** (2), 129 (1976).
56. Karjalainen, et al., in ref. 45.
57. H. Rapson, et al., *Pulp Paper Mag. Can.,* **78** (6), T–137 (1977).
58. G. C. Meyer, *Paper Trade J.,* **157** (50), 31 (1973).
59. T. L. Scroggins, *For. Ind.,* **98** (2), 24 (1971).
60. J. W. Austin, *Pulp and Paper,* **47** (3), 63 (1973).
61. R. M. Lindgren and W. E. Eslyn, *Tappi,* **44** (6), 419 (1961).
62. G. J. Hajny, *Tappi,* **49** (10), 97A (1966).
63. D. E. Allen, *Pulp and Paper,* **42** (8), 33 (1968).
64. E. L. Springer and G. J. Hajny, *Tappi,* **53** (1), 85 (1970).
65. J. K. Shields, "Microbiological Deterioration in the Wood Chip Pile," Canadian Dept. of Forestry Publication No. 1191 (1967).
66. W. C. Feist, et al., *Tappi,* **57** (8), 112 (1974).
67. O. Bergman, *Svensk Papperstidn.,* **77** (18), 681 (1974); *Abstr. Bull. Inst. Paper Chem.,* **45** 10714 (1975).
68. H. P. Dahm, *Norsk Skogind.,* **20** (1), 10 (1966); *Abstr. Bull. Inst. Paper Chem.,* **37** 363 (1966).
69. W. C. Feist, et al., *Tappi,* **56** (4), 148 (1973).
70. E. L. Springer, et al., USDA Forest Service Research Paper FPL–110, Madison, Wis., (May 1969).
71. W. C. Feist, et al., *Tappi,* **56** (8), 91 (1973).
72. M. A. Hulme and J. K. Shields, *Tappi,* **56** (8), 88 (1973).
73. E. L. Springer, et al., *Tappi,* **58** (3), 128 (1975).
74. K. Hunt, *Tappi,* **59** (10), 89 (1976).
75. E. L. Springer and L. L. Zoch, *Tappi,* **53** (1), 116 (1970).
76. W. C. Feist, *Tappi,* **56** (4), 148 (1973).
77. W. J. Bublitz and J. L. Hull, in "Tappi Conference Papers," Forest Biology and Wood Chemistry Conference, Madison, Wis., 177, June 1977, p. 177.
78. E. L. Springer, *Tappi,* **58** (11), 145 (1975).
79. F. L. Schmidt, *Tappi,* **52** (9), 1700 (1969).
80. R. A. Horn, *Paper Trade J.,* **156** (46), 55 (1972).
81. W. M. Newby, *Pulp Paper Mag. Can.,* **74** (10), 80 (1973).
82. R. M. Samuels and D. W. Glennie, *Tappi,* **41** (5), 250 (1958).
83. G. M. Dick, *Tappi,* **35** (5), 212 (1952).

84. *For. Prod. J.,* **6** (9), 19 (1956).
85. J. S. Abel, *J. For. Prod. Res. Soc.,* **2** (5), 113 (1952).
86. G. Ranhagen, *Paper Trade J.,* **158** (3), 20 (1974).
87. F. Pine, *Paper Trade J.,* **155** (1), 32 (1971).
88. N. Hartler, *Svensk Papperstidn.,* **66** (18), 696 (1963); *Abstr. Bull. Inst. Paper Chem.,* **34,** 3963 (1964).
89. *Pulp Paper Mag. Can.,* **67** (C, WR-122), (1966). [Convention issue.]
90. J. V. Hatton, *Tappi,* **59** (6), 151 (1976).
91. W. J. Nolan, *Tappi,* **46** (8), 458 (1963).
92. J. V. Hatton, *Tappi,* **58** (2), 110 (1975).
93. J. R. Erickson, *Tappi,* **55** (8), 1216 (1972).
94. D. J. Miller, *For. Prod. J.,* **20** (11), 49 (1975).
95. R. A. Arola and J. R. Erickson, USDA Forest Service Res. Paper NC-85 1973.
96. A. L. Wesner, *Pulp and Paper,* **36** (17), 61 (1962); U.S. Patent 3032188, (1962).
97. L. E. Hudson et al., *Tappi,* **58** (7), 162 (1975).
98. D. A. Hartman et al., "Conversion Factors for the Pacific Northwest Forest Industry," Institute of Forest Products, U. of Washington, Seattle, Wash., (1976).
99. V. Heiskanen, *Paperi Puu,* **55** (11), 861 (1973); *Abstr. Bull. Inst. Paper Chem.,* **44,** 10842 (1973).
100. J. V. Hatton, *Tappi,* **60** (8), 107 (1977).
101. J. V. Hatton, *Tappi,* **59** (6), 151 (1976).
102. L. H. Blackerby, *Pulp and Paper,* **45** (8), 84 (1971).
103. W. J. Bublitz, Survey of United States and Canadian pulp mills 1977, Oregon State U., Corvallis, Oreg., [Private communication.]
104. J. V. Hatton and J. Hejjas, *Paper Trade J.,* **160** (10), 49 (1976).
105. J. V. Hatton, *Pulp Paper Mag. Can.,* **77** (6), T-99 (1976).
106. J. V. Hatton and J. L. Keays, *Pulp Paper Mag., Can.,* **74** (1), 79 (1973).
107. J. V. Hatton, *Pulp Paper Mag. Can.,* **76** (7), T-217 (1975).
108. J. V. Hatton, *Tappi,* **60** (4), 97 (1977).
109. W. J. Nolan et al., *Tappi,* **34** (12), 529 (1951).
110. W. J. Nolan, *Tappi,* **51** (9), 378 (1968).
111. W. J. Bublitz and R. O. McMahon, *Tappi,* **60** (8), 105 (1977).
112. W. J. Bublitz and R. O. McMahon, *Tappi,* **59** (12), 125 (1976).
113. J. M. MacLeod, *Tappi,* **60** (2), 128 (1977).
114. J. W. Wilson and H. Worster, *Pulp Paper Mag. Can.,* **61** (11), T-524 (1960).
115. J. J. Rettinger, *Tappi,* **58** (12), 94 (1975).
116. H. Magnusson and T. Konradsson, *Svensk Papperstidn.,* **74** (24), 835 (1971); *Abstr. Bull. Inst. Paper Chem.,* **42,** 11391 (1972).
117. L. A. Wilhelmsen et al., *Tappi,* **59** (8), 56 (1976).
118. *Tappi Standard T-14 os-74,* Technical Association of the Pulp and Paper Industry, Atlanta, Ga.
119. J. V. Hatton, *Tappi,* **59** (6), 151 (1976).
120. A. B. Wardrop, *Tappi,* **52** (3), 396 (1969).
121. D. W. Einspahr, *Tappi,* **55** (12), 1660 (1972).
122. B. Zobel, *Tappi,* **59** (4), 126 (1976).
123. R. L. Blair et al., *Tappi,* **58** (1), 89 (1975).
124. J. M. Dinwoodie, *Tappi,* **49** (2), 57 (1966).
125. J. P. Van Buijtenen et al., *Tappi,* **58** (9), 129 (1975).
126. W. J. Bublitz, *Pulp and Paper,* **45** (9), 82 (1971).
127. *Tappi Standard T-234 su-67,* Technical Association of the Pulp and Paper Industry, Atlanta, Ga.

128. J. M. Dinwoodie, *Tappi,* **48** (8), 440 (1965).
129. G. A. Matolcsy, *Tappi,* **58** (4), 136 (1975).
130. A. P. Binotto and G. A. Nicholls, *Tappi,* **60** (6), 91 (1977).
131. R. V. Braun et al., Canadian Patent 840269 (1970).
132. K. W. Hardacker and J. P. Brezinski, *Tappi,* **56** (4), 154 (1973).
133. R. M. Kellogg et al., *Tappi,* **58** (12), 113 (1975).
134. O. J. Kallmes and M. Perez, "New Theory for Load/Elongation Properties of Paper. Consolidation of the Paper Web," in *Transactions of the Cambridge Symposium* 779 (1966); *Abstr. Bull. Inst. Paper Chem.,* **37** 7370 (1967).
135. D. G. Kirk, *Tappi,* **55** (11), 1600 (1972).
136. W. J. Bublitz, *Tappi,* **54** (6), 928 (1971).
137. R. M. Kellogg and E. Thykeson, *Tappi,* **58** (4), 131 (1975).
138. W. A. Côté, Jr., *For. Prod. J.,* **8** (10), 296 (1958).
139. *Tappi Standard T-401 os-74,* Technical Association of the Pulp and Paper Industry, Atlanta, Ga.
140. *Papermaking Fibers,* State University of New York, College of Environmental Science and Forestry, Syracuse, N.Y.
141. J. E. Atchison, *Amer. Paper Ind.,* **51** (1), 25 (1969).
142. F. S. Conklin, "Farmer Alternatives to Open-Field Burning: An Economic Appraisal, Special Report 336, Agriculture Experiment Station, Oregon State U., Corvallis, Oreg., Oct 1971.
143. W. J. Bublitz, *Tappi,* **53** (12), 2291 (1970).
144. "Oregon State University Research on Field Burning," Circular of Information 647, Agriculture Experiment Station, Oregon State U., Corvallis, Oreg., December 1974.
145. W. J. Bublitz, "Pulping Characteristics of Willamette Valley Grass Straws," Station Bulletin 617, Agriculture Experiment Station, Oregon State U., Corvallis, Oreg., November 1974.
146. P. B. Chaudhuri, *Ippta,* **7** (2), 117 (1970); *Abstr. Bull. Inst. Paper Chem.,* **41,** 8284 (1971).
147. J. E. Atchison, *Tappi CA Report,* **40,** 1 (1971), Technical Association of the Pulp and Paper Industry, Atlanta, Ga.
148. D. K. Deshmukh, *Ippta,* **6** (4), 23 (1969); *Abstr. Bull. Inst. Paper Chem.,* **41,** 5403 (1970).
149. M. O. Bagby et al., "Kenaf Stem Yield and Composition: Influence of Maturity and Field Storage," *Tappi CA Report,* **58,** 69 (1975), Technical Association of the Pulp and Paper Industry, Atlanta, Ga.
150. M. O. Bagby et al., "Kenaf Roots, A Pulping Resource," *Tappi CA Report,* **58,** 63 (1975). Technical Association of the Pulp and Paper Industry, Atlanta, Ga.
151. C. A. Moore et al., *Tappi,* **59** (1), 117 (1976).
152. T. F. Clark et al., *Tappi,* **53** (9), 1697 (1970).
153. K. Ward, Jr., and M. H. Voelker, *Tappi,* **49** (4), 155 (1966).
154. M. Nikolova and Z. Litovski, *Tseluloza Khartiya,* **5** (4), 32 (1974); *Abstr. Bull. Inst. Paper Chem.,* **45,** 11948 (1975).
155. *Paper Trade J.,* **158** (32), 26 (1974).

4 PULPING

The importance of pulping and the large number of different pulping processes that are used to produce papermaking pulp make it desirable to have more than one author cover this subject. Consequently, separate authors have covered the subject of pulping under the headings: stone groundwood, refiner mechanical and thermomechanical pulping, semichemical pulping, chemical pulping, silvichemicals, pulping and bleaching of nonwood fibers, secondary fiber pulping, and pulping of rags and cotton linters.

INTRODUCTION TO PULPING

J. N. McGOVERN, Ph.D.

Department of Forestry
University of Wisconsin
Madison, Wisconsin

Pulp is the product of the separation of the fibers in wood or other plant fiber material and is an intermediate product in the manufacture of paper and paperboard. If wood is the starting raw material, it must be reduced before pulping to suitably sized material, either bolts or chips.

Pulping is achieved by chemical or mechanical means or combinations of the two. In mechanical pulping the original chemical constituents of the fibrous material are unchanged, except for removal of water solubles. Chemical pulping has as its purpose the selective removal of the fiber-bonding lignin to a varying degree with a minimum solution of the hemicelluloses and the celluloses. If white papers are to be made, this purification is continued in the bleaching step. Further purification for the production of chemical pulps with a high alpha cellulose content may be carried out in the bleaching phase.

The properties of the end products, the papers and paperboards, will depend on the properties of the pulps used in their manufacture. These in turn will vary with the wood species or nonwood plant fiber used and the pulping process employed. The types of pulping processes, which have proliferated greatly since the early inventions, will be described generally in the following paragraphs and covered in detail later in this chapter. Recent advances and trends in pulping will also be briefly described in the following paragraphs. Repulping of wastepaper, called secondary fiber pulping, which is a source of over 20% of the total United States fiber supply and even more in Japan and Europe, will also be described in this chapter as will the pulping of nonwood plant fibers.

Types of Pulping Processes

The sulfate or kraft process was the last invented (1879) of the strictly chemical pulping processes, but it has in its nearly 100 years of existence come to dominate United States pulp production at a level of about 70% of the total pulp production. The soda process, the first chemical process, is no longer included in the statistics of the industry, although there is still some small production of pulp through this process. An intermediate or semichemical (part chemical and part mechanical) process based on the use of neutral sulfite, the neutral sulfite semichemical (NSSC) process, was developed in 1925. In recent years there have been numerous modifications of the chemical, semichemical, and mechanical pulping processes as shown in Table 4-1[1]. The motivations for this process diversification have been the need for more efficient utilization of the wood resource and the need for special pulps to meet the diversified requirements of the expanding paper and paperboard markets.

Polysulfide-type kraft pulping has been fairly intensively investigated and installed in several mills in Scandinavia and the United States to take advantage of the high-yield, carbohydrate-retaining characteristic of this process. A modification of soda pulping employing an additive, anthraquinone and hence called soda-AQ pulping, has been developed and is being put into commercial practice; the advantages are reported to be improved pulp yield and strength, especially tear index, for specific softwoods.[1a] Nonsulfur processes (sodium carbonate or soda-oxygen, for example) have been researched and installed for environmental reasons. Solvent pulping has made little progress.

The thermomechanical pulping process (TMP) and the refiner mechanical pulping process (RMP) employ chips rather than the bolts of wood required in stone groundwood, which is a distinct advantage in supply and handling. They produce groundwood-type pulps capable of replacing or extending chemical pulps.

Advances and Trends in Pulping

Preparation of Wood. The ever-increasing need for new wood supplies to meet the rising demands of paper consumption has led to a steadily deteriorating quality of chips at the source. Upgrading of chips has been achieved by the introduction of chip-washing systems, screening systems for removal of fines from chips, and systems for the separation of bark from chips when a whole-tree chipping system is used. Upgrading efforts to improve chip quality can be expected to continue.

Improved Utilization of Raw Materials. Improved utilization has been achieved in the last 25 years by increased use of hardwoods, by use of manufacturing and forest residuals, and by use of chips from whole-tree harvesting. Other beneficial developments have been the increased yields from chemical pulping, the increased production of semichemical and chemimechanical pulps, and the upgrading of secondary pulps recovered from wastepaper.

The use of hardwoods (including residuals) has increased to the point where they comprise 25 to 30% of the pulpwood supply in the United States. The use of residual chips from primary wood manufacture, such as sawmills, has increased spectacularly to over 35% of the total wood used in pulping. Some mills in the northwestern part of the United States are based entirely on this resource. The production of chips from the whole, rough trees, using special chipping methods developed for that purpose, is a promising development that makes available a source of wood that would not be available from conventional logging. Whole-tree chips may contribute as much as 10% to the total pulpwood demand in the future[2].

Secondary fiber pulping using wastepaper and paperboard was decreasing until concerns for the environment and the fiber supply became prominent in the early 1970s. From a low of slightly over 20% of the total fiber supply in 1968, its expansion rate is such that it is expected to reach 28% of the total fiber supply in 1985. In 1976, nearly 60% of the United States fiber supply for paper and paperboard came from chip residuals and secondary fibers at a tremendous saving of round pulpwood.

Improved Pulping Technology. Continuous processing of wood chips was invented in the 1930s. The process involved the use of steam treatment followed by defibration (fiberizing) under pressure or atmospheric condition to produce a rough pulp for fiberboard and roofing felt. Semichemical pulping using a horizontal continuous digester tube for the chemical stage appeared about 1950. Continuous chemical pulping was first performed about 1960. A number of digester configurations and systems were proposed during the development period and these appear to have settled down to:

1. **Vertical downflow digesters with impregnation and chip immersion (mostly kraft).**
2. **Downflow with vapor phase pulping (mostly semichemical).**
3. **Inclined tubes with raw-material submergence (mostly sawdust).**
4. **Horizontal tube digesters (semichemical and nonwood plant fibers).**

An estimated 40% of the North American kraft mills employ continuous digestion systems with improved equipment, including reduced temperature blowing and in-digester pulp washing. Countercurrent pulping in these digesters is practiced in a few mills. The continuous downflow digester has also been adapted to prehydrolysis kraft pulping with a vapor prehydrolysis stage.

TABLE 4-1 CLASSIFICATION OF WOOD-PULPING PROCESSES BY YIELD[a]

Form of Wood	Common Name	Chemical Treatment	Mechanical Treatment	Pulp Yield (%)
Mechanical				
Bolts	Groundwood, cold and hot[b]	None	Grindstone	93–98
Chips	Refiner mechanical (RMP)	None	Disk refiner	93–98
Chips	Masonite	Steam	Steam expansion	80–90
Chips	Asplund (coarse fiber)	Steam	Disk refiner	80–90
Chips	Thermomechanical (TMP)	Steam	Disk refiner (pressure)	91–95
Chemimechanical and chemithermomechanical				
Bolts	Steamed groundwood	Steam	Grindstone	80–90
Bolts	Groundwood, Decker process	Acid sulfite (Ca, Na, Mg)	Grindstone	80–90
Bolts	Groundwood, Fish process	Kraft	Grindstone	85–90
Bolts	Chemigroundwood	Neutral sulfite	Grindstone	80–90
Chips	Cold soda	Caustic soda	Disk refiner	80–90
Chips	Alkaline sulfite	Alkaline sulfite	Disk refiner	80–90
Chips	Chemithermomechanical	Steam + chemical	Disk refiner (pressure)	65–85
Semichemical				
Chips	Neutral sulfite (NSSC)	$Na_2SO_3 + Na_2CO_3$, $NaHCO_3$	Disk refiner	65–90
Chips	Kraft	$NaOH + Na_2S$	Disk refiner	75–85
Chips	Green liquor	$Na_2S + Na_2CO_3$	Disk refiner	65–85
Chips	Soda	Sodium hydroxide	Disk refiner	65–85
Chips	Nonsulfur	$Na_2CO_3 + NaOH$	Disk refiner	65–85
High-yield chemical				
Chips	Kraft (sulfate)	$NaOH + Na_2S$	Disk refiner	55–65
Chips	Sulfite	Acid sulfite (Ca, Na, Mg)	Disk refiner	55–70
Chips	Sulfite	Bisulfite (Na, Mg)	Disk refiner	55–70

Full chemical

Chips	Kraft (sulfate)	Na_2S + NaOH	Mild to none	40–55
Chips	Polysulfide, one stage	$(NaOH + Na_2S)_x$	Mild to none	45–60
Chips	Polysulfide, two stage	H_2S pretreatment; kraft	Mild to none	45–60
Chips	Soda	Caustic soda	Mild to none	40–55
Chips	Soda-AQ	NaOH + Anthraquinone	Mild to none	45–55
Chips	Soda-oxygen, two stage	NaOH; oxygen	Disk refiner	45–60
Chips	Acid sulfite	Acid sulfite (Ca, Na, Mg, NH_3)	Mild to none	45–55
Chips	Bisulfite	Bisulfite (Na, Mg, NH_3)	Mild to none	45–60
Chips	Magnefite	Magnesium bisulfite	Mild to none	45–60
Chips	Neutral sulfite	Neutral sulfite	Mild to none	45–55
Chips	Multistage sulfite	—	—	
	Stora	Sulfite-bisulfite; SO_2	None	45–55
	Sivola	Sulfite; alkaline sulfite	None	45–55

Dissolving

Chips	Prehydrolysis kraft	Prehydrolysis; kraft	None	35
Chips	Sulfite	Acid sulfite (Ca, Na)	None	35

[a] Moisture-free pulp and wood base.
[b] Low and high grinder-pit temperature.

The advantages reported for continuous digesters are that they require less energy and have greater environmental suitability (digester gas control); batch systems are favored for their greater flexibility,[3] production reliability, higher turpentine yields, and lower maintenance costs. Depending on the specific situation, either system may be superior in pulp quality and uniformity, chemical and personnel requirements, and capital costs. The trend toward continuous digester installations for alkaline pulping appears to have leveled off.

Chemical pulping control has been improved through new measuring methods and instruments enabling use of computers,[4] but better raw-material measurement is needed, such as solids content and moisture and chip thickness for improved pulp uniformity.

Secondary pulping has been made difficult by the use of additives in modern special purpose papers, but improved processes for asphalt dispersion, refining, screening, cleaning, and flotation systems for ink removal have reduced the problem. Recycled newsprint is now used to a significant extent.

Environmental Demands. The characteristic unacceptable odor of kraft mill emissions is being reduced to the point where it is acceptable under current standards of environmental control. The total reduced sulfur (TRS) has been lowered or eliminated in the mill stacks by digester-relief-collection, black-liquor oxidation, and black-liquor evaporation using indirect heat prior to combustion. This has also improved chemical recovery efficiency. Nonsulfur pulping, another approach toward improving the environment, has applications for soda chemical and semichemical pulping and for research in soda-oxygen pulping. Semichemical and sulfite chemical recovery has been aided by introduction of the fluidized bed furnace, which eliminates sulfite stream discharge. The employment of soluble bases in the sulfite processes has also made chemical recovery possible and effluent disposal unnecessary.

A required adjunct to the modern pulp mill, integrated or not, is a secondary treatment plant to reduce the biological oxygen demand (BOD) of its effluent to meet governmental regulations. This development has produced a solid waste in sludge form whose utilization, other than disposal in a landfill, is in need of technological development.

In order to reduce water usage and the volume of mill liquid effluent, pulp and paper mills are moving toward a closed system. Some pulp and paperboard mills have reached this goal. To the integrated mill making pulps for bleached kraft papers, this requires a countercurrent flow system backwards with ultimate combustion in the chemical recovery system. This has introduced corrosion problems, which are undergoing investigation.

Energy Reduction. The pulp and paper industry is one of the largest consumers of process steam and electrical energy. However, the chemical pulp mill, particularly one using the kraft process and a chemical recovery system, can be self-sufficient in energy and even produce a surplus of energy through combustion of the wood substances dissolved in pulping and of the bark and wood residues generated in wood preparation. The energy efficiency of the chemical pulp mill has increased in recent years and continued efforts in energy conservation can be expected in the coming years. The use of bark and other wood wastes as fuel is not new, but its importance has increased. This energy source can fill an appreciable part of the chemical pulp-mill fuel needs, although bark and combination boilers need to be suitably equipped for air emission control.

Energy can be saved in the pulp mill by indirect steam heating in batch digesters, blow and flash steam recovery, condensate return, minimum liquor-to-solid ratios in cooking, and computer control to reduce operating fluctuations.[5] Recycling of water in pulp washing, recycling of liquor spills, and diffusion washing of pulp help to conserve energy as well as to reduce water requirements. In the recovery department, increased efficiency of vacuum evaporation through control of evaporation stages and boilouts can be important. Indirect heating of the black liquor, control of the solids content of the black liquor to the recovery furnace, temperature drop control, waste-heat utilization, and computer control are important. Heat recovery, water-use reduction, and minimum fuel for the lime kiln are necessary for energy conservation in the causticizing area.

General Trends. Capital investments in pulp and paper mills is a major concern in planning new or expanded capacity. For top economy the size limit appears to have been reached in pulp mills at a unit of about 1000 ton d. Trends that are developing to help reduce costs include product specialization, extending chemical with mechanical pulps, combining operations, using closed cycles, and using nonpolluting processes. Up to 10 to 15% of the capital for new construction now goes for environmental control.[6]

By-products of pulping can have great value. For example, tall oil and turpentine are among the major by-products of the kraft pulping of pines, but studies indicate that the yields are seasonal and are tending to decrease due to chip storage and the increased use of purchased chips that have a lower content of extactives.[7]

The technology for the use of nonwood plant fibers, such as bagasse, cotton linters, rags, manila hemp, and flax straw, is advancing through improved material-preparation, application of horizontal-tube continuous digesters, and better chemical recovery. However, these fibers will probably contribute 1% or less to the United States fiber supply in the near future,[2] although the world production will be more in the area of 6%. Tropical forest utilization appears to be nearing a breakthrough due to developments that make it feasible to use mixtures of species.

STONE GROUNDWOOD

WARREN F. DANIELL

Pulp and Paper Consultant
Hanover, New Hampshire

In chemical pulping, chemicals are used for dissolving and removing the lignin that holds the fibers together, thus causing the wood to disintegrate into its component fibers. It is possible also to use purely mechanical methods for disintegrating wood into a fibrous state without the use of chemicals. The most important of the mechanical processes is that used for making groundwood. This process, which was developed about 1843, involves that wet grinding of wood into a fibrous mass by means of a large revolving grindstone. Logs of wood are held with pressure against the surface of the stone, and as the stone grinds the wood into fibers, a stream of water is sprayed on the stone to carry the pulp away and to absorb the heat generated. Temperature in the grinding zone can be over 150°C. The logs are held against the stone in transverse fashion with the length of the log parallel to the axis of the grindstone so that the fibers in the wood are at right angles to the direction of movement of the stone. In this way the longest-fibered pulp is obtained. Friction forces tear fibers from the logs, and the fibers are carried in grooves in the pulpstone to the grinder pit. Since cheap power is essential to economical operation of a groundwood mill, much of the groundwood production is in Canada where there is abundant cheap hydroelectric power. Approximately 1500 kwh of power is required to produce one ton of groundwood. To grind 500 ton of pulp in a day (24 hr) 32,000 kW of continuous power is required.

The energy required for the production of stone groundwood and for refiner pulping is expressed in the customary commercial units of either horsepower-days/short ton or kilowatt-hours/short ton. SI units of megajoules/kilogram are not used because this unit is so seldom found in commercial practice. The conversion factors are:

Horsepower-days (hpd)/short ton $\times$ 7.10 $\times$ 10^{-2} = millionjoules (megajoules)/kilogram (MJ/kg)

Horsepower-days (hpd)/ton $\times$ 17.9 = kilowatt-hours (kWh)/ton

Horsepower-days (hpd)/short ton $\times$ 19.73 = kilowatt-hours (kWh)/metric ton (MT)

The goal of groundwood manufacturers for a century has been to make a quality of pulp that would, at 100% of the furnish, produce newsprint paper that would have the required strength, printing quality, and brightness to satisfy printing-press operators as well as run safely and efficiently on a paper machine. It has been an elusive goal, but one actually reached in a few instances 50 years ago, when paper machines and printing presses were run at much slower speeds than today and good quality spruce wood was ground. The economics of present-day paper and printing operations require high speed to lower capital and labor costs. This in turn requires a stronger sheet of paper, which is obtained by adding up to 25% of long-fibered chemical pulp to the furnish, resulting in some loss of printing quality. Another factor that requires the use of chemical pulp in the furnish is the limited supply of spruce wood. This has forced manufacturers of pulp to use other, inferior quality woods, such as fir, jack pine, southern pines, hemlock, and poplar, all of which produce weaker groundwood pulp. Stronger groundwood pulp, other factors being the same, requires more energy per unit of production. Sharp increases in the cost of power may make it uneconomical to increase the strength of groundwood by increasing power input per ton.

A great deal has been accomplished over the years in the design and building of large capacity grinders that are operated with much less labor than the old small units. Working conditions in the modern grinder rooms are vastly improved over the early grinder rooms, which were wet, messy, hot, humid and had poor ventilation. Now each grinder is equipped with an air and vapor exhaust system and the room is well ventilated and lighted. Automatic controls maintain the grinders at a uniform temperature, which helps to produce pulp of uniform quality.

This chapter continues with a more detailed description of the groundwood process and brief reports on some of the attempts to improve its quality. Some were successful and are now standard procedure; others failed to accomplish the expected results and were not continued. Research groups are studying the fundamentals of the groundwood process to add to the empirical information accumulated since the invention of the process over a century ago. More progress should come in the future. In reporting on the 1977 International Mechanical Pulping Conference the following statement was made: Although more than 90% of the mechanical pulp is being produced by grinders with considerably less energy required per ton than that required by thermomechanical pulping or refiner mechanical pulping, less than 5% of the program was concerned with the need for broader development and research on stone groundwood.[8] It is important that research on production of stone groundwood continue. A summary of the 1975 International Mechanical Pulping Conference by Underhay,[9] who speaks from vast experience and knowledge on the subject, stressed that more good research will bring us closer to the goal of good quality newsprint paper from a 100% groundwood furnish.

Stone Groundwood Process

Groundwood differs from chemical pulp in that is contains practically all the lignin of the original wood, and the fibers do not exist as individual entities, but rather as fiber bundles and fragments of fibers. The yield is about 90 to 95% of the original wood, compared with about 50% for chemical pulp.[1] About 3.5 m³ of spruce yields about one ton of groundwood.

Water at about 350 kPa (50 psi) is sprayed uniformly across the face of the pulp stone on the down running side and it is also sprayed on the top of the stone in two pocket grinders. The required volume of spray water will vary from 20 to 60 l/k of pulp produced, depending on the consistency desired in the grinder pit. This forms a slush of stock, that is collected in a pit under the stone. The bottom part of the grindstone is immersed in this stock in the grinder pit to a height determined by the height of the dam boards. Immersion of the stone in this way aids in cleaning the stone, lubricates its surface, and acts to cool and to equalize the temperature of the stone. If enough water is sprayed on the face of the pulpstone, grinders can be operated without any immersion in the grinder pit.

Excess pulp from the grinder pit flows over the dam into the stock line. This stock contains coarse wood particles that must be removed before the pulp can be used for making paper. The pulp is first screened through a coarse screen, called a bull screen, to remove the coarse splinters and unground wood. Next, it is passed through knotter screens having 6.4- to 3.2-mm (1/4- to 1/8-in.) holes where the smaller splinters and fragments are removed. Then the stock is passed through primary screens. Rejections from the primary screens are diluted and rescreened in the secondary screens. The accepted pulp from the secondary screens goes back to the first screen system and rejects go to some type of refiner to be ground to a finer size. The screens are generally of the rotary type. Among the screens used are the Cowan, Waterous Mohawk, Voith, and Waterous Hydrofoil. Jönsson screens, with 4- to 6-mm holes, are sometimes used. Stepwise screening is common practice. In some integrated mills the screened pulp from the first step may be taken off to be used in the better grades of paper, while the pulp from the third step is used in coarser grades. Screen plate perforations vary in the first step from 1.5 to 1.8 mm; in the second between 1.3 and 1.6 mm; and in the third, between 1.0 and 1.4 mm.[10] In the Cowan screen, which has two screening zones, the perforations in the first plate zone are about 1.25 mm. The efficiency of screening is dependent chiefly on the consistency of the pulp flowing into the screen and the percentage of rejects allowed. Screening practices differ somewhat in Europe and North America.

Centrifugal separators, or cyclone cleaners, are increasingly used to remove grit, short and thick shives, and pieces of bark from mechanical pulp. The removal of heavy dirt, particularly grinder grit, by means of centrifugal cleaners may extend the life of the paper machine wire by 25%.[11] Because vortex cleaners have a relatively high power consumption and deliver the stock in a very dilute state, they are usually installed before the dilute stock mixture flows to the headbox of the paper machines, but additional cleaners are sometimes installed in the pulp mill stock system to clean the accepts from the secondary screens and refiner stages. It is on the secondary screen accepted stock that the highest concentration of undesirable shives can be trapped most economically.[12] About 50% of the objectionable 0.25-mm-screen-fraction and about 85% of the shives can be removed by this arrangement. This allows the use of coarser screens, which permit a large percentage of long fibers to bypass the refiners and be accepted by the primary screens. The capacity of vortex cleaners is about 2.75 ton/d for a 5.0 cm × 6.6 cm cleaner operating at 240 kPa (35 psi) and 0.4% consistency feed.

The tailings or rejects from the screening operations may be used in making a coarse mill wrapper, or they may be refined in disk refiners and returned to the main stream. Disk refined groundwood stock produced this way is as strong and as free as the groundwood from which the rejects were taken[13] The selective refining of groundwood pulps has been tried. In this process, the grinding would be adjusted to obtain 25 to 40% rejects, instead of the 3 to 5% rejects normally obtained. These rejects would then be selectively refined and screened into coarse fractions suitable for boards, medium fractions suitable for book papers, and fine fractions suitable for tissue. The results were not as satisfactory as the conventional grinding and screening system. Another proposed new method is to produce a very coarse groundwood, and then refine this to an acceptable strength and freeness using a disk mill.[14,15] It was hoped that a pulp equivalent (or better) in quality to that of regular groundwood could be obtained at the same power input. The chief advantages claimed are that production can be doubled at low capital cost and flexibility of the operation increased. The system involves: (1) grinding on a special coarse stone (30 grit) to produce a long, free pulp: (2) screening on graduated vibratory screens with perforations of 0.81 mm, 1.04 mm, and 1.75 mm; and (3) refining the screen rejects (at 3.5 to 4.5% of consistency) in a double-disk refiner. Roughly half the stock passes through the screens and bypasses the refiners, but since the power used in refining is about equal to that used in grinding, the total power input to the system is about equal to that required when the pulp is all ground. In the case of a standard news grade of groundwood, the total power input is about 1100 to 1250 kWh/AD/ton, of which 54 to 61% is applied on the grinders. The strength and fiber distribution of the pulp is comparable to conventional groundwood. This type of operation typifies modern thinking in making groundwood, which is to use grinders, screens, and refiners in combination so that each unit performs its own particular function to best advantage to produce a pulp of the desired properties. Each unit is interrelated to the other and can, in fact, substitute for one or both of the other functions.[16] The optimistic results forecast in the above-described system have not been realized in practice.

A flow chart of a typical mechanical pulp mill is shown in Figure 4-1. Groundwood stock is thickened and stored in slush form or put up in the form of wet laps (67% moisture). If stone groundwood is overdried, it is extremely difficult to redisperse the fibers in water.

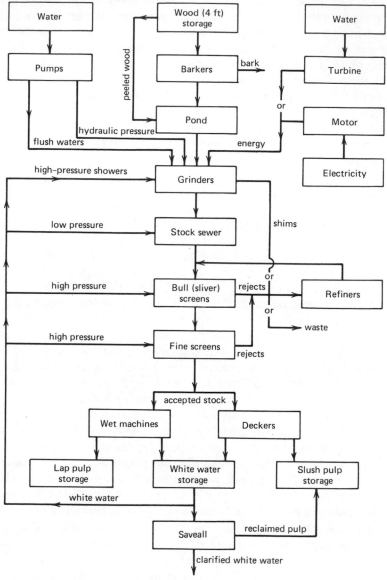

Figure 4-1. Flow chart of stone groundwood process.

Types of Grinders

Modern practice in groundwood operation is to operate at increasingly higher surface speeds, higher pressures, and with longer wood. There are two principal types of grinders: intermittent and continuous.

Intermittent grinders, which were the earlier development, first employed a lever or screw arrangement for producing the necessary thrust of wood against the stone; in modern designs they employ a hydraulically operated pressure foot. The rate of feed is controlled by the pressure employed. There are two principal forms of intermittent grinders: the pocket type and the magazine type (see Figure 4-2A and B).

Pocket grinders may have two, three, or four pockets located about the top part of the grinder. The wood is loaded manually into the pockets from the side

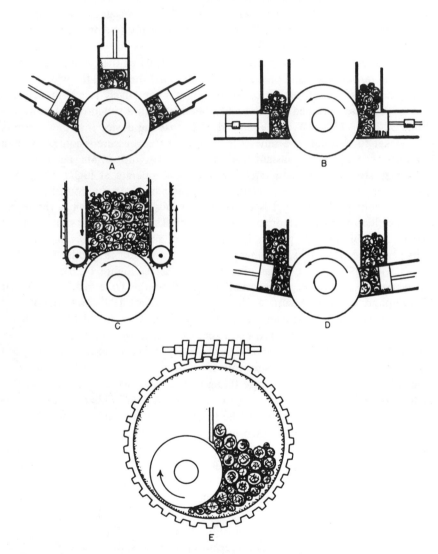

Figure 4-2. Different grinders for producing stone groundwood (A) Three-pocket grinder. (B) Waterous hydraulic grinder. (C) Warren chain grinder. (D) Great Northern grinder. (E) Ring grinder (Roberts).

when the pressure foot is withdrawn for a new charge. An indicator may be used to show the position of the pressure foot. The wood used in pocket grinders is rather short, usually 61-, 81- or sometimes 122-cm long. The production of pocket grinders is relatively low compared with other grinders, and generally varies from 3 to 25 tons of air-dry pulp per stone per 24 hr. They are generally used only in the smaller mills because of their high manpower requirements.

The hydraulically operated magazine grinder was first developed by Voith in 1910 to reduce the labor required in loading the grinder. This grinder has two pockets almost directly opposite each other with a magazine located between them where the logs are stored. When the hydraulically operated pressure foot of the magazine grinder is withdrawn for a new charge, the wood automatically falls by gravity from the magazine into the pocket. When the space if filled with wood, the ram reverses and presses the wood against the stone, the whole operation taking less than 30 sec. In order to maintain a relatively steady load, magazine grinders are installed in pairs and operated with a synchronous motor of about 3000 to 4500 kW and equipped with a timing device that only permits one pocket to be filled at a time. The cylinder water pressure is controlled by a governor to maintain a constant load on the electric motor. Magazine grinders use larger stones, have more grinding area, and operate at higher peripheral speeds than pocket grinders, which results in increased production of pulp per stone. Handling of the wood is easier than in pocket grinders since the magazines are filled from the floor above the grinder. The speed of the stone is about 240 to 250 rpm and the production varies from about 20 to 40 ton of air-dry pulp per stone per 24 hr. Another example of a magazine grinder is the Waterous, which is a highly popular grinder in Canada (see Figure 4-2B).

A variation of the magazine grinder, related to the Waterous, is the Great Northern grinder (see Figure 4-2D), developed about 1929 and widely used since then in North American mills, particularly in the United States. This is a two-pocket grinder that has low magazines located above each pocket. The logs are fed to the magazines from an overhead conveyor, and from there are dropped into the pocket when the pressure foot is withdrawn for filling. The magazines hold just enough wood for a single filling of a pocket, which facilitates loading of the pocket with the correct amount of wood. A gate holds the wood in the magazine and it opens only when the pressure foot is withdrawn to allow the wood to fall into the pocket. Each pocket has its own governor that automatically operates the gate and the piston. A typical Great Northern grinder installation includes a series of grinders arranged in lines, two grinders in a line driven by a 3750 kW-, 11,000 V, 60-cycle synchronous motor. The stones turn at about 257 rpm. Motor loads are controlled by electric load governors. Logs of 122 cm are fed into the grinder by floating in a sluice. The wood is held against the revolving stones at a pressure of 689 to 1720 kPa (100 to 250 psi).

Continuous grinders differ from intermittent grinders in that the wood is fed to the stone without interruption. Continuous grinders were developed at a later date than the intermittent grinder; the primary object in their development was to increase production. The principal continuous grinders are the pocket type (e.g., Kamyr), the chain type (e.g., Warren and Voith), and the rotary type (e.g., Tidmarsh and Roberts). The Kamyr is a two-pocket, hydraulically operated, continuous grinder that is equipped with a spring device for holding the wood under pressure as the pressure foot is withdrawn to receive a new charge. Thus the advance of wood is practically continuous. This grinder is used principally in Europe.

The Warren and Voith chain grinders are widely used in Europe. In these grinders (see Figure 4-2C), the wood is placed in a central magazine over the stone and is continuously pulled downward to the stone by means of specially designed chains running in grooves down the sides of the magazine. Dogs on the chains grip the sides of the outer logs and impart a downward thrust to the logs held in the magazine. The speed of the chains determines the pressure exerted between the wood and the stone. The speed is usually controlled by governors to maintain a constant load on the electric motor driving the grinder or to maintain a constant shaft speed on the hydraulically driven grinder. These grinders have practically a constant load since the chains always maintain a constant pressure of wood against the stone.[17]

The ring grinder (Tidmarsh and Roberts) is a continuous grinder in which the stone is set eccentrically in an inside-serrated cast-iron ring that forms a crescent-shaped pocket at about the seven o'clock position. The ring revolves slowly (about one revolution in 3 to 5 min) in the same direction as the stone, thus forcing the wood under continuous pressure in a gradually constricting pocket. Stone speed in the Roberts grinder is about 1050 m/min or more, and the unit pressure is about 225 kPa (33 psi).[18] The pulp is discharged from both ends of the stone, dams being used to regulate the depth of stock held in the grinder. The load is held constant by regulating the flow of oil to the fluid motor, and the ring speed is maintained to advance the wood at a rate that absorbs the power determined by the governor. In this type of grinder, the pressure is not uniform, but increases as the log approaches the nip formed by the stone and outer ring. Because of the large grinding area, high grinding speeds are used. No coarse splinters leave the grinder, since they are reground. (See Figure 4-2E for the Roberts grinder.)

Theory of Grinding

There have been many attempts to present a theory of grinding. It is accepted generally that there are two processes involved: a defibering process and a refining process.[19] In the first stage, wood fibers and fiber bundles are torn or brushed from the pulp log by the action of the grits on the pulpstone surface. The heat generated in doing this is important, since it softens the lignin. The size of the grit in the pulpstone and the rate of feed of wood to the stone are other important factors influencing the defibering process.[20] In the second process, which goes on simultaneously with the first, the fibers and fiber bundles produced in the first stage are refined and to some extent reground to a smaller size. The amount of refining is largely determined by the smoothness of the stone surface. The amount of regrinding is about 15 to 25%, but in special cases it can be higher. Higher consistency or greater submergence of the stone results in more regrinding, which in turn reduces the freeness.[21]

Of the large amount of work expended in grinding, part is converted into heat and part is consumed in fiber separation. Only an extremely small part of the grinding energy is usefully employed in creating new wood surface because a substantial part of the energy is consumed in viscoelastic deformation of the wood.[22] It has been estimated that for a stone velocity of 1200 m/min, each fiber is deformed by a passing grit for a duration of about 10 μsec after which there is an interval of about 30 μsec before the next deformation.

Variables in Groundwood Process

The principal variables in the groundwood process are stone surface, stone speed, grinding pressure, power input, energy consumption, production, temperature of grinding, consistency of grinding, freeness and strength of pulp, and wood variables. These variables are discussed separately in the following sections, but it should be pointed out that they are interrelated one to another and consequently cannot be completely isolated for individual study. There is a definite relationship between pulp properties, such as freeness and strength, and grinding characteristics, such as energy consumption and production. Pulp quality and power requirements in grinding are almost independent of the type of grinder.

Large amounts of energy are required to convert wood into groundwood. Hence, energy consumption is the most important factor in mill operation, but actually the energy required to make groundwood pulp is less than the total energy required in making chemical pulps. Grinding characteristics are defined by the energy consumption per ton of pulp and by the tonnage produced if the quality of pulp, stone surface, pressure, and pit temperature are held constant. The principal operating variables are stone surface and grinding pressure. In order to maintain production and energy consumption at acceptable levels, it is necessary to resharpen the stone at regular intervals during grinding to reduce the pressure.

Instrumentation in a groundwood mill is likely to consist of temperature controllers and load regulators that have recording meters.

Effect of Stone Surface in Grinding

Grindstones used in mechanical pulping are composed of hard grit embedded in a softer matrix. Stone surface is one of the principal factors in determining the quality of groundwood; grit sharpness, pattern, smoothness, and uniformity are the most important considerations. One of the conditions necessary for the production of good pulp is projection of grit above the surface of the matrix in the stone surface.[23] The grit must be of the proper type for the grade of pulp being produced; stones with coarse grit tend to produce coarse, shivy pulp, whereas stones with fine grit tend to produce fine pulp. The corners of the exposed grit should be rounded off to minimize cutting the fibers. Increased grit coarseness increases the strength at constant freeness, but tends to reduce the production.

Both natural sandstones and artificial pulpstones are used. Natural sandstones were used exclusively up to about 1925, but as the demands for production increased, artificial stones became much more important. In natural sandstone the grit consists of small particles of quartz. Artificial sandstones, made from quartz or silicon carbide grits bonded with cement, are used in Europe.[24] In the United States, artificial stones are made by mixing grit of the proper sizes (either silicon carbide or aluminum oxide) with a vitrified bonding agent (consisting of clay, feldspar, flint, and other materials) in proper amount to obtain the desired texture and hardness of stone. The stones are made in wedge-shaped segments that are jointed together and connected by metal bolts to the reinforced center of the stone. Artificial stones are more uniform, have longer overall life, are available in larger size, and require less frequent dressing than natural stones. They are available in a number of combinations of grit type, size, hardness, density, and type of bonding agent. The most important factors are grit size, ratio of grit to bonding agent, and the hardness of the bonding agent. A medium size grit (50 to 60) is used for newsprint; coarse grit (20 to 24) is used for insulating board stock; fine grit (80 to 100) is used for fine groundwood stock for printing papers. Artifical stones are good for between 30,000 to 160,000 m^3 of wood, depending on the hardness of the stone and species of wood, compared to an equivalent life of about 4000 m^3 for natural stone. The pulp produced from artificial stones is generally longer fibered and has a higher freeness than the pulp produced from natural stones.

Before being used, pulpstones must be dressed, that is, the surface of the stone must be treated lightly with a burr that rotates in contact with the stone and roughens its surface as it moves across the width of the pulpstone. This produces a definite cutting pattern by exposing the grit and provides grooves in the stone surface where the fibers remain protected until they are washed free by the shower water. Thus, the function of the burr pattern is twofold: (1) to increase the unit pressure of grinding through reduction in area of contact and (2) to provide voids in which the fibers can be removed from the grinding area to avoid excessive regrinding. Fiber length and freeness are controlled principally by the width of the grooves in relation to the width of the lands.[25]

Burring can be done by hand or by means of mechanically or hydraulically operated stone-sharpening lathes equipped for controlled penetration of the burr into the stone. The sharpening operation is extremely important in determining groundwood quality, and, although a number of mechanical devices have been used to help, it is still an operation that is dependent on the skill of the operator. The burr must be moved across the pulpstone in the same direction each time it is used.

There are four principal patterns of burrs: the straight-cut, spiral-, thread-, and diamond-shape. The thread burr cuts grooves in the stone perpendicular to the axis of the stone; the straight burr cuts grooves parallel to the axis of the stone; the spiral burr cuts grooves at an angle to the axis of the stone; the diamond burr (a combination of thread and straight burr) cuts little pyramids in the stone. The stock produced with a diamond burr is generally short and stubby, but the production is high because of the large number of exposed grits. The straight burr produces long, shivy, stock. The thread burr tends to produce a short-fibered stock due to regrinding of the stock on the stone face; it is used mostly for light dressing of the stone surface. The most popular patterns are obtained with spiral burrs ranging from 3 to 5 cuts per cm with a lead of 2 to 5 cm. However, practically the same quality pulp can be produced by different burrs, depending on the way they are used; for example, essentially the same grade of newsprint can be made by using a 6-diamond burr or a 4-cut-by-4 cm lead, which is a special burr.

Burrs come in different degrees of fineness, which is usually expressed in terms of the number of ridges, threads, or points per inch. The proper coarseness of the burr depends on the type of stone, species of wood to be ground, the amount of available power and pressure to be used in grinding, and the freeness desired in the pulp. The coarser burrs are generally used on the soft stones with coarse grit. The finer burrs are often used in the summer months to slow up the stock. Sometimes the stock leaving the grinder immediately after burring is too coarse, in which case it may be necessary to dull (knock back) the stone by holding a very fine burr or a brick against the stone. An important factor in burring is the amount of pressure applied, since this determines the depth of the markings on the stone surface. Increasing the depth of burring increases the freeness of the pulp, but excessively deep burring results in a high percentage of shives, which increases the waste. Immediately after sharpening, the production rate is high and the energy consumption low. The temperature in the grinding zone drops and the pulp becomes of poor quality, that is, it is short and chunky and has poor felting properties. As grinding continues, the quality gradually improves.

The profile of the stone surface may be reproduced for observation by pressing a thin aluminum sheet against the stone surface or by using a specially built piece of equipment consisting of a pin fastened to a flexible beam that is placed against the pulpstone surface and held there while the stone is slowly revolved.[27]

As the stone becomes conditioned in use, production decreases; unit energy consumption increases, freeness decreases; and bursting strength increases. Thus, although strength increases with continued wear on the stone, too much wear cannot be tolerated from the standpoint of production and energy consumption. A well-conditioned stone consists of rather prominent grits, which have been worn away to flat surfaces. The conditioning appears to be directional, since reversal of direction of a well-conditioned stone produces results similar to a freshly surfaced stone.[28] To offset wear the stone must be resharpened at intervals, but burring must not be excessive since this lowers the strength of the pulp and increases the amount of screenings. The proper balance between frequency of sharpening and depth of burring should be maintained. The schedule of burring must be so arranged that the interval between sharpenings is as uniform as possible in order that the properties of the pulp do not undergo abrupt and radical change. Even under the most favorable conditions, there is considerable fluctuation in pulp quality from a given grinder. The best operation is one that maintains a constant quality stock from the whole production of the grinder room, which is more easily done in a plant with many stones than in a small plant.

Usually the stone is reburred when the freeness and production have dropped below a predetermined value. Another method for determining when stone resharpening is required is by measurement of the heat produced in grinding. Since this is dependent on the friction between the wood and stone, it reflects the sharpness of the stone. The amount of heat generated can be determined by measurement of the grinder pit temperature, the shower white water temperature, and the grinder pit consistency. Using the following formula, Klemm[29] reports values for S (sharpness) of 11 for sharp stones and 24 for dull stones:

$$S = \frac{\text{Temp. grinder pit} - \text{Temp. shower white water}}{\text{Consistency in grinder pit}}$$

In some mills visual examination of the pulp, or an increase in energy consumption over a predetermined value, is used as the criterion for reburring. Klemm[30] suggests resharpening when energy consumption is 35% higher (with low load) to 25% higher (with high load) than normal. The usual sharpening interval is 75 to 200 hr for vitrified stones and less for cement-bonded stones. Some mills have a simple schedule of reburring each stone once a day.

Magruder[31] has studied the effect of stone surface age on grinding rate, power consumption per ton, and pulp properties, using the Roberts grinder. The results are reproduced in Figures 4-3 and 4-4, from which it can be seen that a

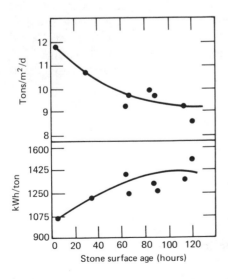

Figure 4-3. Effect of stone surface age on production and power consumption (Roberts grinder).

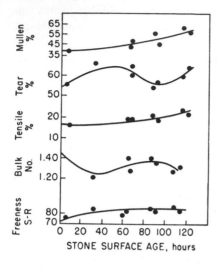

Figure 4-4. Effect of stone surface age on properties of groundwood (Roberts grinder).

burring cycle of about 80 hr is optimum since beyond this point power consumption and production are affected adversely. Natural stones may require sharpening about every eight hours.

Effect of Stone Speed in Grinding

Older mechanical pulp mills were operated with water wheels, but modern grinders are generally driven by constant-speed synchronous motors on which a constant predetermined load is maintained by varying the pressure of grinding. Variations in pressure on grinders driven by water wheels result in a fluctuation in stone speed, but similar fluctuation does not occur on motor-driven grinders.

In most North American mills, the grinders are arranged in lines, with one motor driving two or more grinders. Continuous grinders are generally driven by individual motors. Rotational speeds in pocket grinders vary from 200 to 275 rpm. Surface speeds for natural stones vary from 850 to 1050 m/min. The average speed with artificial stones is about 1200 m/min, but speeds up to 1500 to 1800 m/min are practical.[32,33]

For normal grinding operations, production and power input to the grinder are directly proportional to stone speed. Energy consumption is generally not appreciably affected by stone speed,[34] although in certain ranges energy consumption is decreased by increasing stone speed.[35] Pulp strength is generally reduced by increasing stone speed.

Effect of Grinding Pressure

Grinding pressure has an important effect on the grinding characteristics of the wood and the physical properties of the final pulp. In general, increasing the grinding pressure increases the production, increases the power input to the grinder, and decreases the energy consumption (see Figure 4-5). The effect on

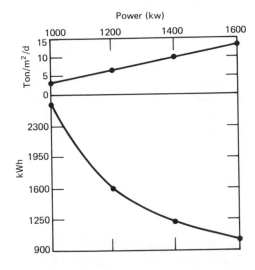

Figure 4-5. Effect of power input on production and power consumption (Roberts grinder).

pulp properties is to increase the freeness, decrease the strength, and increase the number of slivers. As the stone becomes dull through grinding, increased pressure is required to maintain a predetermined rate of grinding and freeness of the pulp. To prevent loss in production, the stone is reburred whenever the freeness falls below a certain level. For a given energy consumption, high pressure will generally produce a stronger pulp than low pressure. In commercial operation, grinding pressure is generally confined to a fairly narrow range, the upper limit of which is established by the available power.

Grinding pressure is a difficult calculation to make because of the variation in pressure between the wood and stone during grinding, resulting from the change in the area of wood in contact with the stone as the wood is ground away. In the case of pocket grinders, unit pocket pressure can be calculated from the length of the wood, width of the pocket, cylinder pressure, and the diameter of the cylinder,[36] but calculation for continuous type grinders is not so simple. Pocket pressures vary from 170 to 690 kPa (25 to 200 psi).[32,37] Average operating pressure is probably in the neighborhood of 345 to 415 kPa (30 to 60 psi) but best quality pulp is obtained at an average pressure of about 241 kPa (35 psi). Newsprint is generally made in the range of low pressure, whereas groundwood for wallboard is made at high pressure.

It has been reported [37a] that increasing the pressure, using pressurized air and steam in a two-pocket grinder at 200 kPa greatly increases the long-fiber content and strength of the pulp and reduces power consumption. A quality of pulp that is comparable to thermomechanical pulp is claimed.

Governors for Electrically Driven Grinders

In the early grinders, which were water-wheel driven, it was necessary to control the speed of rotation to prevent bursting of the stone at high speeds. In modern electric-driven grinders speed control is not a problem, but power control is necessary to prevent excessive power swings. Such control was first attempted by use of current-control governors (ammeters) and was followed by resistance-controlled and wattmeter-controlled automatic load regulators. These control the power load of the grinder room and maintain a constant total power-room load. Governors regulate the speed of the driving chain in continuous grinders and regulate the hydraulic pressure on the pressure foot in intermittent grinders. Thus, in the case of pocket grinders, when one pocket is backed off for filling, more pressure is automatically applied to the remaining cylinders to maintain a constant total load on the motor.The process is then reversed when the pocket is refilled and is again grinding. This method of operation means that pulp quality is being sacrificed to greater power efficiency, but for practical purposes the effect on pulp quality is not important.

Effect of Power Input in Grinding

Power input to the grinder increases directly with grinding pressure. In practice, power input is widely used in place of pressure for controlling grinder operation because it is easier to measure power input than to measure specific pressure. Power input per unit of grinding surface can be calculated from the total power input per line and the projected grinding area of the pockets. For the same wood with a constant condition of the stone face and grinding temperature, power input can be calculated from the applied pressure, peripheral speed of the stone, and the coefficient of friction between the wood and the stone surface, using this formula:

$$\text{Total kW delivered to grinder} = \frac{\mu W \times \pi d N}{60,000}$$

where μ = coefficient of friction; W = total thrust of pressure foot against stone in newtons; d = diameter of stone in meters; N = rpm of stone. In this formula μW equals the force opposing rotation and $\pi d N$ equals the lineal velocity of the stone in meters per minute. The coefficient of friction for grinding varies from about 0.13 to 0.16 for commercial grinders to about 0.22 to 0.38 for miniature grinders.[38] The effect of power input on the production and consumption of power in the Roberts grinder is shown in Figure 4–5.[39]

The power required in grinding is determined by the species and density of the wood, type of pulp being made, condition of the stone surface, grinding pressure, speed of the stone, and temperature and consistency of stock in the grinder pit.

The properties of the pulp vary appreciably with changes in power input for a given stone surface. The properties of pulp can also vary widely with constant power input to the stone, but this is always accompanied by a variation in the production of pulp due to changes in stone surface.

The bursting strength of the pulp decreases rapidly with increasing power in-

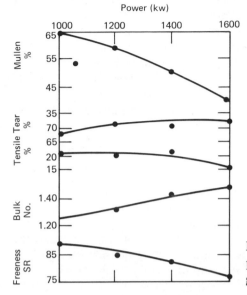

Figure 4-6. Effect of power input on properties of groundwood (Roberts grinder). No change in stone face.

put, while the tear and tensile show little change except at the extremes and bulk tends to increase linearly. Typical results are shown in Figure 4-6 which is taken from work by Magruder[40] on the Roberts grinder. (The bulk number is reported as the caliper of eight sheets divided by 8; the result is divided by the basis weight, and this result is multiplied by 10.) High strength and low bulk are obtained at low-power input, but this is the point where production is low.

Energy Consumption in Grinding

Energy consumption in mechanical pulping is one of the most important factors. With electric motors, the power used in grinding is easily obtained from the wattmeters of the individual grinder motors or from a master wattmeter in the grinder room; the readings are converted from kilowatts to horsepower. Specific energy consumption, which is usually expressed in horsepower-days per ton of air-dry pulp produced, is easily calculated by dividing the horsepower input to the grinders by the tons of pulp produced per day.

Pulp strength and energy consumption are closely related variables. For a given stone surface, the strength of groundwood increases directly with an increase in the amount of energy consumed in grinding. A relatively fixed amount of energy is required to produce a particular grade of pulp from a given species of wood. Sometimes energy consumption and pulp strength are expressed in one factor combining energy consumption per ton of pulp per meter of breaking length.[41]

The energy supplied to the stone in grinding is consumed in overcoming the friction between the stone and the surface of the wood. Practically all the energy consumed reappears as sensible heat in the grinding water; only a very small amount of energy is absorbed in forming new surfaces owing to fibers using separated from the wood.[42] Power consumption runs from about 1000 to 1200 kWh/ton for hanging (wallpaper) stock, 1100 to 1300 kWh/ton for newsprint, and 1400 kWh/ton or over for book-grade groundwood (spruce).

Energy consumption in grinding decreases with an increase in stone sharpness, an increase in pressure, or an increase in stone speed. The power consumption of continuous grinders can be reduced by operating at low pit consistency (e.g., 2 to 2.5%), with low submersion, and by better cleaning of pulp from the surface of the stone.

Effect of Production in Grinding

The production of pulp is a variable of obvious importance in mechanical pulping. Production may be calculated as tons of air-dry pulp produced per grinder per 24 hr, or as tons of pulp produced per square meter of grinding surface per day. The latter is obtained by dividing the tons of pulp produced per day by the pocket area in square meters. Values for the grinding of spruce in pocket grinders vary from 6.5 to 20 ton/m^2/d. The total pulp production is obtained by multiplying the wood consumption in cubic meters per day by the yield in tons (OD) per cubic meter of wood.[43] Wood consumption recording instruments are available.[44]

Production increases with increase in stone sharpness, increase in grinding pressure, or increase in stone speed. With constant power input the only way production can be increased is by sharpening the stone, which increases the production per unit of grinding surface. In general, bursting strength decreases with increasing production, and freeness increases with increasing production.

Temperature of Grinding

The effect of temperature on grinding is of recognized importance. The temperature that is important is the working temperature in the grinding zone at the wood-stone interface. This temperature reaches a level of 182 to 194°C, which is sufficient to plasticize the lignin, and makes it easier to defiber the wood.[45] In commercial practice, it is customary to measure the temperature in the grinder pit or take that of the stock on the discharge side of the stone, where the temperature is lower than the working temperature.[46] Grinder-pit temperature can be controlled at the desired point by regulating the amount of shower water or extra white water (from the deckers) added to the grinder pit. Higher pit temperature, without change in consistency, can be obtained by raising the shower water temperature. Temperature controllers are used on discontinuous grinders, while normal control of shower water is generally satisfactory for continuous grinding.

The amount of heat generated in grinding varies with the pattern on the stone, the density of the wood, and the pressure applied. The temperature is also dependent on the consistency in the grinding zone. It fluctuates considerably according to the condition of the stone surface.[45] Very sharp stones may generate only enough heat to create a temperature in the grinding zone of about 163°C, which is not enough to plasticize lignin.

There are, in reality, two types of grinding processes, known as hot grinding and cold grinding. Nearly all grinding in the United States and Canada is hot grinding, that is, the temperature in the grinder pit is in the range of 66 to 88°C and the consistency is in the range of 2 to 8%. The cold grinding process, which is used in some European mills, employs larger quantities of water, and the consistency of the pulp in the grinder pits is correspondingly lower, about 2 to 4%; the temperature is about 50 to 65°C. Corresponding energy requirements are about 1040 to 1220 kWh for hot grinding and 1250 to 1520 kWh for cold grinding.

Increasing the grinding temperature (at constant consistency) increases the grinding rate, decreases the energy consumption, increases the freeness, and increases the pulp strength.[47,48] Power savings on the order of 2 to 3% can be obtained for each 6°C increase in white-water temperature in a closed system[49] up to about 70°C. Some mills with a well-closed white-water system use heat exchangers to keep the white water from exceeding the desired maximum temperature. In general, hot grinding produces a long, strong fiber, whereas cold grinding produces a fine stock of low freeness. In the grinding of spruce, Davis[50] found that fiber length and freeness increased up to about 77 to 82°C pit temperature, after which both decreased with further temperature increase.

Magruder[51] found, with the Roberts grinder, that temperature had relatively little effect on pulp properties until the temperature was increased over 71°C, after which production rose sharply, tear and bulk increased, and bursting strength slightly decreased. Andrews[52] obtained a progressive increase in bursting strength for constant freeness when the temperature was increased up to 78°C, the highest temperature studied, as shown in Figure 4-7. The effect of

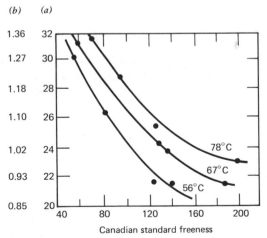

Figure 4-7. Effect of temperature of grinding on bursting strength in groundwood (a) percent Mullen test. (b) Burst index kPa·m²g.

temperature on the quality of pulp appears to be less as the speed of grinding is increased. In practice, temperatures as high as 88 to 93°C are used, but it is generally believed that the most satisfactory all around operation is obtained at temperatures of 68 to 77°C. Klemm states[53] that the temperature should not be higher than 77°C with dry wood, nor higher than 60°C with over-age wood. At least part of the effect of high temperature on grinding can be attributed to the reduced viscosity of the water.[54]

Consistency of Grinding

Uniform temperature is obtained in commercial grinding at the expense of uniform consistency because of the fact that temperature is generally controlled by regulation of the amount of cooling water used in the showers. For this reason, it is difficult to separate the effects due to temperature from those due to consistency; ordinarily no attempt is made to do so. The favorable effects obtained by raising the temperature are generally offset by the unfavorable effects of the increased consistency that follows.

Pit consistencies are generally about 3 to 4%, although pit consistencies as high as 8 to 9% are sometimes used. As a rule, increasing the consistency of the pulp in the grinder pit above 4% results in an increase in the energy consumption and a decrease in the grinding rate.[55] For increases in the pit consistency between 1 to 10%, Andrews[56] obtained these results:

	For Constant Grinding Pressure	For Constant Freeness
Production	decreases	increases
Freeness	decreases	—
Burst	increases	decreases
Energy consumption	increases	increases

Freeness and Strength of Groundwood

Bursting strength and freeness (Canadian Standard and Schopper-Riegler) are the two classical means of measuring the quality of groundwood, although fiber classification is often used. The freeness test by itself means very little, and it must be combined with at least a visual examination of the stock, and preferably with a strength test.

Groundwood pulps produced by the same grinder from the same wood under unchanging operating conditions show a definite relationship between freeness and bursting strength. This relationship is exponential in nature, that is a plot of burst and freeness on semilog paper is a straight line, as shown in Figure 4-8.[57,58] Assuming that the curve in Figure 4-8 denotes average quality, then a shifting to the right signifies improved quality and a shifting to the left signifies lowered quality.

By the use of curves such as those shown in Figure 4-9, groundwood pulps of different freeness and strength values can be compared on a common basis by bringing to a common freeness (usually 100 cc) or to a common bursting

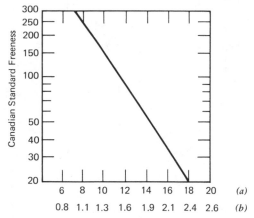

Figure 4-8. Relation of freeness to burst and burst index (a) burst in psi (32 avdp basis weight). (b) Burst index in kPa·m²g.

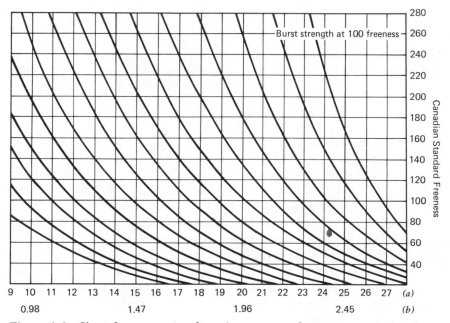

Figure 4-9. Chart for converting from burst at one freeness to equivalent burst at a different freeness (a) burst ratio in g/cm² ÷ g/m². (b) Burst index in kPa·m²/g. (Tappi Data Sheet, courtesy Technical Association of the Pulp and Paper Industry.)

strength. For example, the strength of groundwood of any given freeness may be converted into strength corresponding to any standard freeness, such as 100 cc, by finding the curve that most closely corresponds to the given freeness and bursting strength, and then following that curve to the corresponding strength at freeness 100 cc. The curves in Figure 4-9 are useful for comparing the effect of

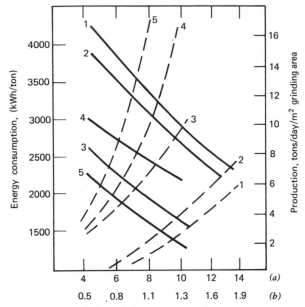

Figure 4-10. Relation between bursting strenth, energy consumption (— —), and production (——) in grinding of different woods: (1) spruce; (2) balsam; (3) shortleaf pine; (4) poplar; (5) cottonwood (a) burst in psi (32 avdp basis weight). (b) Burst index in kPa·m²g.

wood, grinder stones, and process variables on the quality of product. The curves hold fairly well for most conditions, but specific curves may be necessary where a high degree of accuracy is required.

For a given species, bursting strength generally varies in a definite relationship with freeness; both of these properties of the pulp are determined by the conditions of grinding. If increased strength (reduced freeness) is desired, it is necessary to supply more unit energy to the pulp. This in turn reduces the rate of production if all other conditions are unchanged. On the other hand, if the production is increased, keeping the unit energy constant, the pulp strength will decrease and the freeness will increase. These effects are shown in Figure 4-10, where production and energy consumption are plotted against bursting strength for several different species of wood.[59]

Wood Variables Affecting Grinding

The important wood variables affecting the grinding characteristics of the wood and the quality of the mechanical pulp are species, density, and moisture content. The desirable characteristics of wood for mechanical pulping are good color, a high degree of softness (low density), and long fibers. Better quality pulpwood is required for mechanical than for chemical pulping. For best results, each species should be ground separately to permit the optimum grinding conditions for that species.

Softwoods are generally preferred for mechanical pulping although hardwoods, especially aspen, have enjoyed an increasing use in recent years. Spruce is the most desirable species because of its low energy consumption in grinding and high production rate. In the northern states and Canada, spruce is widely used, together with smaller amounts of jack pine, balsam, hemlock, white fir, and tamarack. Black spruce from high land, where the rate of growth is slow, has the highest density and is considered to be the finest wood for mechanical pulp.

Spruce and balsam fir are rather similar in grinding properties, although less power is required with spruce, and the resulting pulp produces a harder sheet than balsam. The quality of the groundwood is good, but the strength and yield are lower than with spruce. Hemlock groundwood has a shorter fiber length than spruce groundwood, and the color is slightly reddish.

In the southern states, increasing quantities of pine are being used, but in some cases, only young trees in which heartwood has not developed can be used. Coarse-grained dull stones are required for southern pines (compared with the stones used on spruce) to produce uniformly long-fibered pulp. The energy consumption per ton of pulp is generally 20% or higher than that required for spruce of comparable quality. Southern pines have a much higher density than spruce or fir and produce paper of lower strength, higher bulk, and less rigidity than paper from northern softwoods.[60] The strength of pine groundwood can be improved by heating the pulp suspension.[61] Jack pine has a limited use in the northern states, but it, too, requires higher energy consumption than spruce, and the wood is knotty. The pulp is resinous, poor in color, and has a lower strength than spruce for a given freeness. It has been stated that more than 15% jack pine in a groundwood furnish increases the likelihood of pitch trouble on the paper machine.[62] On the West Coast, spruce (Sitka and Engelmann) and the true firs are preferred, although hemlock is used. Lodgepole and Monterey pines generally produce better quality pulp than the southern pines.

Under ordinary grinding conditions, hardwoods (aspen) grind at a higher rate and consume less energy per ton than spruce, but the pulp obtained is much lower in strength and is very free and bulky.[63] Strength can be greatly improved by increasing the energy consumption through use of a dull stone surface, but this increases the cost by sacrificing production and power. Aspen is the most satisfactory of the hardwoods; appreciable amounts of native aspen are ground in the United States; a hybrid species grown as a crop is used in Italy for newsprint.[64] The pulp from aspen compares favorably in color with spruce and balsam groundwood. A pulp can be produced that is nearly equal in strength to that of spruce groundwood, but the energy consumption is higher. Typical results obtained on a laboratory grinder (32 in.-diameter, 11 in.-face) are:

Species	Burst Index $(kPa \cdot m^2/g)$	Energy Consumption (kWh/ton)
Spruce	1.19	1164
Aspen	0.93	1343

If a relatively bulky stock is desired, aspen can be ground with a fairly sharp stone at low energy consumption. Some eucalyptus is ground in Australia, but it is interesting that trees (*Eucalyptus regnans*) over 200 years old grind well, whereas younger wood does not make a good pulp.[66]

The grinding characteristics of a few different species are shown in Table 4–2, where it can be seen that spruce and balsam do not require as high an energy consumption as poplar or the pines and that the pulps produced have a higher freeness for a given bursting strength.[59] The values for energy consumption given in Table 4–2 are not exactly comparable with commercial values because of the low grinding pressure used, 138 kPa (20 psi) of pocket area, but the values serve for comparing one wood with another. Klemm[67] rates various woods as follows: At the same freeness, specific energy consumption increases in the order of pine → spruce → fir → poplar, and production decreases in the same order; at the same applied load, freeness increases in the order of poplar → fir → spruce → pine, while production decreases in the order of poplar ← fir ← pine ← spruce.

The freeness requirements for mechanical pulp depend on the grade of paper that is to be produced. Coarse pulps for boxboard, liner, wrapper, and similar grades generally have a freeness (Canadian) of 160 to 250. Newsprint and wallpaper stocks have a freeness of 85 to 95. Fine groundwood for use in book, magazine, and other printing grades has a Canadian Standard Freeness of about 60 to 80.

Density of the wood appears to be the most important factor in grinding since pulps produced from high-density hardwoods are uniformly low in strength and poor in color. High-density hardwoods (e.g., birch, beech, and maple) produce pulps having less than 25% of the strength of spruce groundwood. These high density hardwoods are of interest, since they yield from 25 to 30% more pulp per cord than spruce or the low-density hardwoods; the energy consumption is less, but the low strength and poor color have limited their use. Yield of pulp and density of wood bear an approximate linear relation to each other, as shown in Table 4–3.[68] Fast growth wood is desirable for maximum strength although the yield is low.

TABLE 4–2 GRINDING CHARACTERISTICS OF DIFFERENT WOODS (BURSTING STRENGTH CONSTANT AT 55 kPa (8 psi) AND BASIS WEIGHT 52g/m^2)

Species	Freeness (cc)	Production (ton/m^2/d)	Power Consumption (kWh/ton)
Spruce	115	12.0	1250
Balsam	68	9.6	1430
Hemlock	40	7.4	1935
Jack pine	58	8.0	1880
Shortleaf pine	42	6.4	2240
Poplar	37	6.0	2500
Loblolly pine	25	4.8	2635
Cottonwood	22	3.2	3550

TABLE 4–3 RELATION OF YIELD TO DENSITY IN GROUNDWOOD PULPING

Species of Wood	Density, Oven-Dry Weight on Green Volume	Yield of Screened Pulp per m^3, kg
Loblolly pine	.49	482
Shortleaf pine	.46	457
Cottonwood	.44	433
Poplar	.43	413
Jack pine	.41	401
Hemlock (eastern)	.40	392
Spruce	.38	381
Balsam	.34	329

As a general rule, fresh wood that is still slightly moist grinds more easily and produces longer fibers than wood that is seasoned and dried. Freshly cut wood produces a brighter pulp than wood that has aged; in the case of spruce and balsam fir, brightness may drop off by three points during one year's aging.[69] Moist wood also requires less chemical for bleaching.[70] The moisture content of the wood at the time of grinding should be 40 to 45% or higher for maximum strength, since the strength of the pulp drops off if the moisture content is much below this value. Remoistened wood that was once overdried is believed to be less satisfactory than wood that never was overdried, although if moisture is not reduced below about 18%, the wood can be resoaked to a higher moisture level without any significant loss in strength. Grinding of dry wood, that is moisture of 25 to 30% or less, produces a stubby, soft stock containing a high percentage of fines. This type of pulp may be preferred for some grades of paper (e.g., printing papers), and for these papers grinding dry wood may be advantageous. If the moisture content is increased and the grinding pressure held constant, production is increased and energy consumption is decreased. On the other hand, if the moisture content is increased and the pressure adjusted to maintain constant freeness, production is decreased and the energy consumption is increased,[71] but the strength increases. It is generally desirable to use duller stones and greater loads when the wood is unusually moist. Dahm and coworkers[72] obtained the following results for spruce at different moisture contents when ground to the same freeness (68 to 69°S-R):

Moisture (%)	40.00	28.00
Fineness (% retained on 25 mesh)	23.00	19.00
Breaking length (km)	2.50	2.20
Burst index (kPa· m^2/g)	1.01	0.91
Tear factor	38.90	33.40

To grind dry wood, coarse burring and reduced load should be used. Decayed wood is not satisfactory for grinding. It produces dirty pulp that has low strength, requires more sizing, and foams excessively.[73] Furthermore, the yield is reduced and the pulp is more susceptible to further decay.

Characteristics of Groundwood Pulp

Groundwood pulp does not contain morphological fiber entities, but instead the pulp is composed of individual fibers, broken fibers, fines, and coarse-fiber bundles. The quality of the fiber in groundwood can be determined in qualitative fashion by the blue glass test. A small quantity of dilute stock is placed on a blue glass plate set in a wooden frame and the contents examined visually against a strong light. The glass is covered to a depth of about 2 mm with stock at about 0.05% consistency. Since the light is blue and the fibers white, the nature and uniformity of the fibers show plainly. In the hands of an experienced person, the test will show something about the ease of clotting, number of shives, number of fines, and the average length of the fibers.

A somewhat better technique than the above is to project a slide of diluted fibers on a screen at a magnification of about 50. For a quantitative evaluation, a microscope examination or a fiber classification test to determine the proportion of fiber of different fiber lengths is frequently carried out. The fibrous constituents of mechanical pulp have been classified as long (0.8 to 4.5 mm); short (0.2 to 0.8 mm); and fines, which are further classified into flour (0.02 to 0.2 mm) and mucilage (less than 0.02 mm).[74] The proportion of these ingredients determines the felting properties, the wet-web strength, and the dry strength of the paper. The amount of fiber fines (the portion of the pulp passing a 150 mesh screen) is usually from 40 to 60% of the pulp.

In addition to the time-honored blue glass test and the projection and classification methods, other tests used to characterize groundwood include freeness, drainage rate, and physical tests on handsheets. Ullman [75, 76] proposed strength measurements on undried handsheets at the 75% moisture level as a test that correlates well with machine performance, particularly with the tendency for breaks to occur on the paper machine.

The ray cells in the wood are usually ground finer than the wood fibers because of their orientation relative to the stone during grinding and because of their smaller size and thinner walls. In grinding much of the coloring matter contained in the ray cells is set free in the form of finely divided particles that tend to segregate in the sheet during formation.[77] The presence of coloring matter causes considerable trouble in some groundwood pulps, for example groundwood from certain of the western woods of the United States (e.g., western hemlock) has a decidedly pink color. On the other hand, certain hardwoods, such as basswood and sycamore, produce very bright pulps due to their lack of coloring matter. Sometimes a small percentage of basswood and sycamore is used for the express purpose of increasing the brightness of the pulp.

Groundwood is low in strength compared with sulfite and sulfate pulps. The pulp does not "hydrate" in beating, but tends to break up, thus producing bulky papers. Because groundwood contains practically the whole wood, papers made from groundwood deteriorate in strength and yellow on aging. For this reason, groundwood is used only in relatively impermanent papers, such as newsprint, cheap books, catalogs, rotogravures, magazines, cheap drawings, toilet tissues, towelings, hanging stocks, coating raw stocks, and some grades of wallboard and paperboard. For these grades, groundwood has the desirable properties of low cost, good printing quality, high opacity, and drainage characteristics that can be controlled for high-speed paper machine operation. Printing qualities are good because of the high bulk, high smoothness, resiliency, and good ink absorption. Groundwood papers also have less tendency to curl than papers made from chemical pulps. The particular properties required for a given grade of paper are obtained by selection of the proper species and regulation of the grinding conditions. For example, groundwood to be used in book papers is ground in such a way that the fiber is short and fine, whereas groundwood for catalogs is ground in a way to produce a longer-fibered stock. Coarse groundwood has a higher freeness and produces bulkier paper than fine groundwood. The history of mechanical pulp has been the story of newsprint. Initially, groundwood was developed as a cheap pulp for low-grade papers, but advances in cleaning and bleaching techniques and the general recognition of the good qualities of groundwood have greatly widened the use for this pulp. An article by Mannstrom[78] concerning influence of groundwood quality on runnability and printability summarizes and discusses the relationship of test results on groundwood to the paper made from it: (1) good runnability is dependent on high tearing strength; high breaking-length; and clean groundwood, free of shives and (2) good printability is dependent on high brightness; high scattering coefficient; high porosity; and high oil-absorbtion rate.

The amount of groundwood used in newsprint and hanging stock is about 85% of the fibrous furnish. Groundwood is used in printing papers to the extent of 30 to 85% of the fibrous furnish. Other grades of paper may contain from 10 to 100% groundwood. A small amount of groundwood (up to 10 to 20%) in mixture with long-fibered chemical stock will generally increase the bursting strength of the paper slightly, without lowering the tearing resistance, owing to the groundwood filling between the longer fibers of the chemical pulp. Higher percentages of groundwood materially lower the strength of the paper. Hardwood groundwood generally has a lower strength than spruce groundwood, although this depends on the species and the method of grinding the wood. Aspen groundwood has been used up to about 15% of the furnish in newsprint and up to 50% of the furnish in printing papers without harm to the paper. On the other hand, Sears[79] found that over 10% birch groundwood reduced the strength of newsprint materially.

Groundwood from Pretreated Wood

Ever since the invention of the groundwood process, there has been considerable interest in methods of reducing the power requirements and improving the quality of the pulp. Among the early processes that attracted considerable attention were: (1) the Hall process, in which the wood was subjected to boiling water before grinding and (2) the Enge process in which the wood was heated in water at 110 to 125°C for 5 to 10 hr at a pressure higher than the steam pressure at that temperature. Soluble organic acids are formed when wood is treated in this way, and there is some depolymerization of the cellulose,[80] but the principal effect is a softening or partial solvent action on the lignin in the wood, which renders the wood more easily separable into fibers. The result is a pulp that has a higher freeness and a greater strength and stiffness than regular groundwood. The pulp is dark colored, however, and this limits its use to paperboards and cheap grades of brown paper.

Heating or steaming of wood before grinding is not widely practiced in the United States, but it is a fairly common practice in Central Europe where pulpwood is received in a drier state than in the United States. Because steaming softens the wood, steamed wood can be readily ground "cold." The total specific energy consumption is lower than normal when softwoods are used.[81] At high temperatures and pressures [e.g., 158°C and 462 kPa (67 psi)], a considerable portion of the wood is dissolved—that is about 12% (for pine) to 17% (for spruce). Steaming has some merit, particularly for the hardwood species, which generally do not discolor as much as the softwoods. One beneficial effect of steaming is reduced trouble with pitch. Increasing the degree of treatment by increasing the time or temperature of steaming increases the strength, lowers the yield, and lowers the color of the pulp; the energy consumption may be slightly increased.

In more recent times, attempts have been made to improve the strength and fiber length of groundwood by treatment of the wood with chemicals. Two different processes have been proposed: (1) treatment of the wood at the time of grinding by introducing chemicals in the grinding water and (2) pretreatment of the log at high temperature and pressure before grinding. The former process has little to recommend it, although both sodium carbonate and sodium sulfite have been used this way. Sodium carbonate has been reported to increase production and reduce power consumption,[82] and at least one mill has added sodium sulfite at the grinder showers to reduce the energy requirements and increase the brightness. At an addition level of 0.8% sulfite on the weight of the screened pulp, the energy requirements are reduced by about 9% and the brightness increased by 2.5 points.[83]

Processes for pretreating the log or bolt before grinding were suggested years ago, but they never attained any practical importance, principally on account of the difficulties in obtaining uniform softening of the log and because the process was tried mostly on softwoods, which are not well suited to this type of treatment.[84] One of the early processes (the Hall process) for pretreating wood used alkaline liquors, for example, sodium hydroxide, sodium sulfite, sodium bisulfite, sodium sulfide, and sodium carbonate, but these alkaline liquors caused considerable darkening of the wood and the pulps were not suited to the manufacture of white papers. However, pretreatment of pulpwood in mild caustic at high temperatures (110 to 125°C) is being used for making pulp for insulating board, where color is not a problem.[85] Another process (the Decker process) used sulfur dioxide as a means of improving the color of the pulp,[86-88] but the acid liquors caused trouble with corrosiveness. The most satisfactory pretreatment process has been that using hardwoods and neutral sulfite liquor.

Lougheed[89] has described a commercial process in which the wood in log form is treated in an autoclave with a solution of sodium sulfite and sodium bicarbonate in a ratio of 7 : 1, at a temperature of 115 to 125°C for 2 1/2 to 4 1/2 hr, and at a pressure of about 413 kPa (60 psi) higher than the saturated steam pressure, using about 7lb of liquor per ton of wood or, in other words, enough to keep the final pH of the liquor at 7.2 to 7.4. A four-cycle process for pretreating hardwoods with neutral sulfite liquor before grinding has been proposed by the Forest Products Laboratory.[90] In this process, the wood is submerged in a liquor consisting of 60 g/l of sodium sulfite and 20 g/l of sodium bicarbonate and then subjected to:

1. A half-hour vacuum treatment.
2. Heating to 121°C at a pressure of 964 kPa (140 psi) for 30 min.
3. Vacuum for another 30 min.
4. Finally heating again at a temperature of 121°C and a pressure of 964 kPa (140 psi) for 30 min.

The pressure is then relieved, the liquor is drained, and the wood is ground within two hr.

Extensive work on chemigroundwood pulp has been done at the New York State College of Forestry in an effort to establish optimum operating conditions. This original investigation established the most satisfactory cooking liquor (for aspen, birch, beech, and maple) as one containing a ratio of 6 parts of sodium sulfite to 1 part of sodium bicarbonate. The liquor is conveniently made by sulfiting a soda ash solution with sulfur dioxide until a pH of about neutral is obtained and then buffering with additional soda ash. Higher ratios of sulfite to bicarbonate resulted in lower energy consumption in grinding and higher brightness pulp, but the strength of the pulp was lowered. The optimum concentration of chemical for impregnating 1.22 m-logs of wood is about 184 g/l. This amount of chemical results in a pH of 9.3 to 9.5 during the cooking period. After loading the logs in the digester, cooking conditions are established by evacuating the digester to a vacuum of 68 to 71 cm Hg for about 30 mins, forcing in the cooking liquor without breaking the vacuum, raising the temperature to 140 to 150°C by admitting steam to an outside heat exchanger, applying hydrostatic pressure (826 to 1034 kPa (120 to 150 psi) to raise the overall pressure level to about 1378 kPa (200 psi) and holding under these conditions for about 6 hr. In one large commercial installation,[91] 1.22 m-logs are loaded by means of a rotary turntable into a large vertical digester that is partially filled with a pad of water which is gradually drained as the digester fills with wood. The cooked logs are discharged through a large door in the bottom of the digester. An accumulator can be used to hold the pressure during cooking.[92] Temperatures over 150°C or treating periods over 6 hr tend to reduce the strength of the pulp. Lowering the temperature and pressure in the later stages of the cook has the advantage of producing a lighter colored pulp.

The consumption of chemical by the wood when treated in the above manner is about 10 to 12% of the dry weight of the raw wood. Free liquor recovered from the cook can be fortified with fresh chemical and reused repeatedly without harm to the color or strength of the pulp, which means that no chemical recovery or waste disposal plant is required.

Optimum grinding is obtained at a grinder pit temperature of 54°C and a pocket pressure of 276 kPa (40 psi), using a fairly dull stone. However, the type of stone used in grinding is not as important as with untreated wood, and the power requirements in grinding are about one-half to two-thirds those of spruce groundwood. If regular spruce groundwood requires 1074 kWh/ton, the energy requirement for aspen and birch chemigroundwood would be about 716 kWh/ton, and for beech and maple chemigroundwood about 537 to 572 kWh/ton. For a coarse grade for corrugating, the energy consumption in grinding is about 537 to 626 kWh/ton.[93] The production rate can be about double that of regular untreated spruce. If regular spruce groundwood is produced at a rate of 14 ton/m² of grinder surface per day, then aspen chemigroundwood would be produced at a rate of about 17 ton/m², and birch, beech, and maple between 20 to 23 ton/m²/d. The yield per cord for the high-density hardwoods (birch, beech, and maple) is about one-third greater than for regular spruce, that is about .52 to .55 ton/m³ compared to .41 ton/m³ for spruce. Aspen chemigroundwood yield is somewhat less than regular spruce, about .35 ton/m³. The above advantages, plus the fact that the high-density hardwoods can be used, gave this process the hope of economic advantage over the regular grinding of spruce.

Chemigroundwood prepared by the above process is a unique pulp for groundwood since there is uniform separation of the fibers, the typical bundles and broken fibers of ordinary groundwood being absent.[94] The fibers show fibrillation, which serves to differentiate them from semichemical or full chemical pulps.[95] The strength of the pulp is excellent, being three to four times as strong as regular spruce groundwood. The wet-rub strength of birch chemigroundwood is comparable to that of groundwood from spruce.[96]

One disadvantage of the chemigroundwood process is the low brightness of the pulp (about 45 to 50). Even with aspen, a normally bright wood, dark-colored pulps are usually obtained. The low brightness is believed to be caused by infected areas in the heart of the wood, known as wet wood. These areas are difficult to penetrate because the vessels in the affected region are closed by barriers that prevent the movement of the treating liquor. The pulp responds readily to bleaching, and brightness gains of 7 to 15 points can be obtained with 2% peroxide, or 10% available chlorine added as hypochlorite. However, because of the lower initial brightness, bleached hardwood chemigroundwood is only slightly brighter than unbleached spruce groundwood. Unfortunately, serious problems developed when a large commercial chemigroundwood plant was built in the United States as part of a large integrated newsprint mill. Several million dollars was invested in this plant. After several years of unsuccessful experimenting to overcome the problems the plant had to shut down. The major problem was that chemigroundwood cooked to make strong pulp, bleached, and then used in the furnish for newsprint paper lowered the opacity, brightness, and ink-absorbancy of the paper. In an attempt to overcome these problems, the cooking time and chemical concentration was reduced. These changes reduced, but did not cure, the problems. More power was required to grind the wood and the resulting pulp was weak so that no reduction in the expensive chemical fiber in the furnish was possible. The result was a pulp that cost more than regular groundwood and was of inferior quality. The final problem that could not be easily or cheaply solved was stream pollution. The 10% chemical, which was absorbed by the wood during the cooking and was in the wood when it was ground, could not be washed out and recovered by any practical method. The result was increased stream pollution at a time when efforts were being made to reduce stream pollution. Smaller chemigroundwood plants built in Europe and South America also did not operate as well as had been expected.

REFINER MECHANICAL AND THERMOMECHANICAL PULPING

JOSEPH A. KURDIN

Technical Director
Pulp, Paper and Board Division
Sprout Waldron Operation
Koppers Company, Inc.

In this section four related, but different, pulping processes are described. They are basically mechanical processes and are commonly known by the terms: refiner mechanical pulping, thermomechanical pulping, chemimechanical pulping, and chemithermomechanical pulping.

Refiner Mechanical Pulping

Grinding of wood against a stone to produce pulp, called stone groundwood, was the only method of producing mechanical pulp until the development of the refiner mechanical pulping process during the period 1948-1956. Refiner mechanical pulping involves the mechanical reduction of wood chips in a precision built attrition disk mill, called a refiner. Refiner mechanical pulping became popular because it produced a pulp that was stronger than stone groundwood, it expanded the types of raw materials that could be used to produce acceptable pulp, and it offered the possibility of reducing the amount of expensive chemical fiber required in a newsprint furnish.[97-100] The growth of refiner mechanical pulping has resulted in the development of large refiners and different designs. Disk diameters have increased up to 137 cm (54 in.) with installed drive capacity of 10,500 kW of (14,000 hp) and disk speeds from 1200 to 1800 rpm. The acceptance of refiner mechanical pulp was relatively slow due to a number of problems and shortcomings. Refiner mechanical pulp is bulkier than stone groundwood, which results in a higher sheet caliper. However, additional calendering, (or the use of a breaker stack) on the paper machine will reduce the bulk to the desired levels. Although the pulp is relatively free of shives, it contains a significant amount of short stubby material (cubical debris).

Stages in Refiner Mechanical Pulping

In refiner pulping, chips are fed to the disk refiner by means of a ribbon feeder operated at high speeds. On some of the larger refiners, the feeder rotates with the main shaft. The chips are conveyed into the eye of the refiner, and steam, generated by the refining process, escapes from the refining zone along the shaft or through the center of the feeder. This separation of the generated steam from the incoming chips allows for an even feed to the refiner.

Production of refiner mechanical pulp involves two basic steps.[101] The first step, called fiberization (defibration), converts the original wood structure into single fibers. The second step, called fibrillation, reduces a portion of the fibers into cell-wall fragments. The object of the first step is to produce single, long fibers with a minimum of debris. This separation requires little specific energy. In a multistage refining process the fiberization (defibration) must be accomplished in the primary stage. To accomplish fiberization without reduction of fiber length requires high consistency and large refiner plate clearances. In practice this requires a high degree of fiber-to-fiber refining and the separation of fibers by a rubbing action. Fiber-to-plate contact must be kept to a minimum to prevent shortening of fiber length. Pulp produced in the primary stage must have few fiber bundles, shives, and chop because these cannot be eliminated in the secondary stage of the refining. The second stage in mechanical pulping, called fibrillation, involves the conversion of a portion of the whole fibers into fibrils and cell-wall fragments, which provide the bonding characteristics required in the paper. Fibrillation requires refining at high specific energy consumption. It requires refining at high consistencies to prevent shortening of the fibers.

Disk-refiner-plate design is very important. Generally, there are three sections: breaker bar section, intermediate refiner bar section, and fine bar section. The breaker section has wide bars with deep grooves and serves to break up the chips. The refining bar section has narrow bars and shallow grooves. The bars are parallel to each other and arranged in sections that are fed by feed grooves. As chips are fed into the refiner, the breaker bar section of the refiner plates shreds them into matchstick size. Action of the breaker bar is extremely important in producing high-quality pulp. Shredding will take place only if the chips can tumble between the bars, which is impossible if the space is plugged or overcrowded. More importantly, some of the steam generated within the refining zone vents through the breaker zone and out of the inlet, thereby softening the incoming chips for shredding. A screening action takes place at the transition between the breaker bars and the intermediate refiner bars because plate clearance determines the size of the wood particles passing between these sections. Failure to reduce the particle size sufficiently to permit materials to pass to the next bar zone results in excessive wear at the transition point, which will result in formation of a dam that will restrict the pulp flow. The principal refining treatment takes place in the finer bar section. This section of the refiner plate accomplishes the fibrillation. The overall degree of fiberization and fibrillation will depend greatly on the power input and the throughput rate.

The success of any refining installation depends to a large extent on the handling equipment feeding the raw material. Without uniform feed rate, the refiners cannot be loaded properly, thereby making it impossible to achieve uniform high pulp quality. If a ribbon feeder is used to deliver the chips, some of the steam escapes through the refiner inlet. With plug feeding, however, the steam can escape only around the periphery of the refiner plate. This becomes difficult after the plates have been worn and the height of the bars reduced. If the steam is not allowed to escape, pressure will build up and result in non-uniform loading with consequent reduction in pulp quality.

Production of high-quality refiner mechanical pulps with high-tear and burst values requires high consistencies between the plates, usually between 20 to 30%. Consistencies below 18% will lower the strength of the pulp drastically.[102] High consistency in the primary stage of refining is most critical since production of debris and shives at this point cannot be subsequently corrected. Refining consistency is one of the important variables in the refining process because it affects burst, tear, shive content, and freeness. The effect on bursting strength, tearing strength, shive content, and freeness are shown in Figure 4–11.

Single-Stage versus Multistage Refining. The most popular systems involve two-stage refining, but some systems use a single stage and some use three and four stages.[103] Single-stage refining requires high-power input to each individual refiner.[104] This high-energy input puts excessive strain on the equipment and requires special design.[97] In addition, single-stage refining means lower throughput rates and more difficulty in maintaining a uniform flow; as a result, feeding becomes more critical. The close plate clearances required in single-stage refining present the constant danger of the refiner plates clashing, resulting in the shutdown of the system. However, there are situations where single-stage refining produces high-quality mechanical pulp. It is entirely possible to apply 1600 to 1700 kWh/ton (90 to 95 hpd/ton) in a single-stage refiner to produce a 160 Canadian Standard Freeness pulp that, after screening and centrifugal cleaning to remove the fiber bundles, will have acceptable strength and debris content for such papers as newsprint. A system such as this is shown in Figure 4–12.

In some two-stage installations, two refiners are used in parallel in the primary

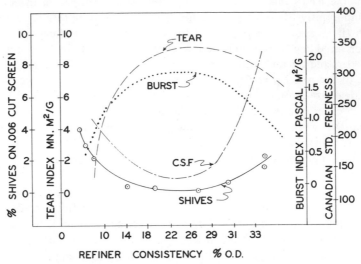

Figure 4-11. Effect of consistency in refiner mechanical pulping on bursting strength, tearing strength, shive content, and freeness. Pulp evaluated directly from refiner before screening.

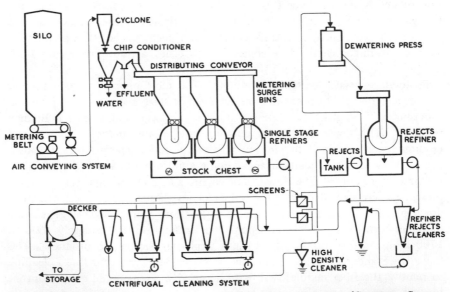

Figure 4-12. Single-stage refiner mechanical pulping system. (Courtesy Sprout Waldron Koppers Co., Inc.)

200

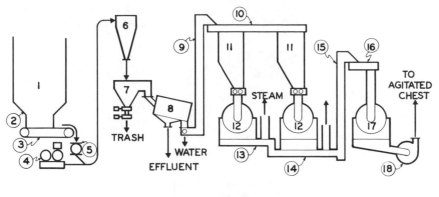

Figure 4-13. Two-stage refiner mechanical pulping system. (Courtesy Sprout Waldron Koppers Co., Inc.)

1. CHIP SILO	7. STONE & METAL TRAP	13. PULP CONV. 14" DIA.
2. SILO DISCHARGER	8. DRAINER-WASHER	14. PULP CONV. 16"
3. BELT CONVEYOR	9. VERTICAL LIFT	15. ROTOR LIFT
4. BLOWER	10. DISTRIB. CONVEYOR	16. PADDLE CONV. 14"
5. ROTARY VALVE	11. SURGE BINS 100 CU. FT.	17. SECONDARY REF.
6. COLLECTOR	12. PRIMARY REFINERS	18. STOCK PUMP

stage to feed a single second-stage refiner. The high bulk of the raw material at the primary stage makes it possible to use relatively large plate clearances. After the primary refining, the bulk density is increased and a considerably smaller plate clearance is required to do a satisfactory refining job. The two primary refiners operating in parallel will provide a throughput that is sufficient to load properly the second-stage unit, which is operated at double the throughput rate. Installing a single primary refiner with more power will achieve similar results, although this will sometimes require a larger diameter refining disk. Twice as much refining energy is generally used in the primary refining stage as compared to the second-stage refiner. A diagram of a two-stage refiner installation is shown in Figure 4–13.

Thermomechanical Pulping

Thermomechanical pulping is a mechanical pulping process that evolved from a combining of the refiner mechanical pulping process and the Asplund process used for the manufacture of fiberboard. It involves the steaming of raw wood chips for a short period at relatively low pressure and refining after the lignin has softened.

Asplund High-Temperature Process

Steam pressurized refining was developed by A. Asplund in the early 1930s.[105] In this process, wood chips are preheated in a steam vessel to 6- to 10-kg/cm^2

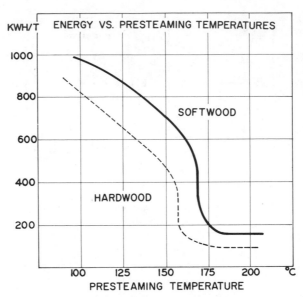

Figure 4-14. Comparison of softwoods and hardwoods for refining energy needed at various softening temperatures in the Asplund process.

guage pressure, which corresponds to 165 to 185°C. Refining is carried out at these high temperatures by close coupling the refiner to the steaming vessel. At this temperature, the lignin is softened to such an extent that the fibers are almost completely separated in undamaged condition with very little refining energy input. As shown in Figure 4-14, softwoods need higher temperatures compared to hardwoods before easy fiber separation can take place. The fibers separated under these high temperatures are not only undamaged, but they are coated with a thin layer of case-hardened lignin. This coating creates a smooth fiber surface, as shown in Figure 4-15, and makes any attempt to fibrillate very difficult.[106] This type of fiber can be used successfully in the building board industry, but it is unsuitable for papermaking. The explanation for this phenomenon is that when refining is carried out at high temperatures, the lignin undergoes a reversible glass transition thermal softening.[107,108] Fibers produced above 690 kPa (100 psi) are separated at the interface between the primary wall and the middle lamella, as shown in Figure 4-16. These fibers are intact, but are coated by softened lignin, which on cooling reverts to a glassy state. The bonding characteristics responsible for strength, which are so important for papermaking, are lost.

Effect of Temperature in Thermomechanical Pulping

When thermomechanical pulping is used to produce pulps for papermaking, it is a compromise. The chips are heated to a temperature that does not exceed the

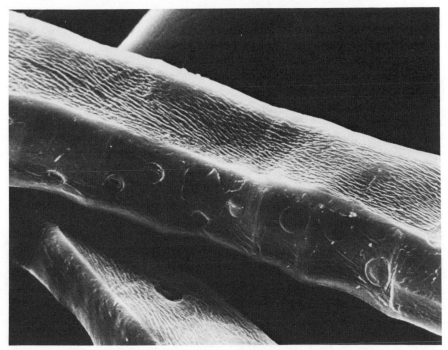

Figure 4-15. Scanning electron micrograph showing smooth surface on fiber produced from eastern black spruce chips by refining at 170°C. (Courtesy Canadian Pulp and Paper Research Institute and Svensk Papperstidning.)

glass transition point of the lignin so that the lignin remains in the glassy state. Refining separates the fiber at its weakest point in the S_1 layer and separation often occurs by complete fracture of the tracheid walls, as shown in Figure 4-17.[106] The middle lamella lignin remains intact on the outer surfaces of the fiber and is not dissolved to any great extent. This is quite unlike what happens when the fibers are refined at temperatures above the glass transition point of the lignin, when rupture occurs in the middle lamella, as shown in Figure 4-18. Distribution and condition of the lignin on the surface of the fibers has an important influence on the papermaking properties of the pulp. Thus, steaming time and temperature are very important variables in the thermomechanical process.

In the early days of thermomechanical pulping, it was believed that refining at elevated temperatures was fully responsible for the improved pulp qualities.[109] It is known today that thermomechanical pulping is a process involving two separate steps. The first step is preheating of the chips with the purpose of softening the lignin. This makes the chips less brittle so that they are reduced in size at the breaker bar section of the refiner, thereby allowing easier fiber separation without fiber breakage or splintering, creating more long fibers, less debris,

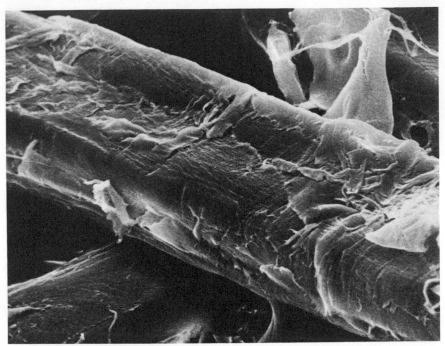

Figure 4-16. Scanning electron micrograph showing separation between primary wall and the middle lamella of fiber produced from eastern black spruce chips by refining at 115°C. (Courtesy Canadian Pulp and Paper Research Institute and Svensk Papperstidning.)

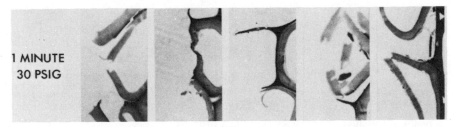

Figure 4-17. Ultraviolet electron micrograph of cross section of thermomechanical pulp refined below the glass transition temperature, showing rupture in the secondary fiber wall. (Courtesy Canadian Pulp and Paper Research Institute.)

and fewer shives. If this chip reduction is carried out without presteaming as it is in ordinary refiner mechanical pulping, the chips are brittle and tend to break under the impact, thus resulting in a large percentage of fines and debris in the pulp. Presteaming alone will not produce high-quality mechanical pulp. It is not sufficient merely to separate fiber bundles, but it is also necessary to fracture the

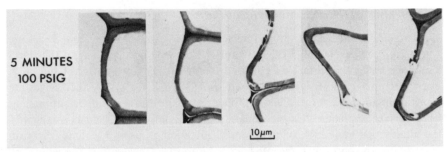

Figure 4-18. Ultraviolet electron micrograph of cross section of thermomechanical pulp refined above the glass transition point of the lignin. Rupture took place in middle lamella and fibers are covered with dark lignin. (Courtesy Canadian Pulp and Paper Research Institute.)

secondary wall of the individual fibers. This second step requires mechanical energy applied by a refiner at high consistencies. In this phase evidence suggests that the fibers are stress cycled by the refiner a large number of times at values less than their breaking point, either as a result of bar-to-fiber contact or of fiber-to fiber contact. This compression and decompression generates heat leading to the production of steam. Eventually, due to fatigue failure, the fibers are released from the wood matrix and, at the same time, ribbon-like wall segments are produced from the S_1 layer. The quantity of this ribbon-like wall substance is critical to the pulp quality. It affects the scattering coefficient of the pulp. The quantity of this substance formed in the pulp depends on the refining conditions and not on presteaming.

It should be noted that the frequency of oscilation during testing in the torsional pendulum is of the order of 10Hz, while the bar crossing frequency in a disk refiner is in the range of 10 to 20 kilohertz (kHz). Furthermore, the strain levels are very low, almost entirely in the elastic region. Tests by Hoglund[110] using frequencies of the order of 10 kHz, showed that the maximum softening temperature shifted from 90°C to about 120°C. If this is representative of what happens inside a refiner, there is an optimum temperature that depends on the amount of the stress cycling in the refiner. This could have a significant relationship to pulp quality in terms of fiber length and bonding potential.

Wood chips that are heated to elevated temperatures will remain soft when the temperature is decreased. The softening behavior can be observed as a function of temperature by measuring the torsional pendulum. The torsional modulus decreases when the temperature is increased to 135° C and does not return to its original value when the temperature of the wood is lowered. For example, chips heated to 135° C and then cooled to 110° C will be almost as soft as at the higher temperature. Also, wood at 50% moisture, that has been heated to 105°C, appears to be sufficiently soft for refining. This all suggests that fiber separation can be carried out at temperatures lower than the preheating temperature. It raises the question whether refining equipment has to be close coupled to the steaming vessel, and whether refiners need to be designed for pressure operation.

Properties of Thermomechanical Pulp

To characterize mechanical pulp, Forgach[111] has developed the L and S factors, which are used extensively. The L factor is fiber-length dependent, while the S factor expresses the surface area related to the bonding properties of the pulp. For thermomechanical pulps, however, the L and S factors by themselves are insufficient to characterize the pulp. Fiber flexibility, fiber size distribution, and fiber surface characteristics are equally or more important, and consequently the microscope evaluation of thermomechanical pulps, using a scanning electron microscope, is becoming accepted as an important evaluation procedure.[112] When one compares thermomechanical pulp with stone groundwood microscopically, it becomes evident that thermomechanical pulp contains a significantly large portion of long fibers and fewer fines. It is the presteaming of the chips prior to refining that results in a greater portion of the fibers remaining intact, causes fewer fiber bundles to be broken into short segments, and results in an overall reduction in the shive and debris content.

Separating the fibers from a typical thermomechanical pulp produced for newsprint by using a Bauer McNett classifier and by evaluating each fraction individually reveals interesting information. The plus 30-mesh fraction contains mostly whole fibers with smooth surfaces as shown in Figure 4–19. This fraction represents 30 to 40% of the total pulp by weight and includes approximately 5% of long, stiff fibers retained on a 16-mesh screen. These long, stiff fibers, which seem to be thick-walled summerwood tracheids, do not form well with other fibers, and, as a result, they are unable to create fiber-to-fiber bonds. The drainage of this fraction is free and rapid, about 700 Canadian Standard Freeness. Handsheets prepared from this fraction are low in both tear and bursting strength. In contrast, the long fibers in stone groundwood have a more uneven surface that aids in fiber bonding, and as a result the long fiber fraction in stone groundwood has better strength properties. Examining under the scanning electron microscope the 30- to 50-mesh fraction of thermomechanical pulp, which

40 μ

Figure 4–19. Plus 30-mesh fraction of thermomechanical pulp.

Figure 4-20. The 30- to 50-mesh fraction of thermomechanical pulp.

Figure 4-21. The 50- to 100-mesh fraction of thermomechanical pulp.

represents 17 to 25% of the total weight, reveals that this fraction is similar to the plus 30-mesh fraction in that it contains shorter but still whole fibers with relatively smooth surfaces, as shown in Figure 4-20. However, in addition, there are some fibers that have been stripped of the fiber wall exposing the S_2 layer. This fraction has low overall bonding properties, but it does contribute to the tear strength of paper. The plus 50-mesh fraction shows especially poor Z-directional tensile strength, Scott Bond, and is responsible for the surface roughness of paper. The 50- to 100-mesh fraction, shown in Figure 4-21, constitutes about 10 to 15% of the pulp weight and contains mostly ribbon-like segments of the fiber wall and the S_1 layer and some short fibers. The ribbons are long and very uniform in length. The higher the proportion of ribbons to short fibers, the better the pulp quality. The 100- to 200-mesh fraction, shown in Figure 4-22, contains practically all ribbons of the fiber wall and S_1 layer. While stone groundwood also contains some ribbons in this fiber fraction, their presence is less dominant than in thermomechanical pulps.

Figure 4–22. The 100- to 200-mesh fraction of thermomechanical pulp.

Figure 4–23. Minus 200-mesh fraction of thermomechanical pulp.

The overall bonding strength of thermomechanical pulp is represented in the 50- to 200-mesh fraction where the ribbon-like fraction is concentrated. It is these flexible, ribbon-shaped lamella that conform well when a sheet is formed resulting in high fiber-to-fiber contact, improved internal bonding, high tensile strength, and smoothness. The specific-surface area of the ribbons is large; this results in a high scattering coefficient. The minus 200-mesh fraction represents the fines. In thermomechanical pulp, fines are seldom more than 20 to 30% of the total pulp by weight. The fines consist mostly of flakes from the middle lamella, thin ribbons from the fiber wall, and short ray cells, as shown in Figure 4–23. The short ray cells have poor bonding properties and could be the main cause of linting in offset printing. The fines in thermomechanical pulp are much more important than in stone groundwood. They greatly affect the fiber-to-fiber bonding properties of the sheet, influence the optical properties, and improve the scattering coefficient. However, the fines contain some lignin and resin and could adversely affect light absorption and brightness. The shive content of thermomechanical pulps is usually very low. Depending on operating conditions, it will be below 0.5% before screening or cleaning. Screening and cleaning can easily reduce it to the 0.02% level. Tests carried out on an optical shive analyzer comparing thermomechanical pulp with stone groundwood shows thermomechanical is much lower in shives that are wider than 125μ compared to stone groundwood. Table 4-4 shows a comparison between thermomechanical pulp and stone groundwood, for (1) shive content over 0.15 mm as measured on a flat screen and (2) number of shives in a one-gram sample in various length ranges. There are far fewer shives in thermomechanical pulp when measured by the latter method.

The Z-directional tensile strength, which is an indicator of the internal bonding properties of thermomechanical pulp, is mostly affected by the 50- to 200-mesh fraction of the pulp. Removal of this fraction from the total pulp will drop the Z-directional tensile strength to about 15% of the original value indicating the importance of this fraction in bonding properties. Fiber flexibility and the ratio of long fibers to ribbon-like lamella affect the density of the pulp as measured by handsheets. Higher-density pulps usually show better bonding properties, primarily due to their high content of ribbon-like lamella. Density is an important and practical test for prediction of quality. As thermomechanical pulps find their way into finer paper grades traditionally produced from chemical fibers, wet-web strength has become a property of increasing concern. Thermomechanical pulp, if produced properly, will contain a high proportion of the long-fiber fraction as well as a high proportion of the fine fraction, which contains mostly lamellae from the fiber walls. Such pulp will have a considerably higher wet-web strength than stone groundwood. Because the fine fraction plays a very important role in the wet-web strength, reliable tests on handsheets cannot be carried out without white-water recirculation when forming the handsheets. At equal density, thermomechanical pulp has about 50% higher wet-web strength than stone groundwood, but it has a lower rupture stretch. Thermomechanical pulp

TABLE 4-4 COMPARISON OF NUMBER OF SHIVES IN STONE GROUNDWOOD AND THERMOMECHANICAL PULP

	Stone Groundwood	Thermomechanical Pulp
Somnerville flat screen (fraction over 0.15 mm)	0.3%	0.3%
STFI shive analyzer (number of shives in length range in one gram of pulp)		
0.5–1.5	46,000	21,000
1.5–2.5	6,100	800
2.5–3.5	1,200	130
3.5–4.5	300	20
Over 4.5	90	3

has a higher modulus of elasticity, which allows processing at a much lower stretch for a given stress without rupture.

Brightness of thermomechanical pulp is usually lower than stone groundwood for the same wood species. This is partially caused by exposure of the wood chips to air during storage and partially by the presence of bark, often found in the chip supply. It is also affected adversely by the presteaming step prior to refining. The lower the presteaming temperature, the less the brightness loss. Brightness of mechanical pulps is an important property in papermaking because these pulps have a significantly larger surface area than chemical pulps and will, therefore, have a more pronounced affect on the sheet brightness. In a furnish where mechanical pulp exceeds 40%, the sheet brightness will be determined by the mechanical pulp; chemical-pulp brightness will have a negligible affect. Thermomechanical pulps, because of their high lignin content, will turn yellow when exposed to ultraviolet light (e.g., sunshine). This property makes the use of thermomechanical pulps unsuitable for some critical fine-paper grades, where resistance to ageing is an important requirement. Thermomechanical pulps have the same chromophore groups as refiner mechanical pulp and stone groundwood. For this reason they can be bleached with the same techniques. For high-brightness levels, peroxide is recommended. Bleaching processes based on delignification, like hypochlorite bleaching, are definitely not suitable since they result in drastic yield loss.

The light-scattering coefficient of paper, which is related to the unbonded surface area, is improved by increasing the density of the pulp, reducing the drainage, and mechanical disintegration of the whole fiber. At the same time, the bonding strength is increased. Thus, the scattering coefficient is not only an important measure of the optical properties in printing-grade papers, but also is an indirect test of the bonding properties and surface strength. Thermomechanical pulp usually has a lower scattering coefficient than stone groundwood, but it is interesting to note that increasing the density of thermomechanical pulp to the same level as groundwood will produce the same scattering coefficient.

The dry strength of paper made from thermomechanical pulp is higher than that from other mechanical pulps. This is especially true for tear strength. The tear index is dependent on the L- factor and the long-fiber content, especially in the 30- to 50-mesh fraction of the pulp, but it should be realized that tear values are influenced by the bonding properties of the pulp and, as a result, the 50- to 200-mesh fiber fraction is important. A comparison of stone groundwood and thermomechanical pulp properties are shown in Table 4-5. Thermomechanical pulps produced from different wood species are compared with stone groundwood produced from spruce in Table 4-6.

In close-coupled thermomechanical pulping the fibers are separated at elevated temperature, while the lignin is in a plasticized state. As a result, a high percentage of the fibers are long and intact without fibrillation. These long fibers do not compress well and show poor conformity to the surface of the sheet. This results in low sheet-surface strength and a tendency for the sheet surface to break up, causing linting in the press room. Evaluation of these long unde-

TABLE 4-5 COMPARISON OF NEWSPRINT GRADE STONE GROUND-WOOD AND THERMOMECHANICAL PULP FROM SOUTHERN PINE

	Stone Groundwood	Thermomechanical Pulp
Refining energy kWh/ton		
Grinders + rejects refining	1750	–
Refiners + rejects refining	–	2500
Pulp characteristics		
Freeness Canadian Standard (ml)	60	100
Shives, Somnerville 0.15 mm (%)	0.10	0.05
Bauer McNett fiber classification % retained on:		
+ 14 mesh	2	11
+ 28 mesh	6	30
+ 48 mesh	15	16
+100 mesh	16	9
+200 mesh	19	7
–200 mesh	42	27
Burst index (kPa $\cdot$ m^2/g)	0.8	1.8
Tear index (mN $\cdot$ m^2/g)	3.1	7.3
Breaking length (km)	2.1	3.3
Bulk (cm^3/g)	2.6	3.0
Stretch (%)	1.2	1.8
Opacity (%)	97.0	94.0
Brightness (%)	58.0	56.0
Fiber-length index	0.072	0.088

TABLE 4-6 TYPICAL CHARACTERISTICS OF THERMOMECHANICAL PULPS FROM DIFFERENT SPECIES COMPARED WITH SPRUCE GROUNDWOOD

	Species			
	Spruce Stone GwD	Spruce TMP	Jack Pine TMP	Hemlock TMP
Energy (kWh/ton)	1460	1900	1950	1960
Freeness Canadian Standard (ml)	88	120	115	80
Burst index (kPa $\cdot$ m^2/g)	1.3	2.8	1.8	2.0
Tear index (mN $\cdot$ m^2/g)	4.5	9.7	8.2	8.3
Breaking length (Km)	2.7	4.8	3.5	3.7
Stretch (%)	1.4	2.1	1.8	1.9
Bulk (cm^3/g)	2.6	2.8	2.9	2.6
Shives, Somnerville 0.15 mm (%)	–	0.07	0.05	0.02
Brightness (%)	95.0	92.0	92.5	97.0

veloped fibers indicate that they are mostly summerwood fibers. Compared to springwood fibers, summerwood fibers normally have thicker and tougher fiber walls and require more mechanical energy to fibrillate. Sufficient energy applied in mechanical pulping to develop all summerwood fibers may result in overrefining of the springwood portion, which would increase the fine content of the pulp. Selective removal of the unfibrillated summerwood fibers by screening followed by additional high consistency refining will develop these fibers for the desired higher strength. Reblending with the original pulp will greatly increase the overall strength properties and will also reduce linting.

The decision whether to install screening and cleaning equipment in a thermomechanical pulping system is dependent on the type of paper made and by the amount of thermomechanical pulp used in the furnish. If used as a reinforcement in a predominately stone-groundwood furnish, the presence of long undeveloped fibers will have little adverse affect on sheet properties. The same is true for board grades when high-drainage pulps are the objective. In the case of newsprint, where thermomechanical pulp represents close to 100% of the furnish, the quality of the pulp is more critical and screening and cleaning are well justified. In some mills centrifugal cleaners have been removed from the pulp mill and cleaning is done by hydrocyclones located near the paper machine. Others have installed single screens with large perforations to function as barrier screens for removal of large fiber bundles present as a result of upset or startup. In the latter case, the reject rates maintained are very low, in the range of 3 to 5%, and the screening is accomplished by the perforation of the screen plates, not by the pulp mat that forms on the screen.

Operating Variables in Thermomechanical Pulping

Chip Capacity. In both refiner mechanical pulping and thermomechanical pulping, uniform chip size is critical to assure a uniform feed rate to the refiner. Metering devices are generally based on volume control, whereas uniformity is dependent on gravimetric control. As a result, changes in chip density will create uneven gravimetric feed even when volumetric feeding is maintained properly. The problem becomes acute if wood species with great differences in density are blended in an unknown ratio or when chips, sawdust, and shavings are blended without ratio control.

In thermomechanical pulping, uniform size of chips is particularly important because it affects the time required for heat transfer to the center of the chips. The chips must be thoroughly softened before entering the refiner and achieving this with all chips is a basic requirement for producing a high-quality thermomechanical pulp. On the other hand, brightness deteriorates with prolonged steaming time or high steaming temperature and for this reason steaming should be kept to a minimum. Tests by Frazier[113] indicate that thermoconductivity depends on the species and on chip thickness. Since the largest chips heat more slowly, they determine the necessary softening time and not the average chip size. Table 4-7 shows the time needed to heat chips at two different steaming

TABLE 4-7 TIMES REQUIRED TO HEAT CHIP CENTERS TO THE SOFTENING TEMPERATURE

Steam Temperature °C	Inches			Time to Reach 117°C in Minutes	
	x =	y =	z =	Western Hemlock	Black Spruce
132	1/8	1	1	0.45	0.62
132	1/4	1	1	0.89	1.6
132	1/2	1	1	1.50	2.7
132	1/2	1 1/2	1 1/2	2.50	4.6
121	1/8	1	1	1.40	2.6
121	1/4	1	1	2.30	4.2
121	1/2	1	1	3.90	7.1
121	1/2	1 1/2	1 1/2	6.90	12.6

temperatures and clearly demonstrates the need for uniform chip size to assure a minimum of stiff, undeveloped fibers and debris in the pulp at the maximum brightness level. By determining the shear modulus in a torsional pendulum, it has been shown that once the chips reach the desired temperature, softening occurs immediately and, contrary to previous belief, no reaction time is required.

Effect of Air in Steaming. Investigations by Frazier[113] indicate that when chips are fed to a steaming vessel they carry a large amount of air into the vessel. Unless this air is removed by steam purging or by compression it will prevent access of the steam to the chips, thereby greatly increasing the time needed for heat transfer. In thermomechanical pulping it is common to control the steaming vessel at constant pressure. The assumption is that the steam is saturated and, hence, the pressure represents a known temperature, but the presence of air changes this relationship, as shown in Figure 4-24. If air is present in the steaming vessel, the temperature will be significantly lower than in the case of saturated steam at a given pressure. Air may also delay the diffusion of steam to the chip surface, causing delay in chip heating. Ordinarily two minutes of steaming time is sufficient, provided the temperatures are on the order of 120°C and no air is present in the steaming vessel. Although steaming time has relatively little affect on physical strength of thermomechanical pulp, it significantly affects debris content.

Effect of Consistency in Thermomechanical Pulping. Consistency is very important.[114] Tear strength drops off both at low and high consistencies. At 15% consistency, the tearing strength value will be equal to that obtained in refiner mechanical atmospheric pulping. At very high consistencies, the generated heat will dehydrate the fibers and make them weak and brittle. Maximum tear strength is achieved at 20 to 25% consistency at the point between the

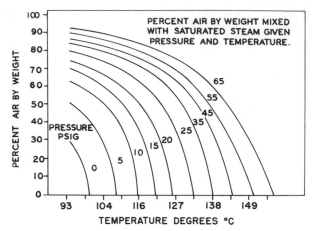

Figure 4-24. Relation of air content to temperature and pressure in stea vessel. (Courtesy Canadian Pulp and Paper Assoc.)

refiner plates. There is an optimum consistency around 25% where maximum burst strength is obtained. Scattering coefficient increases with sistency, but light absorption decreases. Both effects contribute to an inc in brightness with increased consistency. However, thermomechanical nearly always have a 2.0 to 2.5 lower brightness than stone groundwood d the presteaming used. The effect of consistency on tearing strength, burst, brightness are shown in Figure 4-25.

Discharge consistencies, which are often used as a control test by opera will be considerably higher than the consistency between refiner plates du flashing of some of the moisture when pressure is relieved at the blow v The relation between the two is shown in Figure 4-26, based on typical op ing conditions of 1160 kWh/ton (65 hpd/ton).

Effect of Temperature and Steaming Pressure in Thermomechanical Pulp
The effect of temperature on the physical properties of the pulp have alre been discussed in the section on properties of thermomechanical pulp.[11] general, as steaming pressure and temperature are increased, the tear strer first increases and then decreases with a pronounced maximum occurrin about 275 kPa (40 psi), which is close to the glass transition point of the nin.[114] Bursting strength increases similarly, whereas bulk and scattering efficient decrease linearly as the steam pressure is increased. Total refir energy increases linearly with an increase in steaming temperature for a gi freeness value. The curves, which are shown in Figure 4-27, indicate that strength properties of thermomechanical pulp can be altered by changing steaming pressure without change in the freeness.

High-energy refining at high consistencies, as practiced in thermomechan pulping, produces large volumes of steam between the refiner plates. Due to

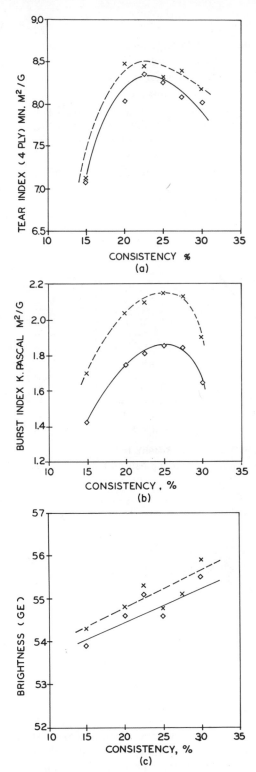

Figure 4-25. Effect of consistency on (a) tearing strength, (b) burst, and (c) brightness of thermo-mechanical pulps. (X = 150 Canadian Standard Freeness, ◊ = 200 Canadian Standard Freeness.)

215

self-pressurization of the refiner plates, not all of the steam will pass with
pulp and a large percentage will move toward the inlet of the refiner whe
can interfere with chip feed.[116] While the volume of steam generated is
than the steam generated in straight mechanical atmospheric refining, it is
sufficient to interfere with the feed to the refiner (see Figure 4-28). It is c
mon to relieve steam-generation problems by adding dilution water to the
of the refiner. This will collapse some of the steam and will reduce the seve
of steam buildup. However, pulping consistency will also be reduced.
adversely affects pulp quality by reducing the coefficient of friction between

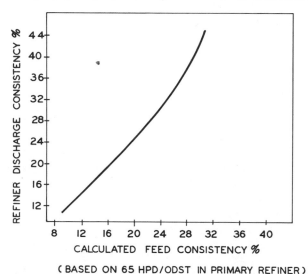

(BASED ON 65 HPD/ODST IN PRIMARY REFINER)

Figure 4-26. Showing change in discharge consistency caused by steam flashir
in thermomechanical pulping.

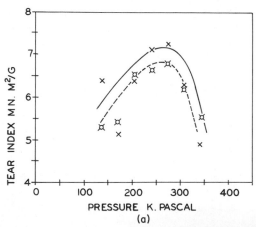

Figure 4-27. Effect of steaming pressure on (a) tearing strength, (b) bursting
strength, (c) scattering coefficient, and (d) refining energy required in thermomechanical pulping. (X = 150 Canadian Standard freeness, ⊠ = 200 Canadian
Standard freeness.)

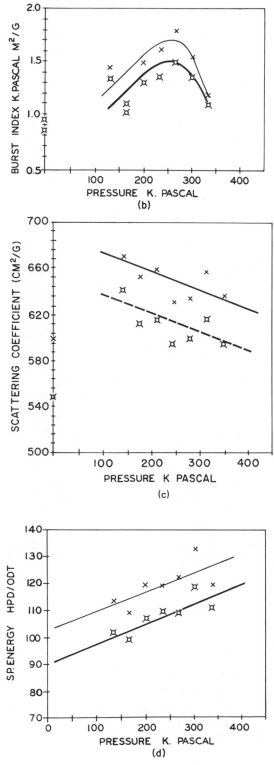

Figure 4-27. Continued

217

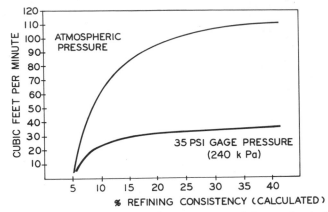

Figure 4-28. Comparison of steam volumes generated in mechanical pulping compared to that generated in thermomechanical pulping at different consistencies.

fibers. Another approach is to release the generated steam through the blow valve with the pulp. This creates a large pressure drop across the refiner plates, results in reduced retention time, and reduces pulp-mat thickness. This approach also reduces pulp quality. Highest-quality pulp is produced when the blow valve opening is kept to a minimum and the amount of steam released is not more than needed to convey the pulp to the receiving cyclone. In thermomechanical pulping systems in which at least 50% of the refining energy is applied in a pressurized first stage, more steam is generated than required for conveying the pulp from the casing and more than is needed to heat the cold chips. To utilize the generated steam for preheating the chips, the steam must be separated from chips and fibers and be delivered to the steaming tube. It is best to relieve the excess steam from the top of the steaming vessel rather than through the blow valve of the refiner. A proper control system will allow for the addition of fresh steam to the steaming tube until equilibrium is reached, although under balanced conditions no fresh steam makeup is needed. The pressure in the steaming tube can be controlled by exhausting excess steam from the tube. Venting exhaust steam carries away the noncondensables, thereby assuring a uniform temperature in the steaming vessel.

In many commercial installations, the refiner is operated with a great pressure drop across the plates. This pressure drop assists feeding the refiner and reduces steam-related problems, but it has an adverse affect on pulp quality.

Effect of Energy Applied in Thermomechanical Pulping. The most important variable affecting pulp quality in mechanical pulping is energy applied in the refiners. Increasing the refining energy will increase the strength of the pulp with the exception of tear strength, which is first increased and then decreased as the energy is increased. Thermomechanical pulping requires more mechanical energy than refiner mechanical atmospheric pulping and considerably more than stone groundwood to produce pulp of the same freeness value. The amount of energy required is increased as the steaming temperature is increased and further increase can be expected by mild chemical pretreatment. In thermomechanical pulping the extra energy applied is not wasted, but returned in the form of a higher degree of fiber development and a higher strength. If the burst factor or the breaking length are plotted against the total refining energy applied, it can be shown that thermomechanical pulp requires no more energy than any other mechanical pulping method. Thermomechanical pulp is capable of absorbing more energy, and this can be done without excessive loss in drainage.

The refining energy can be applied in a single refiner, but it is more common to distribute the power between at least two refining stages. In multistage refining, only the primary refining stage is carried out under pressure; the second stage is atmospheric.[117] Installing two pressurized refiners in series will have an adverse affect on fiber development by producing too many long, unfibrillated fibers. There are no fixed rules on how to distribute the energy between stages, but it is accepted practice to apply more than half the total energy in the primary refiner. It is important that the pulp leaving the first stage be separated mostly into individual fibers. The desired surface area can then be developed by peeling off the fiber walls in the second stage. Loading of the second-stage refiner is always more difficult because separated fibers will support less load than chips. To reduce the chance for plate clashing in the secondary refiner, a thicker fiber mat should be carried between the refiner plates. This can be accomplished by increasing the throughput either by (1) following two primary refiners by a single secondary unit, (2) recirculating some of the pulp from the secondary discharge to the feed, or (3) blending screen rejects to the primary pulp before feeding to the secondary refiner. Single-stage thermomechanical pulping is of interest to simplify the operation and to eliminate the problems related to handling, feeding, and metering the fluffy primary pulp. The high cost caused by the additional steaming equipment required can be overcome by connecting two refiners to a single steaming tube.

Thermomechanical Pulping Systems

There are four major thermomechanical pulping systems: Bauer, Defibrator, Sprout Waldron, and Sund.

In the Bauer system,[118] Figure 4–29, chips are metered, fed through a rotary valve into a horizontal digester, discharged into a leveling conveyor, and fed by a twin-screw feeder into the eye of a double-revolving-disk refiner.[119] The refiner is close coupled to the steaming vessel, and steaming pressure must be higher than the discharge pressure in the refiner casing to assure positive feed. Pulp is discharged through a blow valve to a cyclone, where steam is separated from the fibers. The pulp is quenched with dilution water at this stage. The primary pulp is fed to a second-stage atmospheric refiner for further fiber development.

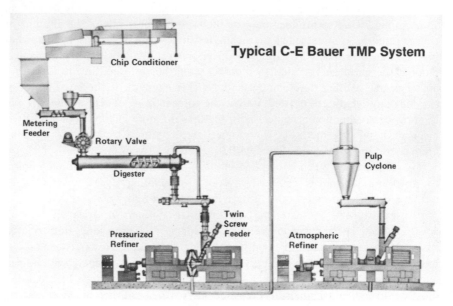

Figure 4-29. Bauer thermomechanical pulping system. (Courtesy C-E Bauer division of the Process Equipment Groups of Combustion Engineering, Inc.)

The Defibrator system, Figure 4–30, is operated on the close-coupled principle.[120] Chips are compressed by a tapered screw into a continuous plug, which is then discharged into a vertical steaming vessel. An air-loaded cylinder prevents blowback in case chip feed is interrupted. The chip level in the steamer is controlled to give the desired steaming time, while the temperature is maintained by constant steam pressure. The steamed chips are discharged at the bottom by horizontal variable-speed-metering twin screws and fed by gravity into the feed screw of the steam-pressurized primary refiner. The steamer and the refiner operate under the same pressure. Some of the steam generated during refining is flowed back to the steaming vessel to heat the chips and the remaining portion is used to convey the refined pulp to the cyclone. After the cyclone, the pulp is diluted and fed to the second-stage refiner, which operates at atmospheric pressure.

The Sprout Waldron system is shown in Figure 4–31. It consists of a vibrating bin from which chips are taken by a plug-screw feeder that compresses the chips and provides a seal on the feed end of the vertical steaming vessel. If the load on the screw feeder should drop, indicating the loss of the plug, a valve automatically seals the opening to prevent steam blowback. The level in the vertical steaming vessel is maintained by a gamma-ray, nuclear-level sensor that sets the time for steaming. The preheated chips are discharged by two individually driven metering screws into the load-sensing conveyors of the twin refiner. The chips are fed to the eye of the hydraulically loaded refiner by a ribbon feeder

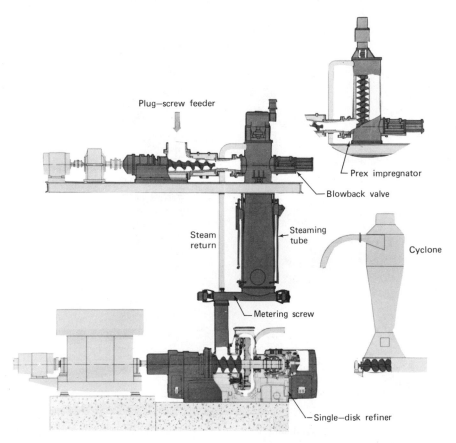

Plug—screw feeder

Prex impregnator

Blowback valve

Steam return

Steaming tube

Cyclone

Metering screw

Single—disk refiner

Figure 4-30. Defibrator thermomechanical pulping system. (Courtesy American Defibrator, Inc.)

mounted on the main shaft. The pulp is discharged through a blow line into a cyclone from where it is fed by a load-splitting conveyor to both sides of the twin-design, secondary (atmospheric) refiner. Some of the steam generated in the primary refiner is used for carrying the pulp from the refiner to the cyclone, but the majority of the steam is returned to the steaming vessel so that no auxiliary steam is needed to maintain the desired steaming temperatures. Excess steam is vented to the atmosphere from the steaming vessel carrying noncondensable gases with it, thereby assuring an accurate temperature in the steaming tube.

The Sund system,[121] Figure 4-32, differs in principle from the other systems in that it uses an atmospheric refiner in the primary refining stage.[122] A metering bin feeds the chips through a rotary valve into a horizontal steaming tube. If the steaming pressure is higher than the feed pressure to the refiner, a rotary valve is used on the discharge end of the tube. The preheated chips are fed

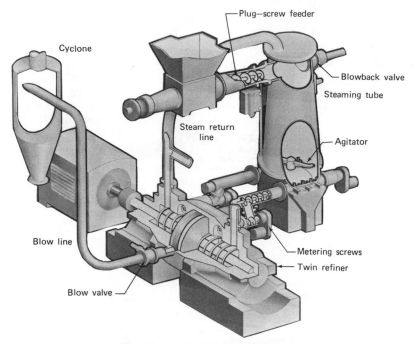

Figure 4-31. Sprout Waldron thermomechanical pulping system. (Courtesy Sprout Waldron Div. Koppers Company Inc.)

through a steam expansion chamber and a twin-screw feeder into the eye of a double-revolving disk refiner. The pressure on the feed end assures uniform feed rate and eliminates blowback so that refining can be carried out at high consistencies. The double-revolving disk refiner is especially built to withstand high thrust created not only by the torque, but also by the steam pressure generated during refining. The steam generated during refining is partially returned to the steaming vessel to preheat the chips in cases where the steaming vessel temperature does not exceed the pressure at the refiner.

A modified system[123] of thermomechanical pulping is based on the observation that chips heated to temperatures below the glass transition point can be cooled and the lignin will remain soft for a long period.[124] This means that refining does not have to be carried out at the same temperatures as presteaming and does not have to be done in a steam pressurized refiner, but can be carried out in an atmospheric discharge refiner. Also chips that are preheated and then cooled to 100°C can absorb more mechanical energy than at 130 to 140°C; this will result in better fiber development. This observation opens the possibility of preheating the chips in a separate steaming vessel and refining in atmospheric refiners, thereby providing the advantage of easier operation and lower capital investment compared to close-coupled, fully pressurized thermomechanical

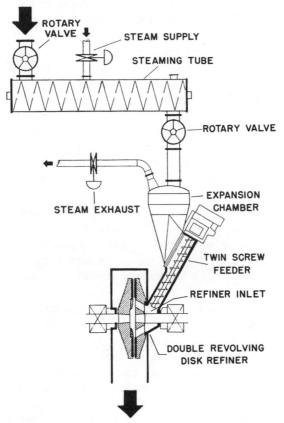

Figure 4-32. Sund thermomechanical pulping system. (Courtesy Svenska Cellulosa Teknik A. B).

pulping. In one case, a single steaming vessel is followed by five atmospheric refiners. Lower production loss and more uniform pulp quality are claimed. One disadvantage is that the steam generated in the refiner cannot be recovered and used for preheating the chips.

Equipment Used in Thermomechanical Pulping

The two principal items of equipment are the presteamer and the refiner. The steamer may be fed with a plug-screw feeder or by a rotary valve. The plug-screw feeder compresses the chips and by doing so expels a good portion of the air trapped between the chips. As previously explained, air in a steaming vessel creates problems. The second advantage of a plug-screw feeder is that it will remove a significant portion of the color-forming material in the chips, including resins and fatty acids. Chips are normally pressed to 60% consistency oven-dry

(OD). The effluent is highly concentrated and if it can be disposed of without mixing with the total mill effluent, it will greatly reduce the overall pollution from the mill. A rotary valve has the advantage of being insensitive to bulk density of the raw material and it will allow the feeding of sawdust and shavings with the chips. Metering, however, must be provided prior to the rotary valve since the pockets cannot be filled without severe wear on the blades.

Steaming of the chips is accomplished in a vertical or horizontal steaming vessel. Horizontal steaming tubes are equipped with a slow-rotating screw to convey the chips from the inlet to the discharge. This creates a fluctuating discharge, and a high-speed leveling conveyor is used before feeding to the refiners. Vertical steaming vessels are equipped with variable-speed-metering discharge screws at the bottom. A slow-moving agitator or scraper keeps the chips above the metering screws in suspension, thereby allowing a uniform metering to the refiner. Steam generated in the refiner is usually returned to the steaming vessel in a close-coupled thermomechanical pulping system and is utilized to heat the raw material. Excess steam is vented at the top removing noncondensables, like turpentines and resins. (See the section on semichemical pulping for a description of digestors and for the general layout of a semichemical pulping plant).

To protect the refiners from damage, foreign material, such as stone, sand, and tramp metal, must be removed from the chips prior to the steaming step. Air cleaners will remove heavy contaminants, but washing the chips is the best method to remove sand. Dewatering the chips after washing is important. Dewatering can be done by a rotating drum equipped with perforated plates or mesh screens, or it can be done by vibrating screens, which have the advantage that they will retain sawdust and small wood particles. Screw feeders, due to their compressing action, will remove excess moisture from the chips.

The disk refiners used in pulping were developed from attrition mills used by the grain industry. They are large, precision built, and have sophisticated, rugged designs to assure plate parallelism and accurate control of plate clearances. There are three types: double disk, single disk, and twin disk, shown in Figure 4–33.[125] The double-disk refiner has two rotating disks mounted on cantilevered shafts, each driven by integrally mounted motors. The rotation, which is in opposite directions, provides large differential speeds normally 2400 rpm in case of 60-Hz current and 3000 rpm with 50-Hz current. The raw material is force-fed by a twin-screw feeder through the spokes of one of the rotating disks. Both shafts are equipped with thrust bearings; loading is accomplished by hydraulic means. Plate clearance is maintained accurately by LVTD (lineal velocity transducers) controls. Single-rotating disk refiners are available with cantilevered or with true shafts that are supported in the middle. Only one of the disks is rotated at a speed of 1800 rpm in case of 60-Hz current and 1500 rpm at 50-Hz current. Feeding the refiner is easily accomplished through the unrestricted opening on the stationary disk. Plate-clearance adjustment is achieved by a hydraulic system or by a gear motor. The twin-disk refiner combines two single-rotating-disk refiners into one unit. It is a major departure in design because it utilizes both

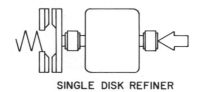

SINGLE DISK REFINER

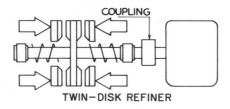

TWIN–DISK REFINER

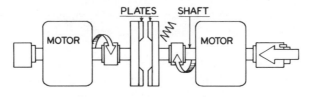

COUNTER ROTATING DISK REFINER

Figure 4-33. The three major refiner configurations used in mechanical pulping.

sides of the rotating disk, thus producing two refining chambers in one refiner. Rotating-disk deflection is nonexistent, even when operating at high speed. Steam generated within the refiner is allowed to exit from both the inlet as well as the discharge end of the unit. The dual set of nonrotating plates are hydraulically adjusted and loaded against the rotating assembly, which is locked into position. This type of loading permits operation without the use of a large thrust bearing. A unique suspension system permits the twin refiner to expand lengthwise as well as circumferentially. The design allows the application of very large horsepowers and 14,000-hp refiners with 127-cm (50-in.) disk diameters are being used.

Disk refiners must be built to withstand a large thrust. In straight (atmospheric) refiner pulping, the thrust load in the refiner is created principally by torque, resulting from the coefficient of friction between the fibers and to a lesser extent from steam generated between the plates. In thermomechanical pulping, where the pressure differential across the plates can be significantly higher than in refiner mechanical pulping, large-thrust loads are developed in addition to the refining loads created by the torque. If not accounted for, this can create problems in bearing life. Periphery speed of the disk is very important, both in atmospheric refiner pulping and in thermomechanical pulping.

Higher disk speed develops a better quality pulp with low-debris content. Higher periphery speeds can be achieved by increasing the rpm or by increasing the disk diameter. Single-disk refiners generally operate at 1800 rpm (60-Hz current) or at 1500 rpm (50-Hz current). Double-revolving disk refiners usually operate at 1200 rpm (60 Hz) and 1500 rpm (50 Hz). Increasing the disk diameter increases the disk area, thereby allowing the use of larger motor loads to operate the refiner. Refiner-disk diameter of 137 cm (54 in.) is common 157 cm (62 in.) is being contemplated. Disks with larger diameter driven at 1500 to 1800 rpm need special metallurgy since the centrifugal forces can deform metals, such as 316 stainless steel. Large disk area is needed when large loads are applied in order to support the load and prevent breakthrough of the chip or pulp mat. Large disk diameter and high rpm together result in more refiner bars crossing each other, thereby increasing the frequency of compression and decompression and resulting in the creation of more of the desired ribbon-like fibrils in the pulp. A large disk diameter develops greater centrifugal force and this can result in difficulties in maintaining a fiber mat between the refiner plates. The problem is less severe in double-revolving-disk refiners. On single-disk refiners plate-design damming effects will compensate for the centrifugal force and allow the pulp to stay between the plates.

Refiner-plate pattern has an important effect on the quality of thermomechanical pulp. The main objective is to find a plate design that will allow the utilization of refining energy without causing the refining surfaces to break through the pulp mat. It must allow steam to escape with the pulp on the outer periphery as well as on the feed end without interfering with the feed itself. As the chips enter the refiner, they are first fractionated by the breaker bar section of the refiner plate. The length of the breaker bars must be sufficient to achieve size reduction before the chips enter the refining section. Dams, located on the outer periphery of the plate, function to restrict the flow of material and thereby maintain a suitable mat of fibers between the plates. The degree of damming depends on the type of refiner with less being required on a double-revolving disk than on a single rotating disk. It also differs for different disk speeds and diameters. Two different refiner-plate designs are shown in Figure 4–34.

The energy applied during refining develops heat due to friction between the fibers. A temperature profile across the refiner-plate surface measured by thermocouples is shown in Figure 4–35.[126] The curve begins with unheated chips entering the breaker bar zone and at this point reflects the temperature of the steam blowback from the refiner plates. The temperature increases rapidly as the chips enter the refining zone, because the fiber-to-fiber friction is increased.[127] From this point the temperature gradually decreases and then sharply drops at the discharge point. The heat developed is very important in refiner (atmospheric) mechanical pulping, but is of less importance in thermomechanical pulping, where softening of the lignin is accomplished by external preheating. Refining consistency greatly affects the temperature profile between the refiner plates.

The life of a refiner plate depends primarily on its metallurgy as well as on operating conditions. High refining consistency promotes longer life in addition

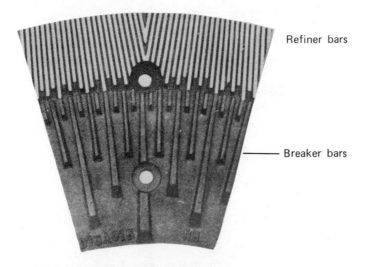

Refiner bars

Breaker bars

Primary refiner plate

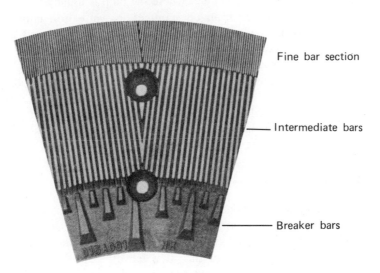

Fine bar section

Intermediate bars

Breaker bars

Secondary refiner plate

Figure 4-34. Typical refiner plate designs used in thermomechanical pulping, showing breaker bar sections and refining section. (Courtesy Sprout Waldron Div. Koppers Company Inc.)

to producing better pulp. In multistage refining, the life of the secondary refiner plates is shorter than that of the primary refiner plates due to the closer plate clearances used in the secondary refiner. The most common metallurgy for refiner plates is Ni-Hard, an alloy of white iron with approximately 4% nickel content. It has excellent abrasion resistance and a high coefficient of friction,

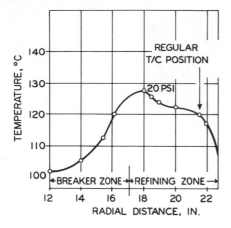

Figure 4-35. Radial temperature profile across the surface of a disk refiner.

which makes it well suited for mechanical pulping. It is brittle and it has a low tensile strength. Its corrosion resistance is poor, and this is its most severe shortcoming for thermomechanical pulping, where the pulp is usually in the 4 to 5 pH range.[128] Increasing the chromium content in white iron improves the corrosion resistance. High-alloy white iron has good corrosion and abrasion resistance as well as good resistance against steam erosion, which makes it well suited for high-consistency, high-energy refining. Heat-treated stainless steel of the 440 series is sometimes used. Although corrosion is nonexistent, the edges will round off due to wear after which pulp quality suffers. If heat treated to a lower hardness, the metal will wear at a greater rate, but the bar edges will remain sharp throughout the life of the plate and good pulp quality will be maintained.

Screening and Cleaning of Thermomechanical Pulp

Whereas stone groundwood contains mostly thick, long, and stiff fiber bundles,[129] and whereas refiner (atmospheric) mechanical pulp contains mostly short, stubby, fiber bundles and cubic debris, thermomechanical pulp contains only a little of each.[130] Because of this, it is felt by some that screening and cleaning can be eliminated for thermomechanical pulp. This would greatly reduce the capital and operating costs.

Thermomechanical pulp does contain a small portion of fiber bundles and unfibrillated, stiff, uncompressed fibers that add nothing to the bonding strength of paper. The long, stiff fibers adversely affect sheet smoothness; the shorter bundles can cause linting during printing. Figure 4-36 is a photograph taken by a scanning electron microscope of a newsprint sheet produced from 100% thermomechanical pulp. The surface contains a high percentage of poorly bonded fibers that show little conformity to the surface of the sheet. The production of these unfibrillated fibers is difficult to avoid. Their removal is bene-

Figure 4-36. Electron micrograph showing undamaged fibers on the surface of newsprint made from 100% thermomechanical pulp. These fibers are stiff and do not conform well to the sheet. (Courtesy Swedish Forest Products Laboratory.)

ficial and, furthermore, if they are refined separately at high consistencies and added back to the pulp, they can add overall physical strength to the pulp.

Screens for mechanical pulps were originally developed for stone groundwood, where the rejects to be removed are mostly shives and fiber bundles. Screens work very efficiently to remove these fractions, and a screening efficiency of 90 to 95% is common in stone groundwood mills. In refiner (atmospheric) mechanical pulping, the shive content is relatively low and the pulp contains high percentages of short chubby fiber bundles or cubical debris. Screens are relatively ineffective in removing these particles, and this makes the use of centrifugal cleaners mandatory for the cleaning of refiner mechanical pulp. Thermomechanical pulp contains little, if any, shives or fiber bundles and it is also low in debris. Therefore, screening and cleaning becomes a fractionation step rather than a cleaning step, with the rejects containing mostly unfibrillated stiff fibers. Screening of thermomechancial pulp is common, but in many cases it is used only as a barrier to protect the system from large wood particles. Screens with perforations of 2.5 mm are used in such applications fed by high consistencies, and the rejects are less than 3%. The screening efficiency of thermomechanical pulp is low, in the range of 50 to 60%. The efficiency can be improved by increasing the amount of rejects. A 50 to 60% efficiency at 20 to 25% rejects can be improved to 70 to 80% at a 40% rejects level, but this will remove close to 50% of the plus 28-mesh fiber fraction resulting in loss of tear strength.[131] While screening efficiency will increase with an increase in rejects, the efficiency will be different for pulps made at different refining energy levels. Pulps produced at a higher specific energy are screened more efficiently. The effect of screening and cleaning on thermomechanical pulps is shown in Table 4-8. Screening reduces the drainage of the pulp. The percentage change in

TABLE 4–8 THE EFFECT OF SCREENING AND CLEANING ON THERMOMECHANICAL PULP QUALITY[a]

	Eastern Spruce and Balsam		Western Hemlock and Fir	
	Before	After	Before	After
Bulk (cm^3/g)	2.92	2.76	2.66	2.57
Burst index (kPa · m^2/g)	1.2	1.7	1.6	1.9
Tear index (mN · m^2/g)	6.2	7.4	7.2	7.8
Breaking length (km)	3.1	3.7	3.3	3.6
Stretch (%)	2.0	2.1	1.6	1.9
Folding Endurance (M.I.T.)	4.2	9.0	7.0	10.2
Freeness Canadian Standard (ml)	117	93	103	75
Drainage (sec/60g/m^2)	13.8	18.0	17.3	26.1
Brightness (%)	59.1	61.5	52.3	53.0
Opacity (%)	96.4	97.3	97.9	98.3
Scattering coefficient (cm^2/g)	580	630	580	650
Shives Somnerville 0.15 mm (%)	0.16	0.08	0.10	0.002
Bauer McNett fiber classification % Retained on:				
14 mesh	4.3	2.8	5.9	5.0
28 mesh	27.4	25.4	30.9	29.6
48 mesh	20.5	22.6	19.7	18.3
100 mesh	17.8	13.2	12.1	11.8
200 mesh	10.6	8.5	7.4	7.9
Fines – 200 mesh	25.4	27.5	24.0	27.4

[a] Data from producing mills.

freeness-drop across the screen is independent of the original freeness, but it is affected by the amount of rejects. This relationship applies to all thermomechanical pulps and allows the prediction of freeness-change for a given reject rate. The amount of screen rejects affects the physical properties of the pulp. Bulk is reduced initially, but not much change is noticed above 15%. Bursting strength and breaking length increase until the rejects reach 30%, after which they may decrease slightly. Wet-web strength is increased only slightly. Experimental data has shown that the shive-removal efficiency is the same in centrifugal cleaners as it is in screens. However, it has been shown that the specific surface characteristics of the fibers in the rejects are quite different. Centrifugal cleaners are more efficient in removing cubical debris and also reject a much higher fraction of low specific-surface material. Since the latter are mostly responsible for linting during printing, increasing the rejects by centrifugal cleaning will greatly reduce this problem. Since linting is one of the most serious shortcomings of thermomechanical pulp, centrifugal cleaning can be quite useful. Most commercial centrifugal cleaning systems operate at an approximately 10% reject rate. This is insufficient to remove a significant amount of lint fibers; an increase of rejects from 20 to 30% is necessary to accomplish this. The rejects can be transformed into lint-free fibers by additional refining at high consistency.

The efficiency of centrifugal cleaners is improved at low consistencies. Feed consistencies should not exceed 0.6% in the primary cleaning stages and should be even lower in the following stages, where debris concentration is high. The size of the centrifugal cleaner determines the size of particles it will remove. With refiner mechanical pulp, 20-cm (8-in.) cleaners have proven most efficient, whereas with thermomechanical pulp, 15-cm (6-in.) cleaners are the most accepted. Smaller cleaners are used in mills where high cleanliness is needed, as in coated-paper grades. The disadvantage of smaller cleaners is their higher capital cost and the increased tendency to plugging. High temperature has an adverse effect on the capacity of centrifugal cleaners. The loss in capacity at temperatures of 80°C is 10 to 15%. Mechanical pulps are typically high in reject concentration ratio, which means that the rejects leaving the cleaner nozzle are high in consistency and as a result tend to plug the opening. To counteract this, larger openings have been designed into the cleaner. This results in increased rejects, and places a heavier load on the secondary and the tertiary cleaning systems, which often do not contain sufficient cleaners to take care of the larger amount of rejects, thus reducing cleaning efficiency. For design calculations it is more or less agreed that primary 15-cm (6-in.) cleaners should reject approximately 25%, the secondary stage should reject 35%, and the tertiary 50%. Rejects are slightly less for the 20-cm (8-in.) cleaners. If possible, a cleaning system where the last-stage accepts are returned for further refining is recommended.

Although screening and cleaning will greatly reduce the linting tendencies of thermomechanical pulps, the most significant improvement can be achieved by applying more mechanical energy in the refining stages. Mechanical pulp is produced in the refiners and its quality can be improved only marginally by screening and cleaning.

In high-consistency refining, a condition known as latency can be introduced.[132-134] It is caused by the twisting action on the fibers during refining. This introduces kinks and curves into the fiber; these are then frozen into the fiber as they are cooled. Removal of this latency before screening is important since the screen cannot differentiate between fiber bundles and the kinks and curves caused by the latency.[135] In commercial operations, latency is removed by agitation of the stock for 40 to 60 min at 4% consistency or less and at temperatures of 70 to 90°C.[136] Latency is normally removed from the pulp by the time it reaches the deckers. Any subsequent high-consistency handling can reintroduce some latency, such as a high-consistency thickener, press, or high-density pump. This, however, will be removed by the pumps and agitators before the pulp reaches the headbox of the paper machine.

Rejects Refining of Mechanical Pulp

All mechanical pulps, whether produced on grinder stones or in disk refiners, contain a fraction of unacceptable fibers, fiber bundles, and debris. The amount and type of undesirables depend on wood species, condition of the raw material, energy consumption, and type and condition of refiners. Debris content will vary between 0.2% for thermomechanical pulps to 5% for some refiner mechanical pulps. The amount to be handled is considerably higher because screen and cleaner rejects always contain some good fibers and rejects can run as high as 20% of production. Years ago these rejects were discarded as waste. Today, raw-material costs dictate maximum utilization, and environmental considerations prohibit the sewering of rejects. Therefore, it becomes necessary to recover and prepare the rejects from mechanical pulps for acceptance in the paper-making process. The majority of the so-called undesirable fibers can be developed into a pulp with high-strength properties, which, when reblended into the original pulp, will improve the overall quality.[137]

Refining Stone Groundwood Rejects

In stone groundwood mills, the first rejects come from the coarse or bull screens. These contain large slivers that must be reduced to small particles before they can be refined properly.[138,139] In most cases a hammer mill or a shredder is used to accomplish this. The hammermilled rejects are usually mixed with other rejects and refined together. However, best results can be obtained by pre-refining the coarse-screen rejects in a separate small refiner prior to mixing with the screen and centrifugal-cleaner rejects. Refining requires consistencies between 15 to 20% for the best results, and due to the coarseness of the material, this consistency can be easily obtained by gravity drainage. Power requirements in the prerefining stage range from 360 to 450 kWh/ton (20 to 25 hpd/ton).

Refining of Rejects from Refiner Mechanical Pulping

Rejects in a refiner mechanical pulping system originate from the screens and from the centrifugal cleaners. Rescreening of rejects from the primary screen is a common practice. This concentrates the undesirable fibers and reduces the overall amount of rejects. Centrifugal-cleaner rejects contain mostly cubical debris, originating from knots in the woods, which cannot be developed into fibers. Such debris should be removed from the system and discarded. A refiner mechanical pulp mill needs an outlet for a small amount of debris; the system cannot be closed up entirely without adversely affecting pulp quality. Reducing the amount of cleaner rejects to be discarded is an important objective, but unfortunately lower rejects mean a compromise in cleaning efficiency. One answer to the problem is to add a fourth cleaning stage to the system. This system is operated with elutriators that allow only the high-density, cubical debris to be rejected. The accepts in such a system will include shives and fiber bundles; these are not cascaded back, but are combined with the screened rejects and sent to the rejects refining system. This arrangement allows the production of high-quality pulp, but requires sufficient dewatering capacity to thicken the low-consistency cleaner rejects before reject refining. Gravity deckers, side-hill screens, and rotary drainers can be used to obtain consistencies in the 5 to 8% range. For higher consistencies, it is necessary to use screw presses (8 to 16% consistency), a tapered-spindle variety of screw press (30 to 35% consistency), a V-notch disk press, or a vacuum thickener with press rolls (22 to 24% consistency).

Feed-rate uniformity stands out as the most important factor in the successful refining of mechanical-pulp rejects. Since metering devices operate on a volumetric basis, it is important that the bulk density or consistency of the rejects be kept uniform and that they be fed at a uniform feed rate. A horizontal, variable-speed-metering screw-feeder will deliver the rejects at a preset rate. Excess rejects are returned to the tailing chest. Another factor in refining mechanical-pulp rejects is sufficient power. The rejects must be reduced in size so that the majority will be accepted by the screening and cleaning system. If satisfactory reduction is not obtained, recycling of the rejects will take place creating a buildup to the point that the system must be purged. A separate centrifugal cleaning system to handle the refined rejects is desirable. Exact sizing of the reject refining system is difficult but, depending on the objective, power applied should be in the range of 800 to 1400 kWh ton (45 to 75 hpd/ton).

Refining consistency is another important factor affecting the refining of mechanical-pulp rejects.[140] Drainage rates are reduced at high consistencies for a given refining energy level. The burst index will be increased in relation to the power applied, but higher burst values will be obtained at a given unit energy level if the consistency is increased, as shown in Figure 4-37. Improvement in the breaking length follows the same pattern. The most important objective in the refining of rejects from mechanical pulps is to reduce the debris content in the pulp. Significantly better reduction of debris content can be accomplished at higher consistencies, and this is reflected in a reduced specific volume of the pulp, as shown in Figure 4-38. One of the most significant advantages of higher consistency in refining is its effect on refiner-plate life, which can be increased

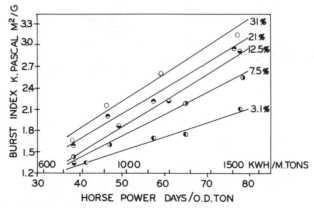

Figure 4-37. Relation between burst index of pulp from the rejects of mechanical pulping versus specific energy applied in refining at different levels of consistency.

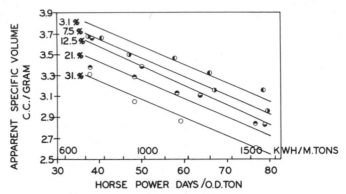

Figure 4-38. Relation between apparent specific volume of pulp from the rejects of mechanical pulping versus the specific energy applied in refining at different levels of consistency.

Figure 4-39. Electron-scanning micrographs showing rejects from thermomechanical pulp (a) before and (b) after refining at high consistency. (Courtesy Sprout Waldron Div. Koppers Company Inc.)

234

threefold by increasing the consistency from 6 to 12% and which will approach a five-times higher life when operating in the 25% range. Refining mechanical-pulp rejects at consistencies above 18% will produce latency in the pulp. It is important that this latency be removed by holding the refined rejects in a well agitated chest at 70 to 85°C for at least 20 min prior to screening and cleaning.

Refining of Rejects from Thermomechanical Pulping

It has been pointed out that thermomechanical pulp, especially when produced close to the glass transition temperature, contains a significant portion of long, unfibrillated fibers that have no bonding properties. These are summerwood fibers that require a high level of mechanical energy for development. By selectively removing these fibers in a screening system and refining at high consistency, they will develop high strength and can be blended with the original pulp to produce a sheet with greatly increased strength and reduced linting tendency. Sheets made from these fibers before and after refining are shown in Figure 4-39.

It has been advocated that the screened rejects be blended with the primary refined pulp and the combination be refined together in the second refining stage. This is not a satisfactory arrangement, since either the tough summerwood fibers, which require more energy to develop than the primary refined pulp, will not receive sufficient energy, or the primary refined fibers will be overrefined, resulting in excessive fines that cause poor drainage.

Environmental Effects of Thermomechanical Pulping

Pollution from a mechanical mill is considerably less than that from a chemical pulp mill due mainly to higher pulping yields. Thermomechanical pulp mills will, however, produce higher pollution compared to stone groundwood or refiner mechanical pulping.[141,142] The higher pollution from thermomechanical pulping is caused by the higher pressures and elevated temperatures used in presteaming which dissolve more solubles from the wood. If combined with chemical pretreatment, as in the case of hardwoods, a further increase in the pollution load will occur. Pulps for publication-grade papers, which are usually produced at low temperatures and without chemicals, will have a pollution load close to that obtained in refiner mechanical pulping. A comparison of different mechanical pulping processes on biological oxygen demand (BOD) is shown in Figure 4-40. Suspended solids from a mechanical pulp mill are usually higher than from chemical pulping plants due to the higher fines content. Thermomechanical pulps, however, contain longer fibers than stone groundwood and are relatively free of fines, which means less suspended solids in the effluents.

Effluents from a mechanical pulp mill are toxic and, if not diluted before entering a stream, will be fatal to fish life.[143] About 60 to 90% of the toxicity comes from the acid extracts in the wood, predominantly resin acids and unsaturated fatty acids. Neutral components, such as pimarol, juvabione, and

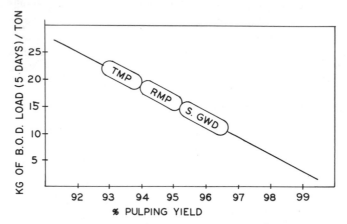

Figure 4–40. Relation between different mechanical pulping processes on BOD (five-day load) and pulping yield — thermomechanical (TMP), refiner mechanical (RMP), and stone groundwood (S. GWD). (Northern softwood species.)

diterpene-alcohols (mainly from pine), represent 25 to 30% of the total toxicity in the effluent. The primary factor governing effluent pollution from a mechanical pulp mill is the wood species used. Most northern softwoods, like spruce, hemlock, and western cedar, are low in resins and fatty acids. As a result, they create little pollution when compared to high resin-content species, like the pines, which can create a pollution load ten-times higher. The amount of resin acids will be reduced in storage, thereby reducing the level of pollution. The BOD from a mechanical pulp mill depends greatly on the system configuration, basically on how well the system is closed up.[144] In a mill producing market pulp, the amount of effluent can be reduced by recirculating the white water from the thickeners. In an integrated mill, paper-machine white water should be used for makeup. A completely closed-up system can seldom be achieved. Excessive recycling of process white water can lead to an increase in the level of iron and manganese with deleterious effects on the optical properties of the pulp, or, if the logs are stored in sea water, to a buildup of sodium chloride leading to accelerated corrosion of equipment. High temperatures and low pH will cause the hydrolysis of carbohydrates.

The major source of pollution in a mechanical pulp mill is the wood room if the logs are debarked and chipped on the mill site; these effluents are high in suspended solids and in toxicity due mostly to the bark. The second source of pollution originates from the chip-washing system. If a plug-screw feeder is used to seal the steaming vessel, close to one-third of the toxic pollutants in the wood chips can be extracted. These are highly concentrated, low in volume, and can be disposed of easily. Steam exhaust from a thermomechanical pulp mill contains some volatile resin components and turpentines.

Postrefining of Mechanical Pulps

Postrefining refers to the treatment of mechanical pulps in a disk refiner, after screening and cleaning, with the objective of improving the printing characterisitcs of the paper by reducing its shive content.[145] Since shive reduction is always accompanied by a freeness drop, the latter frequently becomes the primary objective.

Mechanical pulps are used in printing grades not only because they are less expensive than chemical pulps, but also because they produce a smoother sheet, higher opacity, and less show-through. To meet these requirements, mechanical pulp must possess a large surface area and a significant amount of fines. The type of fines, however, is critical. Fines produced by cutting of fiber bundles are undesirable since they have no bonding characteristics and will result in linting and dusting of the sheet. The types of fines desired are those produced from individual fibers and consist of fiber segments and fibrils. Shives are undesirable in printing-grade papers, particularly in coated grades, because they contribute to blistering, heat-set roughening and fiber puffing, particularly when heat-set inks are used, and they contribute to linting.[146]

There is an optimum level of energy for the postrefining of mechanical pulps to achieve the desired level of shive reduction, as shown in Figure 4-41. As shown in this Figure, 35 to 60kWh/ton (2 to 3 hpd/ton) is sufficient to obtain optimum results in shive reduction. The effect of postrefining on strength is somewhat different. During the no-load condition, when the refining plates are

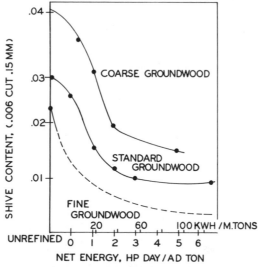

Figure 4-41. Relation between energy applied and shive reduction in the postrefining of stone groundwood in a disk refiner. Consistency 4%.

open, a certain amount of work is accomplished and burst and tear are improved slightly. When the refiner plates are closed and the load increased, the strength begins to decrease. At the optimum point for shive reduction, the loss in physical strength is negligible and will not exceed 10%, as shown in Figure 4-42. The overall effects of postrefining on freeness, fiber distribution and strength are shown in Table 4-9.

The initial pulp quality greatly influences the improvement that can be obtained by postrefining. Clean pulps with low shive content can be further improved and the shive content reduced without any significant change in freeness or loss in physical strength. On the other hand, postrefining of mechanical pulps high in shive content is accompanied by a large freeness change and significant loss in tear and burst strength. The medium- and short-fiber fraction, which represent 60 to 80% of the total fiber present in most mechanical pulps, are little affected by postrefining. Fibers one millimeter and longer are reduced in amount by postrefining and the longest fibers are most affected. Long-fiber bundles are reduced to short-fiber bundles at low-energy inputs. If the power is increased to a higher level (60 to 80 kWh/ton), the short-fiber bundles will be separated into individual fibers.

Consistency of the pulp during postrefining is critical, just as it is in any refining operation. The lower the consistency the more fiber cutting. There is a limit for consistency below which refining becomes unstable and severe erosion of refiner plates will take place, reducing plate life drastically. High refining con-

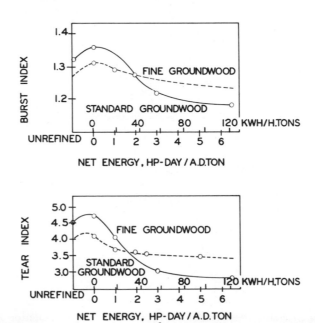

Figure 4-42. Relation between energy applied and strength development in the postrefining of stone groundwood in a disk refiner. Consistency 4%.

TABLE 4-9 EFFECT OF POSTREFINING OF DIFFERENT SPECIES ON FREENESS, FIBER DISTRIBUTION, AND STRENGTH[a]

	North American Spruce		North American Spruce		North American Aspen	
	Before	After	Before	After	Before	After
Freeness Canadian Standard (ml)	63	52	48	33	94	58
Burst index (kPa · m²/g)	3.1	3.3	3.0	3.5	1.0	1.1
Tear index (mN · m²/g)	5.3	5.2	4.6	3.6	4.0	3.5
Breaking length (km)	3.1	3.4	2.9	3.3	3.0	3.2
Brightness (%)	69.4	69.4	69.4	69.4	61.6	61.2
Opacity (%)	92.4	92.3	94.3	94.5	96.7	96.4
Bauer McNett fiber classification % retained on:						
14 mesh	.91	.38	1.1	.11	–	–
28 mesh	8.7	7.2	8.1	2.6	8.7	4.1
48 mesh	18.4	17.8	16.7	14.1	17.2	17.1
100 mesh	19.8	19.7	18.2	19.8	25.2	27.4
through 100 mesh	52.2	54.9	55.9	63.4	48.9	51.4

[a] Data taken from producing mills.

sistency results in less fiber shortening, the upper limit being the highest consistency that can be pumped and will not plug the line to the refiner, usually 4.5 to 5.5%. This means that postrefining is a low-intensity refining operation using a large refining-plate area.

Peripheral disk speed in postrefining is usually in the range of 1525 to 1675 m (5000 to 5500 ft) per minute. Reducing the speed reduces the no-lead energy consumption and allows the utilization of more power to fiber treatment. At the same time, lower speeds increase the intensity of refining and this results in more fiber cutting, produces only a small reduction in shive content, and leaves numerous shorter-fiber bundles. Increasing the speed increases the no-load energy consumption, decreases the intensity of refining, and the gentle refining action minimizes fiber cutting and tends to brush out the fiber bundles. The improvement achieved by using higher than standard speeds is, however, marginal and is recommended only in special cases.[147]

In all refining operations the configuration of refiner plates has a significant effect on the results. The accepted theories applicable to the refining of chemical wood fibers do not apply for postrefining of mechanical pulp since mechanical fibers cannot be properly developed by beating due to the high lignin content. A refiner-plate pattern that allows treatment at low intensities is recommended. The applied power should be supported across a large disk area and many bar edges utilized. There are, of course, restrictions in achieving the optimum plate pattern; a refiner plate with very fine bars will have a short life and a limited production rate.

Postrefining to increase the production from a stone groundwood mill has been advocated. Production increases up to 10% are claimed by grinding to a coarser pulp of higher drainage and then using postrefining to reduce the free-ness to the desired level. This will reduce the grinder-room energy and increase its production, but total energy will not be reduced because there is no evidence that postrefining is more efficient than stone grinding. It is known that mechanical pulp containing highly hydrophobic lignin cannot be fiber-developed at low consistencies with a pump-through refiner, a term used to describe a disk refiner that is used for preparation of chemical pulp fibers in front of the paper machine. Mechanical pulps should be screened, cleaned, and then thickened to a high but pumpable consistency prior to postrefining. This can be done with low-drainage pulps and the freeness reduced by postrefining without reducing fiber strength, but this cannot be done with high-freeness pulps where fiber-bundle content is high. High-consistency postrefining has definite advantages, but the selection of equipment should be made after determining the pulp-quality objective. When strength development plus some shive reduction is the main objective, a high-speed, open-discharge refiner operated in the 15 to 25% consistency range is better suited than a pump-through unit. Postrefining for the sole purpose of shive reduction in stone groundwood will generally result in no strength improvement and more likely result in some loss. The freeness of the pulp is affected to a lesser degree in an open-discharge refiner operated at higher consistencies compared to a low-consistency, pump-through refiner. Open-discharge refiners that do not have the limitation on power applied to them can be used to improve the physical strength characteristics in a straight relation to the applied energy. The long-fiber fraction of the pulp is reduced much more in a pump-through refiner than in an open-discharge refiner, especially when the open-discharge refining is carried out at a high-consistency level. If the major objective is shive reduction, a pump-through refiner is preferred over an open-discharge refiner because of the lower capital and operating expense. Open-discharge refiners usually cost more than twice as much as pump-through refiners, and the operating cost is higher due to the higher refining energy used.

Chemimechanical Pulping

As pointed out in the section on semichemical pulping, there is no essential difference between it and chemimechanical pulping since both involve chemical pretreatment of wood chips followed by mechanical refining. In this section the greater emphasis is placed on the refining step, and the reader is referred to the section on chemical pretreatment in semichemical pulping for more information on the chemical pretreatment step.

In straight mechanical and thermomechanical pulping, softwoods tend to break down to produce fibrilar material that has good sheet forming and bonding properties. Hardwoods, however, because of their more rigid fiber structure, do not form such fibrils.[148] Instead, they break up into short nonfibrilar debris. This debris, and other fines, which come from the high percentage of ray cells present in hardwoods, add little to the bonding properties. Preheating of hardwood chips below the glass transition point prior to refining does not assist in separating the fibers. For this reason, thermomechanical pulping of hardwoods will do little to accomplish improved fiber strength.

Some bright-colored, low-density hardwoods, such as aspen, poplar, and cottonwood, are being processed by thermomechanical pulping,[148,150] and the pulps are being used successfully in the production of coated-paper grades. The pulps have a high degree of cleanliness, but they add little if any strength to the paper. These pulps are considered to be filler grades with high-bulking, high-optical properties, but low strength. Chemical pretreatment of hardwoods weakens the fiber structure so that the subsequent refining process will cause the fibers to fail between the S_1 and S_2 layer, just as in the case of softwoods treated by the thermomechanical pulping process. Pulps made from pretreated hardwoods will contain a large percentage of fibrilar fines that will provide strength properties equal to or better than mechanical pulps produced from softwoods. Furthermore, if sulfonation is used in the pretreatment, the swollen lignin will make the fibers more flexible and will further assist in fiber-to-fiber bonding. The various processes for pretreatment of hardwoods for chemimechanical pulping are discussed below.

Cold Soda Process

This is the oldest chemimechanical pulping process. It is described in the section on semichemical pulping. The process involves soaking the wood chips in sodium hydroxide solution at low temperature prior to refining.[151] The sodium hydroxide causes swelling of the hemicellulose and of the amorphous region of the fiber. Soaking time is normally 30 to 120 min at atmospheric pressure. Increasing the temperature to 80°C will improve the physical properties of the pulp, but high temperatures should be avoided, since these will degrade and partially dissolve the hemicellulose. The absorption of chemical and uniformity of distribution throughout the chips is improved if pressure impregnation is used.

The more alkali absorbed the lower the refining energy that is required to develop the pulp to a high strength. The pulp will have more long fibers and correspondingly fewer fines. On the other hand, increased chemical pretreatment will lower the scattering coefficient and opacity, reduce the yield, and lower the brightness. No reaction with the lignin is likely to take place during the sodium hydroxide pretreatment, but the alkali reacts with the hemicellulose and softens the fiber wall, both within the middle lamella and within the secondary wall, thus assisting in the fibrillation of the S_1 and S_2 layers. The effect of the amount of sodium hydroxide absorbed on pulp properties is shown in Table 4-10.

Sodium Sulfite Treatment

Treatment with sodium sulfite under alkali conditions is frequently used in the chemithermomechanical pulping of softwoods[151] and is discussed in that section. The process is well suited for low-density hardwoods, like aspen, poplar,

TABLE 4-10 EFFECT OF AMOUNT OF SODIUM HYDROXIDE ABSORPTION ON PULP PROPERTIES IN COLD SODA PROCESS (*EUCALYPTUS REGNANS*)

	Sodium Hydroxide Absorbed (%)			
	2	4	6	8
Freeness Canadian Standard (ml)	200	230	180	300
Bulk (cm^3/g)	3.3	3.2	2.7	2.4
Burst index (kPa · m^2/g)	0.4	0.7	1.5	1.8
Breaking length (km)	1.5	2.5	3.5	4.2
Stretch (%)	1.6	2.0	2.5	2.6
Tear index (mN · m^2/g)	2.1	3.0	4.2	4.8
Folding endurance (M.I.T.)	1.0	1.5	4.0	5.0
Brightness (%)	53	55	54	52
Shives (Somnerville 0.15 mm) (%)	1.5	1.2	0.5	0.4
Bauer McNett fiber classification % retained on:				
+ 30 Mesh	3.4	4.4	3.3	7.1
+ 30–50	35.6	39.7	44.0	35.7
+ 50–100	3.4	2.3	3.5	10.0
+100–200	30.2	16.7	15.4	26.3
−200	27.4	37.0	33.8	20.9

and eucalyptus, with which it will produce acceptable pulps in yields above 90%. It is seldom used for high-density hardwoods.

Sodium Bisulfite Treatment

Chemical pretreatment of wood chips with sodium bisulfite is used to sulfonate the lignin.[153] Sulfonation converts the native hydrophobic lignin into the more hydrophilic lignosulfonic acids so that the fiber is capable of swelling and of absorbing water. To produce a high-strength chemimechanical pulp from high-density hardwoods, a high degree of sulfonation is necessary and the chips must be cooked to a yield in the range of 80 to 85%.[154] During chemical impregnation with bisulfite, care must be taken that the liquor remains between pH 4 to 6 and that excess chemical should be left at the end of the treatment. If the pH of the cooking liquor should fall below 4, condensation of the lignin will take place resulting in a significant loss in brightness. The brightest pulp is produced in the range of 6 to 6.5 pH, but the highest strength is obtained at 4.5 to 5.0 pH. Above pH 7, the pulp becomes brownish in color. Cooking time is generally short.

The most important factor is to obtain thorough impregnation of the chips with the liquor prior to heat treatment. Short and thin chips will assist in good impregnation. The chips should be green and have a relatively high-moisture content because this aids the diffusion of chemical into the chip. Diffusion is also affected by wood density and this makes the chemimechanical pulping of mixed hardwoods difficult. Diffusion occurs across the grain of the wood and is effective only over short distances, hence the importance of thin chips. The higher the liquor concentration the more uniform the impregnation. Compressing wood chips prior to impregnation will combine the effect of diffusion with that of forced penetration. The latter reduces the differences between species of different wood densities. To avoid uncooked raw-chip centers it is best to carry out the impregnation under pressure at 490 to 980 kPa (5 to 10 kg/cm^2) in the absence of air.

An important factor in chemimechanical pulping with bisulfite is wood density, which is a reflection of the fiber wall thickness. Low-density species, such as aspen and willow, have thin, flexible wall fibers that collapse easily on refining. As a result, these fibers will produce a well-bonded and relatively strong sheet. Wood species with densities less than 0.45 g/cm^3 will produce a pulp with high breaking length and good opacity. High-density wood species above .060 g/cm^3 are less suited for chemimechanical pulping. They contain thick wall fibers that are stiff and resistant to collapse, tending instead to disintegrate into short fines. Such fibers will produce a coarse, bulky sheet with low strength. These high-density species can be developed into a pulp having acceptable physical strength, but the yield will be low and the opacity will be low.

As discussed previously, good quality mechanical pulp requires the rupture of the fibers in the fiber-wall region. This can be best accomplished at temperatures where the lignin is plasticized, but is below its glass transition point. While untreated softwoods have a glass transition temperature of about 120 to 125°C, chemically pretreated woods (depending on the degree of sulfonation) will have a glass transition temperature in the range of 70 to 90°C. This means that the refining of treated fibers must be carried out at lower temperatures using atmospheric refiners that are not close coupled to the presteaming vessel. In addition, refining should not be carried out in a single stage because the high-energy input will result in temperatures between the plates well in excess of the glass transition point.

The bonding properties of chemimechanical pulps produced from hardwoods can be improved by increasing the degree of sulfonation and increasing the chemical absorption. The higher the degree of sulfonation the more the pulp will resemble chemical fibers. Refining of such fibers will improve their properties, whereas the refining of unteated pulps is useless. If chemical absorption is limited to a pulping yield of 85 to 90%, the pulp will exhibit characteristics of a mechanical pulp; at lower yields the results will resemble chemical pulps. The tear strength will be influenced mostly by the original fiber length of the species. Hardwoods, in general, contain short fibers as compared to softwoods. Special care should be taken during refining to avoid any further shortening of the already short fibers. Opacity and the scattering coefficient are the properties adversely affected by increased pretreatment due to the loss in fines and reduction in surface area.

The brightness of chemimechanical pulps depends first on the color of the wood. When chemical pretreatment is performed under correct conditions, the brightness of the pulp will be higher than the wood itself. Brightness levels of 65 to 75 GE can be achieved without bleaching on bright-wood species like aspen by using a bisulfite chemical pretreatment. One of the shortcomings of chemimechanical pulp is color reversion resulting from exposure to sunlight. The color reversion is caused by the lignin. The tendency toward reversion becomes stronger after the lignin has been sulfonated and as a result is more pronounced in chemimechanical pulps than in mechanical pulps.

Acid Sulfite Treatment

Another approach to producing chemically pretreated mechanical pulps with the objective of obtaining improved strength is chemical impregnation with acid sulfite liquor.[155] The process involves atmospheric impregnation for approximately 60 min with an acid sulfite solution at a pH of 1.5 to 2.0 followed by steaming in a vapor-phase digester at approximately 120°C. The pretreated chips can be refined in an atmospheric refiner or close coupled to the digester under pressure, as in thermomechanical pulping. Chemical pretreatment with acid sulfite liquor reduces the softening temperature of the lignin, the extent of which depends on the degree of sulfonation. Acid sulfite pretreatment provides a high rate of sulfonation; the reaction is about three times faster than with neutral bisulfite. In some cases, acid sulfite pretreatment will provide a higher brightness than that obtained with neutral sulfite.

Softwood pulps pretreated with acid sulfite demonstrate a significantly higher bonding potential but lower tear strength when compared to untreated refiner pulps. Brightness will also be lower, but the opacity will be slightly higher. Acid sulfite treated softwoods are lower in bonding strength than high-yield bisulfite pulps, but they show better printing qualities due to improved scattering coefficient and higher bulk. The mechanical energy required to develop these pulps will be significantly higher due to the higher yield.

Chemithermomechanical Pulping

Thermomechanical pulping of softwoods can produce a stronger pulp than any other mechanical pulping method, but this high strength can be obtained only at lower drainage levels. As has already been shown, thermomechanical pulp contains long, stiff fibers that do not bond properly to each other. Mild chemical pretreatment of the chips in a process known as chemithermomechanical pulping makes it possible to increase the softness of the chips, still maintain a yield above 90%, and to produce a fiber with a very high flexibility that can be refined with a minimum of fiber cutting.[156]

In chemithermomechanical pulping, softening of the chips is achieved by impregnating the chips with sodium sulfite that has been adjusted to a pH of 9 to 10 with sodium hydroxide. Following impregnation, the chips are pretreated in a steaming vessel to 130 to 170°C and then refined in a steam-

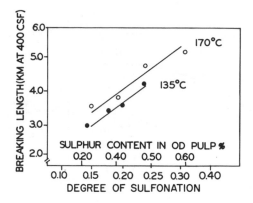

Figure 4-43. Effect of degree of sulfonation and preheating temperature on the strength (breaking length) of chemithermomechanical pulp. (Courtesy Svenska Cellulosa Teknik A. B.)

pressurized or atmospheric disk refiner. The energy consumption in the refining stages is normally higher than that in the thermomechanical pulping process when refining to the same freeness value. Chemithermomechanical pulp is characterized by exceptionally low shive content even at a high-freeness level, and the long-fiber fraction is higher than in thermomechanical pulp.

The properties of chemithermomechanical pulp can be varied over a wide range by varying the main process parameters which are: the preheating temperature, the amount of consumed sodium sulfite, and the energy input in the refiner. The breaking length, as a function of the amount of sodium sulfite absorbed by the chips, is shown in Figure 4-43 at two levels of preheating temperature. The freeness of the pulp is 400 Canadian Standard. To obtain this freeness, single-stage refining is sufficient. If lower freeness values are needed, multiple stages are normally used, the reasons being that chemically impregnated chips require smaller plate clearance compared to thermomechanical pulp, the pulp is unable to support high loads without plate-to-plate contact, and too small a plate gap involves the risk of cutting the fibers.

The brightness of chemithermomechanical pulp is increased with an increase in the sodium sulfite consumption, as shown in Figure 4-44. The relation between the strength of the pulp and the light-scattering coefficient is shown in Figure 4-45. The typical increase in scattering coefficient obtained with mechanical pulps by refining has been changed by the sodium sulfite treatment to such an extent that refining will have no effect at all.

The yield of chemithermomechanical pulp normally is only 2 to 3% lower than that of thermomechanical pulp. At temperatures not exceeding 135°C the yield is about 95%, and it decreases only slightly with an increase in the amount of sodium sulfite absorbed by the chips. At 170°C, the yield is reduced to 90%. It is interesting to note that the yield is lowered by a high presteaming temperature even without the use of chemical impregnation and that the addition of sodium sulfite in the range of 2 to 14% will actually increase the yield by acting as a buffer to prevent the dissolution of lignin and hemicellulose. The optimum yield is obtained in the range of 9 to 12 pH.

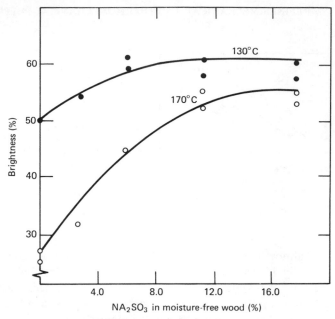

Figure 4-44. Effect of amount of sodium sulfite absorbed on the brightness of chemithermomechanical pulp. Freeness 100 to 200 Canadian Standard (Courtesy Svenska Cellulosa Teknik A. B.)

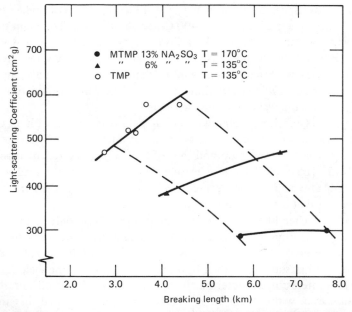

Figure 4-45. Relation between strength (breaking length) and light-scattering coefficient of chemithermomechanical pulps compared to a typical thermomechanical pulp. (Courtesy Svenska Cellulosa Teknik A. B.)

246

A comparison of the papermaking properties of chemithermomechanical pulp (CTMP) with stone groundwood (GWD), refiner mechanical pulp (RMP), thermomechanical pulp (TMP), unbleached sulfite (UBS) and semibleached kraft (SBK) is shown in a series of curves in Figure 4-46.[157] Figure 4-46a shows tensile strength versus density; it shows the progressive improvement made in the mechanical pulping processes and shows that a high-yield chemithermomechanical pulp can produce paper with a tensile strength comparable to that of a relatively low-yield chemical pulp and do this at a lower density. The second curve (Figure 4-46b) shows tear versus tensile strength. The chemithermomechanical pulp approches the value for the sulfite pulp although it is below that of the semibleached kraft pulp. It is interesting to note that the maximum tear strength for both the semibleached kraft and the chemithermomechanical pulp occur at the same tensile strength. The third curve (Figure 4-46c) shows that the light-scattering coefficient versus tensile strength for chemithermomechanical pulp is higher than that of the chemical pulps but is lower than stone groundwood, refiner mechanical pulp, or even thermomechanical pulp.

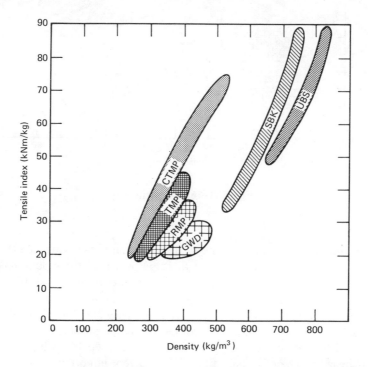

Figure 4-46a. Comparison of sheet strength properties of chemithermomechanical pulp (CTMP), stone groundwood (GWD), refiner mechanical pulp (RMP), thermomechanical pulp (TMP), unbleached sulfite pulp (UBS), and semibleached kraft pulp (SBK): (a) tensile versus sheet density, (b) tear versus tensile, and (c) light scattering versus tensile. (Courtesy Svenska Cellulosa Teknik A. B.)

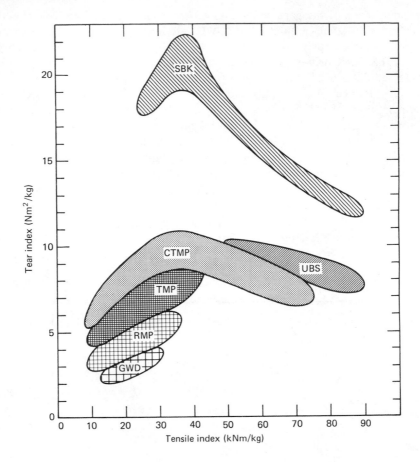

Figure 4-46b. Continued

Typical fiber classifications for the six different pulps mentioned are shown in Figure 4-47 at typical yield values for these pulps. All of these pulps represent samples in a freeness range suitable for newsprint production (100 to 120 for the mechanical pulps and 120 to 150 for the chemithermomechanical pulp). As shown in this figure, chemithermomechanical pulp has about three times the long fibers and two-thirds the fines content of stone groundwood. Because of the large yield difference, chemithermomechanical pulp fibers are about two times as heavy as chemical pulp fibers.[158] It is very important to obtain good treatment of these long fibers in the refiner. If this is not accomplished, and the freeness is reduced mainly by fiber cutting and beating of the finer fiber fractions, the stiff, long fibers will reduce the papermaking potential of the pulp, especially for the finer grades of printing papers. The shive content is very low for unscreened chemithermomechanical pulp even at high freeness levels; actu-

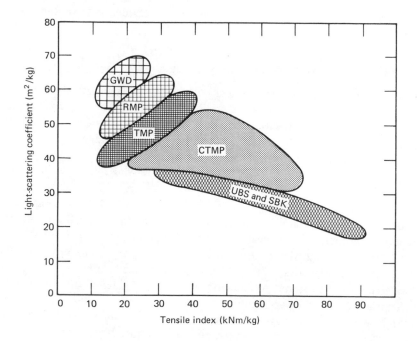

Figure 4-46c. Continued

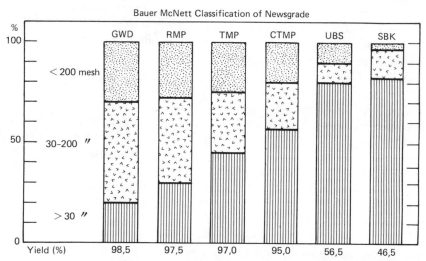

Figure 4-47. Comparison of fiber distribution for stone groundwood, refiner mechanical, thermomechanical, chemithermomechanical, unbleached sulfite, and semibleached kraft pulps at typical yield values. Newsprint grade, Bauer McNett classification. (Courtesy Svenska Cellulosa Teknik A. B.)

ally the shive content is so low that the pulp can be regarded as shive-free for all freeness levels of practical interest.

Thermomechanical Pulps in Papermaking

Thermomechanical pulp is a mechanical pulp, but it has significantly different characteristics from stone groundwood and for this reason a straight replacement should not be considered. It should also be noted that thermomechanical pulp is a general term and a wide variety of qualities can be obtained by changing operating variables. Thermomechanical pulp was originally developed for the newsprint industry,[159-162] but its exceptional characteristics make it suitable for other paper grades and its usage is growing.[163-165] In 1976, about 71% of thermomechanical pulp capacity was used for newsprint, 14% for magazine- and coated-paper grades, and 15% for market pulp and other speciality products. More and more thermomechanical pulp is finding its way into other publication grades, low-weight coated sheets, boards, tissues, and absorbent products. Its major acceptance originates from its high physical strength in both the wet and dry state compared to stone groundwood. This makes reduction of more expensive chemical fibers in the paper furnish possible.

Thermomechanical pulp has a bulk that is lower than refiner mechanical pulp, slightly higher than stone groundwood, and significantly higher than chemical pulps. The higher bulk could represent an advantage for printing, especially when show-through is important as in low-weight paper grades, but it increases the sheet thickness and thus results in less footage in paper roll of given diameter. The amount of thermomechanical pulp in a sheet could be limited by the maximum acceptable thickness of the sheet. Thermomechanical pulp is stronger than stone groundwood and as a result it can replace a significant amount of chemical pulp without adversely affecting runnability on the paper machine or in the press room. The physical strength of thermomechanical pulp depends on wood species and quality of raw material but, as a guideline, semibleached kraft can be replaced by two or three times its weight of thermomechanical pulp. As an example, if a newsprint mill is operating with 24% kraft and 76% stone groundwood, it can use 72% thermomechanical pulp and 28% stone groundwood with no loss in sheet strength. The typical characteristics of thermomechanical pulps are its high content of long fibers and its low content of fines. This kind of fiber distribution is well suited for a newsprint furnish in which thermomechanical pulp represents less than 30% and stone groundwood is used to provide the fines needed to satisfy the printing requirements. In such a case, thermomechanical pulp is considered to be a reinforcement pulp.

If thermomechanical pulp is close to 100% of the furnish, it is important that the pulp be developed by refining to a low-drainage level. The reduction or elimination of the long, unfibrillated fiber portion is of great importance because these fibers have a low-bonding potential and they adversely affect surface roughness and reduce the surface strength of the sheet, the latter creating severe problems in the press room. It is necessary to develop thermomechanical pulp by beating or refining to a very high extent, even when used at low levels if the paper is used for offset printing. If the pulp is well developed into ribbons by splitting the fiber wall and S_1 layer, the scattering coefficient will be high and the surface strength will be improved correspondingly. While the quality of thermomechanical pulp greatly influences surface strength of the sheet, it should be remembered that strength will also be affected by the operation on the paper machine. Twin-wire machines, in general, have a tendency to produce sheets with a higher surface strength. Wet pressing will also improve surface strength and a breaker stack may also be helpful.

How much thermomechanical pulp can be used in newsprint and whether 100% thermomechanical newsprint is feasible are questions that have been raised. No general statement can be made; the decision depends greatly on the location of the mill, its wood supply, and its power costs, all of which affect the overall economy of the operation. Technically, a 100% thermomechanical newsprint is possible, although some of the properties of the sheet will not be all that is desired. For coated papers, thermomechanical pulping of hardwoods offers little advantage. However, since thermomechanical pupling of softwoods produces pulps with lower shive and debris content than is possible by stone groundwood or refiner mechanical pulping, this use is growing.[166] Due to the significantly improved strength properties, larger percentages of these pulps are being used in the furnish to replace expensive chemical pulps. Although in the past maximum mechanical-pulp content seldom exceeded 40%, it is now possible to use 70% TMP in some of the low-grade coated papers. While the strength of the thermomechanical pulp is important to allow the replacement of larger percentages of chemical pulps, the optical properties should not be over-looked. Shive and debris content cannot be tolerated and, in addition, long-fiber content must be kept low. Whereas a typical thermomechanical pulp may have a +28 mesh fiber content close to 40%, this must be reduced to about 10% in pulp for coated grades to assure good smoothness of the sheet. Lower-drainage pulp, obtained by applying more energy in the refiners, will accomplish this, but at considerable additional energy per ton and correspondingly higher cost. Postrefining appears to be less expensive and this is now being practiced in most coated-grade paper mills.

Thermomechanical pulp is a logical supplement in board grades because of its high bulk and stiffness, which are two major requirements for paperboard. In the case of multilayer boards, however, it must be remembered that the bonding, which is very critical, is poor for thermomechanical pulps. The plybond can be improved by additional fiber development through refining, but this will reduce drainage of the pulp and will adversely affect stiffness and bulk. Chemical pre-treatment of the chips will improve the bonding properties (including those of plybond) and maintain a high-drainage rate, but if excessively applied will reduce the bulk and stiffness. Chemical pulps, after beating, will improve the overall properties of a mixture of thermomechanical and chemical pulps and provide adequate bulk and stiffness. Because thermomechanical pulps are fast-draining and low in shive content, their use in tissue grades is expanding. The better bio-degradable properties of all mechanical pulps compared to chemical pulps is an advantage in sanitary tissues and with the added strength of thermome-chanical pulp, 100% use in this grade of paper is not impossible. High quality thermomechanical pulp is now used in toweling and in facial tissue at percentage levels as high as 50 to 60%. Mild chemical pretreatment allows the production of a long-fibered pulp with fast drainage, good absorption properties[167,168] and sufficient softness, thus making it a low-cost raw material for the tissue industry. One of the fastest growing uses for mechanical pulp is in disposable diapers. For this grade, the brightness instability of mechanical pulps is of little importance. The two most important requirements for fluffed-paper products are fiber weight and thickness. Fiber weight is the more critical because it affects water-holding capacity, while thickness is more of a consumer value. Thermo-mechanical pulp meets both requirements and is replacing the more expensive chemical pulps among the disposable fluffed paper products.

Thermomechanical pulp has been slow in entering some of the paper grades traditionally produced from chemical pulps. Fine papers, where resistance to aging is requisite, cannot be made with high-lignin mechanical pulps and, for this reason, they are not used in bond and other fine-paper grades.[169] Papers for photocopying and for computer cards, grades that have a short life span before being recycled, are a logical target for thermomechanical pulp. The good dimensional stability and stiffness properties of thermomechanical pulp make it well suited for computer cards and similar paper grades.

Masonite Process

The Masonite process is a unique process for reducing wood to a fibrous form by heating chips in a digester or "gun" under very high pressure for a very short period and then releasing by means of a quick-opening valve. The explosive effect of the rapid release breaks up the wood chips into a fibrous mass. The fibers contain nearly all the lignin present in the original wood, but the lignin is in a thermoplastic state. Because of this, the fibers can be formed and pressed into a high-density product in which the lignin functions as the binder. The liquor washed from the fibers is high in sugars and organic acids and these can be recovered as valuable by-products.

SEMICHEMICAL PULPING: THE NEUTRAL SULFITE SEMICHEMICAL OR NSSC PROCESS

W. W. Marteny, Ph.D.

Owens-Illinois Inc.
Forest Products Division
Toledo, Ohio

The number and types of pulping processes have proliferated greatly in recent years. In the broad sense, many of these must be classified as semichemical processes because they use a combination of chemical pretreatment followed by a mechanical fiberizing step. These processes were developed with the objective of producing pulp in yields higher than are common in full chemical pulping, thereby conserving wood and reducing cost. A related objective is to produce pulp from hardwoods with properties approaching those from softwoods.

This section is concerned principally with one particular semichemical process known as the neutral sulfite semichemical or NSSC process. This is a two-stage process in which wood chips are given a mild chemical treatment at elevated temperatures to soften and partially remove the lignocellulosic interfiber bonding material. This is followed by mechanical refining to complete the fiber separation. Because the cooking liquors used in the early development of the process contained sodium sulfite that was buffered with sodium carbonate: bicarbonate to reduce corrosion, the process became known as the neutral sulfite semichemical, or NSSC process. For years NSSC and semichemical have been used interchangeably until it was found that other pulping agents could be used to produce high-yield semichemical pulp.

Types of Semichemical Processes

Before describing the neutral sulfite semichemical process in detail it is helpful to review the other broadly related processes (see Table 4-1). These processes have been discussed in the previous section. They are briefly reviewed here:

1. Refiner mechanical pulping (RMP) and thermomechanical pulping (TMP). Refiner mechanical pulping is an outgrowth of the groundwood-from-chips process in which chips are fiberized in disk refiners at atmospheric pressure. In thermomechanical pulping the chips are heated with steam and fiberized under pressure while the lignin is in a plastic state.

2. Chemimechanical pulping (CMP) and chemithermomechanical pulping (CTMP). These are semichemical processes in that they are two-stage processes. However, the chemical treatment is less severe and the mechanical fiberizing is more drastic than that used in the conventional neutral sulfite semichemical process. Both pulping processes convert wood chips into pulp with little breaking or splitting of the fibers and, microscopically, the pulps resemble fully cooked chemical pulps rather than stone groundwood. These pulps, in contrast to fully cooked pulps, contain more than half the original lignin of the wood.

3. Mechanochemical pulping. This process refers to the production of high-yield pulps from nonwood sources, such as cereal straws and reeds. This process differs from conventional semichemical pulping in that it applies chemical and mechanical action simultaneously at atmospheric pressure.

Chemical Treatments Other than the Neutral Sulfite (NSSC) Process

The usual procedure in semichemical pulping is to treat wood chips with a pulping reagent in batch or continuous digesters, adjusting pulping time, pulping temperature, and/or chemical-to-wood ratio to obtain the desired pulp yield.

Complete fiber separation is achieved mechanically under digester conditions or at atmospheric conditions. Reagents most commonly used are: sodium sulfite, sodium hydroxide, sodium carbonate, and kraft green liquor. Treatment of wood with these chemicals is described in the following sections.

There is no essential difference between semichemical pulping and chemimechanical pulping since both involve chemical pretreatment of wood chips followed by mechanical refining. In this section (under the heading of semichemical pulping), emphasis is placed on the chemical treatment, whereas in a previous section (under the heading of chemimechanical pulping) greater emphasis was placed on the mechanical refining step. The reader is referred back to the section on chemimechanical pulping. The greatest emphasis in this section is placed on the neutral sulfite or NSSC process, but first other forms of chemical treatment will be described briefly.

Acid Sulfite Process. In this process the cooking liquor consists of a solution of bisulfite plus free sulfur dioxide. Delignification is the predominate reaction. The mechanism probably includes sulfonation of lignin in the middle lamella in the solid state, followed by hydrolysis to soluble lignosulfonic acids. The hemicelluloses are less dissolved. High-yield acid sulfite pulping is becoming increasingly popular in sulfite mills making newsprint, but pulp yields are generally limited to the range of 60 to 70%. The pulp produced is cleaner, has a lower freeness, higher burst, lower tear, and lower wet-web strength than conventional sulfite pulp. Extensive studies have been made at the Pulp and Paper Research Institute of Canada on ultra high-yield (93 to 95%) pulping of spruce with sodium-base acid sulfite liquors[170] using the chemimechanical (CMP) and chemithermomechanical (CTMP) processes. Pulps were prepared from spruce waferized chips and their properties were compared with those of the corresponding RMP and thermomechanical pulps (TMP) (see the section on sulfite pulping).

Bisulfite Process. Bisulfite pulping liquor contains no free sulfur dioxide. The mechanism of the predominate delignifying reaction is similar to that in acid sulfite pulping. High-yield pulp (53 to 70%) is produced in many Canadian newsprint mills from a spruce and a balsam wood furnish using a magnesium or sodium base. The pulp replaces acid sulfite pulp. Sodium bisulfite pulp from spruce and balsam has been proposed for use in newsprint.[171] Pulp suitable for corrugating medium has been produced from hardwoods.

Alkaline Sulfite Process. In this process, the cooking liquor contains sodium sulfite and any or all of the following: sodium carbonate, sodium hydroxide, and sodium sulfide in such proportions that the pH falls between 9 and 12. The reaction rate is lower than for kraft pulping and consequently cooking times are longer. Pulp with high-strength properties suitable for linerboard has been produced on a pilot-plant scale. The yield of alkaline sulfite pulp will be 6 to 10% higher than kraft pulp (60 versus 54%) at equal burst-strength levels. The alkaline sulfite pulp is considerably brighter than kraft. Some mills produce corrugating medium from hardwoods using sodium sulfite at buffer ratios of 1:1 or less.

Sodium sulfite to which sodium hydroxide has been added is commonly used in chemithermomechanical pulping. The chips are impregnated with chemical, heated to 130 to 170°C and refined in a disk refiner. Pulps have been produced using chips, shavings, or sawdust from softwoods and hardwoods. Pulp strength is a function of the amount of chemical used. In one trial, a CTMP pulp was made by adding 3.4% Na_2SO_3 and 0.2% NaOH based on wood weight, cooking for about 30 min at 93°C, followed by disk refining at high consistency in two stages. The specific energy applied in the first stage was 1098 kWh (61 hpd) and in the second stage 774 kWh (43 hpd) per oven-dry ton to give a pulp of about 180 Canadian Standard Freeness.[172] In comparison, RMP (refiner mechanical pulp) required a total energy of about 1620 kWh (90 hpd) per oven-dry ton of pulp at the same freeness.

Sulfate Process. Semichemical pulps produced by the sulfate processes are darker, lower in strength, and contain more lignin and less hemicellulose than NSSC pulps. Some sulfate mills produce high-yield pulps in the range of 65 to 80%, which are in reality semichemical pulps. These are generally made from coniferous woods; the pulps are used in linerboard and corrugating medium. About one-half the normal amount of chemical is used and this may be weak white liquor from an associated kraft mill. Other cooking conditions, such as temperature, are usually the same as in normal alkaline pulping. A summary of the cooking conditions used for making coarse pulp for paperboard is shown in Table 4-11. The pulps are blown "raw" and refined "hot" in disk refiners with a power requirement about 66 kWh (3.7 hpd) per ton. Screw pressing before refining is becoming popular as a means of fiberizing the chips and recovering the spent liquor in concentrated form. The lignin content of high-yield kraft pulps generally ranges from 8 to 14%.

A very high-strength kraft pulp can be produced from pine by cooking to a high yield with a sulfate liquor of low active alkali and purifying the resulting pulp with acidified sodium chlorite. The semicooked pulp yield is about 74% and the chlorinated pulp yield about 54%. The strength of the chlorinated pulp is superior to a straight kraft pulp of 44% yield. Hagglund[173] obtained promising results with a cooking liquor of 25 to 30% sulfidity and an active alkali

TABLE 4-11 COOKING CONDITIONS USED IN SULFATE SEMI-CHEMICAL PROCESS FOR BOARD PULP

Pulp yield (%)	70–75
Cooking chemicals:	
NaOH + Na_2S [as Na_2O] (% of wood)	4–7
Cooking temperature (°C)	160–185
Time at cooking temperature (hr)	0.3–2

content as low as 11 to 14%. However, the pulp had a fairly high lignin content and was harder to bleach to a high brightness than pulp cooked by the acid or neutral sulfite process.

Cold Soda Process. This is a semichemical pulping process developed at the United States Forest Products Laboratory. It is used to produce coarse pulps for corrugating and finer grades of pulps for printing papers, using hardwoods. This process, which is sometimes referred to as groundwood-from-chips process, is interesting because the pulp can substitute in some papers for regular groundwood. Furthermore, the process is applicable to the high-density hardwoods that do not make acceptable groundwood.

Basically, the cold soda process consists of treating chips with sodium hydroxide solution and fiberizing the softened chips in a disk refiner. Cold soda pulping is usually done on a batch basis draining off the excess liquor before fiberizing, but the process can also be operated on a continuous basis by applying the sodium hydroxide solution at the same time the chips are being fiberized. The cold soda process is not as specific for lignin removal as the neutral sulfite process. Typical conditions for a batch process are steeping for 30 to 120 min at atmospheric pressure or for 20 to 30 min when hydrostatic pressure is used. From 61 kg (135 lb) of caustic soda (for newsprint) to 115 kg (255 lb) of caustic soda (for corrugating) is required per ton of air-dry pulp. Snyder and Premo[174] describe the basic steps in a batch cold soda process as including:

1. Pressure impregnation of 9.5 mm (3/8 in.) hardwood chips at 1033 kPa (150 psi) with cold 2.5% sodium hydroxide for about 20 min.
2. Removal of unabsorbed liquor for reuse.
3. Retention of alkali-soaked chips in a bin for about 40 min.
4. Screw pressing to about 60% solids.
5. Disk refining in two stages at 12 to 15% and 8% consistency respectively.
6. Screening, cleaning, and finally disk refining to the desired strength.

The chips absorb from 2.5 to 4.0% sodium hydroxide (see Table 4-10). The optimum temperature is about 25°C with higher temperatures producing a dark-colored pulp. The power requirements are about 378 kWh (21 hpd) per ton in the primary and 414 kWh (23 hpd) in the secondary refiners. An 80% yield of pulp from poplar suitable for tissues has been claimed[175] by soaking chips in sodium hydroxide at 40 g/l and 0.05% surface-active agent for 10 min at 40°C followed by pressing through a series of screw presses.

Penetration of alkali in the cold soda process is a problem with the dense hardwoods. In the original process, soaking or steeping only was used. This was a time-consuming operation and also resulted in an overtreatment of the outside of the chip while the inside was undertreated. Pressure impregnation can be used to speed up the penetration, but pressure of 1033 kPa (150 psi) for 20 to 30 min is required for complete penetration. Decreasing the moisture content of the wood and removing air helps penetration. Reducing the dimensions of the chips before treatment also improves penetration, but tends to reduce pulp strength. Subjecting the chips to compression and decompression in the presence of sodium hydroxide solution in a roller mill has been shown to improve the penetration. The use of small chips and a two-stage treatment in fiber presses has been proposed for continuous pulping under conditions of rapid chip penetration. The chips are treated in the first press with partially spent liquor from the second press and then treated with fresh caustic and run through the second press. Continuous operation can also be achieved by spraying the chips with caustic solution, followed by screw pressing, and then passing the treated pulp upward in an inclined reactor tube while immersed in caustic soda liquor (retention time 20 min, liquor-to-wood ratio 6:1).

Yields in the cold soda process vary from about 87% for corrugating pulps to 85 to 92% for printing papers. The process is generally cheaper than the groundwood process because the chemical cost is more than offset by the lower energy consumption in fiberizing and, in some locations, lower wood costs. Energy requirements run from about 360 to 450 kWh (20 to 25 hpd) for corrugating to about 630 to 900 kWh (35 to 50 hpd) for printing paper per ton of air-dry pulp.

Cold soda liquor used for impregnating can be reused after fortification. This results in some loss of brightness, which may be as much as 4 to 5 points after reusing 15 to 20 times. Most mills discharge the spent liquor after about 15 to 20 cycles partly because of the loss in brightness, but also because the liquor contains enough solids to inhibit pulping. The spent liquor from the screw press is generally discarded in the case of high-grade pulps although this can be reused in the case of coarse pulps. For small mills, the discharge is not a very serious problem because the volume is small and the solids content is relatively low. If the mill is operating in conjunction with a kraft mill, cross recovery can be practiced.

High-grade cold soda pulps from hardwoods have bursting strengths that are equivalent to those of spruce groundwood and higher than those for pine groundwood. The freeness is generally higher and, unlike groundwood, the pulp has the ability to develop strength in beating. One of the disadvantages of cold soda pulp is the low brightness. It has a yellow color that persists after bleaching by conventional single-stage methods. The opacity of the pulp is also lower than that for groundwood. Unbleached brightness is about 40 to 45, but this can be increased to a brightness of 65 to 70 by bleaching with 10% chlorine at the cost of a bleaching loss of 2 to 10%. The most effective bleaching process appears to be treatment with a combination of peroxide and hypochlorite.

Nonsulfur Process. A process for the production of pulp for corrugating medium for hardwoods has been developed that uses no sulfur. The active ingredients are sodium hydroxide and sodium carbonate.[176] In one instance, the mill uses a Pandia-type, horizontal-tube continuous digester with a Defibrator pressurized-disk refiner at the bottom of the bank of reaction tubes. The defibered pulp is blown to a tank, diluted with spent pulping liquor, prerefined in disk refiners, and washed on vacuum drum washers. The addition of cooking liquor to the digester is adjusted to maintain the pH of the blow stock at the desired level. The steam pressure in the digester is 1275 kPa (185 psi). The retention time estimated from the speed of the timing screws approximates 6 min. There is evidence that pulp made at 1275 kPa (185 psi) has different papermaking properties than pulp made at 1035 kPa (150 psi).[177]

The Neutral Sulfite Semichemical (NSSC) Process

As mentioned earlier this is the principal semichemical process.

History of the NSSC Process

In principle this process dates back to 1874 when A. Mitscherlich proposed treating chips with a solution of sulfur dioxide or of bisulfite followed by grinding or rubbing to make pulp, and to 1880, when C. F. Cross described the advantages of cooking chips in a neutral or slightly alkaline solution of sodium sulfite. Exploitation of the process was delayed by the lack of a demonstrated use for the pulp and the lack of suitable fiberizing equipment.

Workers at the United States Forest Products Laboratory later found it advantageous to add a small amount of an alkaline agent to the sodium sulfite solution to control the pH and reduce corrosion. They also found that cooking to the fiber liberation point was not necessary. Partially pulped chips at 60 to 75% yield could be defibered mechanically without damage to the fibers. The first NSSC mill started operations in 1925 for the manufacture of corrugating medium from chestnut chips from which the tannin had been extracted. The product found ready acceptance and, following the first publication about the process by Rue, Wells, Rawling, and Staidl of the United States Forest Products Laboratory in 1926, several similar mills were soon in operation. During the 1930s total daily capacity remained close to 350 ton, but rapid expansion began in 1944, following the conversion of a kraft mill in the United States to NSSC with a daily production of 200 ton. The same year marked the first preparation of NSSC pulp by a continuous process. Four years later, fully bleached pulp went into commercial production. By the late 1950s, 47 mills were operating in the United States; in 1963 about 2.5 million ton of NSSC pulp were produced. The future growth of hardwood NSSC pulps is assured by the unrivaled stiffness that these pulps impart to corrugated board, the demand for which is expected to increase steadily. Total production of bleached NSSC pulps is small and further growth is restrained by high bleach costs. The NSSC process has reached a state of development where new plants are becoming standardized in design and construction details.

Semichemical pulping was fostered by the development of more efficient refining equipment to replace the older hammer mills, plug refiners, and rod mills. The process is attractive economically because of the high yields attainable, the low chemical consumption, and because the process lends itself to small units and a minimum of plant investment. Yields are generally about 10 to 40% higher than conventional pulping processes because only about 25 to 50% of the lignin and 30 to 40% of the hemicelluloses in the wood are removed, compared with a removal of about 90 to 98% of the lignin and about 50 to 80% of the hemicelluloses in ordinary chemical pulping.

Full bleaching of semichemical pulps produces pulps of low lignin content and exceptionally high hemicellulose content. Such bleached pulps are more like full chemical pulps, the main difference being that a large part of the purification is obtained in bleaching rather than by cooking. However, because the action of the cooking and bleaching is directed selectively on the lignin, bleached hardwood pulps can be produced in a yield of 60%, compared with a yield of 45 to 50% for fully cooked pulps. In other words, a ton of bleached pulp can be produced from 1.3 cords of hardwoods, using the semichemical process, compared with the 2 cords of pine that are needed to produce a ton of bleached pulp by the sulfate process. The semichemical process is best suited for the hardwood species, and is little used for pulping of pine, spruce, and similar softwood species. It produces a stronger pulp than can be obtained by full chemical cooking when the hardwood species are used. The bleaching of semichemical pulps is discussed in the chapter on bleaching.

Chemical Reactions in the NSSC Process

The mechanism of the NSSC attack on the interfiber bond involves primarily the sulfonation in the solid state of a portion of the sulfonatable groups of the so-called A-lignin in the lignin-carbohydrate complex, followed by partial hydrolysis to a soluble lignin sulfonate and soluble carbohydrates. Thus the middle lamella is partially dissolved and the fiber bonds weakened. Application of very high neutral sulfite concentrations and temperatures will bring about nearly complete solution of all the lignin, including the ordinarily acid soluble B-lignin. Although hemicelluloses are fairly resistant to hydrolysis under neutral or slightly alkaline conditions, appreciable quantities are removed by neutralization of easily removed acetyl and other acid groups and by hydrolysis of carbohydrates in the lignin-carbohydrate complex.

Composition of Cooking Liquor in the NSSC Process

When the concept of semichemical pulping originated, the conventional acid sulfite liquor was quickly found to be unsuitable. Acid sulfite liquor required less chemical and shorter cooking time than neutral sulfite, but the strength and brightness of the pulp were lower. The most satisfactory pulping agent was found to be a solution of sodium sulfite containing sufficient buffering agent to keep the pH around 7 during cooking, from which the name, neutral sulfite, was derived. The buffering agent is used to control the corrosion of equipment and increase the yield of pulp. It neutralizes the organic acids that are formed when wood is heated to 120°C or more. Originally the ratio of sulfite to buffering agents was considered to be of great importance. However, Husband[178] was able to obtain strong bleachable pulps from hardwoods at a pH within the range of 3.7 to 5.8 without the use of added buffer, controlling the pH by proportioning sodium sulfite to the amount of wood acids produced during digestion, thus creating an in situ sodium acetate-acetic acid buffer system. The ratio of buffer-to-sodium sulfite is now considered to be a secondary variable with the concentration of bisulfite ions being much more important. Initially, a ratio of sodium sulfite to sodium bicarbonate of 4:1 was recommended for unbleached pulps with higher ratios of about 7:1 recommended for bleached pulps. The whitest pulps were claimed when the pH of the charge at the end of the cooking period was around 7.2 to 7.5. Sodium bicarbonate is the preferred buffer because of its strong buffering action and because it does not darken the pulps as much as sodium hydroxide. At least one mill making pulp for corrugating medium has found that board of good quality can be produced with a sulfite-buffer ratio as low as 1:0.5. This change results in a considerable saving in chemicals.[179]

Cooking liquor for NSSC pulping has been prepared in a number of ways:

1. Purchased sodium sulfite and sodium carbonate are dissolved in water or weak spent pulping liquor.

2. Sodium carbonate or sodium hydroxide or mixtures of the two are reacted with sulfur dioxide in solution.

3. Green liquor is sulfited directly with burner gas in a concurrent flow absorption tower.

4. The liquor in the smelt dissolving tank is a solution of sodium bisulfite.

The liquor at a pH of 8 to 8.5 contains two buffer systems: the sulfite-bisulfite system and the carbonate-bicarbonate system. The alkaline liquors, soda ash or green liquor, can be sulfited either batchwise or continuously. Magruder[180] has described a continuous process for sulfiting sodium carbonate solution for the NSSC process. Sodium carbonate ($Na_2CO_3 \cdot H_2O$) crystals stored in slurry form are extracted continuously with water at 40 to 45°C to produce a uniform saturated solution (30.0 ± 0.5%). Molten sulfur (121 to 135°C) is pumped from storage to a spray-type burner. By controlling the flows of these two materials and the dilution water, a liquor of constant composition can be prepared with little attention by the operator.

The effect of time and temperature in semichemical pulping is similar to that in alkaline pulping, except that the neutral liquor reduces the wood chips to pulp more slowly than sodium hydroxide. Chari[181] made a study of the kinetics of NSSC pulping of aspen chips. He found the rate of delignification can be estimated by the following equation:

$$r_L = \frac{-dL}{d\theta} = (1.6798 \times 10^{11})(\exp -27{,}925/T)(L^{2.1825})(S^{0.4120})$$

where:

$$r_L = \frac{-dL}{d\theta} = \text{rate of delignification}$$

$$= \frac{\text{grams of lignin removed}}{(100 \text{ g OD wood}) (\text{minute})}$$

θ = time minutes

L = lignin content of pulp, $\dfrac{\text{grams of lignin}}{100 \text{ g OD wood}}$

T = reaction temperature °R.

S = concentration of $Na_2 SO_3$ in cooking liquor in grams per liter as $Na_2 O$.

The effect of temperature on the rate of reaction can be computed from the relationship:

$$p = \exp -27{,}975 \left(1/T - 1/672\right)$$

where:

T = reaction temperature °R

and the values of the H-factor can be computed by the relationship

$$H = 1/60 \int_0^\theta p\,d\theta$$

Neutral sulfite cooking liquor is relatively specific for lignin and it has been claimed that lignin is the principal substance removed up to temperatures of 170°C in the pulping of aspen. Some of the more soluble hemicelluloses are removed, but most of the alpha cellulose and more resistant hemicelluloses are left in the pulp. In an 80% yield pulp from aspen, the removal is about 31% of the soluble hemicelluloses (soluble in 5% KOH), 5% of the total hemicellulose, and 50% of the lignin. The pulp contains about 19% lignin. There is some evidence that at temperatures over 180°C the reaction is less selective toward lignin and the lignin remaining in the pulp is less susceptible to bleaching agents.

Cooking Variables in the NSSC Process

The conditions of semichemical pulping vary according to the grade of pulp desired. When making coarse grades of pulp, the digester is charged with unbarked wood. The cooking liquor is added, heated to temperature in 1 hr, and held there for 15 min to 4 hr. The presence of bark lowers the yield and the density and strength of the pulp somewhat, but moderate amounts of bark do not require any significant changes in cooking conditions from that used with unbarked wood. In one mill, pulp suitable for corrugating board is made by pulping a mixture of unbarked hardwoods with a small percentage of softwoods in neutral sulfite liquor using a chemical ratio of 7:1, 13% chemical on the wood, a cooking time of about 4.5 hr, and a pressure of 690 to 825 kPa (100 to 120 psi). The yield is 80 to 90% of the weight of the original wood (on a bark-free basis). A process suitable for the manufacture of corrugating paperboard from New England hardwoods has been described as follows: chemical ratio of sodium sulfite to sodium bicarbonate 5:1; concentration of liquor (calculated as sodium carbonate) of 35 g/l; amount of chemical on the basis of the wood, 14%; temperature of about 170°C; pressure of about 689 kPa (100 psi); time at maximum temperature of 2 to 3 hr. The yield is about 70%, which is low for this type of pulp.

Bleachable grades of semichemical pulp are made by cooking barked wood, using a higher chemical ratio, lower temperature and pressure, and a longer cooking schedule. The original semichemical process described by the United States Forest Products Laboratory involved:

1. Steaming of wood chips for 0.5 hr at atmospheric pressure.

2. Introduction of liquor and application of pressure of 690 kPa (100 psi) at a temperature of 120 to 125°C for 1 hr.

3. Removal of excess liquor not absorbed by chips by blowing back to the liquor tank for reuse in succeeding cooks.

4. Digestion at 140 to 160°C for 1 to 6 hr.

This process is practiced today, except that higher temperatures, up to 170 to 175°C, are used. A typical process for making bleached semichemical pulp using a mixture of southern hardwoods (25% red gum and 75% white and red oak) is based on the following digester charge:

$$22,453 \text{ kg OD wood}$$

$$4,264 \text{ kg } (19\%) \text{ Na}_2\text{SO}_3$$

$$712 \text{ kg } (3.2\%) \text{ Na}_2\text{CO}_3$$

$$21,215 \text{ l. pink liquor} \pm 200.4 \text{ g/l. Na}_2\text{So}_3 \text{ and}$$
$$34.8 \text{ g/l. Na}_2\text{CO}_3$$

$$33,308 \text{ l. brown liquor}$$

$$(\text{liquor ratio, } 3.25:1)$$

Indirect heating is used, and the cooking cycle is 2 hr to 760 kPa (110 psi) and 1 hr and 45 min at 760 kPa (110 psi). The blow liquor contains 10 to 12 g/l of residual sodium sulfite and has a pH of 8.0.

A summary of the cooking conditions used in the NSSC process for board-grade and bleaching pulp is given in Table 4-12. The amount of sodium sulfite

TABLE 4-12 COOKING CONDITIONS USED IN THE NEUTRAL
SULFITE SEMICHEMICAL PROCESS FOR BOARD AND BLEACHING
PULPS

	Board Pulp	Bleaching Pulp
Pulp yield (%)	70–80	65–72
Cooking chemicals		
Sodium sulfite (% of wood)	8–14	15–20
Sodium bicarbonate (% of wood)	4–5	3–4
Chemical requirements		
Sulfur (kg/ton AD pulp)	3.2	52.2
Soda ash (kg/ton AD pulp)	119	209
Cooking temperature (°C)	150–185	160–175
Time at cooking temperature (hr)	0.3–4	3–8
Energy for fiberizing ton of AP pulp (hpd)	12–18	10–15
(kWh)	216–324	180–270

in the NSSC process varies, depending on the yield desired. Typical amounts
used for varying yields expressed as percent sodium sulfite on the oven-dry
wood weight are as follows:

Yield, %	% Sodium Sulfite
70	14–18
75	12–16
80	10–14
85	8–12

As a guide to chemical usage, Schied[182] recommends a ratio of approximately
0.54 kg (1.2 lb) of sodium sulfite per 0.45 kg (1 lb) of lignin in the wood, which
comes out, in the case of aspen, to be 204 kg (450 lb) of sodium sulfite per cord
of wood, or a little less than 272 kg (600 lb/ton) of air-dry bleached pulp. An
average concentration of the liquor is around 120 g/ℓ of sodium sulfite and
20 g/ℓ of sodium carbonate, in the case of aspen, with higher concentrations
used for the more dense woods.

The NSSC process has the advantages of producing light-colored pulps, high
yields for a given degree of delignification, high strength, and no offensive odor.
The economy of wood and chemical consumption and the wide applicability
to different wood species, particularly those readily available but little used
(such as the oaks, gums, aspen, and birch), has made this a very popular process.
The NSSC process produces pulps that make more rigid papers than the usual
alkaline pulps, which is an advantage for making corrugating medium. Ammonium sulfite reacts similarly to sodium sulfite, although it is slightly more
reactive (lower chemical-to-wood ratio and shorter cooking time for the same
yield) and the pulp tends to be darker and slightly lower in strength.

In semichemical pulping, the chemical charge based on the oven-dry wood is reduced below levels for full chemical pulps. It is good practice to test a sample of the liquor with the pulp or softened chips at the end of the cook to be sure that the correct amount of active chemical has been charged with the chips. In some cases, sufficient liquor will be introduced to cover the chips in the digesters. Hydrostatic pressure and heat may be applied to improve liquor penetration of the chips before the liquor is drained away. Liquor so removed can be fortified with chemical and reused. Steam can be introduced to finish the pulping in the vapor phase. Complete impregnation of the chips with liquor is essential to insure uniform pulp with a minimum of fiber bundles. Particular attention should be paid to the dimensions of the chips and to the concentration of the pulping liquor. The rate of penetration of alkaline solutions is independent of the direction; acid liquors penetrate preferentially parallel to the fibers. Penetration of liquor into wood chips can be facilitated by:

1. Submersion of the chips in liquor.
2. Physically reducing chips in a hammer mill or special disk refiner.
3. Crushing chips in a screw press.
4. Presteaming.
5. Preevacuation.
6. Application of hydrostatic pressure.

High-yield pine-kraft pulp produced from chips passing a 12.7 mm mesh and retained on a 6.3 mm (0.25 in.) mesh develops strength properties equivalent to that made from longer chips. Bark should be removed from wood intended for use in bleached pulp; however, small-diameter centrifugal cleaners are very effective in removing fine bark specks.

The objective of cooking in the semichemical process is to put the chips into a condition in which they can be fiberized with a reasonable power consumption. Therefore, it is important to know the extent to which the wood fiber structure has been "loosened" during cooking. The uniformity of commercial operation can be followed, over the short term, by testing samples of pulp collected before and after the refiners, at regular intervals, and noting the flow through the refiners and the load on the drive motors. It is common practice to verify in the laboratory the validity of the tests on the production equipment. This frequently is judged by the power consumed by a bench-scale refiner or by determining the change in freeness or other properties of the pulp when treated under controlled refining conditions. Another technique, described by Sadler and Trantina[183] and others, involves the testing of samples of the cooked wood in a device that measures the shearing strength. The correlation is fairly good with refining power consumption and pulp properties obtained under mill conditions.

Batch Digesters in Semichemical Pulping

Much semichemical pulp was originally produced by the batch method, using either spherical rotary digesters or old stationary vertical digesters. Liquor charged to the digesters usually consists of part fresh liquor and part spent liquor from previous cooks. Maximum cooking pressures vary between 689 to 1033 kPa (100 to 150 psi), corresponding to temperatures of 170 to 186°C. Cooking time ranges from 2 to 10 hr.

Rotary globe digesters are usually 4 to 5 m (13 to 16 ft) in diameter and are equipped for direct steaming. In the early operations the digesters were filled completely with chips and strong NSSC liquor and heated at 120°C for 1 hr. The free liquor was blown from the digester and the cook conducted with direct steam in the so-called vapor phase. At the end of the cook, the chips were dumped into a tank equipped with a leach caster. In other cases existing vertical stationary digesters were adapted for semichemical pulping by installing larger blow valves and blow lines. The digestion is conducted by conventional means. The pulp may be blown into blow tanks or into blow pits fitted with leach casters.

Continuous Digesters in Semichemical Pulping

Continuous digesters are used in a number of mills, and the number of installations will undoubtedly increase in the future. These digesters not only offer the usual benefits of uninterrupted operation, but also have the additional advantage of presenting the softened chips to the refiners at high temperatures. Other advantages are the economy of space required, the uniform boiler-load maintained, and the fact that pulp rather than chips is blown to the blow tank, thus facilitating handling and washing. Although commercial continuous digesters are physically different, they all include equipment that will accomplish three functions: (1) introduction of chips into a pressure vessel, (2) retention of chips in a vessel under steam pressure, and (3) discharge of cooked chips.

Feeding Chips to Continuous Digesters. In the early stages of the development of continuous pulpers various means of introducing chips into the vessel under pressure were considered. A reciprocating plunger was found to be less reliable than a feed screw. The feed screw compressed the chips to form a plug and forced this plug into the digester against the cooking pressure. There is still some question about the effect of the severe mechanical treatment in the screw on the papermaking properties of the pulp produced, especially with softwood chips. There are indications that the liquor, added to the chips as they expand after discharge from the screw, completely penetrates the chips in a very short time.

Rotary pocket valves have been used to introduce chips into the digester as well as to control their discharge from the digester. They are commonly used in situations where the pressure difference across the valve is not great. As each pocket discharges the chips into the digester and continues to turn, it is filled with steam that will be lost to the atmosphere unless precautions are taken to prevent this. The loss can be reduced if vent lines are installed to connect pockets of high pressure to those of lower pressure. Some preheating of the chips undoubtedly takes place. The Kamyr feeder (shown in Figure 4–50) is very effective, does not subject the chips to severe mechanical action, and there is no steam loss.

Horizontal-Tube Digesters. The first method of continuous pulping was to presteam the chips, add chemicals and then pulp continuously in a horizontal tube with a flight screw to convey the chips. The digester consists of from one to several tubes (reaction chambers) mounted one above the other. These units are fabricated by a number of machinery suppliers. One type of steaming tube made by Beloit-Jones is shown in Figure 4–48. Cooking liquor, dispersed by steam, is sprayed on the chips as they leave the feeder. The volume of liquor is insufficient to cover the chips completely. The chips are moved along continuously by the screw conveyors from chamber to chamber until they are discharged from the last tube. Retention time is determined by the speed of the screws. A development in the horizontal-tube continuous digester is to have the top (or first) tube after the feeder inclined at about 45°. The tube is partially filled with cooking liquor so that the chips are totally immersed in the liquor. A considerable portion of the liquor drains from the chips; the cook is completed in the vapor phase in the succeeding tubes. In early installations the diameter of the tube was about 0.6 m (24 in.) or less. The size of the tube has been increased over the years, up to a diameter approaching 1.8 m (6 ft). Since the digester is completely filled with chips only when operating at maximum capacity, it is possible to decrease the production rate considerably without changing the speed of the screws or the retention time.

Inclined-Tube Digesters. The Messing-Durkee digester consists of a tube, divided into chambers by a midfeather, erected at an angle of 45°. A drag conveyor moves the chips down the length of the upper chamber and back up to the high end, where they are discharged. The tube is partially filled with

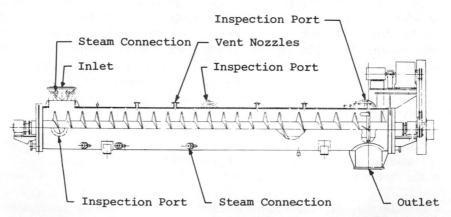

Figure 4–48. Beloit-Jones steaming tube used in semichemical pulping. (Courtesy Beloit Corp.)

cooking liquor so that the chips are submerged during a great part of the time. There are installations in which the chips are discharged into another similar unit or into a vertical reaction tank for additional reaction time. Since the retention time is set by the speed of the drag conveyor, the production rate can be changed without changing the retention time.

Vertical-Downflow Digesters. The vertical-downflow continuous pulpers use relatively tall slender tanks equipped for zone circulation and heating, as shown in detail in Figure 4–49. Chips may be introduced into the digester zone by either: (1) special low- and high-pressure rotary-type feeders with an intermediate steaming vessel, or (2) a compression-type feed screw. The mass of chips in the digester flows downward by gravity. The diameter at the bottom of the tank is greater than at the top in order to promote plug flow. The taper may be either a continuous or a step-wise one. If the digester is used for semichemical pulping, steam is added directly to the digester near the point where the chips enter. In the Kamyr system, which is shown in Figure 4–50; an inclined separation tube at the top of the digester is used so that the liquor can be recycled back to the high-pressure chip feeder. The liquor level in the digester is maintained approximately 1 m below the top so that the chips are exposed directly to steam before being immersed in the liquor.

Some installations have special features to insure thorough impregnation of the chips. Kamyr uses a separate preimpregnation vessel. The Kamyr digester and the modification using preimpregnation are described in the section on alkaline pulping. In the Defibrator digester, shown in Figure 4–51, the chips are presteamed before the screw feeder, which introduces the chips into the bottom of a built-in preimpregnating section. This is a vertical tube inside the digester and has double-screw conveyors that lift the chips and liquor close to the top of the digester and then drop them into the vacant space. Liquor is totally absorbed by the chips and pulping occurs in the vapor phase. If the level of chips in the digester must be held constant, the retention time will vary with the production rate.

Discharge Mechanisms. Pulp from continuous digesters discharging under pressure is blown to a cyclone for reduction of pressure and is then dropped into a blow tank or a live-bottom bin, which is a tank with a bottom composed of a number of parallel screw flights for removing the contents continuously. Digesters of the horizontal-tube type are frequently installed with a variable orifice mounted so that the inlet side is wiped mechanically to prevent plugging. In another common arrangement, the cooked chips fall on a screw that conveys them into a disk refiner operating under full digester pressure. One or two variable orifices are mounted on the shell of the refiner and discharge into blow lines to a cyclone above the blow tank. The consistency in these cases approximates 25%. Most of the inclined-tube digesters use rotary pocket valves as discharge devices. Sufficient steam is carried in the pockets to transfer the pulp to a blow tank. The Kamyr and Esco vertical-downflow digesters use a rotating

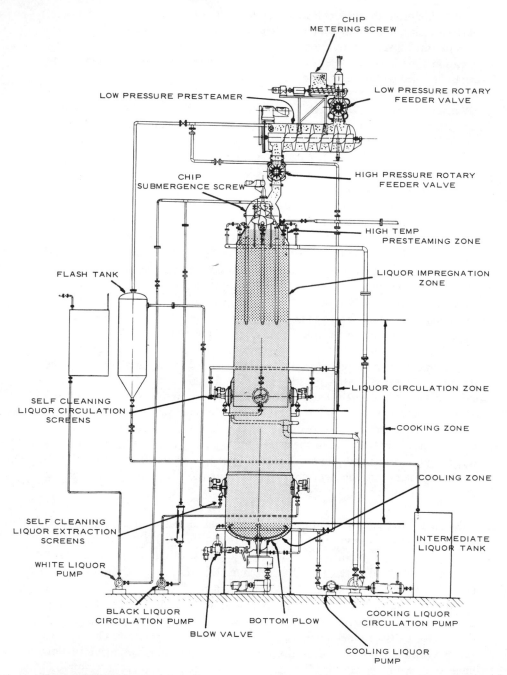

Figure 4–49. Showing details of a continuous digester suitable for semichemical pulping. (Courtesy Esco Corp.)

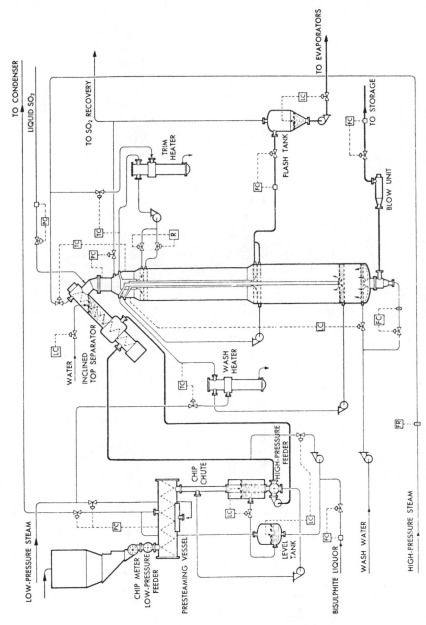

Figure 4-50. Typical Kamyr system for production of NSSC pulp showing how liquor is recirculated from top separator to the high-pressure chip feeder. (Courtesy Kamyr Co.)

269

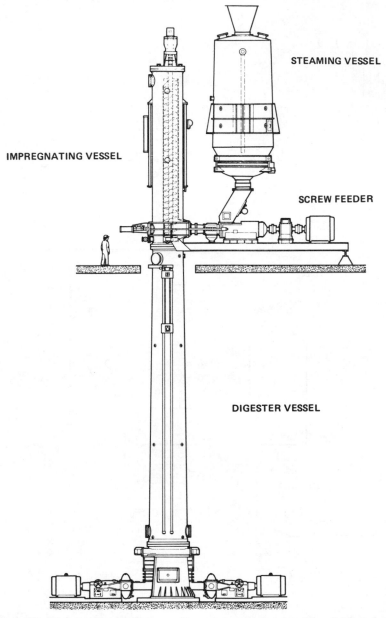

STEAMING VESSEL

IMPREGNATING VESSEL

SCREW FEEDER

DIGESTER VESSEL

Figure 4–51. Showing a Defibrator digester used for semichemical pulping. (Courtesy American Defibrator, Inc.)

arm to collect the pulp from the entire bottom of the digester and to direct it to a central opening leading to the blow line. Flow in the blow line is controlled by a system of valves so that the chip level in the digester remains constant. A disk refiner may be installed in the blow line as the first fiberizing step. The Defibrator digester uses a defibrator and variable orifice in this position.

Cooking in Continuous Digesters. Because the pulping temperature in continuous systems is generally higher (up to 190°C) than in batch systems, the rate of reaction is higher and the total pulping time is reduced. Thorough liquor penetration of the chips is essential if burned chips are to be avoided. Penetration is facilitated by the wringing action of the screw feeder, which opens up the chips along lines of natural cleavage. If the compacting ratio of the screw is too high, the pulp may be degraded. A compacting ratio of 1.4 to 1 has been found to facilitate penetration of cooking liquor. Another explanation for good liquor penetration in chips after a feed screw is that the compressed chips swell after the relief of pressure and soak up the liquor applied at this point, much like a sponge. Excess liquor may be removed by draining and the cook finished in the vapor phase. When pulping at temperatures in the range of 170°C, impregnation is less essential and the chips may simply be sprayed with cooking liquor as they enter the top tube. Steam may be added in the second tube to maintain cooking temperature.

NSSC pulp for corrugating medium is made by spraying chips with sodium sulfite liquor containing enough carbonate to give a final pH of 8.0 to 8.5 at a liquor-to-wood ratio of 2.5 to 1. Representative cooking times range from about 12 min at 190°C to 34 min at 177°C. Pressures in the digester range from about 1033 to 1275 kPa (150 to 185 psi). The consumption of cooking chemical is generally about 9% sodium sulfite on the weight of the wood for a yield of about 80%. Loving[184] reports a yield of 75% with a chemical consumption of 6.75% sodium oxide on the weight of pulp produced.

Bleachable grades of hardwood semichemical pulps are produced in continuous digesters using NSSC liquor. The cooking cycle is generally longer than that used with pulp for corrugating medium, but it is still shorter than the 6- to 8-hr cycle required for batch digesters. The chemical charge, based on the oven-dry wood, ranges up to 20% sodium sulfite (9.84% Na_2O), and the yield is correspondingly lower. Bleachable pulps have been produced in the horizontal-tube (Pandia) digester, using monosulfite liquor in the ratio of 5 to 1 sodium sulfite to sodium carbonate, a cooking pressure of 792 kPa (115 psi), a cooking temperature of 186 to 188°C, and a cooking cycle in the 40- to 45-min range. The unbleached yield is approximately 70 to 73% at a Roe chlorine number of 14 to 15. The bleached pulp yield is in the 55% range based on moisture-free wood. The brightness following three-stage bleaching is 80 to 84.

Several mills producing pulp for corrugating medium started using NSSC liquor with a ratio of sodium sulfite to sodium carbonate of 5 to 1; however, over a period of several years, they gradually reduced the ratio to less than 1 to 1 as part of the normal cost-reducing program. The chemical charge was controlled so that the final pH of the blow stock was between 9.0 and 9.5. Later developments include the use of cooking liquor composed of a mixture of sodium hydroxide and sodium carbonate in order to reduce air pollution resulting from the direct sulfitation of green liquor produced in the recovery furnaces. Under these conditions the smelt produced consists mainly of sodium carbonate. Sodium hydroxide is added as makeup. Pulp for corrugating medium has been produced by pulping hardwoods with 3% Na_2O as kraft liquor; the pH of the blow stock varied between 5.5 and 6.0. The color of the board was chocolate brown. The odor of furfural was noticeable around the digesters. Pulp for corrugating medium has been produced with a 5% sodium hydroxide solution from hardwoods. The consumption of chemical approaches 5 to 8% sodium oxide per ton of pulp. The yield varies between 75 and 90%.

E. Santiago[185] has reported that the reaction of sodium carbonate with hardwood chips at 190°C depends on the chemical charge based on the moisture-free wood. The data have been presented in a Ross diagram as shown in Figure 4-52. Treatment with water alone causes a loss in yield to about 63%, including losses of both lignin and carbohydrate. As the sodium carbonate charged to the digester is increased to 10%, less carbohydrate is removed, while the lignin content remains essentially unchanged. When the sodium carbonate is further increased, lignin is removed preferentially, while the carbohydrate content remains essentially constant.

Energy Conservation. To conserve energy and use waste heat, continuous digesters have provisions for using low-pressure steam to preheat the chips before they are introduced into the high-pressure vessels. This steam is obtained

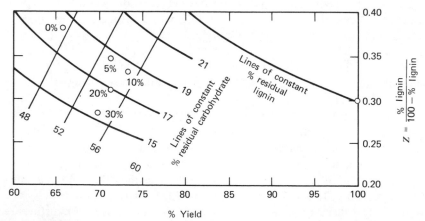

Figure 4-52. Ross diagram showing treatment of hardwoods with sodium carbonate solutions (0 to 30%) in semichemical pulping showing lines of constant percent residual carbohydrate and constant percent residual lignin at various yield levels.

from flash tanks that are used to cool the spent pulping liquor before it is sent to the multiple-effect evaporators or from the blow tanks. The presteaming of chips to 100°C with the low-pressure steam recovered from the process will permit a considerable reduction in the amount of high-pressure steam required to heat the chips to the maximum pulping temperature. Assuming an ambient temperature of 15.6°C, then:

$$\Delta T_1 = 99° - 15.6 = 83.4°C$$
$$\Delta T_2 = 170° - 99 = 71.0°C$$

and total

$$\Delta T_T = 170° - 15.6 = 154.4°C$$

In some cases[186] the steam is used in the multiple-effect evaporators. Rosenblad[187] has designed a system in which blow steam from batch digesters is used to partially concentrate weak black liquor.

Refining Semichemical Pulp

Refining is an integral part of semichemical pulping, and hence must be discussed in relation to the pulping process. Refining can be carried out in two stages referred to as primary and secondary refining, or it can be done by progressive multistage refining with each refiner doing a share of the work on the pulp. Washing may follow refining or it may precede refining if presses are used for washing. A full description of refining is given in the section on mechanical and thermomechanical refiner pulping.

In early installations, refining followed immediately after cooking and the pulp was washed on vacuum-drum washers. Later with the advent of screw presses, cooked chips from the digester were washed, drained, and pressed to a consistency of 60% or more before refining. Foaming problems can be avoided by pressing the pulp to 40% consistency. If especially clean pulp is desired, the cooked chips may be pressed, refined, and then washed on vacuum drums. When pulp for corrugating medium is being made, washing may be eliminated; the paper-machine white water is used for dilution to retain as much as possible of the dissolved solids in the paperboard. This has the effect of increasing the yield to over 80% and improving the stiffness of the board. Washed stock runs better on the paper machine and causes less corrosion. These factors, plus more stringent regulations imposed on mill effluent, have made it desirable and necessary to wash all grades of pulp.

The major variables in fiberizing which affect the quality of the pulp are:

1. The type of raw material fed to the refiner.
2. The type of fiberizing equipment.

3. The design of the fiberizing surfaces.

4. The shear speed between the fiberizing surfaces.

5. The power expended in fiberizing.

6. Consistency.

7. Temperature.

8. Clearance between surfaces.

9. Effective area of fiberizing.

10. The throughput rate.

The principles of refining semichemical pulps have not been thoroughly established, but best results appear to be obtained by the use of high-shear and high-peripheral speeds, high stock temperatures, and high consistencies.

The first refining after cooking is primarily a defiberizing action. It is generally carried out at high temperatures, up to 93°C, using disk refiners with plates set at 0.25- to 0.30-mm (0.010- to 0.012-in.) clearance to break up the chips without excessive power consumption or excessive hydration of the fibers. With batch digesters, where the consistency is 12 to 14%, the chips are allowed to drain to 17 to 20% consistency and then they are diluted with hot, strong spent pulping liquor to a favorable consistency for refining, about 12%. Nolan[188] found that low-consistency refining requires as much as 70% more energy than at high values. High consistency and wide plate clearance produce more of a rubbing action between chips, with the result that the fibers and fiber bundles appear longer. Secondary refining is carried out for strength development.

The total work done on semichemical pulp in fiberizing and refining varies considerably. The range for NSSC pulp is 180 to 360 kWh (10 to 20 hpd) per ton of pulp and about 360 to 540 kWh (20 to 30 hpd) per ton for cold soda pulp. Power requirements for bleachable pulps refined to a freeness of about 625 CS is approximately 190 kWh (10.5 hpd) per ton. Energy consumption is increased with close plate clearance but is reduced by a high-moisture content of the chips. To develop maximum strength, the temperature of the stock should be between 50 and 90°C. Most secondary refining is carried out at consistencies of 4 to 8%, but consistencies up to 30% can be used for very soft pulps. Tensile and bursting strength are increased at high consistencies, while tear resistance is reduced at values over 10%.

One of the problems with high-yield pulps for use in fine papers is shives. The amount of shives can be reduced to acceptable levels by judicious screening and refining.

Washing Semichemical Pulp

Semichemical pulps can be washed by any of the methods used for fully cooked pulps. Generally the process will use two to four stages in countercurrent flow of wash water. The chemical loss with the pulp is dependent on the dilution factor used.

Drum Washers, Vacuum or Pressure. The pulp must be refined so that it can be picked up on the washing drum and a continuous cake of uniform thickness formed. The refining must be controlled so that pulp in the cake is free and wash water will flow through easily. If the pulp is refined too much, it will tend to blind the facing wire and the filter cake will be difficult to remove.

Diffusion Washers. Some pretreatment will be required to eliminate fiber bundles and chips.

Press Washers. No pretreatment of the stock is necessary when using presses for washing because the rather severe mechanical action will open up chips and fiber bundles. The first-stage press will not only remove a considerable amount of spent pulping liquor at a reasonable concentration, but also will exert a fiberizing action on the softened chips. The power consumption in increasing the consistency of the pulp to the 60 to 70% level may be expected to be in the range of 27 to 72 kWh (1.5 to 4.0 hpd) per ton and should permit a reduction of energy in later stages of refining.

The overall layout of a semichemical plant is shown in Figures 4–53 and 4–54.

Preparation of NSCC Cooking Liquor and Chemical Recovery

In the early 1950s it was common practice to prepare NSSC cooking liquor by dissolving sodium sulfite and sodium carbonate or bicarbonate in water or dilute spent pulping liquor. This was a simple operation and required little special equipment. Those mills that had sulfur burners prepared cooking liquor by sulfiting soda ash in a simple absorption tower. There was considerable latitude in the composition of the sulfited liquor.

Since the chemical charge based upon moisture-free wood was considerably less than that used in kraft mills, some considered it permissible from an economic standpoint to discharge the spent pulping liquor to the receiving stream. Several installations use kraft green liquor as the pulping agent for the production of corrugating medium. At least one is not associated with a kraft mill; it concentrates and burns the spent pulping liquor. The smelt solution may be used without further treatment in the digesters or sodium sulfite may be added. Other methods of recovery are discussed below and also in the section on sulfite pulping.

Some semichemical pulp mills preimpregnate the chips by covering them in the digester with cooking liquor. A portion of the liquor is withdrawn, fortified and reused in subsequent cooks. In some mills, more than half the spent pulping liquor is recycled resulting in significant chemical and heat economies. Keller and McGovern[189] found that reuse of spent NSSC liquor when pulping aspen had no effect on pulp strength, increased the yield from 75 to 76%, and reduced the brightness from 51 to 47%.

Some mills discharge the spent pulping liquor to a receiving stream. However the high biological-oxygen-demand (BOD) content, due to the hemicellulose and

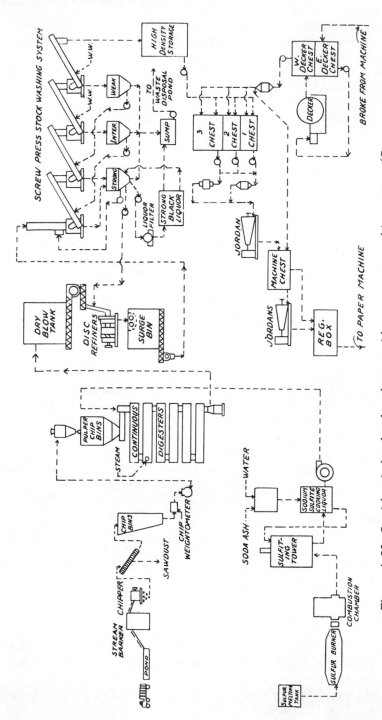

Figure 4-53. Semichemical pulp plant layout with screw press washing system. (Courtesy *Paper Trade Journal*.)

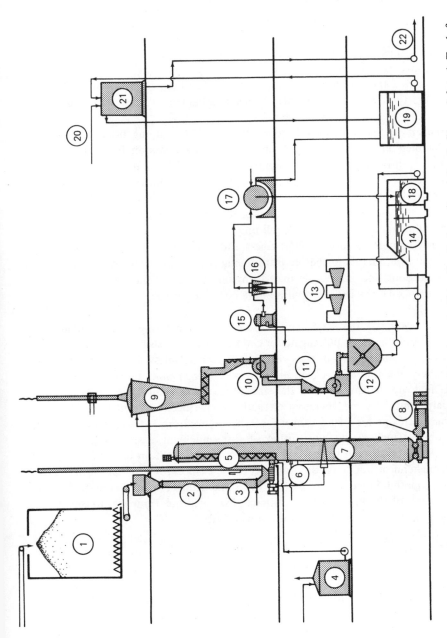

Figure 4-54. Flow sheet of a semichemical pulp plant. 1. Chip silo; 2. Steaming tube; 3. Steam inlet; 4. Tank for cooking acid; 5. Impregnation apparatus; 6. Steam inlet; 7. Digester; 8. Defibrator; 9. Pulp cyclone; 10 and 11. Davenport press; 12, 14, and 19. Pulp chest; 13. Raffinators; 15. Pressure screen; 16. Radiclones; 17. Thickener; 18. White water tank; 20. Water for pumping; 21. Pumping tank; 22. Pipeline. (Courtesy American Defibrator, Inc.)

acetate content, creates a serious stream pollution problem. Some mills impound the spent liquor, releasing it to the receiving stream at times of high water. Others concentrate it for use as road binder. In other cases, the spent pulping liquor is concentrated and burned under oxidizing conditions in a fluid bed reactor to give a mixture of sodium carbonate and sodium sulfate, which is of value to a kraft mill or to a glass plant, but has little value to a NSSC pulp mill.

Evaporation of Spent Pulping Liquor. Spent NSSC pulping liquor can be readily handled in the long vertical-tube evaporators common in kraft mills until the concentration of the liquor approaches 45 to 50% total solids. Forced-circulation concentrators and/or direct-contact flash evaporators must be used for concentration to higher levels. The liquor is extremely viscous and highly thixotropic at concentrations approaching 60% solids. Since hardwoods are generally pulped and some hot stock refining occurs before washing, the filtrate is frequently contaminated with short fibers and debris. It is therefore advisable to screen or filter the liquor before introducing it into the evaporators. Under certain conditions, inorganic scale may form on the liquor side of the tubes. Both calcium oxalate and calcium sulfite have been identified.

Cross Recovery. If the semichemical pulp mill is operating in conjunction with a kraft mill, the spent pulping liquor can be sent to the sulfate mill recovery system. This practice, termed cross recovery, is the simplest recovery system. Unfortunately the improved chemical recovery in kraft mills has reduced the chemical makeup demand so that the size of the semichemical operation may be severely limited. Some operating difficulties have been claimed if the two spent-liquor streams are blended ahead of the evaporators. Attempts to avoid these problems include the substitution of the semichemical spent liquor for weak black liquor charged to the kraft digesters or the addition of the concentrated semichemical liquor to the direct contact evaporator after the furnace. Another proposed procedure is to add the semichemical spent liquor to the dissolving tank of the kraft recovery furnace. (See also the Tampella process in the section on pulping and bleaching of nonwood fibers.)

Wet Combustion Process. The wet combustion processes for chemical recovery are described in greater detail under acid sulfite pulping. These processes cover the oxidation of black liquor in the liquid phase, using compressed air at 5512 to 10,335 kPa (800 to 1500 psi) and high temperatures (over 374°C). Sodium sulfate, steam, and uncondensed gases are produced. No sulfur is recovered. The sodium sulfate may be converted to sulfite. The steam and hot uncondensed gases are used to drive a turbine.

In the Zimmerman process, as practiced in one plant on a pilot scale, the spent liquor is raised to a pressure of 5512 kPa (800 psi), the temperature is raised to 121°C, and air is supplied. Steam at 1033 kPa (150 psi) initiates the combustion, and the reaction becomes self-sustaining at about 207°C. During combustion the organic matter is oxidized to carbon dioxide and water, and the inorganic matter is converted to sodium sulfate and sodium acetate.

Direct Sulfitation of Wet Combustion Spent Liquor. The sodium sulfate produced by the wet combustion of NSSC spent liquor can be converted to sodium sulfite in a chemical recovery process by mixing it with a calcium sulfite sludge (obtained from the process) and sulfiting with sulfur burner gas. The reactions are as follows:

$$CaSO_3 + SO_2 + H_2O = Ca(HSO_3)_2$$
$$Na_2SO_4 + Ca(HSO_3)_2 = 2\ NaSO_3 + CaSO_4$$
$$2\ NaHSO_3 + Ca(OH)_2 = Na_2SO_3 + CaSO_3 + H_2O$$

Up to 90% of the sodium sulfate can be converted to sodium sulfite.

Dry Combustion. Most of the processes for recovery of NSSC spent liquor are based on dry combustion of the concentrated black liquor. This consists of burning the concentrated black liquor in a recovery furnace. Burning can be done in a small furnace or a smelter, which is a smaller edition of the larger kraft-type furnace. The evaporated liquor at about 30% solids is concentrated by flue gas (538°C) to 65%, at which point it is suitable for burning. The reducing conditions in the furnace change the higher valence forms of sulfur to the sulfide form so that a smelt is produced containing principally sodium sulfide and sodium carbonate. Some sodium sulfate and sodium thiosulfate are also produced.

Direct Sulfitation of Green Liquor. The main problem in sodium-base sulfite liquor recovery is to convert the sulfide (either in the smelt from the furnace or in the green liquor) to sulfite without the formation of sodium thiosulfate. Thiosulfate is generally regarded as deleterious to the bleachability of the pulp, although it is possible to use semichemical liquors containing appreciable sodium thiosulfate under alkaline conditions. The presence of thiosulfate has been reported as accelerating corrosion.

The direct sulfitation of green liquor developed by The Institute of Paper Chemistry keeps the thiosulfate production within reasonable limits. Hot sulfur burner gas saturated with water vapor is introduced into the top of an absorption tower concurrent with a stream of hot green liquor without recirculation. The gas separated from the liquor is discharged through a short stack. The sulfited liquor flows to one of two parallel tanks. When one is full the flow is changed to the empty one. A sample of liquor is analyzed. If the composition is within selected limits, the liquor is sent to the digester house. If the analysis falls outside specified limits, the composition is adjusted before sending it on. This system has been used in several mills making corrugating medium with excellent results for quite a number of years. It is outlined in Figure 4-55. An alternate method of sulfiting green liquor that would reduce the formation of thiosulfate has been developed. In this case sodium bisulfite prepared by reacting sulfur dioxide and sodium carbonate together is rapidly and thoroughly mixed with green liquor. The resulting liquor is satisfactory for producing unbleached

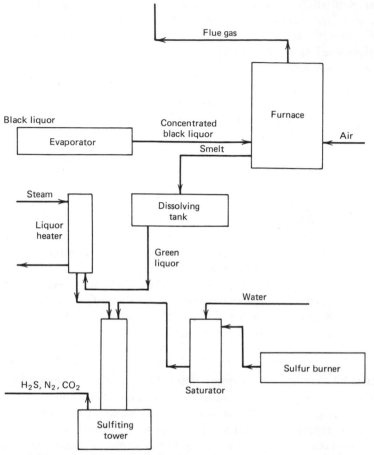

Figure 4-55. Direct sulfitation process for sodium sulfite chemical recovery. (Courtesy The Institute of Paper Chemistry.)

semichemical pulp. In another process[190] smelt from the chemical recovery furnace is contacted with a solution of sodium bisulfite to produce a cooking liquor. The sodium bisulfite solution is produced by scrubbing flue gas containing sulfur dioxide with a solution of sodium carbonate.

In order to overcome the undesirable formation of sodium thiosulfate, a two-stage process of carbonation followed by sulfitation has been proposed. In the carbonation step, the sulfur is displaced and removed and the sodium is converted to carbonate. The separated hydrogen sulfide is burned to sulfur dioxide, which reacts with the carbonate solution to form sodium sulfite. If it is not desirable to recover the hydrogen sulfide as sulfur dioxide, it can be discarded and replaced with sulfur dioxide by burning fresh sulfur.

An early process proposed the use of sodium bicarbonate instead of carbon dioxide for converting sodium sulfide in cooled smelt to hydrogen sulfide as follows:

$$Na_2S + 2NaHCO_3 = 2\,Na_2CO_3 + H_2S$$

The sodium carbonate was dissolved and carbonated to the bicarbonate. The hydrogen sulfide was burned to sulfur dioxide and used for sulfiting a portion of the recovered soda.

A further development was the use of the Sulfox smelt separation system for crystallizing out sodium carbonate from concentrated sodium sulfide, the carbonate being less soluble than the sulfide. Thus only the concentrated sodium sulfide liquor would be carbonated and no carbon dioxide would be consumed in converting sodium carbonate to bicarbonate.

In an American process (Mead), the liquor is evaporated and burned. The green liquor is treated with flue gas so that the carbon dioxide in the flue gas converts the sodium sulfide into sodium carbonate, while the hydrogen sulfide is driven off. The hydrogen sulfide is returned to the furnace where it is oxidized to sulfur dioxide. The carbonated liquor is used to scrub the total furnace flue gases for recovery of sulfur as sulfite. The following steps are used:

1. The evaporated black liquor is scrubbed.

2. The black liquor is burned to produce a smelt.

3. The smelt is dissolved to form green liquor.

4. The green liquor is partially carbonated with flue gas (precarbonation) so that all the sodium sulfide in the incoming green liquor is converted into sodium hydrosulfide and some carbonate is converted into bicarbonate; the hydrogen sulfide in the flue gas is taken off at about 4% by volume and sent to the furnace for burning.

5. The liquor is carbonated again with flue gas during which time the liquor absorbs carbon dioxide and releases hydrogen sulfide.

6. The liquor is then sulfited and sodium makeup is added at the carbonated liquor storage tank as sodium carbonate; sulfur makeup is obtained by burning raw sulfur followed by the introduction of the sulfur dioxide into the flue gases prior to the sulfiting system.

A simplified flow sheet of the process is shown in Figure 4–56. The equations are:

$$Na_2S \rightleftharpoons 2Na^+ + S^{2-} \tag{1}$$

$$Na_2CO_3 \rightleftharpoons 2Na^+ + CO_3{}^{2-} \tag{2}$$

$$S^{2-} + H_2O \rightleftharpoons OH^- + HS^- \tag{3}$$

$$2HS^- \rightleftharpoons S_2{}^- + H_2S(aq) \tag{4}$$

$$H_2S(aq) \rightleftharpoons H_2S(g) \tag{5}$$

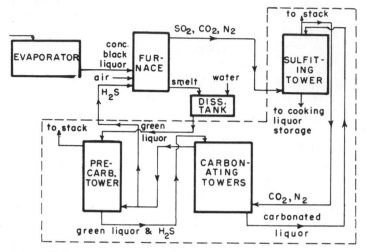

Figure 4-56. Carbonation process (Mead) for sodium sulfite liquor recovery. (Courtesy *Paper Trade Journal.*)

$$CO_3{}^{2-} + H_2O \rightleftharpoons OH^- + HCO^{3-} \qquad (6)$$

$$2HCO_3{}^- \rightleftharpoons CO_3{}^{2-} + H_2CO_2(aq) \qquad (7)$$

$$H_2CO_3(aq) \rightleftharpoons H_2O + CO_2(aq) \qquad (8)$$

$$CO_2(aq) \rightleftharpoons CO_2(gas) \qquad (9)$$

By maintaining a high concentration of CO_2 in the carbonation tower, the equilibrium of equation 9 is shifted to the left. This increases the concentration of dissolved CO_2, which in turn shifts the equilibra of equations 8, 7, 6 to the left with the net that there is a substantial increase in carbonate ions in solution. As the carbonate ions increase, they combine with more of the sodium ions and equations 1, 3, 4, and 5 shift to the right releasing hydrogen sulfide. Since the partial pressure of hydrogen sulfide is low in the gas stream, conditions are such that hydrogen sulfide is stripped out of the liquid phase and carbon dioxide is absorbed. Thus, in the carbonation tower, the CO_2 concentration of the gas is reduced from about 17% to 10 to 13%, while the hydrogen sulfide is raised from about 0 to 7 to 10%. The liquor leaving the bottom of the tower is practically all carbonate and bicarbonate. In the sulfiting tower, sulfur dioxide is absorbed and carbon dioxide desorbed from the cooking liquor. The resulting liquor then contains sodium carbonate, sodium bicarbonate, and sodium sulfite. The efficiency of the recovery is about 97% of the sodium (80 to 90% for the overall pulping and recovery operations) and about 92% of the sulfur (75 to 85% for the overall pulping and recovery operations). The principal objection is the hazard in handling flue gas of high hydrogen sulfide concentrations.

Direct Oxidation Processes. Several recovery processes are based on the oxidation of sodium sulfide directly to sodium sulfite. In the proposed Greenwalt process, molten smelt is solidified, pulverized, and oxidized with air at 225°C. In the Aries-Pollak patent, green liquor is oxidized with air, while being absorbed on a surface material that can be the end product of the oxidizing reaction (sodium sulfite) or sodium carbonate present in the green liquor. The temperature is held below 149°C by controlled addition of water.

In the Western Precipitation process, the sulfur in the spent liquor is fixed by oxidation of a mixture of sodium sulfide and spent liquor before evaporation. The process uses the furnace as a carbonator. By increasing the green-liquor concentration, sodium carbonate is salted out. The sulfide mother liquor, in which the sulfide is stabilized by oxidation to prevent losses during evaporation, is recycled to the multiple-effect evaporators and added to the entering black liquor. A simplified flow sheet is shown in Figure 4–57. The steps include:

1. Steam stripping of the spent liquor to recover sulfur dioxide for reuse.

2. Neutralization, evaporation, and burning of the solids in a recovery furnace.

3. Crystallization of sodium carbonate from the carbonate-sulfide smelt.

4. Sulfiting of the sodium carbonate solution to provide a sodium bisulfite solution for reuse.

5. Return of the sodium sulfide mother liquor to the furnace to be converted in part to carbonate on a continuous basis.

Nugent and Boyer[191] describe some of the details of the Western Precipitation operation as follows:

1. Adding the dissolved smelt, which contains all the sodium sulfide and part

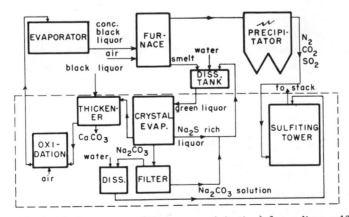

Figure 4–57. Oxidation process (Western precipitation) for sodium sulfite liquor recovery. (Courtesy *Paper Trade Journal.*)

of the sodium carbonate (see point 5 below) from the previously burned liquor, to the incoming spent liquor from the washing operations.

2. Blowing this mixture by air into a foam so that the sulfide from the spent liquor reacts with the sulfite in the lignosulfonates to form thiosulfates and polythionates. (This step is required so that the corrosive sulfite ion or hydrogen sulfide will not be given off in the evaporation step.)

3. Evaporation of the treated liquor to make a high sulfur-content concentrated liquor for the furnaces.

4. Burning the concentrated liquor. This produces sodium carbonate and sulfur dioxide equivalent to the sodium and sulfur in the spent liquor as well as sodium sulfide equivalent to that in the smelt. The important consideration is how the sulfur in the fired liquor is divided between the sulfur that goes into the furnace gases as sulfur dioxide and the sulfur that stays in the smelt as sodium sulfide.

5. Separation of the sodium carbonate from the dissolved molten smelt by crystallization and scrubbing the sulfur dioxide in a tower with this sodium carbonate to produce sodium and sulfite that is returned to the process as cooking acid. One advantage of this process is that no hydrogen sulfide has to be piped from a tower to a burner. If the carbonate solution were used for scrubbing the flue gases from the recovery unit, a high recovery efficiency could be obtained.

In a similar method, the liquor from the evaporators is burned in the furnace and/or carbonated so that the sulfur is converted to sulfur dioxide and the sodium to sodium carbonate. The sulfur dioxide is then absorbed in a scrubbing tower by use of a sodium carbonate solution prepared from the smelt solution. The residual smelt solution, containing sodium sulfide and sodium carbonate, is recycled to the furnace. A simplified drawing is shown in Figure 4-58.

Atomized (Pyrolysis) Process. In the atomized or pyrolysis process, developed by the Pulp and Paper Research Institute of Canada, the spent liquor is evaporated and pyrolyzed under pressure in a jacketed heated furnace (wall temperature $760°C$) into which the liquor is introduced in the form of atomized droplets in the absence of air. In the tower, the droplets pass through successive zones of evaporation, drying, and thermal decomposition to produce sodium carbonate, hydrogen sulfide, and steam. The advantage is that a solid residue is obtained that is very high (about 90%) in sodium carbonate, plus a small amount of sulfate. Unlike other combustion processes, little or no objectionable thiosulfate is produced because the temperature in the reactor is high enough (about $659°C$) to decompose it. The sodium sulfide is decomposed by reaction with steam at the high temperatures prevailing in the reactor. The sodium in the residue is in a ready form for direct sulfitation. The gas obtained, which contains hydrogen sulfide, carbon dioxide, hydrogen, and hydrocarbons, can be burned to make the process thermally self-sufficient. As much as 80% of the sulfur originally present in the spent liquor can be recovered after combustion in the form of sulfur dioxide. This can be used for sulfiting the sodium carbonate residue.

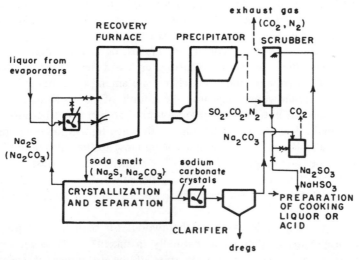

Figure 4-58. Oxidation process (Mannbro) for sodium sulfite liquor recovery. (Courtesy *Paper Trade Journal.*)

Properties of NSSC Semichemical Pulps

Unlike chemical pulps, which are fairly well standardized, semichemical pulp is not a standard grade of pulp; it varies considerably according to the mill that produces it. Semichemical pulps for coarse-paper manufacture cannot be well characterized by their lignin content because so much depends on the location and nature of the lignin. One method of characterizing is to process the pulp in a laboratory beater for a specified time and then screen to determine the amount of fiber rejects.

Species Used. Semichemical pulping is best suited to the hardwood species. It is not so well suited to the softwood species. One disadvantage of softwoods is their high chemical requirement; for example, the pulping of jack pine by the neutral sulfite process requires about 45% more chemical (on a sodium oxide basis) than pulping by the sulfate process. The energy requirement in refining is higher than for hardwoods. Softwood semichemical pulps have a high lignin content and comparatively low strength compared with softwood pulps produced by other pulping processes.

In general, both the low-density, low-lignin-content hardwoods (e.g., aspen) and the high-density, intermediate-lignin-content hardwoods (e.g., birch) produce pulps of high strength, whereas the high-density, high-lignin-content hardwoods (e.g., oak) produce pulps of low bursting strength, high tearing resistance, and high lignin content. The most widely used species in the northern states are aspen, birch, beech, and maple. Aspen and birch produce pulps of particularly good color and strength. In the southern states the species commonly used are the gums, tupelos, extracted chestnut, and oak. Oaks and certain other dense hardwoods produce pulps of low strength and high bulk, although some oaks produce better pulps than others. In spite of relatively low strength, poor brightness, and high lignin content, large amounts of oak are pulped for bleaching grades as well as for corrugating grades. Because of their bulk, these pulps: (1) impart adequate rigidity to corrugated board and (2) bulk and opacity to bleached pulp. Ash, elm, and black cherry are among the other species used. Mills using mixed hardwoods may limit the amount of oak to 10 to 20%. Elm is sometimes excluded, not because of any pulping problem, but because it is so difficult to bark. The color of the pulp is fair when light-colored woods (e.g., birch, poplar, maple, and gum) are used, being close to the color of unbleached sulfite under optimum pulping conditions. The semichemical process can be used for the pulping of mixed tropical woods, but many tropical hardwood species are not well suited to semichemical pulping because of their high lignin or high extractives.

There is a growing tendency to use hardwoods for semichemical pulping in the same proportion as they occur naturally in mixed stands. Some mills use mixtures of as many as 12 species. In spite of the differences in species, no serious trouble in pulping mixtures has arisen, although there is a difference in the color and strength of the pulp produced.

Paper Grades Using NSSC Pulps. The NSSC semichemical pulps have been used successfully to replace a large portion of coniferous chemical pulps in a wide range of papers. The greatest use for semichemical pulps is in corrugated board; about 80 to 85% of unbleached NSSC pulp is used in this grade. The pulps tend to be stiff due to residual lignin; this makes them ideal for corrugating medium. The very high-yield types do not always beat readily, in spite of their high hemicellulose content, because the lignin content is high and the hemicellulose inaccessible for hydration.

When used alone in the furnish, medium- to low-yield semichemical pulps produce papers having a high bursting strength, but the paper tends: (1) to be stiff and rattly owing to the high hemicellulose content and (2) to be low in tearing and folding strengths because of the rapid hydrating properties of the pulp. Because the pulps are fast beating, they are well suited for use in grease-proof papers and bond papers. They are not so well suited for soft tissues, toweling, chart papers, or vegetable parchment. However, when properly prepared and mixed with other pulps, it is possible to use a high percentage of semichemical pulp in bond papers without detracting seriously from the opacity of the paper. Bleached NSSC pulp will produce excellent printing papers that have high strength, low density, and good formation. NSSC pulp can be included in the furnish for almost any product that does not require maximum tearing resistance or folding endurance. It may be necessary to include liberal portions of other fibers to produce papers conforming to commercial specifications established before NSSC pulp was available.

Chemical Properties of NSSC Pulps. In contrast to the 3 to 7% lignin found in fully cooked pulps, semichemical pulps for bleaching contain 8 to 12% lignin, while pulps for corrugating medium contain 15 to 20%. This lignin is largely on the surface of the fibers.

Hardwood NSSC pulps contain more pentosans than pulps prepared by any other common process. This is largely because of the neutral nature of the cooking liquor, which restrains hydrolysis and dissolution of hemicelluloses during pulping. Both yield and strength are benefited by the high retention of hemicelluloses. With light mechanical treatment the pentosans on the surface of the fibers swell rapidly and furnish additional interfiber bonding. The chemical composition of selected pulps made by the NSSC process is shown in Table 4-13.[192] Birch wood and pulps have an exceptionally high pentosan content, 23%.

Physical Properties of NSSC Pulps. Semichemical fibers are heavier than those that are fully cooked because of their high lignin content; consequently

TABLE 4-13 THE CHEMICAL COMPOSITION OF SELECTED SEMICHEMICAL PULPS MADE BY THE NSSC PROCESS

Constituent	Birch		Spruce	
	Wood (%)	Pulp (%)	Wood (%)	Pulp (%)
Analyses of wood and unbleached neutral sulfite pulps:				
Yield	—	77.0	—	81.0
Pentosan	22.5	23.3	7.5	7.9
Uronic acids	4.7	2.8	2.3	1.1
Lignin	20.0	10.8	28.0	20.2
Methoxyl	6.1	4.4	4.9	3.5
Roe number	—	15.0	—	25.0
Carbohydrates, total	75.5	84.0	70.5	75.0
Polysaccharides in wood and bleached neutral sulfite pulps:				
Galactan	1.5	(trace)	3.0	(trace)
Glucan	59.5	67.5	66.0	69.0
Mannan	2.5	1.0	17.0	19.0
Araban	2.5	1.0	1.5	1.0
Xylan	27.5	27.5	9.0	9.5
Rhamnose	0.4	0.0	0.0	0.0
Uronic acid	6.1	3.3	3.5	1.5
Losses in bleached neutral sulfite pulps:				
Pentosans	20.5	—	19.0	—
Uronic acids	54.0	—	61.0	—
Lignin	58.5	—	41.5	—
Galactan	100.0	—	100.0	—
Glucan	3.0	—	10.5	—
Mannan	66.0	—	3.0	—
Araban	65.0	—	33.0	—
Xylan	14.0	—	10.0	—
Rhamnose	100.0	—	—	—

a sheet of standard weight will contain fewer fibers than a sheet made from chemical pulps. As the yield is reduced by extended cooking time, the increasing number of fibers per gram and the increasing flexibility contribute to greater bonding within the sheet and also to increased density. The properties of several corrugating boards made from different hardwoods by the NSSC process are shown in Table 4-14.[193]

As the freeness decreases during the refining of NSSC pulps, tear drops sharply, but burst, tensile, folding endurance, and sheet density increase. A NSSC pulp from one of the better hardwoods, such as birch or aspen, is only slightly weaker than a bleachable-grade spruce sulfite pulp in burst and is equal in tearing resistance. In the high- and middle-yield semichemical range, NSSC pulps have superior papermaking properties than the corresponding kraft pulps. As yields approach the fully cooked stage, kraft pulps begin to show their well-known superiority. Primary advantages of NSSC include much lighter color and the ability to form stiffer corrugating medium. Cooking at too high an alkalinity darkens the pulp.

Hardwood semichemical pulps are stronger than hardwood sulfite and sulfate pulps. This is shown in Table 4-15 where aspen semichemical pulp is compared for strength with aspen pulp produced by the full chemical processes (sulfite, soda, and sulfate). Compared with spruce sulfite pulp, Chidester[194] reports high bursting strengths for birch, aspen, and cottonwood in the yield range of 70 to 80%. These bursting-strength values are about 75 to 80% of the corresponding values for spruce sulfite. In tearing resistance, unbleached semichemical pulps are generally 90 to 120% of spruce sulfite and in folding endurance about 10 to 50%. The semibleached grades rate somewhat higher in strength and full bleached grades have still higher strengths.

Effect of Yield on NSSC Pulps. Yield in pulping has a pronounced effect on the properties of the pulp. The strength is very low at yields over 80%, but increases steadily when yield is decreased to about 65%. In the case of coarse pulps used for making corrugating medium, coarse wrapping paper, linerboard, hardboard, insulating board, and roofing felt, the most desirable combination of high yield and acceptable strength is obtained in the range of 70 to 80%. Such high-yield pulps are coarse and shivy unless heavily refined, but it is possible to produce semichemical pulps in yields as high as 75% having fibers that are fairly well separated. Shives are common but can be removed by centrifugal separation. The pulps are susceptible to pitch and slime troubles, but these can be kept under control by proper measures.

Pulps for the finer grades of paper (book, bond, glassine, writing, waxing, and tissue) are produced in lower yield. With aspen, typical yields for book-grade pulp would be between 70 to 74% unbleached and between 62 to 64% bleached on a moisture-free basis. Yield varies according to the species; oak has a yield of about 10% less than aspen for comparable grades of bleached pulp. The brightness of unbleached NSSC pulps is fairly high, about 40 to 50 GE.

TABLE 4-14 PROPERTIES OF CORRUGATING BOARDS MADE FROM NEUTRAL SULFITE SEMICHEMICAL PULPS

Wood Species	Thickness (mm)	Bursting Strength (kPa)	Tear Strength (mN)	Tensile Breaking Load, (kN/m)	Ring Crush (N)[a]		Flat Crush (N)
					Machine Direction	Cross-Machine Direction	Concora Medium Test
		Commercial boards					
Mixed northern hardwoods	0.223	330	912	6.26	253	191	351
East central hardwoods	0.238	234	755	5.05	262	187	329
Red alder and Douglas fir (3:1)	0.206	283	—	—	—	218	245
		Experimental boards					
White and red oaks (2:1)	0.231	324	726	6.26	280	205	302
Sweet gum	0.208	352	1059	6.96	267	205	378
Mixed northern hardwoods	0.206	372	726	6.79	307	249	338
Mixed Wisconsin hardwoods	0.208	414	794	7.83	338	267	400

[a]15.2-cm specimen; 1.27-cm wide.

TABLE 4-15 COMPARATIVE STRENGTHS OF ASPEN PULPS PRODUCED BY DIFFERENT PULPING PROCESSES

	Burst Index (kPa · m²/g)	Tear Index (mN · m²/g)	Burst Tear (relative %)	
Neutral semichemical, bleached	1.32	10.7	100	100
Neutral semichemical, unbleached	0.96	8.7	73	81
Sulfate, unbleached	1.08	7.6	82	71
Sulfate, bleached	0.96	9.2	73	86
Soda, unbleached	0.78	7.6	59	71
Soda, bleached	0.60	6.1	45	57
Sulfite, unbleached	0.60	6.1	45	57

Bleached NSSC Pulps. In general, NSSC pulps are easily bleached compared to softwood kraft. Bleachable grades are made by cooking to a lignin content of about 10%. The multistage bleaching of semichemical pulps with chlorine compounds is (more than with other pulps) an integral part of the purification process and, therefore, the reader is referred to the bleaching chapter. The brightness values of NSSC pulps before bleaching are in the 35 to 55 range. They can be readily brightened to 75 by using peroxides, dithionates, and other agents normally used on groundwood.

The NSSC pulps produced for full bleaching are cooked to yields between 63 and 75%, depending on the lignin and extractives content of the wood and the preference of the individual mill in proportioning lignin removal between pulping and bleaching. These pulps contain 10 to 14% lignin compared to a lignin content of 6 to 8% for sulfate pulps and 1 to 2% for acid sulfite pulps. The lignin in semichemical pulps is largely residual in the middle lamella and not in the cell wall as in full chemical pulps. Lignin in hardwoods, reeds, or straw is highly reactive and can be easily removed by relatively simple multistage procedures, including chlorination, extraction, and oxidation steps. Chlorine dioxide has been used successfully in one or more stages. Because bleaching is a continuation of the delignification started in pulping, the loss in yield is considerable, approximating 20% of the unbleached pulp. In spite of this shrinkage, the yield of bleached pulp based on the moisture-free wood-weight will be in the range of 55 to 62%, which is considerably higher than in full chemical pulps. The removal of lignin from the surface of the fibers improves the interfiber bonding potential and the strength of the sheet. The high hemicellulose content of the bleached pulp enhances its rapid hydration characteristics. Considerable refining energy can be saved in producing greaseproof and glassine papers if bleached semichemical pulp is substituted for acid sulfite pulp in these grades.

Because bleached hardwood NSSC pulp is similar to bleached softwood acid sulfite pulp in many properties, it can be used to replace it in many grades of paper. Although bond paper can be made from 100% NSSC pulp, general practice limits it to 70 to 80%. In greaseproof and glassine the limit is about 60%.

NSSC pulp use in printing, coating, offset mimeograph, ledger, envelope, tissue, and bristol boards is limited to between 25 and 50% of the furnish.

SULFITE PULPING

J. R. G. BRYCE, Ph.D.
Manager, Pulp, Paper, and Packaging Research
Domtar, Inc.
Senneville, Quebec, Canada

The two principal chemical pulping processes, sulfite pulping and alkaline pulping, are discussed in the following sections.

Description of the Sulfite Process

The sulfite process has had a significant role in the production of chemical pulps for nearly 100 years. That role has declined and, as shown in Table 4-16, growth in chemical pulp production has been almost entirely in the kraft process.[195] In 1935 the two processes were used about equally, but the sulfite process now supplies less than 10% of the total paper-grade chemical pulp production in the United States. Nevertheless, where suitable conditions exist for the use of the sulfite process, it offers distinct advantages over alkaline pulping processes. The

TABLE 4-16 PRODUCTION OF SULFITE AND KRAFT PULPS IN UNITED STATES AND CANADA, 1935-1976

Year	United States		Canada	
	Sulfite[a]	Kraft[b]	Sulfite[a]	Kraft[b]
1935	1,390	1,467	1,025	206
1940	2,281	3,747	1,281	372
1950	2,370	7,502	1,875	1,054
1955	2,555	11,280	2,415	1,471
1960	2,578	14,592	2,291	2,442
1965	2,685	21,509	2,924	3,904
1970	2,344	29,471	2,720	6,844
1972	2,172	31,826	—	—
1973	2,230	32,838	2,488	8,893
1974	2,209	33,010	2,865	9,139
1975	2,025	29,357	2,159	7,051
1976	2,207	33,561	2,539	8,708

[a] Includes both unbleached and bleached.
[b] Includes unbleached, semibleached, and bleached.

sulfite process was originally discovered by an American, B. C. Tilghman, who found that a bright pulp could be produced by treating wood at elevated temperatures and pressures with solutions of sulfurous acid and calcium bisulfite. However, his equipment was unable to withstand the highly corrosive conditions of these solutions. Tilghman was, therefore, not successful in developing a commercial sulfite pulping operation. C. D. Ekman constructed and operated a pulp mill at Bergvik, Sweden in 1874 to manufacture pulp by the sulfite method. The digesters were lead-lined and steam jacketed and, although the process was operated with great difficulty, Ekman was able to produce sulfite pulp continuously on a commercial scale. At about the same time, without knowledge of the work of Ekman, Mitscherlich in Germany developed a similar process employing a ceramic-lined digester containing steam coils for use in heating. A few years later, Rittner and Kellner in Austria developed the direct-steam-heated digester, which allowed the use of larger digesters and established the sulfite process on a viable basis. With the development of commercially suitable equipment, the sulfite process became the dominant method for producing chemical pulps.

The chief reasons for the rapid commercialization of the sulfite process were: the high yield; the low cost of cooking chemicals compared to the alkaline processes; the high brightness of the unbleached pulps, permitting them to be used in many grades without bleaching; and the easy bleachability of the pulps with the relatively simple bleaching agents available at that time. However, the sulfite process suffered from two distinct disadvantages: (1) only a limited number of wood species could be pulped and (2) the pulps produced were distinctly weaker than those produced by the kraft process. With the development of efficient kraft recovery systems about 1930, the kraft process became dominant in the unbleached, high-strength field. With the development of the modern bleach plant, and particularly the use of highly efficient chlorine dioxide as a bleaching agent in the 1950s, the kraft process became the principal process for producing bleached pulps as well.

Outline of the Sulfite Process

The early literature on sulfite pulping used widely varying terms to describe the process, but in later publications the standards adopted by Tappi and the Technical Section, CPPA., have been widely used.[196] These terms will be described and used in the description of the sulfite process that follows.

A flow sheet outlining the main steps in producing low-yield sulfite pulp is shown in Figure 4–59. In the preparation of cooking liquor, or acid, for the sulfite process, sulfur is burned to yield sulfur dioxide. After cooling, this gas is absorbed in the desired base to form storage acid. This liquor is transferred first to a low-pressure recovery tower and then to a high-pressure accumulator where heat, water vapor, and sulfur dioxide (relieved from a previous cook) are absorbed to form the actual cooking liquor. Meanwhile, the digester is filled with chips, presteamed to remove air, and then filled with the cooking liquor.

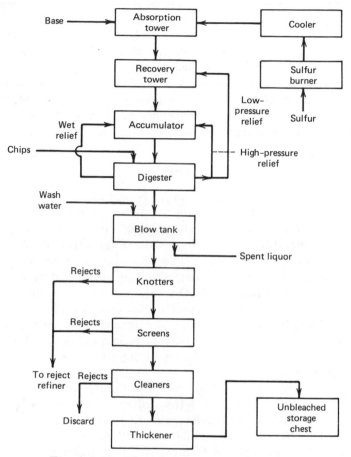

Figure 4-59. Flow sheet of the sulfite process.

The digester is heated to cooking temperature, using either direct steam injection or indirect heating. Both natural and forced circulation of liquor in the digester are used, but the latter is the more common method. At an early stage of the cook, a portion of the digester acid, known as side relief, is withdrawn from the digester and returned to the accumulator. After the side relief has been removed, digester pressure is controlled by relieving gas (top relief) from the top of the digester, which is also returned to the accumulator. Steaming is continued until the digester reaches the desired temperature. Near the end of the digestion period, blowdown is initiated, that is the digester pressure is reduced by relieving gas at a rapid rate to the accumulator. When the pressure has been substantially reduced, the blow valve at the bottom of the digester is opened and the residual pressure is used to discharge the contents of the digester to a blow pit.

Spent cooking liquor is drained from the blow pits through perforated plates in the bottom. Simultaneously, wash water is added by means of sprays at the top of the blow pit and washing is continued until virtually all the spent liquor is removed. Alternatively, pulp and liquor can be continuously removed from a blow tank and washing performed on a washer, either a drum filter or some other design. If a high-yield pulp is being produced, at least one stage of refining usually takes place before the final washing in order to open up the chip structure for more effective washing. After washing, the pulps are screened and cleaned and then thickened for storage prior to further processing, such as bleaching, drying, or inclusion in a paper-machine furnish. In some cases, the spent liquor is sent for evaporation and incineration or for by-product recovery.

Analysis of Sulfite Pulping Liquors

Sulphur dioxide in cooking liquor can be present in various chemical forms depending on the ratio of sulfur dioxide to alkali used in preparing the liquor. The possible forms are: dissolved sulfur dioxide (SO_2) and/or sulfurous acid (H_2SO_3); bisulfite (HSO_3^-); and monosulfite ($SO_3^=$). The bisulfite can be present alone or in combination with either of the other two forms. The conventional method of describing the composition of sulfite cooking liquors is in terms of total, free, and combined SO_2. The total SO_2 is the sum of all three forms listed above; the free SO_2 represents the dissolved SO_2—sulfurous acid plus one half the SO_2 present as bisulfite; the combined SO_2 represents the SO_2 present as monosulfite plus one-half the SO_2 present as bisulfite. Each form is usually expressed as percent concentration of SO_2 in the liquor (more properly as g SO_2/100 ml of liquor).

The total, free, and combined SO_2 in conventional sulfite liquors are determined by the method of Palmrose.[197] This analysis consists of two steps: (1) the oxidation of all the SO_2 in the liquor with KIO_3 to determine the total SO_2 content and (2) titration of the available acid with NaOH to determine the free SO_2 content. Combined SO_2 is then calculated by difference. The reactions involved in the iodate oxidation are:

$$KIO_3 + 5KI + 6H^+ \rightarrow 6K^+ + 3H_2O + 3I_2$$

$$3\begin{bmatrix} H_2SO_3 \\ HSO_3^- \end{bmatrix} + 3I_2 + 3H_2O \rightarrow 3\begin{bmatrix} H_2SO_4 \\ HSO_4^- \end{bmatrix} + 6H^+ + 6I^-$$

It should be noted that the acid stays in balance through this sequence of reactions and that the NaOH titration can be carried out on the oxidized sample. In this way, possible errors in the proportions of combined and free SO_2, due to small differences in sample volume, are eliminated. In addition, a sharper end point is obtained for the acidimetric titration because the highly acidic HSO_4^- ion is being measured instead of the weaker HSO_3^- ion. Acid must be present for the initiation of the iodate reaction. In the original sulfite pulping method, in which cooking liquor was always highly acidic, this did not give rise to any problems. However, with pure bisulfite and monosulfite-containing liquors, it is necessary to acidify the liquor by adding a measured volume of standard acid prior to the first titration and then adjusting the volume of caustic required for the second titration. Standard procedures for this analysis have incorporated this correction.[198]

The total, free, and combined SO_2 terminology traditionally used in commercial operations does not describe the actual composition of sulfite liquors and can at times actually be misleading. The composition is described more accurately by giving the content of sulfurous acid, bisulfite, and monosulfite. These composition terms can be determined from the total, free, and combined SO_2 values. For liquors containing more free than combined SO_2 (no monosulfite present), the sulfurous acid (or "true" free) content is numerically equal to the difference between the free and combined SO_2. The remainder of the total SO_2 content can be considered to be bisulfite. In liquors containing more combined SO_2 than free (no sulfurous acid present), the difference between the two is equivalent to the monosulfite content and the remainder is bisulfite. A solution having equal free and combined SO_2 contains only bisulfite.

The Sulfite System and Its Composition

When sulfur dioxide is dissolved in water, a series of equilibria are established:

$$SO_2 \text{ (gas)}$$
$$\Updownarrow$$
$$SO_2 \text{ (solution)}$$
$$(+H_2O)$$
$$\Updownarrow$$
$$H_2SO_3$$
$$\Updownarrow$$
$$H^+ + HSO_3^-$$
$$\Updownarrow$$
$$H^+ + SO_3^=$$

It has been established by Campbell and Maass,[199] using vapor pressure and conductivity measurements, that in solutions of sulfur dioxide in water relatively little sulfurous acid is formed. This, rather than the limited dissociation of sulfurous acid, is the cause of the low acidity of these solutions. When a base is added to the sulfur dioxide-water system, first bisulfite and then monosulfite is formed. For example, when sodium hydroxide is used, these reactions occur:

$$H_2O + SO_2 \rightleftharpoons H_2SO_3$$

$$H_2SO_3 + NaOH \rightleftharpoons NaHSO_3 + H_2O$$

$$NaHSO_3 + NaOH \rightleftharpoons Na_2SO_3 + H_2O$$

The change in pH at room temperature that occurs during this sequence for both sodium and magnesium base is shown in Figure 4-60. It can be seen from these curves that the bisulfite inflection occurs at about pH 4.5 and the monosulfite inflection at about pH 9. At these levels, the solutions are unbuffered and relatively small changes in composition cause substantial variations in pH. Since the bisulfite pH range is the one predominantly used for sulfite pulping and acidity

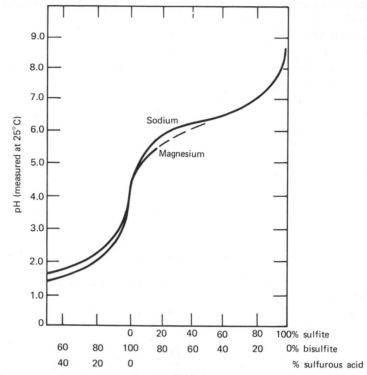

Figure 4-60. Composition of sulfite solutions at various pH levels.

(pH) is of major importance in controlling the rate of sulfite pulping, the pH as well as the composition should be specified to describe these liquors adequately.

The effect of the elevated temperatures used in sulfite pulping on the above equilibria is important for a proper understanding of the process. Rydholm[200] investigated the system, using chemical methods, and found that acidity decreased as temperature increased due mainly to the higher reversion of sulfurous acid into sulfur dioxide and water. The development of a pH electrode suitable for use at elevated temperature[201] has permitted a much more detailed examination of these solutions. It has been confirmed that for solutions of SO_2 in water as well as for solutions containing substantial amounts of bisulfite, the pH increases (acidity decreases) with increasing temperature.[202–204] Figure 4-61 shows this effect for the $NaOH–SO_2$ system over a temperature range from 20 to 145°C. These changes in pH with temperature are different with different bases, and these differences, in turn, influence the course of pulping.

Bases Used in Sulfite Pulping

Calcium Base. From the inception of sulfite pulping in the 1870s until the mid 1950s, calcium was the dominant base used in the process. This was due to

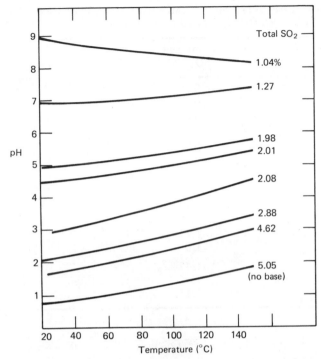

Figure 4-61. Temperature–pH relationship in the system NaOH–SO₂ at a combined SO₂ concentration of 1.0%.

the low cost of this base and its wide availability. The most common source of calcium for sulfite liquor preparation is limestone. Although limestone was normally used in the rock form, in some instances it has been powdered to speed up the rate of dissolution in the sulfur dioxide solution. In addition to limestone, calcium hydroxide (milk of lime, slaked lime) can also be used as a source of calcium base. Dolomite (Ca Mg (CO₃)₂ may also be used when it is more readily available than limestone.

While calcium was suitable for the conventional acid sulfite process, it suffered from some severe limitations that became impediments to further process development. These include:

1. The relative insolubility of calcium sulfite at elevated temperatures, which limits the process to very low pH levels.
2. Severe scaling of the process equipment due to the low solubility of calcium sulfite, sulfate, and carbonate.
3. Lack of a suitable recovery system.

Of these restrictions, the limited solubility of calcium sulfite is the most serious. Precipitation temperatures for a range of compositions in the system CaO–

TABLE 4-17 PRECIPITATION TEMPERATURES FOR CALCIUM ACID
SULFITE LIQUORS OF VARYING COMPOSITION

Free SO_2 (%)	Combined SO_2 (%)	Precipitation Temperature (°C)
0.52	0.59	Below 7°
1.41	0.58	102°
3.03	0.57	120°
5.81	0.55	Above 120°
0.64	0.89	53°
2.00	0.88	93°
2.03	0.88	94°
4.56	0.85	Above 108°
1.07	1.14	28°
2.24	1.12	79°
3.22	1.11	106°
4.24	1.09	Above 108°
1.05	1.47	Below 19°
2.42	1.45	65°
5.14	1.43	Above 105°

SO_2-H_2O are given in Table 4-17.[205] Calcium sulfite exhibits reverse solubility—
it becomes less soluble with increasing temperature. Thus to maintain the required combined SO_2 concentration of 1 to 1.2% (0.9 to 1.1% CaO) in solution
up to the normal cooking temperature of 140°C, a very high SO_2 concentration
in the range of 5 to 7% total SO_2 must be maintained. The resultant low pH of
the cooking liquor limits the flexibility of the process and reduces the quality
of the sulfite pulp.

Magnesium Base. Magnesium base can be used over a wider pH range than
calcium because its monosulfite ($MgSO_3$) is soluble to the extent of about 1%
as SO_2.[206,207] No problems are encountered at the bisulfite pH, but care must be
taken during liquor preparation to ensure that this level is not exceeded. The
limited solubility must also be considered in selecting conditions for applying
magnesium base in multistage pulping. Magnesium has been the most widely
used base where liquor-burning systems are required as part of the mill system.
Combustion of magnesium base liquors produces magnesium oxide and sulfur
dioxide as the inorganic products and these can be readily combined to form
fresh cooking liquors.

Magnesium base is normally obtained in the form of dry or slurried magnesium
hydroxide[208] or powdered magnesium oxide. Magnesite, the naturally occurring
form of magnesium carbonate, does not react with sulfur dioxide at a sufficient
rate to be applicable in conventional liquor-making systems. Dolomite (Ca Mg
$(CO_3)_2$ and brucite (Mg $(OH)_2$ · $CaCO_3$) contain too much calcium to be used as

true magnesium base. For mills with recovery systems, magnesium sulfate can be added to the liquor ahead of the furnace to serve as makeup. In this case, the sulfur dioxide formed from the decomposition of the sulfate becomes part of the sulfur makeup to the system.

Sodium Base. The main attraction of sodium as a base for sulfite pulping is the solubility of sodium sulfite and bisulfite. Both SO_2 absorption and pulping are simplified because composition levels where precipitation might occur are virtually never encountered. The fact that most sodium salts, both inorganic and organic, are more soluble than the corresponding salts of the divalent bases minimizes deposition of scales and pitch. Sodium is at a disadvantage, however, where a base recovery system is required. Although numerous sodium-recovery systems have been proposed (to be discussed later), none has the simplicity of the magnesium-base recovery system. Most of them require the handling of reduced sulfur compounds, which are potential sources of obnoxious atmospheric emissions.

Sodium-base liquors are usually prepared from either sodium hydroxide or sodium carbonate (soda ash). Caustic soda is usually received as a 50% solution, obviating any need for solids handling. Soda ash is delivered in solid form, stored as slurry, and readily dissolved in water prior to use.

Ammonia Base. Ammonia resembles sodium in solubility, in that its sulfites are highly soluble throughout the whole pH range. However, because of the volatility of ammonia and the weak basic nature of ammonium hydroxide, there are distinct dissimilarities between ammonium and sodium base in the pulping and liquor-making processes.

Ammonia can be obtained either in anhydrous liquid form or as aqueous ammonia; the former is more common for permanent installations. An unloading, dissolving, and storage system is required to handle the liquid ammonia, which is usually dissolved to about 15 to 20% concentration for storage. Special precautions must be taken with ammonia-base absorption systems to ensure that ammonia gas does not escape.[209] This loss can be prevented in packed-tower absorption systems by either introducing the aqueous ammonia into a recycled stream of product liquor or by installing a water-scrubbing spray section above the main absorption bed.[210] Since ammonia gas in the presence of SO_2 tends to form an aerosol of ammonium pyrosulfite, the former system, which eliminates the presence of gaseous ammonia in the tower, is preferred.

Delignification is more rapid with ammonia base than with other bases,[211] probably due to the weaker acidity of the ammonium bisulfite and consequent lower pH at cooking temperature. Ammonia-base pulping also produces darker pulps than those from the other bases, and this prevents its use where the pulps are to be used in the unbleached form. However, the darker unbleached pulp does not affect bleaching, and ammonia base has been used mainly for bleached pulps.

Ammonia-base spent liquors are readily incinerated to dispose of the organic material. Sulfur dioxide can be recovered from the flue gas, but the ammonia is decomposed to nitrogen and water under the conditions of the incineration and is not available for reuse. The recovery of sulfur dioxide from the flue gas with the weakly basic ammonia is only achieved with some difficulty. Special precautions must be taken to control the unsightly mist that tends to form and be discharged in the stack gases from this operation.

Pulping without Base. In some respects, sulfite pulping could be distinctly simplified if the reactions were carried out in the absence of base, that is with solutions of sulfur dioxide in water. Under these conditions, much of the sulfur dioxide could be recovered from the spent liquor by gas relief and steam stripping at the end of the cook. Scale and pitch formation should be minimized and liquor burning would be simplified with no ash to handle. However, there are substantial disadvantages to such a system: (1) the solutions of sulfur dioxide are very corrosive when there is no base present; (2) in the absence of base, very low water temperatures would be required to reclaim SO_2 from relief and particularly from furnace flue gases; and (3) very high SO_2 partial pressures are required to shift the sulfite equilibrium

$$SO_2 + H_2O \rightleftharpoons H_2SO_3 \rightleftharpoons H^+ + HSO_3^-$$

to the right to form enough bisulfite ion for the sulfonation reactions to occur. Laboratory studies have shown that pulping in this manner is possible,[212,213] but the practical difficulties in operating this process and the weak nature of the pulps produced leave little doubt that this method is only a laboratory curiosity and not a practical means of producing chemical pulps. However, it has been claimed that a vapor-phase modification of this process could be used in the pulping of eucalyptus to high yield levels.[214]

Comparison of Bases

Numerous comparisons of the effect of using different bases in sulfite pulping have been made.[211,215-218] Each base has slightly different optimum conditions. Once these have been identified, the base does not have a significant effect on pulp properties, at least in the chemical pulp range. However, definite differences exist in a number of important process aspects. These have been summarized for the four commonly used bases (Ca, Mg, Na, NH_4) in Table 4-18. The monovalent bases (Na, NH_4) have the advantage of being soluble throughout the total pH range and this simplifies the SO_2 absorption equipment. All bases, except calcium, are sufficiently soluble to permit pulping at the bisulfite level, but the solubility of the monovalent bases at higher pH levels gives them increased flexibility in multistage pulping processes. The higher solubility of sodium and ammonium bases reduces the possibility of scale and deposit formation in the liquor preparation and digestion system. Magnesium does not give scaling problems in the digester system at bisulfite pulping pH levels. But mag-

TABLE 4-18 COMPARISON OF BASES FOR SULFITE PULPING

	Calcium	Magnesium	Sodium	Ammonium
SO_2 absorption system	Complex	Relatively simple	Simple	Simple
pH range for digestion	Below 2	Below 5	0 to 14	0 to 14
Rate of pulping	Intermediate	Intermediate	Slowest	Fastest
Level of screenings	Moderate	Moderate	Low	Low
Scaling tendency	High	Moderate	Low	Low
Ease of liquor incineration	Difficult, no base or SO_2 recovery	Simple, base and SO_2 both recovered	Complex, base and SO_2 both recovered	Simple, but no base recovery

nesium is much more susceptible to deposits in the SO_2 absorption system, where it is desirable to maintain the pH as high as possible to obtain maximum absorption efficiency. Calcium-scaling problems are encountered in both digestion and liquor-preparation systems, even in the acid sulfite pH range.

The rate of pulping varies slightly with the base used for acid sulfite pulping—but more significantly with bisulfite pulping—in the increasing order $NH_3 >$ $Mg > Na$. The differences in rate have been found negligible if rate comparisons are made on the basis of equal pH at cooking temperature.[218] The monovalent bases give lower pulp-screening rejects than magnesium or calcium, although the differences are relatively small in the 30 to 40 Kappa-number range used in chemical pulp mills, that is where no refining is used to disintegrate the chips.

Recovery is simplest with magnesium base and most difficult with sodium. Ammonia base can be disposed of by burning, but the base is not recovered. Chemical can not be recovered with calcium base, although the liquor has been burned for its heat value in certain cases. Recovery systems will be discussed and compared in more detail later.

Thus the choice of base will to some extent depend on the requirements of the particular system. Where the spent-liquor organics must be destroyed, a situation that is widespread and is likely to become a universal requirement, magnesium base, or in some instances ammonia base, is now preferred. In the case of mills supplying long fiber to newsprint mills, on the other hand, high-yield sulfite pulping using sodium base has been found to produce pulps of highest quality.[217]

Sulfite Pulping Processes

Acid Sulfite Pulping. The traditional method for sulfite pulping since its inception almost 100 years ago has been the acid sulfite process, which is characterized by a high excess of free sulfur dioxide. The liquor has a pH in the 1.2 to 1.5 range and contains excess free SO_2 that is not consumed during pulping. This excess is relieved from the digester and recovered by dissolving in the fresh cooking liquor for reuse. The low pH used in acid sulfite pulping offers two main advantages that were the major reasons for the dominance of this method for a long period of time: (1) since free sulfur dioxide is added in large excess, there is no need for accurate control of the chemical charge if sufficient combined SO_2 is provided and (2) because of the acidity, calcium can be used as the base in the form of the readily available and inexpensive limestone.

A typical time-temperature schedule for an acid sulfite cook is shown in Figure 4-62. Because of the high vapor pressure of the excess free sulfur dioxide, the gas can penetrate the chip faster than the base. If the temperature is elevated under these highly acidic conditions, lignin condensation will occur, producing black, uncooked centers in the chip. The acid sulfite cook is therefore normally started at a relatively low temperature in the range of 70 to 80°C and heated slowly through the low temperature range so that full penetration of base can take place before the temperature reaches 120°C. Maximum cooking tempera-

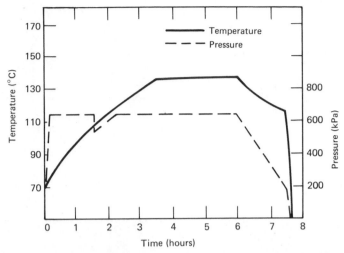

Figure 4-62. Temperature and pressure during a typical acid sulfite cook.

ture is usually in the range of 130 to 145°C. The digester pressure is held substantially above the corresponding steam pressure at all times during the cook to maintain the excess free SO_2 level and avoid calcium precipitation. Relief pressure is normally in the 550 to 700-kPa (80 to 100-psi) range. At the end of the period at cooking temperature, pressure must be reduced gradually so that the excess sulfur dioxide can be reclaimed for a subsequent cook. Soluble bases can, of course, be used for acid sulfite pulping, but at substantially increased cost if there is no recovery system.

The acid sulfite process has a number of serious disadvantages. The high free SO_2 content makes it difficult to pulp resinous woods and chips containing bark. Cooking cycles are long because of the long heating and relief periods required. The high sulfur dioxide partial pressure leads to frequent losses and to atmospheric pollution with unpleasant odors both inside the mill and in the surrounding area. Acid sulfite pulps, although easily refined, are of low strength compared with most other pulps. For most paper grades, acid sulfite pulping has been superceded by other pulping processes. It continues to be used mainly for specific end products, such as rayon and acetate-grade specialty pulps as well as pulps for newsprint, tissue, glassine, and greaseproof papers. The empirical relationships in acid sulfite pulping were thoroughly investigated in a series of publications by Miller, Swanson, and their collaborators.[219]

Bisulfite Pulping. Bisulfite pulping is carried out with a liquor having an equal free and combined SO_2 content (square acid) and having a pH in the range of 3.0 to 5.0. Thus this liquor has the chemical composition of a true bisulfite solution and does not contain the excess free SO_2 that is characteristic of the acid sulfite process. This composition prohibits the use of calcium base because of its low solubility and consequently sodium, magnesium, and ammonia are

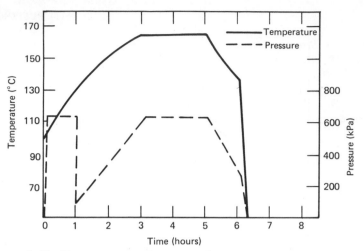

Figure 4-63. Temperature and pressure during a typical bisulfite cook.

the bases employed. A typical cooking curve for a bisulfite cook is shown in Figure 4-63. Since the main liquor components (SO_2 and base) penetrate into the wood simultaneously, higher initial temperature and a more rapid heating cycle can be used than in acid sulfite pulping. The lower acidity of the bisulfite cook necessitates a higher temperature for cooking. However, at the bisulfite pH levels degradation of the cellulose is much less pronounced than at the lower pH levels. In spite of the higher cooking temperature, the total digester pressure can be held at about the same level as with acid sulfite pulping since there is only a slight SO_2 partial pressure. Steam requirement is not significantly increased since most of the heat in the digester is transferred to the incoming liquor for a subsequent cook by means of the blowdown relief. Bisulfite liquor can be introduced into the digester at much higher temperatures than acid sulfite liquor. Normally, the accumulator temperature will stabilize between 90 and 100°C in a bisulfite system. For this reason, it is possible to use a shorter cycle with bisulfite pulping than with acid sulfite.

Although the SO_2 partial pressure in a bisulfite cook is quite small, it is important that it be controlled at the proper level. This overpressure controls the pH of the liquor during the cook, and since pH has an important effect on both the rate of the bisulfite cook and the quality of the final pulp, accurate control is essential. Normally, it is desirable to control the pH slightly on the acid side of the bisulfite point, that is, in the vicinity of 3.5 pH.[220] The higher pH of the bisulfite cook also permits the use of a wider range of wood species since the competing reactions occur to a much lesser extent at the bisulfite pH. Tolerance to bark is higher and the problem of bark-stained wood, which arises from the diffusion of phenols into the wood during water delivery of pulpwood logs, is not as serious a factor with bisulfite as with acid sulfite pulping.

With the substantial advantages of the bisulfite method, it is surprising that the acid sulfite process remained dominant for so long. The bisulfite process was studied in the laboratory for many years,[221-223] but only became commercially successful when the Arbiso (sodium bisulfite)[224] and the Magnefite (mag-

304

nesium bisulfite)[220,225] processes were developed in the late 1950s. In these processes, the importance of controlling the chemical charge and the pH during cooking in order to achieve a good result was recognized for the first time. Following publication of these methods, numerous other workers investigated bisulfite pulping and substantiated its advantage over the acid sulfite process.[216,226,227] Mill trials and conversions were carried out [228–232] and, in the ensuing years, the majority of the sulfite installations producing paper-grade pulps were converted from acid to bisulfite cooking. Although the bulk of the experience on bisulfite pulping has been with sodium or magnesium base, ammonia can also be used for bisulfite pulping.[233,234] However, the unbleached pulps still have low brightness, which is detrimental to their use in newsprint furnish.

Alkaline Sulfite Pulping. The alkaline sulfite pulping process uses a combination of sodium sulfite and sodium hydroxide as the pulping chemicals.[235] As a result it combines features of both kraft and sulfite pulping. Like kraft, the rate of pulping is rapid, there are no species limitations, and the pulps have high strength, as illustrated in Table 4-19. Yield and brightness are low and similar to kraft cooks. On the other hand, this process resembles sulfite in that no significant odor is generated at the digester and the pulps are easily bleachable. Recovery processes for alkaline sulfite pulps have been suggested,[235,236] but they are not very attractive because they combine the complexity of the sodium-recovery systems with the causticizing and lime-burning systems required for alkaline recovery.

Several suggestions have been made to use mixed sulfide-sulfite liquors for pulping.[237–240] Such systems would be extensions of the alkaline sulfite process and, like the latter, have been suggested as alternates to kraft, rather than sulfite pulping. None of these mixed sulfide-sulfite forms of alkaline sulfite pulping have been attempted on a commercial scale.

Multistage Sulfite Pulping. The use of the soluble bases has not only allowed a wider choice in the pH of the initial pulping liquor, but also has permitted changes to be made in the pH during the course of the cook. A variety of these multistage pulping processes have been proposed,[241,242] but most of these can be summarized in two broad categories. In one category are processes in which the first stage is a neutral or alkaline treatment and the second stage is more acidic, either bisulfite or acid sulfite. In the following discussion these processes, as a group, will be referred to as neutral-bisulfite or neutral-acid sulfite pulping. In the other category are processes in which the first stage is carried out at an acid or bisulfite pH and the second stage is at or near the neutral point. These processes are referred to as two-stage neutral pulping methods. These two techniques have significantly different effects on both the delignification process and the properties of the pulps produced.

In the neutral-bisulfite process, the first-stage treatment is normally carried out with a sulfite-bisulfite mixture at a pH in the range of 6 to 8. First-stage

TABLE 4-19 COMPARISON OF ALKALINE SULFITE AND KRAFT PULPS

Type of Cook	Total Yield (%)	Kappa No.	Brightness	Bulk (cc/g)	Strength at 300 C.S.F.		
					Burst Index (kPa · m^2/g)	Tear Index (mN · m^2/g)	Breaking Length (m)
Kraft	47.2	26.8	28.9	1.31	10.9	9.22	14,200
Alkaline sulfite[a]	43.9	43.6	37.9	1.30	10.9	9.62	14,300
Alkaline sulfite[b]	45.5	35.8	33.0	1.29	9.31	11.0	13,000

[a]pH at 175°C—8.0
[b]pH at 175°C—9.5

treatment is usually of about 2-hr duration at temperatures between 125 and 140°C. After completion of the first stage, some liquor is withdrawn from the digester and replaced with either SO_2 gas or low pH liquor for completion of the cook. The liquor withdrawn in the first stage can be reused after the concentration and pH are restored to their original values. This two-stage process was developed by Stora Kopparberg[243,244] and has frequently been referred to as the Stora process. It has been operated commercially in several mills. One of its major advantages is its ability to pulp resinous species, such as pine, because the sulfonation is achieved during the initial high pH stage, where the undesirable side reactions do not occur. It also gives pulp-yield values several percentage points higher than conventional sulfite processes because of higher retention of the glucomannon fraction of the wood.[245,246] (These reactions are described later.) The pulps produced by the Stora process are no stronger than conventional acid sulfite pulps, but they refine very easily and develop their maximum tensile strength with very low-power input. Bases other than sodium sulfite can be used for the initial stage, including ones containing no sulfur dioxide, such as caustic[247,248] or calcium hydroxide.[249] The Kramfors method uses a first stage similar to the Stora process but employs calcium acid sulfite liquor in the second stage.[250] Magnesium base can also be used in a similar two-stage process.[251] Since the solubility of magnesium monosulfite is limited, liquor is pumped to the digester at about pH 5 and the pH then increased to between 5.5 and 6 by addition of magnesium hydroxide directly to the digester. The first-stage liquor is still lower in pH than that used in the corresponding sodium-base process, but this is compensated by increasing the temperature to 150 to 160°C. The second stage is carried out at lower pH in the normal manner by adding either SO_2 water or magnesium bisulfite.

In the two-stage neutral process the first stage is acidic and the final stage is approximately neutral. When sodium is used as the base, this type of pulping is often referred to as the Sivola process (after its inventor) or the two-stage bisulfite-neutral process.[252] The first stage, in which the major portion of the delignification occurs, follows the regular acid sulfite or bisulfite cooking procedure. If it is bisulfite, the pH adjustment to the higher level for the second stage can be made by adding relatively small amounts of base directly to the digester. If, however, the first stage is acid sulfite, it is desirable to either relieve some SO_2 or withdraw part of the liquor before the pH adjustment is made in order to reduce the base consumption. The cooking temperatures and pressures are similar to those used for single-stage bisulfite cooks, although the higher pH levels necessitate somewhat longer cooking times. The temperature can be elevated slightly for the second stage without increasing the pressure since there is no SO_2 overpressure at the higher pH levels. Since pH adjustment for the second stage adds substantially to the base requirement, a recovery system is an economic necessity for this two-stage process. Magnesium base can be used for this process[251,253] by addition of magnesium hydroxide to the digester towards the end of a Magnefite cook. Since the sulfite concentration is low at this point and the dissolved organic solids appear to exert a stabilizing effect, the pH in the second stage can be as high as 6.5 to 7.0 without precipitation. The main advantage of this process is the increase in the strength properties of the pulp, which results from the high final pH.[254,255] This is particularly true of tear strength, which can equal kraft if the proper cooking conditions are used. The yield is slightly lower than that of the bisulfite process, but higher than that of kraft. To some extent, a tradeoff can be made between strength and yield. At second-stage pH levels in the range of 8 to 9, high strength is obtained, but yield is lower than for regular bisulfite. At pH values in the neutral range, there is very little loss in yield, but the strength increase is also not as great. This two-stage process can also be used for high alpha cellulose pulps. In this case, a low pH is used in the first stage for viscosity control and a high pH level and high temperature is used in the second stage to extract the hemicellulose.[256,257] By this method, pulps with up to 96% alpha cellulose can be prepared, and most of the hemicellulose, which would normally have to be removed in the bleach plant and become part of the mill effluent, becomes part of the pulping spent liquor and is disposed of in the recovery system.

A three-stage pulping process has been developed and is in commercial operation for producing high-alpha pulps from pine.[258] The three stages employed consist of a bisulfite first stage for sulfonation, an acid sulfite second stage for viscosity control, and an alkaline third stage for hemicellulose removal.

Dissolving Pulps. The sulfite process is frequently used for the preparation of cellulose for chemical modification, such as xanthation or acetylation, for regenerated fibers and films (rayon, cellulose acetate), and for etherification for methyl and carboxymethyl cellulose.[259] These products require specific properties in the pulp and, as a result, some modifications to the sulfite pulping method are required. The pulps require a low-hemicellulose (high-alpha-cellulose) content and a controlled final degree of polymerization or viscosity. These properties are most often obtained by using the acid sulfite process at elevated temperatures (150°C or higher) and cooking to low Kappa numbers. This reduces the chain length of the hemicelluloses to a point where they are removed in the caustic extraction stage in the bleach plant. The process for producing dissolving grades of sulfite pulp are similar in most other respects to that used for paper production and although the subsequent descriptions are directed mainly at pulps for papermaking, they are also applicable to the production of sulfite dissolving pulps.

High-Yield Sulfite Pulping. When bleaching is not required, a substantial yield advantage can be obtained by stopping the sulfite cook before the defibration point and separating the fibers mechanically. The use of these high-yield pulps brings about a substantial reduction in the wood, cooking chemical, and steam usage. High-yield pulping is particularly suited to mills where the sulfite pulp is being used as the reinforcing pulp for newsprint.

When time, temperature, and liquor composition are at their optimum levels, the maximum screened yield that can be obtained without refining is in the

TABLE 4-20 PROPERTIES OF CALCIUM- AND MAGNESIUM-BASE HIGH-YIELD PULPS

	Roe Chlorine No.	Freeness (°SR)	Tensile (m)	Tear Index (mN · m²/g)	Burst Index (kPa · m²/g)	Brightness (%)
Calcium acid sulfite	21.0	17.0	8290	7.37	4.96	48.4
Magnesium bisulfite	22.8	16.4	8630	8.37	5.77	56.6

range of 52 to 54%.[260] Mechanical refining is necessary to operate at higher yields. The first commercial installations of the high-yield technique were made between 1955 and 1960 and they used the calcium acid sulfite process.[261-263] Pulp yield was usually in the range of 63 to 68%, and although pulp quality was acceptable, the brightness was lower and the pulp contained more dirt and shive than conventional sulfite pulps. A search was begun for other pulping methods that would retain the economic advantages of the high-yield process but would give improved pulp quality. Laboratory studies demonstrated that the then newly developed bisulfite methods would achieve these goals,[224,264-267] and most of the high-yield mills were converted to the higher pH pulping process.[268] The gains in tensile, burst, tear, and brightness experienced by one mill in changing from high-yield calcium acid sulfite to magnesium bisulfite are shown in Table 4-20.[269] The improved quality permitted still further increase in the pulp-yield levels; values as high as 75% were reported by some mills. A major portion of the sulfite pulp used in newsprint is now produced by the high-yield bisulfite process.

Refining conditions play a major role in the quality of the pulp produced.[270,271] Disk refiners are universally used for fiber separation. Sufficient refining must be applied to give good fiber separation, but since paper-machine drainage will be adversely affected if excessive refining is used, the degree of refining must be carefully controlled. A final freeness in the range of 600 to 650 CSF usually meets both of these requirements. The power requirement to achieve this level is in the range of 0.7 to 1.4 MJ/kg (10 to 20 hpd/ton) depending on the yield. Increasing the refining consistency increases the pulp strength and reduces the power requirement; as a result, in some installations, presses have been placed ahead of the refiners to increase the feed consistency to 25 to 30%.

An important reason for adding chemical pulp to newsprint furnish is to maintain adequate wet strength and drainage on the paper machine. It has been found, however, that both drainage and wet strength of high-yield fibers are reduced if insufficient polyvalent ions are present during refining, as is the case in sodium-base mills, where the water supply is very soft. This effect can be overcome by adding calcium or aluminum ions to the high-yield stock before the primary refining stage.[272]

Sulfite Liquor Preparation

The important steps in sulfite liquor preparation are: (1) preparation of the sulfur dioxide gas, (2) absorption of the gas in the base solution, and (3) recovery of digester relief for reuse.

Preparation of Sulfur Dioxide

Sulfur dioxide used for making sulfite cooking liquor is prepared by burning pure sulfur in either rotary or spray burners in the presence of air.[273] Sulfur is received either as a powder or as molten sulfur. It melts at 113°C and care must be taken to ensure that the temperature in the molten state does not rise much above this value since it becomes more viscous with rising temperature until at 200°C it will no longer flow. It is most fluid between temperatures of 120 and 160°C. When sulfur is received as a powder, it is usually melted in a melting tank or a tubular melter prior to feeding into the burner.[274]

Rotary burners vary in size, depending on the production desired, but normally are between 3 and 10 m in length and 1 to 2 m in diameter. The burner is rotated at about 1 to 2 rpm so that the inner surface is completely coated with molten sulfur. Primary air is added at the front of the rotary element so that the sulfur is vaporized and partial combustion occurs. A secondary supply of air is added at the discharge end of the rotary element and combustion is completed in a combustion chamber. Temperatures in the combustion zone are in the range of 900 to 1000°C. Modern installations in sulfite mills employ the spray or jet-type burner.[275-277] In this type of burner, molten sulfur is pumped through a steam-jacketed line to the nozzle of the horizontal burner where it is atomized by compressed air. Preheated secondary air is added in several stages through entry ports in order to create substantial turbulence and thoroughly mix the air and the sulfur. The advantages of the spray burner are:

1. Startup and shutdown are simple. Unlike the rotary burner, in which it is necessary to build up or burn off a pool of liquid sulfur, the spray burner has little sulfur inventory and stable combustion conditions are achieved rapidly on startup.

2. Burning rate is easily varied. A simple adjustment in sulfur and air flow allows the rate to be quickly changed over a wide range.

3. High SO_2 concentration is possible with little SO_3 formation.

4. Maintenance cost is low since there are essentially no moving parts.

The sulfur burner should be operated to minimize sulfur trioxide formation, which not only results in a loss of sulfur and base from the system, but also produces undesirable corrosion and scaling in the subsequent equipment. Sulfur trioxide can be formed from decomposition of sulfuric acid present in the sulfur during the sulfur-burning process, or more likely, from a secondary oxidation of SO_2 by excess oxygen in the gas leaving the burner.[278] Sulfur trioxide formation is minimized by maintaining a high gas temperature in the combustion zone and avoiding a high air ratio in the burner gas.

In the burning of sulfur, one volume of oxygen combines with sulfur to yield one volume of sulfur dioxide. Since air is 21% oxygen by volume, it is theoretically possible to obtain a burner gas containing 21% sulfur dioxide. In actual operation, rotary burners usually yield about 15 to 17% concentration of sulfur dioxide and spray burners about 17 to 19% concentration. A low sulfur dioxide concentration is indicative of high excess air levels, which in turn are conducive to the formation of sulfur trioxide. On the other hand, it is essential to maintain some excess air because if there is no excess, sublimed sulfur will carry through with the product gas (drawing sulfur) and cause plugging and inefficient operation of the subsequent equipment. In most mills the concentration of sulfur dioxide in the burner gas is used as the main control on the sulfur-burning operation. This concentration can be measured manually using the Orsat or other absorption apparatus.[279] For on-line process control, a continuous sensor based on the conductivity of a solution prepared from the gas is normally used.[280] The secondary air to the burner can be controlled by this response. Sulfur trioxide in the gas can be determined by the difference between acidimetric (SO_2 + SO_3) and iodometric (SO_2 only) titrations.[279] Pyrometers are used to measure the gas temperature in the combustion chamber or at the inlet to the gas cooler.

For many years, iron sulfides (pyrites) were used as a source of sulfur dioxide by a number of sulfite mills; however, this practice has largely disappeared. Liquid sulfur dioxide is used by many mills as a supplementary sulfur dioxide source where sulfur burner capacity is insufficient to supply the full mill need. In Canada in particular this is attractive, because liquid sulfur dioxide is available at low cost as a by-product of base metal refining operations.

The gas leaving the combustion zone has a temperature of $1000°C$. For preparation of acid sulfite liquors, it is necessary to cool this gas to 25 to $30°C$ to keep the liquor temperature as low as possible since the solubility of sulfur dioxide decreases as the temperature increases. For bisulfite liquor preparation it is not necessary to maintain such a low temperature and the gas temperature can be between 50 and $70°C$ entering the absorption system. The cooling system must be designed so that the gas passes rapidly through the range of 600 to $850°C$, which is the optimum range for the formation of sulfur trioxide. Both spray-type and indirect surface-type coolers are used, the former being more satisfactory. In the spray-type cooler, the gas is cooled by direct contact with the spray water. In the Chemipulp-KC spray cooler, shown in Figure 4-64,[281,282] there are two stages. The gas enters the primary tower where it is contacted by a concurrent spray of water and the temperature is lowered rapidly through the critical range for sulfur trioxide formation. There is very little reduction in the heat content of the gas in this tower as the temperature change is caused by vaporization of part of the water spray. The small excess of water remaining in liquid form is removed from the bottom of the tower, carrying away sulfur trioxide and any solid impurities that may be in the gas. The wet gas, at approximately saturation temperature, then passes into the bottom of the secondary tower where it meets two sprays that complete the cooling. The lower portion of this tower is packed with resistant stoneware to ensure even distribution of the gas. Because the liquor in the bottom of the secondary tower contains some SO_2, it is cooled by a heat exchanger and recirculated through the spray nozzles. The cooled gas leaving the secondary tower enters the suction side of a fan. This system holds the sulfur trioxide level in the exit gas to very low levels.[277,283] In indirect coolers, the gas is conducted through pipes made of corrosion-resistant material, normally lead, and is cooled by an external flow of water. In the most common configuration the gas flows through horizontal pipes submerged in a vat of water and then through large vertical pipes with fresh cooling water flowing down the outside.

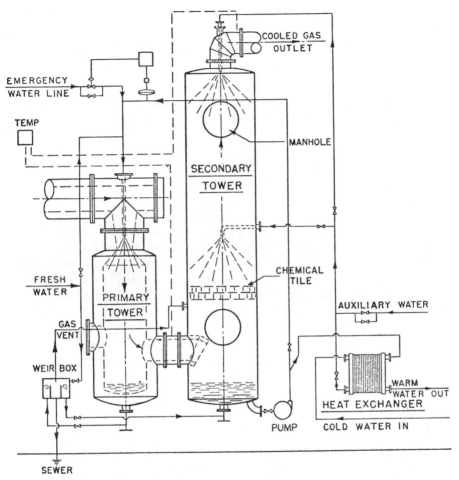

Figure 4-64. Chemipulp–KC spray cooling system.

Absorption of Sulfur Dioxide

Absorption systems for sulfite liquor preparation vary substantially, depending on whether a soluble base or calcium base is used and whether acid sulfite or bisulfite liquor is the final product. The soluble-base bisulfite systems can be of relatively simple design because the liquor does not have the high sulfur-dioxide vapor-pressure characteristic of the acid liquors. Compared to an acid sulfite system, several simplifications to the absorption system are possible, including the following:

1. Higher temperature can be used, permitting more rapid reactions and, as a result, smaller absorption towers.

2. Liquor preparation can take place at ambient pressures and only moderate pressures are required in the accumulator.

3. Since the solution is unbuffered, control is primarily based on pH rather than ratio of free to combined SO_2.

4. The recycled SO_2 makes up only a small portion of the total, whereas a substantial portion of the SO_2 in an acid sulfite liquor must be recycled from a previous cook.

Three types of absorption system are used for preparing sulfite cooking liquor: the Jensen tower, packed towers, and moving bed towers. Jensen towers were usually preferred by North American mills when calcium was the chief base used in the sulfite process. In this system, the cooking acid is prepared by feeding sulfur dioxide gas and water through coarse calcium carbonate (limestone) resting on a grate at the bottom of a large tower. There are normally two towers varying from 2 to 4 m in diameter and 20 to 33 m in height. They are operated in series and the gas flow is countercurrent to the liquid flow, creating a strong and a weak tower. The piping is arranged in such a way that the strong and weak towers can be reversed daily. This allows the level of stone to be maintained by adding fresh material to the weak tower. At the same time, the more acidic liquor dissolves calcium sulfite deposits, which have formed in the weak tower. The following reactions are involved in the preparation of calcium sulfite liquor from the limestone:

$$SO_2 + H_2O \rightarrow H_2SO_3$$
$$2H_2SO_3 + CaCO_3 \rightarrow Ca(HSO_3)_2 + CO_2 + H_2O$$

About 75 to 90% of the sulfur dioxide reacts in the first or strong acid tower. The weak tower serves to complete the absorption so that the exit gas can be discharged to the atmosphere. When operated under favorable conditions a Jensen tower can absorb in excess of 97% of the sulfur dioxide in the inlet gas. The Jensen tower has the advantage of simple operation, but its disadvantages are the high cost of manually loading limestone and the lack of control over the composition of the raw acid. In areas where there is a substantial seasonal variation in water temperature, the liquor analysis shows corresponding seasonal swings. In summer a low free and high combined liquor is obtained, whereas in winter the rate of reaction is too low and the liquor does not contain sufficient combined SO_2. Several mills continued to use their Jensen towers when converting to soluble base by filling them with an inert material, such as pine logs.[208] This permitted the conversion to be made with a very low capital cost, but the system was difficult to control.

When materials other than limestone rock are used as the source of the base, packed towers can be used for absorption of the burner gas. These towers are considerably smaller than the Jensen tower and only one is required for most sulfite mills.[284] The actual size depends on the capacity, but in general they range from 1 to 2 m in diameter and 8 to 14 m in overall length. The packing material may be Raschig rings, cross partition rings, intalox saddles, or other conventional shapes. To withstand the elevated temperatures sometimes encountered in bisulfite systems, the packing material should be made of ceramic or polypropylene. In the operation of a packed tower, the flow is countercurrent with the cooled sulfur dioxide gas introduced into the bottom of the tower and the water and base at the top. Milk of lime $[Ca(OH)_2]$, magnesium hydroxide, caustic, soda ash, and ammonia are typical bases used with this equipment. In order to achieve efficient absorption of the sulfur dioxide and to avoid deposits and plugging in the tower, particularly when using slurried materials, it is important to ensure that the liquor flow is uniformly distributed across the tower cross section. Weir-type liquor distributors are normally installed at the top of the tower for this purpose. Mist eliminators are inserted at the gas outlet to minimize the loss of liquor to the atmosphere. With this system absorption efficiencies in excess of 99% can be achieved.

The Turbulent Contact Absorber (TCA) is a relatively new development in gas-liquid absorption equipment that has found application in the preparation of sulfite liquor. This tower contains a packing of low density, hollow plastic spheres placed between retaining grids sufficiently far apart so that the high gas and liquid velocities maintain the spheres in a turbulent and random motion. This turbulence contributes to intimate contact between the gas and liquid phases, resulting in a high degree of mass transfer and allowing efficient absorption to take place in towers much smaller than would be needed if static packing were used. The tower is essentially nonplugging and is particularly useful when a solid phase is involved, either in the feed material or the product.[285,286] A TCA tower for a typical sulfite mill is about 1 m in diameter and 7 to 10 m in height, with the configuration shown in Figure 4-65. Four stages are usually used, each partially filled with hollow polypropylene spheres 40 mm in diameter. To achieve turbulent motion in the bed, the static volume of the packing should fill less than 50% of the total absorption zone volume. The tower should be sized so that the gas velocity is in the range of 200 to 400 m/min to maintain the proper bed action. The gas is admitted to the bottom of the tower and the base is introduced above the top bed. Alternatively, the base may be added to the third bed and fresh water above the top bed to minimize liquor carryover and to recover heat from the exit gas. An expanded section in the base of the tower (sump) collects the liquor. Product acid and any liquor required for recirculation are taken off at this point.

Venturi scrubbers have also been used to prepare soluble-base liquors, particularly for magnesium base.[228,287] Control is difficult, however, since multiple stages must be used to achieve efficient absorption and base must be added to each stage. Each of these flows must be controlled to maintain the pH at a level sufficiently high to give efficient sulfur dioxide absorption, but low enough to prevent monosulfite-scale formation.

Whereas only limited control can be exercised over the strength of acid sulfite liquor prepared from limestone, the operator has much better control over bisulfite liquor quality. The production rate is controlled by the operating rate of the sulfur burner (sulfur and air-feed rates). The strength of the liquor is controlled by the total water flow to the system. The pH or ratio of free to combined SO_2

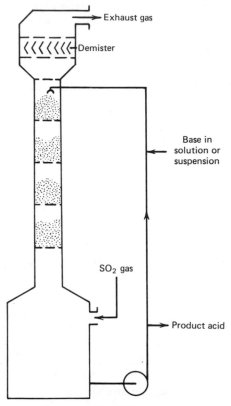

Figure 4-65. TCA tower.

in the liquor leaving the tower is determined by the flow of base. The flow of base should be controlled to maintain a slight excess of monosulfite in the system to ensure good absorption efficiency. Final cooking pH control usually occurs at the accumulator, where a small flow of base is added to compensate for the SO_2 absorbed from the digester relief gases.

Two factors are involved in preparation of sulfite liquors: (1) the equilibrium solubility of the gas in the liquid being used for absorption and (2) the rate at which this equilibrium is achieved, that is the reaction kinetics. Sulfur dioxide solubility is a function of the temperature, pressure, and pH during absorption, while the kinetics are influenced by the degree of turbulence between liquor and gas, the base concentration, and the temperature. The turbulence differs widely between different absorption equipment, depending on the system design. The solubility relationships for sulfur dioxide and water with various bases have been extensively studied.[203,205,206,288-292] The partial pressure of sulfur dioxide is quite low over solutions containing no true free SO_2 (combined equal to or greater than the free). However, when solutions contain true free SO_2, there is a partial pressure of SO_2 above the solution corresponding to

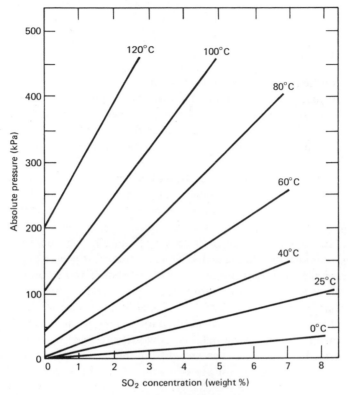

Figure 4-66. Vapor pressure of sulfur dioxide solutions at various concentrations and temperatures.

the true free SO_2 content of the solution, as shown in Figure 4-66.[289] This means that rapid and efficient SO_2 absorption can only be achieved in solutions containing no true free SO_2. Conversely, to build up high true free SO_2 concentration, gas of high SO_2 concentration is required and the liquor must either be at a very low temperature or a substantial external pressure must be maintained.

Recovery of Digester Relief

In carrying out a sulfite cook in a batch digester, liquid and gas are withdrawn from the digester at certain times during the process. The material removed is referred to as relief and is taken in the following forms:

1. Side relief—cooking liquor withdrawn in liquid form during the early stages of the cook to provide a vapor space so that gas can be relieved at the top of the digester to control pressure, and also, in direct steamed cooks, to provide space for the steam condensate formed.

2. Top relief—mainly gas vented from the top of the digester during cooking to limit the pressure to the desired level.

3. Blowdown relief—taken at the end of the cook to reduce the pressure before discharging the pulp.

In an acid sulfite cook, these relief flows contain between 30 and 70% of the sulfur dioxide charged to the digester. They must be recovered both for economic reasons and to meet environmental standards. They are particularly useful for building up the concentration of free sulfur dioxide in the cooking liquor because they contain a much higher concentration of sulfur dioxide than the burner gas, which contains residual air.

In the bisulfite process only about 10% of the charged sulfur dioxide is in the digester gas relief, but a considerable heat economy is achieved by reclaiming these flows. Virtually all mills use some variant of the hot acid (Chemipulp) system of heat and chemical recovery, as shown in Figure 4–67. The system is a countercurrent one, with the fresh cooking liquor flowing from the acid plant to the digester and the relief gas flowing back towards the acid plant. The hot liquors from the side relief are returned to a pressure vessel, or accumulator. High-pressure gas relief is also returned to the accumulator through an eductor, where it contacts the fresh cooking acid. The accumulator usually has a capacity equivalent to 1.5 to 3 digester charges of liquor and is operated at about 275 kPa (40 psi) pressure in acid sulfite pulping. Low-pressure relief from the latter part of the digester blowdown relief and relief from the accumulator go to a pressure or recovery tower, which is usually operated at a slight pressure. The digester relief is first cooled in a heat exchanger with the fresh acid flowing to the ac-

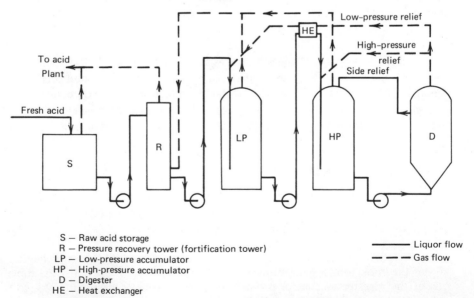

S — Raw acid storage
R — Pressure recovery tower (fortification tower)
LP — Low-pressure accumulator
HP — High-pressure accumulator
D — Digester
HE — Heat exchanger

——— Liquor flow
– – – Gas flow

Figure 4–67. Chemipulp hot-acid system.

cumulator to reduce its temperature and facilitate SO_2 recovery in the recovery tower. Some systems use two accumulators, with the second, low-pressure accumulator being operated at about 70 to 100 kPa (10 to 15 psi). The hot-acid system was developed to handle the high free SO_2 content liquors used in the acid sulfite pulping process. However, the same system is used for bisulfite liquor, mainly to recover heat, but also to recover the small amount of SO_2 in the relief. Since the pH of these unbuffered liquors is affected markedly by even these relatively small quantities of SO_2, it is common to recirculate the accumulator liquor and control the pH of the liquor pumped to the digester by injecting the required amount of base into this stream. The acid system runs at little or no pressure when bisulfite solutions are used.

The composition of sulfite cooking liquor is normally determined manually by the standard analytical methods discussed earlier.[198] Frequently the pH is continuously recorded in both the acid plant and at the accumulator in bisulfite systems. Instrumental methods to analyze liquor continually for chemical composition have been proposed,[293,294] but are not widely used.

Digestion in Sulfite Process

Digestion for the sulfite process can be either continuous or batch, but batch digestion predominates by a wide margin.

The Digester

Batch digesters for the sulfite process vary between 70 and 350 m^3 (2500 and 12,000 ft^3) in volume. Normally, a number of digesters of similar capacity are used in sequence to even out the operation to some extent and to allow the preceding acid-making stage and the subsequent stock-treatment system to be run continuously. Since relief liquors and gases are recirculated from one digester to the following one by means of the accumulator and other reclamation equipment, uniformity in both size of digester and in the scheduling of cooks is important. Digesters are normally cylindrical in shape with a conical bottom and hemispherical or conical dome. The top of the digester has a large, flanged opening with a removable top for filling with chips and for access. A strainer for gas relief is located in the neck of the digester. Connections from this strainer carry high-pressure relief to the accumulator and low-pressure relief to the recovery tower. A strainer near the bottom of the dome, also connected to the accumulator, is used for side relief. Collection and return strainers for the liquor circulation system are located in appropriate positions in the digester. The bottom of the digester has a special fitting that includes steam connections and a specially designed blow valve, which is opened to discharge the digester contents at the end of the cook.

Sulfite digesters are normally built of riveted carbon steel with reinforcing butt straps to provide the required rigidity. Since sulfite cooking liquors are highly corrosive and materials coming directly in contact with the liquor must be resistant to its attack, an inner lining of either ceramic tile or stainless steel cladding is used[295] to protect the main digester structure. Ceramic linings are by far the most common type in North America. For calcium-base pulping, coarse-pored clay brick and litharge-glycerine mortar were originally used as the standard lining materials. Calcium compounds deposited in the brick sealed off the pores. With soluble base pulping, denser brick and a more resistant furan resin mortar are required.[296-298] These linings have become standard for brick-lined digesters, regardless of base. The composition and structure of the lining must stand up to the physical and temperature stresses that occur during the filling, heating, and blowing of the digester. Single-coarse bricks are used on small digesters, double-coarse bricks on large digesters. In Europe carbon brick has been used in a number of installations to increase lining life. Brick linings have the advantage of high chemical resistance and low initial cost. However, they have to be replaced periodically, the average life of a lining being about 10 years. This results in a substantial loss in production, and even routine maintenance and inspection requires substantial down time for cooling. Furthermore, a substantial volume of the digester interior is occupied by the brick. These difficulties have led to the use of an inner lining or cladding of stainless steel.[295,299] These are more costly to install than brick linings, but they are much more versatile as to liquor composition and base, are easier to maintain, and result in much less loss in volume inside the structural shell than in brick-lined digesters. Corrosion problems have been encountered in some instances, however, particularly where liquors contain significant chloride levels due to pulping of saltwater driven wood. It has been found necessary to maintain the molybdenum content of the steel at a minimum of 2.5% to avoid these problems. Extra low carbon content, particularly at welds, and avoidance of scale deposits are also important.[300]

Digester Charging

In order to ensure that the available digester space is filled with the maximum quantity of chips and that these are uniformly distributed through the digester, a chip packer is normally used. These devices impart a tangential motion to the chips so that the chip bed maintains a flat rather than a conical profile and both packing density and uniformity are increased. These chip packers, which may use either mechanical action or steam, can increase the digester charge by from 10 to 40%.[301,302] The steam used in packing heats and partially presteams the chips. Air and other noncondensable gases are removed from a separate connection, usually the circulation line, with aid of an evacuation system. Additional steam to complete the presteaming may be added through the bottom connection.

When the digester is filled and the cover has been securely fastened, hot liquor from the high-pressure accumulator is pumped to the digester. During pumping, the displaced digester gases are allowed to escape from the top of the digester, either to the atmosphere or to the primary absorption system. Pumping is continued until the digester is completely full and then is continued for a few minutes, taking the overflow back to the accumulator. This precirculation displaces the liquor that has been diluted by initial contact with moist chips and ensures a uniform distribution of chemical throughout the digester.

The ratio of the volume of liquid in the digester to the dry weight of wood charged is referred to as the liquor-to-wood ratio. In addition to the liquor pumped to the digester during filling, the liquor-to-wood ratio should include the wood moisture; condensate formed during steam packing and presteaming; and direct steam condensed during the heating period, less the liquid removed by wet and side relief. The volume of liquor that can be charged is strongly influenced by the degree of chip packing. The liquor-to-wood ratio must be known to ascertain the liquor concentration and to supply to the digester the desired chemical charge based on wood weight.

Heating and Cooking in Sulfite Process

When the digester is filled with liquor and chips, it is ready for heating to cooking temperature. Since it is very difficult to eliminate temperature differentials within the digester once they are established, temperature should be maintained as uniform as possible while heating. Either forced or natural liquor circulation may be used. In natural circulation steam is added to the bottom of the digester, and the resulting natural convection currents and the relief of gas from the top of the digester bring about a circulation of the liquor. Where chip packing is used to increase the digester wood charge, it is considered essential to use forced circulation established by a circulation pump. The most common system for forced circulation of sulfite digesters is the split circulation method, in which liquor collection strainers are located in the middle of the digester wall and the return flow is split, one portion being returned to the top of the digester and the remainder to the bottom.[303] Top collection and bottom return is an alternative system. Usually separate side-relief strainers are located above the collection strainers to ensure that sufficient head is maintained after side relief to provide an adequate flow to the circulation pump.

Liquor heating can be either with direct steam injection into the digester or indirectly by means of a heat exchanger. A number of mills have reported improved quality after conversion to indirect heating,[304] but this system is more costly to install and maintain. It is usually preferred by mills making high-quality pulps, such as bleached paper grades and dissolving pulps, and for mills having a recovery system so as to avoid unnecessary dilution of the spent liquor. In direct steaming with forced circulation, the steam is usually added to the liquor circulating line to aid in attaining temperature uniformity. During the early part of the heating period the digester remains full of liquid and a hydrostatic pressure is maintained. However, once the chips at the top of the digester have been impregnated, it is desirable to remove a portion of the liquor to provide a gas space at the top of the digester and to allow the maximum pressure to be set at a preselected value by relieving gas from the digester dome. Side relief should be taken before the temperature level becomes too high, usually between 115 and 130°C for acid sulfite pulping and 125 to 145°C for bisulfite pulping. Taking the side relief at too high a temperature results in increases to the accumulator temperature and carries chemicals back to that vessel, which can initiate undesirable side reactions to be described later. Digester heating is continued until the cooking temperature is reached. In choosing the cooking temperature, the effect on pulp quality must be considered as well as the length of the cooking cycle, since sulfite pulp quality is affected by cooking temperature. Rate of temperature rise during the heating period can be controlled either manually by controlling steam flow to the digester or automatically by using the signal from a temperature sensor to regulate flow according to a predetermined schedule. The temperature sensor is usually located in the liquor circulating line, after the steam mixer or heat exchanger. Other digester instrumentation commonly used includes side temperature sensors, pressure control, a liquor-level recorder and circulation-flow recorders.[305,306]

Digester Relief and Blow

At the end of the cooking period the digester pressure is partially relieved in preparation for blowing. This relief gas is returned to the hot-acid system for heat and sulfur dioxide recovery as described earlier. Once the pressure and free sulfur dioxide level have been lowered by this means, the rate of pulping slows down to an insignificant level and no significant additional delignification occurs. When the appropriate pressure level for blowing is achieved, the blow valve is opened and the digester contents are allowed to discharge into the blow pit or blow tank. Blow pressure is selected on the basis of the blow-pit structure and the degree of sulfur dioxide and heat recovery desired; normally it is between 175 and 275 kPa (25 and 40 psi). If the blow pressure is too high, sulfur dioxide and heat recovery is not complete. If it is too low, a dirty blow may result, that is the full contents of the digester are not discharged. In this case, the remaining material must be removed either by washing or by reblowing with steam pressure. Many mills have installed flushing systems to ensure clean blows. In these systems, spent liquor or hot water is pumped into the digester at the start of the blow to loosen the material in the bottom cone.[307] Flushing appears to be particularly necessary when pulping to high-yield levels.

Although much of the heat and chemical available at the end of the cook is recovered during the gassing-down period, in the case of acid sulfite pulping substantial amounts of sulfur dioxide are released during the blow and discharged through the blow-pit vent. Environmental requirements in some locations have made it necessary to ensure that this gas is not discharged to the atmosphere. Several mills have installed scrubbers to remove SO_2 from these gases. These scrubbers are normally packed towers consisting of cooling and absorption stages, with the absorption medium being either water[308,309] or base.[310] The recovered sulfur dioxide can be returned to the acid-making system. The value of the chemical and heat thus recovered is sufficient to cover the installation and maintenance costs of the system. The sulfur dioxide content of the blow gases from a bisulfite cook are much lower and additional recovery at this stage is not needed.

Determining the End Point of a Sulfite Cook

The start of the pressure relief, after the period at cooking temperature, is the critical time in a sulfite cook. It determines the degree of delignification since the rate of pulping rapidly decreases as temperature and pressure are reduced. Several means are used to select the proper point to begin pressure relief, including time of cooking, liquor color, residual SO_2, and visual examination of pulp samples.

Relief can be initiated according to a standard time schedule if the cooking cycle and other pulping variables are closely controlled. This method is widely used because it keeps the digesters in a uniform sequence. In acid sulfite pulping, where the large excess of sulfur dioxide employed minimizes the effect of chemical charge, this method of control is usually adequate. In bisulfite pulping, where liquor pH and chemical charge have a major influence on the rate of pulping, other means of control are often required.

Liquor color deepens as the cook proceeds and this color change can be used to follow the progress of the cook. The end point can be selected by comparison to preselected standards or by measurements made on a simple colorimeter. These measurements are not very reliable, however, because the color intensity varies with changes in wood species, age, and other variables. A more precise liquor measurement is based on the ultraviolet absorption of lignin at 280 or 205 nm.[311-313] The latter wavelength has been shown to be a particularly accurate measure of the lignin dissolved during the cook.[312]

Residual sulfur dioxide concentration in the circulating liquor can be used to determine the point at which relief should be started because there is a correlation between the degree of delignification and the sulfur dioxide consumed. As an alternative to manual analysis, indirect methods, such as measurement of the electrical conductivity or index of refraction, have been suggested.[314]

Pulp samples removed from the side of the digester can be used to follow progress of the cook. However, there is not sufficient time available to carry out a permanganate test and assessment must be based on visual inspection. It is questionable whether a representative sample is obtained from a single point on the digester shell.

Continuous Cooking

Although most sulfite mills use batch digestion, several continuous digesters have been installed for bisulfite pulping.[315-317] The Kamyr digester unit using an inclined top separator has proven particularly satisfactory.[318] The system is shown in Figure 4-50. A slurry of the chips in the cooking liquor enters the digester through the high pressure feed valve and is then pumped to the expanded lower zone of the inclined separator, where a substantial proportion of the liquor is removed through a screened section and returned to the feeder. Makeup chemical is added to this recirculation flow at the desired rate. The drained chips are carried upward by the screw conveyor and, along with any excess liquor, drop into the main digestion vessel. The chips absorb sufficient chemical during the short impregnation period in the feed zone so that the cooking vessel can be maintained at cooking temperature and the chips rapidly heated without adverse effect. Additional chemical is taken up by the chips as they pass through the liquid-phase zone of the digestion vessel. The retention time in the vessel is determined by the production rate; degree of cooking is therefore controlled by adjusting the digester temperature. Noncondensable gases and excess sulfur dioxide are relieved from the top of the vessel to control pressure. A countercurrent wash zone is maintained in the lower part of the digester by introducing weak liquor near the bottom and withdrawing spent liquor from extraction screens near the middle of the digester.

In the earlier installations the standard Kamyr hydraulic design was used. In this configuration, which is shown in Figure 4–83, there is no vapor-phase zone separating the impregnation and cooking stages. As a result, some back mixing occurs, allowing reaction products to mix with high-bisulfite concentration impregnation liquor and initiating thiosulfate-producing side reactions. Thus liquor decomposition limits the use of this design to the production of newsprint-grade pulps in the 50 to 80 Kappa-number range. The top separator unit is required where a lower Kappa-number target is sought for bleachable grades. The Finnish company, Rauma-Repola, has built and operated a similar digester (different in design, but similar in principle to the Kamyr) to produce high-yield sulfite pulp for newsprint.[317]

Main Factors in Sulfite Pulping

There are a large number of variables that affect the sulfite pulping process. These include the wood species, cooking-liquor composition, and conditions in the digester. The most important of these factors are now discussed.

Impregnation of Wood with Cooking Chemical

The most important requirement in chemical pulping processes is to ensure that the cooking chemical is uniformly distributed throughout the wood. Improper distribution of chemical in the chips leads to uncooked chip centers, high screen rejects, low pulp yield, and high dirt count. Penetration is particularly important in sulfite pulping at low pH levels because sulfur dioxide can penetrate gas spaces in the chips ahead of the base and cause lignin condensation as the temperature is increased.

The impregnation of chemical into wood can be considered to take place by two separate processes: (1) mass flow of liquid and solute into the chips and (2) diffusion of solute through the liquid-saturated chips. These two kinds of movement act under different forces. The movement of liquid into chips is governed by capillary forces and applied pressure, while the diffusion of chemical into liquid-filled wood is dependent on the concentration gradient between the liquor surrounding the chip and that inside the chip.

In softwoods, the flow of liquid into chips takes place through the lumen of the fibers. Movement from fiber to fiber is by way of the bordered pits that connect adjoining tracheids. The membrane in these pits, known as the torus, has a network of small openings that act as capillaries. These contribute the main resistance to the flow of liquid through the chips. The flow of liquid through a capillary is dependent mainly on the diameter of each capillary rather than on the total cross-sectional area, and the morphological characteristics of the wood therefore play a significant role in penetration.

In deciduous woods the nature of penetration is slightly different. There are few pits interconnecting the individual tracheids, and the longitudinal flow takes place mainly through vessels that are connected to the tracheids through pits. Since the vessel elements are substantially greater in diameter than the tracheids, mass flow of liquid can be very rapid in hardwoods. However, the membranes are frequently blocked by tyloses, which are bladder-like growths in the vessel cavities. These growths are particularly noticeable in the heartwood fraction. Measurements by Stone[319] have shown air permeability can differ between sapwood and heartwood samples taken from adjacent locations by a factor of as much as 2000 to 1 due to the presence of tyloses. However, the difficulties encountered in the pulping of the heartwood of these species is much less than would be anticipated from these measurements, and it appears that the chemical pulping reagents attack the tyloses, lessening their effect on the cooking process.

The rate of mass penetration of cooking liquor into wood depends on the physical properties of the cooking liquor, the wood species, the pressure differential, the length of the penetration path, the temperature, whether sapwood or heartwood is being penetrated, and the structural direction in the wood. These factors were originally investigated in detail by Maass and coworkers[320,321] and the fundamental aspects of this subject have been reviewed by Stamm.[322-325] The mass flow of liquid into wood can be accelerated by applying external pressure to the system to augment the natural capillary forces. In a digester this pressure can be applied as hydrostatic pressure by continuing to operate the filling pump after the digester is filled or by adding direct steam to the liquid-filled digester and controlling the pressure through the top relief valve. The rate of penetration of liquids into wood also increases with temperature, as a result of both the reduction of the viscosity of the penetrating liquor and also by the removal of resins and other deposits from the pit membranes, thus allowing freer passage of the liquor. The use of the hot-acid system in sulfite mills is thus an aid to good liquor penetration. The effect of temperature and pressure on the degree of penetration of water into spruce chips as measured by direct weighing on a quartz spiral balance is shown in Table 4-21.[326] It can be seen that, over the ranges reported, pressure has a major effect on the water uptake and temperature has a lesser but significant effect.

TABLE 4–21 EFFECT OF PRESSURE AND TEMPERATURE ON PENETRATION OF WATER INTO SPRUCE CHIPS

Temperature (°C)	Penetration (%)							
	at 207 kPa (30 psi)		at 310 kPa (45 psi)		at 414 kPa (60 psi)	
	1 hr	4 hr	1 hr	4 hr	1 hr	4 hr
25	58	72	70	83	81	91
60	72	82	85	90	89	93
90	81	89	89	93	94	96

Initial mass flow of sulfite liquors is almost entirely in the longitudinal or fiber direction, since the ratio of fiber length to width is high and the resistance to flow within the lumen is less than that through the pit membranes or the cell wall. The actual rate of flow is dependent on the morphological features and physical condition of the particular wood particle; Maass has claimed this flow-rate can be as high as 50- to 100-times greater in the fiber direction than in the transverse directions.[327] The rate of diffusion of solute through the liquid-filled wood does not show as great a difference in the two directions as the rate of mass flow. The ratio of longitudinal to transverse diffusion varies with species and conditions; for sulfite liquor it is normally between 3 and 10 to 1.[324,325] The movement is slightly greater in the radial direction than in the tangential for most species, probably because of the additional diffusion through the ray cells. Sapwood is several times more permeable than heartwood, the actual magnitude of the difference varying between species. This is believed to result from closing off the pit membranes in the heartwood when they are no longer required for the transport of sap in the tree.

The initial penetration will proceed more rapidly if the air is first removed from the chips. Small bubbles of entrapped air will greatly increase the resistance to the flow of liquids through the wood capillaries. Methods to remove air prior to immersing the chips in liquid are therefore frequently employed. The common method is to presteam the chips during the filling of the digester,[328] either as part of the steam-packing operation or by a separate steam addition. It is important that atmospheric-pressure steaming continue until a temperature of 100°C is reached, an indication that all the air has been displaced from both the chips and the surrounding vapor. At moisture contents of about 30% or higher (the fiber saturation point) there is sufficient water in the chip to displace all the air in the chip when it is vaporized. Chip steaming is more effective at or above this moisture level. If the digester is then filled with liquor at a temperature slightly below the presteaming temperature, water vapor will be condensed and a vacuum created that will draw the cooking liquor into the chip. Even after presteaming, some air remains entrapped in the wood structure, however, and a procedure was developed by the Pulp and Paper Research Institute of Canada[329,330] to expel this residual air. In this procedure, known as the Va-purge technique, the steam pressure in the digester containing the chips is rapidly built up to a predetermined level, held for a short period of time, and then rapidly released to a lower level, normally atmospheric. This cycle may be repeated if desired. Normally, the chips are presteamed at atmospheric pressure prior to the purge treatment in order to remove the bulk of the air from the digester. The maximum pressure in the purge cycle is usually 206 kPa (30 psi), although higher levels may be used if advantageous. Although this technique is effective in removing air from the chips and thus aids penetration,[331] it is difficult to carry out on the mill scale, particularly in batch digesters, and it consumes a substantial quantity of steam. It has not been adopted to any extent on a commercial scale.

A comparison of atmospheric steaming, vacuum evacuation, pressure impregnation, Va-purge, and "Vilamo" impregnation[332] (a pressure pulsation technique), has been carried out by the Swedish Forest Products Laboratory.[333,334] Two different criteria were used to assess the effectiveness of the impregnation: (1) direct measurement of the weight increase in individual chips, using a quartz spiral balance and (2) the level of screenings after cooking. Using both methods, it was found that the conventional impregnation technique most commonly used in both batch and continuous digester sulfite mills, that is, atmospheric steaming followed by pressure impregnation, was as good as or superior to the other more complicated procedures.

Effect of Chip Dimensions and Quality

Chip size has an important bearing on the uniformity of impregnation of the chips and on the quality of the pulp produced in the sulfite process. Penetration and diffusion take place at different rates in the three different directions of the wood structure, that is in the longitudinal, radial, and tangential directions. Bulk penetration occurs almost entirely in the fiber (longitudinal) direction, while diffusion takes place in all directions, although still preferentially in the longitudinal direction. Thus both chip thickness and length are of importance in sulfite pulping.

Chip length (the grain direction) affects not only the penetration of liquor into wood, but also the quality of the pulp produced. The mass flow-rate is inversely proportional to chip length and, hence, more rapid penetration occurs with shorter chips. However, the fibers in the wood are damaged during cutting due to both fiber shortening at the point of the cut and to compression damage in the area surrounding it. This fiber damage tends to increase when the chip length is short. When chipping under optimum conditions (hand-fed wood and sharp chipper knives), deterioration of pulp strength properties could be detected only when the nominal chip length was decreased to 12.7 or 6.4 mm (1/2 or 1/4 in.), although some fiber shortening was detectable below 19 mm (3/4 in.).[335] On the other hand, when chips produced on a commercial chipper were separated into fractions, both according to length and to thickness, the main pulp-quality parameters (rejects, brightness, shive content, and strength) were found to correlate more closely with the thickness.[336] The optimum thickness was found to be 4.8 mm (3/16 in.) for calcium acid sulfite pulping and 7.9 mm (5/16 in.) for sodium bisulfite pulping. The high quality obtained from the relatively thin chips is apparently due to their more uniform penetration with the cooking chemical. However, with conventional chippers it is only possible to control chip length. The chipper knife penetrates the wood until the shear force is great enough to split the wood. As the length setting is increased, greater shear is required to bring about this splitting. For this reason, increased length is usually accompanied by increased thickness.[337] However, knots, grain and other gross morphological features, as well as chipper variables have a major effect on these forces, and conventional chips vary widely in thickness.

The effect of compression damage during chipping on the quality of sulfite

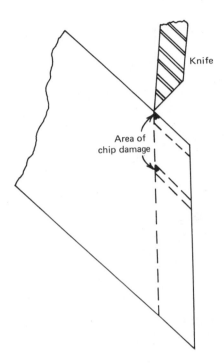

Figure 4-68. Action of chipper knife and cause of wood damage.

pulps was first demonstrated by Green and Yorston,[338,339] who noted that when a chipper knife enters a log one end of the chip is subject to extensive compression, as shown in Figure 4-68. When chips from commercial chippers were cut in half and the damaged and undamaged halves pulped separately, the damaged half of the chips showed a substantial decrease in strength, as shown by the results in Table 4-22. It was also shown that a similar loss in strength could be induced when wood was subjected to severe compression parallel to the grain. Compression in this direction is more damaging than compression perpendicular to the grain and the degradation is greatest at pH 2.0 (acid sulfite), less at pH 5.0 (bisulfite), and least with neutral sulfite and kraft liquors.[339-343] Strength loss is apparent in an acid sulfite pulp when the wood has been subjected to as little as 5% deformation parallel to the fiber. The degradation increases in proportion to the deformation when the crushing load is increased. Since only a portion of the wood is damaged during the chipping operation, the full effect of this strength loss will not be found in the pulp as a whole, but it has been well established that poor chipper operation (inadequate frequency of knife changes, incorrect grind angles, excessive fines production) does result in production of pulps of decreased strength properties. Hartler has conducted a very thorough investigation of the effect of chipper variables on the quality of chips produced.[344] He showed that chip quality is markedly affected by the relationships between knife grind angle, the angle of the feed spout, and the clearance be-

TABLE 4-22 SULFITE PULP QUALITY FROM DAMAGED AND UNDAMAGED ENDS OF MILL CHIPS

	Unbeaten Pulp			Beaten Pulp		
	Undamaged Half	Damaged Half	Ratio	Undamaged Half	Damaged Half	Ratio
Bulk (cc/gm)	1.43	1.44	—	1.36	1.33	—
Burst ratio[a]	52.8	38.6	1.37	86.1	58.4	1.47
Tear ratio[a]	1.48	1.13	1.31	1.19	0.88	1.35
Breaking length (m)	6770	5600	1.21	8680	7340	1.18
Double folds	1210	467	2.59	3940	874	4.51
Retained 28 mesh	—	—	—	5.0	8.7	1.74

[a]Units not given in SI units; to be used for comparison.

tween the cut log and the face of the chipper. In order to produce good quality chips, not only must the chipper knives be kept sharp, but also the correct relationship between these angles must be maintained. The loss in pulp strength when poor chipping techniques are used is apparently caused by the increased accessibility of cellulose to acid attack where dislocations in the fiber wall are created by the mechanical action on the wood.[345] The subject of chip quality was discussed in the chapter on pulpwood.

A number of attempts have been made to design chippers that will minimize the damage to the wood during the chip formation. These include the parallel or wafer chippers, such as the Anglo drum chipper,[346] the Soderhamn H-P chipper,[347] in which the main cutting action is parallel to the wood grain; and the CCL chipper,[348] in which the shear force is exerted perpendicular to the fiber axis. These chippers not only produce chips with minimum fiber damage, but also produce thin chips of uniform thickness, which ensures uniform takeup of cooking chemical and minimum pulp rejects.

Mechanically reducing the size of chips for sulfite pulping by shredding in disk refiners has been suggested[349-351] to increase the uniformity of liquor penetration and to attain higher screened pulp yields. Shredding increases the wood surface area available for penetration by breaking the chips along their natural lines of cleavage. It can be achieved in a refiner having coarse-tooth plates and operated with a relatively open clearance between the disks. The fraction of chips retained on the larger screens (25 and 19 mm, 1 in. and 3/4 in.) is reduced and the portion passing through to the smaller screens is increased, as shown in Figure 4-69. In addition to lowered pulp rejects, chip shredding can increase the amount of wood that can be charged to a digester,[350] thus increasing production in a mill limited by digester capacity. However, the mechanical action of

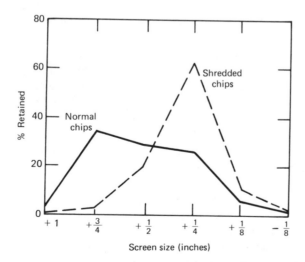

Figure 4-69. Classification of normal and shredded softwood chips.

the refiner reduces pulp tear strength if the degree of shredding is not carefully controlled.

Effect of Temperature, Time, and Pressure

Temperature. Temperature mainly affects the sulfite process in three ways. Increasing the temperature:

1. Decreases the hydrogen ion concentration (increases the pH).
2. Decreases the solubility of sulfur dioxide thereby increasing the SO_2 pressure above the cooking liquor.
3. Increases the rate of lignin dissolution and cellulose degradation.

Two aspects of temperature will be considered: the rate of rise to maximum temperature and the maximum temperature itself. The rate at which the maximum temperature can be safely approached is dependent on both the effectiveness of chip penetration by the cooking reagent and the nature of the cooking liquor. In acid sulfite pulping, free sulfur dioxide can enter the gas phase and move through the wood structure more rapidly than the base, and produce high local acidity in the center of the chip. Under these conditions lignin condensation reactions occur at elevated temperatures, and these condensation reactions consume the active sites in the lignin molecule where sulfonation normally occurs. The condensed lignin can no longer be dissolved by the sulfite liquor and uncooked fragments of wood remain after the completion of the cook.[352,353] The phenomenon of dark centers commonly observed in sulfite pulping is largely a result of such condensation reactions. Thus, for acid sulfite pulping a heating cycle must be used that allows for full penetration of the liquor containing the bisulfite and base before the critical temperature for lignin condensation is reached. This critical temperature is generally found to be in the range of 110 to 120°C. Thus the practice for many years in the sulfite industry was to use a very slow rise to temperature. It was common to hold the digester temperature at 110 to 115°C for periods of 1 to 2 hr to ensure that impregnation was complete before heating to higher temperatures. This led to long, slow cooking schedules. Digester cooking times in the range of 10 to 15 hr were not uncommon.[354,355] With modern penetration treatments the heating schedules can be speeded up to permit the cook to be carried out in a much shorter time. With the use of presteaming, pressure impregnation, and forced liquor circulation, heating times of 2 to 3 hr to maximum temperature can be used without lignin condensation, since the base then permeates all portions of the chip and as elevated temperatures are reached, the sulfonation reaction occurs in preference to the condensation reactions. It has even been suggested on the basis of laboratory cooks that sulfite digestion can be carried out by introducing the liquor to the digester at cooking temperature, provided the chips have been suitably deaerated and presteamed by a series of Va-purge treatments.[356] Although there are a number of technical barriers to applying this system commercially, it does demonstrate the importance of adequate penetration conditions in sulfite pulping.

The maximum temperature used in acid sulfite pulping is determined by the quality of pulp desired, the digester pressure limitations, and the available digester time. The cooking temperature affects the rate of removal of hemicellulose more than that of the other major components, cellulose and lignin.[357,358] Thus cooking at low temperature favors hemicellulose retention, while cooking at high temperature favors dissolution. Pulps for use in glassine and grease-proof grades, which require high hemicellulose content to provide a high level of bonding, transparency, and easy beating, are usually cooked in the range of 125 to 135°C, while pulps for dissolving grades, which must have low hemicellulose content, are cooked at about 150°C. Acid sulfite paper-grade pulps are normally cooked between these extremes.

With the bisulfite process, now predominant in sulfite pulping, the requirements are quite different. Since the cooking liquor contains only a very small partial pressure of SO_2 and the cooking chemical is composed almost solely of bisulfite, the sulfur dioxide cannot penetrate ahead of the base to any extent, and lignin condensation reactions are much less likely to occur. Heating rates can thus be quite rapid, limited only by the digester circulation rate, so that a relatively short cooking time is possible. However, it is still important that good penetration be obtained, since acids generated by the wood can also initiate lignin condensation in portions of the wood devoid of cooking chemical. Because the cooking liquor has lower acidity, bisulfite pulping must be carried out at higher temperature than acid sulfite pulping. Thus cooking temperatures are normally in the range of 160 to 166°C for the production of bleachable pulps. At the high pH levels used, the hydrolysis of hemicellulose is retarded and only minor losses of cellulose occur, so that in spite of the higher temperature, bisulfite cooks exhibit excellent carbohydrate retention.

Decreased maximum cooking temperature gives higher retention of hemicellulose, reduced degradation of cellulose, and results in increased pulp yield in both the acid and the bisulfite pulping processes.[226,359] Figure 4-70 shows that, for a given degree of delignification, higher yields are obtained at three different levels of pH as the temperature is decreased. Since yield is an important consideration in the economics of pulping, this indicates that digesters should be operated at quite low temperatures. However, from the standpoint of productivity of the digesters, it is also important that the cooking time should not be extended too long and so both yield and production rate must be considered in choosing the appropriate cooking temperature in a particular situation. Investigation of the effect of temperature on the kinetics of sulfite delignification[359] show that the reactions obey the Arrhenius equation:

$$Ln\ k = Ln\ Z - \frac{E}{RT}$$

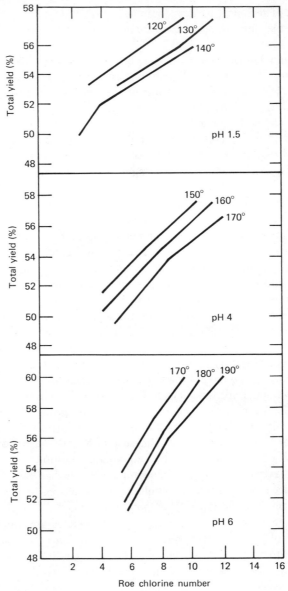

Figure 4-70. Effect of cooking temperature shown in degrees Centigrade on pulp yield.

where

k is the rate constant for the reaction

E is the energy of activation

R is the gas constant

T is the absolute temperature

Z is a constant

These studies how that the energy of activation for sulfite delignification is 20 to 21.5 kcal/mole and that the rate of reaction approximately doubles for each 10°C rise in the maximum cooking temperature.

Time. Cooking temperature and liquor composition are the main factors in determining the cooking time required to achieve a given degree of delignification. It has been found,[360] for example, that the residual lignin content can be expressed by a quadratic equation of the type:

$$Y = a + bX + cX^2$$

where

Y is the residual lignin

X is the cooking time

$a, b,$ and c are constants dependent on the temperature and pH of the cook.

The values shown in Table 4-23 were obtained for the three constants in the above equation for two series of tests, one carried out on very high-yield spruce fiber and the other on thin spruce wafers. The substantial effect of both pH and temperature on cooking time can be seen from these data. Chemical charge,

TABLE 4–23 COEFFICIENTS FOR DELIGNIFICATION RATE EXPRESSION ($Y = a + bX + cX^2$)

Temperature (°C)	pH	a	b	c
135	1.5	21.2	−0.1858	0.000415
	2.0	22.9	−0.1066	0.000143
	3.0	23.5	−0.0303	0.000040
	4.0	23.5	−0.0183	0.000019
150	1.5	25.0	−0.2930	0.000868
	2.0	20.5	−0.1757	0.000386
	3.0	23.4	−0.0504	0.000014
	4.0	22.6	−0.0246	0.000032
165	1.5	24.4	−0.4783	0.001050
	2.0	24.4	−0.3139	0.000778
	3.0	25.3	−0.1449	0.000209
	4.0	24.3	−0.1117	0.000126
180	1.5	21.5	−0.7476	0.005333
	2.0	23.3	−0.5653	0.001858
	3.0	23.8	−0.2930	0.000354
	4.0	24.3	−0.2205	0.000476

on the other hand, has only a minor effect on the reaction rate. A similar relationship can be derived for the rate of carbohydrate removal.[360] When a certain degree of delignification has been reached, usually corresponding to about 95% lignin removal, further lignin removal is very slow. Carrying the delignification process beyond this level will result in severe cellulose degradation and loss in yield.

Two series of cooks, showing the effects of increased cooking time on pulp properties for calcium acid sulfite and magnesium bisulfite pulping of spruce chips, are shown in Tables 4-24 and 4-25.[225,361] These two tables show the

TABLE 4-24 EFFECT OF COOKING TIME AT MAXIMUM TEMPERATURE ON PROPERTIES OF PULP FROM BLACK SPRUCE—ACID SULFITE

	Time at Maximum Temperature (hr)				
	2.0	2.5	3.0	3.5	4.0
Total yield (%)	56.0	52.6	50.6	48.2	47.4
Screened yield (%)	46.3	49.0	47.8	47.1	46.1
Permanganate number	22.5	18.1	13.9	12.9	14.6
Beating time (min)	40	42	43	49	46
Burst Index (kPa · m²/g)	7.35	7.16	7.75	7.16	7.06

Maximum temperature	134°C
Time to maximum temperature	7 hr
Liquor ratio	5 l/kg OD wood
Liquor composition	5.0% free, 1.2% combined SO_2

TABLE 4-25 EFFECT OF COOKING TIME AT MAXIMUM TEMPERATURE ON SPRUCE PULPING AND PULP PROPERTIES—MAGNESIUM BISULFITE

	Time at Maximum Temperature (hr)			
	0.5	1.5	2.25	3.0
Yield—accepted (%)	—	54.2	58.3	51.5
—rejected (%)	—	11.3	0.5	0.5
—total (%)	80.2	65.5	58.8	52.0
Roe chlorine number	27.2	21.1	13.4	7.6
Strength factor[a] —450 freeness	99	129	123	134
—300 freeness	107	123	123	131

[a]Strength factor = burst factor + 1/2 tear factor.

Maximum temperature	166°C
Time to maximum temperature	30 min
Liquor-to-wood ratio	4.2 l/kg OD wood
Liquor composition	4.0% total SO_2
	2.0% free SO_2
	2.0% combined SO_2
	3.6 pH (at 25°C)

effect of time on the degree of delignification measured by permanganate number and chlorine number respectively. The shorter cooking time requirement for the bisulfite process can readily be seen from these data.

Pressure. The main aspect of operating pressure in the sulfite cooking process is its use to regulate the amount of free sulfur dioxide in the system.[362-364] This controls the partial pressure of SO_2, which in turn fixes the free SO_2 concentration and pH in the liquor during the cooking period, both of which are important pulping parameters. Digester pressure limitations generally control the maximum pressure that can be used in a particular mill. Older digesters were built to withstand 480 to 550 kPa (70 to 80 psi), while newer ones have been built for about 700 kPa (100 psi) to take advantage of the faster cooking cycles which can be used with higher temperatures and higher SO_2 concentration. In the calcium base systems, high free sulfur dioxide concentrations were necessary in order to prevent calcium sulfite precipitation (liming) as the temperature was raised.[205] In order to maintain this high free sulfur dioxide concentration throughout the cook, the pressure must be maintained sufficiently above the steam pressure of the liquor to prevent relieving excessive amounts of sulfur dioxide gas. This precipitation problem does not exist with the soluble bases and the same constraints do not need to be applied in selecting the temperature, pressure, or free SO_2 levels.

Effect of Chemical Composition

The pH of Sulfite Liquor. When a base is added to a solution of sulfur dioxide in water, an equilibrium is established between the constituents, as shown below, using sodium as the example:

$$SO_2 + H_2O \rightleftharpoons H_2SO_3 \overset{(+\ NaOH)}{\rightleftharpoons} Na^+ + HSO_3{}^- \overset{(+\ NaOH)}{\rightleftharpoons} 2Na^+ + SO_3{}^=$$

The composition is normally expressed in terms of total, free, and combined sulfur dioxide, as defined previously, and pH. The effects of composition variables over a wide range and at different temperatures on sulfite pulping have been thoroughly investigated by Ingruber.[365]

Probably the most important parameter governing the rate of the bisulfite cook is the initial pH. The pH level varies during the cooking cycle due to the effects of temperature, SO_2 relief, and acids generated during the cooking process. Thus, for a full understanding, it is important to know not only the initial pH of the liquor, but also the changes in pH as the cook proceeds. For many years this could only be achieved by withdrawing a series of cooled samples from the digester. However, Ingruber[201] has developed pH electrodes that can be inserted into the digester liquor-circulating line to detect the pH at digester temperature, giving a finer insight into the role of pH in cooking. It has been suggested that the liquor pH could be controlled during a cook by installing a hot pH sensor in the circulating line to regulate the rate of sulfur dioxide relief.[366,367] The course of pH, measured both hot and cold during a typical acid sulfite cook, is shown in Figure 4-71. The hot pH values are 1 to 2 units higher than the cold values in this range. The temperature coefficient of pH in acid sulfite pulping liquors is about 0.011 to 0.012 pH units/°C.[368] The temperature coefficient is dependent on the wood species, type of base, and concentration. It is less during wood digestion than when the liquors are heated alone, probably due to the buffering effect of the acids released during cooking.[369] The importance of the hydrogen ion concentration in sulfite pulping was recognized many years ago by Hägglund,[370] who postulated that the delignification reactions took place in two main stages—sulfonation followed by hydrolytic cleavage. The hydrolysis stage was considered to be the slower of the two reactions, and thus the overall rate-controlling step. Maass and coworkers[371,372] found that both hydrogen ion concentration and bisulfite ion concentration were significant factors in determining the sulfite pulping rate, and proposed kinetic expressions involving both these terms.

The effect of pH on the rate of sulfite pulping has become particularly important as the emphasis has shifted to the bisulfite range.[373] Because the hy-

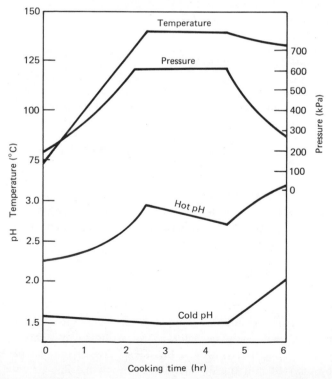

Figure 4-71. Temperature, pressure, and hot and cold pH during a typical acid sulfite cook.

drogen ion concentration available for the hydrolytic reactions is limited at this pH level, pulping can only be achieved by using elevated temperature. Hall and Stockman[226] found that the temperature required to produce a pulp of Roe No. 6 in 3 hr was as follows:

pH 1.5	130°C
pH 4.0	160°C
pH 6.0	180°C

Other workers have confirmed that when the pH of the initial cooking liquor rises much above the bisulfite point (pH 4), the rate of pulping becomes very slow,[374] and operation at these levels would be impractical with existing equipment.

The quality and yield of the pulp produced is also affected by the pH of the cooking liquor. Pulp yields in the bisulfite pH region are higher than those obtained at lower values because more hemicellulose is retained.[226,365] The difference is even greater when screened yield is compared, since the bisulfite liquor penetrates the wood more uniformly and the level of screen rejects is reduced. The defibration point (that is the yield or Kappa number giving 1% screenings after disintegration by a standard defibration treatment) is also higher at the bisulfite pH.[375,376] This permits the system to be operated at higher Kappa-number levels, giving still further gains in yield. Unbleached brightness also reaches a maximum at the bisulfite pH level (pH 4) as shown in Figure 4-72. For a given pH level, the brightness is higher for lower cooking temperatures.[226] For softwoods, pulp-strength properties reach an optimum when the initial pH of the cooking liquor is about 4 (measured at 25°C).[225-228,376] Strengths at this pH level are higher than those obtained at either 1.5 or 6.0 pH. In the case of

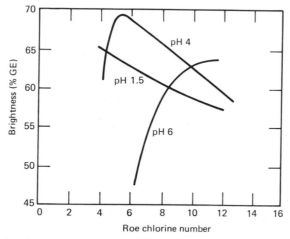

Figure 4-72. Brightness versus Roe chlorine number for sulfite cooks at three pH levels.

hardwoods, the strength from bisulfite pulping is greater than that from acid sulfite cooking, but the strength continues to increase as the pH rises to the 5 to 7 range.[376,377]

Combined Sulfur Dioxide. This is important in sulfite pulping in two different ways—its concentration in the initial liquor and its weight ratio to wood. The concentration mainly affects the penetration, which has already been discussed. The charge on wood, on the other hand, mainly affects the course of the cook. It determines the rate at which the concentration is depleted and thus affects the rate of pulping. The more fundamental parameter is the bisulfite ion concentration, which is twice the combined SO_2 level for solutions containing no monosulfite. Increasing the level of combined SO_2 in a bisulfite cook increases the yield at a specific Kappa-number level and increases the brightness of the resultant pulp. The tensile strength of the pulps increases with increasing combined SO_2 but tearing resistance decreases. The effect of combined SO_2 on some of these properties is illustrated in Figures 4–73 and 4–74,[375] which show that combined SO_2 has a significant effect at the lower levels of charge on wood, but that the effect diminishes at higher charges. This demonstrates the importance of using the optimum charge in pulping operations since the gain from using additional cooking chemical diminishes at the higher addition levels. The use of excessive amounts of combined SO_2 in bisulfite pulping results in substantial chemical costs for systems without recovery, and an overloading of the recovery and liquor regeneration equipment for systems with recovery. In addition, excessive bisulfite enters into autocatalytic decompositions involving thiosulfate, which can deplete the bisulfite concentration and cause burnt or black cooks. The combined SO_2 requirement for pulping varies with wood species and the desired yield level, but for chemical pulps 7 to 10% combined SO_2 on wood is normally the optimum charge. The combined SO_2

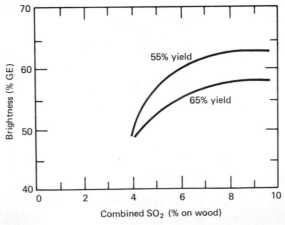

Figure 4–73. Effect of chemical-to-wood ratio on pulp brightness.

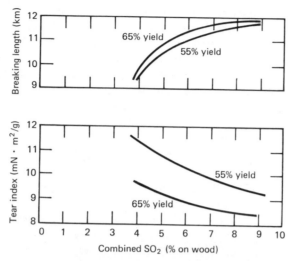

Figure 4–74. Effect of combined SO$_2$ on strength properties.

requirement would be in the lower part of this range for spruce and aspen, while that for pine and dense hardwoods would be at the upper end. High-yield bisulfite systems require less chemical, although care must be taken not to reduce the level too far because this increases the refining-power requirement for a given yield. It has been suggested that, for pulping to the 65% yield level, 5 to 6% combined SO$_2$ on wood (16 to 20% NaHSO$_3$) is satisfactory,[224] but many operating mills find it preferable to operate slightly above this range.

The effect of combined sulfur dioxide is also significant in acid sulfite pulping. In many older publications it has been suggested that the chief role of the base was to prevent lignin condensation by neutralizing the highly acidic lignosulfonic acids as they are formed. However, it is now recognized that the bisulfite ion plays an active part in lignin sulfonation, and that under normal pulping conditions a certain minimum level of bisulfite is required. On the other hand, if the combined SO$_2$ level is too high, the rate of pulping is retarded, probably because of the lower acidity due to the lower ratio of free to combined SO$_2$. In a study of the effect of combined SO$_2$ on acid sulfite pulping,[378] it was found that for pulps cooked beyond the defibration point, brightness was increased by increasing combined SO$_2$ level. Other properties were not affected. In high-yield pulping, both brightness and pulp strength improve, and refining-power requirement decreases with increasing charge of base.

Total and Free Sulfur Dioxide. Of the four composition parameters normally considered in connection with sulfite cooking liquors (total SO$_2$, free SO$_2$, combined SO$_2$, and pH), only two need be specified, the other two are then fixed for all practical purposes. For example, if the total and combined SO$_2$ for a liquor are specified, then the free SO$_2$ is the difference between the two, and that composition will correspond to a particular pH. In the foregoing discussion on the effect of composition on sulfite pulping, a detailed discussion of the effects of pH and combined SO_2 has been given because they can be related to the hydrogen ion concentration and the bisulfite ion concentration, usually considered to be the most important parameters in sulfite pulping. Only a few further points relating to the total and free SO_2 content are of significance for this discussion.

In bisulfite pulping, it is important that sufficient SO_2 is charged to complete the sulfonation, making allowance for that removed as relief and that consumed in side reactions. In acid sulfite pulping, the total SO_2 charge is less important than in bisulfite pulping because the digester is charged with an excess of sulfur dioxide and the unused portion relieved for reuse in subsequent digestions. Total SO_2 concentration of the digester liquor for a bisulfite cook is in the range of 3 to 6%, depending on the wood species, liquor-to-wood ratio, and the target yield level. In acid sulfite pulping, total SO_2 concentration normally varies between 5.5 and 8%. The free SO_2 has been shown to have a significant effect on the rate of acid sulfite pulping, although its influence is probably mainly due to the accompanying increase in hydrogen ion concentration. In bisulfite pulping, if the liquor contains a slight excess of free SO_2 (that is free slightly greater than combined), the rate of pulping is accelerated. However, to achieve the increased pulping rate, while retaining the benefits of bisulfite pulping, it is necessary to set the relief pressure about 35 to 70 kPa (5 to 10 psi) higher than the steam pressure to maintain the slight excess of free SO_2 during the main delignification period.[225,375]

Effect of Liquor-to-Wood Ratio

For calculation of the liquor-to-wood ratio, the weight of *all* liquid in the digester should be used, including the initial liquor charge, wood moisture, and any steam condensed during either presteaming or direct heating of the digester, less any withdrawal in the form of relief. The liquor-to-wood ratio varies at different points of the cook as direct steam is added and wet relief is withdrawn. In practice, the important value is the net liquor-to-wood ratio after side relief, and an approximate value for this can be obtained by adding the wood moisture to the average weight of the liquor pumped to the accumulator per cook.

The liquor-to-wood ratio during impregnation must be sufficient to immerse all the chips; the amount of liquor required to do this will depend on the wood density and chip packing, but will usually be about 5 to 1. Once impregnation has been completed, the ratio can be lowered by side relief so that there is less liquor to be heated to the cooking temperature, thereby reducing the steam requirement. Sufficient liquor must remain in the digester to maintain circulation; the liquor-to-wood ratio, after side relief, is usually 3.0 or 4.0 to 1 depending mainly on the level of the collection strainers in the digester. For a given chemical charge on wood, the best results can be obtained by filling the digester

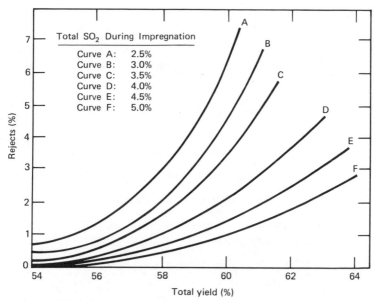

Figure 4-75. Effect of chemical concentration during impregnation on the relationship between rejects and total yield. Chemical charge on wood during cooking maintained constant.

with a relatively concentrated liquor during the impregnation period and then withdrawing down to a low final liquor ratio. The high liquor concentration during the impregnation stage gives a high driving force for diffusion, resulting in excellent penetration of chemical and a lower level of screen rejects. Figure 4-75[379] shows that a substantial reduction in rejects level can be achieved by this technique. Pulp dirt is also decreased, brightness and other properties improved, and bleach demand is reduced. In some mills the side relief strainers have been lowered in order to obtain these benefits.

Kinetics of Sulfite Delignification

A number of investigations have been aimed at developing kinetic equations that can be used to simulate and optimize the sulfite process. The significant variables in predicting cooking rate are the hydrogen and bisulfite ion concentrations, temperature, and pressure. Swedish workers have developed a series of mathematical expressions for the rates of dissolution of lignin and hemicellulose and the rate of formation of strong acids in acid sulfite pulping:[380]

$$-\frac{dL}{dt} = k_L(T)L^a \cdot [HSO_3^-]^\alpha \cdot [H^+]^\beta$$

$$-\frac{dC}{dt} = k_c\,(T)C^d \cdot [\text{H}^+]^\gamma$$

$$\frac{dSA^-}{dt} = \frac{k_{SA}\,(T)}{V} \cdot [SA^-]^n \cdot [\text{HSO}_3{}^-]^b \cdot [\text{H}^+]^c$$

where

L = lignin content

C = hemicellulose content

SA^- = strong acid anion concentration

$\text{HSO}_3{}^-$ = bisulfite ion concentration at cooking temperature

H^+ = hydrogen ion concentration at cooking temperature

(T) = average value of the absolute temperature

t = time

V = liquor-to-wood ratio at the start of the cook

The a, b, c, d, n, α, β, and γ are exponents; the k values are rate constants for the reactions that depend on (T).

Values for the exponents and rate constants were determined from extensive laboratory pulping data, and were generally found to be constant over the range of cooking conditions employed in acid sulfite pulping. However, for the rate of dissolution of lignin, an adequate fit to empirical data could only be obtained by assuming a transition point in the process at about 12% lignin content, with different values for k_L and a on either side of this point. It has been found possible to predict lignin content and yield of acid sulfite cooks based on this type of model with sufficient accuracy to allow process optimization.

Other workers in Finland[381] have used a simplified model of form:

$$\log t = b_0 + b_1 T + b_2 S + b_3 K + b_4 C$$

where

t = cooking time

T = cooking temperature

S = percentage of total SO_2

K = Kappa number

C = percentage of base

b_0, b_1, b_2, b_3, and b_4 are constants.

The values for these constants were determined from extensive analysis of mill cooking data and an equation was then developed to determine optimum conditions for mill operation. In this program, the desired production rate was used to determine the time available for each cook. The minimum temperature that would give the required Kappa number in this time can then be predicted from the kinetic expression. Savings result from the higher yield obtained by cooking at minimum temperature.

Yorston[363,373,382] has also studied the kinetics of the pulping process in detail. In acid sulfite pulping[363] the rate of pulping was found to be proportional to $K(P_{SO_2})^n$ (P_{SO_2} = partial pressure of SO_2), and was best satisfied with a value of n = 0.75. The activation energy was found to be 20.2 kcal. For bisulfite pulping[373] the rate was proportional to the pH and was not substantially influenced by the bisulfite ion concentration. In this case, the activation energy was found to be about 36 kcal.

Effect of Wood Species

Spruce, balsam fir, Western hemlock, and poplar have traditionally been the preferred species for acid sulfite pulping. The limitations in regard to other species and the intolerance to bark have been mentioned. These limitations apply particularly to those species containing phenolic constituents, which can interfere with the delignification reactions. It was known for many years that pine heartwood, for example, gave a substantial proportion of uncooked wood when cooked by the sulfite process.[383] Pine sapwood is usable, as evidenced by the fact that chips from sawmill slabs can be pulped satisfactorily. It was found by Erdtmann[384-386] that two substances in pine heartwood, shown in Figure 4-76, termed pinosylvin (I) and pinosylvin monomethyl ether (II) inhibited acid sulfite pulping. When these materials were removed from the heartwood by extraction with alcohol or acetone, the wood pulped normally. Conversely, when spruce was impregnated with pinosylvin or other phenols, such as resorcinol or phloroglucinol, it could not be pulped.[387] Phenolic materials found in other wood species can play a similar role. For example, it has been found that a flavonone (III) present in Douglas fir heartwood is responsible for the resistance of this species to acid sulfite pulping,[388] although a different mechanism appears to be involved in this case.[389] Similar phenolic materials in bark, called flavotannins, have been shown to inhibit the pulping of both the inner bark and the outer wood of logs that had been left floating for considerable periods in water before peeling.[390,391] Under the acidic conditions of the normal acid sulfite cook, these phenols condense with the lignin[392] in the wood, causing it to become cross-linked and blocking many of the reactive sites normally avail-

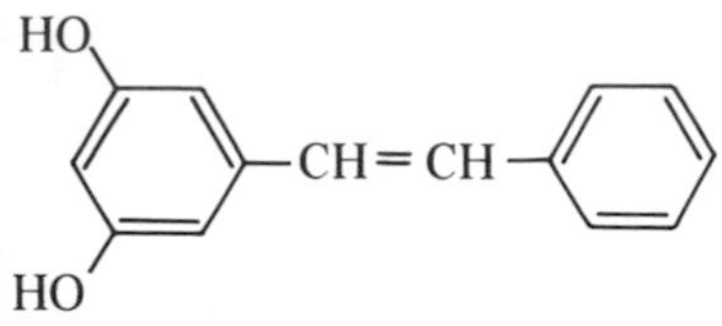

Pinosylvin (I)

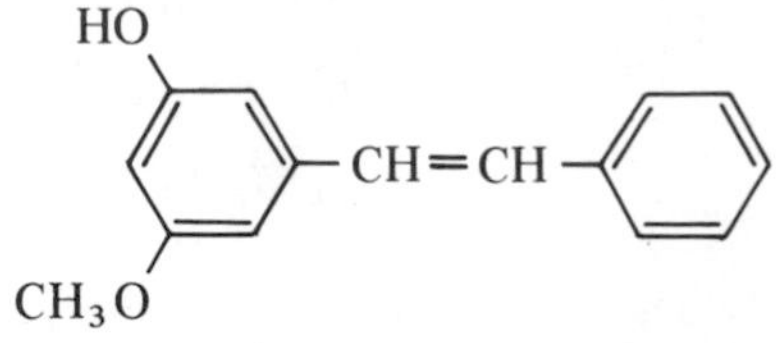

Pinosylvin monomethylether (II)

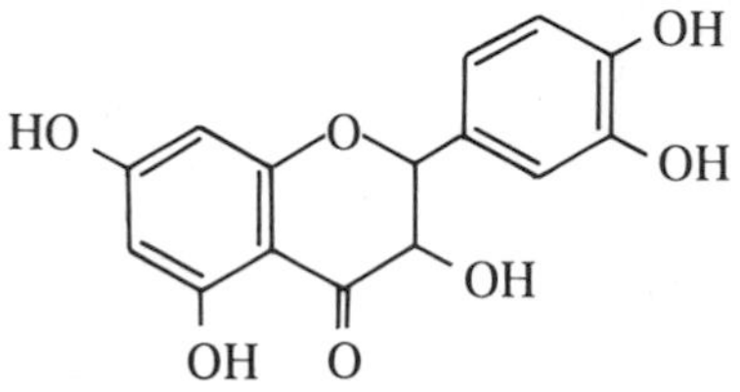

Dihydroquercitin (III)

Figure 4-76. Wood constituents that inhibit sulfite pulping.

able for sulfonation. The resultant complex high-molecular-weight polymer is no longer solubilized by the acid sulfite pulping solution.

Some improvement in the pulping of pine heartwood can be obtained by replacing calcium with a soluble base, either sodium, magnesium, or ammonium, and using a liquor of higher than normal combined SO_2 concentration.[393,394] However, the results are still inferior to those obtained on spruce or even pine sapwood. A more satisfactory solution to the problem of cooking phenolic-containing woods has been the use of bisulfite pulping processes, such as the Magnefite and Arbiso. At the pH levels used in these processes, the condensation reactions between the phenols and lignin occur much more slowly, while the initial sulfonation reaction is accelerated because of the higher bisulfite ion concentration. For example, it has been shown that pulping of pine and Douglas fir by these processes proceeds quite satisfactorily.[226,228,395,396] Inner bark and bark-stained wood are also handled much better at the higher pH level, and less problem with dirt from these sources is encountered with bisulfite pulping.[397] Outer bark is not pulped by any of the sulfite processes and relatively bark-free wood is still a requirement for sulfite pulping. The problems originating from the phenol-lignin condensation reactions are even less of a factor in the two-stage processes having high initial pH, such as the Stora process and the two-stage acid magnesium-base system.

Dense hardwoods are difficult to pulp with calcium acid sulfite because the density of the wood is very high and the volume available for liquid within the chip structure is limited. High combined SO_2 concentrations are required in the initial liquor in order to carry out the sulfonation in the early stages. These concentrations are difficult to achieve with calcium; dense hardwoods are pulped much more efficiently using the soluble-base liquors of high combined

SO_2 concentration, such as the bisulfite and multistage processes.[263,398-400] Another difficulty encountered in cooking hardwoods is that their defibration point is only reached at a relatively high degree of delignification compared to softwood. This is due to the fact that in hardwoods a higher proportion of the lignin is in the middle lamella, and correspondingly less in the cell wall. Since bisulfite liquors tend to decompose if cooking is carried too far, the range of Kappa number in which a chemical pulp can be produced is very small for hardwoods.[376]

Yield in Sulfite Pulping

Since the largest single component making up the cost of producing sulfite pulps is the cost of the wood, it is important that the maximum practical yield of pulp from wood is achieved. However, since other factors, such as bleaching cost, shive content, unbleached brightness, and strength also depend on the yield, it is important to select a yield level appropriate for the particular end product. In the range of 50 to 65% yield, the bursting strength increases with yield, while the tear, beating time, sheet density, and brightness decrease.[401] Bleach consumption increases with increasing yield, and, for bleached pulps, the increased bleach chemical costs counteract the reduced wood cost so that the combined cost of these two items is at a minimum at about 50% yield for Magnefite pulping and a 47 to 48% for calcium acid sulfite.[402] For unbleached pulps, such as those used for newsprint, bleaching costs are not a factor and the yield can be increased as high as pulp quality requirements will allow, provided adequate refining power is available.

There are two different ways to describe yield:[403] (1) the *digester yield*, which is the yield of pulp obtained from an individual cook and which should be used, for example, when evaluating changes in the wood supply or cooking variables and (2) the *mill yield,* which is the tonnage of usable pulp produced in the mill as a percentage of the wood delivered to the mill. The latter value takes into account losses in addition to those in the digester, such as those during barking, chip screening, and pulp spillage, and is used to evaluate the overall economic performance of the pulp mill. The measurement of yield in a mill is quite difficult to achieve;[403,404] yield is not usually used directly as a means of process or quality control. For a particular process, operating under properly controlled conditions, a good correlation exists between Kappa number and yield. Variations in digester yield are detected most readily from lignin-related tests. Yield can also be estimated by a material balance, that is, by determining the quantity of dissolved organic solids in the spent liquor in relation to the weight of pulp produced. Mill yield is normally measured directly by weighing the wood gravimetrically over a period of time and comparing this to the weight of pulp produced over the same period, correcting for inventory changes.

Yield values vary considerably, depending on species, process conditions, and properties required in the end product. In the acid sulfite process, unbleached yields in the range of 45 to 50% are usually obtained for bleachable grades, whereas values of 48 to 53% are obtained with bisulfite pulping. Further shrinkage in yield of several percentage points occurs during the bleaching. High-yield pulps for newsprint are normally produced in the range of 65 to 75% yield.

Chemical Reactions in the Sulfite Process

The reactions that take place during sulfite pulping are: (1) the reactions in connection with delignification, (2) reactions with the carbohydrates in the wood, and (3) a variety of miscellaneous reactions. These are now discussed.

Lignin Sulfonation and Dissolution

The early work on the mechanism of sulfite pulping has been thoroughly reviewed by Hägglund.[405] It was shown that, stoichiometrically, in the sulfonation reaction, the elements of bisulfite are added to the lignin molecule. Klason suggested that sulfonation consists of an addition of bisulfite to ethylenic double-bond units, but since such units have never been found to exist in lignin, this does not appear to be the mechanism. Freudenberg[406] postulated a two-step process in which first an oxygen linkage between two carbon atoms in the lignin molecule was ruptured to produce a phenolic and an aliphatic hydroxyl group, and then the latter was replaced by a sulfonic acid group. This mechanism was found to be inconsistent with the chemical reactions of methylated lignin, which can be sulfonated but not dissolved.[407] Hägglund found that partial sulfonation of the lignin in wood occurred readily in neutral or weakly alkaline solutions, but that these sulfonated lignins were not soluble in the cooking liquor. Erdtmann[408] showed that dissolution occurred only when the sulfur content of the lignin reached a value corresponding to about one sulfonic acid group per two methoxyl groups, the level usually found in technical sulfite lignosulfonic acids.[409] Brauns[410] had shown earlier that in order to dissolve lignin with sulfite solutions, the sulfur content of the lignin had to be at least 3.5%, equivalent to one sulfonic acid group per lignin unit of molecular weight 840. Earlier, Hägglund[411] had found that if a sulfite cook was interrupted while the pulp still contained a considerable proportion of lignin, the sulfonated lignin could be dissolved by heating with acidic solutions containing no sulfite. It was postulated that lignin dissolution occurred in two stages—sulfonation of the lignin followed by an acid-catalyzed hydrolysis of a covalent linkage. The latter was viewed as the rate-determining step. Since the same reactions occurred with hydrochloric acid lignin, which contains very little carbohydrate, it was concluded that the hydrolysis involved lignin-lignin bonds rather than a lignin-carbohydrate linkage.[412] On the other hand, based on the kinetics of the sulfite process, Maass and his associates[382] concluded that the rate of sulfonation was the rate-governing step in sulfite pulping. More recently it has been established[413] that both sulfonation and hydrolysis are probably involved in the reaction; either one can govern the reaction, depending on sulfur dioxide content and pH of the pulping liquor. It was shown by Rydholm[414] that, in the presence of sulfite cooking liquor, sulfitolysis occurred more rapidly than hydrolysis and was the more likely second step. On the other hand, if the correct mechanism involved hydrolysis in the second step, a further sulfonation would be necessary to reach the methoxyl-sulfur ratio observed in lignosulfonic acids from technical cooks. The two mechanisms shown in Figure 4–77 were considered by Rydholm as being possible routes to the sulfite dissolution of lignin. The information available does not clearly differentiate between the two mechanisms, but kinetic and analytical data on technical sulfite cooking indicate that sulfitolysis plays at least some role in delignification.

Considerable work has been done using structural models of lignin functional groupings to determine the nature of the reactions of lignin in the sulfite process. Lignin is made up of substituted phenyl propane units and it is evident that the initial reaction takes place at the α-carbon atom of this structure, which is activated by the phenolic grouping in the adjacent benzene ring:

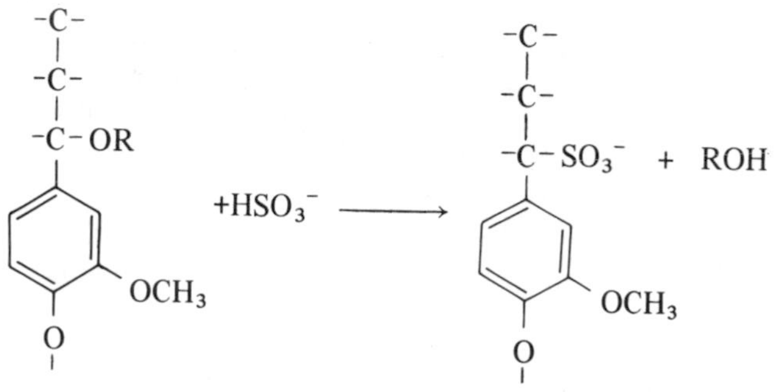

The R group in the structure can be hydrogen (benzyl alcohol), carbohydrate, or another lignin unit (benzyl ethers). Treatment of a wide number of model compounds of this type has indicated that, in all cases, these materials can be sulfonated in the α position.[415–418] Based on studies using many different lignin-model compounds, Gierer[418] has summarized the reactions that occur in the dissolution of lignin in the sulfite process in three general categories:

1. *Sulfonation.* This reaction occurs predominantly, if not exclusively, in the α position by a carbonium ion mechanism. All benzyl alcohol, benzyl alkyl ether, and benzyl aryl ether structures can be sulfonated in the α position under sulfite pulping conditions.

2. *Degradation.* Bonds joining lignin units are mainly of the β-aryl ether type; these are quite stable to acidic sulfite conditions. Thus complete separation of two adjacent lignin units does not often occur. This explains the fact that acid sulfite treatment of lignin does not increase its content of phenolic hydroxyl groups.

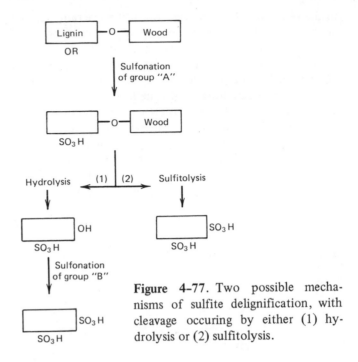

Figure 4-77. Two possible mechanisms of sulfite delignification, with cleavage occuring by either (1) hydrolysis or (2) sulfitolysis.

3. *Condensation.* These reactions compete with the sulfonation reaction for the reactive sites in lignin and counteract any degradation which occurs.

Goring[419,420] has studied the nature of successive lignin fractions removed during the course of a sulfite cook, using a continuous flow digester. There was no difference in the methoxyl content, ultraviolet absorption, sulfur content, refractive index increment, infrared spectra, or neutralization equivalents of the lignosulfonates from the different fractions. The average molecular weight, on the other hand, increased from 10,000 for the first fraction to 140,000 for the final fraction in the case of acid sulfite pulping, and from 10,000 to 77,000 for bisulfite pulping. Some of the values obtained for the bisulfite cooking series are shown in Table 4-26. Since recooking of the fractions with sulfite liquor did not produce any substantial increase in molecular weight, it was concluded that the increasing molecular weight of the successive lignin fractions was not due to condensation of lignin as the cook proceeds. The results appear to support a concept of pulping in which a three-dimensional network of phenyl-propane-type monomer units, having various reactive sites, is simultaneously sulfonated and degraded. The low-molecular-weight fragments become soluble first, while the larger fragments are released only as the slower degradation reaction proceeds to a sufficient extent to permit their solubility.

The topochemical aspects of lignin removal at the fiber level have been studied using the ultraviolet microscope. In an early study Lange[421] reported

TABLE 4-26 PROPERTIES OF SUCCESSIVE FRACTIONS OF LIGNOSULFONIC ACID REMOVED DURING A SULFITE COOK[a]

Fraction	Sulfur (%)	Methoxyl (%)	UV Absorption at 280 nm (cm^{-1}/g^{-1})	Refractive Index (m/g^{-1})	Weight Average Molecular Weight
1	6.2	11.8	13.3	0.185	10,000
2	5.6	12.8	12.1	0.181	13,600
3	5.2	12.9	12.8	0.191	31,000
4	5.5	12.6	14.1	0.198	77,000

[a]Values obtained after separation of carbohydrate material by electrophoresis-convection.

that first middle-lamella lignin is removed, followed by attack of the lignin in the cell wall later in the cook. Using more precise equipment, Goring[422] has shown that, in the early stages of the cook, lignin is removed preferentially from the cell wall. Later, the middle-lamella lignin dissolves rapidly, leaving the residual lignin in the secondary cell wall. A similar mechanism occurs in kraft pulping and will be discussed more fully in the section describing the reactions in that pulping process.

Reactions of Carbohydrates in Sulfite Pulping

Although the pulping of wood with sulfite liquor is primarily a delignification process, hydrolytic reactions that affect the polysaccharide constituents occur simultaneously; this results in dissolution of a significant amount of carbohydrate. Three types of polysaccharides enter into these reactions—cellulose, xylan, and glucomannan. The reactions of these carbohydrates in chemical pulping processes have been reviewed by Hamilton.[423]

Xylose occurs in both deciduous and coniferous woods in the form of its polymer, xylan, a series of 1,4-linked xylose units. In softwoods, O-methyl-glucuronic acid units and also occasional arabinose units are attached to one out of every five or ten xylose units:

$$X-X-X-X-X-X-X-X-X-X$$
$$\underset{\text{Glu}}{|} \qquad \underset{\text{A}}{|} \qquad \underset{\text{Glu}}{|}$$

where

$$X = \text{xylose unit}$$

$$\text{Glu} = \text{glucuronic acid unit}$$

$$A = \text{arabinose unit}$$

Hardwood xylans are similar in structure, but do not contain arabinose, and about seven out of ten xylose units contain an acetyl group. The linkages between the pentose units are particularly susceptible to acid hydrolysis, and in sulfite pulping, the arabinose units are eliminated at an early stage. A considerable number of xylose-xylose bonds are also broken as evidenced by the fact that xylose is the predominant sugar in sulfite spent liquors[424] and the degree of polymerization (DP) of the xylan is decreased during sulfite pulping.[425] For example, in acid sulfite pulping of birch, only 45% of the original xylan remained in the pulp after 20 min cooking, and its original DP of 200 was reduced to between 73 and 98.[426,427] Similar effects were observed during the sulfite cooking of spruce.[425] Acetyl groups and glucuronic acid-xylose linkages, on the other hand, are relatively stable to sulfite cooking conditions.[428,429]

Glucomannan and galactoglucomannan are the major hemicellulose components of coniferous wood, but do not occur to any significant extent in decid-uous species. Glucomannan has a backbone of glucose and mannose units in about a 1:4 ratio with side units of glucose and galactose (in some species):

$$M\text{–}G\text{–}M\text{–}M\text{–}M\text{–}G\text{–}M\text{–}M\text{–}G\text{–}M\text{–}M$$
$$\underset{\displaystyle Gal}{\,\vert\,}\qquad\qquad\underset{\displaystyle G}{\,\vert\,}$$

where

$$M = \text{mannose}$$

$$G = \text{glucose}$$

$$Gal = \text{galactose}$$

The acetyl units of softwoods are mainly located in the glucomannan portion.[429] The glucomannans are more stable to acid hydrolysis in sulfite pulping than xylans,[426,430] but a substantial amount is nevertheless dissolved in both the acid sulfite and bisulfite pulping processes.[427] This fraction of the wood can be stabi-lized to a large extent by treating the wood under neutral or alkaline conditions at elevated temperature prior to the main sulfite digestion. The two-stage neutral-bisulfite (Stora, magnesium two-stage acid) processes[243,251] give higher yield values than other sulfite processes due to this stabilization.[244,431] This stabiliza-tion is accompanied by deacetylation of the glucomannan.[432,433] In a detailed study of this process, Rydholm[434] has shown that the degree of stabilization could be directly related to the decrease in acetyl content of the wood in the pretreatment stage, as shown in Figure 4-78. It appears that the presence of acetyl groups in the glucomannan units does not permit the close contact of

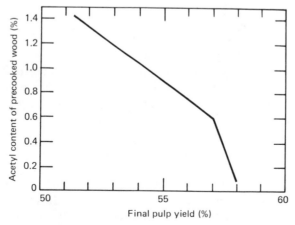

Figure 4-78. Acetyl content of precooked wood versus final pulp yield in two-stage cooking.

the molecules required to assume a crystalline form. After deacetylation, however, the glucomannan units can reorient themselves on the cellulose molecules in a crystalline form that is less susceptible to acid hydrolysis than the original amorphous material. If cotton is included in the digestion along with the wood, some of the glucomannan is deposited on the cotton, and both this material and the corresponding stabilized wood glucomannan are remarkably resistant to hydrolysis.

The glucosidic bonds in cellulose also suffer cleavage during sulfite pulping. Although the proportion of the original wood cellulose which is actually dissolved during the cooking process is relatively small, there is a substantial reduction in its degree of polymerization. One study[435] has shown that the average DP of spruce cellulose is reduced from an original value of 2400 to about 1400 when the normal acid sulfite cook is used. When the milder Mitscherlich cook was used, the DP was only reduced to 2000. Later work[436] has shown that the reduction in viscosity (and therefore also in DP) is greater in the bisulfite process (pH 3.7) than in the acid sulfite process (pH 1.8). Normally the cellulose in the wood is protected from attack by the presence of lignin and by the highly oriented crystalline structure of the cellulose. However, damage to the wood during chipping exposes additional areas of the cellulose microfibrils to attack by the cooking chemical in the early part of the cook and leads to more extensive loss in viscosity and, ultimately, in yield.[344]

Hydrolysis of wood polysaccharides during acid sulfite pulping leads to the formation of sugar monomers, which are in a suitable form for fermentation for the production of alcohol or yeast.[437] In addition, substantial quantities of aldonic acids and smaller amounts of uronic acid are formed.[425] With bisulfite cooking, the solubilized sugars are still in a polymeric form, and it is necessary to carry out a hydrolysis (either with acid or with enzymes) before they can be consumed by the microorganisms.

Other Reactions in Sulfite Pulping

The condensation reactions of lignin have already been mentioned in the sections dealing with lignin reactions, chip impregnation, and rate of heating. These reactions, in which lignin is converted to a form no longer soluble in normal sulfite pulping, occur when lignin is subjected to elevated temperatures and low pH before it is sulfonated.[212,438] These condensations occupy the sites in the lignin molecule at which the sulfonation would normally occur. It has been suggested that carbohydrate degradation products, such as furfural, also play a role in the lignin condensation reactions in the final stages of cooking.[439] It has also been suggested that the unreactivity of lignin is due to a physical coalescence rather than formation of chemical cross-links.[440] This coalescence occurs when the carbohydrates, which form a supporting matrix for the lignin, are removed and the interface available for the heterogeneous sulfonation reaction is considerably reduced. Condensation reactions occurring between phenolic materials other than lignin which are present in certain wood species and in bark, have already been discussed. These substances, which include pinosylvin, flavanones, and tannins, can also prevent lignin dissolution by occupying sites where sulfonation would normally occur.

Another type of reaction that can materially affect the course of a sulfite cook is the autocatalytic decomposition of bisulfite. Such reactions involve the formation of thiosulfates from bisulfite in the presence of reducing agents. Hägglund[441] found that sugars, formed by hydrolysis during cooking, could be oxidized to aldonic acids by bisulfite, forming thiosulfate as the reduction product:

$$2HSO_3^- + 2C_6H_{12}O_6 \rightarrow S_2O_3^= + 2C_6H_{12}O_7 + H_2O$$

Stockman[442] showed that, during sulfite cooking, wood released small amounts of formic acid, which underwent a similar reaction:

$$2HSO_3^- + 2HCOOH \rightarrow S_2O_3^= + 2CO_2 + H_2O$$

The sugar concentration does not become significant until near the end of the cook, and spruce, for example, contains only about 0.3% formic acid, so that these reactions ordinarily consume only minor amounts of bisulfite during a cook. However, it is known that under certain circumstances, sulfite liquors decompose rapidly producing a burnt or black cook. This phenomenon occurs when, for example, spent liquor is recycled to a subsequent cook, or when the sugars or thiosulfate are present in the cooking liquor at the start of the cook. Stockman[442] attributed these effects to a series of reactions involving thiosulfate:

$$5S_2O_3^= + 6H^+ = 2S_5O_6^= + H_2O$$

$$2S_5O_6^= + 2HSO_3^- = 2S_4O_6^= + 2S_2O_3^= + 2H^+$$

$$2S_4O_6^= + 2HSO_3^- = 2S_3O_6^= + 2S_2O_3^= + 2H^+$$

$$2S_3O_6^= + 2H_2O = 2SO_4^= + 2S_2O_3^= + 4H^+$$

or, overall,

$$5S_2O_3^= + 4HSO_3^- + 6H^+ = 6S_2O_3^= + 2SO_4^= + 8H^+ + H_2O$$

These reactions demonstrate that the formation of thiosulfate from bisulfite is autocatalytic, and, once a certain amount of thiosulfate is formed, the reaction proceeds rapidly and consumes bisulfite that would normally be available for pulping. Increasing the bisulfite charge at the start of the cook does not help because bisulfite is also a reactant and kinetic studies have shown that excess bisulfite accelerates the decomposition.[443] Once the combined sulfur dioxide is consumed in a sulfite cook, the pH of the liquor drops rapidly, causing the remaining lignin in the pulp to turn very dark and decomposing the thiosulfate to form sulfur:

$$S_2O_3^= + H^+ = HSO_3^- + S$$

The formation of sulfur in the pulp and the deterioration in pulp properties that accompany this reaction detract seriously from the pulp quality, and it is important to ensure that cooking is terminated before reaching this point. Sodium thiosulfate addition to pulping liquor retards the rate of delignification, so that more time is required to achieve a given degree of cooking. Since carbohydrate dissolution is also retarded, there appears to be little loss in yield.[444]

Bisulfite cooks are particularly susceptible to liquor decomposition because the concentration of bisulfite ions is high and the cooking is done at a temperature that is higher than that used in an acid sulfite cook. Precautions that should be taken with bisulfite pulping to control these side reactions include charging only the required amount of bisulfite and completing the side relief before the temperature reaches too high a level. Recycling of evaporator condensates to the liquor-making system had been suggested as a cause of the buildup of formic acid early in the cook. However, it has been shown[445] that very little formic acid is recycled in this way, particularly when the liquor is neutralized before the evaporation. Recycling of spent liquor, on the other hand, is permissible only if cooking is carried out at very low temperatures.[446] The recovered liquor from some sodium-base recovery systems contains thiosulfate and these systems should not be used with bisulfite pulping, although they may be suitable for neutral sulfite pulping.

The use of sulfur that contains selenium promotes the formation of selenosulfate ($SeSO_3^=$) in the liquor, which has an even stronger effect on liquor decomposition than thiosulfate.[441] Concentrations as low as 0.5 mg of selenium/l can cause serious problems in commercial liquor. Sulfur with minimum selenium content should be selected for sulfite pulping.

Sulfite pulping produces less thiosulfate than would be expected from the amounts of sulfate and aldonic acids formed because thiosulfate is not only produced, but also is simultaneously consumed in reactions with lignin. Analysis of lignosulfonic acids from technical sulfite cooks demonstrated that these materials contained more sulfur than could be accounted for by the sulfonic acid content of the liquor.[447] This "organic excess sulfur" was believed to arise from reactions between lignin and thiosulfate. Studies with lignin-model compounds demonstrated that this was very likely the case. When vanillyl alcohol is heated with thiosulfate at pH 5, vanillyl sulfide is formed:[448]

Studies of the reactions between thiosulfate containing radioactive sulfur and lignin have confirmed that the sulfur enters the lignin molecule.[449] The consumption of thiosulfate in reactions of this type plays an important role in controlling its level in sulfite pulping. Simultaneously, however, the rate of delignification is lowered, probably because the thiosulfate sulfur occupies some of the reactive groups that would otherwise be sulfonated with bisulfite.

Characterization of Sulfite Pulps

The major objective in chemical pulping is to separate the fibers, while leaving the cellulose itself intact. Unfortunately, all known processes for removal of lignin from wood also attack cellulose to some extent. In sulfite pulping, carbohydrate linkages are cleaved by the acidic cooking liquors, particularly where physical disorientation of the cellulose microstructure occurs due to damage during chipping or to the ray-cell crossing in the wood structure. Although very little cellulose is degraded to the extent that it is dissolved in the sulfite process, these attacks reduce the degree of polymerization and produce weak points in the fiber structure that yield to physical stresses.[450,451]

Chemical Composition of Sulfite Pulps

In addition to cellulose, sulfite pulps contain three main constituents—lignosulfonic acids, hemicellulose, and extractives. The ratio of these constituents to the cellulose will depend on the composition of the original wood, the degree of pulping, and the pulping conditions used. The chemical composition of these constituents plays a significant role in the papermaking properties of the pulp.

The lignin content of unbleached sulfite pulps vary over a wide range. The high-yield pulps used in newsprint contain 10 to 15% lignin. Pulps for bleaching are delignified to a much greater extent: softwood pulps to a lignin level of 3 to 5% and hardwoods, because of their low defibration point, to 1 to 2.5% lignin. The undissolved lignin in sulfite pulps is partially sulfonated, but to a lesser degree than the soluble lignosulfonic acids isolated from the liquor. Lignosulfonic acids isolated from the wood residue of a 73% yield sulfite cook, for example, had a sulfur content of 2.13% and a S/OCH_3 ratio of 0.2 compared to a sulfur content of 5.8% and a S/OCH_3 ratio of 0.48 for the lignosulfonic acid isolated from the spent liquor of the same cook.[452] Nearly all of the sulfur present in softwood lignin is present as sulfonic acid; in hardwoods, however, up to 30% of the sulfur is in a nonacidic form,[453] apparently as a result of side reactions not occurring in softwoods. The sulfonic acid groups act as ion-exchangers in chemically bonding inorganic cations to the unbleached pulp. They render the lignin more hydrophilic, probably the reason that a high degree of bonding can be achieved in unbleached sulfite pulps with very little mechanical action.

A substantial portion of the wood hemicellulose is removed during sulfite pulping and modifications take place to those that remain. Softwood sulfite pulps contain glucomannan (no longer containing galactose) and glucuronoxylan (without any arabinose).[454] Hemicelluloses are partially dissolved during the cook, but the fraction of these constituents remaining in a particular wood species for a given degree of delignification is increased by lower cooking temperature and by increased liquor pH.[357] Table 4-27 shows the proportions of the sugars in the hydrolysates from Western hemlock sulfite pulp and holocellulose respectively.[454] The instability of hemicellulose relative to cellulose in sulfite pulping is shown by the higher concentration of glucose and lower proportion of the other sugars in the sulfite pulp compared to the original holocellulose. The two-stage pulping systems described earlier affect the reactions of hemicellulose during pulping, and these differences are reflected in the hemicellulose content of the pulps.[455,456] Table 4-28 shows the carbohydrate composition of spruce pulp using four different sulfite pulping processes.[455] There are only minor differences between the acid and bisulfite pulps, but the two-stage cook with a neutral first stage shows a distinctly higher glucomannan content. Conversely, in the two-stage cook with an alkaline final stage, the hemicellulose content is significantly lower.

Carbohydrate content of pulps is sometimes described in terms of the alpha cellulose content, that is, the fraction of material insoluble in strong alkali. This term is useful in characterizing dissolving pulps where the initial step in the subsequent processing is often swelling in strong alkali. The term is also used to describe papermaking pulps in cases when sugar analysis is considered to be too time-consuming. However, the alpha cellulose content does not describe the chemical composition in unequivocal terms since alpha cellulose contains sugars other than glucose, and the extracted carbohydrates contain some glucosan. The alpha cellulose, in fact, describes a combination of chemical and morphological features of the pulp, and chemical composition is much more clearly described using the content of individual sugars after hydrolysis, since methods for these analyses are now well developed.

Wood resins are only partially removed from pulp by sulfite pulping, since they are not acid soluble. The amount remaining in the pulp is only about 1% of the weight of the pulp, but even this small amount can cause serious problems in pulp and papermaking operations by forming sticky deposits on equipment, such as screens, wires, and presses, and by producing sticky specks in pulp and paper sheets. These deposits are commonly referred to as pitch. In the original

TABLE 4-27 CARBOHYDRATE COMPOSITION OF SULFITE PULP FROM WESTERN HEMLOCK COMPARED TO HOLOCELLULOSE

	Calcium Acid Sulfite Pulp (%)	Chlorite Holocellulose (%)
Galactose	0	4.6
Glucose	89.8	69.4
Mannose	7.6	20.3
Arabinose	0	1.0
Xylose	2.6	4.7

TABLE 4-28 COMPOSITION OF SPRUCE PULPS FROM VARIOUS SULFITE PULPING PROCESSES

	Acid Sulfite (%)	Bisulfite (%)	Neutral Sulfite-Bisulfite (%)	Bisulfite-Carbonate (%)
Lignin	5	5	5	3
Glucomannan	11	10	21	6
Glucuronoxylan	5	6	5	4
Cellulose	79	79	69	87

wood, the resins are concentrated in the resin ducts, ray parenchyma, and epithelial cells.[457] These cells usually survive the sulfite cook, but subsequently their walls become fractured by mechanical handling during pulp processing thus allowing their resin content to be released to form colloidal particles in the liquid surrounding the fibers.[458] These colloidal particles under unfavorable conditions can coagulate to form pitch deposits. Rydholm has given a thorough review of the chemical nature of the extractives in wood and pulp.[459] They consist mainly of resin acids (abietic and other cyclic terpene acids); straight chain fatty acids (oleic, linoleic); esters of these acids; and nonsaponifiable materials, such as waxes, sterols, and fatty alcohols. The unsaponifiable fraction appears to give particularly high levels of deposited material in simulated pitch deposition studies.[460] There is more resin in softwoods than in hardwoods and in heartwood than in sapwood. Seasoning of wood before pulping promotes oxidation and hydrolysis of resin components, rendering them generally less tacky and troublesome during processing.[461-463] Several methods of measuring the extractive content have been suggested. Total extractive content in pulp can be determined by solvent extraction of the pulp, using alcohol or alcohol-benzene extraction of the pulp.[464] Use of ether or dichloromethane has been suggested to extract only troublesome pitch. Deposition of the pitch on metallic surfaces, such as stainless steel or copper, has been recommended as a means of determining whether pitch will be troublesome in subsequent processing.[460] It has been found that in troublesome pitch the concentration of colloidal particles is much higher than in nontroublesome pitch, and direct measurement of these particles by microscopic observation has been suggested as a means of assessing the pitch-depositing tendency of pulp.[465]

Physical Characteristics of Sulfite Pulps

Although sulfite pulps do not possess the overall high strength of kraft pulps, they do have a number of physical properties that make them particularly attractive for certain paper grades, including newsprint and other groundwood printing papers, tissues, glassine and greaseproof, and many high-grade printing papers. The physical characteristics that make sulfite pulp attractive for these uses can be divided into optical and strength properties.

Optical properties include brightness and opacity or, alternatively, light absorbance and light scattering. Lignin in wood is relatively colorless, but when it is treated chemically in pulping processes, it becomes colored. However, sulfite lignosulfonic acids are less highly colored than the lignin degradation products produced in alkaline pulping, so that unbleached sulfite pulp can be used in many printing grades. Even sulfite pulps in the range of 70% yield, which contain 10 to 15% lignin, are sufficiently bright to be used in nonpermanent printing papers. Pulp brightness is higher when pulping at the bisulfite pH than it is when pulping at either higher or lower pH levels.[224,466] In the bisulfite cook, brightness decreases in the early part of the cooking cycle as the undissolved lignin becomes more highly colored. A minimum value is reached at about the level of 70% for spruce and the level of 80% for birch. As cooking proceeds to lower yields, the pulp brightness increases as the discolored lignin is dissolved. Darker pulps are obtained when wood is cooked under conditions which cause the lignin to undergo condensation reactions.[467] In addition, the light absorbance of lignosulfonic acids changes reversibly with pH[468] and reflectance is highest in the range of 4 to 6. Unbleached sulfite pulps also have a tendency to redden under certain circumstances. This color reaction occurs in the presence of the ferric ion with both solutions of lignosulfonates and with pulps containing solid lignosulfonates. This color formation is prevented by the presence of sulfur dioxide, a reducing agent, while it is promoted by addition of limited amounts of hydrogen peroxide. The color is apparently due to the formation of an iron complex with the lignosulfonic acid, which is then stabilized by an internal oxidation-reduction.[469,470]

The strength characteristics of sulfite pulps have been referred to earlier in connection with the various process conditions. They are inferior to those of kraft pulps, as shown in Tables 4-29 and 4-30, and, as a result, the competitive position of sulfite pulp has suffered. However, the sulfite process is quite versatile and can be employed in a variety of modifications to produce the desired pulp properties for particular end products. Bonding strength (tensile and burst) increases as the yield is reduced to the range of 60 to 65%, and then decreases slightly as cooking proceeds to still lower yields. Tear strength, on the other hand, increases continually as the yield is decreased.[401,471] Pulps produced at the bisulfite pH are stronger in all properties than acid sulfite pulps. As the pH rises still higher into the neutral range, strength properties either level off or decrease slightly.[472] Cooking with strongly alkaline sulfite solutions produces pulps of higher strength in the range of kraft pulps.[235] Tensile breaking length and bursting strength of bisulfite pulps decrease with a rise in cooking temperature, while tear increases slightly.[473] Sulfite pulps are readily beaten and reach their maximum tensile strength with much lower refining-power input than that required for kraft pulps.[474] This is a distinct advantage to mills that have limited refining power.

The use of multistage pulping processes in which the pH is modified during the course of the cook produces substantial variations in pulp properties. These changes are related to the retention or removal of hemicellulose during the cook, as described earlier. The processes with a neutral or slightly alkaline first stage (two stage neutral-bisulfite) develop their maximum bursting and tensile strength with even lower refining-power input than the other sulfite pulps; however, the tear strength is lower than that of single-stage bisulfite pulps. Because of their high transparency and excellent greaseproof properties and their rapid hydration during beating, two-stage neutral-bisulfite pulps are particularly useful for glassine and greaseproof grades. Normally, acid sulfite pulp cooked at very low maximum temperature (usually about 125°C) have been used for this purpose. Pulps from the other type of two-stage process, those having a high pH in the final stage, are less easily beaten, but give better strength properties (particularly tear) and higher opacity than other sulfite pulps. A comparison of the pulps produced

TABLE 4-29 STRENGTH PROPERTIES OF UNBLEACHED MULTISTAGE SODIUM-BASE JACK PINE PULPS[a]

	Permanganate No.	Total Yield (%)	Beating Time (min)	Tensile (m)	Burst Index (kPa · m^2/g)	Tear Index (mN · m^2/g)
Bisulfite-neutral sulfite	22.2	48.5	18	11,400	7.26	10.2
Bisulfite-acid sulfite	18.0	49.8	26	10,000	6.38	11.8
Neutral sulfite-acid sulfite	22.1	54.8	7	9,000	5.00	7.35
Bisulfite	18.4	50.8	15	10,400	6.38	9.22
Kraft	17.4	45.5	34	12,900	9.81	13.7

[a]Strength properties at 500 Canadian Standard Freeness.

TABLE 4-30 STRENGTH PROPERTIES OF BLEACHED MULTISTAGE MAGNESIUM-BASE SPRUCE PULPS[a]

	Unbleached Roe Chlorine No.	Bleached Yield (%)	Beating Time (min)	Opacity (%)	Breaking Length (m)	Burst Index (kPa · m^2/g)	Tear Index (mN · m^2/g)
Two-stage Magnefite	7.6	48.1	35	58	11,500	8.14	8.63
Two-stage acid	7.4	52.7	19	51	10,700	7.16	6.28
Magnefite	8.7	48.9	31	56	11,500	7.84	8.43
Kraft	5.7	43.2	39	60	12,900	9.71	10.40

[a]Values at 450 Canadian Standard Freeness.

by the two types of multistage pulping with single-stage bisulfite and kraft pulps is given in Table 4–29, which shows unbleached properties for sodium-base jackpine pulps,[255] and in Table 4–30, which gives similar results for bleached pulps prepared from spruce with magnesium base.[475]

Hardwoods produce sulfite pulps that are considerably weaker than those from softwoods, primarily because of morphological factors. Hardwood fiber length is in the range of 1 to 1.5 mm, much shorter than softwood fibers which are in the range of 2.5 to 3.5 mm. Furthermore, the principal hemicellulose in hardwoods, xylan, is to a large extent lost as a result of hydrolysis during sulfite cooking, leading to low fiber bonding. Hardwood sulfite pulps cannot be used in papers where high strength is required. However, their short, narrow fibers contribute to excellent paper formation, high opacity, and a smooth sheet, all desirable properties for printing papers; hardwood sulfite pulps, therefore, find some application in these grades.

Processing of Unbleached Pulps

After completion of the sulfite cook, the contents of the digester are either blown or sluiced into blow pits or a blow tank. The cooked chips in the digester are still in chip form, but when the digester is blown under pressure, the chips entering the blow pit hit a stainless steel target plate on the wall opposite the blow line. The combination of the mechanical action in falling through the digester into the blow line, the sudden release of steam as the pressure suddenly drops at the blow valve, and the impact against the target plate completely disintegrates the softened chips into fibers. In other cases, the steam pressure in the digester is relieved to atmospheric to increase the recovery of heat and sulfur dioxide, and the digester contents are removed by flushing. The chips, in this case, are not completely disintegrated, and subsequent mild mechanical disintegration is necessary.

Washing of Sulfite Pulps

Separation of the spent liquor from the pulp can take place either in the blow pit or on a filter. When washing in the pit, there is normally a separate blow pit for each digester in order to allow adequate time for washing. These blow pits are usually constructed of wood or of tile-lined concrete and are fitted with a perforated stainless steel false bottom through which the spent liquor and wash water can be drained off. Following the blow, the spent liquor is drained and then the contents of the blow pit are washed by displacement by flooding the pit from the top, while draining spent liquor from the bottom. When the digester is relieved to atmospheric pressure, a considerable volume of spent liquor can be drained from the digester before the chips are discharged. Where it is important to maintain a high level of solids in the extracted liquor for subsequent evaporation, the initial washing can be made with weak spent liquor and only the final displacement with fresh water, thus minimizing dilution.

When spent liquor is not being recovered, it is possible to achieve good washing in blow pits if sufficient time is available and large volumes of water can be used. However, in most instances, regulatory agencies require some form of treatment to spent liquor before it can be discharged. Under these circumstances, it is important to minimize the volume to be handled by minimizing dilution during washing. With blow pit washing, only about 70% of the spent liquor can be recovered at a concentration suitable for evaporation and recovery. This has led many mills to convert to multistage, countercurrent washing on rotary drum filters or diffusers to increase the washing efficiency.[476] The use of drum filters and diffusion washers will be discussed in detail in the section on alkaline pulping. Using a three-stage washing system, spent-sulfite-liquor recovery efficiencies of 97 to 98% can be achieved, while maintaining the solids content high enough for efficient evaporation.[476] Spent sulfite liquors are very corrosive and high-quality stainless steel containing at least 2.5% molybdenum should be used in the fabrication of the washing equipment. Wash water should have a temperature of between 65 and 70°C for efficient washing and to produce hot liquor to minimize steam requirement for evaporation.

Screening and Cleaning of Sulfite Pulps

After washing, sulfite pulp contains visual impurities of various types, which must be removed if a clean pulp is required. These particles include sand and grit, uncooked wood particles, and agglomerated pitch and bark. Different methods are used to eliminate these contaminants from the pulp:[477] chiefly screening, which is based on separation by size, and centrifugal cleaning, which separates by difference in either specific gravity or specific surface.

The coarse material is removed on vibratory knotters, which have screen plates with perforations of 4.8 or 6.4 mm (3/16 or 1/4 in.) in diameter. Where rotary drum filters are used for washing, the knotter is usually placed ahead of the filter in order to achieve more satisfactory washer operation. In high-yield systems, the total digester charge is passed through disk refiners; coarse screening is usually not required. The accepted pulp from the knotters is next treated in fine screens. The most common type is the rotary centrifugal screen.[478] These screens operate at about 2% feed consistency and the perforations of the screen plate are in the range of 1.5 to 2 mm. They operate at high throughput and, within their normal capacity range, they remove a rejects fraction containing very little good fiber, particularly if two stages are operated in a cascade manner. However, the screen-plate openings are substantially larger than the shives, which is necessary in order to attain the high throughput, consequently some shives pass through the screen plate with the accepted stock. In pulps where pitch is a problem, a fractionating screen may be used to remove a portion of the fines fraction from the pulp. The filtrate from this operation contains a high concentration of ray cells and small pitch particles, and its separation decreases the resin content of the pulp substantially.[479] Since discarding this fraction represents a 1 to 2% loss in pulp and presents a disposal problem, this practice has not been widely adopted.

Centrifugal cleaners, or hydrocyclones, separate much of the contaminant material that is not rejected by the fine screens.[480,481] The centrifugal action of these devices will cause high-specific-gravity material to separate at the bottom outlet of the cone. Material with a specific gravity similar to that of the fiber, but with a low specific surface, such as shives and bark, will also accumulate in the bottom reject outlet. The rejection rate is controlled by the size of this bottom aperture relative to the accepted-stock outlet. A cleaning installation usually consists of several stages operated in a cascade manner. By operating the primary stage at a moderately high reject rate, accepted pulps with very low dirt levels can be produced. At this reject level, a substantial portion of good fiber is rejected also, and the subsequent stages are designed to reprocess this material until it is essentially free of good fiber. A disadvantage of centrifugal cleaning is that the cleaning efficiency is high only when the feed consistency is in the range of 0.6 to 0.8%. A substantial thickening capacity is then required to remove this additional dilution. Other parameters affecting the efficiency of cleaners include the pressure differential between the feed and discharge, the size of the unit, and the relative sizes of the inlet, outlet, and discharge orifices.

In order to minimize fiber loss from the screening system and to obtain maximum fiber yield, the rejects from the knotters and the secondary fine screen are often combined and fiberized in a refiner. These rejects should then be treated in additional cleaning equipment before being returned to the main pulp stream. The final-stage cleaner rejects contain a substantial concentration of nonfibrous dirt and should be discarded from the system.

Recovery Processes in Sulfite Pulping

The use of low cost calcium base used in the original sulfite process made it economically unattractive to install systems to recover the cooking chemical. Relatively low fuel costs provided little incentive to burn spent liquor for heat recovery, and, in the absence of stringent control regulations, disposal of spent liquors directly to neighbouring water bodies was an accepted practice. The first attempts at burning spent sulfite liquors were made in Sweden during World War I, when, due to high fuel prices, the combustion of calcium-base spent liquors was studied. Commercial installations were not installed, however, until the early 1940s, when the availability of fuel was again limited during World War II. Further development of recovery processes took place in the following years, first due to the increasing use of soluble bases and then due to the necessity of meeting the much stricter pollution-abatement regulations that have been established. A thorough review describing the different sulfite recovery processes has been published.[482]

Although most of the recovery systems begin with evaporation and incineration of the spent liquor, there are a wide variety of methods for sulfite spent-liquor disposal. The choice of a particular system for a given installation depends on the type of base used, whether recovery of base or sulfur (or both) is sought, and on the importance of heat recovery and economic factors. Recovery of both sulfur and the base, while simultaneously reclaiming a substantial portion of the heat available in the spent liquor, is possible for magnesium base by means of a well-established and relatively simple process. A large number of processes that recover heat and chemicals from sodium-base liquor have been proposed and several of these are being used commercially. However, they all involve complex chemical conversion systems and none have received general acceptance. Ammonia-base spent-liquor burning and heat generation have been practiced with and without sulfur recovery, but recovery of ammonia is not possible from a combustion process. Calcium-base liquor-burning systems have also been installed commercially, but no chemical is recovered and operating difficulties in both pulping and recovery areas have made this process essentially obsolete, leaving only a few installations still in operation.

A number of studies have been carried out in which the equilibria involved in the combustion process have been investigated in considerable detail.[483–487] Although the understanding developed in these investigations has been useful in the development of the pyrolysis-based processes and in improving the standard combustion methods, they are exceedingly complex and a detailed discussion is not within the scope of this work.

Calcium-Base Recovery

Before the changes from calcium to soluble base, more than 30 mills in Scandinavia and North America practiced calcium-base liquor burning. Although most of these mills have abandoned the practice, a brief description of this system is worthwhile because some of the problems encountered are common to all systems.

Because a significant concentration of sulfate ion is formed in the spent liquor during the cooking process and because calcium sulfate has an inverted solubility-temperature relationship (that is, a lower solubility as temperature is increased) early attempts to evaporate sulfite spent liquors encountered serious scaling problems. In addition, suitable acid-resistant materials of construction were not available at that time. Several methods were developed to avoid the scaling problem, including periodic backwashing of heating surfaces with acid condensate on an alternating basis,[488,489] seeding with fine gypsum crystals,[490] and the use of other types of evaporation, such as the horizontal-spray film evaporator.[491] The development of highly acid-resistant stainless steels (types 316 and 317) has provided a suitable material of construction for evaporators and other recovery equipment.

In calcium liquor-burning systems the liquor is atomized by air or steam into a drying and precombustion chamber prior to final burning. The first successful unit and most commonly used one is the Loddby furnace, named after the mill where it was originally installed.[492] In this unit, liquor at a solids content of about 50% is sprayed into a brick-lined cyclonic burner outside the furnace, where drying, ignition, and initial combustion take place. Final combustion takes place in the furnace proper. The cooking chemicals leave the furnace as an ash (composed of calcium sulfate, calcium carbonate, and calcium oxide) and as sulfur dioxide in the flue gas.[493] Conversion of the calcium sulfate ash to calcium oxide and sulfur dioxide can only be accomplished at temperatures well above those normally attained in conventional liquor-burning operations. The low value of the calcium makeup chemicals do not justify such an installation. Similarly, although a portion of the sulfur present in the flue gas as sulfur dioxide could be reabsorbed and reused, its low concentration and the limitations of calcium-base absorption systems make it impractical to do so.[494]

Magnesium-Base Recovery

In 1937, Howard Smith Paper Mills and the Babcock & Wilcox Co. cooperated in a pilot plant to evaluate magnesium-base recovery.[495] At the same time, a similar pilot plant was being operated by Weyerhaeuser Co. in Longview, Washington.[496] These three companies subsequently cooperated in the development of the process. The first full-scale magnesium-recovery installation was installed at the Longview, Washington, mill of Weyerhaeuser in 1948. Since that time, over 30 mills in various parts of the world have adopted the magnesium-base process. Three separate steps are involved in the recovery of chemicals from magnesium-base pulping liquor: spent-liquor concentration, spent-liquor burning, and cooking-liquor preparation or secondary recovery.

Evaporator scaling during concentration is less of a problem with magnesium-base liquor than with calcium-base liquor, although some calcium salts are still deposited due to the calcium introduced into the system by the wood.[497] Scaling is most serious in multiple-effect evaporator systems and a spare unit is commonly added to allow alternate washing of these effects. Multiple-effect evaporation is normally carried out to about the 55% solids level; higher solid levels can be obtained by direct-contact evaporation.[498] The spent liquor is usually neutralized with recycled magnesium oxide prior to evaporation to minimize loss of sulfur dioxide and other acids during the concentration.

The concentrated liquor is burned to recover the heat and chemicals. In the magnesium process, the chemicals are not converted to a smelt in the furnace, but rather are swept through the boiler with the flue gas in the form of magnesium oxide ash and sulfur dioxide gas. Combustion must be complete to produce an ash essentially free of carbon; on the other hand, the time-temperature relationship in the furnace must be controlled to prevent formation of periclase, an unreactive dead-burned magnesium oxide, which would be unsuitable for the subsequent liquor-generation system. Between 90 and 98% of the magnesium compounds are converted to magnesium oxide, the remainder being chiefly in the form of magnesium sulfate. The air supply to the furnace must be carefully controlled to prevent too much excess oxygen, which would carry over into the liquor regeneration system and oxidize sulfite to sulfate. The Babcock and Wilcox magnesium-base recovery units consist of a furnace in which evaporation and combustion take place, followed by a heat-absorbing zone in which the boiler and the superheater are located. Temperature is controlled by liquor concentration so that the flue gases leave the combustion chamber above 1300°C to ensure an essentially carbon-free ash. The magnesium oxide is separated from the flue gas by cyclonic collectors. In addition to chemical recovery, this system is capable of generating a substantial portion of the steam required by the pulp mill. In mills that have converted from calcium to magnesium base, the Loddby furnace has been successfully applied to the magnesium-base liquor.[499] The Copeland fluidized-bed combustion process[500] is an alternative method of converting magnesium-base liquors to magnesium oxide and sulfur dioxide. Liquor at about a solids concentration of 40% is introduced into the freeboard area of the fluidized-bed furnace. This furnace, which operates at 930°C, contains a bed of granular magnesium-oxide pellets fluidized by the combustion air entering at the bottom. The magnesium-oxide ash remains in the reactor becoming part of the bed, while the sulfur dioxide leaves with the flue gas. The flue gas is passed through a waste-heat boiler to recover heat. Magnesium-oxide pellets are discharged continuously from the bed to maintain the desired bed volume and the separated pellets are then crushed for use in the liquor regeneration system. Because of the relatively low solids content of the liquor fed to this system, the evaporation can be simple and easy to operate, but the quantity of steam generated is much lower than with the standard recovery boiler.

The ash collected from the recovery system is slurried and passed to a retention tank where soluble salts, such as those of potassium and calcium brought in by the wood, are extracted. These salts are removed by filtration to prevent their buildup in the system, and the magnesium oxide is sent to slaking tanks where it is converted to magnesium hydroxide. This base is then contacted with cooled, water-saturated flue gas to form magnesium bisulfite. The combination of the low SO_2 concentration in the flue gas and the relatively low solubility of magnesium monosulfite limits the range in which the absorption system can be operated efficiently. The equilibria involved in the dilute SO_2-water-MgO system have been investigated in considerable detail[206,500-502] to define suitable temperature and composition ranges for SO_2 absorption at these low concentrations. The ranges defined in these studies have been used in the absorption-system design. Three types of reactor have been used: packed towers,[503] Venturi scrubbers,[504] and moving bed towers.[505] The first two require multiple-stage systems, each stage equipped with separate pH control and magnesium-addition facilities. The more recent installations are of the third type, using a single moving-bed tower, which needs only a single pH control and magnesium-addition point. Make-up sulfur is provided by an auxiliary burner and is introduced into the product acid in a fortification tower. Magnesium makeup can be added as either magnesium oxide, hydroxide, or sulfate. A well-designed and well-operated magnesium-recovery system will return to the fresh cooking liquor about 90% of the base and 80% of the sulfur leaving the digester in the spent liquor. A schematic diagram showing the simple cyclic nature of the magnesium process is shown in Figure 4-79.

Ammonia-Base Liquor Recovery

Concentrated ammonia-base liquor can be burned either as a secondary fuel in a power boiler or in a separate unit specifically designed for the purpose.[506]

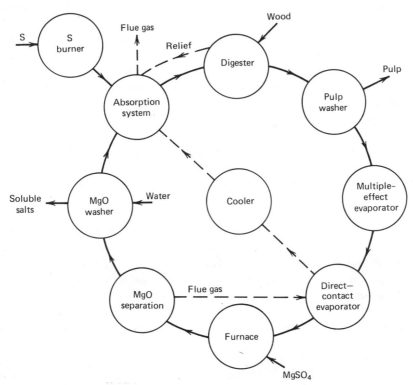

Figure 4-79. Schematic flow diagram of the magnesium-recovery process.

A minimum of excess air should be used in order to reduce sulfur trioxide formation and minimize corrosion in the cooler zones of the system. The small quantity of ash formed during the combustion can be removed by cyclone separators. A fluidized bed combustion system has also been proposed for ammonia-base liquor burning[507] in which liquor at 30 to 35% solids content is sprayed into a fluidized bed of magnesium pellets. The operation is similar to the magnesium fluid-bed recovery system already described for magnesium-base recovery, except that no additional solids are formed. If no supplemental fuel is used in the furnace, the sulfur dioxide concentration in the flue gas is high enough to allow its recovery for cooking-liquor regeneration. In one installation[508] the cooled flue gas is passed through two tile-packed absorption towers where it contacts the absorbing liquor. Bisulfite-sulfite liquor is recycled in the bottom portion of these towers and fresh ammonia is added to the recycle stream. Water is added to the top of the tower to absorb any ammonia stripped from the scrubbing system.

Base recovery is not possible from an ammonia-burning operation. However, it has been suggested that ammonia could be recovered by adding a base and steam stripping before combustion.[509,510] Suggested bases include lime, caustic, magnesium hydroxide, and kraft liquor.

Sodium-Base Liquor Recovery

The number and complexity of the sodium-base recovery systems prohibits a detailed description of each process. This discussion will be limited to a general description of the nature of these methods. A more detailed review of these processes has been published.[511] A description of some processes is given in the section on semichemical pulping.

The evaporation of sodium-base spent liquor is not entirely free of scaling problems because some calcium is extracted from the wood during the cooking process. Occasional boilouts are necessary to remove the resultant deposits. Many of the processes for sodium-base sulfite recovery involve a burning of the spent liquor in a kraft-type furnace to produce a carbonate-sulfite smelt. This is then converted to fresh cooking liquor by a variety of chemical sequences. A serious problem that must be avoided in the conversion of the dissolved smelt to sodium bisulfite is the formation of thiosulfate. As discussed earlier, the presence of thiosulfate in the cooking liquor leads to catalytic decomposition of bisulfite during the sulfite cook, forcing termination of the cook before delignification has proceeded to the desired degree. Thiosulfate is formed during the smelt conversion, both by direct oxidation of sodium sulfide in alkaline solutions and by the reaction between sodium sulfide and sodium sulfite. The other major disadvantage of these recovery processes is that they involve separation and handling of hydrogen sulfide gas, a highly odoriferous and toxic material.

Simple, direct methods for separating sulfide from the smelt, such as the IPC process of direct sulfitation,[512,513] direct oxidation of smelt,[514] and carbonation by flue gas (Mead process),[515] generally lead to high thiosulfate levels. In the more complex sodium-base recovery systems, such as the ones proposed by Stora,[516] Sivola,[517] Tampella,[518] Western Precipitation,[519] and Rayonier,[520] the conversions are arranged to avoid contact between sulfide and sulfite by employing either carbonation with pure CO_2 or bicarbonate to strip H_2S, by carbonate crystallization, or by smelt leaching—or a combination of these processes. Hydrogen sulfide is then converted to elemental sulfur by reaction with sulfur dioxide in the Klaus reaction. Sulfur is burned in the conventional manner and reacted with the previously separated base. Most of these processes have been proven on a mill scale, but they are costly to install, complex to operate, and none has received broad acceptance as a means for spent-sulfite-liquor recovery.

Another approach proposed for sodium-base liquor regeneration is based on the pyrolysis of the spent liquor. In the pyrolysis processes, virtually all the nonsulfate sulfur leaves the reactor with the gas, mainly in the form of H_2S, and none remains as sulfide in the solid residue. The gas is burned to convert the H_2S to SO_2, providing heat for the process. Sodium carbonate is leached from the residue and used to scrub SO_2 from the flue gas to form fresh cooking liquor. The residue from the leaching is mainly carbon and it can be burned for heat. The two main pyrolysis processes are the Atomized Suspension Technique (AST)[521] and the SCA-Billeruds[522] process. The AST process uses very high reactor temperatures and produces highly corrosive products, which require specialized and very costly materials of construction. This process has never achieved a commercial stage for sulfite liquor disposal. The SCA-Billeruds process, which has been operated commercially, avoids this problem by operating at lower temperature. This increases the amount of carbon in the residue very substantially, which makes it difficult to leach out all the sodium salts. A third process, the Sunoco method,[523] includes some features of the smelting methods, but like the pyrolysis methods, it produces a solid ash, free of sulfur. In this method, the liquor at a solids content of about 50% is mixed with alumina and fed to a kiln. This kiln is maintained at about 1000°C with a reducing atmosphere in the bed and an oxidizing atmosphere above it. The sodium compounds in the liquor form high-melting sodium aluminate, while the gases are oxidized to sulfur dioxide in the oxidizing zone of the furnace. The sodium aluminate is dissolved in water and used to scrub the SO_2 from the flue gas, forming an aluminum hydroxide precipitate that is filtered and returned to the kiln feed. The system is in use at an NSSC mill and could be adapted to other sulfite systems.

Fluidized bed combustion can be used for the destruction of the organic content of sodium-base liquors. This method is used in a number of NSSC mills, but has only been used to a limited extent in other sodium-base sulfite processes. It produces a mixture of sodium carbonate and sodium sulfate as the solid residue, and since this mixture can not be used for the regeneration of cooking liquor in a sulfite mill, it is usually sold to a kraft mill. The two fluid-bed systems that have been applied on a mill scale are the Copeland method[524] and the Dorr-Oliver FluoSolids process.[525] A major feature of these processes is that the liquors can be fired at low solids (40%) and the evaporation requirements are therefore reduced. In particular, the high-solids evaporator effects that are most subject to scale and corrosion can be omitted. However, because of the low-firing solids concentrations, steam generation is limited.

A system has been developed at the Broby mill[526] in Sweden and modified by Marathon Engineering Inc.[527] in the United States in which all the heat from the combustion is used to evaporate the spent liquor. In this manner, there is no need for expensive multiple-effect evaporators, and a simple oxidizing-combustion unit with no boiler section can be used. The direct-contact evaporation is carried out in two or three stages and is capable of raising solids from 12 to 60%. The product is a carbonate-sulfate smelt that can be solidified and sold as kraft makeup. The system has also been used for calcium base.[528]

Other Sulfite Recovery Methods

Systems for the recovery of base from sodium- and ammonia-base systems by ion exchange have been developed but not used commercially. In these systems, the base is removed from the spent liquor by passing through an ion-exchange resin that is in the hydrogen form. The other liquor constituents pass through the column and must be disposed. When the resin is saturated, the base can be reclaimed by regeneration with sulfurous acid solution. A very high concentration of SO_2 (in the range of 20%) is required in order to achieve efficient elution of base at the concentration required for pulping. The excess SO_2 must then be stripped from the liquor to leave an acid suitable for bisulfite pulping. Ion-exchange processes allow recovery of base, but do not recover sulfur and do not destroy the organics in the liquor. Concentration and combustion of the effluent solution containing these materials is unattractive because of the severe corrosiveness of the acidic liquors after the base has been removed in the ion-exchange column. The main processes involving ion exchange are the Abiperm process,[529] the Pritchard-ORF process,[530] and the Pritchard-Fraxon process.[531] In the latter process, the concentration of SO_2 required for the elution is reduced by addition of acetone, which increases the acidity of the eluting solution.

Wet combustion of spent sulfite liquor has been developed by Zimmerman[532] as a means of destroying the organic material. The oxidation takes place in the wet state at high temperature (250 to 300°C) and high air pressure (30 to 50 atm). About 90% of the organics in the liquor fed to the system are destroyed and converted into heat with an efficiency at least equal to that of other recovery systems. However, neither base nor sulfur is recovered and the system appears to be costly to install. A few trial installations have been made, but it has not received wide acceptance.

Separation of the spent liquor into several streams according to molecular weight by electrodialysis has also been suggested for treatment of sulfite spent liquors.[533] The lowest molecular-weight fraction would contain predominantly the base that would be available for recycling. However, until a profitable use is found for the organic streams, particularly the lignosulfonic acids, this process does not offer a solution to sulfite-mill pollution problems.

Biological oxidation of spent liquor can be used to destory the oxygen-consuming constituents in high-yield sulfite pulping liquors.[534] A two-stage treatment with aeration in the presence of activated sludge will remove over 90% of the five-day BOD from 73% yield sodium bisulfite liquor in 30 to 50 hr. However, color is increased and toxicity is not reduced to an acceptable level. This process offers the possibility of improving the effluent quality at relatively low capital cost,[535] but a significant operating cost is incurred and there is no possibility of credit for steam or recovered chemical. The use of oxygen instead of air in an activated sludge system substantially reduces the retention time and consequently the space requirement.[536]

Recovery of By-Products from the Sulfite Process

The chemicals dissolved from wood during a sulfite cook offer a wide array of interesting opportunities for the recovery of useful organic chemicals. Several of these, including lignosulfonates, vanillin, alcohol, and yeast, have been produced commercially from spent sulfite liquors for many years. Separation of chemicals not only provides valuable by-products, but also will remove undesirable organic materials from the mill effluent, reducing the pollution load on the receiving water. Unfortunately, the markets for the products are not sufficient to justify their production by all sulfite mills as a means of eliminating the effluent problem, but a number of mills have successfully used this approach.

Whole Spent Liquor and Lignosulfonates

The whole spent sulfite liquor has found a number of uses and the material has been sold in the form of the dilute liquor, a liquid concentrate, and dry solids. These products are used for a wide range of applications[537-539] that are based on their adhesive, dispersant, and surface-active properties. These applications include spraying roads to keep dust down, for use as a cement binder, and as a binder for animal-feed pellets. In the latter application, the sulfite liquor concentrate has proven particularly useful since in addition to functioning as a binder, the carbohydrate and inorganic constituents add to the nutrient value of the feed and the lignosulfonic acid salts lubricate the die used to form the pellets.

For other applications, it is preferable to use only the purified lignosulfonates; in this case the carbohydrates must first be removed. Several means have been used to accomplish this separation. Some mills ferment the liquor to produce alcohol or protein prior to lignosulfonate recovery, substantially reducing the carbohydrate content of the lignosulfonate product. In a process developed by Marathon Paper Mills, now a division of American Can Co., the lignosulfonates are precipitated in a two-stage lime treatment to produce a range of products marketed under the trade name Marasperse. Other techniques for reducing the sugar content of lignosulfonates include ion exchange,[540] ion exclusion,[541] and ultrafiltration.[542,543] Lignosol Chemicals, a division of Reed Paper, produces a series of lignin concentrates and dry powders, using vapor-recompression evaporation, flash drying, and ion exchange to produce the various products.[544] These lignosulfonates find their largest market as dispersants in oil-drilling muds, where they prevent flocculation by contaminants encountered in the drilling. Other uses include use in leather tanning, as a binder for foundry cores and briquettes, as dispersants in pesticide and agricultural sprays, as an ore flotation agent, and many other miscellaneous applications. Lignosulfonates have potential application in resins and plastics, particularly laminated products.[545] An example of the latter application is the use of lignosulfonates as a binder for exterior waferboard.[546] The wafers are sprayed with a sulfuric acid solution and then mixed with the powdered lignosulfates. The board formed from these treated wafers, using the normal press cycle, has both strength and weathering properties similar to those produced using the normal phenol-formaldehyde resin.

Vanillin from Sulfite Spent Liquor

Many attempts have been made to produce pure organic chemicals from the lignin residues in sulfite spent liquors. In one instance, the isolation of vanillin, these efforts have resulted in a successful commercial process. It had been known for some time from lignin-structural studies[547] that small amounts of vanillin were formed when lignin was treated with certain oxidizing agents. Processes with yields and chemical costs sufficiently attractive to justify commercial production of vanillin were not developed, however, until the period between 1935 and 1940. At that time, two processes, differing significantly in detail, were put into commercial production almost simultaneously and without knowledge of each other. In the Marathon process,[548] precipitated lignin was heated with alkali and the sodium salt of vanillin was then extracted from the reaction products with butanol. Free vanillin was produced by acidification and purified by crystallization. In the Howard Smith process,[549] the spent liquor was concentrated, and then made highly alkaline with caustic soda. It was heated to an elevated temperature and pressure to form vanillin, which was then released from its sodium salt by carbonation and extracted with benzene. Purification was by conventional means, using vacuum distillation and recrystallization. It was subsequently shown that addition of oxygen or air during the alkaline digestion gave higher vanillin yields,[550] and this improvement is used in commercial vanillin plants. Several large vanillin plants using sulfite spent liquor are in operation,[551] and lignin-based vanillin has almost completely replaced the natural material in North America and Europe. In addition to food flavoring, it appears to be useful as a starting material for pharmaceutical compounds, such as L-dopa, which is used in the treatment of Parkinson's disease.[538] Future expansion of the vanillin market will probably depend on development of applications of this type in organic synthesis. Other organic compounds that might also be valuable intermediates in organic synthesis, such as vanillic acid, acetovanillone, and syringealdehyde, can also be isolated from spent sulfite liquors.

Alcohol from Sulfite Spent Liquor

Fermentation of the hexose sugars in sulfite spent liquor to form ethyl alcohol is perhaps the oldest use of sulfite waste liquors and for many years was the most widely practiced. This process is still a source of commercial alcohol, although many plants discontinued the operation because of their inability to compete with ethanol produced by the petrochemical route. Only the hexose sugars are fermented to alcohol and so the process is not applicable to sulfite liquors from hardwood pulping, which contain mainly the pentose sugar, xylose.[552] For utilization in alcohol production, the sugars should be in monomeric form and spent liquors from the bisulfite process require acid hydrolysis prior to fermentation because the solubilized sugars are still in the form of oligomers.[553] Pulp yield and alcohol yield are inversely related since the amount of sugar dissolved during pulping is reduced substantially as the pulp yield increases. For this reason, dissolving pulps produce a spent liquor that is a particularly attractive source for the fermentation process, as shown in Table 4-31.[553] Thus compared to regular papermaking pulps, the hexoses, which can be fermented to alcohol, increase by about 40% for dissolving pulp, while the

TABLE 4–31 DISSOLVED SUBSTANCES IN SPENT LIQUORS FROM
CALCIUM ACID BISULFITE PULPING OF SPRUCE AT TWO
DIFFERENT YIELD LEVELS

	Paper Pulp (50% yield)	Dissolving Pulp (44% yield)
Hexoses		
Mannose	110 kg/ton pulp	156 kg/ton pulp
Galactose	26	22
Glucose	26	45
Total hexoses	162	223
Pentoses		
Rhamnose	–	3
Arabinose	9	16
Xylose	46	78
Total pentoses	55	97
Other substances		
Aldonic acids	44	76
Acetic acid	57	72
Polysaccharides and uronic acids	60	72
Lignosulfonates	622	735
Total other substances	783	955
Total dissolved organic matter	1000	1275

total carbohydrate, which can be used in protein fermentation, increases by about 45%.

An interesting combination involving the reclamation of chemical products in the spent liquor, destruction of residual organics, and regeneration of inorganic chemicals is practiced by the Ontario Paper Co. at its Thorold, Ontario, mill.[554] The spent liquor is first fermented for alcohol production. The residue from this treatment is then used to produce vanillin. After the alkaline oxidation stage in the vanillin operation, the product solution containing the vanillin is passed through ion-exchange columns to recover a portion of the excess sodium hydroxide, reducing the amount of acid required for neutralization prior to vanillin extraction. The ion-exchange column is regenerated using $NaHSO_3$-H_2SO_3 solution from the pulp mill acid plant to displace the sodium hydroxide, reducing the sodium requirement for the pulp mill. Finally, the residue from the vanillin operation is incinerated in a Copeland fluid-bed reactor, destroying the remaining organics and providing a sodium carbonate-sulfate product for sale to kraft mills as makeup. In addition to providing a system for disposing of at least 90% of its waste products, valuable chemicals are recovered that add substantially to the economic viability of the sulfite mill. Although this particular solution is not applicable to the sulfite spent-liquor problem in general because of market size limitations for some of the products, it is an excellent example of how sound technical solutions can be developed to solve an apparently critical problem.

Fermentation of Sulfite Spent Liquor to Produce Protein

The other fermentation process based on sulfite spent liquor that has achieved commercial status is the growth of yeasts. These materials can be marketed for their protein and vitamin values, either as animal fodder or in certain types of food for human consumption. The most common strain is Torula yeast (*Torula utilis,* reclassified as *Candida utilis*), which was developed commercially in Germany during World War II.[555] Torula yeast plants are easily maintained in continuous operation, the microbial growth is rapid, and all growth requirements are synthesized from simple organic compounds.[556] The yeast can utilize both the hexose and pentose sugars as well as aldonic acids and acetic acid. The conversion yield is about 45%.[557] Still higher yields can be obtained if the liquor is diluted prior to the fermentation,[558] but this decreases the output of the plant because less solids can be processed. To prepare the liquor for the process, sulfur dioxide must be removed either by stripping or by lime precipitation. Nutrient addition is required, particularly nitrogen, phosphorus, and potassium along with trace elements. Since the reaction is aerobic, air must be supplied continuously during the process, and the heat generated must be removed by a cooling system. The product is recovered by filtration, washed and then dried with heat to kill the yeast organism. The process usually operates best between 4.5 to 6.0 pH and in the temperature range of 34 to 37°C.[556,559] The product, besides having a high protein content, is an excellent source of the B-complex vitamins.

Pekilo protein is produced commercially in a similar operation in Finland, using spent sulfite liquor as the substrate.[560] The product has been thoroughly tested as an animal feed ingredient and is a beneficial source of protein compared to other sources. It contains an excellent balance of amino acids, including the essential amino acids, lysine and methionine. The product is fibrous in nature and is therefore readily filtered and dewatered. The Pekilo organism will consume all the carbohydrates in the spent liquor; in addition it will consume the acetic acid, methanol, and furfural, thereby reducing the BOD of the spent sulfite liquor by about 50%. The residual liquor from the Pekilo treatment is readily concentrated by ultrafiltration to produce concentrated lignosulfonates.

ALKALINE PULPING

J. R. G. BRYCE, Ph.D

Manager, Pulp, Paper, and Packaging Research
Domtar, Inc.
Senneville, Quebec, Canada

General Description of the Alkaline Pulping Processes

The two principal alkaline processes used for the pulping of wood are the soda and the sulfate process. In both, sodium hydroxide is the principal cooking chemical; in the sulfate process, sodium sulfide is also involved. Soda pulping is the older process, but it is now used only occasionally for pulp production. Although environmental problems have brought about some renewed interest in sulfur-free alkaline pulping, the superior pulp quality and other advantages of the sulfate method make it unlikely that cooking with sodium hydroxide alone will be widely used commercially in the foreseeable future.

Development of the Soda and Kraft Processes

The sulfate process is an offspring of the soda process, having been discovered by Dahl, a German chemist, in 1879. Dahl found that when the alkali lost in the soda process is replaced by sodium sulfate instead of sodium carbonate, the sulfate is reduced to sulfide during incineration of the spent liquor. The term sulfate process is therefore misleading since it implies that sulfate is the active cooking agent, whereas the active agents are actually sodium hydroxide and sodium sulfide. The name kraft (from the German and Swedish words for strong) has been applied to this process because the pulps produced are very strong, particularly when the cook is terminated while the lignin content is still at a relatively high level. Initially, the term kraft was used only for unbleached pulp used in high-strength papers; the term sulfate was used for bleachable grade pulps. Today the two terms are used interchangeably. The kraft process is superior to the soda process with respect to the rate of pulping, pulp yield, pulp quality, and production cost, and is currently predominant in the alkaline pulping industry. The kraft process also has several advantages over the sulfite process including:

1. Any wood species can be used, allowing wide flexibility in wood supply.
2. Substantial amounts of bark can be tolerated in the chips.
3. Cooking times are short.
4. Pitch problems are less prevalent.
5. The pulp has excellent strength.

6. The recovery process for the spent liquor is well established.

7. Valuable by-products are produced in the form of turpentine and tall oil from some species.

These factors are responsible for the rapid growth since the early 1930s of the kraft process relative to sulfite.

Terms Used in Alkaline Pulping

In North America, the widely accepted practice is to express all sodium compounds on the basis of the equivalent amount of sodium oxide (Na_2O). This practice will be adhered to in our discussion. In Scandinavian and other European literature it is more common to select NaOH equivalent as the basis. In the laboratory, chemical concentrations are usually expressed as g/l. In North American mills, the most common concentration term is pounds per cubic foot. The generally accepted terms used in sulfate pulping are given below.[561] Similar terms are used in soda pulping, but simpler definitions apply because of the absence of sulfur compounds.[562] The terms are:

1. *Total chemical:* all sodium salts, expressed as Na_2O.

2. *Total alkali:* $NaOH + Na_2S + Na_2CO_3 + 1/2Na_2SO_3$, all expressed as Na_2O.

3. *Active alkali:* $NaOH + Na_2S$, expressed as Na_2O.

4. *Effective alkali:* $NaOH + 1/2Na_2S$, expressed as Na_2O.

5. *Activity:* the percentage ratio of active alkali to total alkali, both expressed as Na_2O.

6. *Causticizing efficiency:* The percentage ratio of NaOH to the sum of $NaOH + Na_2CO_3$ in the white liquor, both chemicals expressed as Na_2O, and corrected for the NaOH content of the original green liquor in order to represent only the NaOH produced in the actual causticizing reaction.

7. *Causticity:* the percentage ratio of NaOH to active alkali, both expressed as Na_2O.

8. *Sulfidity:* in white liquor, the percentage ratio of Na_2S to active alkali, both expressed as Na_2O; in green liquor, the percentage ratio of Na_2S to total alkali, both expressed as Na_2O.

9. *Reduction:* in green liquor, the percentage ratio of Na_2S to the sum of $Na_2SO_4 + Na_2S +$ any other soda-sulfur compounds, all expressed as Na_2O.

10. *Unreduced salt cake:* Na_2SO_4 in the green or white liquor, expressed as Na_2SO_4.

11. *Makeup chemical consumption:* the number of pounds of new Na_2SO_4 (or other sodium compounds expressed as Na_2SO_4) per ton of AD pulp production that are added to the recovery process to maintain a constant sodium level in the cooking-liquor system.

12. *Chemical recovery efficiency:* the percentage ratio—total chemical fed to the digesters minus total chemical in the new Na_2SO_4 (or other makeup chemical) divided by total chemical fed to the digesters. In calculating the recovery efficiency over a given period, due allowance should be made for any change in liquor inventory during that period.

13. *Chemical losses:*

   (a) *Total loss:* the percentage ratio—total chemical in the new Na_2SO_4 (or other makeup chemical) divided by the total chemical fed to the digesters.

   (b) *Loss in cooking and pulp washing:* the percentage ratio—total chemical fed to the digesters minus total chemical fed to the evaporator divided by total chemical fed to the digester.

   (c) *Loss in evaporator and furnaces:* the percentage ratio—total chemical fed to the evaporators minus total chemical in the green liquor divided by total chemical fed to the digester.

   (d) *Loss in recausticizing and mud washing:* the percentage ratio—total chemical in the green liquor minus the total chemical in the white liquor divided by total chemical fed to the digesters.

   In calculating the above chemical losses, due allowance should be made for any change in liquor inventory and for any new chemical added to the part of the system under consideration.

14. *Green liquor:* the name applied to liquor made by dissolving the recovered chemicals in water and weak liquor preparatory to causticizing. In the calculations based on green liquor, quantities should usually be corrected for the chemical content of the weak liquor used for dissolving the smelt.

15. *White liquor:* the name applied to liquors made by causticizing green liquors. White liquors are ready for use in the digester.

16. *Black liquor:* the name applied to liquors recovered from the digesters up to the point of their incineration in the recovery plant.

Composition and Analysis of Alkaline Pulping Liquors

Liquors used in the soda process are composed primarily of sodium hydroxide and sodium carbonate. The sodium carbonate is present because the causticizing reaction, which produces the sodium hydroxide, does not go to completion:

$$Na_2CO_3 + Ca(OH)_2 \rightleftharpoons 2NaOH + CaCO_3$$

Actually, this equilibrium and others shown later are more correctly expressed in an ionic form. For simplicity, however, the molecular form will be used in most cases. At equilibrium, this reaction is substantially to the right because the solubility of calcium carbonate is relatively low. However, lime $[Ca(OH)_2]$ itself has a limited solubility and the reaction is not complete; in practice 80 to 85% of the sodium is in the hydroxide form and 15 to 20% is as carbonate. These liquors can be analyzed by conventional acidimetric techniques. Sodium hydroxide is determined by titrating to neutral or weakly alkaline pH after precipitating the carbonate with barium chloride. The carbonate can then be determined by adding excess acid, boiling to expel carbon dioxide, and back titrating the excess acid with alkali. Alternatively, the total alkali can be measured by titrating a separate sample to the lower carbonate inflection point without barium chloride addition. The sodium carbonate can then be determined by the difference between the titrations with and without barium chloride addition.

Sulfate liquors contain, in addition to sodium hydroxide and sodium carbonate, a number of sodium-sulfur compounds. The principal compound present is sodium sulfide. However, unreduced sodium sulfate, sodium thiosulfate (and possibly higher thionates), polysulfides, and sodium sulfite can also be present. Sodium sulfide hydrolyzes in water according to the equation:

$$Na_2S + H_2O \rightleftharpoons NaOH + NaHS$$

Because this equilibrium is established rapidly, a solution containing both sodium hydroxide and sodium sulfide will show a single inflection point in conventional acidimetric analysis equivalent to the term $NaOH + 1/2Na_2S$. This is the effective alkali, as described in the definitions above, and it is a key parameter in sulfate pulping. The actual position of the equilibrium in the above equation is believed to be substantially to the right, although only limited experimental data is available. The solubility of mercuric sulfide in white liquor under various conditions has been used to determine the ratio of sulfide to hydrosulfide.[563] Since mercuric sulfide reacts with $S^=$ ions but not with HS^-, the solubility of mercuric sulfide in a solution is taken to be a measure of the degree of hydrolysis. With this method it was concluded that in the range of temperature and concentration used in kraft pulping (approximately 165°C and 1 M), the solution would be hydrolyzed to a degree of 50 to 60%. This method may give low values for the degree of hydrolysis because of the shift in equilibrium position as Na_2S is reacted, but pH measurements have indicated this shift to be no more than 10%. The equilibrium has been further studied by measuring the ability of sulfide-containing solutions to promote the alkaline hydrolysis of cellulose compared to solutions of sodium hydroxide.[564] It was found that, if effective alkali was used as the alkali concentration term, the presence of sulfide had no effect on the rate of cellulose hydrolysis, that is, the sodium sulfide reacted as if it were completely dissociated into hydroxide and hydrosulfide ions. The fact that pulping rate and pulp properties in alkaline pulping can be correlated better with the effective alkali charge than with active alkali is further evidence that sodium sulfide is readily hydrolyzed.[565] Any sulfide not hydrolyzed in the initial liquor will be hydrolyzed as the hydroxide is consumed by the wood.

Several different procedures have been recommended for the analysis of sulfate cooking liquors.[566,567] One method, which determines the main constituents rapidly and accurately, uses only acidimetric titrations. The volume of acid to reach three inflection points, corresponding to the effective, active, and total alkali, is determined:

1. Enough barium chloride is added to the sample to precipitate all the carbonate. The sample is then titrated with acid. This gives the first inflection point at pH 10.5 (Titration A), where the sodium hydroxide is neutralized and the sodium sulfide is converted to hydrosulfide. The acid consumed is equivalent to the effective alkali content.

2. Formaldehyde is then added. This forms a complex with the hydrosulfide and releases an equivalent amount of sodium hydroxide, which can then be titrated to a pH equal to, or slightly below, that of the original inflection point. The total titration to this second inflection point (Titration B) corresponds to the active alkali.

3. Titration to a third inflection point at pH 4 will redissolve the barium carbonate, evolving CO_2. The total titration to this point (Titration C) is equivalent to the total alkali.

An example of these titrations is shown in Figure 4–80. In addition to effective, active, and total alkali, these titrations can be used to determine the concentration of sodium hydroxide, carbonate, sulfide, and the percentage sulfidity and causticity.

In order to accelerate the analysis procedure and minimize the chances of operator error, several suggestions have been made for automating the effective alkali analysis. Techniques suggested for this analysis include the use of automatic titrators,[568] calorimetric measurement of the heat generated during neutralization,[569] and conductimetric analysis.[570] A rapid, on-line method of liquor analysis can be an important factor in automatic process control, a subject to be discussed in more detail later.

For liquor samples taken during or following the cook, residual effective alkali can still be determined acidimetrically, but the titration should be terminated at pH 11.[571] Below this value, sodium salts of weak organic acids dissolved from the wood are titrated, and the results are no longer valid. The interference

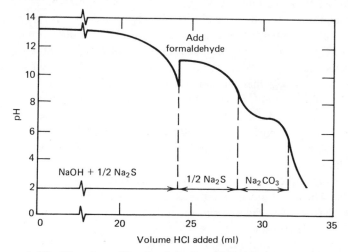

Figure 4–80. Titration of kraft white liquor with hydrochloric acid.

of these substances is even more serious in the case of the sulfide and carbonate determinations. Thiosulfates, polysulfides, and sulfite also interfere with sulfide determinations, and potentiometric methods,[572-574] which differentiate between these substances, must, therefore, be used. Carbonate is most accurately determined by evolution of CO_2.[575,576] Methods have been developed for the determination of other inorganic constituents of the black liquor including sulfate, thiosulfate, sulfite, chloride and potassium.[575,577]

Process Flow in an Alkaline Pulp Mill

An outline of the process flow in a typical kraft mill is shown in Figure 4–81. Chips and liquor are charged to the digester, either by batch method or continuously. The white liquor is charged to give the required chemical-to-wood ratio, and in the case of batch digesters, recycled black liquor from the washers is also added to make up the desired liquor-to-wood ratio. The digester is heated by either direct or indirect steaming to an elevated temperature in the range of 160 to 180°C, and held there until the desired degree of cooking has been achieved. For bleachable pulps, about 90% of the lignin is usually removed in the digester. After the digestion is complete, the cooked chips and spent liquor are discharged to a blow tank. The larger uncooked particles (knots) are then removed on perforated screens (knotters) and are usually returned to the digester for recooking. The spent liquor is separated from the pulp by countercurrent washing. The pulp is then diluted to low consistency and small-sized contaminants are removed by fine screens of either the centrifugal or pressure type, followed in some cases by centrifugal cleaners. The cleaned stock is then thickened and stored for subsequent processing. Pulps in the yield range of 55 to 60%, frequently used for linerboard, require mild refining to separate the fibers; this is normally performed prior to washing in order to facilitate spent-liquor removal.

The filtrate from the unbleached pulp washers is evaporated, first in multiple-effect evaporators and then usually by direct contact with furnace flue gas to achieve a solids content of 60% or higher. To minimize odor problems caused by loss of hydrogen sulfide, the black liquor is often oxidized, either at weak or strong concentration. Makeup sodium sulfate is usually added to the strong black liquor, which is then incinerated in the recovery furnace. This produces an ash consisting mainly of sodium carbonate and sodium sulfide, which leaves the furnace as a smelt. Most of the heat of combustion is carried from the furnace in the flue gas, which passes in sequence through a boiler section to generate steam, an economizer to heat the boiler feed water, and finally through the direct-contact evaporator.

The dissolved smelt is clarified and then causticized by reaction with slaked lime, after which the precipitated calcium carbonate (lime mud) is removed in a second clarifier. The calcium carbonate is then burned to regenerate lime for reuse in the causticization.

Although a wide variety of raw materials are pulped by the sulfate process

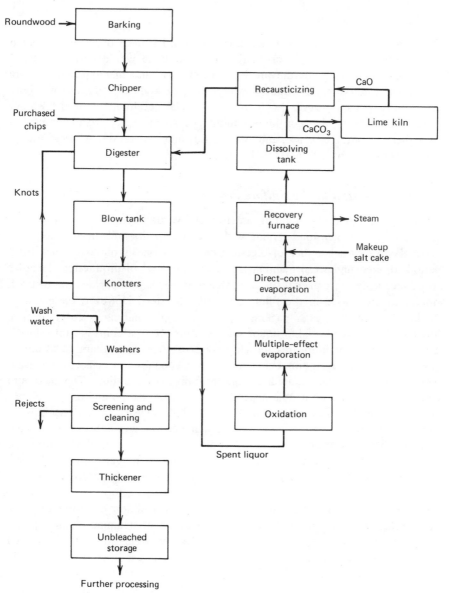

Figure 4-81. Schematic flow diagram of the kraft process.

and a broad range of pulps with differing quality requirements are produced, the above basic process is followed to a remarkably uniform extent. Many variations can be found in both the equipment used and in the pulping conditions applied, but unlike the sulfite process, in which a number of different process variations are used commercially, kraft pulping follows the same general sequence of steps in most of its applications.

Alkaline Digestion

Throughout most of its history, alkaline digestion has been operated as a batch process. It was recognized for many years that there would be a substantial advantage to carrying out the process continuously and many proposals were put forward describing continuous equipment for kraft digestion. It was only after many years, however, with the commercial development of the Kamyr continuous digester, that continuous pulping became a reality. Many companies, nevertheless, continue to use, and even prefer, batch digestion. A brief description of both systems will therefore be given.

Batch Digestion in the Sulfate Process

Batch digesters used in kraft mills are normally vertical cylinders constructed of steel plate, the dimensions varying with the size of the mill. In any case, it is desirable to have a number of digesters so that if one is out of service for a period, the remainder can maintain an acceptable level of production. In order to achieve a high chip-packing density, steam or mechanical packers can be used, similar to those used in the filling of sulfite digesters. However, the common method of achieving high packing density in kraft pulping is to commence introduction of the cooking liquor shortly after the first chips are introduced. The alkaline liquor lubricates the chip surfaces and allows them to assume a closely packed configuration. Using this method, good packing density is achieved, while keeping the time required for digester filling to a minimum. The maximum chip-packing density attainable depends on the wood density and on the relative chip dimensions. Thick chips have a higher bulk density than thin chips.[578] Certain types of chippers, such as the chipper headrig, give chips of lower bulk density compared to other types of chippers.[579] The critical dimension in determining the packing characteristics of chips is the ratio of the longest diagonal to the thickness. The packing density is inversely proportional to this ratio.[580]

Digesters can be heated either by direct steam injection or indirectly by a heat exchanger in the liquor circulating system. Direct steaming of the digester is simpler, but it dilutes the liquor with condensate, which must be evaporated later. Dependence on natural circulation in digesters can lead to variations in temperature in different parts of the digester, resulting in nonuniform cooking and a high level of screen rejects. Heat and chemical will be more uniformly distributed throughout the digester if forced liquor circulation is used, but it is essential that the system be designed so that the liquor distribution is uniform and that an adequate rate of circulation is maintained, particularly in the early part of the cooking period. The resistance to liquor flow through the chip bed increases during the cook; to maintain flow, the collection strainers should be positioned in the lower part of the digester.[581]

A surging technique has been suggested to obtain more uniform distribution of temperature and chemical throughout the cook when liquor circulation is inadequate.[582] In this method, the pressure in the top of the digester is rapidly reduced during the heating period either by gas relief or by injecting cool liquor into the top of the digester. The resultant boiling action at the bottom of the digester initiates a mixing action throughout the vessel. The reduction in pulp rejects achieved by applying this technique at one mill is shown in Figure 4–82.[583] The decrease in screen rejects at constant degree of delignification using this technique confirms the importance of uniform distribution of heat and chemical during batch digestion.

Sulfate liquors are much less corrosive than sulfite liquors, and good quality mild steel is normally a suitable material of construction. However, corrosion does occur and digesters have to be replaced periodically. A life cycle of at least 20 yr is normal for kraft digesters, and longer times have been recorded where special care has been taken. The corrosiveness of sulfate liquors was originally related to concentrations of sodium hydroxide, sodium sulfide, and sodium thiosulfate in the cooking liquor. An empirical relationship describing liquor corrosiveness in terms of these quantities was developed.[584,585] However, the polysulfide content of alkaline liquors has an even more significant bearing on corrosiveness.[586,587] Corrosion rates increase rapidly with increasing polysulfide concentration at low levels. At concentrations above 2 to 5 g/l, however, poly-sulfide has a passivating effect, and the rate of corrosion decreases with increasing concentration until at 15 g/l polysulfide, the rate is negligible. The change

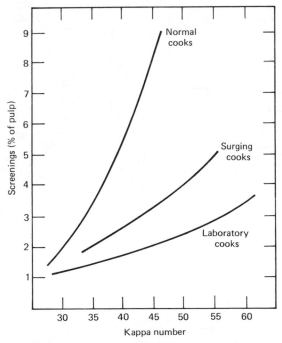

Figure 4–82. Reduction in pulp rejects using the surging technique.

from the active to the passive state can be observed from a polarization curve obtained by measuring the potential between the metal and liquor.[588] At constant polysulfide levels, sodium hydroxide concentration is the most critical factor.[589] For this reason, the most severe corrosion takes place in the early part of the cook when the sodium hydroxide concentration is high. The method of charging liquor to the digester is important to avoid erosive flow of liquor by impingement on the vessel wall. The use of deflector plates on the filling line and properly designed nozzles on return circulation lines can avoid these problems.[590] High silicon-content metal, such as killed steels, have a higher rate of corrosion than rimmed steels.[589] Addition of sulfur to the white liquor to bring the liquor to a passive state can be used to combat digester corrosion.[591] Application of an anodic potential to the digester shell can also be used to maintain the digester in a passive state.[592]

The Kamyr Continuous Digester

Several different designs of continuous digester have been employed for sulfate pulping, but the Kamyr system, developed in Sweden by Johan Richter,[593] is by far the most widely used. The digester used for kraft pulping in most mill continuous-pulping installations is shown in Figure 4–83. An excellent review of the use of the Kamyr digester in alkaline pulping has been published by Rydholm.[594]

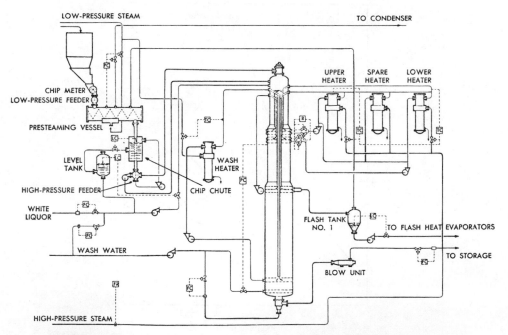

Figure 4–83. Typical Kamyr continuous cooking system for sulfate pulp.

Chip flow into the system is controlled volumetrically by a rotary metering valve. The chips enter the presteaming vessel, which operates at about 100 kPa (15 psi), through a low-pressure star feeder. Steam is introduced into the bottom of the vessel, and noncondensable gases and turpentine are vented from the top. The presteamed chips pass through a chip chute where they are mixed with white liquor before passing into the high-pressure feeder. This unit is a submerged rotary feeder having a series of transverse passages. These pockets become filled with chips and liquor in the vertical position and then, in the horizontal position, the contents are displaced into the digester by a high-pressure liquor flow. This flow consists of liquor extracted from strainers at the top of the digester in combination with the fresh white liquor added to the system. The chips settle to form a plug, which descends through the digester as cooked material is continuously withdrawn from the bottom. The chips pass successively through an impregnation, a cooking, and a washing zone. The impregnation takes place at about 120°C for a duration of 45 min. Recirculated liquor is used to heat the mass to the digestion temperature in two stages: the first to provide the bulk of the heating and the second to provide precise temperature control. Cooking time is usually about 1-1/2 hr at a temperature in the range of 160 to 170°C. At the bottom of the cooking zone, the pulping reaction is terminated by contacting the chips with the relatively cool quench flow recirculation liquor. In the wash zone, the spent liquor is displaced by weaker liquor from the external washing, flowing countercurrent to the chip mass. Chemical diffuses from the chips into the washing liquor and is removed near the middle of the digester through extraction screens. The liquor removed at these screens is reduced to atmospheric pressure through a two-stage flash system; steam from the higher-pressure flash tank provides the steam for presteaming. Washing in Kamyr digesters will be discussed in more detail later. The flow of stock from the digester to the blow tank is controlled by ball valves to maintain a constant chip level in the top of the digester.

Several liquor-recirculation streams, shown in Figure 4–83, play a major role in controlling the cooking process in the Kamyr digester. These streams include the top circulation, upper and lower heater circulation, the quench circulation, and the wash liquor flows. In these flows, the liquor is collected through screens on the digester periphery, circulated through a heat exchanger, and returned to the digester by means of a center pipe. This method of recirculation creates a cross flow through the downward-flowing chip mass, ensuring uniform temperature distribution. The liquor flow into the bottom of the digester provides cooling prior to the blow. This is essential for the production of high-quality pulp. In the early Kamyr installations, which did not have this cooling, pulp was found to be of inferior strength to that produced in batch digesters. This loss in strength was attributed to mechanical damage occurring in the high-temperature discharge; this was confirmed in simulation experiments.[595] The adoption of the cold-blow technique, in which the temperature around the bottom scraper in the discharge zone of the digester was cooled to below 100°C, resulted in an improvement of 10 to 15% in pulp strength.[596,597]

The Kamyr digester gives a very low level of screen rejects even when relatively large chips are cooked. The digester is full of liquid at all times and operates at a hydrostatic pressure substantially above the vapor pressure of the impregnating liquid. This pressure and the relatively long impregnation period at elevated temperature are the main factors that bring about the uniform penetration of cooking chemical attained in this system.

Once the dimensions of the vessel and the locations of the different screens are fixed, retention times in the different zones of the Kamyr digester are set by the rate of production. Typical values are 2 to 5 min for presteaming, 45 min for impregnation, 1-1/2 hr for cooking and between 1-1/2 and 4 hr in the diffusion washing zone. The actual times depend not only on the design of the digester and on the production rate, but also on the degree to which the chips compact as they descend in the digester. Tracer studies have been used to measure the chip flow pattern in the digester[598,599] by adding radioactively tagged chips to the digester and following their progress through the vessel. These studies showed that surprisingly little compaction occurs as the chip mass descends in the digester. At the same time, it was established that there was very little vertical channeling of the radioactive chips or rotation of the chip mass as it moved down the digester; the measured retention time was similar to that calculated from the digester geometry. When the production rate is increased, the time in the cooking zone decreases, and in order to maintain the same degree of delignification, it is necessary to increase the cooking temperature. Since most mills work to a constant degree of delignification (permanganate number level), the H-factor concept (to be described later) can be used to prepare a schedule of cooking temperatures that will produce the desired permanganate number at various chip-feed rates.[600] If off-standard pulp quality is to be avoided during changeovers, the appropriate temperature and flow changes must be made in the proper time sequence. This requires knowledge of the time delays at the different levels in the digester. In order to facilitate these changes in temperature and flow rate and to assist in dealing with other unexpected changes in digester operation, the required adjustments may be programmed into a computer that will follow through the proper sequence of changes on initiation by the operator.

Kamyr digesters, particularly the larger-sized units, are sensitive to the presence of chip fines and pin chips, which tend to plug the screen plates through which liquor is extracted for the different circulation flows. These small chips also increase the resistance to flow in the countercurrent wash zone, which impedes the downward flow of chips and can cause the digester to become plugged. To minimize problems of this type, the chip specifications for a Kamyr digester normally impose a limit of 5 to 10% by weight on the proportion of chips passing through a 7 or 10 mm (1/4 or 3/8 in.) Williams screen. A number of changes have been made to the digester design and to the methods of operation that have reduced the sensitivity to small wood fragments, but the Kamyr continuous digester remains more sensitive than batch digesters to these fine materials. On the other hand, because of the excellent liquor penetration, the

Kamyr digester produces less rejects from the larger chip fraction than batch digesters. Mills having both batch and continuous digesters frequently screen selectively, using the larger chips in the Kamyr and the finer fractions in the batch digesters.

Scale deposition is another problem that can affect the operation of continuous digesters more than batch digesters. Two types of deposit have been identified: inorganic deposits (mainly calcium carbonate) and organic deposits (black-liquor residues). The organic deposits are usually formed behind liquor extraction screens, probably as a result of evaporation caused by a drop in pressure across these screens.[601] Calcium carbonate scale is the most common type and it most frequently occurs on the heat exchanger surfaces. The Kamyr digester includes a spare heater that can be brought into service without interrupting the digester operation, permitting the fouled heater to be cleaned. Heater scale is thus a nuisance, but not a severe problem. Scale deposition on the digester screens is a far more serious problem, particularly when scaling occurs at the top separator and interferes with digester feeding. This scale is usually the result of calcium carbonate precipitation from the liquor rather than accumulation of lime mud particles carrying over from the white-liquor system. The calcium originates mainly from the wood, while the carbonate can arise either from inefficient causticizing or from carbon dioxide generation during the cook. Precipitation occurs more frequently in the earlier stages of the cook because the organic materials dissolved later in the cook can act as complexing agents and thereby produce an apparent supersaturation effect. The sources of calcium (wood, water, and white liquor) are difficult to eliminate, and, therefore, efficient operation of the causticizing system to minimize carbonate in the cooking liquor is a major defense against these deposits.[602]

Corrosion is not as serious a problem with continuous digesters as it is with batch digesters[603] because continuous digesters are not subject to the frequent temperature cycles of batch digesters, and also because there is no exposure of the metal to air. The corrosion problems that do occur with continuous digesters are mainly at the welds on the vessel shell, and periodic inspections should be made to identify any deterioration as soon as possible.

Continuous Digester Modifications

An extensive research program covering adaptation to a wide variety of pulping processes was carried out on a 10-ton-per-day, pilot Kamyr digester installed by Billerud AB at its Jössefors mill in Sweden. A number of different digester operating modes were developed as a result of this program and the results have been reported in considerable detail by Rydholm.[594] In one of the systems, the Mumin tube, the top separator and a screw conveyor to lift the chips from the liquor are located in an inclined tube, which is external to the main digester. Preliminary impregnation occurs in the Mumin tube, and the main digester can be operated partially in the vapor phase, that is, with the liquid level lower than the chip level. In this case, the system is no longer under hydrostatic pressure, and the cooking pressure is essentially the vapor pressure of the liquor. The Mumin tube can also be built inside the main digester structure, allowing construction of much larger units. The Mumin-tube arrangement is useful for processes that require a separate chemical pretreatment at lower temperature than the main delignification temperature, since the vapor zone at the top of the digester avoids any mixing of the cooking liquor back into the feed system. However, the time is very short and the pressure is low during impregnation, leading to higher rejects than obtained with the hydraulically pressurized digester. To overcome these problems, a two-vessel system has been developed that makes use of a preimpregnation tube ahead of the main digester. The preimpregnation tube can operate either at low pressure (requiring a high-pressure feeder between the two vessels) or at high pressure (high-pressure feeder before the preimpregnation vessel). With the low-pressure unit, the presteaming can be done in the upper part of the preimpregnation vessel. The effect of the various digester configurations on pulp-reject levels in standard kraft pulping is shown in Figure 4-84. For pulp with Roe chlorine number of 6 (about 40 Kappa number), the Mumin system gives the highest level of rejects at a given alkali level, and the two-vessel, high-pressure impregnation system gives the lowest rejects level. The standard hydraulic digester and the two-vessel, low-pressure impregnation system both give intermediate reject levels.

The use of a preimpregnation digester is particularly helpful in the production of high-lignin-content base stock for linerboard. Because the time cycle for cooking these pulps is very short, there is little time available for redistribution of the chemical, and uniformity of penetration before heating to cooking temperature is very important. Inadequate chemical distribution leads to high shive content and increased refining power requirement. By using the two-vessel unit to achieve thorough penetration, a satisfactory pulp can be produced at higher yield levels with a substantial reduction in wood cost. Other alkaline pulping processes that can benefit from the use of the two-vessel system include

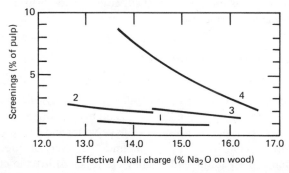

Figure 4-84. Pulp screenings at Roe chlorine number 6 versus white liquor charge on wood for various continuous digester systems: (1) high-pressure impregnation plus Mumin tube, (2) standard hydraulic system, (3) low-pressure impregnation plus Mumin tube, and (4) Mumin tube without preimpregnation.

prehydrolysis kraft pulping and polysulfide pulping. In prehydrolysis kraft cooking, acids generated by the wood at elevated temperatures hydrolyze the hemicelluloses and render them soluble in the subsequent alkaline cook. This hydrolysis is performed by using water in the feed circuit of the Mumin system and allowing the reaction to occur in the vapor-phase zone in the upper portion of the digester vessel. The kraft liquor for the alkaline cooking stage is added in the middle of the digester and weak liquor from the washers for countercurrent washing is added at the bottom of the vessel. Dissolving grade pulps of very low hemicellulose content can be produced using this system. No liquor circulation is required in the prehydrolysis zone, and deposition of pitch on the strainers is thus averted.[604] In polysulfide pulping (to be described later) it is important to fully impregnate the chips with the polysulfide-containing liquor and to carry out the stabilization reaction at moderate temperature because polysulfide is unstable in the presence of alkali at full cooking temperatures. The two-vessel system with high pressure in the first unit provides suitable conditions for this to take place.[605]

A system of countercurrent cooking in the Kamyr digester has been developed in Australia.[606] In this method, the white liquor is added to the digester near the middle. It flows upward countercurrent to the chips and is finally removed as black liquor at the top of the digester. Washing liquor is added at the bottom of the digester and it becomes part of the cooking liquor in the upper portion of the digester. This countercurrent-liquor-flow method uses less alkali than conventional cooking and gives excellent diffusion washing in the digester. Only minimum additional washing is required outside the digester. Brightness is excellent and screen rejects are low. However, there is a slight loss of pulp yield due to the extraction of hemicellulose by the action of the strong white liquor on the already delignified chips.

The displacement impregnation technique[607] is a variation of the full countercurrent method in which the white liquor is added at the upper heating circulation stream, and a portion of the top circulation is withdrawn and transferred to the causticizing plant for reuse. Since less liquor flows down the digester into the cooking zone, the heat and chemical requirement are reduced.

Other Types of Continuous Digesters

Other systems have been developed for continuous pulping, although none has achieved the wide acceptance of the Kamyr system. The Esco digester[608] (see Figure 4-49) is a downflow unit similar to the Kamyr, but with a star-valve feeder and a different liquor-collection strainer design, which makes this unit less sensitive to chip fines. The Impco digester[609] is an upflow unit, which has found only limited application. The Bauer M. & D. digester[610] is an inclined-tube digester through which the chips are carried on a drag conveyor. The feeding unit is a conventional rotary valve. The Bauer digester is quite widely used as a sawdust digester, but it is only used to a limited extent for chips. Digesters having tapered-screw feeders and horizontal-screw conveyors, widely used in high-yield NSSC pulping, are not suitable for chemical pulping of wood chips because of the damage to the fibers during the process. These units are used to some extent, however, in the alkaline pulping of agricultural residues.

Comparison of Batch and Continuous Pulping

Continuous digesters have generally been chosen in preference to batch digesters in new sulfate pulp mill installations and major expansions since they became a proven unit. However, some batch digesters continue to be installed, and both types have particular merits that may apply in any particular situation. There does not appear to be any significant difference between the two types in the yield or the strength of the pulps obtained from a particular chip supply.[611] Initial investment and operating costs usually favor the large single continuous digester. The multiple-unit batch system, however, offers greater reliability and continuity of pulp supply, which is especially important when the pulp mill is integrated with a papermaking operation. Manpower requirements once favored continuous digesters to a significant extent, but with the highly automated batch systems that are available, this advantage is relatively insignificant. The continuous digester should have lower energy requirements because the liquor-to-wood ratio is lower and because the digester shell is not alternatively heated and cooled. A continuous system is much more amenable to control of emissions from relief and blow gases. Control of continuous digesters should also be easier, leading to a more uniform product. The batch digester system, on the other hand, is more flexible where multiple products are being manufactured or frequent quality changes are required. Batch digesters have been mechanically more reliable than continuous digesters, and the maintenance costs are lower. Turpentine yields are usually higher on batch digesters. Both types of digesters require chips of good quality for efficient operation, but the consequences of using off-quality chips are less serious in batch than in continuous digesters.

Impregnation in Alkaline Pulping

In chemical pulping, the fibers are separated from each other by removing as much as possible of the middle lamella lignin that cements them together. Simultaneously, the lignin content in the fiber cell wall must be reduced sufficiently to render the pulp suitable for subsequent processing, which can involve drying (in the case of commercial pulp), bleaching, or direct conversion into unbleached paper. This lignin-removal will take place only if the wood is thoroughly and uniformly impregnated with cooking chemical. The cell wall is swollen by the high pH of the alkaline pulping liquors, substantially altering its structure and, as a result, alkaline liquors penetrate into the wood in a different manner than the acidic sulfite liquors discussed previously.

The penetration of cooking chemical into wood occurs by two different processes: bulk penetration and diffusion.[612] Bulk penetration is the movement of the cooking liquor through the wood capillary structure by means of either the natural surface tension forces or by externally applied pressure. Diffusion, on the other hand, refers to the movement of dissolved ions through the liquid-saturated chip.

Bulk penetration is governed by the rate of flow of the liquid through the capillary structure of the wood, and follows the Poiseuille equation:

$$\frac{V}{t} \propto \frac{nr^4 \, \Delta P}{L\eta}$$

where

V is the volume of liquid passing through the capillary in time t,

n is the number of capillaries

r is their radius

L is the length of the capillary

η is the viscosity of the fluid

ΔP is the pressure drop.

As the flow is proportional to the fourth power of the radius of the individual capillaries, this radius is of prime importance in determining the bulk-flow through the wood structure. Capillary size varies widely between species, and even within a particular wood sample. For example, sapwood is penetrated more readily than heartwood, earlywood more readily than latewood.[613] Dry chips, particularly if the air is removed by presteaming, can take up a much greater volume of liquor through bulk penetration than water-saturated wood. The mechanism of mass transfer of liquid is different for coniferous and de-ciduous woods. In conifers, vertical movement is through the cell lumen, and lateral movement through the bordered pits in the cell wall. In deciduous species, vertical transport occurs through the vessels, while lateral movement occurs mainly through ray tracheids and, to only a minor extent, through the cell wall.[614] Because there are wide morphological differences between species, the rate of chemical uptake will differ between species when they are cooked together, and will result in differences in the degree of pulping.

In contrast to bulk penetration, diffusion of chemicals into wood can only take place efficiently when the wood is saturated with liquid. The rate of diffusion is dependent on the total cross-sectional area available and not on the size of the individual capillaries, as is the case with bulk penetration. Diffusion is also very dependent on the concentration gradient between the liquor surrounding the chips and that within the chips. If the moisture content of a chip when it enters the digester is above 25 to 30% (the fiber saturation point), it will be partially filled with liquid water, even after presteaming. In the liquid-filled portion of the chip, bulk penetration is not possible, and cooking chemical can enter only by diffusion. As the chemical within the wood structure is consumed during pulping, diffusion is the only means of transferring additional chemical into the chip from the surrounding liquor.

In contrast to sulfite liquors, alkaline liquors penetrate into wood at about the same rate in all three structural directions: longitudinal, radial, and tangential. The effective cross-sectional area available for diffusion in aspen wood has been measured in each direction by a conductivity technique. The results are shown in Figure 4-85.[615] It can be seen that the area available for diffusion in the longitudinal direction does not change with pH. In the transverse directions, on the other hand, a distinct change in the diffusion rate occurs at high pH due either to swelling of the cell wall or to dissolution of cell-wall components, or both. This change only begins at around pH 12.5 and does not exhibit its full effect until the pH level is well in excess of 13. Since the rate of penetration of alkaline liquors into wood is not dependent on wood structure, it will take place mainly in the shortest dimension, that is, the chip thickness.[616,617] The effect of chip size on pulping will be discussed in more detail in the next section.

Forced penetration by means of superimposed pressure can be of assistance, particularly in the bulk-penetration phase. In order for pressure treatment to be effective, it is important to purge the chip structure of air prior to its immersion in liquid. This can be done most readily by presteaming the chips at either atmospheric or slightly elevated pressures. Care must be taken to ensure that the conditions during the presteaming are not too drastic, since loss in pulp quality and yield may result.[617] As alternatives to conventional atmospheric presteaming, Va-purge (a series of pressure treatments, each followed by a rapid release of the pressure), or vacuum impregnation have been suggested for removal of air from the chips prior to alkaline pulping.[618] The use of pressure to accelerate the rate of mass transfer of the impregnating fluid into the chip structure assists in counteracting the morphological differences

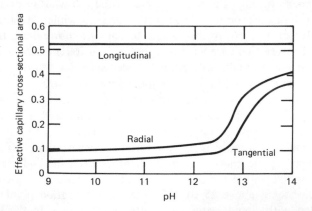

Figure 4-85. Effect of pH on effective capillary cross-sectional area of aspen.

TABLE 4-32 EFFECT OF HYDROSTATIC PRESSURE ON EFFECTIVE
ALKALI UPTAKE BY WOOD[a]

Pressure (kPa)	Ratio of Weight after Penetration to Initial Weight	Effective Alkali Retained (% on wood)
207	2.7	8.3
414	2.7	8.3
621	2.8	9.2
827	2.9	9.8
1034	2.9	10.1

[a]Liquor temperature$-130°C$. Two steam purges [207 kPa (30 psi), 2 min each] were used prior to the experiment for air removal.

in the wood.[614,619] Table 4-32 shows that even after two Va-purge treatments both the total liquid and the effective alkali uptake are increased by increasing the hydrostatic pressure on the system during the impregnation period.

Increasing the temperature of penetration increases the rate of diffusion of chemicals into the wood, and, as a result, the uniformity of distribution of cooking chemicals throughout the wood is improved.[619] However, as the temperature is increased, chemical reaction of the alkali with the wood components is also accelerated, and if too high a penetration temperature is used, chemical concentration in the impregnating liquor is reduced as it moves into the chip. The physical sorption of alkali from the cooking liquor onto the wood also has an important bearing on the distribution of alkali within the chips. The extent of alkali sorption on wood has been measured by Kleinert[620] by determining the decrease in alkali concentration of sodium hydroxide solutions after contact with wood particles for different periods of time. The sorption of alkali on wood occurs quite rapidly, as shown by the fact that the concentration reaches a constant level after only a few minutes contact. The alkali sorption at different concentration levels calculated from these studies is shown in Figure 4-86. At commercial pulping concentrations (4 to 6% NaOH), the wood sorbs about 5% of its weight in alkali. This causes a substantial drop in alkali concentration as the liquor proceeds into the chip and results in an alkali concentration gradient across the chip. This observation further emphasizes the desirability of using thin chips in alkaline pulping to minimize this gradient. At high liquor-to-wood ratios during penetration, the concentration of chemical in the surrounding liquid will decrease to a smaller extent, and the driving force for diffusion into the wood will be maintained at a high level. Figure 4-87 shows the effect of temperature, alkali concentration, and liquor-to-wood ratio on effective alkali uptake. At any temperature and alkali concentration, the highest uptake of effective alkali, and presumably the most uniform distribution of alkali within the chip, occurs at the highest liquor-to-wood ratio. The liquor-to-wood ratio levels shown in this figure are not practical in batch digestion,

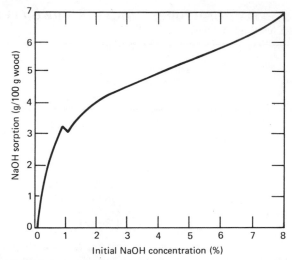

Figure 4-86. Sorption of alkali by spruce sawdust from sodium hydroxide solutions of increasing concentration. Time, 15 min; temperature, 20°C.

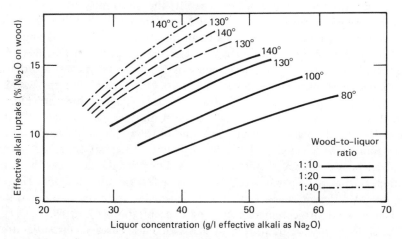

Figure 4-87. Influence of liquor-to-wood ratio and temperature during impregnation on alkali uptake at different liquor concentrations.

but, as pointed out by Kleinert,[619] these high levels can be simulated in continuous pulping by maintaining a high ratio of recirculation rate to fresh liquor addition in the impregnation zone.

It has been shown that lignin will strongly absorb alkali.[621] If, therefore, the proper conditions for impregnation of the wood are maintained so that uniform distribution of chemical is achieved, the alkali sorption process becomes an advantage by concentrating the cooking chemical at the site of the delignification reaction. The pulping reactions, as a result, become less dependent on diffusion of chemical during the cook and are significantly accelerated. Corresponding sorption studies using either sodium sulfide or kraft liquors gave no evidence that either sulfide or hydrosulfide ions were absorbed on wood in a manner similar to sodium hydroxide. However, the effective alkali formed from hydrolysis of sodium sulfide solution does exhibit sorption, although to a lesser degree than sodium hyroxide itself.[620]

Chip Quality in Alkaline Pulping

For many years, the nature of the chips was not considered to be an important factor in alkaline pulping. However, it is now recognized that chip quality affects a number of key aspects of the process:

1. Overthick chips result in nonuniform penetration of cooking chemical and high screen rejects.

2. Fines and pin chips cause a high level of resistance to liquor flow and blockage of liquor-collection screens in both batch and continuous digesters.

3. The ability of chips to flow in bins and feeders depends on their geometry and structure.

4. Chip-bulk density, (which depends partially on chip shape) affects digester production.

5. Over- and under-sized chips result in a loss of material on screening and add to mill wood costs.

As discussed in the previous section, once initial bulk penetration has taken place and the chip is essentially saturated with liquid, additional chemical can only enter the wood by diffusion. In highly alkaline media, such as kraft pulping liquors, diffusion takes place at approximately equal rates in all directions in the chip, and the most critical dimension is the smallest, that is the chip thickness. However, the conventional method of chip analysis, which uses the Williams screen, gives no information on chip thickness.[616] Slotted screens, which do classify chips by thickness, have only been used for chip analysis relatively recently.[622-625 a] Even these screens only measure the maximum thickness, and the thickness of irregular chips will be overestimated. Liquor can enter through fissures in the chip structure, and there is presently no method to measure the degree of fissuring. Visual inspection of the overthick fraction is always advisable to determine if there is a significant difference between the measured chip thickness and the effective thickness from the point of view of alkaline pulping.[626]

Several studies using handmade chips have confirmed the critical nature of chip thickness in alkaline pulping.[626-629] Figure 4-88 demonstrates the pronounced effect that chip thickness has on the level of rejects in the pulp, particularly at the higher Roe chlorine number levels. The chips of 7-mm thick-

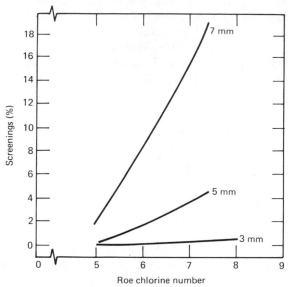

Figure 4–88. Increase in pulp screenings at different Roe chlorine number levels for handmade chips of three different thicknesses.

ness give rapidly increasing reject levels as the Roe number is increased, indicating that the distribution of cooking chemical within the chip is not uniform. The chips with 3-mm thickness, on the other hand, do not produce a significant amount of rejects even up to the 9 Roe-number level, an indication of the good distribution of cooking chemical that is achieved with chips of this thickness. In chemical pulping, the cooking is usually terminated when the amount of rejects in the pulp is reduced to an acceptable level. Substantial yield and quality advantages can often be obtained when chips are cooked to a higher lignin content. As a result, increase in Kappa-number target by use of thinner chips can achieve a substantial cost saving. The thickness of chips produced on commercial equipment will always be quite variable, and a major portion of the total rejects frequently originate from a relatively small fraction of the chips having the greatest thickness. Table 4–33[630] shows the proportion of rejects originating from different chip fractions that were cooked in separate compartments of the same laboratory digester. The high contribution to the total rejects of the larger fractions is quite apparent. For example, the 8- to 10-mm fraction, which makes up only 14% of the weight of the wood, contributes over one-third of the rejects in the pulp. The material above 10-mm thickness, which was not present in this evaluation, generates rejects to an even higher degree. On the basis of results such as these, Hatton recommends that the maximum thickness for acceptable softwood chips should be set at 10 mm. For deciduous woods, chip thickness is even more critical because the defibra-

TABLE 4-33 REJECT LEVELS FROM DIFFERENT CHIP THICKNESS FRACTIONS[a]

	Thickness Classification					
	2 mm	2–4 mm	4–6 mm	6–8 mm	8–10 mm	Total Rejects
Fraction as percent of original sample	2.0	26.6	40.1	21.6	9.6	—
Percent rejects obtained from fraction	1.5	0.5	1.1	3.1	6.8	—
Rejects as percent of original wood	0.03	0.13	0.44	0.67	0.65	1.92

[a] Average values for white spruce, Douglas fir, and Western red cedar.

tion point is at a much lower lignin level than for coniferous species. A maximum of 8-mm thickness is, therefore, recommended for hardwood chips.

Control of the quality of the chip supply to a mill is complex because a substantial portion of the chips are often produced at sawmills and log-handling centers remote from the mill. Wide variations in chip quality occur depending on the wood species, chipper type, moisture content, season, and chipping temperature.[631,632] However, once the importance of chip thickness is understood, proper chipper maintenance and selection of proper chipper operating conditions can be applied to give the highest quality possible under any set of conditions.[633]

Shredding in disk refiners has been suggested as a method of reducing the effective thickness of chips prior to alkaline pulping. Chip-shredding in disk refiners has been investigated in the laboratory,[634,635] and at least one mill has practised shredding commercially for many years with beneficial results.[636] Crushing of chips to produce fissuring in the longitudinal direction has also been suggested[637] and is being carried out at one mill,[638] although primarily for a different purpose. Although these studies show that the use of shredded and crushed chips can lead to distinct advantages when the chip supply contains a substantial fraction of overthick chips, chips treated in this manner have not been used widely in mill systems because of uncertainties regarding their packing and handling properties. Chippers have been developed that will cut chips to the desired thickness by making a controlled cut parallel to the grain of the wood. These machines include the Anglo chipper, the Soderhamn H-P chipper, the CCL chipper, and the Domtar Axial Feed Waferizer. These have not been widely adopted commercially because of their high capital and maintenance cost, and the poor handling characteristics of the chips.

Other chip-quality parameters have been studied in considerably less detail than chip thickness. The bulk density, which has a significant effect on production capacity in the wood-handling and digestion systems, relates inversely to the ratio of the diagonal of the chip face to thickness (d/t).[639] Thus if the chip thickness is decreased to improve the pulping characteristics of the chips, a corresponding reduction must be made in the other dimension of the chip to avoid a decrease in the chip-packing density. Pin chips and fines impede the flow of liquor through the chip mass, reducing the circulation rate and causing a loss of pulp quality in forced circulation digesters; in addition, diffusion washing efficiency in continuous digesters is reduced and the possibility of hangups is increased. Some pin chips and fines are always formed during chipping, but their level increases when chipping cold or frozen wood, when incorrect chipper settings are used, when the chipper is poorly maintained, and when chip-handling systems are improperly designed.[632] A method of classification has been developed to separate the smaller chip fractions into thin chips, pin chips, and true fines.[640] Poor chip-flow properties are caused by curved and extra long chips, which form bridges at critical points in the system. Compression damage to the chips is a much less serious factor in alkaline pulping than in sulfite, and only affects pulp properties when the fibers have been subjected to extreme forces.[641]

Cooking Liquor for Alkaline Pulping

The charge of alkali and its concentration during the cook are the most important of the chemical factors affecting the kraft cook. Alkali charge on wood has a major influence on the degree of delignification, pulp yield, and quality. Alkali concentration is the most important factor controlling uniform penetration of chips and the rate of delignification. In kraft pulping, alkali can be expressed as either active alkali (NaOH + Na_2S) or effective alkali (NaOH + $1/2Na_2S$). It has been established[565] that, of these two expressions, effective alkali correlates more closely with the main pulping responses, and this term will therefore be used in most instances in this discussion.

There are three parameters to be considered in relating to liquor, wood, and chemical in an alkaline cook: chemical charge on wood, liquor-to-wood ratio, and alkali concentration in the liquor. However, when two of these are specified independently, the third will then be fixed. To ensure bulk penetration of all the chips, it is important that sufficient liquor be charged to the digester to immerse the chips completely. The liquor-to-wood ratio is thus to a large extent set by the degree of chip packing and the digester dimensions. The operator can then select the chemical-to-wood ratio to control the degree of cooking based on the residual alkali and pulp permanganate number obtained in previous cooks.

Liquor-to-Wood Ratio. In a batch digester, the white-liquor requirement will normally be substantially less than the volume needed to immerse the chips completely. The normal charging technique is to calculate the white-liquor requirement based on its concentration, the desired alkali-to-wood ratio, and the expected digester wood charge. The volume is then made up to the required liquor-to-wood ratio by addition of weak black liquor from the brown-stock washers. Since uniform chemical distribution throughout the chip charge is more easily achieved if all chips are in contact with the same alkali concentration during their bulk-penetration stage, white and black liquor should be added simultaneously at approximately their final ratio. Dilution of the white liquor with black liquor produces a number of advantageous results:[642]

1. The white-liquor concentration is reduced, thereby minimizing cellulose degradation, which occurs at excessive alkali concentrations.

2. Dilution is achieved without adding to the evaporation requirement of the spent liquor.

3. Residual chemical in the spent liquor is utilized.

4. The higher liquor-to-wood ratio ensures all chips will come in contact with the impregnation liquor.

Some investigations[642-644] have shown that recycling spent liquor can cause adverse effects, such as loss in pulp yield and brightness, increased permanganate number, higher chemical usage, and increased bleach requirement. However, others have found no significant disadvantages from the use of black liquor as a diluent.[645] It appears that as long as the proportion of black liquor used is not too high (normal chip packing should limit the requirement to about 50%) and an adequate residual alkali level is maintained in the liquor until the end of the cook, any adverse effects of black-liquor recycle are very small.

Liquor-to-wood ratio has to be viewed from a slightly different standpoint in continuous digesters. In the normal operation of a continuous digester, no black-liquor recycle is required, and the apparent liquor-to-wood ratio is quite low. White liquor is normally introduced to the system by addition to one of the recirculation flows. However, the residence time of liquor in a continuous digester is often longer than that of the chips, and this has much the same effect as black-liquor recycle in batch digesters. The effective liquor-to-wood ratio in the cooking zone is actually higher than that based on the white-liquor flow alone. In the continuous system, the volume of liquor heated to cooking temperature per unit of wood is less than that for a batch system, thus giving a saving in energy. Furthermore, the liquor has essentially plug-flow characteristics through the digester, and the fresh chips do not come in contact with spent liquor.

Alkali Charge. The amount of alkali charged in an alkaline cook varies with the wood species, cooking conditions, degree of delignification required, and other conditions. Some approximate ranges of chemical usage in commercial processes for the principal grades of alkaline pulp are:

	Normal Kappa-Number Range	Effective Alkali Charge (% Na_2O on wood)
Hardwood soda	10 – 12	15.0 – 17
Bleachable hardwood kraft	13 – 15	13.5 – 16
Bleachable softwood kraft	28 – 35	14.0 – 17
Linerboard kraft	75 – 100	11.0 – 13

For a particular wood species, there is a relationship between the amount of NaOH consumed and the amount of wood material dissolved.[646-648] This relationship for spruce and pine sawdust, treated at cooking temperatures with sodium hydroxide, is shown in Figure 4-89. A similar relationship exists for alkali consumption in kraft pulping. In order to complete the cook in a reasonable time, a small excess of chemical is usually charged, and the chemical

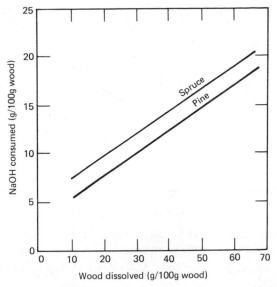

Figure 4-89. Relationship between the amount of wood dissolved and the amount of sodium hydroxide consumed in the soda process.

actually fed to the digester is about 10% higher than that consumed. The residual alkali is important if optimum results are to be obtained from the cooking process. If insufficient alkali is charged, then the pH will drop towards the end of the cook to a level where redeposition of undesired lignin-derived materials onto the fibers can occur.[649] Under these conditions the pulp lignin content, instead of decreasing, actually increases with time, and the redeposited material is very difficult to remove during bleaching. This redeposition, which is apparently a sorption effect, begins when the cooking liquor pH falls below 12, and it continues to increase as the pH becomes lower until a maximum level of deposition is reached at pH 6. Excess of alkali at the end of the cook has a noticeably beneficial effect on the pulp brightness, which is important for pulps to be used in unbleached grades. The lower diagonal line in Figure 4-90[650] shows the minimum effective alkali charge that can be used to provide a desired permanganate number, while the contour lines above the base curve indicate the magnitude of the brightness gain that can be attained by using additional alkali beyond this minimum level.

Alkaline cooking dissolves not only lignin, but also a substantial amount of carbohydrate as well, particularly the hemicellulose. The excess alkali charged should, therefore, not be too much above that required for the delignification process. The amount of hemicellulose that dissolves depends on the alkalinity of the cook, and thus on the effective alkali charge on wood. Figure 4-91[651] shows the decrease in pulp yield from Swedish pine as the alkali charge is increased. An increase of alkali charge from 17.5% to 22.5% as NaOH on wood

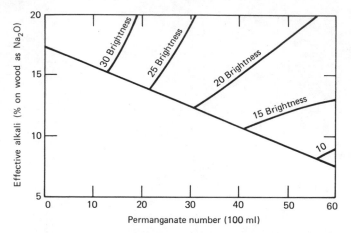

Figure 4-90. Minimum effective alkali required for different permanganate number levels. Contour lines above the diagonal minimum line show the magnitude of the brightness gain that can be attained by using additional alkali above the minimum level.

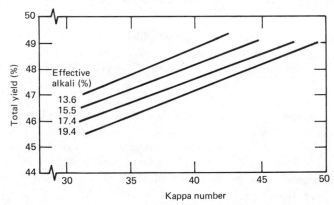

Figure 4-91. Total yield versus Kappa number for four different effective alkali charges on wood.

(13.6% to 17.5% as Na_2O) reduces the yield by about 1% on wood. The xylan-type hemicellulose in hardwood is even more sensitive to extraction by excess alkali than the glucomannan of softwood. Figure 4-92[652] shows the effect of increasing the alkali charge from 15 to 20% as NaOH on wood for Swedish birch. In this case, an increase of 5% in the alkali charge decreases the pulp yield by 2% on wood, double the loss measured in the case of softwoods. The right-hand scale and the dotted curve in Figure 4-92 show that the loss in xylan accounts for all of the additional loss in yield when the higher alkali charge is used.

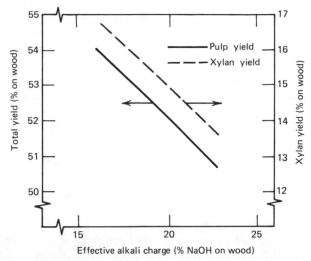

Figure 4-92. Pulp and xylan yield of birch chips cooked to 19 Kappa number with different charges of effective alkali.

In commercial operation, control of chemical-to-wood ratio offers an attractive means for controlling the cooking process.[653] Reliable measurement of this ratio is, however, difficult to make in practice. The chemical charged to the digester can be measured quite accurately, using an integrating flowmeter and titrimetric analysis of alkali concentration. The volume and residual alkali content of black liquor charged to the digester can also be taken into account quite readily. Measurement of wood charge, on the other hand, cannot be achieved with the same level of accuracy. Wet-wood weight can be measured by continuous weighing devices (weightometers), but these devices are both imprecise and inaccurate and significant errors are introduced. Moisture measurement to determine the dry-wood weight can be made using either automatic or manual methods, but both methods have serious shortcomings. Thus it is extremely difficult to obtain representative wood samples by manual methods and the operation is both time consuming and costly. An on-line moisture meter avoids both these disadvantages. A number of different types have been suggested, but none of these has been adopted very widely on a commercial scale, an indication that none has yet proven satisfactory in practice. Where wood density and chip-packing characteristics do not vary widely, a reasonably accurate estimate can be obtained by assuming a constant wood weight per unit of digester volume.

Although removal of lignin is the main goal in alkaline pulping, the dissolution of lignin consumes only a minor part of the total alkali charged in the alkaline cook. The remainder is used up in carbohydrate degradation and other reactions. It has been estimated that in a cook that consumed 16% alkali on wood as NaOH, only 3 to 4% was required to dissolve the lignin.[654] A further 2 to 5%

was used to neutralize acids, mainly acetic and formic, released by the wood. Of the remaining 11% alkali, the major part (over half the total charged) was consumed in dissolution of carbohydrates, while a small amount remained absorbed on the fibers at the end of the cook. The alkaline degradation of carbohydrates leads to formation of sugar acids, and substantial amounts of alkali are needed to neutralize these acids. These statements are confirmed by the observation that sodium hydroxide consumption during the cook is high during periods of high carbohydrate dissolution and much lower at times when dissolution of lignin is the main reaction.[655]

Alkali Concentration. The rate of removal of both lignin and carbohydrate during alkaline pulping is dependent primarily on the concentration of sodium hydroxide. Thus the concentration of alkali in the pulping liquor throughout the cooking period is one of the most important factors in alkaline pulping. Since, in most mill installations, liquor-to-wood ratio is largely determined by digester size and chip-packing properties, and can only be varied within narrow limits, the alkali concentration will be determined by the alkali-to-wood ratio charged at the start of the cook. Figure 4–93[656] shows that for sodium hydroxide alone (soda pulping) the rate of decrease in pulp yield, and consequently the rate of dissolution of wood substance, is accelerated with increasing alkali concentration. In the higher alkali-concentration range, the dissolution becomes less selective towards lignin. In this example, a constant ratio of 30% NaOH (23.3% as Na_2O) was used, based on wood weight, but at two different concentrations of NaOH in the cooking liquor (60 and 90 g NaOH/l). By selecting different cooking times, the alkali consumption and the overall yield are about

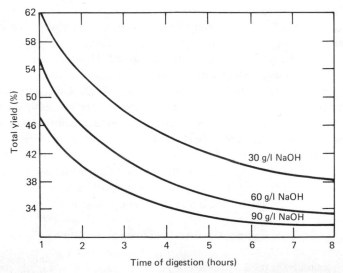

Figure 4-93. Effect of alkali concentration on the rate of dissolution of Loblolly pine.

the same for the two cases. However, the pulp from the cook at lower alkali concentration contains more total cellulosic material (93.3% compared to 93.8%) and less lignin (5.8% compared to 8.1%) than the corresponding pulp from the higher concentration cook. Thus, although the higher alkali concentration can be used to reduce the cooking time (2 hr compared to 4 hr in this case), there is a corresponding penalty in lost yield. This effect of concentration on constituent removal is most apparent when a substantial excess alkali charge is used. At a much lower alkali charge of 15% NaOH on wood, changing the concentration had little effect on the relative dissolution of lignin and carbohydrate in sulfate pulping.[657] Thus, although alkali concentration is the most important factor from the kinetic standpoint, control of chemical charge on wood at the required level is also quite important to avoid yield losses.

The sodium hydroxide concentration decreases as the reaction proceeds, and the concentration profile during the cook is the overall rate-determining factor. A very substantial decrease in concentration occurs in the early stage, when alkali is consumed in neutralizing acids released by the wood and in reactions involving the more labile carbohydrate constituents of the wood. Woods of different origin often have quite different initial alkali demand due to variations in wood composition, bark content, and other factors. The degree of change in the initial alkali consumption is frequently the cause of variations in the total alkali requirement and in the pulping rate.

The fact that high alkali concentration causes excessive carbohydrate degradation led to the suggestion that only part of the liquor should be added when the digester is initially charged and that additional alkali should be charged to maintain an optimum concentration level as the cook proceeds. This cooking procedure is generally referred to as injection cooking.[658,659] It was further suggested that a conductivity cell installed in the digester circulating line could be used to measure the alkali strength of the cooking liquor for control of the injection rate.[660] Since spent liquor from an injection cook has relatively high sulfidity levels, gives pulps of better color and 10 to 20% higher physical strength at lower alkali consumption than conventional kraft cooking. The advantages It has been found that optimum cooking results are obtained in injection cooking when the alkali level is maintained at about 15 g/l Na$_2$O.[662] Below this level of alkali, pulp rejects are high and the rate of delignification is too slow; while above this level, pulp strength and viscosity are reduced, although brightness is increased. Injection cooking, using either the soda or the kraft process at low sulfidity levels, gives pulps of better color and 10 to 20% higher physical strength at lower alkali consumption than conventional kraft cooking. The advantages gained by use of the injection cooking method are reduced at higher sulfidities.[662] The process has not been practiced commercially because of the complexity of its operation and control.

Sulfidity. The presence of sulfide has a strongly beneficial effect in alkaline cooking. This effect occurs mainly as a result of more rapid lignin dissolution, while the rate of attack on the carbohydrates is relatively unaffected. Figure 4-94[663] shows the relative rates of dissolution of lignin and carbohydrates from spruce chips when treated with soda liquor and with 30% sulfidity kraft liquor at equal effective alkali charge. The amount of lignin removed in a given time is much greater with the kraft process and the normal target of 90% delignification is reached in about half the time required for the soda process. On the other hand, the rate of carbohydrate removed plotted on a time scale is essentially identical for the two processes. The rate of removal of pentosans and alpha cellulose and the reduction in the degree of polymerization of the alpha cellulose fraction, as indicated by viscosity measurement, are also similar for the two processes. The strength properties of the pulp improve with increasing sulfidity, particularly strength of the beaten pulps, since the kraft fibers retain their length to a greater degree than soda pulps when subjected to mechanical action.[664] It is apparent from these results that the principal effect of sulfide is an enhancement of the lignin reactivity rather than the protection of the carbohydrates. The reduced dissolution and degradation of carbohydrates in a kraft cook results from the less drastic conditions required for lignin dissolution, namely shorter times and lower alkali charges. The chemical changes in the lignin structure resulting from the presence of sulfide in alkaline cooking liquors will be discussed later.

The optimum sulfidity for any particular mill operation will depend on the type of wood used, the alkali charge, the cooking temperature, and the properties desired in the final pulp. In commercial kraft mills, sulfidity varies from about 15 to 35% with the common range being between 20 and 30%. The sulfidity level cannot be readily controlled by the operator at will, since it is

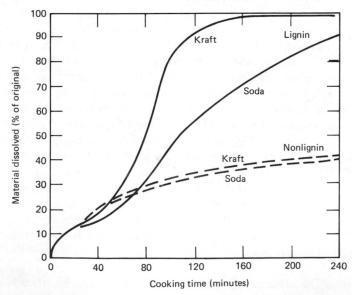

Figure 4-94. Removal of lignin and nonlignin from wood by kraft and soda liquors.

dependent on the makeup chemical used and the sulfur losses in the digestion, pulp washing, and liquor recovery systems. Some sulfur is lost from these systems in the form of gases, such as SO_2, H_2S, and methyl mercaptan, and since these losses do not involve a simultaneous loss of sodium, they determine the equilibrium sulfidity at which the entire system stabilizes. In addition, the ratio of sodium to sulfur in the makeup chemical will have a marked influence on the sulfidity. Caustic soda, sulfur, and chlorine dioxide generator effluent are commonly used to supplement the salt cake makeup. The proportion of the total makeup from these sources will affect the final sulfidity, since they have a different sodium to sulfur ratio than salt cake.

Because of the increased odor levels encountered at higher sulfidities, many mills prefer to keep the sulfidity as low as practical by using caustic soda as part of the makeup supply. For this reason, it is important to know the minimum sulfide level at which the desired benefits of kraft pulping can still be achieved. When comparing results at different sulfidities, the comparison should be made at equal effective alkali charge rather than at equal active alkali charge, since effective alkali includes only that fraction of the sodium sulfide actually available as alkali:

$$Na_2S + H_2O \rightleftharpoons NaOH + NaHS$$

The optimum level of sulfidity for cooking varies with wood species. For black spruce cooked at constant effective alkali charge, pulp strength shows a marked improvement with sulfidity in the 0 to 20% range, a smaller rate of improvement in the 20 to 40% range, and almost no further improvement at higher sulfidity levels.[665] When active alkali is held constant, many of the pulp properties are at an optimum at 20% sulfidity, but are adversely affected above this level due to depletion of alkali before cooking is completed. For Douglas fir, the pulping rate and pulp quality reach an optimum at 34% sulfidity. Further increases in sulfidity result in only slight additional quality improvements.[666] With Loblolly and shortleaf pine, the optimum sulfidity is about 25%, although acceptable pulps are produced in the 17 to 19% sulfidity range.[667] The change in properties with sulfidity is gradual and can only be seen when changes are of sufficient magnitude, for example, 5 to 10 percentage points in sulfidity. Another study[668] examined in detail the effect of sulfidities in the range of 20 to 32% on the pulping of mixed western Canadian wood species. It was found that, although there was a gradual improvement in pulp quality with increasing sulfidity within this range, the effect was small, and no greater than the changes likely to occur as a result of normal variations in species mix in a commercial mill operation. With hardwoods the optimum sulfide charge is usually lower than with softwoods.[669]

Since the role of sulfide is to modify the lignin, it is apparent that the sulfide charge on wood (or on lignin) is actually the critical requirement, rather than the sulfidity level in the liquor.[669,670] At constant sulfidity, sulfide charge on wood will vary with the alkali charged. However, the sulfide-charge variations resulting from the normal variation in alkali charge would be small and should not significantly affect the cook. Figure 4-95[671] shows the rate of consumption of sodium sulfide and sodium hydroxide during a kraft cook. About one-third of the sulfide charged to the digester is consumed early in the heating period, and no significant additional consumption takes place during the balance of the cook. This is in contrast to the sodium hydroxide, which continues to be consumed throughout the course of the cook. Thus the optimum level of sulfidity, which is about 20%, is not determined solely by the amount of sulfur required to form thiolignins during the cook. Since the sulfide consumption occurs very early in the cooking cycle, only that part of the sulfide that is inside the chip after the initial impregnation period is available to take part in these reactions. Any sulfide in the liquor surrounding the chips does not penetrate the wood structure in time to be used in these reactions and remains unconsumed at the end of the cook.

Wood can be pulped with sodium sulfide alone, that is, using an initial cooking liquor of 100% sulfidity.[665,666] In this case the full alkali charge required for pulping is supplied by the hydrolysis of the sodium sulfide. The main advantage of pulping with sodium sulfide alone is the elimination of the need for causticizing and lime reburning in the recovery system. With conventional cooking processes, it is impractical to cook in this manner because the total sodium charge would be much higher than at conventional sulfidity levels, placing a heavy load on the recovery system and making the liquor very difficult to burn. Furthermore, the quantity of sulfur compounds released to the atmosphere would be very substantial, giving unacceptably high odor levels. However, a method known as the Alkafide process has been developed[672] in which sodium

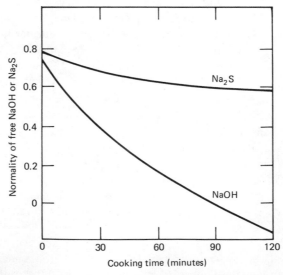

Figure 4-95. Consumption of sodium sulfide and sodium hydroxide in a sulfate cook.

sulfide is the sole pulping chemical and the pulping is carried out very rapidly in a steam atmosphere. Under these pulping conditions,[673] both alkali requirement and mercaptan formation are reduced, and it has been suggested that pulping with sodium sulfide could become practical. The process has been demonstrated at the pilot-plant scale, but is not practiced commercially.

Cooking Cycle in Alkaline Pulping

Time and Temperature. The batch cooking cycle can be divided into two phases: the rise to temperature and the period at temperature. In heating a digester, it is important to ensure that alkali has penetrated to the center of the chip before the temperature rises above 140°C, since undesirable lignin condensation reactions can occur at high temperature in the absence of alkali. Once enough chemical has penetrated the chip to maintain the alkalinity, the digester can be rapidly heated to cooking temperature. Whether natural or forced circulation is used, the rate of heating of the digester contents must be compatible with the circulation rate. Too rapid a heating rate will lead to nonuniformity of temperature throughout the digester. Since the rate of circulation decreases markedly once the digester has reached cooking temperature, any temperature differences, once established, are very difficult to eliminate and can result in variations in the degree of delignification in different parts of the digester.[674] In continuous cooking, the initial part of the cook is usually an impregnation period at a temperature below the level at which significant delignification occurs. Uniform temperature is achieved by maintaining adequate flow through the circulation system in which the liquor-heating takes place.

The effect of increased temperature on the pulping process has been studied for both soda[675] and sulfate[676] pulping by Bray, who found that within the normal pulping temperature range of 160 to 190°C, temperature had no major effect other than on the rate of pulping. However, when high temperatures above 200°C are used, the process becomes less selective for lignin, and the pulp yield at constant Roe chlorine number is reduced, as shown in Figure 4-96.[677] The actual temperature at which this yield loss occurs depends on the wood species and conditions applied, particularly the alkali charge, since the presence of excessive alkali near the end of the cook leads to substantial carbohydrate degradation when pulping at very high temperatures. Some degradation of the carbohydrate component takes place even at temperatures below the level where significant yield loss occurs, as evidenced by the reduced viscosity observed for pulps produced at 190 to 195°C.[678] However only small changes in pulp strength properties have been observed below a 200°C cooking temperature.[677]

The H Factor. The delignification reaction has essentially the characteristics of a first-order homogeneous reaction, even though the reaction is actually heterogeneous. In other words, the lignin dissolution rate at any time is proportional to the amount of undissolved lignin still remaining in the wood at

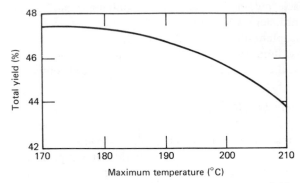

Figure 4-96. Effect of maximum cooking temperature on yield at Roe chlorine number 6.

that time.[679] The lignin concentration remaining is usually corrected for that portion of the lignin that is difficult to remove and is left undissolved at the end of the cook, giving the following rate expression:[680]

$$\frac{dL}{dt} = k\,(L - L_D)$$

where

L is the lignin remaining undissolved at time t

L_D is the difficult-to-remove lignin

k is the reaction velocity constant

The value k is affected by temperature according to the Arrhenius equation:[679]

$$\ln k = B - A/T$$

where

T is the absolute temperature and A and B are constants.

Based on these relationships, Vroom[681] has developed the concept of *H factor*, a numerical value that represents time and temperature as a single variable in the alkaline cooking process. By assigning a value of unity to the rate of reaction at 100°C, relative rates can be assigned to all other temperatures by substituting empirical data into the Arrhenius equation. This concept has proven of considerable value in predicting end points for cooks with different rise-to-temperature schedules or different maximum temperatures. Integrating devices have been developed to calculate H factor during the course of a cook.[682] The

H factor does not take variations in chemical charge into consideration and assumes that this charge will be constant relative to the wood requirement. The Tau factor, a modification to the H factor, has been proposed to account for these variations. The use of these factors in digester control will be discussed later.

Rapid Pulping Processes. Two pulping processes have been developed that use modified conditions for the alkaline cook so that it can be completed in a much shorter time than with conventional cooking procedures. Because of their rapid heating cycles and short cooking times, these processes are primarily applicable to continuous operation. These rapid processes could lead to savings due to the reduced capital cost of the digestion equipment. The reduced time lag between charging and blowing should also permit better process control and less variation in pulp quality.

One of these processes, developed at the University of Florida, is based on the use of matchstick chips split along the grain so that the maximum possible wood surface is presented to the reacting chemical. The path the chemical must travel to reach the center of the chip is thus considerably shorter than with whole chips.[683,684] As a result, the chips can be heated to digestion temperature very rapidly by introducing them into liquor already at the digestion temperature. High liquor concentration and relatively high cooking temperature are used to speed up the reaction further. With this process, shredded pine chips can be cooked to a yield of 47 to 48% in 20 to 30 min using a liquor concentration of 20 to 35 g/l of Na_2O and a temperature of 196°C. Pulp yield and quality are similar to those obtained by conventional pulping of normal mill chips, provided the liquor concentration, temperature, and time are properly selected to avoid excessively drastic reaction conditions.

In a rapid alkaline pulping process developed by Kleinert,[685] chips that have been previously steamed are impregnated at about 140°C with sufficient liquor for subsequently completing the cook. The chips are then removed from the excess surrounding liquid and heated rapidly to a relatively high digestion temperature with direct steam. Using this method, chemical pulps in the 50% yield range can be produced by cooking for 15 min at 185°C. Alkali consumption is lower than with conventional, liquid-phase batch pulping and the generation of mercaptan is also significantly reduced.

End-Point Prediction in Sulfate Pulping. The rate of decrease in the lignin content of wood as alkali is consumed during kraft cooking is shown in Figure 4-97.[686] Based on this relationship, the reaction can be divided into three phases: initial reaction, bulk delignification, and residual delignification. Different reactions occur in these phases and each has its own reaction velocity. In the initial reaction phase, alkali is consumed in deacetylation reactions; in neutralization of wood acids; and in dissolution of readily soluble, wood-carbohydrate components; but very little actual delignification occurs. Most of the lignin is removed in the bulk-delignification phase, which is more rapid

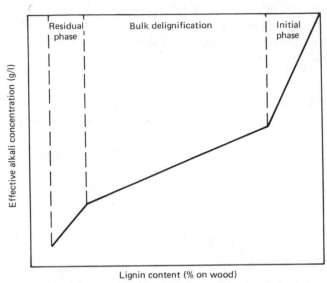

Figure 4-97. Idealized plot of effective alkali concentration in cooking liquor versus residual lignin content.

than residual delignification and evidently less destructive toward carbohydrates, since the decrease in pulp yield per unit weight of lignin removed is relatively lower. The cook should normally be terminated immediately prior to the transition point between the bulk- and residual-delignification phases. The lignin content at this transition point varies with wood species, liquor composition, chemical-on-wood charge, and cooking temperature, but it is usually in the range of 1 to 4% lignin, based on the original wood weight. Once the end point of the cook has been selected, it is necessary to be able to predict the point in the cooking period that it is achieved. For control purposes, it is particularly desirable that this prediction be based on information available while the cook is in progress so that the feed-forward control concept can be applied.

Lignin removal during the bulk-delignification stage is approximately first order with respect to the remaining lignin concentration in the wood. However, this is only true when there is enough excess alkali present at the reaction site for the alkali concentration to remain relatively constant throughout the cook.[680,687,688] In commercial cooking, it is not practical to use excess chemical to maintain the alkali concentration constant; the concentration is depleted as the cook proceeds. Under these circumstances, alkali concentration plays an important role in delignification kinetics and it is therefore necessary to include an expression for this parameter in the rate expression. In the lower sulfidity range, where changes in sulfidity level will affect the rate, a term to compensate for changes in sulfide concentration is also required. The H factor was one of the earliest methods used to predict the end point of an alkaline cook.

This factor has proven very useful in accounting for time-temperature variations in alkaline pulping, but it does not take into account the chemical factors affecting the rate of pulping, such as alkali concentration and sulfidity. The Tau factor (T) is a modification of the H factor developed to allow for variations in chemical concentration[689] and is defined as:

$$T = \left[\frac{S}{2-S} \right] \cdot \left[\frac{EA}{L:W} \right]^2 \cdot H$$

where

S is the sulfidity (expressed as a fraction)

EA is the effective alkali charge on wood

$L:W$ is the liquor-to-wood ratio

H is the H factor

The Tau factor has been shown empirically to be related to Kappa number. Several more general relationships, in the form of kinetic models, have been developed to express the rate of delignification in terms of the main variables in alkaline pulping. As an example, one of the proposed expressions[690] takes the form:

$$- \frac{dL}{dt} = k_1 L \ [OH^-] + k_2 \ [S^{-2}] \ ^\alpha \cdot [OH^-]^\beta$$

where   L is the lignin content of the partially digested wood

$[OH^-]$ is the hydroxyl ion concentration

$[S^{-2}]$ is the sum of the HS^- and $S^=$ ion concentrations

k_1 and k_2 are rate constants

α and β are exponents

The standard Arrhenius equation can then be applied to the expression to account for temperature effects. A number of other models similar in principle, but different in detail have been proposed[680,691-694] to express alkaline pulping reactions in terms of chemical kinetics. These models can be programmed into computer-control systems[692,694] as predictors of Kappa number. These control systems will be described in more detail in the next section.

There are other parameters affecting the kraft cook that are not generally included in these equations and that are not readily controlled by the operator. These include wood quality variations, such as species, bark content, and degree of decay, which affect the alkali demand, and also chip size, which can affect the cooking rate.[693] Furthermore, the application of the equations described to practical situations is usually dependent on accurate measurement of alkali-to-wood charge, which is often difficult to attain in practice. These variations in both wood quality and alkali charge will influence the alkali concentration at the beginning of the bulk delignification stage. For this reason, it has been suggested[695,696] that the end point of a cook can be predicted based on an alkali concentration measurement in the circulation system at the beginning of this phase. By substituting this concentration value and empirically determined constants for the particular wood being pulped into an appropriate mathematical relationship, a suitable target H factor can be calculated for each cook. The model can be further refined to incorporate the detailed rate expressions described above by analyzing for alkali concentration at two or more points in the early portion of the bulk-delignification phase of the cook.[697] On-line alkali analyzers are essential to obtain the data necessary for this control without an excessive amount of manual testing.

Digester Control Systems. The use of the dedicated computer to perform many of the routine control functions in mill-scale alkaline digestion is a development that has been used to increase the efficiency and reduce the variability of the operation. In their simpler form, these units receive signals either from conventional digester control instruments or by operator input, perform any computational functions required, and either initiate a control action themselves or alert the operator to the need for such action. In their more complex forms, the computers are programmed with kinetic process models, such as the ones described above for the delignification reaction, and are linked with specific sensors, such as alkali analyzers, providing advanced system control. These systems have been applied to both batch and continuous digesters.

Some of the functions performed by computer systems for batch digestion are:

1. *Digester charging.* From the wet weight and moisture content of the wood and the effective alkali concentration of the white and black liquors, the computer can control the charging of liquor into each digester, once the sequence is initiated by the operator. The desired alkali-to-wood and liquor-to-wood ratios are controlled, and the white and black liquors are mixed in the correct proportion throughout the filling period to ensure that all chips are contacted with liquor of the same concentration.

2. *Heating-up period.* The rate of either direct or indirect heating can be controlled to a preset program. In indirect heating systems, if the desired heating profile cannot be achieved (due to heater fouling, for example), supplemental, direct-steam addition can be automatically initiated. By controlling the digester pressure close to its saturation level as the temperature is increased, the computer system can operate the gas relief system in a manner that will ensure that noncondensable gases are purged and that the maximum yield of turpentine is recovered.

3. *Steam demand.* To balance the high steam flows required by batch digesters during relatively short peak periods, which periodically put an excessive load on the steam-generation system, the computer can be programmed to arrange digester scheduling and adjust the heating rate to minimize the peak steam demand to the full digester system for any desired production level.

4. *H factor.* The computer can be programmed to calculate continuously the H factor or any other similar value on a cumulative basis and alert the operator prior to the attainment of the target H-factor level so that he can initiate the blowing operation.

5. *Digester scheduling.* The computer can be programmed to assist in selecting optimum digester scheduling for particular circumstances, such as changes in production rate, limited supply flows (white liquor, steam) and downstream holdups (usually indicated by high blow-tank level). Digester schedules can also be arranged to cook at the lowest practical temperature to minimize steam usage.

6. *Alarms.* The system can notify the operator when any unusual conditions occur, such as a main parameter falling outside the normal operating range. This helps to anticipate potential problems in time to make compensating adjustments.

The computer can be programmed to display a variety of status reports by accumulating data over short time spans or by issuing reports in a wide variety of formats for more permanent records. A number of systems are commercially available for computer control of kraft batch digesters.[692,698-702] In order to justify the cost of installing and maintaining these systems, savings in operating cost arising from four main sources have been cited:

1. Reduction of cost through selection of the optimum mode of operation for a given set of circumstances. In digester operation, for example, these savings can result from increased pulp production or reduced consumption of steam or chemicals.

2. Quality improvement through better product uniformity, as shown in the graph on the left side of Figure 4-98. If, for example, permanganate-number variation is reduced through better control, chemical savings should be realized in the bleach plant as a result of the more constant chemical demand.

3. Increased pulp yield by shifting to higher Kappa target, as shown on the right side of Figure 4-98. When the variability is high, the targets are normally set away from the optimum in order to minimize the occurrence of particularly undesirable results, which will happen from time to time. For example, digester permanganate-number targets are usually maintained at a relatively low level to avoid occasional hard cooks with very high levels of screenings, which can then only be processed by reducing the production rate in the screen room. With the improved uniformity achieved through better process control, the permanganate-number target

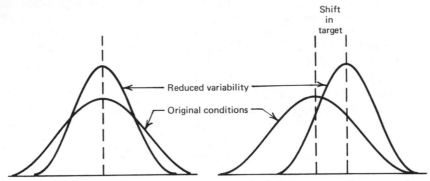

Figure 4-98. Improved product quality and shift in target due to reduced variability.

> can be shifted upward, increasing the average screened yield of pulp without increasing the probability of exceeding the upper permanganate number limit.
>
> 4. In conjunction with other automated equipment, such as digester capping valves and other remote control devices, computer control can contribute to manpower reduction in the operating unit.

Computer-control systems are also in use for continuous digesters.[703-705] Although these systems differ in their details from the batch-digester control systems, they have similar objectives, namely the reduction of variability and improved pulp quality. Changes in wood species, pulp grade, or production rate in a continuous digester require a series of changes in operating targets at the correct time intervals through the digester as the modified conditions reach the different sections. Computer systems can be programmed in advance to make these changes in the proper time sequence, reducing the amount of off-quality pulp produced during the changeover period. In normal operation, the computer can assist in controlling the cooking temperature-time relationship, alkali-to-wood ratio, wash-zone extraction flow, and other process parameters to reduce the variation in permanganate number and to reduce consumption of chemicals and steam. One mill reports[705] that, after the installation of a continuous-digester computer-control system, the average Kappa number was increased by almost three units (from 19 to 21.8) and the standard deviation was reduced from 5.6 to 4.8. This increase in Kappa number produced savings that should justify the installation cost of the computer-control system.

Wood Composition and Quality in Sulfate Pulping

Wood Species. In contrast to sulfite pulping, there are no side reactions that inhibit the dissolution of lignin in alkaline pulping liquors, and, as a result, all known wood species can be delignified. Resins, phenolic constituents, and other acid materials in wood and bark are soluble in the alkaline reagent and are largely removed with the liquor at the end of the cook. Although the kraft process has been most widely applied to the traditional northern softwood and hardwood species, it has also been used successfully on a wide range of tropical and subtropical species, bamboos, reeds, grasses, annual plants, and agricultural waste. Rydholm[706] has listed many of the wood species that have been used or studied as raw materials for alkaline pulping. The wide applicability of the alkaline method makes it practical to pulp similar species as mixtures, providing there is flexibility in wood procurement and simplification of the storage and handling of the wood prior to pulping. Woods of dissimilar nature, such as hardwoods, softwoods, or mixtures of low- and high-density woods, can also be delignified together. However, this practice should be avoided if possible because of the difficulty in controlling both the pulping process itself and the quality of the pulp produced. Where it is necessary to pulp dissimilar woods together, careful blending is required to ensure a relatively constant feed to the digester.

Wood density plays an important role in the operation of the pulping process and in the production capacity of the pulping equipment. Densities of softwoods are generally lower than those of hardwoods, varying from low specific gravities of 0.32 to 0.35 for cedars and balsam firs to high values of 0.45 to 0.50 for Douglas fir and some of the southern pines. Hardwood specific gravities range from 0.35 for some of the aspens and cottonwoods to 0.55 to 0.60 for hard maples, oaks, and some eucalyptuses. Some of the tropical hardwoods have even higher densities. Wood density is the more important factor in determining the chip-bulk density (the other being chip shape) and thus plays a major role in wood charge per unit of digester volume. Wood-handling facilities and storage capacity of chip silos and bins are also related to the bulk density, and the productivity of a mill can be markedly increased where wood of higher density can be used. The rate of wood supply to most continuous digesters is regulated by volumetric feeders, and variations in density will cause variation in feed rate on a weight basis.

Yield. Because resins are dissolved in the alkaline cook, pulp yields from resinous species, such as pine and Douglas fir are slightly lower than those from the low-resin-content wood species. Hardwoods give higher yields than softwoods, since hardwood xylan is more resistant to alkaline degradation than the glucomannan of softwoods. Yield values from the sulfate pulping of hardwoods are at least equal to those from sulfite pulping and usually slightly higher. For this reason, and because the dense hardwoods are more readily penetrated by alkaline liquors, the kraft process is preferred over the sulfite process for the pulping of dense hardwoods.

Wood cost is a major factor in overall pulp costs, and it is therefore important to know the yield obtained during digestion and to be able to determine how it is affected by process changes. Studies on four single species[707,708] have led to the derivation of an empirical expression for relating total pulp yield to pulping-process parameters. The general form of this equation is:[709]

$$Y = A - B \,(\log_{10}H)\,(EA)^n$$

where

        Y is the total pulp yield

        H is the H factor (defined previously)

        EA is the applied effective alkali charge

        A and B are constants for a particular species

        n is a value less than 1.00, also species dependent.

Values for the constants A, B, and n for the four different wood species evaluated are shown in Table 4-34. This table includes the correlation coefficient r^2 for these equations and a good fit between actual and predicted pulp yield is indicated in all cases. Similar equations can be developed to predict permanganate or Kappa number of the pulp from the same process parameters. These relationships are valid in the 15- to 30-permanganate-number range for softwoods and in the 10- to 20-permanganate-number range for hardwoods. However, they are not useful in mill situations where significant variations in wood-species mix or wood quality occur. Furthermore, they cannot be used to predict screened yield since this value is very dependent on chip size.

Effect of Bark and Whole-Wood Chips. Most of the nonpolysaccharide components of bark are dissolved by alkaline pulping liquors; kraft mills can therefore tolerate a higher bark content than sulfite mills. Specifications for bark content of chips to be used for kraft mills are usually set at 1 to 2%; no problems are encountered at these levels. It is difficult to remove bark from cold or frozen wood[631] and the specified levels are often exceeded in winter in the northern climates. The elevated levels are, however, accommodated by the kraft process with only minor difficulty. Laboratory studies[710,711] have suggested that even higher bark contents (up to 24%) can be pulped satisfactorily, provided sufficient alkali is added to dissolve the extraneous bark components. Cooking of bark with alkali produces some fiber, but the yield is very low and the alkali requirement is high. Pulp yields from bark are in the range of 15 to 20% for softwood bark and the range of 20 to 25% for hardwood bark. Shortleaf pine bark, for example, when cooked with kraft liquor to a bleachable pulp, produces only 18% of screened fiber,[710] while northern oak bark yields 22.9% of screened pulp at 16.5 permanganate number.[712]

As a means of extending limited wood resources, it has been suggested that unbarked wood[713,714] or whole-tree chips[715,716] be used as the raw material for pulping. Chipping of the total aboveground portion of the tree in the bush without the prior removal of bark appears to be particularly advantageous. In

TABLE 4-34 VALUES OF CONSTANTS IN THE YIELD EQUATION $Y = A - B (\log_{10} H (EA)^n$

Species	A	B	n	r^{2a}	Total Pulp Yield Range (% on OD wood)
Western hemlock	96.8	4.84	0.41	0.94	40.8–60.0
Western red cedar	84.1	4.68	0.35	0.96	43.9–49.2
Jack pine	94.3	5.88	0.35	0.87	43.7–51.0
Trembling aspen	85.5	1.37	0.76	0.96	54.9–63.9

[a]Correlation coefficient.

the case of one hardwood operation,[717] it is estimated that the use of whole-tree chips increases wood recovery over conventional logging by 40%, while at the same time reducing the manpower requirement. The gain in wood harvested with softwood is less, but it is still substantial. Equipment has even been developed to extract and chip the full tree, including the stump and main root system.[718] There are, however, several disadvantages to the use of whole-tree chips, which affect their application in some instances. The branch wood contains a high concentration of juvenile wood and compression wood, whose carbohydrates are relatively unstable to alkali. As a result of the high content of bark and juvenile wood in whole-tree chips, pulp yields are slightly lower and alkali requirements correspondingly higher. The proportion of the different wood components for spruce and pine and their pulping yield values are shown in Table 4–35.[719] It can be seen that the yield from branches, twigs, needles, and bark is considerably lower than that from the bole, while the yield from the top is slightly lower. Fortunately, these extraneous materials make up less than 25% of the tree weight, and the overall effect on the yield of the mixed chips is not a large one. However, it can be significant where digester or recovery capacity are limiting factors. Branches and twigs are usually not completely chipped and remain in the chips as long thin fragments that make the chips difficult to handle and give them very poor packing properties. Some models of whole-tree chippers are equipped with a split discharge spout to remove these fragments during the chipping. The chipper produces small particles from much of the outer bark; screens can be used to remove these fines and to reduce the bark content of the chips, either directly at the chipping site or as the chips enter the pulp mill. In some terrains, sand and grit are picked up by the tree as it is skidded along the ground and these cause considerable wear problems on equipment, both at the chipper and in the subsequent mill equipment. The use of forwarding equipment that carries the trees off the ground will avert this problem.

The use of whole-tree chips or unbarked wood increases the wood supply, simplifies the operation in the woods and results in lower wood costs. This is attractive for a mill that is limited by its wood supply or by its debarking system, but that has surplus pulp mill capacity available. On the other hand, the low pulp yield per unit of digester volume from bark, twigs, and branches, would reduce production in a digester-limited operation. Furthermore, the additional dissolved organic material in the spent liquor must be incinerated in the recovery furnace and the additional chemical required for pulping must be produced by the recausticizing system. Thus whole-tree chips or unbarked wood would not be an attractive option in mills limited by digester, recovery, or recausticizing capacity.[720,721] Adequate cleaning equipment must also be available to handle the potentially higher dirt level. In a new mill, the pulping and recovery facilities could be sized to deal with the modified wood supply and the increased capital investment and operating cost would then have to be balanced against the savings in wood cost.

TABLE 4-35 PROPORTION AND YIELD OF TREE COMPONENTS[a]

Material	Proportion (%)		Kappa Number	Unbleached Yield (%)	Bleached Yield (%)
	Spruce	Pine			
Bole (>50 mm)	58	59	30	51.5	48.0
Stump	5	5	27	48.4	46.2
Roots (>25 mm)	10	7	26	48.9	45.7
Branches (>5 mm)	3	3	37	40.8	36.3
Top (<50 mm)	2	2	32	47.7	45.1
Twigs (<5 mm) [unbarked]	4	3	34	21.2	18.4
Needles (spruce)	5	–	60	26.4	24.2
Needles (pine)	–	3	37	23.7	22.9
Bark (inner + outer)	13	18	28	21.4	19.3
Mixture 1[b]	–	–	30	47.0	42.5
Mixture 2[c]	–	–	28	48.5	44.7

[a] Average of results of cooks for spruce and pine except where otherwise noted.
[b] Mixture 1 –above materials in listed proportions.
[c] Mixture 2 –above materials assuming half the bark, twigs, and needles have been removed.

Wood Deterioration. The presence of decay in wood chips used for kraft pulping has a detrimental effect on both the cooking process and the pulp properties. The extent of the decay, which can vary widely from an incipient to an advanced stage, will determine the magnitude of the effect on the pulping operation. Brown rot appears to be more deleterious than white rot.[722] The presence of decay results in more rapid pulping, but it reduces the pulp yield substantially, and if the decay is in an advanced state, it will decrease pulp strength. Yield losses have been reported to vary from a few percent to over 30% of the wood, depending on the type and stage of decay. Staining and mild decay give only a slight loss in yield and an insignificant loss in pulp quality.[723] Other losses occur, however, during chipping and chip screening, since decayed wood produces a much higher proportion of fines than sound wood. Wood density is also decreased, causing a loss in digester capacity.

Deterioration occurs when the wood is stored after cutting, particularly when the storage is in the chip form. Heat is generated in chip piles as a result of continued respiration of the living cells in wood and the temperature rises to a level where attack by microorganisms occurs, resulting in substantial loss in wood weight.[724,725] If the storage period is prolonged, this biological attack severely degrades the quality of the chips for pulping.[726] As in the case of decayed wood, lower yields are obtained in kraft pulping and the pulp produced has reduced strength properties. The magnitude of the effect is dependent on the chip storage time, the temperature developed in the pile, and the particular organisms involved.[727] Substantial losses in the yields of valuable by-products of kraft pulping, such as tall oil and turpentine, also occur during extended storage of chips.[728]

Sawmill Residues. In the operation of a sawmill, the outer portions of the log cannot be cut into lumber, and this residual material is chipped for use in pulping. These sawmill chips make up a substantial portion of the wood used for pulp production, comprising 35% of the supply to United States pulp mills in 1973.[729] In addition, the sawdust and shavings produced during the cutting and planing operations are used to some extent as a raw material for pulping. Because they are quite different in handling and pulping characteristics compared to normal chips, sawdust and shavings are normally pulped in a separate digester. Since circulation of liquor through sawdust is extremely difficult, noncirculating continuous digesters are used rather than batch digesters or conventional continuous digesters. The commonly used units are the Pandia horizontal-screw type,[730] the Bauer M. & D. digester,[610] a special design of the Kamyr digester, and the Esco digester.[731] Sawdust is pulped rapidly, but produces lower yield than chips and requires a higher alkali charge. The extra alkali is required to avoid the presence of small uncooked particles in the pulp, often referred to as "birdseed." Typical conditions for continuous operation are 25 to 30 min cooking at 1035 kPa (150 psi) pressure, using 16 to 18% effective alkali as Na_2O on wood.

TABLE 4-36 COMPARISON OF CHIP AND SAWDUST PULP[a]

	Chip Pulp	Sawdust Pulp
Beating time (min)	93	56
Burst index (kPa · m²/g)	8.83	5.88
Tear index (mN · m²/g)	9.32	6.86
Fold (MIT)	1,635	280
Breaking length (m)	10,750	8,500
Sheet density (gm/cc)	0.74	0.73
Fiber classification (Bauer-McNett)		
Retained 12 mesh	47%	1%
Retained 28 mesh	34	45
Retained 48 mesh	8	24
Retained 100 mesh	3	12
Passing 100 mesh	8	18

[a]Values at 250 Canadian Standard Freeness.

Pulp strengths from sawdust are lower than from regular chips. A comparison of typical strength values of pulps produced in one commercial operation from chips and from sawdust using the wood from the same source are given in Table 4-36.[732] Most of the sawdust pulp-strength properties (except fold, which is very low) are about 70 to 80% of those from the chip pulp; however, they are still equal to or better than a good hardwood kraft or a softwood sulfite pulp. The sawdust kraft pulp has other characteristics that are of value in many paper grades, such as lower refining-energy requirement, reduced porosity, and improved opacity. The strength properties of sawdust pulp are highly dependent on the particle size of the raw material, which, in turn, is a function of the coarseness of the saw blades used in the sawmill.[733] In general, pulp strength properties decline with decreasing particle size of the raw material. Sawdust pulps have been blended with other pulps for use in such paper grades as linerboard, corrugating medium,[730] and numerous printing papers.[732,734]

Chemical Reactions During Alkaline Digestion

The reactions that take place with the lignin and with the carbohydrates in wood during alkaline pulping are discussed in the following sections. (See also the chapter on cellulose and hemicellulose and that on lignin).

Reactions of Lignin in Alkaline Pulping

The reactions of lignin during alkaline pulping are complex and are not completely understood. It is known that the presence of sulfide accelerates lignin dissolution without increasing cellulose degradation[663] and that the attack on the lignin molecule involves formation of groups that render the lignin more soluble in alkali. Some understanding of the reaction mechanism has been achieved by examination of dissolved lignin residues from kraft cooks. However, only limited progress has been achieved by this method since a wide variety of chemical entities are present, and many of the primary degradation products, particularly those involving sulfur, are transitory in nature. As in the case of sulfite delignification reactions, much of the knowledge of the delignification reactions during alkaline pulping has been gained by studying the reactions of lignin-model compounds.

In early investigations, particularly those of Hägglund and Enkvist, it was established that in kraft delignification sulfur reacts chemically with the lignin and becomes organically bound to it.[664,735,736] The resultant thiolignins contain an average of about 2.5% of bound sulfur. However, the sulfur is not uniformly distributed. The low-molecular-weight, dialyzable fractions contain more sulfur than the high-molecular-weight fractions.[737] When the model compound vanillyl alcohol is heated in alkaline solutions containing sulfide, vanillyl monosulfide is the main product, as shown in Figure 4-99. The rate of increase in the yield of this product with increasing ratio of Na_2S to NaOH is very similar to the increase in the rate of alkaline pulping with increasing sulfidity.[738] It was suggested that this type of reaction accelerates alkaline pulping by blocking reactive groups in lignin, which would otherwise enter into condensation reactions, thus inhibiting the dissolution of the lignin.

The cleavage of ether linkages on the lignin side chains also appears to play an important role in alkaline delignification. According to Gierer,[739] hydroxyl ions, acting as nucleophilic agents, bring about the cleavage of certain types of ether linkages in alkaline pulping. In sulfate cooking, the combination of hydrosulfide and sulfide ions present is less alkaline but more strongly nucleophilic than hydroxyl ions alone. This characteristic is thought to be mainly responsible for the differences between soda and sulfate pulping. Studies with lignin-model compounds have identified three cleavage reactions that probably occur in the alkaline degradation of lignin during sulfate pulping:

1. Cleavage of α-aryl ether bonds in phenolic units by way of quinonemethide intermediates.[740] The additional phenolic groups formed in these reactions increase the solubility of the lignin and make it more susceptible to other degradation reactions.

2. Cleavage of β-aryl ether bonds in phenolic units by way of episulfide intermediates.[741] The quinonemethide intermediate reacts more rapidly

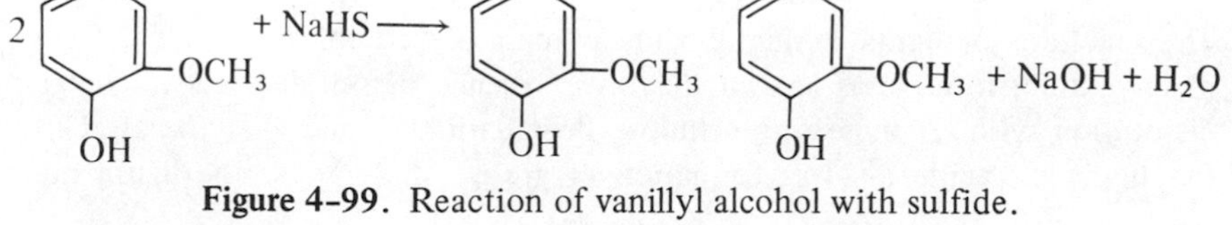

Figure 4-99. Reaction of vanillyl alcohol with sulfide.

with sulfide ion than with hydroxyl ion. This explains the more rapid and extensive degradation of lignin by kraft pulping liquors than by soda liquors.

3. Cleavage of β-aryl ether bonds in nonphenolic units by way of epoxide intermediates.[740] By this reaction phenolic and glycolic groups are liberated, resulting in complete separation of neighboring units in the lignin structure and the formation of more soluble, lower-molecular-weight fragments.

These reactions are discussed in the chapter on lignin.

The degradation of the lignin in alkaline pulping is much less than might be expected from the rates of the above reactions, because condensation reactions occur simultaneously. In these condensation reactions, various carbanions, formed during the lignin degradation process, compete with the nucleophilic anions ($S^=$, SH^-, and OH^-) from the cooking liquor for the reactive sites in the lignin, particularly the quinonemethide intermediates. The use of model compounds has indicated the occurrence of this type of reaction.[742] The role of sulfide in inhibiting these condensation reactions, while simultaneously promoting the degradation of lignin, has been demonstrated using gel filtration of lignins that had been isolated and then treated with sodium hydroxide alone or in combination with sodium sulfide.[743] The dissolved product was found to contain substantially less high-molecular-weight material when the sulfide was present during cooking than when caustic alone was used. The molecular weight of the lignin dissolved during kraft pulping increases as the cook proceeds. By using a continuous flow digestion reactor, material dissolved during the delignification process can be rapidly removed from the digester and cooled before extensive condensation of the dissolved lignin can occur. Under these conditions, the molecular weight of the lignin passing into solution shows a marked increase as the cook proceeds, whereas the other properties of the lignin vary to only a minor extent, as demonstrated in Table 4–37.[744] The distribution of molecular sizes was measured using Sephadex gels and the size of the lignin polymers removed corresponds quite well to the diameters of the pores in the fiber wall.[745]

Lignin in wood is located both in the middle lamella between the fibers and also within the fiber cell walls. Although the lignin concentration is highest in the middle lamella, the greatest portion of the lignin is within the cell wall, since the volume of the cell wall is much greater than that of the middle lamella.[746] During kraft pulping, lignin is preferentially removed from the secondary cell wall during the initial stages of the cook. When the delignification is about 50% complete, however, the rate of dissolution of the middle-lamella lignin becomes dominant and continues to predominate until the end of the bulk-delignification stage. The residual lignin at the end of the cook is mainly located in the cell wall.[747] The rate of lignin removal from the secondary cell wall parallels the rate of hemicellulose dissolution during the early stages of pulping.[748] It appears that the size of the pore openings in the cell wall are increased by this hemicellulose removal and that the size of these openings governs

TABLE 4-37 PROPERTIES OF SUCCESSIVE FRACTIONS REMOVED BY ALKALINE DELIGNIFICATION

Fraction	Delignification (%)	Carbohydrate Content (%)	Methoxyl[a] (%)	Sulfur (%)	Weight Average Molecular Weight
1	5.5	22	12.1	2.5	1,780
2	17.2	9	11.2	2.1	11,700
3	31.0	16	12.5	1.7	13,100
4	45.0	11	12.2	1.5	15,500
5	60.0	14	13.1	2.1	32,800
6	75.0	12	12.5	1.5	30,900
7	88.0	3	11.8	1.7	51,000

[a]Calculated as percentage of lignin.

the rate of lignin removal. The rate of hemicellulose removal is, therefore, an important factor in determining the rate of lignin removal in pulping processes. Since hemicelluloses in the secondary cell wall are removed quite rapidly during the early stages of kraft pulping, lignin located in this portion of the fiber is also rapidly removed.[749] The slower rate of removal of middle-lamella lignin during this phase of the cooking is apparently due to an intrinsic chemical difference in the lignin rather than to topochemical effects.[750]

Reactions of Carbohydrates in Alkaline Pulping

The reactions of the carbohydrate constituents of wood (the cellulose and hemi-cellulose) with the alkaline cooking liquors have an important effect on the alkali consumption and pulp yield and also affect the physical characteristics of the final pulp. The polysaccharides in wood can respond to the alkaline cooking conditions in several ways: they may be dissolved into the liquor as polysaccharides; degraded to soluble low-molecular-weight products; or remain in the fiber, either in their original form or as insoluble degradation products having a reduced degree of polymerization. The degradation reactions are quite complex, and include both alkaline hydrolysis and end-group peeling reactions. Acetyl groups, which are mainly linked to the hemicellulose portion of the wood, are unstable to alkali and are cleaved at a very early stage of the cook. It was pointed out previously that carbohydrate reactions in alkaline pulping are essentially independent of the sulfide content in the cooking liquor, although by accelerating the rate of lignin removal, the presence of sulfide reduces the time of exposure of the polysaccharides to alkaline degradation.[663]

In the alkaline pulping of hardwoods, part of the xylan is removed from the wood by the alkaline liquor. A substantial portion of this xylan is dissolved at, or close to, its original degree of polymerization (in the range of 100 to 200 xylose units) without degradation, and it remains essentially in this form until the temperature reaches at least 150°C.[751] Above this temperature, depolymerization of the dissolved xylan takes place, and the degree of polymerization of the xylan at the end of the cook is only about 40 xylose units. The addition of polysulfide to the cooking liquor stabilizes the xylan so that less is removed from the wood, and the portion dissolved in the spent liquor retains a higher degree of polymerization at the end of the cook than the xylan isolated from standard sulfate cooking liquor.[752] In the case of softwoods, some of the hemi-celluloses are also dissolved during the early stages of the cook, but only the xylan portion remains in polymeric form in the cooking liquor.[753,754] The glucomanan is apparently rapidly degraded as it is extracted from the wood, even at relatively low temperatures, and does not appear as a polysaccharide in alkaline pulping spent liquors. The addition of reagents, such as polysulfide or borohydride, however, will stabilize the dissolved glucomannan to a limited extent.[755]

Xylan is not only dissolved during alkaline cooking, but also some of the dissolved material can be redeposited on the fibers under favorable conditions.[756,757] Several methods have been suggested to enhance this sorption in order to increase the pulp yield in alkaline digestion. If a substantial portion of the pulping liquor is withdrawn after the main uptake of cooking chemical has occurred and before high temperature is reached, it will contain significant amounts of the xylan in polymeric form. The addition of this liquor back into the digester near the end of the cook will result in sorption of some of this xylan onto the wood fiber, giving an increase in yield of about 1% on wood.[758] In cooks having a high alkali charge, the redeposition can be increased by adding acid near the end of the cook to lower the pH.[758,759] Similar results can also be obtained by reusing the withdrawn liquor at the beginning of a subsequent cook in place of the black liquor normally added as diluent.[760] Since the re-cycled liquor already contains a significant concentration of hemicellulose of the type normally dissolved from the wood during the early stages of cooking, the dissolution of additional xylan is inhibited and a gain in yield of about 1.5% on wood is achieved, along with a significant reduction in alkali requirement.

The main reaction contributing to the dissolution of hexose polysaccharides during alkaline pulping occurs at the reducing end group of the chain and is usually referred to as the peeling reaction, because it removes these terminal end groups one at a time. The mechanism for this degradation has been worked out, using glucosides and disaccharides as models.[761,762] For example, when cellobiose is treated with strong alkali at elevated temperatures, the final prod-ucts are glucose and isosaccharinic acid. The reactions, which are believed to occur during alkaline degradation of cellulose, are as follows.[763] Initially, the reducing end group of the cellulose molecule isomerizes to the keto form, which then forms the ene-diol in the presence of strong alkali. Cleavage of the glycosidic bond in the 4-position frees the end group from the cellulose chain, leaving the chain shorter by one glucose unit, but still with a reducing end group. The other product of the cleavage is the diketo unit, which rearranges to the isosaccharinic acid unit in the alkaline medium. The shortened chain can then undergo the same reaction again with the loss of another reducing unit from the end of the chain. This peeling reaction is repeated one step at a time, dissolving glucose units from the cellulose chain and consuming alkali to neutralize the acidic groups formed. It is eventually halted by a much slower reaction, the stopping reaction. This occurs on the end unit without prior isomerization. Initially a hydroxyl group is eliminated from the C-3 position of the enol forming the dicar-bonyl, which rearranges to form a metasaccharinic acid as the end unit. Since the chain no longer contains a reducing end group, it is not possible for the peeling reaction to continue. As mentioned, this stopping reaction occurs at a much slower rate than the peeling reaction, perhaps because of the inhibiting mass-action effect since it involves expulsion of a hydroxyl ion into the already highly alkaline solution. The steps involved in the peeling and stopping reactions are shown in the chapter on cellulose and hemicellulose. It has been estimated that an average of approximately 65 glucose units are dissolved by the peeling reaction before the stopping reaction terminates the process.[764] Some process modifications that have been suggested to prevent degradation resulting from the peeling reaction are discussed in the next section.

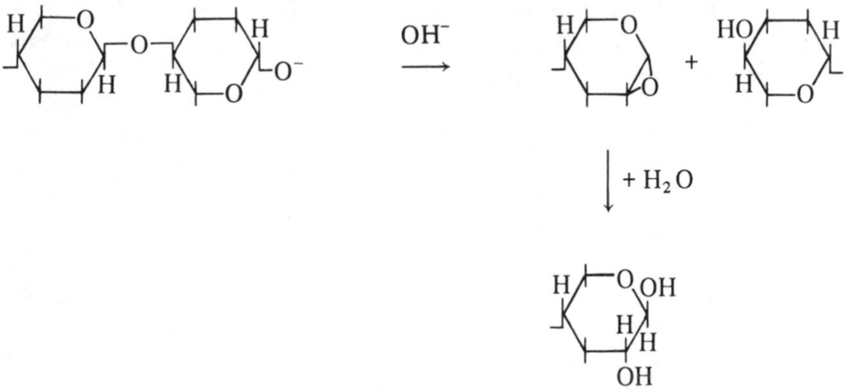

Figure 4-100. Alkaline hydrolysis of cellulose.

Hemicelluloses of both the xylan and glucomannan type are degraded by the alkaline peeling reaction, but at quite different rates.[765,766] The xylans are more stable to alkaline reagents than the glucomannans. In the case of birch xylan, this stability has been attributed to the presence of a galacturonic acid side group in the xylan chain adjacent to the reducing xylose end group. Since this galacturonic acid group is in turn linked to a rhamnose unit through its C-2 position, the normal peeling reaction is blocked once the xylose end group has been removed.[767,768] On the other hand, the glucomannan in softwood hemicellulose is quite unstable in alkaline liquor, as discussed earlier in this section, and the dissolution and decomposition of this constituent is the main reason for the lower yields obtained with the kraft process compared to sulfite in the pulping of softwoods.

When cellulose is treated with alkali at high temperature, the degree of polymerization decreases rapidly, and this decrease can not be accounted for by the peeling reaction alone. It has been shown that at temperatures of 170°C or higher, alkaline hydrolysis by cleavage within the cellulose chain occurs, producing new reducing end groups that can then participate in the peeling reaction.[769,770] The mechanism probably involves the formation of a 1, 2-epoxide that is readily hydrolysed to a reducing end group, as shown in Figure 4-100.[763]

Improvement of Carbohydrate Retention in Alkaline Pulping

Since the reducing end group plays a major role in the loss of carbohydrates in alkaline pulping processes, various treatments to modify this group have been used to decrease the degradation and increase the pulp yield. Reagents used for this purpose include polysulfide, hydrogen sulfide, anthraquinone, and sodium borohydride. These chemicals react in particular with the glucomannan fraction of the wood, and in the case of softwoods the increased retention of glucomannan in the final pulp is usually a good measure of the effectiveness of the treatment.

Polysulfide in Alkaline Pulping. Polysulfide can be generated in white liquors by a number of different methods. The most direct method is to add elemental sulfur to the white or green liquor: [771]

$$Na_2S + S \rightarrow Na_2S_2$$

$$Na_2S_2 + S \rightarrow Na_2S_3$$

and so on.

The elemental sulfur can combine with the sulfide solution in varying proportions. The sulfur in the zero oxidation state is called the excess sulfur or polysulfide sulfur and is available to react with the carbohydrate end groups. Sodium polysulfide can also be formed by adding sulfur to sodium hydroxide alone, but the reaction is slower than with sodium sulfide and the conversion efficiency is lower. The formation of polysulfide by elemental sulfur addition increases the inorganic solids load on the recovery system and increases the atmospheric sulfur emissions. For this reason, methods that generate the polysulfide from the sulfur already in the system are much more attractive. This can be accomplished by air oxidation of the sulfide:

$$4Na_2S + O_2 + 2H_2O \rightarrow 2Na_2S_2 + 4NaOH$$

However the formation of polysulfide is only an intermediate stage in the total oxidation process since the reaction can proceed further to form thiosulfate:

$$2Na_2S_2 + 3O_2 \rightarrow 2Na_2S_2O_3$$

Three catalysts have been suggested that accelerate the rate of the first reaction without affecting the second and, by this means, increase the maximum concentration of polysulfide that can be attained. These catalysts include black liquor,[772] manganese oxides,[773] and granular activated carbon.[774]

The gain in pulp yield achieved in polysulfide pulping is proportional to the polysulfide sulfur-to-wood ratio,[775,776] as shown in Figure 4-101, if other pulping conditions are optimum. The optimum cooking conditions include a slow heating-up period and a relatively low maximum temperature, preferably 160°C or less, to avoid forming new end groups through alkaline cleavage reactions. Under these conditions, a polysulfide sulfur charge of 3% on wood results in a pulp-yield increase of 6% on wood. The increased polysaccharide retention is attributed to the oxidation of reducing end groups to aldonic and metasaccharinic acid structures, which can no longer participate in the peeling reaction.[777] However, in a competing reaction, polysulfide decomposes to thiosulfate and hydrosulfide at elevated temperatures in the presence of alkali:

$$Na_2S_5 + 3NaOH \rightarrow 3NaHS + Na_2S_2O_3$$

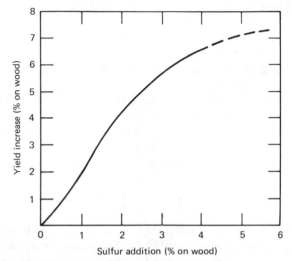

Figure 4-101. Yield increase at Kappa number 36 versus amount of sulfur added, using Norwegian spruce.

This decomposition reaction becomes a significant factor at temperatures above 100°C, and hence the importance of slow heating to permit carbohydrate stabilization to take place before polysulfide decomposition occurs.[778] It has been shown that a multistage process in which the stabilization stage is separated from the digestion stage will give a greater yield increase for a given charge of polysulfide.[779] In the first stage, the chips are treated with a liquor, high in polysulfide content but low in alkali, at a relatively low temperature of about 130°C to promote carbohydrate stabilization. More alkali is then added and the temperature increased for the delignification period. In a further refinement, the impregnation and polysulfide stabilization are also carried out separately from each other, allowing each to be performed at its optimum temperature. In this case, the impregnation is carried out at about 100°C and the stabilization at about 130°C.

Polysulfide cooking is used commercially,[771,776] but the increased emphasis on odor control and the consequent trend to lower sulfidity in alkaline pulping leave less sulfur available for polysulfide formation and the potential benefits from the process are correspondingly reduced.

Hydrogen Sulfide in Alkaline Pulping. The hydrogen sulfide process is an alternative method of end-group stabilization that can produce substantial gains in pulp yield.[780] The process involves pretreatment of the chips with hydrogen sulfide gas under mildly alkaline conditions. The pretreatment stage is usually about 40 min in duration at 125°C with sufficient hydrogen sulfide gas to produce a total pressure of 690 to 965 kPa (100 to 140 psi). When this pretreatment is followed by a kraft cook, a pulp yield increase of about 6% on wood is obtained compared to cooking to the same permanganate number without the pretreatment stage. In addition, there is a reduction of about 2.5% on wood in the active alkali requirement as a result of the reduced formation of sugar acids from the peeling reaction. This is an advantage over polysulfide pulping, where the alkali charge must be increased to compensate for that consumed in the decomposition of polysulfide.

The stabilization of polysaccharides when treated with hydrogen sulfide is due to a reductive thiolation at the end unit, as shown in Figure 4-102.[781] This modification in the end group prevents its participation in the peeling reaction. The glucomannan hemicellulose appears to be the wood constituent most affected since the retention of glucose and mannose in the pulp is increased.[780] Xylose retention, on the other hand, is not increased in softwood pulp and it therefore appears that the xylan is not stabilized by the pretreatment. Yield can also be increased in hardwoods by a hydrogen sulfide pretreatment, but more drastic pretreatment conditions are required than for softwoods, either increased time (85 min) or temperature (140°C).[782] The hydrogen sulfide for the pretreatment stage can either be purchased from an outside supplier or generated on site from the green liquor by carbonation, as practiced, for example, in the Tampella recovery process.[783] It may also be generated internally within the chip pretreatment vessel[784] by modifying the green liquor system to supply the pretreatment vessel with a high sulfidity liquor that has been precarbonated with flue gas. The higher pH and lower hydrogen sulfide pressure, occurring in the pretreatment stage when the internal generation system is used, necessitate the use of higher temperature and longer time in this stage (50 min at 140°C) to accomplish the desired carbohydrate stabilization.

The major advantage of the hydrogen sulfide process over the polysulfide process is the fact that substantial yield benefits can be achieved without adding substantially to the sulfur content of the system. Only about 1 to 1.5% sulfur on the wood weight remains with the pulp after the pretreatment stage; even this need not affect the overall sulfidity if the hydrogen sulfide is generated

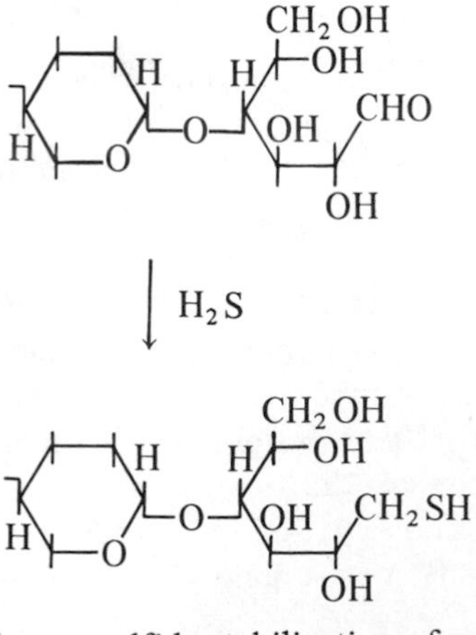

Figure 4-102. Hydrogen sulfide stabilization of cellulose end group.

within the liquor system. On the other hand, the need to handle hydrogen sulfide gas at elevated pressures creates both a safety hazard and a potential source of air pollution should any leakages occur.

Other Treatments for Carbohydrate Retention. Reductive treatments using sodium borohydride have also been suggested as a means of stabilizing polysaccharides during the alkaline pulping process.[785] The addition of sodium borohydride to the cooking liquor prior to sulfate pulping increases the pulp yield, the extent of the increase depending on the quantity of borohydride added. For example the addition of 2.1% of this reagent on wood increases the pulp yield from pine chips from 46.8% to 52.1% at Roe number 5. The increase in yield is largely due to the retention of more of the original glucomannan hemicellulose in the pulp. At the same time, the alkali requirement to reach a desired degree of delignification is reduced. The borohydride is more effectively used if it is applied in a pretreatment stage with the aid of pressure impregnation to assist in the liquor penetration.[786] Under these conditions, good results are obtained with about 1% sodium borohydride on wood. Even at these low addition levels, however, it appears likely that the high cost of borohydride would outweigh the advantage of increased pulp yield.

Another family of additives that has been shown to be quite effective in stabilizing wood polysaccharides includes polynuclear quinones, such as anthraquinone[787] and anthraquinone-2-sulfonic acid.[788] Although anthraquinone is not as soluble in the alkaline liquor as its sulfonic acid derivative, this does not appear to hamper its effectiveness. It costs much less than the sulfonic acid and is much more effective in preserving pulp yield. Anthraquinone is effective at extremely low levels, a dosage of 0.05 to 0.10% on wood giving good results in most cases. The precise mechanism by which anthraquinone stabilizes wood polysaccharides has not been completely clarified, but both glucomannan and xylan yields are increased. Aldonic acids have been isolated from the pulp hydrolyzates, indicating that stabilization takes place through conversion of the end group to the acid form by an oxidation reaction.[789] Polarographic measurements have shown that reduction of lignin by the anthraquinone reduction product formed as a result of these carbohydrate reactions plays a key role in the accelerated delignification rates attained using additives of this nature.[789a] Besides increasing yield, anthraquinone has the additional benefit of increasing the rate of delignification, possibly by blocking lignin cross-condensation reactions. A given degree of delignification may be reached with up to 33% less cooking time than in the absence of anthraquinone.

The increased hemicellulose content of the pulps produced by the polysulfide and hydrogen sulfide processes changes their strength properties relative to those of conventional kraft pulp.[790,791] These pulps develop their maximum strength with a relatively low power input, and the maximum burst and tensile strength level attained are about equal to those for kraft. The tear strength is lower, however, because of the smaller number of cellulose fibers present per unit weight of pulp. The strength deficiency of the polysulfide and hydrogen sulfide pulps compared to kraft pulp decreases with increasing lignin content.[790]

Characteristics of Alkaline Pulps

The composition of the carbohydrates remaining in pulps following alkaline cooking have, to a large extent, been described in the discussion of the carbohydrate reactions during pulping. The most significant differences in carbohydrate composition between the pulp and the original wood occur in the hemicellulose fraction.

Chemical Composition of Alkaline Pulps

In softwoods the two main hemicellulose components are glucomannan and 4-0-methylglucuronoarabinoxylan; in most softwoods the glucomannan is accompanied by galactoglucomannan. The glucomannan fraction suffers the most severe attack by the peeling reaction, and yet it is still a significant component of the pulp produced in a kraft cook. Table 4–38[792] compares the sugar analysis of pine kraft pulp with that of the original wood. From these analyses, the proportions of the original xylan, glucomannan, and cellulose remaining in the final pulp can be calculated. Only about 25% of the glucomannan content of the original wood remains in the final pulp. The xylan, on the other hand, is comparatively stable and more than half of this material is still present in the pulp. The stability of these two hemicellulose components in kraft pulping is thus opposite to that in sulfite pulping, where the glucomannan is relatively stable

TABLE 4–38 SUGAR ANALYSIS OF PINE KRAFT PULP WITH AND WITHOUT BOROHYDRIDE ADDITION COMPARED TO THE ORIGINAL WOOD

	Wood	Kraft Pulp	Borohydride Kraft Pulp
Pulp yield (% of wood)	100.0	48.6	51.6
Lignin (% of wood)	27.3	3.3	3.0
Carbohydrate composition (% of total sugars):			
Xylose	10.2	10.9	7.9
Arabinose	2.9	1.0	0.2
Mannose	18.9	7.1	13.8
Glucose	62.6	79.7	76.9
Galactose	6.1	1.3	1.2
Polysaccharide content (% of wood):			
Xylan[a]	8.7	5.6	4.4
Glucomannan[b]	15.3	4.1	8.5
Cellulose	32.3	34.3	34.8
Galactan	4.0	0.6	0.6
Araban	1.1	0.0	0.0

[a] Assuming a xylose-to-arabinose ratio of 8 to 1 in the wood and 10 to 1 in the pulps.
[b] Based on a mannose-to-glucose ratio of 3.5 to 1.

and the xylan quite sensitive to the acid hydrolysis.[454] The 4-0-methylglu-curonic acid side units of the xylan chain are largely eliminated during kraft pulping, since this group is also less stable in alkaline liquor than in acid.[793,794] When borohydride is added to the cooking liquor, more of the glucomannan is retained in the pulp, but the xylan retention is actually slightly lower than in the standard cook. This is apparently due to a reduced tendency of xylan to resorb on the fibers as a result of the higher residual alkali level when the liquor contains borohydride.[792] In hardwoods, xylan is the major hemicellulose component, occurring in the form of a 4-0-methylglucuronoxylan. As in the case of softwoods, this xylan is relatively stable to alkaline pulping conditions, and a substantial proportion remains in the kraft pulp.[795] The glucuronic acid side unit in the hardwood xylan is, however, not stable to alkali and a substantial portion of it is lost during the cook.[796]

Cellulose is the main product of the pulping of wood. Conditions used for alkaline pulping have, to a large extent, been selected to produce the maximum yield of cellulose without degrading it substantially. Nevertheless, some cellulose degradation does occur in the kraft process as evidenced by a substantial decrease in the viscosity of cellulose that has been treated with alkali at pulping temperature.[797] When cotton cellulose is treated under alkaline conditions similar to those encountered in kraft pulping, its degree of polymerization (DP) decreases very rapidly at first from its initial value of 2000 to 4000, until it reaches a level of about 500, where it remains relatively stable.[770] Wood cellulose behaves in a similar manner, although the initial rate of decrease in viscosity is less rapid and the terminal viscosity slightly lower than that for cotton. These differences between cotton and wood cellulose appear to be due to differences in the fine crystalline structure of the cellulose rather than to differences in the chemical mechanism of the reactions involved.[797] The decrease in the DP of cellulose is accompanied by an increase in its carboxyl content,[798] probably as a result of the formation of stabilized saccharinic acid end groups by the stopping reaction. The decrease in the degree of polymerization of cellulose during alkaline pulping is thought to be caused primarily by the alkaline hydrolysis reaction discussed in the preceding section, since this large effect can not be accounted for by alkaline peeling alone. The attack of alkali on wood cellulose appears to occur at random points in the fiber, producing a relatively uniform effect throughout the fiber and resulting in a minimum loss in fiber strength. This is in contrast to degradation of cellulose during sulfite cooking, where the acid attack is concentrated at points where previous fiber damage has occurred and, therefore, results in severe strength loss. However, when the viscosity of alkaline pulp is reduced below a critical level, pulp strength can be reduced. Care should therefore be used in selecting both the alkali charge and the maximum cooking temperature in order to minimize these degradation reactions.

The residual lignin in pulp influences its physical properties including strength and color. It also affects the chemical behavior, particularly in the bleaching process. When the pulp is to be used in the unbleached state, delignification should be stopped at the highest lignin content at which acceptable pulp properties can be achieved. If this results in a high level of screen rejects in the pulp from the digester, mechanical defibration may be used. When the pulp is to be bleached, the decision on the degree of lignin removal involves additional factors. Pulping and bleaching costs, wood cost (which is determined by the lignin-free yield), and the required pulp characteristics must all be considered. For bleached pulp production, hardwoods are usually cooked to a lignin content of about 2% on pulp compared to 3 to 4% for softwoods. The hardwoods are cooked to a lower lignin content because hardwood fibers are only completely liberated when cooked to low lignin levels, and difficulties are encountered in pulp screening if the pulp is not well defibered at the end of the digestion.

The amount of pitch remaining in alkaline pulps is much lower than in the original wood, and also much lower than in sulfite pulps from the same wood species. This is because resin and fatty acids and their esters are converted into soluble soaps in alkaline pulping. These soaps in turn emulsify a large proportion of the unsaponifiable fraction of the resin. If the pulp is adequately washed, these objectionable materials are removed with the black liquor and should not cause problems in the pulp system. When woods of high extractive content are pulped, the resulting soaps can be separated from the black liquor during evaporation and recovered for sale, as described later. Although pitch deposition problems are less common in kraft pulp mills than in sulfite mills, they do occur from time to time. The deposited pitch can cause major operating difficulties, including plugging of centrifugal-cleaner tips and fouling of consistency- and magnetic-flow meter probes. Pitch can also appear as undesirable dirt specks in the final pulp.[799] In contrast to sulfite pulps, aging of wood in either log or chip form increases pitch problems in alkaline pulps. Apparently, as the wood ages, its resin and fatty acid content decreases, and the dispersant action of these materials is lost, causing more of the unsaponifiable material to remain with the pulp.[800] When the pH of the washed pulp suspension is above 6, the tendency for pitch to deposit on metallic surfaces in the presence of calcium ions is increased.[801] Pitch problems in a kraft pulp mill can be minimized by reducing chip-pile storage time through pile rotation and by maintaining a uniform white water pH of about 6 throughout the pulp screening system.[799] The judicious use of dispersants can also reduce pitch deposition,[799,802] but this can be expensive. In addition to resinous materials arising from the wood, pulp-mill pitch deposits often contain a large proportion of inorganic materials, particularly calcium carbonate particles, resulting from incomplete white liquor clarification.[803] These materials add substantially to the volume of the deposit and also promote scavenging of further organic materials in the system, including wood extractives, black-liquor solids, defoamers, retention aids, and dispersants. For this reason, pitch problems are usually more severe when the white liquor is not adequately clarified.

Physical Characteristics of Alkaline Pulps

The high physical strength of kraft pulps makes them attractive for use in papers with a high-strength requirement, such as linerboard, bag and wrapping papers, and solid bleached boards. However, alkaline pulps are very dark in appearance and in the unbleached state are not suitable for the production of printing papers. Chromophoric groups formed from lignin during pulping are responsible for most of the color in kraft pulps.[739] It has been estimated that about 90% of the color of kraft pulps originates from the lignin, about 10% from the carbohydrates, and less than 1% from the extractives.[804] However, in a few special cases, such as eucalyptus pulps, the extractives contain functional groups sensitive to alkali. In these cases the extractives make a significant contribution to the pulp color. Although the concentration of the lignin remaining in the pulp decreases as the cook proceeds, this lignin is relatively highly colored, so that the pulp has a strong color even at low residual-lignin levels. Redeposition of lignin towards the end of the cook has a significant effect on pulp color[649] since the precipitated thiolignins contain a relatively high proportion of chromophores compared to the original wood lignin, and when conditions are used that promote this redeposition, such as low residual-alkali or high-solids concentration, dark pulps are produced.

The contribution of redeposited lignin to the color of kraft pulp can be demonstrated by comparing bleached kraft pulp placed in a regular batch cook with the same pulp placed in a digester in which the liquor is continuously displaced. The lignin reabsorbed on the pulp in the batch cook has considerably more color than that from the displacement cook and as a result produces a pulp of lower brightness.[805] The redeposited lignin appears to be chemically bonded to the carbohydrate since it cannot be removed by thorough washing or by alkali extraction.[806] Lignin redeposition increases when lignin concentration in the black liquor becomes high towards the end of the cook. Pulp brightness can therefore be increased by using more dilute liquor. The continuous liquor-displacement method of pulping can be used to produce unbleached kraft pulps as bright as sulfite pulp, but of higher strength. More xylan is, however, dissolved during these cooks, giving lower pulp yields than a conventional kraft cook.[807] The high alkali charge and excessive volumes of liquor used in the continuous displacement method along with the increased hemicellulose dissolution appear to make it impractical for commercial use. When the liquor withdrawn during this cooking procedure is recycled to subsequent continuous-flow cooks to reduce this high alkali usage and low yield, the gain in brightness is relatively small.

High pulp strength is characteristic of the kraft pulping process; the process derives its name from this feature. The strength of kraft pulp is dependent on many process parameters already discussed, including cooking time and temperature, alkali charge on wood, sulfidity, degree of delignification and chip size.[808,809] However, the most important parameters in determining pulp strength are the fiber characteristics and chemical constitution of the particular wood species being pulped.[455] The fibers of different wood species vary considerably in such morphological features as fiber length, fiber diameter, cell-wall thickness, and the relative proportions of springwood and summerwood. These morphological differences persist during pulping and the fibers making up the resultant kraft pulps show corresponding variations. The chemical properties of greatest importance are the hemicellulose content and, in some instances, the degree of polymerization of the cellulose, as described in the preceding section. Kraft pulps from five different wood species (Western hemlock, Douglas fir, cedar, Eastern spruce, and Loblolly pine) cooked to a standard permanganate number level of 30 are compared in Table 4-39.[810] To a large extent the strength results can be correlated with fiber length as measured by the Bauer-McNett fractionation. Douglas fir and Loblolly pine, which have the highest fiber length, show the highest tear, but the lowest bursting strength; spruce has the shortest fiber length and shows the lowest tear, but the highest burst strength. Western hemlock is intermediate in fiber length and in pulp properties. Cedar is exceptional in that, although its fiber length is quite high, it has very high burst strength and low tear strength. More precise prediction of pulp strength properties can be made if additional fiber characteristics, such as fiber coarseness, fibril angle, and cell-wall thickness, are taken into consideration.[811]

The biggest factor affecting the strength of alkaline pulps is the difference between the long-fibered coniferous species and the short-fibered deciduous species. The short hardwood fibers do not have sufficient strength to be used to any extent where high strength properties are required. Nevertheless, the high density and relative ease of pulping of hardwood species make them attractive from a process standpoint, and their abundance and rapid growth rates, particularly in tropical climates,[812,813] give them a cost advantage over softwoods in many locations. Hardwoods can be used up to a limit of about 10 to 20% in grades where high strength is required, such as linerboard.[814] Since proportions greater than this have a noticeably adverse effect on pulp strength, uniform blending of the hardwood and softwood species is very important to maintain satisfactory pulp quality, whether the pulping of the two species is carried out together or separately.[815] The main use for hardwood kraft is as bleached pulp in printing papers, since the thin, short hardwood fibers form a smooth, uniform sheet with excellent surface properties and good opacity. Hardwoods can make up as much as 70 to 80% of the fiber furnish of these grades with the remainder being softwood to provide the required paper-machine operating efficiency and paper-strength properties.

The degree of pulping is an important parameter affecting the strength qualities as well as the color of kraft pulp. In general, pulp strength decreases with increasing yield due to the increasing content of lignin. For softwoods, burst and tensile strength reach a maximum at about the 50% unbleached-yield level, while tear strength decreases throughout the whole yield range.[816] At yield levels above 50%, mechanical refining must be used to defiber the pulp, and the method of mechanical refining has an effect on the strength results. Properties will be least affected if the pulp is refined at high consistency and elevated temperature; under these conditions fiber shortening is at a minimum.[817] The decrease in bonding strength with increase in pulp yield is in part due to the increased lignin content in the cell wall, which reduces the fiber flexibility and therefore allows less fiber-to-fiber contact when pressed into sheets. However, this fiber rigidity is an advantage when the pulp is used for linerboard for cor-

TABLE 4-39 PHYSICAL PROPERTIES OF KRAFT PULPS FROM FIVE DIFFERENT WOOD SPECIES

	Western Hemlock	Douglas Fir	Cedar	Black Spruce	Loblolly Pine
Retained 20 mesh–Bauer-McNett (%)	79.0	81.6	80.9	75.2	82.5
Burst (%)[a]–400 CSF	200	154	231	214	152
Tear resistance[a] –400 CSF	1.74	2.65	1.53	1.36	2.20
Beating time to 400 CSF (min)	48	47	53	55	42

[a]Not given in SI units; used for comparison only.

441

rugated containers, since stiffness is an important property for this grade. Pulps for linerboard (hard stock) are generally produced in the yield range of 54 to 57%, using refining to defiber the pulp; some kraft pulp mills producing stock for linerboard operate at yield levels as high as 62%.[818]

Washing and Screening of Unbleached Pulp

Following the cooking stage, a series of physical separations and mechanical treatments is required to prepare the pulp for its end use. The functions performed by these treatments include separation of the cooked chips into individual fibers, removal of soluble solids from these fibers, and elimination of particles of uncooked wood and of nonfibrous impurities. The equipment used for these stages has been adequately described elsewhere.[819] The present discussion will be limited mainly to the process aspects.

Defibration (Fiberization) of Sulfate Pulp

When the desired degree of cooking in the digester has been reached, the contents are discharged into a blow tank. Normally the digester is blown at or near full cooking pressure. In the case of well-delignified chips, the fibers are effectively separated by the expansion that occurs as steam is released within the chips. For pulps of high lignin content, such as linerboard hard stock, the chips do not disintegrate during the blow and additional mechanical action is required, either in the blowline or subsequent to it. When discharging from continuous digesters, the chip mass is cooled in the digester to prevent fiber damage, and a mild action is imparted to the stock by the outlet device to complete the defibration (or fiberizing) process. The blow tank serves as a temporary storage vessel to even out the flow of pulp from batch digesters and also to provide reserve stock and storage space during short shutdowns of equipment, either upstream or downstream from it. The top of the blow tank is designed as a cyclone with a tangential entry to separate the steam and noncondensable gases from the pulp. These vapors are then processed for heat recovery and removal of odorous substances, while the pulp and black liquor drop into the blow tank storage zone. The bottom of the blow tank is a dilution zone. It has an agitator and connections for weak black liquor addition to allow dilution of the brown stock to a pumpable consistency for further treatment.

Pulps cooked to or beyond the defibration point do not require mechanical disintegration and the stock is usually passed directly over a coarse screen or knotter to remove the large rejects. These knotter screens may be of an open vibratory, centrifugal, or pressure design and usually have plate openings of about 1 cm. Operating consistency is usually about 3 to 4%. The coarse reject fractions can be returned to the digester for repulping[820] or mechanically refined and then recombined with the main stock,[821,822] while the screened pulp (accepts) passes to the washers.

In high-yield kraft systems, mechanical disintegration of the cooked chips is required. This may take place either in the line to the blow tank or between the blow tank and the washers.[823,824] The alternative of refining after washing is not as satisfactory, because washing is not efficient on pulp that has not been defibered (fiberized). For refining in the blow line, a defibration unit is placed between the digester and blow tank; this method is used only in conjunction with continuous digestion, since it would not be practical to operate a refiner with the intermittent flow characteristics of batch digester operation. For batch digesters, and frequently also with continuous digesters, the stock is defibered between the blow tank and the washers at 3 to 4% consistency. This method is commonly referred to as hot-stock refining.[825] The power requirement for this operation depends on the pulp yield, wood species, and the temperature of refining. In the yield range of 54 to 56% (70 to 80 Kappa number), about 0.1 to 0.2 MJ/kg (1.5 to 3.0 hpd/ton) are required. The stock from the refiners goes to screens before washing, and the screen rejects are again refined, either by returning to the primary refiners or in a separate unit. In order to minimize foaming and air entrainment in the stock, both the refiners and the screens should be of the pressurized type. Trapped air in the pulp lowers the brown-stock washing efficiency, and the washing will be significantly improved if the stock does not contact air between the blow tank and the washers. Hot-stock refining is suitable for stock up to 80 Kappa number, but for stock at higher Kappa-number levels, additional refining is usually required to achieve satisfactory defibration. This additional refining should preferably be at high consistency, since low-consistency refining tends to reduce fiber length and tear strength. The additional refining can be performed following the washers at the washer-discharge consistency of 10 to 15%. Refining at this consistency gives excellent disintegration of the shives without fiber shortening and is usually referred to as deshive refining. Refining at still higher consistency (20 to 30%) is even more efficient, but has not been widely adopted because the additional investment for pulp pressing equipment has not been justified by the advantages achieved. The power requirement in deshive refining is dependent on the yield and the power in the hot-stock refining stage; it has been shown that for a 60% yield pulp, a low screen-reject level can be achieved with a combination of 0.2 MJ/kg (3 hpd/ton) in the hot-stock refining stage and 0.4 to 0.57 MJ/kg (6 to 8 hpd/ton) in the deshive refiner.[824]

Brown-Stock Washing

Efficient removal and recovery of the soluble materials from cooked pulp is of importance to minimize the cooking chemical makeup requirement, the bleaching chemical demand, and the pollution load in the mill effluent stream. In addition, incomplete washing can lead to foam problems in the screening and papermaking operations and to excessive consumption of chemicals used for pH control in the stock preparation system. To achieve high washing efficiencies, multiple-stage, countercurrent washing systems are employed, including washing within continuous digesters, in separate diffusers, and on drum filters.

Washing Efficiency. The washing system has two main objectives: to remove the black liquor solids from the pulp as efficiently as possible and to deliver the total black liquor to the recovery system with as little dilution as possible. Although the problems arising from underwashing are obvious, it is also important to recognize that overwashing can be equally serious, since it results in overloading of evaporators and consequent sewering of the excess black liquor. The two factors—degree of dilution incurred and level of residual chemicals left in the pulp—are normally used as a measure of the washing efficiency.[826]

In an ideal washing operation, the thickened pulp mat on the washer would be contacted with an amount of washwater exactly equal to its own liquid content. This washwater would enter the mat and completely displace the original liquor without any of the washwater passing through the mat with the displaced liquid. The original liquor would be completely removed without any dilution and the level of dissolved solids in the washed pulp would depend only on their concentration in the washwater. In practice, this ideal situation does not occur; usually, some of the original solids remain in the pulp and some of the washwater passes through into the displaced liquor. These nonideal conditions arise because: (1) there is channeling and mixing at the interface between the original liquor and the washwater, (2) some of the original liquor remains trapped inside the chips or fibers and does not have time to diffuse into the main liquor stream, and (3) some of the solids remain chemically or physically sorbed on the fiber.[827] The effectiveness of a single washing stage in removing solids from the pulp can be expressed in terms of the displacement ratio, which is defined as the ratio of the actual reduction in soluble solids in the stage compared to the maximum possible reduction.[828] This value is determined by the solids content of the three main liquid flows around the washer. It can be expressed as follows:

$$D.R. = \frac{S_V - S_E}{S_V - S_W}$$

where
 $D.R.$ = the displacement ratio
 S_V = the fraction of dissolved solids in the liquor entering the stage with the pulp
 S_E = the fraction of dissolved solids in the liquor leaving the stage in the washed pulp
 S_W = the fraction of dissolved solids in the washwater supply

In the ideal case, the solids content of the liquor leaving the washer would be the same as that in the washwater feed, that is $D.R. = 1$. In practice, the displacement ratio is a measure of the effect on washing of all the operating variables, independent and dependent, including such variables as wash-liquor distribution, uniformity of the fiber mat, wash-liquor temperature, and variables dependent on pulp characteristics. However, washing efficiency is most affected by the amount of wash liquor applied.

The quantity of water used for washing pulp is normally expressed as dilution factor, which is defined as the weight of washwater introduced into the black liquor per unit weight of pulp washed.[829] The water added to the black liquor is the difference between the washwater added to the washing unit and the water remaining in the pulp after washing. A negative dilution factor represents the case where less washwater is added to the system than the amount of water leaving with the pulp. Mathematically, the dilution factor can be expressed as:

$$D.F. = \frac{L_W - L_E\,(1 - S_E)}{1000}$$

where

$D.F.$ = dilution factor in kg of water per kg of air-dry pulp
L_W = washwater flow to the last stage, kg per ton
L_E = liquor leaving the last stage in the pulp mat, kg per ton
S_E = fraction of dissolved solids in the liquor with the pulp leaving the last stage

Norden has viewed washing as a series of countercurrent mixing and rethickening stages in series. He has analyzed the system mathematically on this basis and has suggested an alternative method for measuring the efficiency of a washing stage based on the Norden efficiency factor.[830] Essentially, the Norden efficiency factor is the number of mixing stages that will give the same results as the washing equipment under consideration when operated at the same dilution factor. The Norden method of calculating multistage washing efficiency has the advantage of assigning an efficiency factor to equipment that is independent of dilution factor. However, this method of calculating the washing efficiency of a system becomes increasingly complex as different types of washing equipment with different consistencies and dilution factors are connected in series. The Norden efficiency factor has therefore been redefined[831] as the number of equivalent mixing stages operated at 10% consistency and the same dilution factor. Table 4–40 shows a range of modified Norden efficiency factors for a

TABLE 4–40 MODIFIED NORDEN EFFICIENCY FACTORS

Equipment	Modified Norden Efficiency Factor
Drum washer (one stage)	2.5– 4
Decker converted to washer	2 – 2.5
Single-stage diffuser	3 – 5
Double-stage diffuser	6 –10
Kamyr digester wash zone:	
1 hr 30 min	4 – 6
3 hr	7 –11
4 hr	9 –14
Diffusion/extraction	1.5– 2

number of commonly used washing units to be described later in this section. The dilution-factor range for each value is from +2 to -3. The total modified Norden efficiency factor for a series of washing units can be found by adding the factors for the individual components. The anticipated washing efficiency for a system can then be found from the desired dilution factor and the total system Norden efficiency factor.

A number of methods are used to express the chemical losses due to incomplete washing. The substance-removal efficiency of a wash system is the ratio of the dissolved solids in the spent liquor leaving the system to that in the liquor entering the system with the pulp. In practice, however, sodium-removal efficiency during washing is of particular interest, and washing losses are usually described in terms of the sodium content of the washed pulp leaving the final washing stage, expressed as kilograms (or pounds) of salt cake (Na_2SO_4) per ton of pulp. The sodium in the pulp exists in two different forms—total sodium and washable sodium. The washable sodium refers to the portion that is water soluble, and that can be determined by thorough washing of the pulp with water. The total sodium includes, in addition, the sodium sorbed (bound) by the pulp; this quantity is determined by acidifying the pulp to a low pH value and then removing the sodium by washing. The washable sodium is the more practical value in determining the effectiveness of a washing system. The value for the total sodium lost with the pulp, on the other hand, is an important factor in determining the amount of makeup chemical required in the cooking operation. The difference between the two is the sodium bound to the acidic groups in the kraft pulp. This sodium is bound primarily to carboxyl groups in the carbohydrates and to phenolic groups in the residual lignin; in mills with efficient washing systems this can amount to half the washer sodium losses.[832;833] The actual level of sodium bound in this way is dependent mainly on the Kappa number and pH of the pulp. The effect of pH at two different Kappa-number levels for pine kraft pulps is illustrated in Figure 4-103.[832] The reduction in the quantity of sorbed sodium as the pH approaches the neutral point from the alkaline side is caused by the release of sodium from the phenolic groups of the lignin; the rate of release is quite rapid. This desorption occurs to some extent at the end of a normal washing cycle.[834] The sodium released on the acid side of the neutral point is derived from carboxylic groups, but this release does not occur in normal washing because the pH remains alkaline.

Although washing efficiency is normally defined on the basis of sodium loss, it is important that the organic material dissolved during pulping be removed from the pulp as well. The black liquor in pulp is a homogeneous solution and during most of the washing process the various components are removed simultaneously. However, at low solids concentrations at the end of the washing, physical effects play an important role, and the lignin appears to be removed more slowly than the sodium. A substantial part of the solubilized lignin remains within the fiber wall following the cook, and measurements using an ultraviolet microscope have indicated that removal of this lignin requires a substantially longer time than is available in commercial pulp washing systems.[835] To make

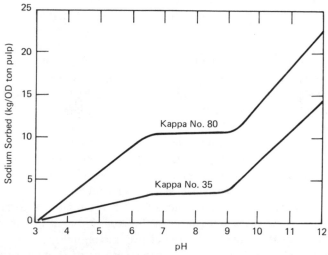

Figure 4-103. Effect of pH on sodium sorbed by slash pine kraft pulp at Kappa numbers 80 and 35.

allowance for these sorption effects, it would be more appropriate in highly efficient washing systems to use total dissolved solids removal as a measure of washing efficiency rather than sodium losses alone.[836] In order to determine the solids removal efficiency in a washing operation, it is necessary to measure the solids content of the liquor in the pulp leaving the final stage of the system. In practice, this quantity has been measured by manual sampling and analysis. However, there is a considerable time lapse between the sampling time and the reporting of the result; the result is therefore of limited practical value to the operator. The use of electrical conductivity to measure the concentration of the residual black liquor is more rapid and is more readily automated. There is an excellent correlation between the conductivity of kraft black liquor and its sodium content[837] and a conductivity probe combined with an adequate sampling system can be used for on-line evaluation of the washing operation.

Rotary Vacuum Filters. The most common method of washing kraft pulp is on a rotary-vacuum-filter washer. The filter consists of a drum partially immersed in a vat that is fed with a low-consistency slurry of the stock to be washed. A mat of pulp is formed on the outer surface of the drum, and some of the liquid is removed during this thickening process. As the drum rotates, showers spray wash liquor onto the sheet to displace additional liquid from the pulp mat. The filter is installed in an elevated position so that the liquor flowing through the drum and down a drop leg to a seal tank creates the vacuum within the washer drum. A washing system usually consists of two to four such units in series, with the filtrate flowing countercurrent to the pulp. Repulpers between the stages ensure proper mixing of the stock with the weaker liquor from the succeeding stage. The washwater entering the system is sprayed onto the sheet on the final washer immediately prior to discharge of the pulp sheet from the system.

Many factors influence the operation of this type of washer. Mat formation must be uniform, both end-to-end across the face of the drum and also around its surface as it rotates. This requires that the pulp consistency in the vat be uniform. The showers must deliver the washing liquid evenly across the surface without disturbing the mat. The choice of operating temperature is important since higher temperature reduces the liquor viscosity and thus improves liquor drainage; however, too high a temperature increases the vapor pressure and thus reduces the vacuum in the drop leg. Optimum washing temperature is about 60 to 70°C and this temperature is maintained by using hot water on the final stage showers. Finely divided air entrained in the stock decreases the drainage rate of filtrate through the pulp sheet on the drum.[836] The minute bubbles of air do not pass through the sheet, but collect in the voids between the fibers and prevent the free drainage of the liquor. The diffusion rate of washwater through the sheet is decreased, and there is also a greater tendency for the shower liquor to channel through the sheet, so that it does not contact all fibers to the same extent. In older systems, the stock is frequently exposed to air prior to the filter, leading to foam formation, particularly when the pulp has been produced from resinous wood species. It is usually necessary to add defoamer to the washer vat to achieve satisfactory washing efficiency.

The effect of dilution factor on dissolved solids loss for three and four stages of drum washing is shown in Figure 4-104.[838] It can be seen that quite low

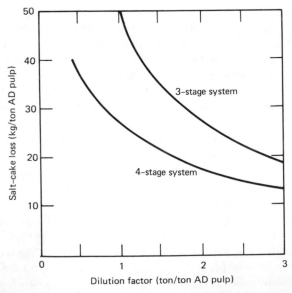

Figure 4-104. Washing efficiency versus dilution factor for three- and four-stage filter washing systems.

losses can be achieved with either system, but that any particular solids removal efficiency is achieved at a lower dilution factor in the four-stage system than in the three-stage system. The mass-flow dynamics of brown-stock washer operation are extremely complex, particularly for rotary filters, involving slurrying, thickening, and displacement operations in combination with the extensive recycle flows required to convey the pulp at low consistency levels. In order to develop control methods and to predict the different efficiency factors for different washing systems, a number of mathematical models have been developed.[839-841] These models are too complex for detailed discussion here, but in the future they may form the basis for much more sophisticated control of the complex washer operation.

Digester Washing. Washing inside batch digesters is only rarely employed because it consumes valuable production time and also because difficulties are encountered in establishing a uniform liquor flow through the bed of cooked chips. However, it has been standard practice for many years to include a countercurrent wash zone in Kamyr digesters,[842] and, as a result, the solids load to the external washers is substantially reduced. A typical internal washing zone for a Kamyr digester is shown in Figure 4-105.[831] At the end of the digestion period the chips, which flow downward in the digester, enter the wash zone from the cooking zone. Strong liquor from the external pulp-washing system is introduced into the bottom of the digester, is heated in an external circulation loop to about 130°C, and then moves up in the digester countercurrent to the chip flow. This liquor is removed, along with the spent cooking liquor, through extraction screens located at the bottom of the cooking zone. The digester can be sized to give the desired retention time in the wash zone to control the degree of washing. The retention time in the washing zone of kraft digesters varies between one and four hours. Washing in a continuous digester is basically a diffusion process and is governed by the factors that normally affect diffusion: time, temperature, chip size, ratio of wash liquor flow rate to pulp flow (dilution factor), and concentration difference between the chip and the surrounding liquor. Longer times are required in digester wash zones compared to rotary vacuum filters because appreciable time is required for chemicals to diffuse out of the undisintegrated chips. Diffusion of dissolved solids from chips is dependent on the smallest chip dimension, which is its thickness,[843] and doubling the thickness will reduce the rate of diffusion by a factor of four. Thus both the average thickness and the uniformity of thickness are important factors. However, it is relatively easy to provide for long wash times at elevated temperatures by appropriate design of the vessel. A favorable concentration gradient is provided by the countercurrent operation, as long as the liquor flow is uniformly distributed through the chip mass.

Mathematical models to predict the efficiency of countercurrent washing systems have been developed using physicochemical and mass-flow principles.[839,844] Based on these theoretical considerations, a typical countercurrent digester wash with 1.5-hr residence time should be approximately equivalent to two stages of rotary filter washing at the same dilution factor.

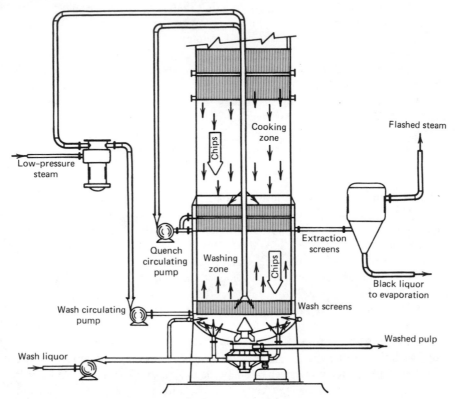

Figure 4-105. Typical internal wash zone in a Kamyr digester.

Diffusion Washing. Batch diffusers were widely used for pulp washing before rotary filter washers became generally adopted, but the large cumbersome units available at that time were difficult to operate and they performed inefficiently. More recently, a continuous diffusion washer that overcomes the problems of the earlier designs has been developed by Kamyr.[845] A single-stage diffuser of this design is shown in Figure 4-106. The diffuser is a series of concentric screen rings mounted on structural arms that radiate from the center of a tower towards the periphery. Each ring consists of two sets of screens through which the wash liquor is collected and removed from the unit. The pulp flows upward through the tower and the screen ring is lifted upward at a rate that approximately equals the flow rate of the pulp. When the assembly reaches the top of its stroke, the extraction flow is stopped and the screen drops rapidly to its bottom limit, clearing the screen openings of fiber as it drops. The wash liquid is added to the annular space between the screen rings through rotating nozzles connected to the top scraper assembly. The flow to each wash nozzle is proportional to the volume of the pulp between the annular rings. The presence of large particles in the stock interferes with wash-liquor flow, and diffusers are more efficient if knots are removed prior to washing.

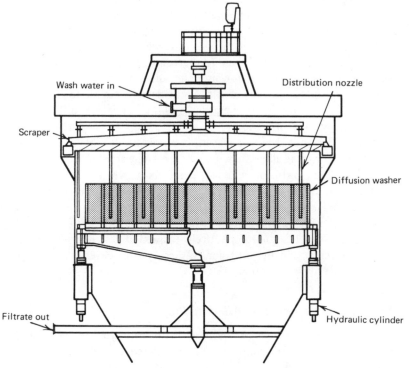

Figure 4-106. Single-stage diffusion washer.

Diffusion washing offers a number of advantages over filter washing. The residence time in the diffusion zone is about 5 to 10 min, several times greater than that in the filter system, allowing more time for residual chemicals to diffuse out of the fiber wall. The system design ensures uniformity of stock flow and provides excellent distribution of the wash liquor. There is no exposure to air so that air entrainment is avoided and defoamer is seldom needed. The system is very simple to operate and control. It can be readily handled by an operator at a remote location. It is possible to erect more than one diffuser in a single tower and consequently the overall space requirement for a diffuser system is relatively small compared to that of a filter system of equal washing efficiency. The diffuser units are enclosed in a tower, which can be outdoors, and the system requires very little building space. These factors are attractive to older mills contemplating installation of improved washing facilities but having limited space available in the existing washing plant. Diffusers use less pumping energy because no dilution is required between stages. The enclosed design allows diffusers to operate with minimum vapor discharge, and the release of odorous substances to the atmosphere is much lower than that encountered in vacuum-filter operation.

Diffusion washers can be used in multiple stages, and commercial installations of two and three stages are in operation.[846] Diffusers are frequently used after continuous digester wash zones and the combination can produce very efficient washing. One system[847] combining a three-hour digester wash with a single-stage diffuser has a total soda loss of only 7.5- to 10–kg salt cake/ton of pulp (15 to 20 lb/ton). A double-stage diffuser following 65 min of in-digester washing results in losses of no more than 7.5 kg/ton (15 lb/ton) at a dilution factor of 3.0.[846] A three-stage unit following batch digesters gives losses consistently less than 12.5 kg/ton (25 lb/ton) of salt cake. It is not possible to obtain samples from within a diffuser during its operation, and it is, therefore, difficult to assess by process analysis the influence of many of the operating parameters on the performance of the unit. For this reason, dynamic models have been developed for diffuser systems that allow these parameters to be studied from a theoretical standpoint.[848–850] In diffusion washing, particularly when it follows a digester wash, the concentration of sodium and other soluble materials is very low, and sorption phenomena and diffusion rates play an important role. These factors must therefore be included when the process models are being derived. These dynamic models have been used effectively to predict design modifications that improve the efficiency of the diffusion washers and to develop more advanced systems to control their operation.

Screening and Cleaning of Alkaline Pulps

Unbleached pulp contains many impurities arising both from the wood and from external sources. These contaminants include incompletely cooked wood particles, fiber bundles, bark, pitch particles, grit, metallic substances, scale, and fly ash. Since most paper grades must be free of these impurities, pulp cleaning systems are required for their removal. Methods for separation of these contaminants from pulp are based on the difference in size, shape, specific gravity, and specific surface compared to individual pulp fibers. When bleached pulp is being produced, there are normally cleaning stages both before and after the bleach plant. The cleaning equipment prior to the bleach plant is aimed mainly at removing particles that will cause operating difficulties in the bleaching process or that will be more difficult to remove after bleaching. For example, sand, grit, and metal particles in pulp cause severe wear on equipment; these should be removed as early as possible. Outer bark and large uncooked wood particles are reduced in size by the mechanical and chemical treatments in the bleach plant. Although they remain large enough to be visible as dirt in the bleached pulp, it is desirable to remove these materials at the unbleached stage where they can be more efficiently separated because of their larger size at this point in the process. On the other hand, small shive, fiber bundles, and fibrous inner bark are largely fiberized during bleaching, and do not need to be removed prior to the bleaching operation. Consequently, the screening and cleaning steps at the unbleached stage should focus on the removal of the largest and the heaviest particles, but do not need to be as efficient in removing the smaller fibrous wood particles. Coarse screening to remove the large knots and uncooked chips that would interfere with pulp washing is carried out prior to the washers. This has already been described. Following washing, fine screens and, in some cases, centrifugal cleaners are used to remove the above described undesirable materials prior to bleaching.

The two main designs of fine screens are the free discharge centrifugal screen[478] and the enclosed pressure screen.[851,852] Although open screens have been widely used in the past because of their simplicity of operation and maintenance, there is a trend toward the use of pressure screens as screen room water systems are closed. These sealed units eliminate air entrainment and, therefore, minimize foaming and increase the capacity of the subsequent thickening equipment. Fine screens may be fitted with either perforated or slotted screen plates, but the perforated design with openings of about 2 to 3 mm (0.080 to 0.110 in.) is widely used for unbleached kraft pulp screening. Since these openings are larger than many of the reject particles to be removed, the separation depends on the rigidity of the particles as well as their size. The feed to open discharge screens is usually of a consistency from 1.25 to 2.0%, depending on the type of fiber. Pressure screens can operate at somewhat higher consistencies, up to 2.5% in some cases. The rejected material from the screen will contain some good pulp, and the rejects are normally rescreened in a secondary unit with smaller perforations to recover this material. The rejects from the secondary screen are sometimes discarded but frequently are either refined and returned to the screen feed[822] or recycled to the digester for recooking, usuable fiber being produced from these rejects in either case. Both the principles of the screening operation[853] and the equipment normally used for fine screening of unbleached pulp[854] have been described in the literature.

In some mills centrifugal cleaners are used in addition to screens to remove contaminants from the unbleached pulp. Since centrifugal cleaners separate impurities primarily on the basis of specific gravity and specific surface, whereas screens separate by size and shape, the two types of equipment eliminate different contaminants and complement each other in their action. The hydrocyclones will readily remove heavy contaminants, such as sand, grit, and scale. They are also quite effective in removing bark and partially disintegrated fibrous material, which have similar specific gravity to pulp fiber but higher specific surface. Large diameter units (usually about 30 cm in diameter) are preferred for unbleached pulp installations, because this allows the use of a relatively large-diameter reject nozzle, which will, in turn, minimize plugging. The use of combined screening and centrifugal cleaning at the unbleached stage permits the mill to produce clean pulp while using a minimum of chemical in the digesters and in the bleach plant.[855] At the same time such use allows the mill to accept wood of relatively high bark content. There are, however, disadvantages to the use of centrifugal cleaners, which have limited their use at the unbleached stage. They operate efficiently only at very low-consistency levels (0.5 to 0.8%) so that substantially higher unbleached-pulp thickening capacity is required. The pressure drop between the cleaner inlet and the accepted stock outlet should be about 275 kPa (40 psi) for efficient cleaning. This pressure drop combined with the high dilution results in a high power requirement for pumping. Centrifugal cleaners give high rejection rates when operated at elevated temperatures, and for this reason do not fit into newer concepts of operating with a closed white water system and high temperature in the brown-stock area.

The efficiency of contaminant removal is determined visually by estimating the number and area of the foreign particles in pulp sheets.[856] These measurements are quite subjective and provide only limited information regarding the size and shape of the contaminating materials. An automated optical method has been developed[857] that measures the dimensions of these particles in two different planes and classifies them according to these dimensions. This method can be used either on-line or as an off-line laboratory device. The classification of particles according to size in this manner permits optimum use of the screening equipment so that it eliminates the most undesirable particles.

Liquor Recovery in Alkaline Pulping

Chemical recovery has been an integral part of alkaline pulping almost from the outset because the cost of the chemicals is too high for economic operation unless they are recovered and reused. This is in contrast to sulfite pulping, where the burning of spent liquor has been of only marginal economic interest, and has only been extensively adopted when effluent quality regulations made it necessary.

The alkaline recovery system fulfils four important functions in the process:

1. It destroys the organic matter dissolved during the pulping stage, eliminating a major potential source of stream pollution.
2. It generates substantial quantities of heat, which can supply a major portion of the heat requirement for the digestion process.
3. It regenerates the sodium hydroxide consumed in the cook.
4. In the case of the kraft process, it converts the sulfur compounds in the black liquor into sodium sulfide.

The main steps in the recovery of alkaline pulping chemicals are:

1. Concentration of the spent liquor to a high solids content by evaporation.
2. Burning the concentrated liquor in a specially designed furnace.
3. Causticizing the resulting sodium carbonate to sodium hydroxide with lime.
4. Regeneration of the lime from the precipitated calcium carbonate by burning.

The kraft liquor system is thus cyclic in nature, as outlined in Figure 4-107.

Some losses of sodium always take place in the pulp mill, mainly at the washers, through the recovery furnace stack and from any liquor spills that occur in the operation. These losses must be restored by adding makeup sodium to the system. The makeup chemical may be added as: purchased salt cake (anhydrous sodium sulfate); chlorine dioxide generator effluent (a solution of sodium sulfate in sulfuric acid) added before the recovery furnace; or purchased caustic

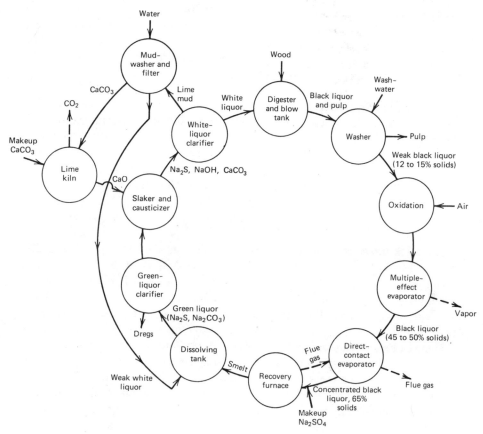

Figure 4-107. Cyclic nature of the kraft recovery process.

soda added to the white liquor. The sodium losses and makeup in a kraft mill are normally expressed as the equivalent amount of salt cake, regardless of whether the actual chemical is $NaOH$, Na_2CO_3, or Na_2SO_4. This permits direct comparison of losses from different parts of the system. The sodium within the pulping system, on the other hand, is expressed as sodium oxide. The extent of the sodium losses varies considerably from mill to mill, but with good operation in older mills, losses can be as low as 50 kg/ton pulp produced. New mills are being designed for chemical losses of 20 to 25 kg, or less, salt cake/ton pulp. Losses of calcium occur in the lime circuit. These are made up by adding either limestone to the kiln or fresh lime to the reburnt product from the kiln.

Evaporation of Kraft Liquors

The black liquor leaving the washers normally contains about 14 to 18% solids, and it must be concentrated to about the level of 60% or higher before burning in the recovery furnace. Evaporation practices differ in detail between North

America and Scandinavia. The most common sequence in North America is evaporation to about the solids level of 50% in multiple-effect evaporators, followed by a final concentration by direct contact of this liquor with the hot flue gases from the furnace. In Scandinavia the complete evaporation is normally performed in multiple-effect evaporators.

Black Liquor Properties. The physical and chemical properties of the black liquor have a significant effect on the performance of the evaporation system.[858] Black liquor normally enters the evaporators at about 15% solids and a temperature of between 70 and 98°C, depending on the dilution and temperature during washing. Temperatures in the upper part of this range are usually encountered only when the final pulp washing takes place in the Hi-heat zone of a continuous digester. The solids content and temperature of the black liquor entering the evaporators will have a significant effect on the throughput rate and steam consumption.

The black liquor contains some residual alkali from the cooking and normally has a pH of at least 12. If there is insufficient residual alkali to prevent the pH falling below this level, there is an increased possibility of lignin precipitation when high solids concentration is reached. A substantial portion of the sodium in black liquor is in the form of sodium salts of the organic acids formed from carbohydrates and lignin during the alkaline cook. Inorganic components include sodium carbonate, sodium sulfate, sodium thiosulfate, and, if the weak black liquor was not oxidized, sodium hydrosulfide. The sodium salts (soaps) of the resin and fatty acids from the wood are also important constituents of kraft black liquor. With resinous wood species, the concentration of these soaps is sufficiently high that they are "salted out" as their concentration increases during the evaporation, and they can be collected and sold as an important by-product of the process.

The most important physical properties of black liquor that affect the evaporation process are its specific gravity, viscosity, specific heat, and boiling point characteristics. The specific gravity depends mainly on the solids content of the liquor, particularly the inorganic components, which have a greater effect on specific gravity than the organics. For a particular type of liquor, there is a good correlation between specific gravity and solids content. The specific gravity, or density, of the liquor is usually used as a measure of the solids content for routine control purposes. However, changes in liquor composition can affect the solids-specific gravity relationship, since specific gravity is more sensitive to inorganic solids than to organic. When substantial changes are made to the process that can affect liquor composition (e.g., wood supply or cooking yield), a new relationship must be developed to satisfy the new operating conditions. The viscosity of kraft liquors is dependent on both the temperature and solids content. In this case, organic solids have a greater influence than the inorganics. It is necessary to keep strong black liquor at elevated temperatures during storage to maintain it in a fluid state. The viscosity of the liquor affects both the flow characteristics and heat transfer rate in the heat exchange section of the evaporator,[859] and it is an important factor in the design of this section of the unit. The specific heat of black liquor solids is less than that of water, and therefore the specific heat of the liquor decreases as the solids content of the liquor increases. The boiling point of the liquor increases with solids content, and this is a major factor determining the temperature differences between stages of the evaporation system.

Multiple-Effect Evaporators. A multiple-effect evaporator consists of a series of evaporation stages in which the vapor formed by boiling the liquor in one stage is used as the source of heat in another stage. For kraft black liquors, either five, six, or seven effects are usually employed. Five-effect units are the most common in Scandinavia, whereas six effects are commonly used in North America. The effects in a multiple-effect evaporation system are always numbered in the direction of decreasing steam temperature; that is, the fresh steam enters the first effect and the lowest pressure steam leaves from the final effect. Several different liquor flow configurations are used; a common system is to feed weak liquor simultaneously to the two final steam effects and then proceed counter-current to the steam flow with the strong product liquor leaving at the first stage. A flow pattern often used with five-effect evaporation is to feed liquor at effects three and four, and then proceed in sequence to stages five, one, and two. This system is often used where evaporation is to a very high solids level so that the liquor at strongest concentration is not subjected to the highest temperature, avoiding deposition as a result of lignin condensation. With the multiple units, it is possible to use the latent heat of vaporization in the inlet steam several times over. With a six-effect evaporator, for example, it should theoretically be possible to evaporate 6 kg water/kg steam fed. In fact, due to heat losses and other inefficiencies, the actual ratio is usually 4.8 to 5.0. The ratio of weight of water evaporated to weight of steam entering the evaporators is referred to as the steam economy of the system.

The pressure and temperature of the steam decreases in successive evaporator effects, due to changes in boiling point and in heat-transfer requirements. As the solids level of the liquor increases, its boiling point at constant pressure rises (vapor pressure decreases) so that the flash vapor from each successive stage will condense at a lower temperature than the previous one. Furthermore, to transfer heat from the steam to the liquor at an acceptable rate, a minimum temperature differential must be maintained between the two. The rate of heat transfer (Q) is given by the expression:

$$Q = U A \, \Delta T$$

where A is the heating surface area

ΔT is the temperature differential across the heating surface

U is the overall coefficient of heat transfer.

Thus the temperature differential that must be maintained between any two effects is the sum of the required heat-transfer differential (ΔT) and the difference in boiling point of the liquor at the two stages. The maximum permissible temperature for the strong liquor is about 120°C. Above this temperature, condensation reactions occur, which cause liquor deposits to form on the heating surface, thus requiring shutdown of the unit for cleaning. Since the pressure in the final effect is set by the temperature of the water used in the surface condenser, which provides the vacuum to the system, there is a limit to the total temperature range available to the system. This means that a compromise must be achieved between capital cost and steam economy. If a very high steam economy is desired, then a very large heating-surface area must be provided in order to minimize the ΔT value required and permit the maximum number of effects. On the other hand, if the capital costs are to be kept low, heating surface must be reduced, permitting a smaller number of effects and consequently lowering steam economy. One method of overcoming this problem is to use vapor-compression evaporation.[860] However, this alternative also involves additional capital costs and has not been widely adopted in black liquor evaporation.

The main cause of loss in heat transfer in black liquor evaporators is formation of deposits (scale) on the heating tubes. These deposits must be removed to restore a satisfactory operating efficiency. This usually requires a shutdown of the unit for washing and results in a loss in production. Various sources of this scale have been identified.[861] The scale is generally deposited on the black liquor side of the evaporator and is usually most prevalent in the high-temperature, high-concentration stages. Calcium carbonate is the most common component of evaporator scale.[862] It has an inverted solubility-temperature relationship. It usually deposits in high-temperature effects through the reaction between calcium (from the wood) and sodium carbonate (present in the black liquor). The latter arises from incomplete causticizing or is formed during the cook. Calcium carbonate scale can be removed by washing with acid containing corrosion inhibitors. However, care must still be taken to avoid corrosion of the tubes. Similarly, sodium sulfate has an inverted solubility curve. If its concentration is high from inefficient reduction in the recovery furnace, it will precipitate. It can be removed with a water boilout. Aluminum silicate scale forms by reaction of alumina (from refractory equipment linings) with silica (present in the liquor). Acid cleaning, mechanical treatment, and thermal shock have all been used to remove silicate scale.

When processing black liquor from high-resin-content woods, a gummy soap deposit will appear on the tubes unless these materials are previously removed from the liquor by a soap-skimming system when the solids concentration reaches the 25 to 30% level, usually between the third and fourth effects. This deposit can be removed by a water boilout. Pulp fiber should be removed from the liquor before evaporation because it can build up on other scale and foul the tube with extreme rapidity. On the vapor side of the heat-exchanger tubes, the main source of scale is iron sulfide, formed by reaction of hydrogen sulfide in the flash vapors with mild steel tubes.[863] This scale does not form when stainless steel tubes are used. The addition of a polymethacrylate preparation to the black liquor has been recommended to minimize deposition on the black liquor side of evaporators and to improve evaporator thermal efficiency.[864]

The most common type of evaporator unit used for black liquor is the long-tube, vertical (LTV) (or rising film) evaporator.[865,866] In these units the increase in volume, as vapor is formed, carries the liquid up the tube in the form of a film or foam, giving intimate contact with the heating surface. Because of the relatively low hydrostatic pressure in this type of unit, the liquor boils at minimum temperature, thus providing the maximum temperature differential for heat transfer. After passing through the tube section, the liquor passes into the vapor head, which separates liquid and vapor. A condenser is used at the low-pressure end of the evaporator sequence to produce the required vacuum. The condenser may be of either the jet or surface type, although surface condensers are more common because they provide warm water for use elsewhere in the mill.

Direct-Contact Evaporation. In order to provide strong black liquor in the solids concentration range of 62 to 65%, additional evaporation is required after the multiple-effect evaporators. Direct contact of the black liquor from the multiple-effect evaporators with the recovery furnace flue gases is one method that has been used to achieve this concentration range. Several types of equipment are used in direct-contact evaporation, including cascade evaporators, cyclone evaporators, and the venturi evaporator-scrubber. The cascade evaporator consists of a series of steel tubes arranged cylindrically in a rotating wheel that is partially immersed in a vat of the black liquor. The flue gas passes through the space above the vat and contacts the liquor adhering to the wheel. A cyclonic evaporator is cylindrical in shape with a tangential entry for the flue gas. Liquor is sprayed into the gas as it enters the vessel, is carried to the walls, and is collected from the bottom in concentrated form, while the gas leaves through the top. In a venturi scrubber, the black liquor is sprayed into a turbulent flow of the flue gas as it passes through a venturi throat.[867] The concentrated liquor is subsequently separated from the gases in a cyclone.

In a typical direct-contact evaporator operation, the hot flue gas enters the scrubber at 300 to 400°C and is cooled to near its saturation temperature at 80 to 90°C. In the process, black liquor is evaporated from the 50% solids range to the 62 to 65% level. The chief disadvantage of direct-contact evaporation is that, unless the black liquor is completely oxidized, the sulfur dioxide in the flue gas releases a substantial quantity of hydrogen sulfide and methyl mercaptan from the black liquor. These released gases constitute a major portion of the total kraft mill odor. This subject will be discussed in more detail later.

To eliminate direct-contact evaporation, evaporation by indirect heating must be extended to the 60 to 65% solids level. To maintain the overall heat economy of such a system, more of the heat in the flue gas must be converted into steam in the boiler and economizer sections of the recovery furnace to supply the additional evaporation steam requirement. The additional evaporation can be supplied either by simply extending the operation of the multiple-effect evaporation unit to the higher solids level[868,869] or, alternatively, by adding an additional evaporation unit, referred to as a concentrator, which uses a separate steam flow[870] for the final evaporation. The former system will have the highest steam economy because the entire evaporation is now multiple effect. The latter approach will reduce the overall economy of the system, but will entail a lower capital investment because less overall heating surface will be required.

Recovery Furnace

The chemical recovery unit receives concentrated black liquor from the evaporators, burns it in a furnace, and discharges it as a molten smelt into a tank where it is dissolved to form green liquor. A substantial amount of heat is generated during the combustion process, and this heat is converted into steam in a boiler following the furnace. Early models of the recovery furnace were batch operations, which burnt the liquor in rotary kilns, producing a black char from which the inorganic chemicals were recovered by leaching. Heat recovery was either nonexistent or, at best, inefficient. The evolution of the modern, efficient alkaline recovery system began about 1930 when G. H. Tomlinson conceived, designed, and installed the first successful continuous furnace for the combustion of alkaline pulping liquors. The unit had water-cooled walls, removed the completely burned smelt continuously from the bottom of the furnace, and generated steam from the combustion gases with high thermal efficiency.[871] Many design improvements have since been made to the unit. A modern furnace can handle, reliably and efficiently, the amount of spent liquor obtained from a pulp mill well in excess of 1000 ton/day of pulp production.

Recovery Furnace Process Description. The liquor from the direct-contact evaporator passes into a small mixing tank where the required salt cake makeup is dissolved in the liquor. It is then heated indirectly to about 125°C and sprayed into the furnace in the form of coarse droplets, which are first evaporated to dryness and then partially incinerated by the hot gas in the furnace. The solid residue accumulates in a char bed on the hearth of the furnace. Air for combustion passes through a heat exchanger, where it is preheated indirectly by flue gas, steam, or hot water, to about 150 to 175°C. It is then divided into two or three streams, which are injected at different levels of the furnace. The primary air is admitted immediately above the char-bed level and burns the carbon and any organic material remaining in the dry mass. The inorganic sodium salts, principally sodium carbonate and sodium sulfide, form a melt, which flows out of the furnace through a spout into a tank where it is dissolved in weak wash liquor from the causticizing area. The gases from the initial combustion are contacted in the upper section of the furnace with secondary and, in some cases, tertiary air to complete the combustion. It is important that the ratio of primary to total air be carefully controlled to maintain a reducing atmosphere in the furnace bed, while maintaining an oxidizing atmosphere in the upper zone of the furnace. In this way, most of the sulfur in the smelt remains in the reduced state as sodium sulfide, while the sulfur and carbonaceous material in the combustion gases are oxidized to sulfur dioxide and carbon dioxide, and consequently release their maximum heat value to the flue gas. It is also important that the total air flow to the furnace be maintained at the optimum level.[872] A slight overall excess of oxygen is required to complete the combustion and avoid reduced sulfur compounds in the flue gas. However, large excesses of air in the exit gases must be avoided because the additional gas volume in the furnace stack increases the heat losses from the system. Furthermore, excess air increases the gas velocity in the furnace, resulting in higher solids carryover into the heat-recovery section.

The hot gases from the combustion zone pass through to the steam generation zone of the furnace, which includes the screen tubes, superheater, boiler, and economizer. A certain quantity of ash is carried with the gas, as a result of both sublimation of sodium salts and mechanical entrainment of fine ash particles in the gas stream.[873] Near the ash fusion temperature (about 750°C), these materials are very sticky in nature and will subsequently adhere to the tube surfaces, reducing their heat exchange capacity. The water-cooled walls of the furnace and the screen tubes are designed to reduce the temperature of this carryover material well below its fusion point before the gases enter the boiler section. The screen tubes and the superheater section which follows them are located directly above the furnace hearth so that material dislodged from them, either by its own weight or by routine soot-blowing, will fall back into the hearth. The main steam generation in the furnace takes place in the boiler section, where the feedwater, which has been preheated in the economizer, is converted into steam by the hot flue gases. The temperature of this steam is increased further as it passes through the superheater tubes. From the boiler, the flue gas passes through the economizer where its heat content is further used to heat the boiler feedwater. In Scandinavian and low odor North-American systems, sufficient economizer surface is used to cool the gas to the range of 120 to 150°C.[874] In older North American systems, gases leave the economizer at about 300°C; further cooling takes place in the direct-contact evaporator. In the latter system, the capital cost of equipment is less, but the system is less efficient in steam generation efficiency and more likely to produce unpleasant odors. The cost of fuel at the particular location will be an important factor in determining the relative economics of the two systems.

Recovery Furnace Heat Balance. The efficiency of the recovery furnace from an energy standpoint can be expressed as the ratio of heat output in useful form (such as steam or electrical power) to the total of the heat inputs to the furnace, including sensible heat and the potential heating value of the black liquor. The efficiency of the furnace in recovering the potential heat for reuse involves both chemical and heat-exchange considerations. Chemically, the combustion must effectively convert the various components of the spent liquor to the desired end products, Na_2S and Na_2CO_3 in the smelt and CO_2, H_2O, and SO_2 in the flue gas. The formation of these materials has been studied from a thermodynamic standpoint[484],[486] and the conditions that favor their formation have been identified. The heat-exchange equipment, through which the gases pass after combustion, is designed to make the most efficient use of the heat content of the gas.

The heat input to the furnace is the sum of the following:[875]

1. The sensible heat in the strong black liquor leaving the storage tank.
2. The heat added in the liquor heater, which follows the direct-contact evaporator.
3. The gross heating value of the black liquor solids going to the furnace.
4. The sensible heat in the combustion air entering the furnace.

Some of this heat is lost to the system as a direct consequence of:

1. The sensible heat in the dry flue gas.
2. The latent and sensible heat in moisture lost from the system, including water originally in the liquor fed to the furnace, water formed from the hydrogen in the liquor solids, and moisture in the air to the furnace.
3. Heat in the smelt.
4. Heat of reduction of the makeup salt cake. [The heat of formation of sodium sulfide is 5000J/g (2118Btu/lb) less than that of sodium sulfate, and this heat absorbed must be supplied by other combustion heat to allow the reduction to proceed.]
5. Radiation losses.

If these various heat losses, plus a small allowance for other unaccounted losses, are deducted from the heat inputs, the difference gives the heat available for steam generation.

The heating value of the black liquor plays a very important part in determining the heat input to the recovery system. This value is normally determined by direct measurement in a bomb calorimeter. However, corrections must be made to the value measured in this way because the products formed in the bomb calorimeter are not the same as those from the furnace. In the calorimeter, all of the sulfur in the liquor is converted to sulfate. In the furnace, on the other hand, most of the sulfur in the smelt is in the sulfide form; in addition, some leaves the furnace in the flue gas as SO_2 and H_2S. To correct the bomb calorimeter value, an elemental analysis of the black liquor is made. The heats of formation of the actual (or assumed) products of the furnace combustion can then be calculated. The heats of formation of the products of the calorimeter combustion can be calculated in a similar manner. The difference in the total heat of formation of the two sets of products is termed the "heat of reaction correction," and it is the value that should be deducted from the calorimeter heat value to correct it for the actual combustion products.[876] The heating value of the liquor depends on its composition but usually varies between

14,000 and 16,000 J/g solids (6000 to 7000 Btu/lb solids); a value of 15,000 J/g solids (6600 Btu/lb solids) is frequently used as an average working figure.

The heat input to the furnace from the black liquor depends on the quantity of solids fired as well as the heat value. The weight of solids produced during pulping varies with the pulp yield, washing efficiency, and chemical charge: for example, linerboard-grade kraft pulp production generates about 1300 kg solids/ton pulp (2600 lbs/ton), while bleachable kraft pulp generates 1500 to 1750 kg solids/ton (3000 to 3500 lb/ton). Gross steam generation in an average kraft mill recovery furnace is in the range of 4500 to 6000 kg/ton pulp (9000 to 12,000 lb/ton pulp).[875] Additional heat can also be recovered in the form of hot water if it can be used elsewhere in the mill.[877] The net heat available from the recovery system after providing for soot blowing, air heating, and black liquor evaporation, can supply a substantial portion of the pulp mill steam requirements. The recovery furnace can also be operated at a sufficiently high temperature and pressure to drive a turbine for the generation of electrical power, supplying a significant portion of the mill requirement.[878]

Particulate Carryover and Removal. Some of the inorganic material in the black liquor enters the gas stream of the furnace instead of leaving with the smelt. This inorganic material is of two types: fused particles, which are carried mechanically from the furnace, and true fume, which consists of sodium salts volatilized in the high-temperature zone of the furnace and which makes up the major part of this material. The amount of these substances in the gas is dependent on the temperature in the combustion zone, the relative positions of the black liquor nozzles and the secondary air ports, and the gas velocity through the furnace. If the combustion temperature is too high, there is excessive volatilization of inorganic material in the gas stream. Overloading the furnace causes higher gas velocities, adding to the mechanical carryover and making particulate removal more difficult. Under normal operating conditions, about 9% of the total sodium charged to the furnace leaves in the gas stream.[879] Fume particles are very small in size, mainly in the submicron range, and are difficult to remove from the gas stream.

The main component of the furnace dust is sodium sulfate (normally above 90%), with much lower contents of other sodium salts, such as sodium sulfide and carbonate. When saltwater-borne wood is used for pulping, the spent liquor contains a substantial amount of sodium chloride,[880] which is also relatively volatile and concentrates in the furnace dust. Where salt buildup in the liquor system is a problem, removal of chlorides from the particulate carryover before returning it to the liquor cycle has been suggested as a means of control.[881]

Solids carried in the gas leaving the furnace should be removed from the flue gas prior to its discharge to the atmosphere. Two methods are used to remove this material: wet scrubbing and electrostatic precipitation. The venturi scrubber can serve both as a direct-contact evaporator and as a solids-removal device.[867] The high turbulence in the venturi throat brings the particles into contact with the liquor droplets, which coalesce and are removed from the gas stream in the cyclone separator. The efficiency of fume removal depends on the pressure-drop in the venturi. Scrubbing efficiencies in the range of 90 to 95% can be achieved when the pressure-drop across the venturi is 7.5 kPa (30 in. of water). The electrostatic precipitator[882] consists of wire electrodes placed in the gas stream after the direct-contact evaporator. The particulate material in the gas stream becomes electrically charged as it passes between the electrodes and is then deposited on a collecting electrode of opposite charge. The particles are removed from this latter electrode by vibration and fall to the bottom of the unit where they are removed by either a dry conveying system (dry bottom) or with strong black liquor from the multiple-effect evaporators (wet bottom). The efficiency of these units is normally between 95 and 99%, depending on the design of the unit. There would be no need to install facilities for higher than 96% efficiency if economics only were to be considered. However, environmental constraints require that the particulate discharge to the atmosphere be maintained at very low levels; the newer precipitator installations are capable of achieving 98 to 99% particulate-removal efficiency.[883]

Accumulation of solids on the heating surface reduces the efficiency of heat absorption and increases the resistance to gas flow in the system. In order to maintain these surfaces clean at all times, it is necessary to perform regular cleaning routines while the furnace is in operation. At one time, a considerable amount of manual lancing was required to carry out these functions, but in the modern furnace, this cleaning is performed automatically by retractable soot-blowers employing superheated steam. In the lower-temperature zones, falling shot pellets have also been used occasionally to loosen adhering deposits from heating surfaces.

Recovery Furnace Safety. The conventional recovery furnace has performed efficiently and safely for many years in the vast majority of installations in spite of the high operating temperatures and pressures frequently used and the highly reactive chemicals handled during its operation. However, there have been a number of explosions during the operation of recovery furnaces, and because of the resulting risk of injuries to personnel, property damage, and loss of profit, these incidents have been a cause of major concern to pulp and paper mill operating and supervisory personnel, furnace manufacturers, and insurance company investigators. A committee has been formed from all interested groups (Black Liquor Recovery Boiler Advisory Committee—BLRBAC) to investigate the cause of these explosions, identify safe operating practices, and recommend procedures to be followed during emergency situations. The work of this and other industry groups has made considerable progress in reducing the hazards of recovery boiler operation. This committee has issued a number of publications on the subject, which are available to the general public.[884] The committee identified the occurrence of 92 recovery boiler explosions and found that 62 were caused by interaction between smelt and water; most of the remainder were caused by problems related to the firing of auxiliary fuel.

The smelt-water interaction is entirely physical in nature and it does not involve chemical reaction. The explosive effects are produced by the volume change during the rapid vaporization of water to steam. Water can enter the smelt bed from several sources. Firing of black liquor that is too weak, tube failures, and accidental entry of water into the furnace by other means have all been identified as causes of smelt-water explosions. Auxiliary fuel firing can lead to problems if the fuel is improperly ignited or if there is an inadequate air supply to ensure complete combustion. Emergency procedures to deal with these hazardous situations include immediately shutting off all fuel to the furnace and rapidly draining all water from the system.

Because of the safety measures required with the conventional kraft black liquor recovery furnace and because of its high capital cost and undesirable environmental effects, alternative methods of recovery have been investigated.[885] The methods suggested employ mainly known concepts, such as wet and dry pyrolysis and fluid-bed combustion, but considerable development work remains to be done in all cases before these can be considered as practical alternatives to conventional kraft recovery. These processes either operate without smelting of the sodium salts (wet pyrolysis, fluid-bed combustion) or they separate the smelting process from the heat generation operation so that the chances of contact between smelt and water are materially reduced.

Recausticizing

After the smelt from the recovery furnace has been dissolved, the sodium carbonate in the resultant solution (green liquor) must be recausticized to sodium hydroxide before it can be reused in the digestion process.[886] This conversion is achieved by treating the green liquor with calcium hydroxide (slaked lime):

$$Na_2CO_3 + Ca(OH)_2 \rightarrow 2NaOH + CaCO_3$$

The calcium carbonate produced in this conversion is separated, washed, and burnt to regenerate active lime for reuse in the recausticizing reaction. The sodium sulfate and other minor constituents of the green liquor do not play an active part in these reactions and carry through this operation without change. A generalized flow diagram of the recausticizing operation is shown in Figure 4-108.

Green Liquor Clarification. The first step in the recausticizing process is the removal of the insoluble material from the green liquor. The flow from the furnace spout contains impurities consisting mainly of heavy metal compounds and fine carbon particles, as well as silicates and aluminates from the furnace refractory lining. These materials do not dissolve during the smelt dissolution process, and are referred to as the green liquor dregs. These dregs are bulky and gelatinous in nature and can be removed most effectively by sedimentation in a clarifier. The filtrate from the lime-mud washing, referred to as weak wash (Figure 4-108), is used to dissolve the smelt from the furnace. This weak wash

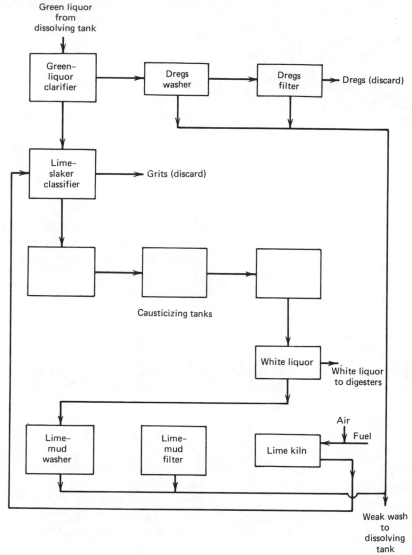

Green liquor
from
dissolving tank

Green-
liquor
clarifier

Dregs
washer

Dregs
filter

Dregs (discard)

Lime-
slaker
classifier

Grits (discard)

Causticizing tanks

White liquor

White liquor
to digesters

Lime-
mud
washer

Lime-
mud
filter

Lime kiln

Air

Fuel

Weak wash
to
dissolving
tank

Figure 4–108. Schematic flow diagram of the recausticizing system.

usually contains calcium carbonate particles, and these particles are separated with the dregs in the green liquor clarifier. In a modern mill with well-chosen materials of construction, the quantity of dregs is only about 5 kg/ton of pulp. However, if these dregs are not removed prior to the causticizing reaction, they retard the subsequent lime-mud settling and washing stages.[887,888] They also accumulate in the lime recovery circuit, reducing the reactivity of the reburned lime. Since the dregs do not settle to a very high-solids content (usually about

10% solids) the underflow from the green liquor clarifier contains a considerable amount of dissolved soda that must be recovered. The dregs are, therefore, washed with water and rethickened before being discarded. The washings are either used to wash the lime mud or added directly to the weak liquor storage tank.

Slaking, Causticizing, and Clarification. The clarified green liquor is pumped to a slaker where it is mixed with the quicklime (CaO) from the lime kiln plus the required amount of fresh makeup lime. The calcium oxide is slaked to the hydroxide by reaction with the water in the green liquor:

$$CaO + H_2O \rightarrow Ca(OH)_2$$

Usually the slaker is built in two compartments: a mixing section, where the green liquor contacts and slakes the lime, and a classifying section, in which the slaking is completed and gritty material is separated by rakes and discharged from the system. The major part of the slaking reaction actually occurs in the mixing section. There is no sharp division between the slaking and causticizing processes since as soon as some of the lime is hydrated, the causticizing reaction begins. Proper operation of the slaker is important because its performance has a significant bearing on both the causticizing efficiency and the settling properites of the final lime mud. The two main operating variables that require attention at this stage are the temperature in the slaker and ratio of lime to green liquor fed to the unit.

It is important that the slaking temperature be neither too high nor too low. The slaking reaction is highly exothermic and if too much heat enters the system with the liquor and the lime, the heat generated will cause the liquor in the slaker to reach its boiling point of about 110°C, and will result in a boilover. On the other hand, if the temperature is too low in the slaker, the lime does not slake completely. The causticizing efficiency is then reduced, and the quality of the lime mud also deteriorates. As a result, the subsequent clarification stage is adversely affected. Temperature can be maintained in the desired range in this system by controlling the temperature of the incoming green liquor and by keeping the flow of liquor and lime in the required proportion at all times. In order to provide time for the causticizing reaction to reach equilibrium, the process is completed in suitably equipped causticizing tanks. These tanks are agitated to keep the solids in suspension and usually provide about 90 min detention. Since the tanks must be agitated, three tanks are commonly used in series to reduce short circuiting and provide more uniform retention. A combined slaker-multitray causticizer has been used to simplify the equipment and reduce the space requirement in the causticizing plant.[889]

The ratio of lime to sodium carbonate in the green liquor has a significant effect on both causticizing efficiency and settling characteristics of the calcium carbonate produced. The addition of a stoichiometric amount of lime will give a causticity of about 80 to 82% in a commerical causticizing system. When excess amounts of lime are added, there is a noticeable decrease in the settling rate of the lime mud with no compensating gain in causticizing efficiency. A causticizing efficiency of 80%, or slightly higher, is usually considered satisfactory. The relationship between the proportion of lime added and the causticizing efficiency is shown in Figure 4–109.[890]

After the liquor has passed through the causticizers, the suspended solids must be removed to produce a clear white liquor for use in cooking. This is usually achieved in a sedimentation-type clarifier.[891] Belt filters[892] and centrifuges [893] have also been suggested, but the clarifier-type separator is still the most commonly used by a wide margin. The more modern design is the unit clarifier in which the mud-settling and liquor-storage functions are combined with the clarification taking place in the lower portion of the tank and the storage in the upper portion.[894] To ensure that the clarifier runs efficiently, the precipitation of the calcium carbonate in the slaker and causticizer must be under conditions that produce an easy settling, compact lime cake.

The effect of different variables on the causticizing equilibrium has been studied in the laboratory by Kinzner,[895] using the degree of causticizing, the settling rate of the mud, and its settled volume as the criteria for evaluation. Increasing the reaction time increased the degree of causticizing (to a maximum of 86% in about 1-1/2 hr), but the settling rate was decreased. Increasing the temperature also increased the degree of causticizing, but again decreased the settling rate and increased the settled volume. This effect of temperature on mud quality is contrary to the usual experience from mill operation.[896] If an overdose of lime is fed, there is a slight increase in the degree of causticizing but a substantial reduction in the quality of the mud. Lime-mud quality is best when the full liquor flow is used in the slaker; it is adversely affected when only partial flows are used (i.e., when some liquor bypasses the slaker). Both the causticizing efficiency and settling properties of the mud are improved if the concentration of the green liquor is decreased. This is generally not prac-

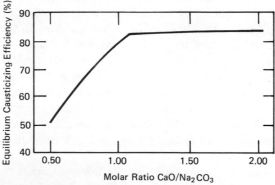

Figure 4-109. Equilibrium causticizing efficiency at varying ratios of CaO to green liquor.

ticable because the use of a weaker green liquor increases the volume of liquor that must be processed by the equipment in the recausticizing area and adds to the subsequent black liquor evaporator load. However, excessively strong green liquor should be avoided because of the adverse effect that the resultant higher liquor viscosities would have on the causticizing system, settling rates, and the increased chance of boilovers at the slaker.

Mud Washing. The lime mud from the clarifier must be washed to recover sodium compounds dissolved in the entrained liquor. If the soda content of the lime sent to the kiln for reburning is too high, balls and rings will form during calcining and interfere with the operation of the unit. The mud-washing is usually carried out in a sedimentation washer. The mud-washings are combined with those from the dregs washer to form the weak wash solution, which is utilized for dissolving the smelt from the furnace. The permitted washwater volume is limited since the total volume of water used to wash the lime mud and dregs establishes the upper limit of the white liquor concentration. To conserve water, the mud-washing is usually carried out in two stages. The efficiency of this type of washing is dependent on the amount of liquor entrained in the solids in the washer underflow; mud that will settle to a high solids content will give higher washing efficiency. Good quality lime mud will normally settle to a solids content of 40 to 43%. Final washing and dewatering of the lime mud takes place on the mud filter, which may be either a standard vacuum filter or a precoat filter, the latter type giving a slightly higher solids level. Solid-bowl centrifuges are also used occasionally for final thickening. Good operation will give a final mud of 60 to 70% solids content for feeding to the lime burning operation.

Lime Burning. The incineration of calcium carbonate to generate lime is one of the oldest chemical processes known. The burning of the lime mud generated in a kraft mill recovery cycle presents some special problems, however, because of the wet feed, its residual sodium content, and the need to produce a lime that will exhibit high reactivity in the causticizing reaction and good settling properties in the subsequent clarification. Lime is most frequently burned in a rotary kiln although fluid bed calciners are used in a few installations. The rotary lime kiln is a long, cylindrical, refractory-lined vessel, sloping from feed to discharge end and rotating at about 1 rpm. Fuel, usually oil or natural gas, is burned to maintain the kiln temperature between 1100 and 1250°C. Heat is required both to evaporate the residual water in the feed and to provide the heat of reaction for the dissociation:

$$CaCO_3 \rightarrow CaO + CO_2 \; ; \; \Delta H = 43,500 \text{ cal/mole}$$

Chains suspended in the feed end of the kiln dip into and out of the solid mass and assist in the heat transfer by exposing a much greater surface area of the lime to the hot gases. With the use of these devices, the gases leave the kiln at the feed-end drying stage at about 200°C. Under these conditions, heat consumption in the kiln is 8 to 10 $\times$ 10^6 kJ/ton (7 to 9 $\times$ 10^6 Btu/ton) of lime. The gases leaving the kiln are normally passed through a scrubber to remove entrained lime dust.[897] In the fluidized-bed calciner,[898] the wet mud is mixed with hot flue gas in a cage mill where it is dried and disintegrated to a fine powder. This material is then separated from the gas in a cyclone and fed to the calciner, which is a two compartment unit. Dried mud and fuel are fed to the upper bed where calcination of the carbonate takes place at about 900°C. The fine particles agglomerate to pellets ranging from 6 to 65-mesh in size. The pellets formed in excess of those required to maintain the upper bed at the desired level are removed to the lower bed, where they are cooled by the incoming combustion air. The pellet size is controlled in the desired range by maintaining the residual sodium content of the feed in the range of 0.2 to 0.4% on a dry basis.[896] The fluid-bed calciner offers the advantage of reduced emissions, low fuel requirement, high quality product, low maintenance, high flexibility, and minimum space requirement. However, the system is more costly to install and its operation requires more electrical power to operate than a conventional rotary kiln.

It is important in the operation of a causticizing system that lime of high quality be produced for the subsequent slaking and causticizing stages, and this quality is largely determined by the operating conditions in the lime reburning system. The degree of burning can be determined by measuring the CO_2 generated on acidification of the product.[899] Lime with optimum properties for causticizing is obtained when the product from the kiln is not completely burned, but still contains a small amount of residual calcium carbonate. The optimum residual calcium carbonate level varies with the particular conditions for individual systems, but should normally be less than 5%.[899,900] Calcining to an excessive extent produces lime with poor settling characteristics.[900,901] Furthermore, too rapid a rise to temperature as well as excessive temperature and time during burning produce lime with low porosity and reduced reactivity in the slaking process.[902,903]

Kraft Odor Emissions

The public concern over the obnoxious odors produced during kraft pulping has grown considerably. Although progress has been made in reducing this problem, the goal of the completely odor-free kraft mill has not become a reality. Several reviews and bibliographies on the subject have been published.[904-907] They outline both the nature and the scope of the problem and review the steps that have been taken to alleviate it.

Sources and Causes of Kraft Mill Odors

The main four constituents of kraft mill emissions that contribute to odor emission are hydrogen sulfide (H_2S), methyl mercaptan (CH_3SH), dimethyl sulfide (CH_3SCH_3), and dimethyl disulfide (CH_3SSCH_3). Of these, hydrogen sulfide and methyl mercaptan are usually present in the highest proportions.

The odor thresholds of both hydrogen sulfide and methyl mercaptan are extremely low, on the order of one part per billion.[908,909] Obviously, it would be very difficult to operate a kraft mill under such close control that its gaseous emissions never exceeded this level. Nevertheless, steps can be taken in the mill to reduce the amount of these materials produced and to eliminate them as much as possible from the discharge streams. Although kraft mill emissions cause unpleasant odors, they do not have other harmful effects at the concentration levels encountered in the atmosphere in and around mills. Hydrogen sulfide is lethal at high concentrations, but neither it nor methyl mercaptan have any known deleterious effect on either animal or plant life in the parts-per-billion concentration range.

Hydrogen sulfide is always present to some extent in the gas phase over sulfide solutions; the concentration is governed by the following equilibrium:

$$H_2 S(g) \rightleftharpoons H_2 S \text{ (solution)} \rightleftharpoons HS^- + H^+ \rightleftharpoons S^= + 2H^+$$

This relationship predicts that the concentration of hydrogen sulfide in the gas phase over sulfide solutions will increase appreciably as the pH decreases.[910] The empirical data for $H_2 S$ vapor pressures over 0.01M sulfide solutions shown in Figure 4-110[904] illustrates this relationship. This figure also shows the effect

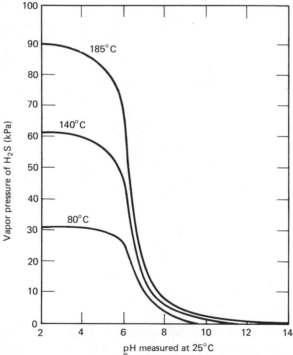

Figure 4-110. Vapor pressure of 0.01M $H_2 S$ versus pH at 25°C at various temperatures.

of temperature, which, although not as great as that of pH, still has a significan
influence on the hydrogen sulfide vapor pressure. Although these data sugges
that hydrogen sulfide vapor pressures are very low in the range of 11 to 12 pH
which is usually encountered in spent liquor, these very low vapor pressures are
still significant from an odor standpoint at the higher concentrations of sulfide
used in alkaline pulping.

Methyl mercaptan also dissociates in alkaline solution and consequently its
volatility also decreases with increasing pH. It is a much weaker acid than hy
drogen sulfide, however, and thus retains much higher vapor pressures at ele
vated pH levels. The effect of pH and temperature on the vapor pressure of
methyl mercaptan is shown in Figure 4-111.[904] In the range of 10 to 12 pH
encountered in black liquor systems, there is a substantial vapor pressure of
methyl mercaptan, particularly in the higher temperature range. As a result
significant quantities of methyl mercaptan can be lost to the atmosphere even
at very high pH levels. Methyl sulfide and dimethyl sulfide do not form sodium
salts; their volatilities are therefore not dependent on the pH.

Methyl mercaptan and dimethyl sulfide are formed during pulping as a result
of bimolecular, nucleophilic, substitution reactions with the methoxyl groups
of lignin as shown below:[911]

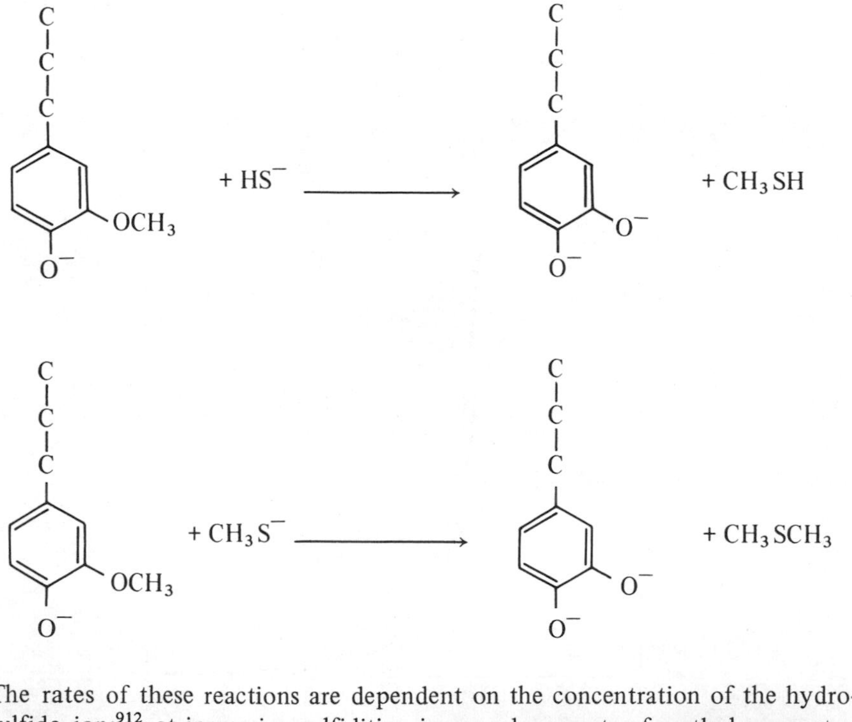

The rates of these reactions are dependent on the concentration of the hydrosulfide ion;[912] at increasing sulfidities, increased amounts of methyl mercaptan
and dimethyl sulfide are formed. For softwoods, the combined yield of these
two products increases in an essentially linear manner with time. The mercaptan

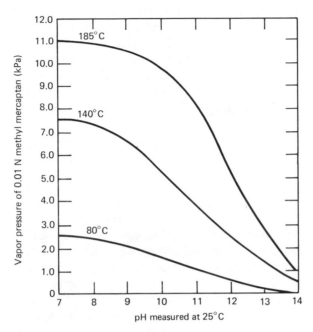

Figure 4-111. Vapor pressure over a 0.01N solution of methyl mercaptan as a function of pH at different temperatures.

forms more rapidly at first; dimethyl sulfide only begins to form when the mercaptan concentration reaches a significant level. Even at the end of the cook, the concentration of mercaptan is substantially higher than that of dimethyl sulfide. Hardwoods generate both methyl mercaptan and dimethyl sulfide at a more rapid rate than softwoods, consequently the odor level is particularly high when hardwood is used in the kraft process.[913] The activation energy for formation of methyl mercaptan and dimethyl sulfide is 22.5 and 19.0 kcal/mole respectively.[912] Since this is substantially less than that of kraft delignification (32 kcal/mole), an increase in cooking temperature will decrease the amount of these two contaminants formed at a constant level of delignification.

Dimethyl disulfide is the least odorous and the least volatile of these four odorous materials. It is formed in side reactions by oxidation of methyl mercaptan:

$$4CH_3SH + O_2 \rightarrow 2CH_3S\text{-}S\text{-}CH_3 + 2H_2O$$

Even though precautions are taken to keep the concentration of these four contaminants as low as possible, they are always present to some extent in spent liquors. A considerable portion of the sulfide charged at the beginning of the cook remains in the spent liquor, and becomes a potential source of hydrogen sulfide emission. Formation of methyl mercaptan and dimethyl sulfide is unavoidable during kraft cooking. The vapors from kraft spent liquors must, therefore, be kept contained and these contaminants removed before the gases are discharged if odor levels are to be kept to a minimum. The steps that can be taken to control and eliminate these materials in the different departments of the pulp mill will be discussed in the following sections.

Digester Relief and Blow

In batch digestion, some relief is necessary during cooking to remove air (entrained with the chips) and liquor during digester charging as well as to release other volatile materials formed during the cook. With many wood species valuable terpenes from the wood are included with this relief, and these materials are condensed and removed. Some of the above sulfur compounds condense and pass into the aqueous phase (underflow) from the turpentine separation. The remainder of the sulfur compounds are discharged with the noncondensable gases, which also contain air, methanol, water vapor, carbon dioxide, and low-molecular-weight terpenes. During the blowing of batch digesters, similar volatile contaminants are released. Because of the cyclic nature of batch digestion, large volumes are released in short time spans, making the problem of contaminant removal particularly difficult. Compared to relief gases, blow gases contain much less turpentine and are essentially free of noncondensables, since these materials have already been removed with the relief during cooking. The blow gases are first cooled with direct or indirect heat exchange to condense most of the steam. To deal with the surges in flow encountered in batch-blow systems, a vaporsphere or gas accumulator is often used to collect the gases.[914] These units usually have a floating diaphragm that moves up and down as the surges of gas flow enter the vessel. A constant flow, about equal to the average gas flow rate over the full digester cycle, is withdrawn from the vaporsphere and fed to the foul-gas disposal system.[915]

In continuous digesters, a much smaller volume of gas is released than in batch digesters, and these gases flow from the continuous system at a relatively steady rate. The highly efficient presteaming systems, which precede most continuous digesters, remove most of the noncondensable gases from the chips before they enter the digestion zone. The Kamyr digester is liquid-filled and operates at a substantial hydrostatic pressure so that little or no gas relief is necessary. Before the pulp and black liquor leave the digester through the blow line, they are cooled to below the flash point of the liquor and the volume of gas released is consequently relatively small. The combined black liquor from the digestion and countercurrent wash zones leave the digester at elevated temperature through a two-stage flash system. The vapors formed during this flashing contain both steam and foul gases. The vapor from the first of these flash tanks is used for presteaming. The gas vented from the presteaming vessel is combined with that formed in the second digester flash tank and this combined stream is fed to a heat exchanger to produce clean hot water. The gas discharged from this heat exchanger contains most of the odorous substances released in the digestion system. The other significant source is the gas released by the vents from the low pressure valve, which feeds chips to the presteamer. A modification to the flash-tank system, which would generate the steam for presteaming indirectly from the flash steam, has been proposed.[916] In this sytem, the total gas release is limited to a single point of emission. Thus, although the continuous digester does not eliminate odor problems, its design simplifies the steps required to collect the undesirable emissions for disposal in subsequent equipment.

Treatment of Digester and Evaporator Condensates

Condensates from the relief and blow system, the multiple-effect evaporators and the turpentine-decanter aqueous phase in mills collecting turpentine contain substantial amounts of hydrogen sulfide and methyl mercaptan with smaller amounts of dimethyl sulfide.[917] Hydrogen sulfide is present to a much lesser extent in the evaporator condensates when the weak black liquor is oxidized, as discussed in the next section. When the liquor is not oxidized, additional methyl mercaptan is formed in the high-temperature evaporation effects by reaction between sulfide and the methoxyl groups in lignin,[918] leading to increased concentration of this compound in the condensate. A portion of the odorous constituents of these condensates is released to the atmosphere when they are discharged, causing unpleasant odors in and around the mill. The sulfur-containing materials are highly toxic and the portion remaining in the condensate constitutes a significant fraction of the toxicity of the mill effluent. Both air and steam stripping have been used to eliminate these undesirable materials from the condensates. Air stripping is less costly and simpler to operate,[919] but it is less flexible when condensates of variable loading must be handled.[920,921] The stripped gas is disposed by further treatment described below, while the clean condensate constitutes an excellent source of hot water for pulp washing and other process uses.

Several methods have been proposed to treat the gases released during the digester relief and blow, from condensate stripping, and from other sources, once the condensable vapors, such as turpentine and water, have been removed. These methods include absorption in white liquor, scrubbing and oxidation in the black liquor oxidation tower, scrubbing with chlorine, and incineration. The latter two are the ones usually employed in commercial practice. Chlorine converts methyl mercaptan and dimethyl sulfide into methanesulfonyl chloride (CH_3SO_2Cl), methanesulfonic acid (CH_3SO_3H), and other oxidized derivatives. Although some of these materials still have an unpleasant odor, they are relatively nonvolatile,[922] and chlorination in the gas phase is effective in reducing the odor from kraft mill relief-gas streams. Either purchased chlorine[923] or effluent from the chlorination stage[924] in the bleach plant can be used as the source of chlorine for this system. Bleach liquor is less costly, but when it is used, the reaction must take place in the liquid phase. The odorous substances are not very soluble in the spent bleach liquor at the low pH required for the reaction between chlorine and the sulfur compounds, and the conversion is consequently less efficient.[923] The most widely used method of disposing of the odorous gases is to burn them so that their sulfur content is converted to SO_2. Incineration may take place in the lime kiln,[915,925,926] in the recovery furnace,[927] or in a separate incinerator.[920] The lime kiln is most commonly employed, because there is less explosion hazard than in the recovery furnace, and the gas replaces purchased lime-kiln fuel, reducing operating cost. Burning in the recovery furnace may also recover heat values, but in many cases furnaces are already operating at their maximum steam-generating capacity. Regardless of whether combustion takes place in the kiln or the furnace, care must be taken to dilute sufficiently with air to avoid an explosive mixture of gas and air.[922,928] The gas in the vaporsphere should be protected by installation of a flame arrester in the line to the lime kiln or other incineration equipment.

Black Liquor Oxidation

The residual sulfide in black liquor can be oxidized with air so that the quantity of hydrogen sulfide released during evaporation is reduced to a minimum. The oxidation of sulfide proceeds through a complex series of reactions, involving the intermediate formation of unstable polysulfides, and producing sodium thiosulfate as the stable end product.[929] The overall reaction can be summarized in the following simplified form:

$$2Na_2S + 2O_2 + H_2O \rightarrow Na_2S_2O_3 + 2NaOH$$

Thus, in addition to converting sodium sulfide to a potentially less harmful form, the oxidation increases the alkalinity of the black liquor, helping to retain in solution other acidic materials that might be released, such as methyl mercaptan. Methyl mercaptan can be oxidized to dimethyl disulfide by air, according to the reaction shown in the section on kraft odors. This reaction occurs to a limited extent in black liquor oxidation systems since there is some dimethyl disulfide in the air leaving the oxidation tower.[930] However, the reaction conditions in the oxidation tower do not appear to be favorable for this reaction because most of the methyl mercaptan is not oxidized, but remains in the form of its sodium salt in the oxidized liquor. Most mills that have weak black liquor oxidation systems observe some reversion to sulfide during evaporation and storage, the greatest reversion usually occurring when the oxidation takes place at low temperature.[931] If the liquor temperature is not at least 60°C during the oxidation reaction, some elemental sulfur is formed; this sulfur can later form sulfide during the storage and evaporation of the spent liquor:[932]

$$4S + 6NaOH \rightarrow 2Na_2S + Na_2S_2O_3 + 3H_2O$$

Reversion also takes place when the oxidation is carried out using times and temperatures that completely oxidize the sulfide,[933] although to a much more limited extent. It is suggested that this reversion is due to a disproportionation reaction of methyl mercaptan:

$$2CH_3 SNa \rightleftharpoons CH_3 SCH_3 + Na_2 S$$

The main requirement in the design of an oxidation system is to allow sufficient contact time between air and liquor for the required amount of oxygen to dissolve and react.[934] The temperature must be maintained above 60°C throughout this reaction period to minimize subsequent reversion to sulfide. Although an excess of air should be used, too large an excess is undesirable because it causes an unnecessary loss of heat, and can result in the stripping of odorous organic sulfur compounds, such as methyl mercaptan and dimethyl sulfide. Several designs have been employed for weak black liquor oxidation. The B. C. Research Council design[929] is a packed tower using a corrugated asbestos-cement packing plate. The pressure drop through the tower is very low, minimizing power costs. The Trobeck and Collins systems use bubble-cap towers, which have similar capital costs to the packed tower, but a higher power requirement. A long contact time between liquor and air is ensured with this system because it creates a foam; however, the foam must subsequently be broken mechanically.

In mills where liquor foaming makes weak black liquor oxidation impractical, the oxidation can be performed on strong black liquor from the multiple-effect evaporators, which has less tendency to foam.[935] In this case, the benefit of reduced sulfide loss during multiple-effect evaporation is lost and the evaporator condensates therefore require stripping before release or reuse. However, strong black liquor oxidation does reduce sulfide emission to the flue gas in the direct-contact evaporator. In one strong black liquor oxidation system, air is added through a series of sparger nozzles in an oxidizer tank. The efficiency of the oxidation is higher when two separate oxidizing stages are used.[936] An alternative system employs an in-line oxidizer in which finely dispersed air is mixed with the liquor in a baffled section of pipe fitted with high-speed agitators.[937] The optimum air flow for a strong liquor oxidation system is about 3.5 times the theoretical amount to oxidize all the sulfide. Air-liquor contact is the most critical parameter in ensuring that oxidation is complete, and it is, therefore, important that either a relatively long retention time or a high degree of agitation be used.[938] A frequently employed dual oxidation treatment is one in which the main oxidation takes place on the weak liquor and a final treatment is carried out on the strong black liquor to reoxidize any sulfur formed through reversion reactions.[939]

Another means of oxidation, which can be applied to particularly foamy liquors, is to use oxygen instead of air.[940] When suitable conditions are used, about 95% of the sulfide can be oxidized with almost complete consumption of the oxygen, minimizing the opportunity for foam formation. The oxidation equipment can be quite simple in this case, usually an in-line reactor. By running the system at 95°C and by allowing the remaining dissolved oxygen to react in the storage tanks, reversion to sulfide can be kept to a minimum.

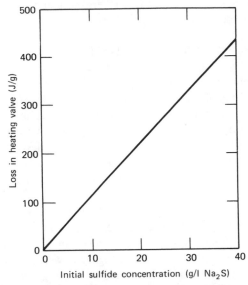

Figure 4-112. Decrease in heat value of black liquor during oxidation to 0.2 g/l of residual Na_2S.

In assessing the net benefit to be gained from black liquor oxidation, the energy balance must be considered as well as the effect on odor. Some evaporation occurs as a result of the saturation of the air passing through a black liquor oxidation system, but since this evaporation uses sensible heat already in the liquor, it does not result in any net energy gain or loss. Since this evaporation is only single effect, the heat is not used in an efficient manner.[941] The increase in temperature of the air as it passes through the tower removes heat from the liquor and is a loss to the system. Heat is also generated in the oxidation since the reaction is exothermic to the extent of 5.40×10^6 J/kg (2320 Btu/lb) of sodium sulfide oxidized.[942] On the other hand, the thiosulfate produced must be reduced to the sulfide form in the furnace; the heating value of the liquor to the furnace is consequently decreased. The extent of this reduction in heat value for two different liquors at various levels of sodium sulfide concentration is shown in Figure 4-112. At 30 g/l sodium sulfide concentration, it has been estimated that the net loss in the heating value of the black liquor due to oxidation is 3.26×10^5 J/kg (140Btu/lb) of original solids, equivalent to about 2% of the original heating value of the liquor.[942]

Odor Control during Direct-Contact Evaporation and Recovery

The flue gas discharged from the kraft recovery boiler is one of the main sources of odor from the kraft mill system. The odorous materials arise from two main sources: incomplete oxidation during combustion (often due to overloading of the furnace) and to the release of hydrogen sulfide, methyl mercaptan, and dimethyl sulfide to the flue gas in the direct-contact evaporator.

Furnace Operation. There is considerable evidence to show that the flue gas entering the direct-contact evaporator from a kraft recovery furnace, operating with an adequate air supply and with good turbulence in the upper oxidizing zone, will contain only a negligible amount of malodorous sulfur compounds.[934] The total air flow to the furnace should be sufficient to provide about 3% of oxygen in the flue gas. When the weight ratio of total air to solids in a recovery furnace falls below a value of about 4.5, there is a marked increase in the total reduced sulfur (TRS) compounds emitted by the furnace.[904] However, the distribution of air is important as well as the total flow (discussed in the section on the recovery furnace). The primary air flow mainly affects the level of reduction in the smelt, but within the normal operating range it has little effect on the hydrogen sulfide emission from the furnace. On the other hand, when the weight ratio of the secondary air flow to black liquor solids fired to the furnace falls below 3.1, the hydrogen sulfide emission level increases rapidly; at secondary air ratios above this level only insignificant amounts of hydrogen sulfide are released, as shown in Figure 4-113.[943] Measurement of the levels of the various flue-gas components under different furnace operating conditions has shown that there is a relationship between the amounts of hydrogen sulfide and carbon monoxide generated and that no hydrogen sulfide is present if the carbon monoxide level is less than 250 ppm.[944] Monitoring of the carbon monoxide level has therefore been suggested as a means of controlling furnace emissions. It is also important,

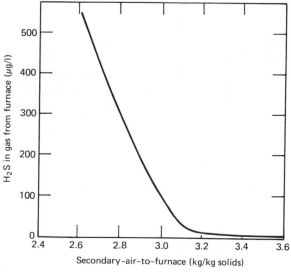

Figure 4-113. Relationship between secondary air flow to the furnace and hydrogen sulfide emission level.

however, that the oxygen levels not be allowed to rise too high because excessive oxygen in the flue gas leads to SO_3 formation; a sticky ash will then be produced, increasing fouling of tubes in the boiler and economizer.

Much of the gaseous sulfur emitted from the furnace is generated during the pyrolysis of the dried solids in the lower zone of the furnace;[945,946] this generation is particularly high when the pyrolysis occurs at low temperature in "cold spots" in the furnace, for example near the liquor sprays.[879] The most frequent cause of these low-temperature zones in the furnace is a low solids concentration in the fired black liquor.

Even when the air supply to the furnace is adjusted to minimize generation of reduced sulfur compounds, the flue gas will contain sulfur dioxide. Although this material has a much less objectionable odor than the reduced sulfur compounds, it is also subject to control regulations for a variety of environmental reasons. Hydrogen sulfide generated in the lower bed zone of the furnace is oxidized to sulfur dioxide in the oxygen-rich upper portion of the furnace. Conditions that minimize the generation of hydrogen sulfide in the bed will also reduce the SO_2 emission from the furnace.[947] These conditions include:

1. High solids content of the black liquor.
2. High temperature of the combustion air.
3. Low moisture content in the combustion air.

Sodium compounds in the fume produced by the high-temperature combustion absorb SO_2 from the gas, thereby reducing the emission level. Addition before firing of makeup chemical that does not contain sulfur (such as caustic or soda ash) increases the sodium available for this absorption and may be a means of reducing furnace SO_2 emission.

Direct-Contact Evaporation. The other main cause of odor in the flue gas is the release of hydrogen sulfide and methyl mercaptan during the direct contact evaporation. The main factors determining the hydrogen sulfide content of the gas leaving the direct-contact evaporator are the hydrogen sulfide content of the inlet gas and the pH and sodium sulfide content of the black liquor entering the unit.[948] At pH 12.5 hydrogen sulfide can be either absorbed or given off in the direct-contact evaporator, depending on the sodium sulfide content of the black liquor. Figure 4–114 shows the equilibrium inlet and outlet (exit) levels of hydrogen sulfide at pH 12.5 for different sodium sulfide concentrations. The dotted line indicates the conditions where inlet and outlet concentrations are equal (i.e., where neither absorption nor emission of hydrogen sulfide occurs). When the operating conditions are above this line, emission of hydrogen sulfide will occur, while below it absorption takes place. In a normal system, conditions that favor absorption will prevail only when the black liquor has been efficiently oxidized or when the inlet gas has a very high hydrogen sulfide content. Similar data at pH 11.5 indicate that H_2S emission occurs at all significant sodium sulfide concentrations, demonstrating the importance of maintaining a sufficiently

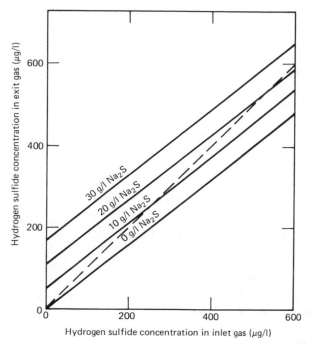

Figure 4–114. Hydrogen sulfide content of the inlet and outlet gases of a direct-contact evaporator at pH 12.5 and various sodium sulfide concentrations calculated from equilibrium data.

high pH level in the black liquor. In view of these results, it is evident that the direct-contact evaporator can either improve or degrade the quality of the flue gas from a recovery furnace. When unoxidized black liquor is used, hydrogen sulfide is emitted under almost all conditions. On the other hand, with fully oxidized black liquor and a sufficiently high pH level, the liquor will absorb hydrogen sulfide in operations where there is a high level of H_2S leaving the furnace. However, a black liquor oxidation efficiency of about 99% must be maintained to achieve a substantial benefit from this absorption.[949] The direct-contact evaporator is effective in removing sulfur dioxide from flue gases under most conditions, provided there is adequate contact between gas and liquor.

Low-Odor Recovery Systems. When a new recovery boiler is being installed, two options exist to minimize odor from this system: efficient black liquor oxidation or elimination of the direct-contact evaporator. If the latter option is selected, alternative means are then required for completion of the evaporation and for extraction of heat from the flue gas. Extension of multiple-effect evaporation to over 60% solids and the use of forced-circulation concentrators for the final evaporation have already been discussed. These systems are in use to evaporate liquor for combustion in commercial furnaces.[950] Additional economizer surface is used to cool the gases to the desired outlet temperature of 175 to 200°C. Some of the heat generated in this manner can be used in a hot-water or steam-coil preheater for the combustion air.[951] The air-contact evaporator is an alternative system for reclamation of flue-gas heat in which the flue gases are used to heat air in an indirect heater. This air is then used for evaporation in a Cascade evaporator. The exit air from the Cascade unit is used as supply air for the recovery furnace.[952] These systems, generally referred to as low-odor recovery systems, eliminate the need for black liquor oxidation and thus give a slightly more favorable heat balance in the recovery system, as explained earlier. They should also give slightly lower emissions than the oxidation direct-contact evaporation method if the emissions in the exit air from the oxidation tower are taken into account. However, modifying an existing furnace to this low-odor mode of operation would entail a considerable expense, and for recovery systems operating within their capacity level, installation of an oxidation system may be the more attractive option.

Wet scrubbing of flue gases has also been suggested as a method of controlling emissions from a kraft furnace. In one system the hydrogen sulfide is absorbed in a solution of sodium carbonate-bicarbonate. This scrubbing solution is also circulated through a separate oxidation unit, which oxidizes the sodium sulfide to thiosulfate, in order to maintain the sulfide concentration at a low enough level to permit efficient absorption of hydrogen sulfide.[953] In a second system[954,955] the absorption of H_2S is carried out in a sodium carbonate solution containing suspended activated carbon particles, which increases the absorptive capacity of the solution. This system is capable of removing methyl mercaptan and dimethyl sulfide as well as sulfide. The technique is particularly applicable to mills already using wet scrubbing for heat recovery and particulate removal.

Sulfur-Free Alkaline Pulping

The problems arising from the release of sulfur compounds during kraft pulping and recovery have led to the investigation of sulfur-free pulping methods as possible alternatives. A variety of approaches have been considered, including the soda process, the application of oxygen to the delignification process, and the use of organic additives to perform a role similar to that of sulfide in the kraft process. These additives may be used to allow complete removal of sulfide from the system or to reduce the sulfidity level during digestion to a point where odor from the black liquor system is more readily controlled.

Soda Pulping. Sodium sulfide was widely adopted as a constituent of alkaline pulping liquor because it produced substantial improvements in the process, including higher pulp yields, shorter cooking time, and stronger pulp. In certain locations where sensitivity to odors is particularly high, it has been necessary to forego the advantage of the kraft process and convert the operation back to soda pulping.[956] The soda process is particularly applicable to hardwoods where the advantages of sulfidity are of a lower magnitude than for softwoods and where the generation of odorous sulfur compounds is also more pronounced than with softwoods. In addition, hardwoods are often used in paper grades where pulp strength requirements are not demanding and can be supplied by the softwood kraft component of the furnish. If care is taken to avoid overcooking, soda pulp may be adequate for the hardwood component of a mixed furnish. Even though caustic is used as the chemical makeup in the soda process, it is difficult to maintain the system completely free of sulfide since sulfur introduced with the wood, water, and fuel oil accumulates in the system and is converted to sulfide in the recovery furnace. In most softwood pulp grades, it is important to achieve maximum physical strength; soda pulping does not produce pulp of acceptable quality in this case.

Oxygen Pulping. In the bleach plant the use of oxygen to remove residual lignin has become a commercially established process. Broadening the scope of oxygen delignification to include the pulping process has received considerable attention as a possible means of eliminating sulfide from alkaline pulping. The proposed oxygen pulping methods can be grouped into two main types: the two-stage soda-oxygen method and single-stage oxygen pulping.

In two-stage soda-oxygen pulping[957,958] the initial stage is a soda cook that is interrupted at a relatively high lignin content, before the defibration point is reached and before substantial cellulose degradation has occurred. The fibers are then separated mechanically and the delignification is continued with oxygen at an alkaline pH, using the techniques established for oxygen bleaching. The quality and yield of the pulp is dependent primarily on the point at which the first-stage soda cook is terminated. Final yield is usually at the maximum value when the first-stage yield is quite high, for example 60 to 65%.[959] However, if the soda cooking stage is terminated too soon, refining-power requirements are high and fiber damage occurs during refining. For maximum tear strength, the first-stage yield should be reduced to the range of 50 to 55%, which is slightly above the defibration point and lower than the level that gives maximum overall yield.[960] The tear strength can also be increased by inclusion of small amounts of magnesium salts in the oxygen stage.[958] Unbleached pulp yield from two-stage soda-oxygen pulping can be slightly higher than that from regular kraft pulping, but the bleached yield is about the same as from kraft. Pulp strength values, particularly tear, are usually slightly lower than those of kraft pulps. The chief advantage of the two-stage process over other oxygen methods is that it uses established equipment: the normal kraft mill digestion and recovery system with the addition of refining equipment and an oxygen delignification system.

In single-stage oxygen pulping the oxygen must penetrate into the wood chips in order to come in contact with the lignin and take part in its dissolution. The oxygen is transferred into the wood structure in solution in alkaline liquor, but its solubility is low. To increase this solubility, oxygen pulping is usually performed at high pressures in the range of 2000 to 2755 kPa (280 to 400 psi) total gauge pressure.[961,962] Most of the proposed single-stage processes require the use of very thin chips (usually no more than 1.5 mm in thickness) to further minimize the oxygen penetration problem.[961] A relatively low pulping temperature in the range of 140 to 150°C is used in order to minimize cellulose degradation

TABLE 4-41 EFFECT OF ANTHRAQUINONE (A.Q.) ADDITION ON KRAFT AND SODA PULPING

Type of Cook	Kappa Number	Yield (%)	PFI (Rev · X10^{-3})	Bulk (cc/gm)	Strength Values at 300 CSF			
					Burst Index (kPa · m^2/g)	Tear Index (mN · m^2/g)	Breaking Length (m)	
Kraft	25.2	44.2	10.7	1.20	8.63	8.63	11,700	
Kraft + 0.13% A.Q.	29.2	49.4	9.7	1.20	10.59	9.42	13,100	
Soda + 0.25% A.Q.	30.5	48.7	9.0	1.30	9.52	9.03	12,300	

and to allow more time for additional oxygen to enter the chip by diffusion. A variety of bases has been used to maintain the pH in the preferred neutral or slightly alkaline range in single-stage oxygen pulping. These bases include sodium hydroxide, sodium bicarbonate, and borax. The best results are achieved when the final pH level is in the 7 to 9 range.[961] A base charge equivalent to 9 to 11% Na_2O on wood is usually sufficient to maintain this pH level. Because a substantial amount of heat is generated during oxygen pulping, it is preferable to use a low consistency for single-stage oxygen pulping to minimize the rise in the reaction temperature. The single-stage method gives excellent pulp yields, several percentage points higher than the kraft process. The major disadvantages of single-stage oxygen pulping is the strength of the pulp, which, particularly in the case of softwoods, is substantially lower than that of kraft pulp. Properties approaching those of kraft can be achieved by using potassium iodide as a protective reagent during cooking and by extracting hemicellulose from the pulp with sodium hydroxide following cooking,[963] but costs are prohibitive if the iodide is not recovered.

Recovery of cooking chemicals from single-stage oxygen pulping could be carried out in conventional systems, but equipment for digestion is not commercially established. It is not likely that present batch or continuous-digestion equipment could be used because of the high pressure and small chips required. The requirement for special thin chips for the process is also a disadvantage. It appears that further technical breakthroughs are required before single-stage oxygen pulping becomes a commercial reality. A proposed design for an oxygen-bicarbonate pulp mill has been published.[963a] A two-stage digester with high liquor circulation was used.

Additives to Alkaline Liquors. Many studies have been conducted to find a sulfur-free additive for alkaline pulping liquors that would have an effect similar to that of sulfide in kraft pulping. Some reagents that partially fulfill this role, such as hydrazine, hydroxylamine, and sodium borohydride, have been found, but they are too costly to be used in the quantities required to make them effective.

One type of reagent, namely anthraquinone and compounds of a related structure, have been found to have a marked effect on kraft and soda pulping, even when added in very small quantities. Both anthraquinone[787] and the sodium salt of anthraquinone-2-sulfonic acid,[788] for example, have been found to function very effectively in this application. The sulfonate has the advantage of higher solubility in alkaline liquors. However, anthraquinone itself is much more effective than the sulfonate at low dosages, in spite of its limited solubility in the pulping liquor, probably because it is rapidly reduced to the more soluble hydroquinone at an early stage of the cook. Addition of anthraquinone to soda pulping liquor in amounts as low as 0.05 to 0.10% on wood will produce delignification rates, pulp yield, and strength properties similar to those obtained by kraft pulping. Anthraquinone can also be added to kraft pulping liquors to give improvements in yield and pulp properties. The effect of adding anthraquinone to kraft and soda liquors is shown in Table 4–41. In the case of kraft pulping, most of the benefits of anthraquinone have been confirmed in mill scale trials.[964] The increase in carbohydrate retention in soda-anthraquinone pulping is due to the oxidation of the reducing end group of polysaccharides to the corresponding aldonic acids.[789] The role of anthraquinone in accelerating delignification, on the other hand, is not yet clear. The H-factor concept was shown to be applicable.[964a]

If the benefits of anthraquinone addition in soda and low sulfidity kraft pulping can be confirmed at the commercial level, the use of this or related compounds could become the solution to the kraft odor problem. It is not clear what the future role of these additives will be, but it seems possible that they could bring about a major change in alkaline pulping.

By-Product Recovery

Lignin can be recovered from spent alkaline cooking liquors more simply than lignosulfates can be recovered from sulfite liquors because alkali lignin will precipitate when the pH is lowered.

Alkali Lignin

Alkaline lignin contains two types of acidic groups: one weakly acidic, probably phenolic, and the other more strongly acidic, probably carboxylic or enolic. Mild adjustment of the liquor pH releases the more weakly acidic sodium salts to form insoluble lignin acid salt. In one process for lignin precipitation[965] the black liquor is carbonated with flue gas to reduce the pH and precipitate the lignin in the form of lignin acid salt. Lignin separation begins at a pH of about 9, but the pH is generally reduced to about 8.2 to separate additional lignin. The liquor is then heated to 90°C, allowing the lignin acid salt to separate as a tar; it is then removed from the black liquor by decanting. The recovered material is then dissolved in hot water and reprecipitated, using hot, dilute sulfuric acid. The stronger acid allows the pH to be taken to lower levels so that the precipitated lignin separates as a solid and is readily filtered and washed. The residual sodium content of the recovered product will depend on the pH of the final precipitation. In the precipitated form these lignins are soluble only in alkaline solutions; this limitation restricts their use. This problem can be overcome by sulfonating the alkali lignins to produce synthetic lignosulfonates similar to those produced from sulfite spent liquor. Because the level of sulfonation can be varied, a range of properties suitable for particular end uses can be obtained.[966]

Alkali lignins from both softwoods and hardwoods are useful in many products. They are used as: stabilizers for asphalt emulsions, modifiers and extenders for latex emulsions, compounding agents in vinyl plastics, soil conditioners, binders in printing inks, wood stains, protective colloids in soap emulsions, dispersing agents for clays, fire foam stabilizers, drilling mud additives, insecticide dispersants, absorbers in storage batteries, and foundry sand binders. Alkali lignin coprecipitates with protein and this property is utilized in the purification of water and clarification of sugar juices. It is an effective metal sequestering agent in the neutral or slightly alkaline pH range. Considerable investigation has been made into the incorporation of lignin into phenol-formaldehyde resins for plywood and particle-board adhesives, for the preparation of molding powders, and for foamed insulating resins.[967] Alkali lignin can also be coprecipitated with latex to serve as a reinforcing agent for rubber.[968,969] Oxidation of the lignin prior to precipitation enhances its value in the latter application.[970] Precipitated alkali lignin has been incorporated into decorative and industrial laminates by applying it in the sheets prior to pressing. The lignin is melted under the temperature and pressure of the press and is bonded into the final product, adding to its strength and rigidity.

Sulfate Turpentine

The term turpentine refers to the volatile oil present in certain coniferous species. Traditionally, it was isolated from the wood by distillation or extraction; however, another source of turpentine is based on the fact that when resinous woods are pulped by the kraft process, the turpentine is volatilized and discharged along with noncondensable gases and steam during regular relief. Whereas in the early years of the twentieth century all commercially available turpentine was isolated directly from wood, today most of the turpentine processed in North America is produced as a by-product of sulfate cooking.[971]

Chemically, turpentine consists mainly of a mixture of unsaturated bicyclic hydrocarbons with the empirical formula $C_{10}H_{16}$. The principal components are alpha and beta pinene; other components, such as camphene, dipentene, and 3-carene, are present in minor amounts, the actual amount depending on the particular wood species, age of the tree, and geographical location where the growth occurred. The yield and composition of the volatile fraction from a wide range of North American wood species have been determined. The highest turpentine yields are obtained with the pines and some of the firs. Other species, including spruce and most hemlocks, give very low turpentine yields, and recovery is not feasible in these cases. In general, it is economical for a mill to recover turpentine when the yield level is in the range of 3 to 4 1/ton (0.75 to 1 gal (U.S.)/ton) of pulp. At lower yield levels, recovery is marginal and is more dependent on mill size and selling price. It has been found that treatment of the wood inside the bark with the herbicide, paraquat, can substantially increase the turpentine yield in Southern pines.[972] The individual components (terpenes) in a turpentine mixture can be determined by gas-liquid chromatography and infrared analysis.[973] In almost all cases, alpha pinene is the major constituent, usually making up about 50% of the total turpentine fraction; the beta pinene content is variable, constituting from less than 10% in some species to as much as 30 to 40% in a few particular cases.[974] The chemical structure of some typical turpentine constitutents are shown in Figure 4–115.

Substantial losses in the extractive components of wood occur when wood is

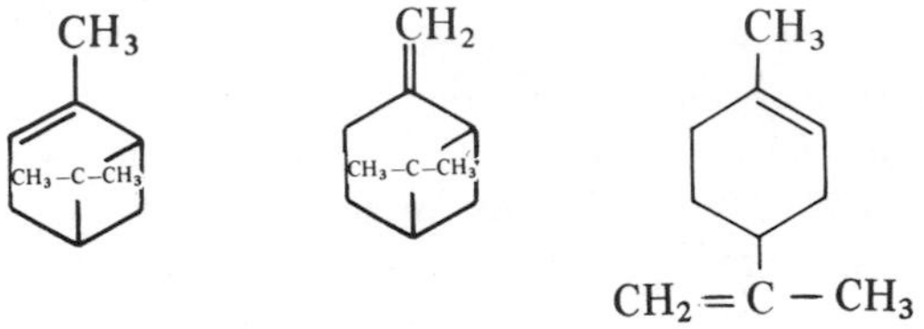

CH₃ CH₂ CH₃

Alpha Pinene   Beta Pinene   Dipentene

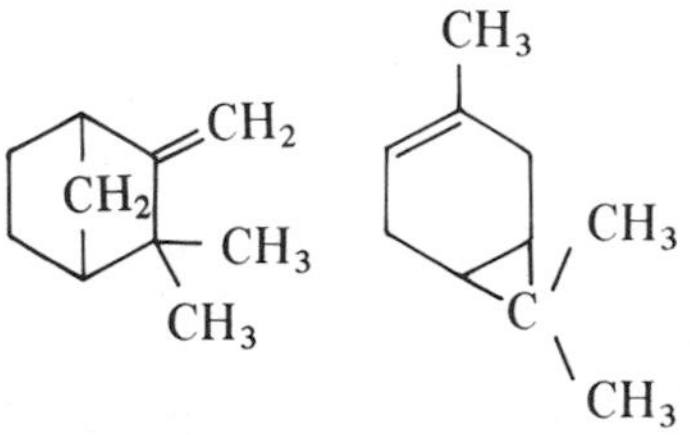

$CH_2 = C - CH_3$

Camphene   3 - Carene

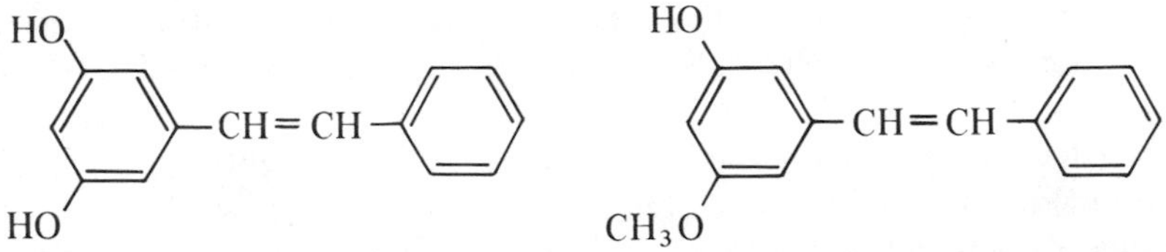

Pinosylvin (I)   Pinosylvin monomethylether (II)

Figure 4-115. Some turpentine constituents.

stored. Maximum turpentine yields can only be obtained if the wood fed to the digester is in a relatively fresh condition. The magnitude of this yield loss is dependent on the conditions during storage. Chip storage results in a greater loss in turpentine than roundwood storage. A typical relationship between loss in turpentine yield and chip storage time is shown in Figure 4-116.[728] Losses are

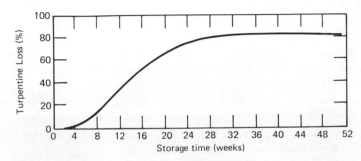

Figure 4-116. Loss in turpentine yield with chip-storage time.

relatively small during the initial storage period (2 to 6 weeks) while the chip pile temperature is increasing, but rapid loss then occurs during the next few weeks of storage. After 18 weeks of storage, the turpentine yield may be decreased by 60% and after 30 weeks by 80%.[975] Treatment of chip piles with chemical reagents to control microbiological action will reduce these losses.[976] The microorganisms do not attack the turpentine directly to any extent, but control of their activity limits the increase in chip-pile temperature and reduces the resultant losses of turpentine by evaporation.

In batch cooking,[977] turpentine and steam are relieved from the digester along with air and other noncondensable gases during the regular digester relief. A cyclone is used to remove fiber and cooking liquor which have carried over in the gas stream. The vapors are then cooled in a condenser. The condensate from this condenser separates into a turpentine phase and an aqueous phase. These are separated either by gravity decantation or with the aid of a centrifuge.[978] Continuous digesters yield less turpentine per unit of wood processed than batch digesters. The Kamyr digester operates completely liquid filled so that there is no gas relief during the heating period from which turpentine can be recovered. However, the turpentine is flashed from the spent cooking liquor in the first digester flash tank.[979] This flashed vapor is conducted into the presteaming tube, where the steam is used to preheat the chips fed to the digester. The gases vented from the presteaming tube are combined with the vapor from the second flash tank, cooled in a condenser, and turpentine is separated from the resulting condensate. Turpentine yields from continuous cooking are about 50% of those obtained from the batch process. These yields can be improved to some extent by collecting additional turpentine from the vapors formed during steam-stripping of digester and evaporator condensates.[980]

For many years, the main use of turpentine was as a solvent, but this application has declined and use of turpentine constituents in chemical applications has become the major use. Polymerization of pinenes with aluminum chloride or similar catalysts leads to the formation of terpene resins,[981] which are used in pressure-sensitive and hot-melt adhesives.[982] Alpha and beta pinene can be converted by hydrolysis to the corresponding terpineols, which are used in synthetic pine oil. The insectide, Toxaphene, is produced by heating alpha pinene to produce camphene, which is then chlorinated. A variety of flavoring and perfume chemicals can be derived from the constituents of sulfate turpentine.[983,984] These include spearmint and peppermint oils, menthol, geraniol, citronellol, and many others. Other minor uses for sulfate turpentine and its derivatives include insect repellants, flotation agents, lubricating-oil antioxidants, and road-surfacing emulsions.[971]

Tall Oil Recovery in Kraft Mills

Resin and fatty acid esters in the wood are saponified during alkaline cooking and to a large extent form colloids and emulsions in the black liquor. As the liquor is concentrated during evaporation, the sodium salts of these materials, referred to as tall oil soaps, rise to the surface of the liquor and can be separated by skimming. It is desirable to remove the soaps from the liquor as they form in order to reduce foaming and to minimize the formation of deposits in the subsequent stages of processing. With highly resinous woods, the soap separation begins in the weak-liquor storage and foam tanks, and it is necessary to provide for the removal of the soap that collects in these vessels. The main separation or skimming equipment is, however, installed following either the third or fourth evaporation effects, in which the solids concentration is between 25 and 30%.[985] About 45 min of retention time in the skimming tank is usually sufficient for complete soap separation.

To produce crude tall oil, the skimmed soap is acidified to a pH of about 3.5 to 5 to convert the fatty and resin acid salts to the free acid form. The acidulation process can be either continuous[986,987] or batch.[988] The acid used may be either sulfuric or waste acid from chlorine dioxide generation. The sulfuric acid and soap must be thoroughly mixed to ensure good contact between the two.[989] The liquor is heated to near the boiling point to assist subsequent separation of the tall oil. In the separator, the mixture settles into three phases: a crude tall oil phase; an aqueous phase, which contains residual acid and any black liquor entrained in the original soap; and an amorphous solids phase of precipitated lignin. The tall oil is separated from the other two phases, which are then returned to the black liquor system. The tall oil produced can be sold either in the crude form or after further refining by vacuum distillation to yield higher-grade products.[990,991]

Tall oil contains constituents of three main types: resin acids (chiefly abietic), fatty acids, and unsaponifiable materials. The latter fraction contains a number of neutral constituents, including hydrocarbons, high-molecular-weight alcohols, and sterols. The composition of the tall oil fraction varies with wood species and geographical location; tall oil from Southern pine contains about 50% resin acid, 40% fatty acid, and 10% neutral constituents. A typical analysis of the three main fractions of crude tall oil is shown in Table 4-42.[990] The acid constituents have the greatest value commercially and the quality of tall oil is often quoted in terms of its "acid number," which is the number of milligrams of potassium hydroxide that is required to neutralize one gram of the particular sample.[992] More detailed analysis of the individual tall oil constituents can be obtained by gas chromatography after preparation of the methyl ester derivatives of the acids.[973,993-995]

The yield of tall oil depends on the wood species and the pulping process. Southern pine gives yield values in the range of 30 to 50 kg/ton (60 to 100 lb/ton) of pulp if efficient recovery conditions are used.[996] Other species give lower yields, but in many cases recovery is still feasible, even in western and northern mills. There is a seasonal effect on yield, the highest yields being obtained from wood harvested in January and February and the lowest yields in July to September. The tall oil components are more concentrated in heartwood than in sapwood and consequently sawmill chips, which are derived mainly from the outer wood, give lower yields than whole roundwood chips.[997] Tall oil yields

TABLE 4-42 APPROXIMATE COMPOSITION OF CRUDE TALL OIL FRACTIONS

Resin acids	
Abietic acid	20–25%
Palustric acid	8–10
Neoabietic acid	22–27
Dihydroxyabietic acid	4–5
Isodextropimaric acid	3–4
Unidentified	29–43
Fatty acids	
Oleic acid	46–48%
Linoleic acid	43–45
Linolenic acid	1–2
Saturated acids	6–8
Unsaponifiables	
Phytosterols	25–35%
Higher alcohols	5–15
Hydrocarbons	33–60

also decrease with storage time in chip piles, and it has been recommended that pile treatments be used to reduce biological action in the pile.[976] In the pulping and recovery process, tall oil yields are increased by improved pulp-washing techniques and by skimming at the weak liquor tanks as well as at the evaporators.[998]

Crude tall oil is sold as a core oil, a flotation agent in mining, for the production of surface-active agents, and for the manufacture of linoleum. However, the main value from tall oil is obtained from use of its separated fractions in specific products. The resin acid fraction is utilized mainly in preparation of rosin sizes for the paper industry. Other applications for the resin acid fraction include its use as a component in synthetic resins, as a plasticizer in the manufacture of paints and varnishes, and in the synthesis of chemicals and pharmaceuticals. Tall oil is the largest source of unsaturated fatty acids in pulp-producing countries. These acids are used to produce alkyd resins for paint and varnish and in the manufacture of soap and other chemical products. Phytosterols can be used in the manufacture of Vitamin D and certain hormones.

Other Alkaline Pulping By-Products

A commercial process for the production of dimethyl sulfide from spent kraft pulping liquors has been developed. In this process, sulfur is added to the 50% solids concentrated liquor, which is then heated to about 200°C for 1 hr. Lignin is thereby demethylated to form first methyl mercaptan and then dimethyl sulfide. About 30 kg dimethyl sulfide/ton kraft pulp (60 lb/ton) can be produced in this way.[999] Dimethyl sulfide can be oxidized to dimethyl sulfoxide, a powerful solvent and reaction medium.

Vanillin can be produced from the lignin in alkaline spent liquors, but in lower yields than those obtained from sulfite liquors. The yield of vanillin based on lignin is equivalent to that from sulfite liquor if the alkaline liquor is removed from the digester during the heating period (e.g., 140°C) while the lignin is still in an undegraded condition.[1000] Vanillin is not produced commercially from alkaline liquors.

Two routes to produce phenols from the lignin in kraft liquor have been proposed—hydrogenation[1001] and high-temperature alkaline degradation.[1002] The principal products of the hydrogenation of lignin are phenol, ortho- and para-cresol, and other alkylated phenols. In the alkaline degradation process, the spent liquor is heated with sodium sulfide and sodium hydroxide at about 290°C for 10 to 20 mins. A significant portion (up to 50%) of the organic content of the black liquor is converted to low-molecular-weight phenols, such as pyrocatechol and its homologs, protocatechuic acid, homoprotocatechuic acid, and other phenol-carboxylic acids. The relatively low price of phenols from petrochemical sources up to the present time has, however, made it difficult to justify the commercial production of these products from lignin.

SILVICHEMICALS

J. N. McGOVERN, Ph.D.

Department of Forestry
University of Wisconsin
Madison, Wisconsin

The paper industry is interested in wood primarily as a source of papermaking fiber. There is, however, a growing interest in wood as a source of chemicals, energy, and even food. As petroleum supplies have diminished, wood has taken on a new importance as a potential source for chemicals and energy. It is conceivable that the paper industry could at some future time become an important supplier of organic chemicals obtained from wood. For the present, however, wood is a more expensive source of energy than coal and a more expensive source of chemicals than petroleum.[1003]

The subject of chemicals from wood is reviewed under four different categories:

1. Primary chemicals from wood and bark that can be obtained by simple extraction or by mild chemical treatment.

2. Chemicals obtained by more drastic treatment of wood, using the pro-

cesses of thermal degradation, destructive distillation, hydrolytic degradation, oxidation, or bioconversion.

3. Chemicals derived as by-products of the pulping operation.
4. Foliage chemicals.

Examples of primary chemicals obtained from wood commercially are: (1) alpha cellulose, one of the major components of wood, which is used to produce dissolving pulp to the extent of about 1.5 million tons per year in the United States (1977) for rayon, acetate, and other cellulose derivatives and (2) oleo-resinous products, turpentine, and rosin, which are obtained from the exudation of pine trees. The conversion of wood components into chemicals by drastic chemical treatment, which was once an important businesss, is no longer of significance, but is receiving renewed attention. The main source of wood chemicals is the by-product streams from the pulping of wood.

Wood and Bark as a Chemical System

Wood and its associated bark form a complex chemical system. The principal components are: cellulose, hemicellulose, lignin, and extractives. Softwoods (gymnosperms) and hardwoods (angiosperms) have different proportions of chemical components and polysaccharide compositions, as shown in Table 4–43.[1004] The hexoses that are present in softwood hemicelluloses can be used in the production of ethyl alcohol, whereas the pentoses present in hardwood hemicelluloses are employed in the growth of yeast using by-product sulfite spent liquor. Hardwood and softwood lignins differ in both quantity and in chemistry: the former are classed as guaiacyl lignins, which yield vanillin on alkaline treatment; the latter are classed as guaiacyl-syringyl lignins. Possible mechanisms for the conversion of cellulose, hemicellulose, lignin, and extractives to other useful chemicals is illustrated in the following diagram for *Tsuga heterophylla.*[1005,1006]

TABLE 4–43 CHEMICAL ANALYSIS AND POLYSACCHARIDE COMPOSITION OF HARDWOOD AND SOFTWOOD

Chemical Analysis (%)			Polysaccharide Composition (%)		
	Spruce[a]	Birch[b]		Spruce[a]	Birch[b]
Extractive-free wood					
Cellulose	43	40	Glucose	65.5	58.5
Hemicelluloses	28	39	Galactose	6.0	1.5
Lignin	29	21	Mannose	16.0	0.5
Extractives			Arabinose	3.5	0.5
			Xylose	9.0	39.0
	1.8	3.1			

[a] *Picea abies.*
[b] *Betula verrucosa.*

MECHANISMS FOR CONVERSION OF WOOD AND (BARK) TO USEFUL CHEMICALS

Bark differs from the wood itself in structure and chemical composition; bark from hardwoods and softwoods also differ characteristically. As shown in the diagram on page 494, Western hemlock bark is typically high in extractives and also in the alkali-soluble lignin fraction, phenolic acids. A distinct feature of the bark in some species is the cork which in the case of the cork oak is used as stoppers for bottles and other purposes. In some bark-heavy species, such as Douglas fir, the bark can be extracted to produce adhesive extenders. Waxes and tannins are other bark ingredients that are extracted for useful products.

Primary Chemicals from Wood and Bark

Primary chemicals are those products that are obtained directly from wood, such as the oleoresins. A discussion of some of the most important primary chemicals follows.

Naval Stores

The oleoresin is collected after scarification of the sapwood of certain southern pine trees and converted into turpentine and rosin. These products are termed naval stores due to their application to sailing vessels in colonial times. There are three sources of naval stores today: gum exudate, wood, and the sulfate process. The distribution from these sources in 1975 was 6, 27, and 67%, respectively.[1007] Gum naval stores are primarily obtained from the wounding of the slash and longleaf pines and are processed into turpentine and rosin. Wood naval stores that contain turpentine, a pine oil fraction, and a crude resin are obtained by the distillation of pine stumps. Sulfate-process naval stores are a by-product of kraft pulping of the pines and other conifers and are called tall oil (pine oil); they will be discussed under by-product chemicals.

Naval stores represent the largest of the silvichemicals in volume and value after chemical cellulose pulp. Their three major components—rosin, turpentine, and fatty acids (from tall oil only)—have the following uses:

1. *Rosin:* chemical intermediates in rubber, paper size, resin-ester gums, coatings, and other minor applications. Rosins from the sapwood of pine trees consist mainly of diterpene resin acids of the abietic and pimaric types.

2. *Turpentine:* pine oil, resins, insecticides, fragrances, and so on. Turpentine from gum and wood sources have varying percentages of α- and β-pinene and other terpenes.

3. *Fatty Acids:* intermediate chemicals, protective coatings, soaps and detergents, flotation compounds, and so on. These compounds are discussed under by-product sulfate tall oil.

Extractives

Extractives are important components of the wood and bark systems, especially the bark. These chemicals include waxes, tannins, carbohydrates, and lignans.

Waxes. Softwood and hardwood barks contain waxes that are mixtures of long-chain fatty acids and esters. In thick-barked species, such as Douglas fir, the waxes may be present in commercial quantities. Although there appear to be limitations to the complete suitability of bark waxes, an industrial installation was started in 1975 for extracting a wax in connection with the production of a plywood adhesive extender from Douglas-fir bark. The wax is extracted with organic solvent, but aqueous alkaline extraction has also been investigated. Aspen bark has been proposed for wax production.

Tannins. Condensed polymeric polyphenols of the catechin type have been important for tanning in leather manufacture. The tannins have also been utilized in preservatives and medicinal products. The present commercial sources of tannins are quebracho wood and wattle bark; formerly an important source was chestnut wood. Extracted chestnut wood chips were the basis for the original semichemical corrugating-grade pulp. Some tanning agent is produced from sulfite spent liquor as described under by-product chemicals. Production of bark tannins and polyphenols by extraction with hot water and solvents from Pacific-Northwest species was of industrial interest in the 1950s. High yields of dihydroquercitin were obtained. Chemicals of this class have properties that make them suitable as antioxidants, dyes, and medicinals, but the markets were small.

Carbohydrates. Extraction of wood with water at temperatures up to 170°C removes 10% or more of carbohydrates. As the extraction proceeds, acids are released by hydrolysis and the hemicelluloses are depolymerized. Mannose predominates in the extract from softwoods; xylose predominates in hardwood extracts. This phenomenon is important mostly in by-product hydrolysates. Larch wood is extracted commercially for its high content of arabinogalactan, a polysaccharide gum which is used as an additive in printing inks. In a review of wood-derived chemicals, Bratt[1007a] includes: 1) animal feed molasses containing 65% solids content and 55% digestible carbohydrate obtained by evaporation of the extract of wood obtained as a by-product in the Masonite process and 2) xylitol derived from the hydrogenation of xylose extracted from birch.

Lignans. Lignans are a category of extractives with a phenolic nature, having a structural relationship to lignin (phenylpropane units). They are extractable with hot water, polar and aromatic solvents, and alkaline solutions. Hemlock, spruce, and Western red cedar have been investigated for recovery of lignans for possible application in fungicides, insecticdes, and antioxidants. The technology of extraction has been developed, but markets have not been established.

Chemicals from Wood by Special Processes

Conversion of wood to chemicals is an ancient process, for example, the destructive distillation of wood to produce charcoal, wood tar, and methanol (wood alcohol). Later, the use of wood for conversion to chemicals was supplanted by more efficient and less costly processes. However, with petroleum becoming more scarce and costly, wood is again receiving attention as a source of chemicals. The renewable nature of wood is an important factor in the long-range potential.

Several reviews have been published on processes for producing chemicals from wood. A study by the Pulp and Paper Research Institute of Canada[1008] concludes that some chemicals from wood can be competitive with fossil chemical sources. Goldstein[1009] concluded that the production of chemicals from wood for plastics, fibers, and rubbers is technically feasible and may some day be economically feasible. The conversion of wood chips into fuel oil in a catalytic process involving high temperature and pressure has been evaluated; a barrel of oil can be obtained from about 900 lb of chips. Some of the technically feasible processes for producing chemicals from wood are described below.

Pyrolysis Products. Pyrolysis, or dry distillation, of wood and bark, through condensation of the gases of combustion, produces charcoal and a pyroligneous acid by-product. The latter is refined into calcium acetate, methanol, and tar. In the process the tar and the noncondensable gases are used as fuels. This process was started in the early nineteenth century and continued through the beginning of the twentieth. The methanol was used as a solvent; the acetic acid derivatives found many uses. Charcoal production was begun in 1645 to furnish material for the processing of iron into steel. Charcoal lost this market when blast furnaces became larger and needed a stronger material to support the overburden. Wood alcohol also lost its place with the development of synthetic methanol. Acetic acid compounds were replaced by acetone recovered from the fermentation process used to make butanol. Charcoal is now used mainly for briquettes and outdoor grills, and its production is about 750,000 tons yearly and growing. A new development is the employment of the fluidized-bed furnace.

The production of methanol from wood and bark by a gasification process has been under consideration.[1010] The methanol would be used to supplement gasoline in motor fuels. The process involves the following steps:

1. Partial oxidation of the wood.
2. Cleaning, cooling, and compressing to remove carbon dioxide, nitrogen, and hydrocarbons.
3. Catalytically shifting the gas to two-parts hydrogen to one-part carbon monoxide.
4. Converting into methanol.
5. Refining the methanol.

Economic feasibility appears to be unfavorable with foreseeable manufacturing costs, wood-residue cost, and conversion efficiencies in comparison with synthetic methanol.

Acid Hydrolysis (Wood Saccharification) Products. The conversion of the carbohydrates, which make up two-thirds of the wood substance, to sugars for further processing into ethyl alcohol or into yeast can be accomplished by a combination of acid hydrolysis and fermentation. The technology has been developed, but the economics of the process in the United States has been unfavorable,[1011] even in the face of the escalating costs of ethanol produced by the hydration of petrochemical-based ethylene. The steps involved in the production of ethanol by the acid hydrolysis route are:

1. The cellulose is converted into sugars by heating in acid medium at high temperature and pressure.
2. Furfural is recovered from the condensate.
3. The sugar solution is neutralized and the calcium sulfate separated.
4. The sugars (C_6-type) are fermented to ethanol and carbon dioxide using yeast, which is then separated for reuse.
5. The ethanol is stripped, purified, and concentrated.

A pentose concentrate is a by-product. The use of enzymes in the place of acid for the conversion of flourlike wood, paper mill residual sludges, or wastepaper into glucose, for subsequent fermentation into alcohol, has also been under development. One proposed process is based on the use of *Trichoderma viride* to hydrolyze paper mill sludge to glucose with simultaneous fermentation to alcohol.

Hydrolyzates from the prehydrolysis stage used in the kraft pulping process for chemical pulp or from the processing of wood chips for fiberboard manufacture contain carbohydrates that are useful for cattle feed. The spent liquor from the acid sulfite pulping process also contains wood sugars capable of fermentation to either ethanol or to yeast, but these wood-sugar sources are in reality by-products.

The pentose-rich hemicellulose of hardwoods can be hydrolyzed to furfural, a chemical intermediate, by:

1. Acid hydrolysis using steam in a digester.
2. Condensation of the furfural-steam vapor.
3. Stripping of the furfural.
4. Dehydration and refining of the crude furfural.

Furfural is produced industrially by the hydrolysis of agricultural residues, including sugarcane bagasse, corn cobs, and oat and rice hulls. The reactions involving the acid hydrolysis of the hemicelluloses to sugars and the conversion of these to furfural proceed concurrently.

Hydrogenolysis Products. Hydrogenation of the lignin in wood or in pulping spent liquors in a hydrogen atmosphere with a catalyst or in the presence of a hydrogen donor (e.g., tetralin) results in heavy and light oil fractions and a residue. The light oil separates into catechol, a phenolic fraction, and a neutral oil. The catechol and phenolics are commercial chemicals. Hydrogenation of acid-hydrolyzed lignin in a benzene solvent medium, using a nickel catalyst, results in a substantial yield of monomeric phenols. Inexpensive metal sulfides have been investigated as catalysts, starting with ligninsulfonates. This route to phenols from lignin has not been competitive with synthetic phenol from petrochemical sources.[1006]

By-product Wood Chemicals

The largest volume, value, and variety of silvichemicals are derived as by-products of chemical, semichemical, and steam pulping. The production of fatty acids and rosin from tall oil and sulfate turpentine accounted for 80% of the total naval stores produced in 1975.[1007] Other important products are the ligno-sulfonates, alcohol, and yeast from sulfite spent liquor. The total annual value of silvichemicals produced in the US in 1977 was over 500 million dollars; this was derived 49% from sulfate mill by-products, 18% from sulfite mill by-products, 17% from naval stores and 16% from miscellaneous sources.[1007a] An outline of the by-products obtained from pulping is given on page 500.

Sulfite Pulping By-product Chemicals

The use of lignosulfonates from calcium-base sulfite spent liquor for tanning agents and adhesives was proposed as early as 1880. The spent liquor at that time was disposed of in adjacent streams. Early in this century the use of appre-ciable amounts of crude and modified calcium lignosulfonates was developed by J. S. Robeson. Sulfite mills in Europe early fermented the wood sugars in sulfite spent liquor to alcohol and yeast, the latter being a product of World War II which has considerable nutrient value. The conversion of sulfite pulp mills from calcium base to soluble bases (Mg and Na) for improved inorganic-chemical recovery and pollution-reduction has changed the amount of sulfite spent liquor available for wood-chemical recovery.

 Fermentation. Sulfite spent liquor contains hexose and pentose sugars from the hemicelluloses in proportion to the softwood (high hexose) or hardwood (high pentose) used. The hexoses are converted into ethyl alcohol by fermen-tation. The alcohol is distilled from the fermented liquor (concentration about 1%) and the yeast recycled. The trend toward producing sulfite pulp in high yields reduces the hexose content of the sulfite spent liquor.

 Fermentation of the pentose-hexose in sulfite spent liquor directed at the production of yeast, *Torula utilis,* is practiced in the United States with an annual production of about 9000 tons. The torula yeast has a high content

OUTLINE OF SOURCE AND TYPE OF BY-PRODUCT WOOD CHEMICALS

Sulfite Pulping	Kraft Pulping	Semichemical Pulping	Steam Pulping (Masonite, prehydrolysis)
Spent liquor	Turpentine	Spent liquor—acetic acid, salt cake	Hydrolysates—hemicellulose extracts
Fermentation—alcohol, yeast	Tall oil—fatty acids, rosin		
Vanillin	Black liquor—DMS, DMSO, kraft lignin, and modified lignins		
Lignosulfonates—dispersants, adhesives, additives, tanning agents			

of vitamin B and protein and is used as a food additive and flavor enhancer. The desugared spent liquor is used for lignosulfonate applications.

Lignosulfonates. The lignosulfonate component of sulfite spent liquor is derived from the lignin and comprises a major class of silvichemicals with a production of 1160-million lb/year and a value of $54 million in 1977 from 11 producers.[1007a] About 60% of the lignosulfonates is from calcium-base pulping. The lignosulfonates are water-soluble, anionic, surface-active chemicals with a multitude of uses, including use as a dispersant, binder, concrete admixture, tanning agent, animal-feed supplement, water-treatment additive, scale inhibitor, feed-pellet binder, sequestering agent, emulsifier, resin extender, drilling-mud thinner, and agricultural-mineral supplement. The spent liquor is used in its dilute form of 8 to 15% solids as a road binder, as a concentrated solution of 50% solids, or as a dried solid.

Vanillin. This organic chemical, 3-methoxy-4-hydroxy benzaldehyde, is a by-product derived from softwood lignin by alkaline oxidative hydrolysis. The reaction is conducted on desugared spent liquor in a high temperature and pressure oxidation with injected air or oxygen and the use of catalysts. The vanillin is extracted from the reaction product with butanol and refined. There are several producers in North America and Europe. Vanillin is used in confectionary and flavorings and as a raw material in medicinal preparations (see also the discussion on the recovery of by-products from the sulfite process). Oxalic acid was formerly marketed as a by-product of vanillin production, but this practice has been discontinued.

Other Chemical By-products. Acetic acid was formerly produced from sulfite spent liquor, but this practice was discontinued for economic reasons. This chemical is being considered for production from the condensate of spent liquor evaporation by adsorbtion on activated carbon and desorbtion as ethyl acetate.

Kraft Pulping By-Product Chemicals

The pulping of pines, especially the southern pines, and other resinous woods by the kraft process results in the presence of turpentine in the digester relief gases and the formation of tall oil in the black liquor.

Turpentine. Turpentine is removed from the digester during the initial steaming and is condensed from the relief gases. Sulfate turpentine contains about 60 to 80% α-pinene, 20 to 30% β-pinene and 5 to 12% other terpenes.[1007] Over 80% of the total production of turpentine in the United States is from kraft pulping.

Tall Oil. Tall oil is composed of two major components: fatty acids and rosin. These chemical products have great commercial value and are recovered as part of the pulping operation. Their recovery is described in the section on kraft pulping (see also the discussion of by-product recovery in the section on alkaline pulping). Their production is given below:[1012]

		Refined Tall Oil	
Year	Sulfate Turpentine (1000 gal)	Fatty Acids (short tons)	Rosin (short tons)
1966	—	168,620	176,254
1970	29,085	200,209	207,769
1974	25,583	181,608	198,458
1976	24,212	186,864	202,202

The production figures above show the static nature of the sulfate naval-stores industry. This is due more to supply than demand because of the increasing use of hardwoods and purchased chips (high in sapwood); outside chip storage; and, in the case of turpentine, the use of continuous digesters—all tending to reduce the unit production of turpentine and tall oil. The development of paraquat treatment of the growing pines to increase the formation of oleoresin may restore the supply. In this case the extraction of chips before pulping may be involved.

Tall oil, which is a product of the kraft pulping of resinous woods, is recovered in the early stages of black-liquor handling as a soap of sodium salts of organic acids. The tall oil soap is treated with sulfuric acid to form crude tall oil, comprising tall oil fatty acids and rosin, which is subsequently fractionated to useful products. The fatty acids are a mixture of oleic and other unsaturated fatty acids; the rosin is similar to abietic acid.

Kraft Black-Liquor Chemicals. Black liquor contains lignin compounds, saccharinic acids, organic sulfides, and many other compounds. Because of the economic necessity for the recovery of the inorganic sodium and sulfur—for recycling in the cooking-liquor preparation system—as well as the heat in the organic constituents (especially with high-cost alternative fuel), very little kraft liquor is converted to chemical by-products. Two products are, however, recovered: dimethyl sulfide (DMS) and kraft lignin. The DMS is present, along with mercaptans, in small amounts, but the DMS-content is enhanced by treatment of the concentrated black liquor with sulfur in proportion to its methoxyl content and the DMS recovered by flashing the liquor. About 9 million lbs are produced annually. The DMS and methyl mercaptan are used as odorants in natural gas and as solvents in chemical synthesis. The DMS is also converted through a liquid-phase oxidation to dimethyl sulfoxide (DMSO), a highly useful solvent in synthetic-fiber manufacture and has potential as a medicinal.

A small quantity (9000 tons annually) of kraft lignin products is produced as such, sulfonated for use as dispersants and emulsifying agents, and converted to lignin amines. Activated carbon for clarification purposes was at one time an important alkaline black-liquor lignin product, but wood waste and coal are now used as the raw materials to produce this product.

Semichemical Pulping By-product Chemicals

The spent liquor from the neutral sulfite semichemical (NSSC) pulping process is evaporated in one mill to 40%, acidified with sulfuric acid, and the acetic and formic acids formed from the hemicelluloses extracted with butanone-2. The separation of these products involves a unique azeotropic distillation procedure. The acetic acid is marketed as the glacial grade. The formic acid production has been discontinued. Sodium sulfate (salt cake) is a product of the reactions and this has a market for kraft recovery makeup. Salt cake is also the by-product of a fluidized-bed combustion of concentrated NSSC spent liquor; this by-product also is utilized as kraft recovery chemical makeup.

Steam Pulping By-product Chemicals

Subjecting wood chips to steam at the high temperatures and pressures used in producing pulps for fiberboard (e.g., Masonite) or in the prehydrolysis stage of the kraft process for dissolving pulp manufacture results in the solution and acid hydrolysis of hemicelluloses. The extract is concentrated by evaporation and can be used as a molasses substitute or supplement in cattle feeds. These materials are regarded as having potential. Hardwood hydrolysates are high in xylose and could be a starting material for furfural manufacture in competition with agricultural residues. A related chemical, levulinic acid, which was produced commercially in the 1960s and then discontinued for want of markets, might start from a prehydrolysate. The levulinic acid was used in foods as an acidulent and as a carrier for calcium in diet foods.[1006] Hardwood prehydrolysates are used commercially in Finland as the basis of the manufacture of xylitol, a wood sugar polyol used as a dietetic sweetener.[1006] The Finnish "Pekilo" process uses an aerobic fermentation for production of a protein-rich cellulose from wood hydrolysates.[1013]

Foliage Chemicals

The foliage of trees has become of interest in connection with the development of whole-tree harvesting and chipping and the consequent separation of the leaf-needle-twig and chip fractions. Conifer needles contain terpenes, which could contribute to naval stores.[1006] Foliage has been of interest historically and prehistorically, and the variety of chemicals in this material is exceedingly large. Nutrients, such as vitamins, protein, and minerals, are present, as are pharmaceuticals, essential oils, chlorophylls, and carotenoids.[1014] *Muka,* a Russian term for an animal feed and vitamin supplement, has become of great interest.[1015] Hardwood foliage contains more nutrients than foliage from conifers.

PULPING AND BLEACHING OF NONWOOD FIBERS

D. K. MISRA, Ph.D.

Director General
"Thessalian"
Pulp and Paper Industries Ltd.
Athens, Greece

Although softwood, temperate hardwood, and some selected species of tropical hardwood constitute the major bulk of the world pulp and papermaking material, nonwood plant fibers are the potential source of raw material in many countries where pulpwood supplies are limited or nonexistent. Until the 1930s, pulp and paper manufacture based on these fibrous materials was confined to very small mills having production capacity of 5 to 25 tons/day. Since World War II there has been a phenomenal development in the utilization of these fibers and many large-capacity, integrated pulp and paper mills have been built all over the world. This has been made possible due to the rapid progress made in pulping technology, equipment design, and process engineering for treating different raw materials to yield various grades of pulp. Based on indigenous annual plant fibers many developing countries have considerably expanded their pulp and paper industries in their efforts to achieve self-sufficiency and to stimulate their national economy. Even many advanced, industrialized countries in Europe have benefited economically from using these fibrous raw materials for pulp production. In 1972, according to FAO,[1016] the estimated paper-grade nonwood plant-fiber pulp production capacity worldwide was 7,850,000 metric tons. According to Atchison,[1017] although pulp production from nonwood fibers constitutes only 5% of the world's total pulp output on a percentage basis, it has been increasing at a faster rate than wood pulp production. In Latin America alone, nonwood pulp production has doubled in the last decade and in Africa and the Middle East almost tripled. It is expected that the growth rate will be compounded in future years in view of a rapidly rising demand for paper and paper products among many populous nations, where the prospects of increased pulp production lie in the exploitation of vast resources of agricultural residues, such as cereal straws, bagasse, and other fast-growing annual plant fibers.

Nonwood plant fibers for pulp are classified in three main categories, in terms of their source, availability, and fiber characteristics:

1. Agricultural and agro-industrial waste, which include cereal and rice straws and sugarcane bagasse.

2. Natural-growing plants which include bamboo, reeds, esparto, and sabai grass.

3. Cultivated fiber crops, which include leaf fibers, such as abaca (Manila hemp), sisal, and henequen; and bast or stem fibers, such as jute, true hemp, sunn hemp, flax-tow, and their salvaged products.

Bagasse, bamboo, cereal straw, esparto grass, and reeds are the dominant raw materials among the nonwood fibers for the production of pulp and a wide range of papers and boards. These pulps are used either wholly or blended with kraft long-fiber pulp or wastepaper to produce corrugated medium, linerboard, insulating boards, sack paper, imitation kraft, M. G. wrappings, greaseproof, and many other unbleached varieties. Similarly bleached pulps produced from these raw materials are used as a portion of the furnish for the manufacture of high-quality writing, printing, airmail, white-poster, glassine, sanitary-tissue, towelling, and other allied grades. Other nonwood pulps produced out of leaf, bast, or stem fibers are used for the manufacture of specialty grades. These fibrous materials have high harvesting-cost and limited availability. These expensive pulps do not find wide application for the manufacture of normal grades of paper and board on a large scale. Advantage is taken of their exceptional fiber characteristics to manufacture specialty papers, such as tea bags, carbon tissue, Bible paper, insulating cable paper, condenser tissue, cigarette paper, and other thin papers.

Further information on the availability and properties of nonwood fibers is given in the chapter on pulpwood.

Pulping Processes

The pulping chemicals used for annual plant fibers and other nonwood materials are basically the same as used in the pulping of wood. The pulping processes include:

1. Straight soda or modified soda process.

2. Sulfate process with varying sulfidity.

3. Neutral or monosulfite process.

4. Acid sulfite process.

5. Soda-chlorine (Celdecor-Pomilio) process.

6. Mechano-chemical process.

7. Cold soda process.

Selection of a particular process is determined by techno-economic and ecological factors, which vary with the geographical location of a mill.

Soda and sulfate processes are being used more and more because of their potential integration with chemical- and heat-recovery systems. Strength properties of pulps produced by the soda and sulfate processes show only marginal differences for most nonwood fibrous materials with the exception of bamboo, particularly when the pulp is to be bleached. In some parts of the world, salt cake is an expensive makeup chemical compared to sodium hydroxide. Pulping chemicals and energy requirements are important cost factors and have a considerable bearing on the economy of the pulping process. Processes lacking thermal- and chemical-recovery arrangements are being challenged from the point of view of both economy and ecology. Most countries have enacted legislative measures and imposed severe restrictions on waste disposal. Therefore, processes that do not include a provision for waste disposal are limited in their commercial applicability.

Pulp mills using the neutral sulfite process can incorporate the Tampella recovery system, although its economics are questionable for small or medium-sized mills. However, in some cases where dual processes, such as soda or sulfate and neutral sulfite, are employed to pulp two different raw materials, a cross-recovery system is feasible.[1018] The waste liquor from both processes is concentrated in multiple-effect evaporators separately and fed to the Tampella recovery boiler by separate feedlines. The smelt, which consists mainly of sodium carbonate and sodium sulfide, is discharged from the furnace to a dissolving tank to prepare the green liquor. The green liquor is treated with lime in a precausticizing stage, where a major portion of the silicate is precipitated and removed along with the calcium carbonate. The precausticized and clarified green liquor is divided into two streams. One stream is treated in a conventional recausticizing system to produce white liquor for use in sulfate pulping; the other portion is treated in the Tampella recovery process to be converted into cooking liquor for NSSC pulp production. The Tampella recovery process has the following steps: precarbonation, carbonation, and bicarbonation/crystallization. In the precarbonation stage, the green liquor is treated with flue gases in an absorption tower, where the carbon dioxide of the flue gases reacts with the sodium sulfide. Most of the sodium sulfide is converted into sodium bisulfide, with a very small quantity of sodium carbonate being produced. The precarbonated green liquor contains sodium sulfide, sodium carbonate, sodium bisulfide, and sodium bicarbonate—in varying proportions, depending on the regulation of pH and temperature. When the solution is heated, sodium carbonate and hydrogen sulfide are produced. If the evaporation is extended, sodium carbonate crystallizes in the form of the monohydrate. These crystals are easily separated from the mother liquor. Part of the crystallized sodium carbonate is used in the carbonation stage. Carbonation is carried out in a special tower provided with perforated floors. The carbonate is introduced at the top floor and the flue gas is introduced into the empty bottom section. The carbonate solution is introduced at a concentration that allows the major part of the resulting bicarbonate to crystallize. The bicarbonate goes into the bicarbonation stage; the hydrogen sulfide produced in the bicarbonation stage leaves the reactor along with steam. The hydrogen sulfide content is approximately 5 to 12 mol percent of the total gas and 92 to 98 mol percent of the dry gases. Water is condensed and separated from the gaseous mass by cooling, producing hydrogen sulfide gas at 90%. The water-free gas is burned along with the makeup sulfur to produce sulfur dioxide. The sulfur dioxide gas formed is washed and is transferred to the absorption tower, where it is reacted with the sodium carbonate. The sodium sulfite produced is used in the NSSC pulping. The neutral sulfite process is employed for the production of semichemical pulps with relatively higher brightness. Fine, bleached pulps can be produced with low-chlorine demand at high-brightness level. Neutral sulfite pulps are comparatively freer in drainage characteristics. Strength properties of neutral sulfite pulps are somewhat low compared to pulps produced by the soda or sulfate processes.

The acid sulfite process is not suitable for pulping nonwood plant fibers. There is appreciable strength loss and in some cases the yield figures are low compared to pulps produced by alkaline pulping processes. Soda, sulfate, and neutral sulfite processes are applicable, for both batch and continuous-pulping operations.

Batch Digestion System

Spherical rotary digesters, or tumbling digesters, are generally used in batch pulping. Operating efficiency of this type of pulping is low. The cooking liquor-to-material ratio must be high in order to achieve complete submergence of the bulky nonwood fibrous material. Direct steam is applied to bring the digester pressure and temperature to the required level. While charging the digester, a considerable amount of air is entrapped and therefore frequent venting is necessary to remove the air and noncondensable gases from the digester during the early stages of cooking. If an alkaline cooking process is used, the presence of oxygen will have a degrading effect on the cellulose and therefore removal of oxygen is essential. These procedures extend the cooking cycle and increase the steam consumption. The total cooking cycle varies from six to ten hours and consequently the pulp-output capacity is low. Steam requirements are high because of the long cooking period and also because of the large amount of steam required for evaporating the dilute black liquor to a concentration that is suitable for spraying into the recovery furnace. The chemical requirement is also high. Stationary digesters equipped with forced-liquor circulation and preheating arrangements are extensively used for bamboo pulping.

Continuous Pulping Systems

Since 1930, continuous pulping has undergone many developments with respect to equipment design and material construction to ensure trouble-free operation. Continuous-pulping methods include: Bauer, Celdecor-Pomilio, Celdecor-Kamyr, Defibrator, Huguenot, Morley, Pandia, and Peadco. Some of these pulping systems are applicable to only a particular raw material or group of raw materials; others have a wider application. The open structure and easy penetrability of most of the annual plant fibers and agricultural residues makes them particularly suitable for continuous pulping.

The advantages of continuous pulping compared to batch pulping are many and include:

1. Easier handling of bulky raw materials, the bulkiness being overcome by mechanical compaction when the raw material is forced to the cooking zone.

2. Greater process flexibility due to the quick response to changes in pulping variables.

3. More uniform pulp quality due to thorough and continuous mixing of chemicals and steam with the fibrous materials all through the pulping zone.

4. Avoidance of sudden surges in cooking-liquor requirements.

5. More uniform steam- and power-requirement, which reduces peak-load demands on utility services and permits operation at high load factors.

6. Lower steam consumption due to the low fiber-to-liquor ratio and efficient heat recovery from the flash steam released at the blow tank.

7. Higher black-liquor concentration to the evaporating station.

8. Greater compactness of the pulping unit and capability to instrument the unit for a single operator.

9. Increased output of pulp.

10. Appreciable saving in labor cost because of increased productivity.

11. Easier expansion of the unit at minimum cost without long shutdown of the unit and without increasing the labor force or building space.

Some of the prominent continuous pulping systems are discussed below.

Celdecor-Pomilio Process. This was the first continuous pulping process, which was originally developed in the 1930s by the Pomilios and later modified and improved with the collaboration of the Cellulose Development Corporation. A number of pulp mills using this process have been built in Algeria, Argentina, Brazil, Chile, China, England, France, Holland, Italy, the Philippines, Spain, Uruguay, and Yugoslavia. The process involves mild digestion of the material with caustic soda, followed by chlorination in the gas phase. Originally the process was intended to use caustic soda and chlorine in the stoichiometric proportions in which they are produced by the electrolysis of brine. It was considered that this process would have economic advantages over other processes by using cheap salt and electricity for electrolysis. Nevertheless, some mills operate with this process without any integration of the electrolytic plant, using purchased chemicals. The pulping process has been successfully used for producing bleached pulp from cereal straw, bagasse, and esparto. A flow diagram of this process is shown in Figure 4-117.

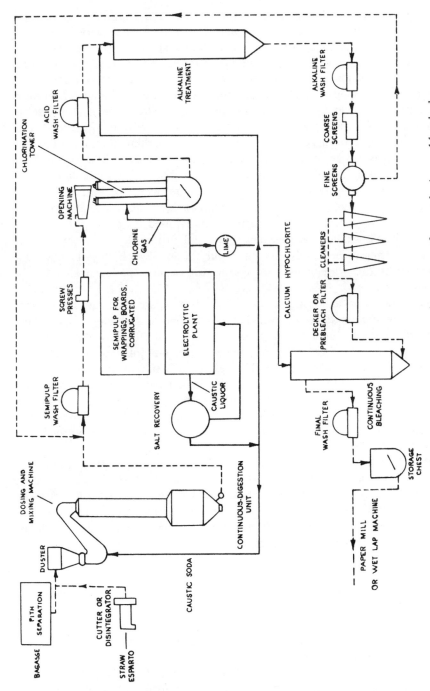

Figure 4-117. Flow diagram for Celdecor-Pomilio process for producing unbleached and bleached pulps from annual plant materials: bagasse, esparto, and straw. (Courtesy of Cellulose Development Corporation.)

509

Upgraded, cleaned fibrous material and caustic soda solution are fed in predetermined proportions to a mixing unit. The alkali-treated material is transported through a screw conveyor to a cylindrical reaction tower. The height of the tower varies, depending upon the raw material and the total dwelling-time required. The material is fed to the top of the tower and travels downward by gravity. Low-pressure steam is injected through a central duct for uniform distribution through the mass of raw material, or alternatively, the vessel is steam-jacketed for indirect heating. Temperature at the bottom of the tower can be maintained between 105 to 130°C without any flashing of steam at the top. Tower designs can be suitably altered to increase the volume both for the purpose of retention as well as for increased throughput. The semipulp or crude pulp is continuously extracted at a regulated rate, which is controlled by the operator from a control panel. The dwelling-time, temperature, and input of caustic soda vary depending on the raw material used for pulping. The mildly cooked pulp is washed over vacuum filters and further dewatered by a screw press to increase the consistency to about 25 to 35%. High-yield raw pulp produced at this stage is used for corrugating medium or low-quality board-manufacture.

The semipulp passes through a shredding unit to expose maximum surface area for effective chlorination. The fluffed pulp is discharged to the top of a chlorination tower, similar in design and construction to the caustic-reaction tower, except that the chlorination tower is lined with acid-proof material to resist the corrosive action of moist chlorine. The descending semipulp comes into contact with the gaseous chlorine, which enters the tower through multiple pipes and nozzles for uniform gas distribution. The alkaline semipulp rapidly absorbs the chlorine gas, which acts as a pulping and bleaching agent. The chlorinated pulp is continuously removed from the bottom of the tower, diluted, washed, and thickened over a vacuum filter. The semiprocessed pulp is subjected to caustic extraction in a high-density tower. The pulp is again removed at the bottom of the caustic extraction tower and washed over a vacuum filter. The pulp is then screened and cleaned by rotary screens and a three-stage centrifugal cleaning system, after which the pulp is thickened and bleached with hypochlorite. Following hypochlorite bleaching the pulp is washed and stored in a chest, either for use in papermaking or for conversion into wet laps for shipment.

Pandia Continuous Pulping System. This system is in use in many mills and has proved very effective and versatile as a pulping unit. A host of nonwood fibrous materials, such as bagasse, cereal straws, esparto, bamboo, reeds, and cornstalks are pulped by alkaline and neutral sulfite processes to produce a wide range of pulp grades.

The Pandia continuous digestion system is shown in Figure 4-118. It consists of a screw feeder and impregnation chamber, a pair of horizontal tubes fitted with screw conveyors joined by vertical connecting necks, and a rotary discharger. The raw material is conveyed continuously either to the mixer-impregnator or

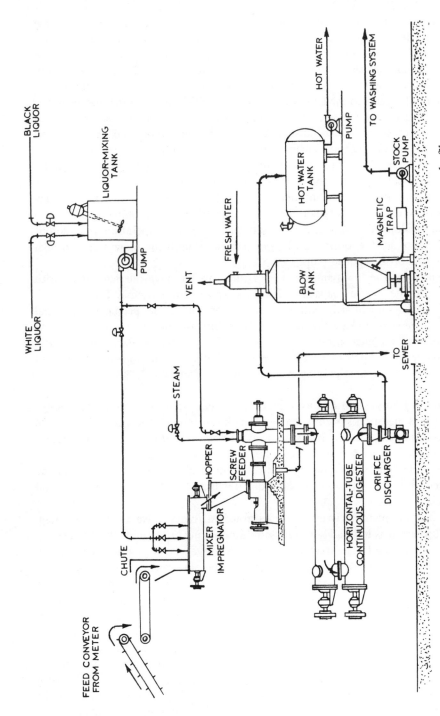

Figure 4–118. Pandia continuous digestion system for pulping various nonwoody fibrous materials. (Courtesy of Parsons & Whittemore, Inc.)

511

alternatively directly to the chute of the screw feeder by means of a suitable metering device. A small amount of cooking liquor is added to the fibrous material at the mixer-impregnator to soften the mass for subsequent treatment. The incoming material is compressed through the screw feeder and the density of the mass is increased sufficiently to withstand the internal pressure of the digester. The plug formed by the compacting effect of the screw feeder serves as a seal against any escape of steam or blowbacks. The compressed mass loses its compactness as it comes into contact with the incoming cooking liquor and the steam at the reaction chamber before it drops into the first tube of the digestion zone. Retention time within the horizontal tubes is regulated by controlling the speed of the internal screw conveyors. Pulping of the fibrous material continues due to intimate mixing of chemicals and steam as the raw material is conveyed through the horizontal tubes. The liquor-to-solids ratio can be maintained as low as 3:1 for satisfactory pulping operation. Injection of black liquor at the orifice discharger facilitates uniform blow and thorough defibration of the cooked pulp.

Diameter of the tubes and number of tubes are the major design criteria with respect to the pulping capacity of the Pandia digestion system. These two criteria and the variable drive arrangement for controlling the speed of the internal screw conveyors provide the necessary flexibility to control the retention time required for pulping various raw materials. The unit can be sized for pulp production ranging from 15 to 200 tons/day. Cooking conditions using different raw materials in a rapid pulping cycle are given in Table 4-44.[1019]

Celdecor-Kamyr Continuous Digestion System. This is a modified Kamyr continuous digestion system suitable for pulping annual crop fibers. In the case of wood pulping, the cooking liquor is heated by preheaters and is introduced to the pulping unit by a forced circulation system, whereas direct steam is used in the modified units for pulping nonwood materials. Figure 4-119 shows the Celdecor-Kamyr continuous pressure digester used for the pulping of nonwood fibers.

The fibrous material is introduced through a rotary valve to the preimpregnator, where it is intimately mixed with hot cooking liquor and with flash steam from the blow tank. The treated material is conveyed through a secondary multipocket valve (LP feeder) to a low-pressure prepulping vessel, where low-pressure steam is provided to maintain an operating pressure from 103 to 240 kPa (15 to 35 psi). The material then passes through a three-port single-pocket valve (HP feeder). When the valve rotor is in the vertical position, material from the low-pressure zone falls through the top port into the pocket. When the valve rotor is moved from this position, the pocket is then aligned with the remaining two ports, one of which connects to an asthma valve and the other to the digester. Steam released from the asthma valve forces the material from the rotor into the digester and 80% of the total steam is introduced into the system by the asthma valve. This system contributes to the overall economy of steam consumption. The preimpregnator and the low-pressure steaming vessel provide thorough

TABLE 4-44 COOKING CONDITIONS USING DIFFERENT RAW MATERIALS IN A RAPID PULPING CYCLE IN A PANDIA CONTINUOUS PULPING SYSTEM

Raw Material	Pulping Chemical Process	Chemical Applied as Na_2O (%)	Dwell Time (min)	Steam Pressure (kPa)	(psi)	Pulp Yield (%)	Permanganate No.
Wheat straw[a]	Soda	4.6	8	515	75	67	—
	Soda	10.0	8	550	80	50	—
Rice straw[a]	Soda	9.8	5.5	690	100	39	4.9
Sugarcane bagasse[b]	Sulfate	12.0	10	895	130	52	7.5
Reeds[a]	Soda	13.8	20	855	130	48	15.0
Esparto	Soda	12.4	20	825	120	52	6.0
Napier grass[a]	Soda	15.0	30	1035	150	44	15.0
Bamboo (Bambusa arundinacea)	Sulfate	18.2	30	855	130	45	10.0
Reeds[c]	Neutral-Sulfite	19.1	25	1035	150	53	12.3
	Neutral-Sulfite	10.7	10	1035	150	62	24.4

[a]Uncleaned raw fiber.
[b]Depithed or cleaned.
[c]Cleaned reeds: chemical applied as Na_2SO_3.

513

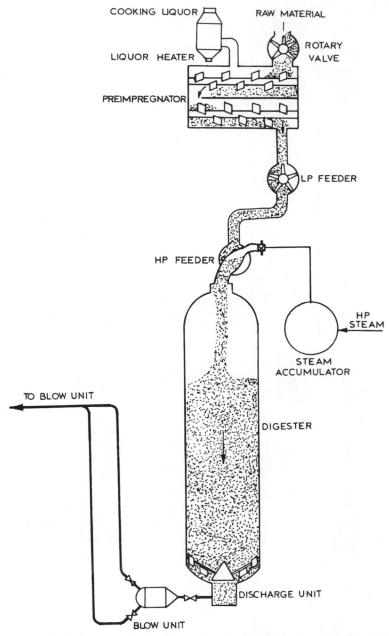

Figure 4–119. Celdecor-Kamyr continuous pressure digester for annual plant materials. (Courtesy of Cellulose Development Corporation.)

mixing of cooking chemicals and uniform preheating so that the final pulping can take place in the digester at optimum temperature. The degree of pulping can be regulated by controlling the input of cooking chemicals, feed rate of raw material, and retention time in the digestion tower. The digestion tower is sized to suit the requirements. Cooked pulp is discharged from the bottom of the tower through an adjustable orifice. This digestion system can be used for pulping a variety of raw materials employing the soda, sulfate or neutral sulfite processes.

The advantages of this pulping unit are:

1. Efficient utilization of flash steam.

2. Multistage arrangement for bringing the fibrous material from liquor impregnation through low-pressure cooking to high-pressure cooking, thereby bringing about delignification with minimum damage to fibers.

3. Positive control of cooking time, from a few minutes up to three hours or more.

4. Positive mechanical-sealing effect provided by the rotary valve and the LP and HP feeders, which prevent blowback problems.

Mechano-Chemical Process

This process was invented and developed for pulping straws and bagasse by Lathrop and Aronovksy[1020,1021] at the Northern Regional Research Laboratory, Peoria, Illinois. In this process either caustic soda or a combination of caustic soda and sodium sulfite is used as the cooking chemical. Pulping is carried out at atmospheric pressure and at a temperature ranging from 95 to 100°C. A hydrapulper equipped with an oversized rotor, a solid bottom plate, and a quick-opening valve form a compact pulping unit. Chopped straw or depithed bagasse is fed to the hydrapulper by means of a conveyor. Cooking chemicals and spent liquor are added to the pulper to maintain a solids-to-liquor ratio of 1:16 to 1:10, and direct steam is added to raise the temperature. Rapid pulping proceeds by the impact action of the rotor in forcing the alkali into the intercellular spaces of the raw material. Delignification proceeds at an accelerated rate as a result of the increased surface area being exposed by the mechanical action of the rotating vanes. Thus a rapid rate of pulping is achieved under relatively mild cooking conditions. Amount of chemical, concentration of chemical, consistency, temperature, and the amount of mechanical action from the pulper rotor are the important variables in determining the type of pulp obtained, which can range from high-yield coarse pulps to bleachable grades. The advantages of the mechano-chemical process are simplicity in operation and modest capital requirement for high production capacity.

Washing, screening, and cleaning operations are similar to those used in other pulping processes. The yields of mechano-chemical pulp, produced by the soda or kraft process, exceed by six to eight percent those obtained by cooking with conventional methods in pressure digesters. The pulp produced by either the soda or sulfate process is stronger than the pulp produced by the neutral sulfite process. The mechano-chemical process has been adopted by mills in Holland, Mexico, Pakistan, Portugal, and other places.

Cold Soda Process

This process is similar to the one used for dense hardwoods (see discussion on semichemical pulping). The process involves soaking or steeping the raw material in caustic soda solution and fiberizing the softened material in a disk refiner with regulated clearance between the plates. Steeping time varies from 15 to 90 min at atmospheric pressure and is reduced to about half when hydrostatic pressure ranging from 517 to 689 kPa (75 to 100 psi) is applied.

Bamboo is sometimes pulped by the cold caustic soda process. In this process, bamboo chips are sprayed with caustic soda solution at a liquor-to-chips ratio of 2:1, allowed to stand for 5 to 10 min, and the unabsorbed liquor then removed by means of a press. The compressed, alkali-treated chips are again sprayed with alkali solution at a liquor-to-chip ratio ranging from 6:1 up to 10:1. In the second stage of caustic treatment the chips are more receptive to liquor impregnation and the impregnation is completed in 10 min time, after which the chips are squeezed to 60% solids content before defiberization. Fiberization is carried out in a disk refiner after dilution to a solids content of 30%. Temperature is raised to 160 to 170°C during the fiberizing operation, which hastens the process. The pulp is further treated in another disk mill at 3 to 5% consistency before washing and screening. The concentrated liquor removed by pressing or by drainage is recycled in the process. The resultant pulp possesses the characteristics of chemigroundwood pulp and can be used as a supplementary furnish in newsprint production. Table 4–45[1022] summarizes the pulping conditions and the characteristics of cold soda pulp produced from *Bambusa polymorpha*. High yield of straw pulp rich in pentosan content and

TABLE 4–45 PULPING CONDITIONS AND CHARACTERISTICS OF COLD SODA PULP PRODUCED FROM *BAMBUSA POLYMORPHA*

	Condition of Pulp Tested	
	Defibered	Fully Refined, Screened Pulp
NaOH concentration (g/l)	31	31
Treatment time (min)	20	20
Pulp yield (%)	89	89
Power consumption (kWh/ton)	197	197
Freeness, Canadian Standard (ml)	705	410
Burst index (kPa · m²/g)	3.3	5.9
Tear index (mN · m²/g)	11.2	15.8
Tensile strength, breaking length (km)	1.2	2.7

suitable for the manufacture of glassine or greaseproof papers can also be produced by the cold soda process.

Chemical Recovery System

The chemical recovery system in an alkaline pulp mill using nonwood fiber materials is very similar to the kraft recovery system employed in wood pulp mills. However, there are significant differences in the quantity and quality of the spent liquor available from the pulping and washing operations. The differences are:

1. Greater volume of liquor per unit of pulp production.
2. Lower concentration of chemical in the liquor.
3. Less free residual alkali in the liquor.
4. Extremely high viscosity at increased concentration.
5. High mineral content, notably silica.
6. Low calorific value of the dry solids.

Nonwood fibers are high in ash content, low in lignin, and rich in pentosans compared to coniferous wood, as shown in Table 4-46.[1023] In the digestion process a high portion of the less resistant hemicelluloses and low-grade carbohydrates are dissolved in the cooking liquor along with the lignin. The inevitable presence of a high concentration of these carbohydrates in the spent liquor makes the liquor much more viscous compared to kraft softwood liquor; the viscosity dramatically increases at higher concentration and lower temperature. The amount of free alkali present in the black liquor affects the viscosity. Table 4-47[1024] shows the effect of free alkali in soda, wheat-straw liquor of 60% solids content at 50 and 90°C temperature, and as can be seen, increased free alkali reduces the viscosity considerably. Maintenance of optimum quantity of free alkali in the weak black liquor has proved a valuable operating technique in overcoming some of the problems of scaling, plugging, and lignin precipitation while concentrating the black liquor to 60 to 65% total solids with multiple-effect and direct-contact evaporators, particularly in straw and bagasse mills where these problems are most acute. Free alkali in the liquor softens the scale in the evaporator tubes for easy removal.

Silica (a major portion of the ash content) is present to the extent of 12 to 18% in rice straw, 6 to 7% in esparto, 4 to 6% in wheat straw, 4 to 5% in reeds, 2 to 3% in bamboo, 1.5 to 2% in well-depithed bagasse, and 2 to 2.5% in upgraded bast and leafy fibers. A high portion of the silica is dissolved in alkaline pulping and appears in the spent liquor, where it creates severe operating problems at various stages of the recovery operation. Silica deposits on the inner walls of the evaporator tubes and forms a hard scale in combination with calcium ions, which seriously affects heat transfer. Frequent cleaning by mechanical or a combination of mechanical and chemical means is imperative to

TABLE 4-46 CHARACTERISTICS OF NONWOODY FIBROUS RAW MATERIALS COMPARED TO PULPWOOD

Group	Fiber Average length (mm)	Fiber Average diameter (μ)	Ash (%)	Lignin (%)	α Cellulose (%)	Pentosans (%)	Raw Material Texture
1. Straws and grasses	1.1– 1.5	9–13	6–8	17–19	33–38	27–32	Open
Rice straw	0.5	8.5	14–20	12–14	28–36	23–25	Open
2. Stalks and reeds	1.0– 1.8	8–20	3–6	18–22	33–43	28–32	Open
Sugarcane bagasse depithed	1.7	20	2	19–21	40–43	30–32	Open
3. Woody stalks with bast fibers:							
woody stems	0.2– 0.3	10–11	2–3	23–27	31–33	15–22	Dense
bast fibers	20.0–25.0	16–22	1–2	1–6	60+	2–6	Open
4. Leaf fibers	6.0–9.0	16–18	1	7–10	53–64	17–24	Open
5. Bamboo	3.0–4.0	14	1–3	22–30	50+	16–21	Dense
For Comparison							
Temperate coniferous woods	2.7–4.6	32–43	1	26–30	40–45	10–13	Dense
Temperate hardwoods	0.7–1.6	20–40	1	18–25	38–49	19–23	Dense

TABLE 4-47 SHOWING THE EFFECT OF FREE ALKALI ON THE VISCOSITY OF BLACK LIQUOR FROM STRAW PULPING (SOLIDS CONTENT 60%)

	Total Solids Content in Weak Black Liquor (%)	Free Alkali in Weak Black Liquor (g/l)	Viscosity at 50°C (centistokes)	Viscosity at 90°C (centistokes)
Wheat straw liquor	13.5	4.48	20500	1430
Wheat straw liquor	13.5	8.00	8000	660
Wheat straw liquor	13.5	12.80	6000	570
Wheat straw liquor	13.5	15.83	5700	550
Kraft pine liquor[a]	16.5	18.35	1400	75

[a]Kraft pine liquor viscosity values are reported for comparison.

maintain the evaporation rate with minimum steam consumption. In some kraft bamboo pulp mills silica forms a hard encrustation in the shape of a "beehive" pattern,[1025] which has to be chiselled and removed in order to keep the hearth and boiler tube surfaces clean. Some silica is carried over to the superheater banks and economizers, where it fouls the tube surfaces and obstructs the passage of flue gases. The adverse effect of silica continues in the recausticizing operation by retarding the settling rate in the clarifiers. Silica builds up in the recovery circuit cumulatively, making sludge burning difficult and nonuniform. This results in the formation of lime with an overburnt outer core and an underburnt inner core. Silica combines with calcium to form calcium silicate, which gives rise to a glass-lining effect on the kiln wall. Excess alkali in the lime mud promotes "ring formation" in the kiln, reducing its capacity and necessitating frequent shutdowns for mechanical cleaning.[1026] Because of this problem a good portion of the sludge must be discarded from the recovery circuit and replaced with limestone in addition to the normal lime makeup.[1027] Silica can be removed by precausticizing the weak black liquor with quick lime prior to evaporation, which selectively precipitates the silica in the form of calcium silicates at 80°C,[1024] while the lignocompounds remain unchanged. The particle size of the precipitated silica is very small and colloidal in character; suitable clarifying and filtering equipment is necessary to separate it from the treated black liquor. Another process involves treatment of the black liquor with CO_2, which is quite effective for the removal of silica. The carbon dioxide is absorbed in the alkaline liquor and the pH is shifted to a point where about 25% of the lignin coprecipitates.[1028] Removal of the lignin in this manner considerably reduces the burning property of the liquor and consequently the thermal efficiency of the chemical recovery system is adversely affected.[1024] However, the carbonation process is well suited for desilication of black liquor derived from pulping of rice straw.

Due to its low-lignin content and unfavorable ratio of organic to inorganic solids the black liquor obtained from the pulping of nonwood materials has a lower heat value than the black liquor obtained from softwood kraft pulping. The calorific value falls within the range of 2150 to 3500 Kcal/kg dry solids. Table 4-48 shows the composition and heat value of spent liquors produced in the pulping of various nonwood fibers. The composition explains the poor burning characteristic of these liquors and the necessity of using a certain amount of auxiliary fuel to support combustion and sustain furnace conditions for satisfactory operation. In earlier installations, triple- or quadruple-effect, short-tube, natural-circulation evaporators, rotary incinerators, and/or smelters, lixivating tanks, batch causticizers and decanters were used for chemical recovery. Operating efficiency and equipment performance were poor and only 50 to 60% of the total chemicals used for pulping were recovered. However, large capacity modern mills using well-designed equipment have achieved chemical recovery efficiency exceeding 90%. The heat value of the black-liquor solids provides 70 to 75% of the total fuel requirement for process steam generation.

In addition to the conventional recovery process, two special processes are

TABLE 4-48 COMPOSITION AND HEAT VALUE OF SPENT LIQUOR PRODUCED FROM NONWOOD, FIBROUS RAW MATERIALS

Raw Material	Ash (including SiO$_2$) (%)	SiO$_2$ (%)	C (%)	H (%)	S (%)	O (%)	N (%)	Na (%)	Calorimetric Heat Value (Kcal/kg.)	Ref.
Bagasse	44–48	1.2	41.50	4.10	0.40	34.50	0.20	15.8	2550	–
Bamboo	–	2.0	–	–	–	–	–	–	–	1031
	48.0	2.2	31.63	2.78	2.01	15.04	0.25	–	3340	1029
	43.2	–	–	–	–	–	–	–	3200	1030
Esparto	35.0	2.0	–	–	–	–	–	–	–	1031
	–	–	–	–	–	–	–	–	3800	1030
Reed	–	2.8	–	–	–	–	–	–	–	1031
	41.3	3.4	–	–	–	–	–	17.2	3620	1032
	39.4	4.7	–	–	–	–	–	–	–	1033
Rice straw	–	16–30	–	–	–	–	–	–	–	1031
	38.9	11.1	–	–	–	–	–	–	3180	1031
	42.5	14.8	–	–	–	–	–	–	2800	1031
Rye straw	–	1–3	–	–	–	–	–	–	–	1031
	41.9	1.9	–	–	–	–	–	–	–	1031
	36.6	2.9	32.50	3.80	–	–	–	–	3200	1035
Wheat straw	–	4–8	–	–	–	–	–	–	–	1031
	54.5	–	–	–	–	–	–	16.0	2550	1030
	51.8	4.0	31.80	3.00	3.20	41.40	–	–	1310	1034
Pine	44.0	–	41.70	4.30	3.60	–	–	–	4130	1036
Birch	–	–	37.50	3.60	4.40	28.50	–	26.0	3660	1037

worth describing. (See also recovery processes described in the sections on chemical and semichemical pulping.)

Copeland Process. This process[1038] can be used in soda, kraft, or neutral sulfite pulp mills for chemical recovery operation. It is based on thermal oxidization of organic matter in a fluidized-bed reactor. The process operates autogenously with black liquor concentrated to about 35% solids. The installation is relatively inexpensive and easy to operate and problems related to high viscosity and high-silica content become less apparent compared to a conventional recovery system. The overall thermal efficiency of the system is somewhat lower than in conventional recovery, but to some extent this is offset by the reduced steam consumption in evaporating the black liquor to lower concentration compared to that required in conventional recovery.

Wet Oxidation Process. Based on eucalyptus wood,[1039] the wet oxidation (Zimmermann) process is in use for chemical recovery in a soda pulp mill in Australia.[1039] Experience gained in operating this recovery system indicates that it could be adapted to pulp mills using nonwood fibers for pulping. The significant advantage of this process is that black-liquor evaporation and incineration of solids in the furnace is eliminated. Weak black liquor, as it comes from the brown-stock washing system, is oxidized in the aqueous phase in a specially designed high-pressure reactor using compressed air. The organic substances are completely oxidized and the treated liquor is practically colorless and free of any carbon particles. The liquor is clarified and a considerable amount of silica, along with other inorganic impurities, is removed with the dregs. The process is unique and is well suited for soda pulp mills. Recovery efficiency is exceptionally high (up to 99%) with almost no chemical losses. Another advantage of the process is the protection of the environment since there is no emission of flue gases to the atmosphere.[1040]

Design of Plant for Alkaline Pulping of Nonwood Materials

Planning, designing, choice of equipment, and coordination of various unit operations for a pulp mill based on nonwood materials require a thorough understanding of the morphology and the physical and chemical characteristics of the raw material.

Material handling arrangements—from the storage yard through the fiber preparation system—differ considerably from one raw material to the other. Straws, grasses, and reeds are stored in the form of bales and are transported to the fiber-preparation station either by trucks or by a series of conveyors. Dry methods are used for chopping, dusting, and cleaning in the case of grasses and reeds. Chopped and uncut straws are cleaned by either dry or wet methods. Bagasse requires a different storage and handling arrangement and it must also be depithed as described in the section on pulping of bagasse. Bamboo is stacked in the form of sticks of varying lengths and is transported directly to the chipper house by trucks or by floating in long flumes. The latter method is preferred because it removes the adhering soil and dirt. Washed bamboo sticks are picked up by submerged conveyors and fed straight to the chippers and the chipping, screening, and rechipping operations are similar to those used for wood. Bast and leaf fibers are treated and upgraded in an entirely different way. Abaca must be decorticated, or stripped (manually or by a spindle system) to separate the long fiber strands from the parenchyma, spongy material. Flax seed straw, sisal, and hemp require crushing, decorticating, shredding, and dusting before they are ready for cooking.

Several factors have to be taken into consideration in choosing the pulping process and the equipment for a particular nonwood fibrous material. These factors are:

1. Bulk density of the raw material.

2. Degree of flexibility required to produce different grades of pulp.

3. Degree of automation and sophistication desired to control pulping variables.

4. The production rate required.

Drainage characteristics of cooked, but unrefined pulps from nonwood materials differ considerably as shown by the freeness values reported in Table 4-49.[1041] Short-fibered pulps are difficult to wash since they form a very close mat on the washing drums, thereby impairing liquor filtration. The specific loading factor of pulps on brown-stock washers ranges from 1360 to 2180 kg/m^2/day (1.5 to 2.4 ton) depending on the freeness value. Three- or four-stage washing provides high washing efficiency, low-soda loss, and high-liquor concentration for the chemical recovery operation. In some cases the brown-stock washers are hooked up with a vacuum pump to increase the filtration rate. In order to obtain pulp free of specks, shives, and dirt, it is worthwhile to carry out screening and

TABLE 4-49 FREENESS VALUES OF COOKED NONWOOD PULPS IN COMPARISON TO WOOD PULPS[a]

	Unbleached	Bleached
Bagasse	20	22
Bamboo	14–16	16–18
Esparto	16	18
Reeds	18–19	22
Rice straw	40–42	42–45
Wheat straw	30	35
For comparison purposes		
Softwood pulp	12	15–16
Hardwood pulp	15–16	18–19

[a] Freeness in °SR

cleaning tests on each particular pulp, using screen plates of different-size perforations or slots and cleaners of different sizes and designs to determine the best combination of screening and cleaning equipment for optimum efficiency. In this way designing of the screening and cleaning system with appropriate equipment can be accomplished.

Each pulp from nonwood material responds differently to bleaching and therefore the retention time, consistency, and operating temperature used in bleaching need to be established to determine the sizes of towers and auxiliary equipment necessary. Number of bleaching stages and bleaching sequences are dictated by economic considerations and the final brightness- and strength-requirements. The economic considerations in designing a multistage bleachery should take into account the capital investment and the marketability of the product.[1042]

The design criteria normally used for a multiple-effect evaporating unit are based primarily on the kilos of water evaporated per kilo of steam used. High viscosity and high-silica content of black liquor are two major adverse factors that greatly influence the total heating-surface area to be provided for a particular installation. Rapid scaling of the tubes necessitates frequent cleaning to maintain proper heat transfer and for this reason a spare evaporator should be included in many cases to ensure uninterrupted plant operation. Forced liquor-circulation, particularly at the critical stage where the viscosity reduces the velocity of liquor-flow through the tubes, should be incorporated in the system to achieve high evaporating efficiency. Due to the poor burning quality and the relatively low-heat value of the concentrated black liquor, adequate surface area should be provided in the furnace to dehydrate the sprayed liquor before the dried solids fall into the combustion zone of the hearth. A number of other modifications to a conventional, stationary furnace are necessary, such as location of the liquor firing points in relation to the smelt spout, amount of hot-air supply, and distribution of the air for stable furnace operation. Some silica is carried along with the fumes and it forms a hard encrustation on the tube surfaces. Adequate spacing should be provided between the boiler tubes, superheater banks, and economizers to permit easy access for periodic cleaning and unhindered passage of flue gas. Provision should be made for additional soot-blowers and airlancing devices at critical points to keep the tube surfaces clean.[1024,1029] The amount of silica in the green liquor determines the overall dimensions of the recausticizing system. The slaking unit and the causticizers should be suitably sized to provide adequate retention time to achieve maximum causticizing efficiency. Oversized white-liquor clarifiers and mud-washing equipment are necessary to compensate for the slow settling rate caused by the presence of silica.[1024] Silica in the lime mud presents problems in reburning and it is, therefore, desirable not to include a lime sludge burning arrangement in the chemical recovery system if waste-solids disposal does not present any problem and good quality burnt lime is available at reasonable price. However, if silica is substantially removed by carbonation of the green liquor prior to causticizing, reburning of desilicated lime sludge is technically and economically feasible.[1028]

Pulping of Straw

Cereal straw is an important source of raw material for pulp and papermaking where grain is grown in vast quantities. Historically, straw is the oldest paper-making material and dates back to the early part of the second century when the Chinese invented papermaking. It remained the major source of fibrous raw material for pulp production in Europe and North America until the wood pulp industry was firmly established in the early 1920s. The decline of straw pulping actually began when straw collection, handling, and storage became increasingly difficult in the face of rising labor costs and changes in harvesting methods. Straw pulping, which once thrived in the United States, gradually lost ground against rapidly expanding wood pulp production and it is now virtually abandoned. On the other hand, straw as a base for the pulp and paper industry has made steady progress in many European countries especially Bulgaria, Denmark, Greece, Holland, Hungary, Italy, Rumania, Spain, and Yugoslavia where pulpwood supplies are extremely limited and the purchase of wood pulp from outside sources is too expensive to support local paper production. Growth in the utilization of cereal straws has also taken place in many developing countries particularly Algeria, Argentina, China, Egypt, India, Indonesia, Mexico, Pakistan, Sri Lanka, Syria, and Turkey. In these countries corrugating medium, boards, and packaging paper are produced from high-yield unbleached straw pulps; bleached straw pulps are used as a major furnish for fine quality writing, printing, and other paper grades.

The availability of straw is closely linked with grain production. Of the huge quantity of straw available after grain collection, a very small amount is used for cattle fodder and bedding. Most of the straw is burned or plowed back to supply humus and soil nutrients. The price of straw in the field is almost nothing compared to the expenses for baling, transportation, and storage. The extensive use of combine harvesters has greatly reduced the harvesting period and the baling takes place simultaneously with the harvesting. Because grain harvesting is the primary objective of combine operation, the crop is cut high from the ground and a good portion of the stem, which is rich in cellulosic materials, is left uncut. However, specially designed units with a scissor-like cutting arrangement are available that simultaneously cut and bale[1043] and with these the straw yield per hectare can be considerably increased.

In the Orient, rice is usually cultivated and harvested manually. The crop is cut close to the ground and vast quantities of straw are available after separation of the grain. In spite of the extensive use of straw for animal fodder, domestic fuel, and roof thatching, a huge surplus is wasted or is ultimately plowed back into the land. Abundant quantities of straw are therefore available for pulp production. In rural areas, straw collection for industrial use provides additional earnings for landowners and employment opportunities for many unoccupied farm hands. In continental Europe, straw collection and storage varies from country to country. In some countries pulp and paper companies buy all their annual straw requirements in the form of bales from the farmers. The entire quantity may be stored in the mill yard or a portion may be stacked in farmlands at a reasonable distance from the mill and shipped from there on a daily basis the year round. In some places where straw supplies cannot be handled by farmers, the paper company maintains a fleet of tractors and balers and other equipment to collect straw and store it at the mill site.[1043] Straw bales are stacked to a height of 12 to 13 m, a length of 150 to 160 m and a width of 20 to 22 m.

Moisture content of the straw bales has an important bearing on the deterioration during storage; acceptable moisture content is between 8 to 14%. Higher moisture content causes the straw to undergo microbial attack and makes it susceptible to fire hazards due to intense heat developed in the stacks. Weathered straw consumes more chemical in pulping, produces pulp of lower strength, and has a lower yield. Chemical preservatives, such as borax and boric acid, are effective in protecting straw against deterioration during storage, but these chemicals are rarely used because they are too expensive. In some mills the stacks are left uncovered, whereas in other cases the stacks are covered with thick polyethylene sheets, canvas, or sheet metal as a protection against adverse weather conditions. In tropical countries straw stacks are covered with palm leaves or are thatched to protect them against rotting.

Physical and Chemical Properties of Straw

Straw stems are elastic and tubular in structure and connected at intervals with nodes. The top portion of the stem where the seeds are attached is called the rachis. Leaves originating from the nodes form a sheath enclosing part of the stem. The head of the plant contains chaff, leaf sheath, blade, glumes, bristles, and rachis. Wheat straw grows to a height of about 1 to 1.5 m depending on the strain, climate, and soil conditions.

Straw has a low cellulose content, but the total holocellulose content is approximately equal to that of wood. Straw contains a high percentage of pentosans and a low percentage of lignin compared to wood. The ash content of straw is invariably higher than that of wood. Analyses of European straws and rice straw from Sri Lanka are given in Tables 4-50[1044] and 4-51[1045] respectively. Wheat and rye are the preferred straws for pulping. Rye straw has a higher cellulose content and more long fibers than other straws. Oat straw is easily pulped but yields more short fibers, whereas barley straw requires somewhat drastic cooking conditions. Papers produced with wheat and rye straw pulps are stronger and stiffer than papers made from other straw pulps.

The bast cells, which exist predominantly in the internodes, are the principal source of fiber on pulping. The bast fibers are associated with epidermal cells, platelets, serrated cells, and the spirals, which are very small and have no fibrous characteristics. Bast fibers are derived principally from the pith or the inner part of the stem and to a lesser extent from the nodes, sheaths, and chaffy materials of the straw. Rydholm[1046] reports that wheat straw contains 50% bast and sclerenchyma fibers, 30% parenchyma, 15% epidermal, and 5% vessels. Straw

TABLE 4-50 AVERAGE COMPOSITION OF EUROPEAN STRAW
(%)

	Summer Barley	Winter Barley	Oats		Summer Wheat		Winter Wheat Heavy Soils	Rye	General Average
			Heavy Soils	Light Soils	Heavy Soils	Light Soils			
Total ash	10.5	8.7	11.9	7.4	11.2	5.1	11.1	3.8	8.7
Ether extract	1.5	1.2	1.2	1.5	1.2	1.4	1.0	1.4	1.3
Total protein	2.5	2.9	2.5	2.3	3.3	3.5	2.5	3.5	2.9
Water extract	12.0	12.3	14.9	14.8	12.0	13.4	12.5	8.8	12.6
Insoluble ash	2.7	1.9	2.3	0.7	4.4	0.9	4.9	0.6	2.3
Insoluble protein	1.4	1.4	1.3	1.3	1.6	1.8	1.5	1.6	1.5
Lignin	16.2	16.9	15.5	16.7	16.3	17.0	15.7	17.8	16.5
Pentosan	27.0	26.4	26.0	26.4	26.2	27.0	27.1	27.9	26.8
Holocellulose	73.9	72.8	71.7	70.2	74.0	72.1	75.0	75.5	73.2
Ash in	2.6	1.8	2.3	1.2	4.2	1.2	4.7	0.9	2.4
Pure	71.3	71.0	69.4	69.0	69.9	70.9	70.3	74.6	70.8
Alpha cellulose	39.4	39.5	39.4	38.0	39.8	38.0	40.5	42.1	39.6
(as % of holocellulose)	53.4	54.4	55.0	54.2	53.7	52.8	54.0	55.8	54.2
Ash in	1.4	1.1	1.3	0.4	2.8	1.2	3.0	0.4	1.5
Pentosan in	2.0	1.7	1.8	2.3	2.4	2.4	2.4	2.5	2.2
Pure	36.0	36.7	36.4	35.4	34.6	34.4	35.2	39.3	36.0
Remainder	3.2	3.2	2.4	3.2	3.7	4.1	2.1	2.6	3.0

TABLE 4–51 ANALYSIS OF SRI LANKA RICE STRAW

Moisture (%)	8.8
Ash content (%)	24.8
SiO_2 (%)	14.3
Alcohol/benzene extract (%)	2.1
Hot water extract (%)	17.7
Lignin (%)	12.0
Pentosan (%)	21.0

fibers are much more heterogeneous than wood pulp fibers. Fiber dimensions of cereal straws are: length 3120μ to 680μ, with an average of 1480μ and a diameter 24μ to 7μ with an average of 13μ. The corresponding figures for rice straws are: length 3480μ to 650μ with an average of 1450μ and a diameter 14μ to 5μ, with an average of 8μ. Ratio of fiber length of diameter is 111:1 in the case of cereal straws and 170:1 for rice straw.[1047] The good paper formation characteristics of straw fibers are attributable to their relatively high ratio of average length to diameter.

Preparation of Straw for Pulping

A considerable portion of the chaff, nodes, rachis, adhering soil and impurities are removed in the chopping, screening and cleaning operations. Good cleaning is essential for fine quality bleached pulp production. Batch digesters can be charged with more straw if it is chopped. Partial cleaning of straw may be adequate for semichemical or coarse pulp production. A typical layout of a straw cutting system consists of a pair of cutters located side by side with a common discharge end. Each cutter is connected with a slow moving conveyor, which feeds straw bales to the cutter. Before the bales approach the cutter the baling wires are removed manually. The straw cutter consists of two rotating spike drums with spring-loaded mechanism and rotating drums fitted with spiral knives. Straw is grabbed between the chain conveyor and the rotating, spike drums and moves forward to the edge of the dead knife where the rotating knives cut it into lengths of 4 to 6 cm. Cut straw is pneumatically transferred to the screening and cleaning unit. Two sets of such cutting stations are needed in order to supply a continuous flow of straw to the digester room without interruption for knife changing and repair work.

Screening is done with two semicircular screen plates arranged in a dome-like fashion and a two-deck horizontal screen located underneath. The semicircular screens have perforations of 1.5 mm and the flat screen has 15-mm holes. An exhaust fan sucks the cut straw from the cutters and delivers it to the screening system. The semicircular screens remove the dust and other suspended impurities; the grains, nodes, and other heavy substances are removed by the flat screen. Two hoppers are provided under the flat screen: the first hopper receives the grain and nodes and the second hopper receives the clean straw. There are star feeders located at the discharge end of both hoppers. The grain and nodes removed by the first star feeder are further treated by vibrators to separate the grain from the nodes. The grain is sold to flour mills and the nodes are mixed with a small amount of grain and sold for animal feed. Broken fibers and other lighter substances are released to a cyclone or dust separator. Heavier material is collected at the bottom cone of a cyclone; the dusty air escapes from the top. A rotating, magnetic drum is located after the screening system to catch broken baling wire or metallic trash before the cleaned straw is transported to the pulping unit. A second-stage screening and cleaning arrangement, similar to the one described above (but without the grain separation arrangement), can be used to ensure well-cleaned straw.[1048]

Wet cleaning is very effective, particularly when the straw is contaminated with dirt and soil during harvesting. Either chopped or whole straw is fed to a continuous or semicontinuous pulper fitted with a heavy duty rotor. Fresh or recycled water from the pulp mill is used for dilution. The vigorous action of the rotor causes partial defibration. The treated straw is then transferred to a chest fitted with a perforated plate and from there a drag conveyor transports the straw to a drainer-conveyor. Dirt and a high portion of the siliceous material embedded in the straw are removed through the perforated plate in the chest as well as through the dewatering conveyor, while the straw is transported to the digester.[1049] This wet-cleaning system is particularly well suited for rice straw, but the power, repair, and maintenance costs are high compared to the dry cleaning system.[1050]

Pulping Processes Used on Straw

Straw is pulped by chemical processes and by the mechano-chemical process. Soda, sulfate, and neutral sulfite processes are used for high-yield semichemical as well as for bleachable grades of pulp. In terms of strength properties the soda process is the best, followed by the sulfate, and then by the neutral sulfite process. Yield-values are in the reverse order.[1051] The acid sulfite process is unsuitable for straw pulping; the pulp produced is brittle and poor in strength. The Celdecor-Pomilio process (soda-chlorine) is mostly used for bleached pulp production, but the process can be interrupted after the alkali predigestion stage to produce a high-yield pulp for corrugating medium and board manufacture. Lime or lime-soda processes are used for the production of coarse pulps for the manufacture of straw boards, packing and wrapping papers, and corrugating medium.

In batch pulping, rotary digesters are used with the lime, lime-soda, and other chemical processes, although continuous digesters are more commonly used for the soda, sulfate, and neutral sulfite processes. Pulping conditions for batch pulping are given in Table 4–52, which shows the various chemicals and pulping variables used to produce high-yield coarse pulps as well as bleachable pulps. Yield-values for wheat and rye straws for the full range of pulp production are

TABLE 4-52 PULPING CONDITIONS FOR BATCH PULPING OF STRAW

	Lime	Lime-Soda	Soda	Sulfite	Neutral Sulfite
Chemical (% on moisture-free straw)	CaO 6–12 or MgO 12–18	5–10 CaO + 3–4 Na$_2$CO$_3$	NaOH 6–15	NaOH 10–12 + Na$_2$S 2–3	10 Na$_2$SO$_3$ alone or 4–10 Na$_2$SO$_3$ + 2–5 NaOH
Liquor-to-straw ratio after steaming	4:1	5:1	4:1–7:1	4:1–6:1	3:1–7:1
Cooking time	6–10 hr	6–8 hr	2.5–4 hr	2.5–4 hr	2–4 hr
Cooking temperature	125–140°C	120–140°C	150–170°C	150–170°C	160–170°C
Variety of pulp produced	Coarse	Coarse	High-yield semi-chemical and bleachable grade	High-yield semi-chemical and bleachable grade	High-yield semi-chemical and bleachable grade
Yield (crude yield) from:					
Wheat + rye straws	70–85%	70–80%	48–70%	50–70%	53–65%
Rice straw	70–80%	65–70%	40–45%	42–47%	45–48%

TABLE 4–53 PROPERTIES OF BLEACHABLE-GRADE, WHEAT STRAW PULP PRODUCED BY SODA PROCESS IN A CONTINUOUS PANDIA DIGESTER

Permanganate number	12.1–12.5
Freeness (°S-R)	30.0
Ash content (%)	1.5–2.8
Alpha cellulose (%)	68.16
Lignin (%)	5.1
Pentosan (%)	21.2
Viscosity (0.5% CED, cp)	30.0
Classification by Bauer McNett fractionater:	
retained on 28 mesh	30.8
retained on 48 mesh	17.2
retained on 100 mesh	14.7
retained on 200 mesh	15.9
passing thru 200 mesh	21.4

higher than the yield-values for the corresponding grades of pulp produced from rice straw.

Pandia and other continuous digesters are used for pulping cereal straws, using 5 to 6% active alkali in the case of the soda or sulfate processes and 4% sodium sulfite plus 2% sodium hydroxide in the case of the neutral sulfite process for the production of coarse pulp for corrugating medium. Bleachable grades are produced with 12 to 14% active alkali in the case of the soda and sulfate processes and 8% sodium sulfite plus 4% sodium hydroxide in the case of the neutral sulfite pulping process. Liquor-to-straw ratio varies between 3 to 4:1. Total retention time in the digesters is 14 to 18 min and cooking temperature ranges from 160 to 170°C. Yield-values obtained in the continuous process are similar to those obtained in batch pulping. However, pulp produced by continuous pulping is more uniform in quality and it is well defibrated.[1048] Rice straw pulping in a continuous system presents problems due to clogging of the blow lines and the piping network.[1052] Table 4–53[1048,1053] provides data on the properties of bleachable-grade wheat straw pulp produced by the soda process in a continuous Pandia digester.

Hot-stock refining of high-yield pulp in alkaline medium prior to washing improves the strength properties of pulp intended for the manufacture of corrugating medium and wrapping grades. Breaker beaters, equipped with washing drums, flat screen, rifflers, and vortraps, are the traditional equipment used for washing, screening, and cleaning of pulp. However, this arrangement has been replaced by a continuous countercurrent three- or four-stage washing system followed by vibrating knotters, rotary screens, and three-stage centrifugal cleaners.

Straw Pulping by Celdecor-Pomilio Process. This process has been described in a previous section (see also Figure 4-117). Straw pulping by the Celdecor-

Pomilio process has undergone several modifications over the years to reduce chemical consumption and to improve yield and strength characteristics of the pulp, particularly the concora value in the case of high-yield pulp for corrugating medium.[1054] Cooking conditions for coarse-grade pulp are 6 to 7% sodium hydroxide based on moisture-free straw, cooking temperature from 110 to 130°C, and retention time in the predigestion alkali tower from 1 to 3 hr. The yield is about 60 to 65%.

For bleached-pulp production, the straw is impregnated with hot caustic soda liquor (85 to 90°C) containing 25 g/l NaOH. The impregnated straw is cooked in the alkali tower for about two hours at a temperature of 115 to 125°C. The tower is of steel construction and steam-jacketed for indirect heating. Total caustic consumption is about 8.5 to 9% on the dry weight of clean straw. Total chlorine used in the gas phase in the chlorination tower is 4 to 4.5% on the dry weight of clean straw. The washed chlorinated pulp is alkali treated in the extraction tower with a retention of 40 min after which the pulp is screened, cleaned, and thickened. Bleaching is completed in a two-stage hypochlorite bleaching sequence, using 1 to 1.5% available chlorine on the weight of clean dry straw. Bleached yield varies from 40 to 45% and the brightness from 80 to 82.[1055]

Straw Pulping by Mechano-Chemical Process. The mechano-chemical process, which is covered in the discussion on pulping processes, can be used to produce a coarse pulp in the yield range of 70 to 75% which is suitable for corrugating medium. Cooking chemicals used are: (1) 10% lime or 6% lime plus 1.5% caustic soda or (2) 6 to 8% caustic soda; based on moisture-free straw. Pulping is carried out at 98°C in a hydrapulper for 60 to 80 min.

Bleachable-grade pulp is produced by mechano-chemical process using 12 to 13% caustic soda with a cooking time of 30 to 35 min.[1020] In this process the nodes and rachis of the straw are only swollen and not defibrated, making it easier to separate them by screening and cleaning before the pulp is bleached. Brightness is limited to 70 in a single-stage bleaching using 10 to 12% chlorine. In a three-stage bleaching sequence (CEH) the pulp can be bleached to 80 brightness with 8 to 9% chlorine. Extremely coarse-grade straw pulp, suitable for board, can be produced by cooking in a hydrapulper at a temperature near 100°C without chemical.[1020] The pulping time ranges from 30 to 90 min and the consistency is about 10%. The pulp can be blended with a portion of chemically cooked straw pulp to provide the necessary structural properties required to produce insulating boards.

A process called the H. F. method (Højbygaard Papir Fabrikker process)[1056] has been developed for producing high-yield pulp from wheat and rye straws for the manufacture of corrugating medium. Cut straw is continuously fed in regulated quantity to an inclined-horizontal diffuser, which works on a countercurrent principle. In the diffuser the straw first passes through an impregnating zone, taking up residual chemicals and heat; then through a digesting zone at a temperature up to 100°C, where the cooking liquor is introduced; and finally through a washing zone, where the dissolved materials, including lignin, silicates, and a portion of the hemicelluloses, are removed with the filtrate. Consumption of steam is exceptionally low since impregnation, cooking, and washing are performed in the same unit. The filtrate is recycled and the concentration of the liquor increases up to 20% by repeated use. At this concentration it would be possible to evaporate the liquor to higher solids content for chemical recovery. The pulp, which does not require any further washing, is removed from the diffuser by a scraper mechanism at 12 to 18% consistency. The semipulp from the diffuser is defibrated in a pulper and refiner. The special feature of this process is that all the steps of pulping are carried out in a single unit.

Bleachable-grade rice straw is produced using 10 to 12% caustic soda in a hydrapulper at 90 to 100°C. The pulp is screened, bleached in a CEH sequence, and refined for glassine or parchment paper manufacture.[1057]

Straw Pulp Bleaching

Cereal-straw pulps are usually bleached in a three-stage (CEH) bleaching sequence. Soda and sulfate pulps with a permanganate number of 12 to 14 can be bleached with 8 to 10% total chlorine to a brightness of 80 plus. Neutral sulfite pulp can be bleached to the same brightness level with 6 to 7% total chlorine.

TABLE 4-54 STRENGTH PROPERTIES OF UNBLEACHED PULP AND BLEACHED WHEAT STRAW PULPS IN CEH AND CEHD SEQUENCES COOKED BY SODA PROCESS

	Unbleached Pulp	Bleached CEH	Bleached CEHD
Permanganate number	12.8	—	—
Ash (%)	5.0	1.92	1.0
Viscosity (cp)	35.8	15.0	16.2
Solubility in 1% NaOH	5.3	5.6	5.6
Initial freeness (°S-R)	30.0	32.0	31.0
Final freeness (°S-R)	50.0	50.0	50.0
Beating time in Jökro beater (min)	4.0	3.0	3.0
Basis weight (g/m^2)	78.1	77.1	78.0
Breaking length (km)	7.6	7.3	7.5
Tear strength (mN)	314.0	275.0	314.0
Tear index (mN · m^2/g)	4.23	3.74	4.23
Bursting strength (kPa)	304.0	314.0	373.0
Burst index (kPa · m^2/g)	4.1	4.2	5.0
Folding endurance, double folds	65.0	40.0	90.0
Brightness	37.0	80+	86.0
Yield of screened pulp based on moisture-free cleaned straw (%)	45–46	40–41	41–41.5
Bleaching shrinkage based on moisture-free unbleached screened pulp (%)	—	11–12	10–11

TABLE 4–55 STRENGTH PROPERTIES OF BLEACHED RICE STRAW PULP

Freeness ($^{\circ}$S-R)	42	51	58	65.5
Breaking length (km)	4.8	5.3	6.1	5.8
Burst index (kPa $\cdot$ m^2/g)	3.4	3.8	4.6	4.5
Tear index (mN $\cdot$ m^2/g)	6.76	6.27	6.03	6.03
Opacity (%)	88.9	87.9	87.9	89.2
Brightness	78.2	77.9	77.2	72.5
Folding endurance	96	149	171	153

Color reversion occurs and is more pronounced when the brightness levels are increased beyond 82 to 83. Sulfur dioxide treatment in the final stage stabilizes the brightness and acts as an antichlor. Chlorine dioxide is used following the CEH sequence to increase the brightness level to 85 or more and some mills in Denmark, Holland, and Rumania bleach wheat straw pulp using chlorine dioxide as the final stage of bleaching. Bleached yield varies from 38 to 45% depending upon the cleanliness of the straw and the extent of degradation during storage.[1041]

Rice straw pulp can be bleached with 8% chlorine to a brightness of 68 in a single-stage hypochlorite treatment using bleaching hollanders. It can be bleached by three-stage (CEH) bleaching to 75 to 80 brightness at a total chlorine consumption of 6 to 8% for pulps in the permanganate number range 6 to 8. It is difficult to bleach the pulp over 82 brightness in a CEH sequence and any attempt to do so results in tremendous strength loss.[1058] Bleached pulp yield is around 35%. Table 4–54[1042] gives the strength properties of unbleached and bleached wheat straw pulp (bleached in CEH and CEHD sequences). Table 4–55[1045] gives the strength properties of bleached rice straw pulp.

Pulping of Bagasse

Sugar cane (*Saccharum officinarum*) is grown in tropical and subtropical countries. Bagasse is the fibrous residue of the sugar cane left after the crushing and extraction process. The quality of bagasse is dependent on the variety of cane, age at cutting, agronomic and soil conditions, and the extent of crushing and milling operations carried out on the cane in the sugar mill. Before cane harvesting the field is set afire to burn the leaves, which helps to get rid of undesirable materials. Cane is harvested by mechanical harvesters and piled up by bulldozers or in the less-mechanized parts of the world, cane is harvested manually. Normally bagasse is used by the sugar mills as fuel for generating process steam and electrical energy. Surplus bagasse, to the extent of 20 to 25%, can be available from sugar mills that have a high thermal efficiency in their boiler operation and properly conserve heat and energy in their process. However, no large-tonnage pulp mill can rely on surplus bagasse alone and a considerable portion of the bagasse used for pulping is released from the sugar mills in exchange for alternative fuel.

Bagasse, being a by-product of the sugar industry, enjoys a unique position among the nonwood fibers for pulp manufacture, mainly due to its availability in large quantities at a central collection point. Its distinct advantage over other nonwood plant fibers is that the costs of collecting, crushing, and cleaning the material are borne by the sugar mill. The crushing season normally lasts from 4 to 6 months but extends over 6 to 9 months in Hawaii, Peru, and Mexico. Adequate quantity of bagasse must be collected and stored during the crushing season to ensure continuous pulp-mill operation until the next grinding season.

Storage of Bagasse

The traditional method of storing bagasse is in the form of well-compacted bales of varying sizes and weights. Where handling and stacking operations are mechanized, large bales weighing up to 750 kg (moist basis) are made for efficient and rapid collection. In the Orient, small-sized bales weighing 30 to 35 kg are used for storage. These involve the use of more labor, but small-sized bales have the advantage of undergoing less fermentation and deterioration during storage.

Wet or moist storage in the unbaled bulk form is increasingly used by many mills to reduce handling and storage costs. Bulk storage has multiple advantages over the baling method, mainly in reduced cost of labor needed for handling bales, elimination of baling wire, and reduced fire hazard and dust problems. The Ritter biological pretreatment process for wet bulk storage[1059,1060] has been widely adopted. In this process, shown in Figure 4-120, whole or partially depithed bagasse is mixed with biological culture liquor, conveyed to an elevated channel, and then flushed down to a large concrete slab. The treated bagasse is piled up and the liquor is drained and recycled. Bagasse is removed from the storage area for further depithing and pulping by heavy-duty handling equipment. The pith is loosened by biochemical action during the storage period and separated in the secondary depithing operation. Bagasse treated by the Ritter process produces pulp of superior quality in high yield.

Physical and Chemical Properties of Bagasse

Whole bagasse is composed of three principal components:

1. The rind, including the epidermis, cortex, and pericycle.
2. The vascular fiber bundles, comprising thin-walled conducting cells associated with relatively thin-walled fibers with narrow lumen.
3. Ground tissue (parenchyma) or pith with fiber bundles distributed irregularly.

Crude bagasse contains 70 to 75% useful fiber and about 30 to 35% pith, dirt, and other water-soluble materials. The pith is an undesirable fraction because it

TABLE 4-56 ANALYSIS OF WHOLE BAGASSE FROM SEVERAL SOURCES[a]

	Louisiana, Fresh (Houma, 1941)	Florida, Fresh, Dry Screened (Clewiston, 1948)	Hawaii, Variety 8560 (1952)	Hawaii, Variety 1933 (Ewa Plantation)	Puerto Rico (Aguirre, 1951, 1952)	Mexico[b] (San Cristobal, 1953)	Philippine (Negros Island, 1952)
Moisture (%)	4.9	7.3	13.2	7.7	7.5	6.2	10.2
Extractives in							
Alcohol-benzene (%)	6.0	10.8	3.2	3.6	5.4	2.3	3.0
Hot water (%)	8.8	11.2	5.7	4.0	8.0	7.6	2.8
1% NaOH (%)	35.9	39.9	33.9	31.3	27.3	40.1	31.3
Ash (%)	2.4	2.2	5.4	2.6	3.9	4.9	2.3
Lignin (%)	18.9	18.1	21.3	19.3	18.1	22.4	22.3
Pentosans (%)	30.0	27.9	27.7	31.3	29.6	29.9	31.8
Cross and Bevan cellulose (%)							
Ash free	53.3	52.0	50.2	55.0	50.9	46.0	56.8
Pentosans in	27.9	26.9	27.4	33.4	31.1	25.5	30.6
Alpha as run							
Ash free	67.3	68.0	69.6	62.6	63.7	61.3	70.2
Pentosans in	7.0	4.7	8.8	8.2	7.0	5.5	12.5
Alpha-basis original, ash and pentosan free	33.4	33.7	31.8	31.6	30.1	26.6	34.9

[a](All values, except moisture, on oven-dry basis).
[b]Pith.

contains nonfibrous cells, which are below 0.4 mm in length. Pith reduces the drainage characteristics of the pulp and increases press stickiness on the paper machine. An approximate analysis of whole bagasse from various sources is shown in Table 4-56.[1061] Dry, fresh bagasse is bright and varies in color from grayish-white to light green depending on the variety and age of cane. The pith is white and is associated with the fiber bundles. Stored bagasse ranges in color from yellowish-brown to dark gray. The pith is relatively free and loose and becomes separated during the storage and fermentation process. Microorganisms grow on the pith cells during storage because of the residual sugar content. Separation of pith from the fiber is easier if the bagasse has been stored.

The chemical composition of the different fractions of bagasse, including whole bagasse, separated fiber, and pith is given in Table 4-57.[1062] The alpha cellulose content is higher in depithed fiber and ash content is high in the pith fraction. Pentosan content is almost 50% higher than in typical temperate hardwoods and exceeds by three to four times that present in coniferous softwoods. Lignin content is about 75% that of softwoods and about equal to that of hardwoods. The holocellulose content is almost equal to that of hardwoods, indicating the possibility of producing high-yield pulp from depithed bagasse under optimum pulping conditions.

Depithing of Bagasse

Depithing is an imperative step to upgrade bagasse for the production of high-grade pulps. Crude bagasse or partially depithed bagasse may be used for low grade corrugating medium, insulating boards, and similar varieties. Well-depithed bagasse requires less chemical in cooking and bleaching to produce high-quality pulp. Depithing also increases the yield and improves the brightness and strength properties of pulp. Basically there are three methods for depithing: dry, moist or humid, and wet.

Dry depithing is carried out on stored bagasse and consists of a bale breaker followed by dusting and screening arrangements to remove pith, dirt, soil, and other extraneous materials. Along with the pith, a considerable amount of useful fiber is separated, which affects the economy of pulping.

Humid or moist depithing, which is shown in Figure 4-121, is normally done at the sugar mill. The pith can be used in the sugar mill boilers, thereby reducing the cost of replacement fuel. Separated pith is also processed for animal feed and briquettes. Moist depithing involves the processing of bagasse as it comes from the sugar mill crushers at about 50% moisture. Depithers are sized to the maximum available bagasse released from the sugar mill in order to avoid any interruption in the sugar-mill operation during the peak crushing season. Several types of depithing machines, such as the Horkel, Rietz, Peadco, Gunkel, and others, are in commercial use, all of which utilize the same basic principle with variations. These machines are designed to break open the fiber bundles and to dislodge the embedded pith by mechanical rubbing and mild disintegrating action. Separation of fiber from pith takes place at the same time. Almost two-

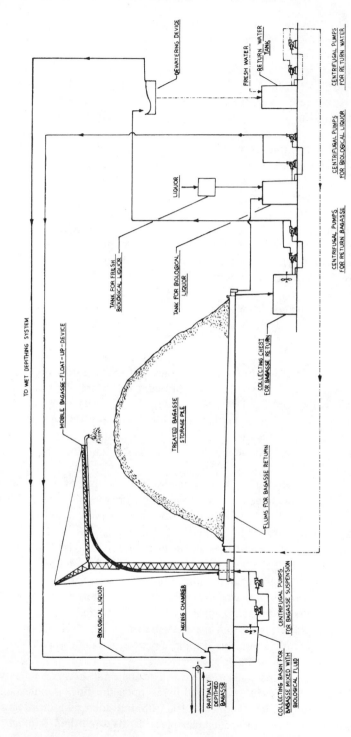

Figure 4-120. Flow sheet for bagasse storage area using the Ritter system after "moist" depithing.

TABLE 4–57 CHEMICAL PROPERTIES OF FRACTIONS OF HAWAIIAN BAGASSE

	Whole	Fiber	Pith
Ether solubility (%)	0.25	0.12	2.5
Alcohol-benzene solubility (%)	4.1	1.8	2.8
Hot water solubility (%)	2.5	0.9	1.9
Lignin (%)	20.2	20.8	20.2
Pentosans (%)	26.7	27.9	28.4
Holocellulose (%)	76.6	77.8	77.7
Alpha cellulose corrected (%)	38.1	42.4	34.8
Ash (%)	1.67	0.7	2.29

thirds of the pith is removed. Essentially the units consist of a rotor to which are attached either swing or rigid hammers that are enclosed fully or partially by perforated screen plates through which the pith fraction is discharged. Feeding of bagasse to the depithers is along either the horizontal or vertical axis of the unit, depending on the construction. The depithed fiber is discharged at the end of the rotating axis. A depithing machine developed by S. P. M. (Swiss Puerto Rican Metallurgical Corporation) differs from the models mentioned above in that it utilizes two horizontal rotors mounted side by side with curved perforated screen plates under each rotor. The rotors are equipped with swing hammers, which rotate in the same direction. Bagasse is fed through a side chute and is carried from one chamber to the next chamber in series. Two material streams are created, which move in opposite directions within the depithing zone located between the rotors. The loosened pith and dirt are discharged through the screen plates and the depithed fiber is discharged through louver bars. The separated fiber and pith are removed by separate extractor fans, which are aided by incoming air through the duct system provided.

Wet depithing, which is shown in Figure 4–122, is carried out exclusively at the pulp mill as the final stage of pith separation to further upgrade the fiber prior to pulping. This process is adaptable for either baled bagasse or bagasse delivered from bulk storage. A typical wet depithing system consists of:

1. A regulating feeding conveyor.
2. A continuous hydrapulper fitted with powerful rotor and an extraction plate.
3. Settling tank.
4. Depithing unit fitted with swing hammers and water jets.

Bagasse is fed to the hydrapulper, where it is thoroughly wetted and the mass partially disintegrated, thereby opening the fiber bundles. The consistency is maintained around 2 to 2.5% by continuous recirculation of process water. The suspended slurry is pumped to the depithing machine either directly or by way

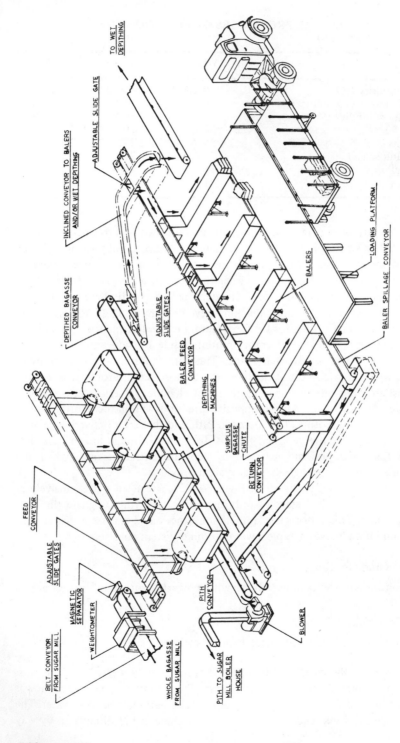

Figure 4–121. "Moist" bagasse depithing system and baling station for location at sugar mill.

TO WET DEPITHING

ADJUSTABLE SLIDE GATE

INCLINED CONVEYOR TO BALERS AND/OR WET DEPITHING

DEPITHED BAGASSE CONVEYOR

ADJUSTABLE SLIDE GATES

BALER FEED CONVEYOR

DEPITHING MACHINES

SURPLUS BAGASSE CHUTE

RETURN CONVEYOR

BALERS

LOADING PLATFORM

BALER SPILLAGE CONVEYOR

FEED CONVEYOR

ADJUSTABLE SLIDE GATES

MAGNETIC SEPARATOR

WEIGHTOMETER

BELT CONVEYOR FROM SUGAR MILL

WHOLE BAGASSE FROM SUGAR MILL

PITH TO SUGAR MILL BOILER HOUSE

PITH CONVEYOR

BLOWER

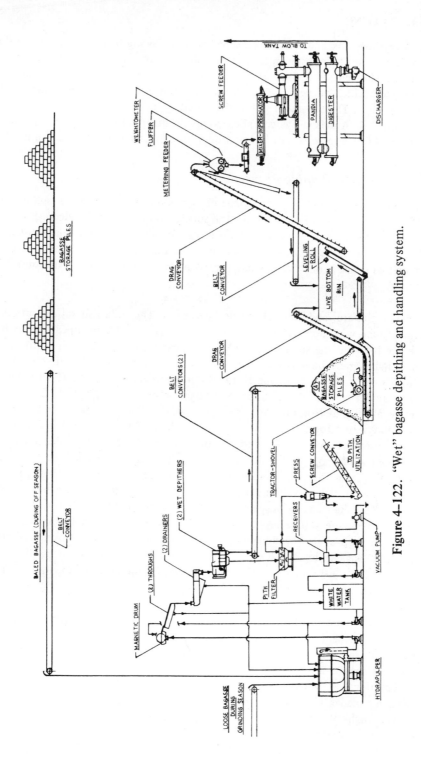

Figure 4-122. "Wet" bagasse depithing and handling system.

541

of a settling tank and a magnetic separator. The swing hammers in the depither complete the defibrating operation and the pith passes through the perforated screen. The separated fiber is discharged at 5 to 6% consistency to a drainer-conveyor, which delivers the depithed bagasse at about 20% consistency to the pulping unit. The separated pith is thickened by a dewatering press or by a filter for disposal. Atchison has compiled complete information about various methods used for bagasse storage[1063] and depithing machines in commercial use.[1064]

Methods for Pulping Bagasse

Bagasse can be pulped by any of the conventional processes: soda, sulfate with 15 to 20% sulfidity, lime-soda, neutral sulfite, soda-chlorine (Celdecor-Pomilio), or mechano-chemical. Two other processes are the Peadco and Cusi-San Cristobal, which have been developed especially for bagasse pulping. These processes, which are based on alkaline pulping with short cooking time, are described below.

Soda, Sulfate, and Neutral Sulfite Processes. The soda and sulfate processes have found wide application in the pulping of bagasse. There appears to be little choice between the sulfate and soda processes when bleached pulp is produced. Bleachable-grade pulp is produced by the soda process in a Pandia continuous digestion system: cooking time 10 to 12 min; temperature 165 to 170°C; digester pressure 6.5 to 7 kg/cm^2; liquor-to-dry-bagasse ratio 1:3.5; NaOH 12% of total active alkali based on moisture-free raw material. Permanganate number of the pulp is 9.5 to 10 and the unscreened pulp yield is 50 to 52%.[1053] Lime-soda process is used in some Latin American countries mostly for the production of unbleached grades of paper, although the pulp can be bleached to a medium brightness level for the manufacture of low-grade writing and printing papers. Cooking with neutral sulfite produces pulp of lower strength, but the yield and brightness values can be respectively 2% and 4 points higher than that of pulp produced by the sulfate process.[1065] Unbleached coarse pulp at 65% yield and bleached pulp at 45% yield and brightness of 80 or above are produced from depithed bagasse by the Celdecor-Pomilio process.[1066] This process has been adopted for the production of bleached pulp in the Philippines, India, Brazil, South Africa, Spain, and other places.

Peadco Process. The Peadco process,[1067] developed by the Process Evaluation and Development Corporation, a subsidiary of W. R. Grace and Company, is a fully integrated process, which includes the depithing and pulping operations. Bales of bagasse are broken up by a bale breaker and the material is run through a depithing unit, where it moves parallel to the rotor axis of the depither. The swinging blades attached to the rotor disintegrate the incoming material and provide high-velocity impact on the mass. Due to the rubbing action, fiber and pith are separated effectively. The pith is removed through the screen pneumatically and the depithed fiber is discharged at the end of the depither. From there it is taken by a compacting screw feeder to the pulping system, where the compacted fiber enters a horizontal, high-pressure digester vessel. Cooking chemicals and steam are injected into the digester at the same entry point as the raw material. Pulping is carried out at (860 kPa) 125 psi in the vapor phase with a total retention time of 18 to 20 min. Cooking conditions are controlled to reduce the lignin content of the fiber to 5%, while the xylans are prevented from being reabsorbed. The pulp is blown from the digester at high consistency to a disk refiner at 145°C. Black liquor is added at the discharger to reduce steam flashing. The semirefined pulp is transferred to the blow tank. The pulp from the blow tank is washed, screened, and centrifugally cleaned. The rejects from the screening system are recycled in the digestion system. This process is in use in mills in Mexico, Venezuela, Peru, and Taiwan.

Cusi-San Cristobal Process.   The Cusi-San Cristobal process[1068] was developed by Cusi in a bagasse pulp and paper mill, Cia Industrial de San Cristobal, in Mexico. Whole bagasse, as it is received from the sugar mill, is screened to remove fines and free pith which have been loosened during the milling operation at the sugar mill. Depithing is carried out in a specially designed depither that operates at high speed and is closely regulated to dislodge the pith from the fiber bundles by mechanical scraping. The pith is mixed with molasses for animal feed. The fiber fraction is given a mild cook either by the soda or by the sulfate process. The raw cooked pulp is screened, employing special techniques, and separated into two fractions, corresponding to the fibers derived from the central and the rind-vascular bundles respectively. The accepted fiber from the screening operation, which consists of the central vascular fibers, is pentosan-rich and has the characteristics of a well-refined wood pulp. The diameter of these fibers is small and the pith cells are swollen so that the pulp needs little refining and has excellent fiber-bonding characteristics, although paper produced from this pulp has poor tear strength and tends to be transparent. The second fraction, which consists of fibers from the more resistant rind-vascular bundles, is in a relatively raw state and it is subjected to more drastic pulping. The pulp produced from this fraction is long fibered and it possesses good tearing resistance and has strength properties that are similar to those of sulfite wood pulp. This pulp is refined to the desired degree of hydration and recombined with the first fraction to produce a wide variety of papers. The salient feature of this process is that it reduces the dependence on long-fiber wood-pulp requirements in the furnish and maximizes bagasse-pulp utilization. The concept of fiber fractionation, on which this process is based, may be used for pulping other nonwood fibrous raw materials.

Mechano-Chemical Process.   This process, using either caustic soda or lime alone or in combination with other alkalis, is employed mostly for the production of high-yield coarse pulp for the manufacture of corrugating medium and other unbleached varieties. Table 4-58[1069] shows the pulping conditions and strength properties of pulp produced by this process.

TABLE 4-58 PULPING CONDITIONS AND STRENGTH PROPERTIES OF PULP PRODUCED FROM HAWAIIAN BAGASSE COOKED FOR ONE HOUR AT 98°C

NaOH (%)	Yield, Screened Pulp (%)	Initial Freeness (S-R, ml)	Strength Characteristics (500 ml S-R)				
			Burst Index (kPa · m²/g)	Tensile Breaking Load (kN/m)	Tear Index (mN · m²/g)	Riehle Ring Crush (N)	Flat Crush (kPa)
8	74.2	700	3.86	8.50	6.40	156	496
10	71.9	730	4.06	9.55	5.98	165	455
12	69.9	740	4.35	9.88	6.12	178	482
14	58.3	820	3.96	9.81	4.87	—	1,723
16	62.4	800	4.01	10.14	5.01	—	1,929
18	65.4	795	3.96	10.33	5.01	—	2,205

Other Processes. Several other processes have been suggested for the pulping of bagasse, including the nitric acid process[1070] and the hydrotropic process.[1071] These processes have not found commercial application.

Bleaching of Bagasse Pulps

Bagasse pulps are easily bleached. With single-stage hypochlorite bleaching, brightness up to 70 can be obtained. With three-stage (CEH) bleaching up to 85 to 86 brightness can be readily obtained with 5% total chlorine on the dry weight of unbleached pulp.[1072] However, under carefully controlled bleaching conditions, pulp of the same brightness level can be obtained with about 3.5% chlorine, as shown in Table 4-59.[1042] Tables 4-59 and 4-60[1042] show bleaching conditions employing the CEH and the CEHD bleaching sequences. The bleaching sequence CEH is satisfactory for the production of bleached pulp for writing and printing grades of paper. The CEHD bleaching sequence will produce pulp up to 90 plus brightness with better strength characteristics. Chlorine dioxide bleaching in the final stage is essential to produce high brightness pulp, particularly when it is intended to be sold as market pulp.

Properties of Bagasse Pulp

Bagasse pulps are of heterogeneous fiber length and their average fiber length depends largely on the degree of depithing used in fiber preparation as well as the pulping process employed. Table 4-61[1053] gives typical fiber classification data. The length-to-width ratio of the fiber is comparable to that of softwood fibers, and therefore bagasse pulp can be used for the manufacture of a wide range of high-quality papers. Table 4-62[1042] shows strength properties of unbleached and bleached bagasse pulp cooked by the soda process. Coarse bagasse pulps are used in the manufacture of insulating board, wrapping paper, and corrugating medium. Bagasse pulps produced by the neutral sulfite process are used in a high proportion for the manufacture of sack paper. Multiwall bag papers with high-tear and energy-absorption value are produced with 60% unbleached chemical bagasse pulp blended with kraft softwood pulp, using the Clupak extensible system.[1073]

Dissolving Grade Pulp from Bagasse

High-grade dissolving pulp can be produced by steam prehydrolysis and cooking by the kraft process. Acid treatment is essential for removing divalent salts and minerals. The pulp is bleached in a CEHD bleaching sequence to 90 to 93 brightness. Alpha cellulose content is 95%, pentosan content 4 to 6%, and viscosity 80 to 85 cp. A more highly purified pulp can be produced with alpha cellulose content 96 to 98% and pentosan content of 2 to 4%, but the viscosity drops to 60 to 80 cp.[1074]

TABLE 4-59 BLEACHING CONDITIONS FOR BAGASSE PULP USING THE SEQUENCE CEH

Reagent	Moisture-Free Unbleached Pulp (%)	Stock Consistency (% Moisture-Free)	Temperature (°C)	Retention Time (hr)	End (pH)	Residual Chlorine Left After Bleaching: (% on Moisture-Free Unbleached Pulp)	Brightness (%)	Yield (%)
Cl_2	2.4	3.5	23	1.0	1.7	0.12	—	—
NaOH	2.5	10.0	75	1.5	11.7	—	59–60	—
$CaOCl_2$[a]	1.0	10.0	38	2.5	8.6	0.21	85–86	94.9

[a]0.6% NaOH on moisture-free unbleached pulp, added as buffer.

TABLE 4-60 BLEACHING CONDITIONS FOR BAGASSE PULP USING THE SEQUENCE CEHD

Reagent	Moisture-Free Unbleached Pulp (%)	Stock Consistency (% Moisture-Free)	Temperature (°C)	Retention Time (hr)	End (pH)	Residual Chlorine Left After Bleaching: (%) (on Moisture-Free Unbleached Pulp)	Brightness (%)	Yield (%)
Cl_2	2.4	3.5	23	1.0	1.7	0.12	—	—
NaOH	2.5	10.0	75	1.5	11.7	—	59–60	—
$CaOCl_2$[a]	0.5	10.0	38	2.5	8.3	0.06	84	—
ClO_2	0.3	12.0	60	2.0	5.7	0.09	90–91	94.5

[a]0.3% NaOH added as buffer.

TABLE 4–61 TYPICAL BAGASSE PULP FIBER CLASSIFICATION DATA

	Retained on				Passing Through 100 Mesh
	20 Mesh	35 Mesh	48 Mesh	100 Mesh	
Unbleached pulp	0.30	13.6	16.5	29.50	29.50
Bleached pulp	0.10	22.1	17.4	31.40	28.10
Bleached beaten pulp (40°S-R)	0.10	12.6	13.5	32.40	41.50

TABLE 4–62 STRENGTH PROPERTIES OF UNBLEACHED AND BLEACHED BAGASSE PULPS—COOKED BY THE SODA PROCESS AND BLEACHED IN CEH AND CEHD SEQUENCES

Item	Unbleached Pulp	Bleached CEH	Bleached CEHD
Permanganate number	8.2	—	—
Ash (%)	1.0	0.34	0.34
Viscosity (cp)	61.5	34.6	46.6
Solubility in 1% NaOH (%)	5.4	4.6	4.6
Initial freeness (°S-R)	18.0	19.0	20.0
Final freeness (°S-R)	50.0	50.0	50.0
Beating time in Jökro beater (min)	17.0	10.0	10.0
Basis weight (g/m^2)	77.6	78.8	77.8
Breaking length (km)	7.9	7.0	8.6
Tear strength (mN)	373.0	373.0	392.0
Tear index (mN · m^2/g)	5.05	4.97	5.30
Bursting strength (kPa)	387.0	343.0	427.0
Burst index (kPa · m^2/g)	5.24	4.58	5.77
Folding endurance, double folds	345.0	470.0	700.0
Brightness	47.5	85–86	90–91
Yield of screened pulp based on moisture-free well-depithed bagasse (%)	48–50	45.6	45.5
Bleaching shrinkage based on moisture-free unbleached screened pulp (%)	—	6.0	6.0

A patented process[1075] for the manufacture of chemical pulp from bagasse consists of:

1. Prehydrolysis with saturated steam at 170 to 200°C or with water at 165 to 170°C in a ratio of 2.5 : 1 (water to moisture-free bagasse), at a pH of 3.5 to 4.

2. Cooking with 22 to 24% alkali as NaOH on the dry weight of fiber, sulfidity 15 to 20%, maximum temperature 165 to 168°C, and time at maximum temperature 60 to 70 min.

3. Washing and screening.

4. Bleaching with two-stage chlorination (20 to 22°C, pH of 2 to 4, and consistency 3 to 3.5%), alkali extraction at 75 to 80°C, pH 10 to 10.5, and sodium hypochlorite at 38 to 40°C with an initial pH 9.5 to 10.

5. Final bleaching with chlorine dioxide at 70 to 80°C to produce pulp of 85 to 87.5 brightness.

Further purification can be carried out by alkali extraction and a second chlorine dioxide bleaching to produce pulp with a brightness of 87.5 to 92. This bleached pulp contains 90 to 94.5% alpha cellulose, 3.5 to 5.5% material soluble in 5% caustic soda, 0.1 to 0.2% ash, 0.0015 to 0.004% iron, and a degree of polymerization (DP) from 700 to 900.[1075]

High-Yield Bagasse Pulp for Newsprint

A number of pulping processes have been developed to produce bagasse pulp for newsprint manufacture. These processes are classified into mechanical and thermomechanical processes and chemimechanical processes.

Mechanical and Thermomechanical Processes. Crown Zellerbach Corp. in conjunction with the Hawaiian Sugar Planter's Association developed a process for producing refiner pulp from well-depithed bagasse. The bagasse is ground and refined in double-disk refiners with refining being carried out at 6 to 7% consistency. The partially refined pulp is thickened in a screw press to almost 30% consistency. The refined stock is screened and cleaned before bleaching. Sodium hydrosulfite is used for bleaching to a brightness of 60 to 63.[1076] The freeness of the screened pulp is about 36°S-R and the pulp produces paper of high opacity and ink absorption.

A thermomechanical process has been developed by Asplund-Defibrator A.B., Sweden. Depithed bagasse is defibrated in a disk refiner with preheating arrangement at 130°C temperature and final refining is carried out in the second stage with reduced disk clearance. The pulp is screened and cleaned. The freeness of the accepted stock is around 63°S-R. The pulp possesses the required strength properties to be used in high proportion for the manufacture of newsprint.[1077]

Chemimechanical Processes. There are several chemimechanical processes used for pulping bagasse.

The de la Rosa process was developed by de la Rosa in Cuba. It was an early attempt to produce pulp suitable for newsprint manufacture. The process involves two-stage digestion of dusted whole bagasse, the first stage being a water prehydrolysis at 150°C and the second stage being a standard sulfate digestion with 12% active alkali on the moisture-free material. The yield of the prehydrolyzed bagasse is 70 to 75% of the weight of the whole bagasse and the crude cooked, but unbleached, pulp yield is 65%. The semibleached pulp has an overall yield of 48% at 65 brightness. Approximately 5% chlorine is used for bleaching the pulp.[1077]

The Simon-Cusi process is similar to the Cusi-San Cristobal process except that for newsprint pulp, the cooking conditions are much milder. Caustic impregnation is effected at low consistency and about 7 to 8% caustic soda is absorbed in a rapid pulping cycle. After digestion the pulp is subjected to coarse screening. The rejects, which are more than 50% of the starting material, are defibrated in a disk mill. Power consumption is less and the strength properties of the pulp are considerably improved if the defibration is done under pressure. The refined rejects are recombined with the accepts from the coarse screening prior to washing, screening, and cleaning. The yield of unbleached pulp is 70%, based on the dry weight of depithed bagasse. The pulp is bleached in a two-stage hypochlorite process to a 62 brightness. The average overall yield of bleached pulp is 62% on depithed bagasse and approximately 46% on whole bagasse. Satisfactory newsprint can be manufactured using up to 95% of this pulp in the fiber furnish.[1078]

The Aschaffenburger-Zellstoffwerke process was developed by Aschaffenburger-Zellstoffwerke in West Germany. Partially depithed bagasse is subjected to water prehydrolysis at 160°C. The prehydrolyzed bagasse is pulped, employing the neutral sulfite process. The crude pulp is refined at high temperature, washed, and screened to produce unbleached pulp of 55 brightness.[1077]

The Grace-Peadco process was developed by the research group of W. R. Grace Paper Division as a modified Peadco pulping system. The accepted cleaned fiber after wet depithing is rapidly prehydrolyzed at 175°C at pH of 4 to 5. The prehydrolyzed bagasse is treated with 2% sodium bisulfite and 1% sodium silicate based on the moisture-free material and cooked in the continuous digester in the vapor phase at 175°C for a period of 10 min at 8 pH. The pulp is blown at high consistency to a disk refiner, where the fiber bundles are defibrated at 145°C. The pulp has high-strength properties with few fines. Presence of sodium silicate in the cooking medium prevents discoloration of the pulp. Pulp yield and brightness values are 85 and 55% respectively. Overall physical properties of the pulp are good and the opacity and brightness of the pulp meet newprint standards.[1079]

Pulping of Bamboo

Bamboo belongs to the grass family and grows in warm tropical and subtropical regions. It is native to Burma, China, India, Japan, Thailand, Vietnam, and the Philippines. Certain species also grow in Africa (Ethiopia, Kenya, Sudan, Tanzania, and Uganda) and in South America (Argentina, Brazil, Ecuador). Attempts have been made to grow bamboo in experimental plantations in the southern part of the United States. Bamboo is the major source of raw material for pulp and papermaking in India. Bamboo is also used for papermaking in Bangladesh, China, Kenya, the Philippines, and Taiwan. *Bambusa arundinacea* and *Dendrocalamus strictus* are the main species commonly used in Indian pulp and paper mills. *Ochlandra travancorica* and *Oxytenanthera monostigma* are used in limited quantities. Other species of bamboo are gradually being exploited in the northeastern region of India.

The bamboo plant has only one culm and every year new culms spring up from the rhizome, forming a clump. Bamboo grows rapidly and reaches a height of 35 to 45 m with a basal diameter of 15 to 20 cm in 6 to 8 yr in the case of certain species. Flowering takes place gregariously in a cycle of 25 yr with *Dendrocalamus strictus* and about 40 yr with *Bambusa arundinacea.*[1080] In the year following flowering and seeding, fresh regeneration normally starts and is complete in 2 yr. In India the cutting cycle ranges from 4 to 6 yr. Care must be taken in harvesting not to damage the rhizomes from which the new culms sprout. The yield from natural forest ranges from 2 to 4 ton/ha/yr. However, with proper plantation and silvicultural practices, the yield can be considerably improved up to 15 ton/ha/yr on a sustained basis.[1081]

Chemical and Physical Properties of Bamboo

Chemically bamboo contains four principal substances: starch, pectins, lignins, and celluloses. The bamboo stem consists of (1) the woody portion or hollow tube referred to as the culm and (2) ground tissue, or parenchyma. In the ground tissue there are fibrovascular bundles. Table 4–63 shows the approximate chemical composition of some bamboo species. The length of bamboo fibers compares favorably with that of coniferous wood pulps. The average dimensions of Philippine bamboo fibers have been given as:[1082] length 2.16 to 3.78 mm; width 0.014 to 0.019 mm; width of lumen 0.006 to 0.007 mm. The average fiber length of *Dendrocalamus strictus* is about 3 mm and that of *Ochlandra travancorica* about 4 mm. Considerable variation in composition and fiber length occurs among bamboo of different species. These differences are due to a number of factors, such as soil, climatic conditions, growth rate, and species. Average length of fiber ranges from 1.5 to 4.4 mm, although the majority of the fibers fall within the range of 2.2 to 2.6 mm. Fiber width varies widely from 7 to 27 μ, with an average of 14 μ.[1083] The thick cell wall with narrow lumen makes bamboo fibers stiff.

The woody portion, or culm, of the bamboo stem is jointed at nodes along the length. These nodes constitute a problem in pulping since they are denser than the rest of the stalk and hence are penetrated with great difficulty. Specially designed crushers are used for crushing bamboo. The nodes and the siliceous epidermis are crushed and the fibrovascular bundles are loosened from the ground tissue with considerable reduction in their rigidity. This renders the fibers more accessible to cooking liquor by providing channels for easy penetration. However, the power consumption in crushing is high, the output from the crushers is limited, and the crushed chips are of irregular size. Hence, the modern trend is to use multiknife chippers with a positive feed device to ensure chips of uniform size. A combination of chipping and crushing may be the most effective fiber preparation to obtain uniform pulp quality with low chemical consumption.[1084]

TABLE 4-63 APPROXIMATE CHEMICAL COMPOSITION OF SOME BAMBOO SPECIES

| Species | Location | Solubility in | | | | Lignin (%) | Pentosans (%) | Ash (%) | Cellulose by Monoethanolamine Method, (%) |
		Cold Water (%)	Hot Water (%)	Alcohol-Benzene (%)	1% NaOH (%)				
Bambusa arundinacea	India	4.6	6.0	1.2	19.4	30.1	19.6	3.3	57.6
Bambusa polymorpha	India	2.9	6.9	1.7	18.4	22.0	21.5	1.7	61.8
Bambusa spinosa	Philippines	–	7.0	3.1	39.5	20.4	19.0	4.8	67.4
Bambusa tulda	India	2.6	5.0	1.9	21.8	24.2	18.4	2.0	64.4
Bambusa vulgaris	Philippines	–	7.8	4.1	27.9	21.9	21.1	2.4	66.5
Dendrocalamus hamiltonii	India	2.5	4.4	0.3	20.8	26.2	21.5	1.8	63.3
Dendrocalamus strictus	India	4.2	5.9	0.2	15.0	32.2	19.6	2.1	60.8
Melocanna bambusoides	India	3.3	6.5	1.4	19.0	24.1	15.1	1.9	62.2[a]
Ochlandra travancorica	India	3.6	5.1	2.2	20.0	26.9	17.8	2.6	61.8[a]
Oxytenanthera abyssinica	Ethiopia	–	–	5.4	29.5	–	–	–	56.7

[a]See reference 1081.

Pulping Processes

Alkaline pulping is the best process for bamboo pulping, particularly the sulfate process. In spite of the dense texture of the bamboo chips, sulfate pulping provides satisfactory delignification as well as high yield. The acid sulfite process produces pulp of lower strength properties and the high cost of sulfur, instability of bisulfite liquor in tropical countries, and the lack of an inexpensive recovery system make this process unsuitable for commercial application in medium sized mills.

At the Forest Research Institute at Dehra Dun in India, Raitt developed a two-stage digestion method for bamboo pulping on a commercial scale.[1085] The process is based on removing starches, degraded products of sugars, gums, tannins, and coloring matter by mild cooking in the first stage and completing the delignification under somewhat drastic conditions in the second stage. In the first stage the cooking conditions are:

1. Six to ten percent total alkali based on moisture-free bamboo.
2. Two hours cooking time.
3. Temperature 115 to 125°C.
4. Liquor ratio 4:1, using spent liquor from the second-stage pulping, after which the liquor is drained and sent to the chemical recovery plant.

The partially cooked chips from the first stage are cooked with 16 to 18% active alkali for three hours at 140 to 153°C. Alkali concentration of the liquor is about 6% in the second stage compared to 2% in the first stage. The first stage is termed "pectin digestion"; the second stage "lignin digestion". Overall chemical requirement for pulping and bleaching is relatively low and pulp of 10 to 12 permanganate number can be produced for bleaching. However, a two-stage cooking considerably reduces the output from a stationary digester due to the longer cycle, and steam consumption is higher than in overhead or single-stage digestion.

Overhead pulping (single stage) is now widely practiced in bamboo pulp mills. The following cooking conditions are normally used:

1. Active alkali 20 to 22% on dry weight of chips.
2. Sulfidity 20 to 25%.
3. Total cooking cycle 4 hr.
4. Cooking time 2.0 to 2.5 hr at 170°C.

The permanganate number and yield are 13 to 16 and 44 to 46% respectively. If milder cooking conditions are used[1086]—active alkali 17 to 19%, sulfidity 22%, cooking time 1 to 1.5 hr, and 150 to 160°C—the unbleached pulp yield is 50 to 52% with a permanganate number of 24 to 28. The milder cooking condition is conducive to the redisposition of hemicelluloses on the fibers. The screening rejects increase and can be used for repulping.

Neutral sulfite semichemical pulping can be used for bamboo. Unbleached yield varies from 55.6 to 69.1%. The pulp can be bleached to a brightness of 72 to 88 employing a CEH sequence with a total chlorine consumption of 18%. Considerable loss occurs in bleaching and the bleached yield is 48.3 to 53.2%. On bleaching, the tear and burst factors are improved. The strength properties are comparable to bleached softwood and the pulp can substitute for long-fiber pulp in various grades of paper in mixtures with pulp from other annual plant fibers.[1087] Stationary digesters with forced-liquor circulation through preheaters are in extensive use for bamboo pulping. However, the trend is toward continuous digestion, using the Kamyr continuous digester or the Pandia continuous digester equipped with a pressurized preimpregnator to provide the necessary flexibility for controlling the cooking variables to produce pulp of uniform quality.

Philippine bamboos, which have higher ash content and lower lignin content than Indian species, are easily cooked under mild cooking conditions to a permanganate number of 13 to 18 and a screened pulp yield of 41 to 48%.[1088] The pulp can be bleached with low chlorine consumption.[1089] Taiwan bamboo species are pulped by the sulfate process with 25% sulfidity using these cooking conditions:

1. Alkali as Na_2O, 13%, 15%, 17%.

2. Liquor ratio 4:1.

3. Digestion temperature 160 to 165°C.

Yields vary from 42 to 48%. Optimum pulping conditions are 15% active alkali as Na_2O on the dry weight of bamboo chips and cooking temperature 160°C. The pulp is bleached in CEH sequence. Pulp from *Dendrocalamus latiflorus* can be bleached to a brightness of 86 using 6% chlorine in the chlorination stage and 2% available chlorine in the hypochlorite stage.[1090]

Bleaching of Bamboo Pulp

Bamboo pulp is more resistant to bleaching than pulps from other annual plant fibers. Pulp of 12 to 14 permanganate number can be bleached in a CEH or a CEHH sequence to 76 to 78 brightness, using 8 to 10% chlorine. Pulp of 16 to 18 permanganate number can be bleached to 75 to 76 brightness, using 13 to 15% chlorine.[1091] Pulp of higher permanganate number (24 to 26) can be bleached to a brightness 78 to 80, using 15 to 17% chlorine in a CEH sequence. Bleaching conditions for such a high permanganate pulp are:

1. Chlorination: consistency 3%, temperature 30 to 35°C, retention time 45 min, chlorine consumption 10%.

2. Alkali extraction: consistency 5%, temperature 55 to 60°C, NaOH 4 to 6%, retention time 60 min.

3. Hypochlorite: consistency 5%, temperature 40 to 45°C, available chlorine consumption 5 to 7%, retention time 120 to 180 min, final pH 8 to 8.5.[1086]

Bleached yield ranges from 42 to 47%. High permanganate number pulp needs an expanded bleachery to bleach pulp to the 80 to 85 brightness level employing a CEHEH,[1092] CEHED, or CCEHD sequence. Sulfamic acid can be used with advantage in hypochlorite bleaching, especially at lower pH conditions, because of the elimination of the buffer.[1092] The bleaching sequences that include chlorine dioxide in the final stage result in an economy in the consumption of chlorine and minimize degradation of the pulp. A CCEHD sequence is suggested for a high permanganate pulp to remove lignin progressively in two stages of chlorination and to reduce the bleaching costs.[1042] Use of sulfamic acid, diammonium hydrozine phosphate, and sodium silicate as bleaching additives is effective in controlling viscosity, preserving strength properties, and improving brightness of the pulp.[1086]

Dissolving Grade Pulp from Bamboo

Dissolving grade pulp is produced by the prehydrolysis-kraft process. Prehydrolysis can be carried out in acidic steam or water medium. The steam or water prehydrolysis-kraft process followed by multistage bleaching gives an average yield of 33%. The pulp contains 96.5% alpha cellulose, 0.7% beta cellulose, 0.1% ash, and 3.46% pentosan. The degree of polymerization ranges from 750 to 900. The pulp can be converted into rayon yarn, staple fiber, and tire cord.[1093] Dissolving grade pulp is bleached in a CEHEDH bleaching sequence with final SO_2 treatment to a brightness level exceeding 90 in a viscosity range of 16 to 20 cp.[1094] Gupta[1095] suggests the applicability of Vroom's H-factor to the prehydrolysis-kraft pulping of bamboo as an effective measure to control pulping variables. Commercial production of dissolving grade pulp from bamboo is well established in India.

Pulping of Reeds

Phragmites communis and *Arundo donax* are the two principal species of the reed family that are used for pulping. *Phragmites communis* grows abundantly in swampy lands and in the delta areas of Russia, Rumania, Egypt, northern China, North Korea, and Iraq. Stalks grow to a height of 3 to 5 m or taller and are about 2.5 cm in diameter. The leaf sheaths contain 8 to 10% silica and are low in cellulose content. *Arundo donax,* or giant cane, is a perennial plant native to the Mediterranean area but now quite dispersed. The clumps grow 2 to 8 m high and 1 to 4 cm in diameter. The wall thickness ranges from 2 to 7 mm and the node lengths vary from 12 to 30 cm. The outer rind is hard and brittle and contains siliceous materials. The cane has a golden-yellow luster at maturity.

During the winter season cane is harvested by mechanical harvesters on the frozen delta areas of Russia, China, and Rumania. Due to low soil-bearing capacity of these terrains, the harvesting and handling equipment requires wide treads.[1096] In Rumania, specially designed amphibious vehicles are used in the Danube delta to cut and haul the reeds during the spring and summer months. In Iraq and Egypt, small boats are used for manual harvesting in the salty marshes at the deltas of the Nile and the Tigris and Euphrates rivers. Freshly cut stalks contain over 30% water and should be stacked in a fashion permitting air draught for natural drying and to avoid rotting. When the moisture content is reduced to 14 to 15%, the reeds are baled and transported to the mill site.[1096] During chopping of the cane at the mill the leaves, nodes, and tops are separated from the stalk and removed in the screening operation. A considerable amount of silica embedded in the sheaths is also removed.

Both batch and continuous digestion are used for pulping of reeds. A wide assortment of pulps ranging from semichemical to easily bleachable grades are produced by a rapid sulfate process in the yield range of 65 to 70% to 42 to 48% respectively,[1097] using these cooking conditions:

1. Active alkali 15 to 20%, with 15% sulfidity.

2. Cooking time 20 min.

3. Cooking temperature 160 to 170°C.

Bleachable sulfate pulp is produced in the Pandia digester,[1098] using:

1. Active alkali 17% based on the moisture-free fibers.

2. Cooking time 20 min.

3. Temperature 180°C.

4. Liquor-to-fiber ratio of 3:1.

5. Breaking length of the pulp, 6.0 to 6.3 km.

Bleachable monosulfite pulp containing 1.6% lignin and 18% pentosans is produced with cooking conditions:

1. Sodium sulfite, 10% on moisture-free material.

2. Cooking time 30 min.

3. Temperature 180°C.

4. 3:1 liquor ratio.

Reed pulp can be bleached with CEH or CEHH sequences to a brightness range of 80 to 85. Pulps from the soda and sulfate processes require more chemicals for bleaching than neutral sulfite pulp. Bleached pulp yield varies from 38 to 42%, depending on the extent of cleaning in the fiber-preparation stage. Reed pulp contains 43.5% long fiber, 26.2% short fiber, and 20.3% fines.[1099] Fiber classification of reed pulp[1041] cooked by the soda process is given below:

	Retained on				Passing Through 100 Mesh
	+20 Mesh	+30 Mesh	+40 Mesh	+100 Mesh	
Unbleached pulp	15.5	8.4	16.9	13.8	45.4
Bleached pulp	20.8	9.0	12.3	11.9	46.0

Reed pulp from *Phragmites communis* shows the highest burst and tensile factors among the commercial nonwood fiber pulps.[1100]

High-yield semichemical pulp is used for the manufacture of container boards and other unbleached grades. Bleached pulp contributes substantially to the bulk and opacity properties of high-quality writing and printing papers.

Pulping of Esparto

Esparto grass grows wild in North Africa and in the Mediterranean areas of southern Spain. *Lygeum spartum* and *Stipa tenacissima* are the two species. *Lygeum spartum* grows in higher interior plains, whereas *Stipa tenacissima* grows around the sea coast. Esparto is a coarse grass, growing in large clumps. The grass is grayish-green in color and reaches a height of 1 to 1.25 m. The plant has strong branched rhizomes and is not creeping. The plant does not propagate vegetatively, but regenerates from seed. Esparto varies widely in quality, depending on the area in which it is grown and the soil conditions. The Spanish is superior to the African varieties because it yields more fiber and responds well to bleaching. Harvesting is done manually in such a way as to leave the root intact for regrowth. Esparto pulping started in England as far back as 1856 as a substitute for rags. It thrived all through the years until World War II, after which transportation of esparto from the Mediterranean to England became increasingly expensive due to high freight costs and gradually esparto pulping in England ceased. However, a number of pulp mills are in operation in the areas where esparto grows in Algeria, Morocco, Tunisia, southern Spain, and France.

Esparto pulp is especially suitable for the manufacture of fine-quality printing papers, where dimensional stability, softness, high opacity, and good printability are of prime importance. Because of its distinctive properties, paper made from esparto is sold at a premium price. The fiber length of esparto pulp varies from 0.4 to 2 mm, with an average of 1.5 mm. The fiber thickness ranges from 4 to 11 μ, with an average of 7 μ. The fibers may be either thick- or thin-walled; an identifying characteristic of esparto is the presence of small, pear-shaped trichomes (17 $\times$ 30 μ) that are hooked at the apex. Esparto grass is composed largely of cutocellulose or pectocellulose rather than lignocellulose. This cutocellulose is broken up by hydrolysis, while the noncellulosic substances, fats, and waxes are solubilized.

To prepare esparto for pulping, a conical screen is used for dusting. For increased output, a specially designed duster with a revolving rotor fitted with spikes is used. Opened-up bales are added to the duster, where sand and dust are separated through the grill, while the clean grass discharges from the other end. Cleaned grass is cut to a length of 4 to 5 cm if a continuous pulping process is used, but uncut grass is used in batch pulping. The soda process is best suited for esparto pulping.

A batch digester consists of a vertical cylinder with a domed top and an internal perforated false bottom, as shown in Figure 4–123. Grass is fed to the digester from the top and the cooked fiber is removed from the side door located close to the false bottom. Cooking liquor and washwaters are introduced to the digester through the manifold. Liquor circulation is effected during the cooking period through two vomiting pipes located on opposite sides inside the digester. When the digester is half full, cooking liquor and a small amount of steam are introduced for wilting and better packing of the digester with grass. Five to six tons of grass can be charged per cook. After charging, the lid is closed and firmly secured and steam is introduced at the bottom of the digester until the pressure reaches 3 to 3.5 kg/cm^2 or about 6 to 7 kPa. Cooking time varies from 2 to 2.5 hr and the pressure is relieved after 30 to 45 min. The liquor is drained through the perforated plate for chemical recovery. Pulp washing can be carried out in the digester by countercurrent diffusion or alternatively the cooked grass is pumped to a blow pit from where it can be transferred to rotary wash filters. If the pulp is washed in the digester, the final wash is with hot water and the drainings are stored for reuse. An improved method for emptying the digesters uses a high-pressure water jet. The digester dome has a sliding gate that is closed during cooking and washing but is opened when the digester is to be emptied. A high-pressure water jet cuts through the mass and converts it into a pulpy condition, facilitating its discharge. A pump delivers the pulped grass to the bleachers, where it receives a final washing on a drum washer. The pulp is bleached at 37 to 40°C in a tower equipped with a centrifugal pump for recirculation. Calcium hypochlorite is used for bleaching. Once the pulp has reached the optimum brightness, it is transferred for screening, cleaning, thickening, and storage in a high-density chest.

Continuous pulping has been incorporated in mills in Tunisia and Algeria that employ the Pandia continuous digester. Pulp is washed in a three-stage countercurrent washing system. The cooking conditions are:

1. Active alkali 14 to 16% based on moisture-free material.
2. Solids-to-liquor ratio 1:3.5.
3. Retention time 16 to 18 min.
4. Temperature 165°C in the digester.

The unbleached pulp yield and the permanganate number are 45 to 47% and 12 to 13 respectively. The pulp is bleached to 83 to 84 brightness in a three-stage continuous bleaching, employing CEH sequence with a total chlorine application of 5 to 6% on the moisture-free unbleached pulp. High-brightness pulp exceeding 90 is produced employing a CEHD bleaching sequence. Use of sulfur dioxide is necessary for controlling color reversion. Bleached-pulp yield varies depending

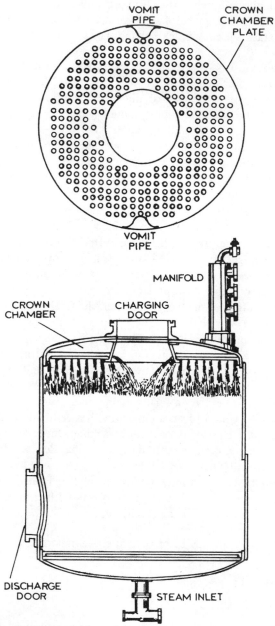

Figure 4–123. Grass digester used for cooking of esparto.

558

TABLE 4-64 CONDITIONS AND RESULTS FOR RAPID CONTINUOUS PULPING OF ESPARTO IN A PANDIA DIGESTER

Digester Pressure (kPa)	Caustic Soda		Dwell time (min)	Residual NaOH (%)	Pulp Yield (%)	Permanganate No.	Viscosity (cp)
	Amount Applied (%)	Concentration (g/l)					
552	16.3	40.0	20	1.6	52.2	5.8	741
828	16.0	47.0	20	5.8	56.9	6.0	401
1035	16.0	47.0	20	5.9	52.7	4.8	149
828	20.4	41.0	20	6.8	52.3	5.0	468
828	16.0	51.3	20	8.0	53.3	5.3	450

TABLE 4-65 STRENGTH PROPERTIES OF UNBLEACHED AND
BLEACHED ESPARTO PULPS AND PULPS FROM MIXED HARDWOOD
AND SOFTWOOD

Freeness Canadian Standard (ml)	Beating time (min)	Burst Index (kPa · m²/g)	Tear Index (mN · m²/g)	Folding Endurance (MIT double fold)	Breaking length (km)	Sheet Density (g/cm³)
		Unbleached esparto pulp				
590	0	1.04	6.77	1	1.68	0.31
450	10	3.08	12.36	17	4.14	0.42
330	20	3.73	12.26	59	5.74	0.48
230	30	4.25	11.37	64	6.65	0.52
		Bleached esparto pulp				
580	0	1.61	9.29	4	2.71	0.33
460	10	2.84	10.84	21	5.33	0.43
450[a]	11	2.94	10.79	24	5.40	0.43
325	20	4.14	11.08	88	6.69	0.50
210	30	4.31	10.44	352	6.87	0.55
		Mixed hardwood (aspen, maple, beech) pulp[b]				
450[a]	—	3.53	8.43	40	7.0	0.69
		Mixed softwood (jack pine, hemlock, fir) pulp[b]				
450[a]	—	7.36	10.39[c]	1600	11.4	0.79

[a]These values at Canadian Standard freeness 450 ml are interpolated.
[b]U.S. Forest Products Laboratory, Madison, Wis.
[c]Tear index for unbeaten pulp.

on the maturity of grass and the extent it is cleaned during dusting; bleached-pulp yield is generally 45% on Spanish esparto and 38% on African varieties.[1101] The conditions and results for rapid continuous pulping of esparto in a Pandia continuous digester on a semicommercial scale are shown in Table 4-64.[1102] The pulp was bleached employing a CEH bleaching sequence at medium density in upflow bleaching towers to 85 brightness. The total bleach consumption was equivalent to 3% chlorine, but in commercial practice the chlorine consumption is almost 6% to develop the same brightness.[1089] Bleached esparto pulp is also produced by the Celdecor-Pomilio process.

Table 4-65[1102] shows strength data on unbleached and bleached esparto pulps. Strength properties of mixed hardwood and mixed softwood pulps are given for comparison.

Pulping of Sabai Grass

Sabai grass (*Eulaliopsis binata*) is a supplementary source of raw material for production of bleached pulp in India and Pakistan. The grass is tough, strong, and durable. Fiber length is maximum 4.9 mm, minimum 0.5 mm, and average

4.1 mm. Widths range between 9 to 16 μ. Fiber length-to-width ratio is 231:1. The fibers are thick walled with a narrow lumen and pointed ends. The chemical properties of sabai grass are:

1. Ash 6%.
2. Solubles in hot water 9.5%.
3. Solubles in 1% NaOH 39.7%.
4. Solubles in ether 1.2%.
5. Solubles in alcohol-benzene 4.1%.
6. Pentosans 23.9%.
7. Lignin 22.0%.
8. Cross and Bevan cellulose 54.5%.[1103]

Sabai grass is pulped in batch digesters by a process similar to that described in esparto pulping. Pulp is bleached to 75 to 80 brightness in single or double stages, using calcium hypochlorite as the sole bleaching chemical. The quality of sabai grass pulp is similar to esparto pulp and it is used for high-grade book and printing papers.

Pulping of Hemp

Bast fibers are present in hemp in the form of fiber bundles underneath the bark. The majority of these fibers occur as strands, filaments, or threads. They are separated from the stalk by retting and mechanical treatment to produce fiber for the manufacture of textiles, ropes and, similar materials. Some hemp straws are used directly for paper making, but the principal source of hemp fiber is rope and cordage collected by junk dealers from seaports, inland shipping points, and other sources where these materials are used in large quantities.

The principal sources of bast fibers are: Manila hemp (*Mussa textilis* or abaca); true hemp (*Cannabis sativa*); sisal hemp (*Agave sisalana*); henequen or Mexican sisal (*Agave fourcroydes*); sunn or Benares hemp (*Crotalaria juncea*); New Zealand hemp (*Phormium tenax*); and Mauritius hemp (*Furcraea gigantea*). Manila hemp, or abaca, is available for direct use in pulp manufacture in the Philippines and in Central and South America. The *Mussa,* or abaca stalk, consists of a central true stem enclosed by leaf stalks or sheaths. Fiber preparation involves several different cleaning and upgrading processes. In the Gocellin process, a process patented in the Philippines,[1104] the whole stalk is used. The process consists of cutting the green stalk into 1-meter lengths and then slicing and crushing these in a specially designed chipper to remove water and to produce chips about 5 cm in length. Following this, the chips are sun dried to prevent rotting. The chips are steam heated at 90 to 100°C for 10 to 15 min, pressure cooked and mechanically defibrated in a hydrapulper at 120 to 130°C for 30 min. The defibrated stock is screened on a Jönsson vibrator and oven dried. The semitreated material is transported to pulp mills for final pulping.

A higher quality material for pulping is obtained by mechanical decortication of the whole leaf sheath separated from the true stem. This gives the highest yield of fiber. The pulp from the decorticated sheaths contains long fiber strands and minimum short fibers, about 80 to 95% long slender fibers (4 to 6 mm in length and average width 17 to 21 μ) and 15 to 20% parenchyma cells and thread-like spiral thickenings of the vessels. The spiral thickening threads show up as cluster-like knots if the pulp is used in fine tissues, such as tea-bags and stencil tissues.[1105] The most satisfactory material is obtained by stripping the tuxy, or outer portion, of the leaf sheath, which contains most of the fiber strands. The tuxy can be peeled or stripped manually from the undesirable portion of the leaf sheath by a knife blade pivoted on the end of a block and adjusted to different heights by a foot pedal. Longitudinal strips of the tuxy are produced by drawing the material through the knife blade, while the parenchyma cells are scraped aways from the fiber strands. Spindle stripping is an improved technique over hand stripping[1106] in that the tension of the knife is controlled. By spindle stripping a uniform raw material is obtained for pulping and consequently the fiber and fines ratio of the resultant pulp has minimum variation compared to hand-stripped fibers. The quality of fiber depends on the particular leaf sheath. The cleaned fibers are graded according to color, luster, fines, and shive content. Fibers upgraded by the tuxying process are not suitable for porous tissues, but they are ideal for dense tissues, such as onionskin. The pulp can be easily refined and high-quality onionskin tissues are produced with high-tear strength.

Abaca fibers are pulped by the soda and the alkaline sulfite processes, the latter process being widely used. Spherical or globe digesters are used for cooking. In the soda process, cooking conditions are:

Caustic soda 12 to 16%, on the moisture-free weight of raw material.

Temperature 140 to 150°C.

Time 5 to 6 hours.

Pulp yield is around 50%. In the alkaline sulfite process, 16 to 18% high-purity sodium sulfite is used as the cooking chemical. Cooking temperature ranges from 150 to 170°C; cooking time varies with the cooking temperature as well as with the quality of the fiber used. With coarse hand-stripped abaca fiber, yield varies from 58 to 62%, whereas the yield from spindle-stripped abaca varies from 67 to 71%.

Hemp rope requires cutting, dusting, and cleaning before cooking. The rope is cut to 5-cm lengths by a heavy-duty cutter. The cut rope is passed through a rotary duster, where the yarns are opened up and the dirt removed. Cooking is carried out in rotary boilers similar to the type used for cooking rags. Caustic soda or a mixture of about 10% lime an 5% sodium carbonate on the moisture-free raw material is used as cooking chemical. Normally the cooking is at 172 kPa (25 psi) for 8 to 10 hr. During the cooking process, the dirt and grease are dissolved and the material is softened for subsequent refining. The yield of fiber is in the range of 50 to 65%. In a patented process developed by Bidwell,[1107] cut manila hemp rope is cooked in an open tub using 10 to 14% lime plus 0.1% or less sodium hydroxide. Following cooking, the pulp is washed and beaten in the same unit.

The washing and bleaching of hemp pulp are carried out in a breaker beater, using the same method as that for rags. Unbleached hemp pulp is used for many grades of paper, but for high-quality specialties the pulp is bleached with hypochlorite to a brightness of 80 in either a single stage or in two stages with intermediate washing. In the first stage the pH should be maintained at about 8 and in the second stage at about 10. Two-stage bleaching reduces the chlorine consumption by approximately 2%. The CED bleaching sequence may be used for producing pulp of high brightness and excellent strength.[1108] Viscosity of unbleached and bleached pulp is 52 and 17 centipoises respectively.

Manila hemp pulp beats readily and produces paper with exceptional strength and pliability. Hemp pulps are used for the manufacture of high-strength bag and wrapping papers, insulating cable paper, emery paper, tags, and gaskets. Bleached pulps are used for thin specialty papers, such as Bible paper, cigarette and carbon papers, and condenser tissues. Papermaking based on hemp pulps requires extremely high-quality water for process use, particularly free from iron and manganese.

Plantings of sisal hemp (*Agave sisalana*) have been started in Brazil to supply raw material for an annual production of 64,000 tons of high-quality, bleached long-fiber pulp exclusively from sisal.[1109] The sisal plant has an eight-year growth cycle, at the end of that time the mother plant flowers and dies. The land is cultivated and the bulbs are uprooted and plowed back to the soil as fertilizer and the new plantings are begun. The sisal leaves are harvested twice a year. The leaf is 1 to 1.5 m in length with a sharp end tip. The leaves are cut to 5-cm lengths and defibrated in a hammermill. The mass is crushed by a specially designed crusher to extract the juice before the solid material is dried and baled for transportation to the pulp mill. The juice is a valuable by-product for the chemical and pharmaceutical industries. The Kamyr continuous digester with a hi-heat diffusion washing and preimpregnation system, similar to that for wood pulping, is used for pulping and primary washing. After cooking, the pulp is washed, screened, and cleaned prior to multistage bleaching. The pulp is bleached in a CEDED bleaching sequence to a brightness of 92. The pulp has exceptionally high-tearing strength and the other strength properties are comparable to softwood pulp. High porosity is a characteristic of sisal pulp. The pulp is spongy and drains rapidly; its bulk density is 20% less that that of conventional wood pulps.

Pulping of Flax

Flax (*Linum usitatissimum*) is grown extensively in Europe for use in the textile industry for the manufacture of linen fabrics. It is also cultivated in India and Argentina, primarily for the seeds, which are used to produce linseed oil for the paint industry. Flax straw is treated by retting and mechanical means to separate the long, strong bast fibers, which are used for weaving. The leftovers are known as seed-flax-straw tow. Considerable seed-flax is grown in the United States, primarily for the manufacture of linseed oil; the by-product straw is used for pulping. Flax fibers in the form of rags, spinning waste, and threads are also collected and used for pulp and papermaking. This material is cooked with sodium carbonate or sodium hydroxide to yield high-grade pulp suitable for fine papers.

Although seed-flax straw contains the same long fibers as those present in waste linen fabrics, it also contains undesirable plant components, such as woody cells, epidermal cells, and small vessels. Unbleached pulp produced from raw straw can be used where the presence of shives can be tolerated, as in bag and wrapping papers and where the high strength is advantageous. However to produce a high-grade pulp from flax-seed straw requires substantial treatment to remove the unwanted woody fractions. The raw tow contains 20% bast fiber and 70% wood fiber. The chemical analysis of seed-flax tow used for papermaking is:

Ash 3 to 4%.

Lignin 13%.

Pentosans 16.5%.

Alpha cellulose (ash and pentosan-free) 45.5%.

To produce high-grade pulp, the raw straw is decorticated by passing through breakers consisting of fluted rollers, which break up the woody portion. After decortication, the material is treated by a combination of mechanical action and screening to produce about 17 to 20% fine tow on the weight of the original straw. A considerable amount of the woody material and fiber fines are separated by this screening and cleaning operation. It is conceivable that raw straw could be disintegrated by a wet cleaning system and screening arrangement similar to that employed for wet depithing of bagasse.

Seed-flax straw is cooked in globe digesters or diffusers. Caustic soda alone, a combination of caustic soda and sodium sulfite, or caustic soda and elemental sulfur are used for cooking. The conditions used to produce pulp suitable for cigarette-grade paper from tow fiber are:

1. Caustic soda 20 to 25%, plus 3% sulfur.

2. Cooking time 5 to 6 hr.

3. Temperature 150 to 160°C.

The pulp is thoroughly washed and bleached to 70 to 80 brightness, using chlorine and hypochlorite. Other bleaching sequences that can be used for bleaching flax pulp are CEP and CED. In the former process 0.5 to 0.6% H_2O_2 may be required, whereas the latter bleaching sequence is limited by the high chlorine dioxide generation cost in small mills. However, chlorine dioxide can be produced from sodium chlorite by the reaction of chlorine or hypochlorite with sodium chlorite. Another bleaching sequence used is CDE, where the pH at the chlorination stage is maintained at 2 with the aid of sulfuric acid. After chlorination, the chlorine dioxide bleaching is carried out by the addition of sodium chlorite and the reaction is completed in about 2 hr. At the final stage of bleaching, the pulp is treated with caustic soda solution and the pH adjusted to 10 to 10.5. The reaction time is 2 hr at about 50°C. The pulp is washed at the end of the bleaching operation.[1108] Bleached yield ranges from 25 to 60%, depending on the extent to which the tow has been upgraded prior to pulping and bleaching.

Indian linseed fiber has been pulped using 10% alkali and cooking at 150°C for 4 hr. The unbleached pulp yield is 70% and the bleached pulp yield is 64.5% on the weight of the raw material, with a chlorine consumption of 0.64%.[1110]

Flax pulp is used for the manufacture of cigarette paper,[1111] condenser tissue, airmail paper, and other expensive thin papers.

Pulping of Jute

Jute (*Corchorus capsularis* and *Corchorus olitorius*) is an annual, dicotyledonous, fast-growing plant cultivated exclusively in the hot and humid climate of India and Bangladesh. The stalk grows to a height of 2.5 to 3.5 m and is cut at maturity about 25 cm above the ground. Cut stalks are retted in ponds of stagnant water. Fermentation occurs and the bast fibers are separated from the bark and woody portion of the plant. The bundles, or strands, of bast fiber have considerable length and are suitable for cordage, sack, and burlap manufacture. The fibers average 2 mm in length and about 20 μ in diameter. The fiber cavity is wide at places, but it is constricted by joints at intervals, resulting in the cell cavity being blocked. The fibers are slender and tapered at the ends.

Whole jute is rarely used for pulp and papermaking. Salvaged products, such as old jute sacks and burlap, are the materials available to the paper mills. Waste jute is cut into small pieces and dusted before cooking. Rotary digesters are employed for cooking, using 15 to 18% lime on the dry weight of the material. Jute is cooked at low pressure ranging from 138 to 207 kPa (20 to 30 psi) and the cooking time is 8 to 12 hr. During cooking, the natural waxes, dirt, and other impurities are dissolved and removed from the fiber. Due to the mild cooking conditions, no appreciable delignification occurs and consequently the fibers remain essentially as bundles at the end of the cooking period. The yield varies from 60 to 65%. The cooked stock is washed and bleached in the same manner as rag pulp. Calcium hypochlorite is used for bleaching with a total chlorine demand of 5 to 10%. The pulp is bleached to a brightness of 50 to 60 and it has a characteristic buff color. Jute pulps are used for the manufacture of high-strength bags, wrappings, drawing papers, and tags.

Pulping of Kenaf

Kenaf (*Hibiscus cannabinus*) grows in slender straight stems to a height of about 3.5 to 4 m and a diameter of 4 cm at the base. The plant attains maturity in 120 to 130 d from the time of seeding. It was originally native to East Central Africa, but it grows extensively in India and other parts of Southeast Asia. It is also grown to a certain extent in Argentina, China, Cuba, Egypt, Guatemala, Italy, Mexico, Morocco, and South Africa. The crop is ready for harvesting in 4 to 5 mo. The yield per hectare varies considerably; when properly cultivated and grown, the average commercial yield per hectare can be 15 to 20 ton. Kenaf is grown as a source of bast fiber for twine and rope. The bast fiber is about 20% of the dry weight of the stalk. After retting, washing, and processing the yield of fiber is around 10%. The average bast fiber length is 2.6 mm, whereas the fiber length of the woody material is 0.6 mm.[1112]

Attempts have been made to pulp the whole stalk. The pulp tends to be slow in drainage characteristics and limits the paper machine speed. Bast kenaf fibers, however, yield pulp with excellent strength characteristics, which can be used as a long fiber substitute. In fact, in Sri Lanka[1113] kenaf bast fiber has to some extent replaced imported long-fiber wood pulp for blending with short fibers produced from agricultural residues. The soda, sulfate, and neutral sulfite processes are suitable for pulping kenaf. Chipped and screened whole stalk kenaf is cooked in a rotary digester with 10% caustic soda on the dry weight of the material to produce bleachable grade pulp. The cooking time is 3.5 hr at a temperature of 170°C. The permanganate number ranges from 14 to 17. In the pulp washing stage, severe foaming occurs due to the presence of saponin in kenaf stems, particularly when the soda process is used for pulping. Following washing and cleaning, the pulp is bleached to 70 brightness using two-stage hypochlorite in a bleaching hollander. Total chlorine demand is 8% and the bleached yield varies from 45 to 48%.[1113] In a CEH bleaching sequence the pulp can be bleached to 80 to 85 brightness. Freeness of the pulp is 19 to 24 °S-R. Since the pulp produced from whole-stalk kenaf consists of long bast fibers from the bark and short fibers from the woody core, fiber fractionation may be used to separate them. The bast fibers can be used as a long-fiber pulp to be blended with other short-fiber pulps for fine quality papers. The short-fiber pulp from the woody core can be used as a furnish in the manufacture of heavy boards, where drainage characteristics are not critical.

Extensive research has been carried out by the Northern Regional Research Center, U. S. Department of Agriculture, Peoria, Illinois, on kenaf as a potential source of raw material for pulp and paper manufacture. These investigations have indicated that kenaf can be cultivated to give the highest yield per hectare compared to all other fibrous raw materials. On experimental plantations in the southern United States, stem yields of 27 MT/ha have been obtained in Georgia,[1114] 34 MT/ha in Texas,[1115] and 45 MT/ha in Florida.[1116] The bast fibers are approximately 20% of the total weight of the stalk and are similar to jute fibers. The crop can be harvested manually or mechanically. Reel-type harvesters can perform the dual function of harvesting and chopping kenaf into small pieces about 8 cm in length. Specially designed decorticating equipment can be used for effective separation of bark and core. Cultivation of kenaf in warmer climates can give a continuity of raw-material supply over an extended period of time in a one year cycle by scheduling the seeding pro-

TABLE 4-66 TYPICAL ANALYSIS OF BARK AND WOODY MATERIAL FROM MARYLAND KENAF

| | Original Unfractionated Stalks | | | |
	Fresh	Dried	Bark	Core[a]
Cellulose:				
Crude MEA[b] (%)	52.0	54.4	57.4	51.2
Alpha (%)	35.5	37.4	42.2	33.7
Extractives in:				
Alcohol-benzene (%)	7.2	3.1	3.5	3.0
Hot water (%)	13.0	10.2	14.0	7.8
1% NaOH (%)	34.8	32.2	·33.7	32.1
Wax[c] (%)	0.9	0.8	1.4	0.5
Pentosans (%)	21.7	19.7	16.1	19.3
Lignin[d] (%)	11.4	13.2	7.7	17.4
Nitrogen[e] (%)	3.4	4.6	6.5	4.2
Ash (%)	4.5	4.1	5.5	2.9
Total sugars, as glucose (%)	3.1	—	None	None

[a] Bark-core ratio = 42:58.
[b] Monoethanolamine procedure of Nelson and Leming. See ref. 1119.
[c] Material soluble in alcohol-benzene and dissolved in warm benzene.
[d] Sulfuric acid procedure of Bagby, et al, (See ref. 1120) pretreated with pepsin to remove protein.
[e] Kjeldahl nitrogen expressed as crude protein; N × 6.25.

gram. The harvesting season can thereby be extended over several months[1117] and storage facilities can be kept to the minimum at the mill and still maintain a continuous supply of raw material throughout the year. Green stalks can be treated in a crusher-screw-press arrangement to remove juices and other extractives prior to pulping. This treatment reduces the chemical requirements for pulping and bleaching. Field-dried kenaf, however, can be chopped and screened to remove surface contaminated materials and used directly in cooking. A typical analysis of bark and woody material from Maryland kenaf is given in Table 4-66.[1118] The bark fibers give higher yield on pulping and possess better strength properties. Pulp from the core or woody material contains relatively more pentosan and lignin.[1118] Soda kenaf pulp exhibits better strength characteristics than kenaf pulp produced by the sulfate process.[1121] Pulp can be bleached in a four-stage bleaching sequence to 90 brightness,[1122] using chlorine dioxide as the final oxidant.

Pulping of Other Fibrous Raw Materials

Apart from the raw materials discussed, which are in commercial use for pulp and papermaking, there are a number of other fibrous materials that have been tested to determine their suitability for industrial exploitation. Agricultural residues, such as cornstalks,[1123] castorstalks and jute-sticks,[1124] can be used as supplementary fibrous materials. Rye grass straw (*Lolium multiflorum Lan.*), which is available in abundant quantities in the Willamette Valley, Oregon, has been found suitable by Bublitz[1125] for use as a supplementary fibrous material for papermaking in conjunction with wood pulp. Various types of grasses such as Lemon grass (*Cymbopogon citratus*),[1126] Ulla grass (*Anthistiria gigantea*),[1127] thatch grass (*Imperata cylindrica*),[1128] and Elephant grass (*Themeda cymbaria*)[1129] have been found suitable for pulping and are used for paper and board manufacture. Papyrus (*Cyperus papyrus*), which grows on the banks of the river Nile in Sudan and Egypt, is known from ancient times as a papermaking raw material. The rind, separated from the stalk of papyrus, contains fibrous material comparable to depithed bagasse. Harvesting, collection, and handling of these fibrous materials require extensive study to make the pulp and paper producers interested in their utilization.

SECONDARY FIBER PULPING

A. J. FELTON

*The Black Clawson Company
Middletown, Ohio*

Secondary fiber pulping involves the repulping of wastepapers and paperboards. There are two basically different methods: (1) a purely mechanical system involving the use of pulpers, screens, and centrifugal separators and (2) a combination chemical and mechanical system in which chemicals are used in the pulping stage to remove ink and other contaminants. For many years secondary fiber pulping did not keep pace with the overall growth of the paper industry,[1130] but recent economic factors and environmental considerations have caused it to expand greatly. Secondary fiber is the second largest source of fiber for paper and paperboard in the United States,[1131] and the percentage of reuse of fiber is greater in Europe and Japan than it is in the United States.

Grades of Wastepaper

The pricing structure of wastepaper in the United States is built around a large number of different grades, but the most common grades are divided into six classes,[1132] shown in Table 4-67. The volume of wastepaper used on an average daily basis from each class in the United States and the value of each, as wastepaper and as finished pulp, is shown in Table 4-68.

Mixed wastepaper is a particularly difficult grade to repulp because of its high

TABLE 4-67 GRADES OF WASTEPAPERS

Mixed wastepaper—contains various qualities of paper not limited as to type or fiber content. Prohibitive material cannot exceed 2% and throwouts not to exceed 10%. Mixed wastepaper is used in roofing and bituminous asphalt shingles, molded articles, center ply in multiply board for boxes, structural dry-wall and common low-cost board.

Corrugated Waste—contains double-lined kraft outer surfaces and a fluted medium center. Also includes double-lined corrugated cuttings, corrugated cuttings, new kraft corrugated cuttings, and used corrugated containers, or boxes. Prohibitive material cannot exceed 1% and throwouts cannot exceed 5%. This grade is used in the production of linerboard, corrugating medium, dry-wall board, and roofing. It comprises the largest tonnage in the secondary-fiber field, with annual consumption well in excess of 5 million ton in the United States

Direct Entry (also identified as pulp substitute)—consists of white paper having no printing, of reasonable uniform brightness, and of no prohibitive material. Throwouts may not exceed 0.5%. This grade is used instead of virgin, bleached pulp in fine papers and publication papers.

Deinking Grades—consists of papers having printing, color, or groundwood content that can be treated in a deinking process that will remove the color, printing ink, and impurities. Should contain no prohibitive material and throwouts may not exceed 0.25%. This grade is used in the production of fine papers, book paper, envelope, and all types of tissue consumer products.

News—consists of baled, sorted, fresh, dry newspapers—not sunburned, and free from magazines, white blanks, pressroom overissue, and paper (other than news), containing not more than the normal percentage of rotogravure and colored sections. Packing must be free from tar. No prohibitive material allowed. Throwouts may not exceed 0.25%.

Prohibitive Material—any material that by its presence in the bale in excess of the amount allowed will make the bale unsuitable or unusable for the grade specified. Any material that may be physically damaging to the equipment.

Throwouts—all papers that are processed or treated in such a manner as to make them unsuitable for consumption in the grade specified.

TABLE 4-68 WASTEPAPER CONSUMPTION: DAILY USAGE IN THE UNITED STATES AND DOLLAR VALUE

Wastepaper Grade	Tons Daily Consumption	$ Value as Wastepaper	$ Value as Pulp
Mixed waste	7,532	75,320	542,340
Corrugated waste	15,506	620,240	3,821,910
Direct entry	498	99,600	136,800
Deinking grades	7,291	583,280	1,691,280
No. 1 news	6,217	248,680	475,470
	37,044	1,627,120	6,667,800

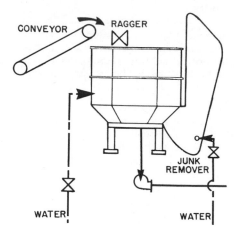

Figure 4-124. Conventional waste-paper pulping system.

degree of contaminants, such as metal particles, stones, bottles, tapes, rags, strings and plastic materials, polystyrene, polyethylene, blown styrene, foamed styrene, hot melts, and bituminus asphalt. It requires a recovery system that will continuously remove these contaminants to a degree that allows paper to be made from the recovered fibers. It can be repulped by the system shown as Figure 4-124, which is used extensively in the United States. The pulping of mixed wastepaper and corrugated waste is expected to increase in the future due to government pressure, economic conditions, tax incentives, and improved pulping systems. A substantial proportion of this increase will come from newly designed plants that will produce better, cleaner grades of secondary fiber for use in linerboard, corrugating medium, tissue, writing paper, and multiply board. The increased consumption of wastepaper will result in a shortage and a deterioration in quality of the overall raw material supply. However, those mills that have the proper equipment, designed to clean and screen plastics and other light-weight contaminants, will be able to use this waste material without problems.

Multiply Cylinder Board

Much of the secondary fiber is used in multiply cylinder board, which is made on a cylinder machine, where the paper is built up to the desired thickness in separate layers. There are many grades of cylinder board and these are listed in Table 4-69. Secondary fibers are almost always used in the filler plys, whereas virgin pulp may be used in the outer (liner) plys. It is generally necessary to have separate pulp-preparation systems for the liner stock and the filler stock.

Secondary Fiber Pulping Systems

The mechanical systems used for the repulping of wastepaper and paperboards are described in this section. There are a number of differently designed systems

TABLE 4-69 GRADES OF CYLINDER BOARD

Chipboard—Chipboard is a grade of cylinder machine board that contains the same stock in all plys and is always wastepaper stock. This board is made in a wide range of thicknesses and both "bending" and "nonbending" grades are made.

Mill and Bogus Bristols—Mill bristols, which are called by this name to distinguish them from index bristols made on a fourdrinier machine, are used mostly for poster cards and advertising. Bogus bristols are generally made from a number of different grades of stock, such as mixed waste, news, blank news, kraft waste, or virgin pulps. They are used for cheap cards, tickets, and colored tickets.

Folding-Box Board—If bending qualities are necessary, as in the case of folding-box boards, the two outside plys of the board must be made from long-fibered stock, such as bleached sulfite or bleached sulfate. Often direct entry grades are suitable. The stock must be refined moderately in order to develop the necessary folding endurance and reduce the amount of fuzz on the sheet, but jordaning of the liner stock is held to a minimum in order to preserve the bending qualities. In comparison, the filler stock, which is usually derived from mixed wastepapers, is customarily refined quite heavily in order to eliminate lumps of undefibered stock or fiber bundles, which may have passed the cleaning and screening system. The principal grades are bleached manila, white-patent coated, and on-machine clay-coated. They are used for packaging cereals, soap powders, cigarettes, wearing apparel, and similar items. Specifications may include weight, caliper, bursting strength, brightness (top liner), sizing, tensile, dirt count, moisture content, and color. Printing qualities are of great importance, and, because most of the board is used in packaging, the gluing properties are also very important.

Combination Manila Board—Another grade, known as combination manila board, designates a board wherein one or both of the outer plies are of a different raw material and/or color than the middle ply, whereas a plain or solid board has the same material throughout. The filler can be composed of approximately 40% unbleached sulfite and 60% mechanical pulp. Clay coating may be used. It is generally used for cartons with high-gloss printing.

Container Board—This board is made with a filler of mixed wastepaper and the liner made of kraft fiber, which may be all virgin kraft or a mixture of virgin kraft and waste kraft. If waste kraft is used, the board is sometimes referred to as jute liner board. Jute liner board may be given either a dry finish or a water finish, and in some cases the board is calender sized with starch, carboxymethyl cellulose, and so on. Filled kraft liner board is similar to jute liner board, except that virgin kraft stock is used in the liners. These grades are made in the standard thicknesses of 12, 16, or 30 points.

Setup Board—This board differs from folding-box board in that bending quality is not required. It is made of short-fibered stocks, such as groundwood, news, straw, or mixed papers. In some cases, the liners may be made from bleached white stock in order to improve the appearance. The board is used for rigid boxes and may be a solid or combination board, depending on the style of box; it ranges in thickness from 0.405 to 1.650 mm (0.016 to 0.065 in.) and weighs 290 to 1005 g/m^2 (60 to 206 lb/1000 ft^2) Stiffness, rigidity, and resistance to abuse are essential qualities.

(continued)

TABLE 4-69 Continued

Other Grades of Board—Other grades of cylinder board include laundry board, calendar board (usually a white-patent-coated board or a clay-coated board), cracker caddies (usually a jute-lined board), bottle-cap board (usually lined with sulfite stock), and matchbook board (either a bleached manila-lined board or a white-patent-coated board). In the case of milk-bottle-cap board, it is desirable to have a weak bond between two of the plys so that a tab can be readily raised on the cap when removing it from the bottle. Weakening of the bond can be achieved by adding wax emulsion to one of the plys at the point where the split is to occur. It is made of bleached prime pulp and normally has a thickness of 16 to 30 points, the actual caliper depending on the type and size of the container, which may be half pint, quart, or half gallon.

Duplex Paper—Duplex paper is a special grade of two-ply paper that is made on either a two-cylinder machine or on a combination cylinder and four-drinier machine. The product is used for bag papers, for example duplex flour sacks that have a white ply on the outside and a dark blue ply on the inside. The blue ply is made from 100% rope and the outside ply from part rope stock and part sulfite pulp. Virgin kraft can be substituted for the rope if adequate refining is available. The inside ply is refined to a lower freeness than the stock for the outside ply and considerable trouble is often experienced in obtaining a good bond between the plys. It is necessary to have the plys very wet at their point of contact and to use graduated pressure on the wet presses in order to reduce the danger of crushing.

Backliners—On all the above grades of board it is possible to use a backliner of different stock between the filler and the top liner. This backliner helps improve the cleanliness, brightness, and formation of the board, and hence is effective in improving the printing surface. The object is to cover up the dark filler stock and produce a better foundation for the liner stock. News stock is a common grade for use as a backliner, and a board made in this manner is said to have a "skim news backing." Manila has also been used as a "skim."

---

used in the United States and European mills.[1133,1134] All systems involve the use of a pulper to break up the bundles or bales of wastepaper; a device to remove heavy "junk"; a device to remove rags, strings, and metallic wires; a screening system for removal of oversized particles; and a centrifugal separator. The objective is to remove contaminants of all types and to accomplish this with a minimum expediture of energy.

Secondary fiber-pulping systems must have a specified tonnage output at a specified consistency and produce pulp having good runability on the paper machine. They must be able to operate without contaminant buildup in the pulper. The amount of power consumed in secondary fiber pulping is one of the most important factors to be considered and it depends on the type of raw material to be pulped. The quality of pulp obtained from a well-designed system can replace virgin fiber for many uses. High-density-type secondary-pulping systems that are available at this time include:

1. Black Clawson Lo-Intensity Pulping (LIP)
2. Voith-Morden Turbo Separator System
3. Beloit-Jones Belcor System
4. Escher Wyss Fiberizer System.

They are described in the following pages.

Black Clawson Lo-Intensity Pulping System. This is a[1135,1136] relatively new system for secondary fiber processing. It requires minimum energy input, about 36 kWh (2 hpd) per ton, which makes for low operating cost. It results in minimum degradation of contaminants, thus making their removal more effective. The pulping system, which is shown in Figure 4-125, operates continuously, depends on extraction through 1.59- to 2.54-cm (5/8- to 1-in.) holes and requires only 18kWh (1 hpd) per ton for defibering. The wastepaper normally used in this pulper contains about 5% contaminants, so even if the system is 100% efficient, there will always be 5% contamination in the stock slurry. The ragger performs the function of continuous removal of strings and rags, and the junk box removes the heavy contaminants that are too large to pass the extraction-plate holes.

The extracted stock is pumped through centrifugal cleaners, which remove the heavy contaminants. The pressure-drop across the cleaner should be 69 to 103 kPa (10 to 15 psi) and, because of the high volume of rejects, an automatic reject dumping system is provided. The stock then enters a pressurized power screen equipped with 0.2- to 0.3-cm (0.079- to 0.125-in.) holes powered by 225 Kw (125 hp) at 4% consistency. The defibered stock accepted by the screen moves downstream for additional cleaning and screening.

The reject flow from the pressure screen amounts to 40% of the fiber weight and contains a certain amount of defibered stock, which acts as a vehicle to carry the contaminants. The reject flow at about 5% consistency passes to a deflaker, which is an adjustable clearance machine designed to handle the large undefibered flakes of paper and contaminants. The deflaker has no bar-to-bar contact so hydraulic shear alone acts on the paper flakes but leaves the contaminants intact. The deflaker uses 18 to 27 kWh (1 to 1.5 hpd) per ton. The reject flow continues to a vibrating screen equipped with 0.47- to 0.95-cm (3/16 to 3/8 in.) holes at 3% consistency. The accepts from the screen are returned to the pulper and the light rejects are rejected for disposal.

Voith-Morden Wastepaper-pulping System. This system consists of a pulper, turbo separator, high-pressure cleaner, and vibrating screen.[1137] It is designed for continuous operation and it removes both heavy and light contaminants with minimal loss of usable fibers. It slushes and cleans the fiber, so that little or no additional screening or cleaning is needed downstream. It operates at 3 to 4% consistency and requires about 36 kWh (2 hpd) per ton.

As shown in Figure 4-126, the pulper receives the baled wastepaper from the

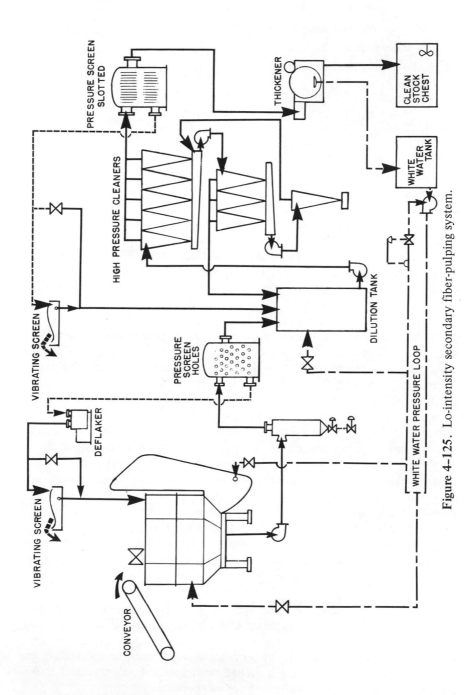

Figure 4-125. Lo-intensity secondary fiber-pulping system.

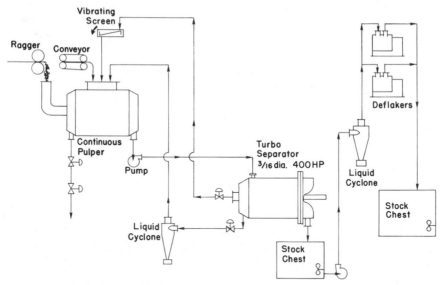

Figure 4-126. Turbo-separator system by Voith-Morden Co.

conveyor. Lightweight contaminants collect in the turbulence zone, form a rag rope, and are removed by a time-controlled ragger. Heavy contaminants collect in the junk trap. Partially defibered secondary fiber plus floating contaminants not removed by the ragger leave by way of the extraction plate. They are pumped into the turbo separator for further defibering and cleaning. A large cup on the rotor stabilizes the vortex, entraining the remaining lightweight contaminants. Accepts pass through the rotor-extraction plate, with heavy contaminants being removed continuously (tangentially). A cleaner receives the light contaminants and return the accepts to the pulper. Lightweight contaminants are periodically discharged and pumped to the vibrating screen, which returns accepts to the pulper. Downstream deflaking and screening completes the system.

Beloit-Jones Belcor System. The Belcor System, as shown in Figure 4-127[1138] is a combination cleaning and defibering system arranged for continuous extraction. The stock from the pulper is pumped into a Belcor unit continuously and tangentially. Its function is to screen out intermediate-size pieces of plastic and wet-strength paper; to serve as a secondary pumping unit after virtually all large pieces of plastic and heavy junk have been removed; and to remove any heavy metal, such as paper clips and staples, that have not been separated in either the pulper or cleaner. Referring to Figure 4-127, the light rejects are discharged at the central outlet (2) opposite the rotor. Usually a flow rate of 10% by volume is adequate to prevent an excessive buildup of rejects in the tank and to insure good cleaning efficiency. The reject flow may be controlled by a hand valve (3) or inlet pressure. The flow is adjustable to the amount of rejects in the waste. The rejects are screened on a vibrating screen (4) and returned to the

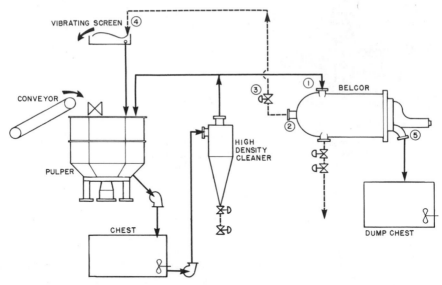

Figure 4-127. Beloit-Jones Belcor system.

pulper. The accepted stock is discharged through a perforated extraction plate behind the rotor (5). Hole sizes range from 0.317- to 0.472-cm (1/8 to 3/16 in.) at stock consistencies of 2 to 3 %.

The Belcor unit serves as a second pulping stage. Its defibering capacity is approximately the same as a conventional pulper and it uses 9 to 18 kWh (0.5 to 1.0 hpd) per ton. The primary pulper, which has large diameter holes, requires low power because some of the pulping power is applied at the Belcor. The cleaning efficiency depends on particle size and concentration of the contaminants and ranges from 75 to 90%.

The Escher Wyss Fiberizer System. This system, which is shown in Figure 4-128,[1139] was designed to separate out contaminants present in wastepaper stock as it leaves the pulper. Within the Fiberizer the functions of centrifugal separation, screening, and deflaking are combined.

Stock enters the Fiberizer through a tangential inlet in the conical housing (1). Rotational movement of the stock is caused by the rotor, which is adjusted close to the screen plate. Fixed defibering bars are positioned in a circle around the rotor to improve the deflaking action and to keep the screen clean by turbulence. Accepted stock leaves through the screen plate, which has 0.3- to 0.4-cm holes (2). Large, undefibered flakes are retained along with the plastic film and heavy contaminants. Stock containing a high percentage of light contaminants is drawn off intermittently through an outlet in the center of the cover (3). Heavy contaminants are collected by centrifugal force in the junk trap at the bottom of the unit (4) and are dumped automatically at selected intervals. A percolating flow of water washes back any fibers. The adjustable gap between

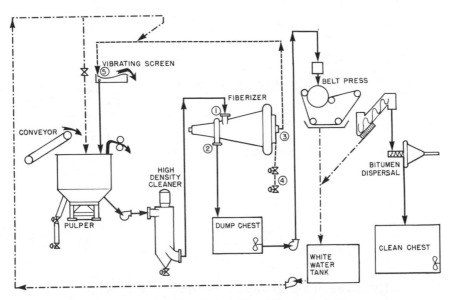

Figure 4-128.   Escher Wyss Fiberizer system.

the rotor and fixed plate as well as the defibering bars defiber the large flakes, but the cutting of thin plastic foil is prevented. The lightweight contaminants, such as foil and polystyrene, are expelled intermittently through a control valve to a vibrating screen (5).

Bituminous Asphalt Dispersion.   Bituminous asphalt, which is commonly used as a laminating adhesive and as a water-vapor barrier in sack papers, fiber barrel containers, and kraft shipping bags has a long history for causing problems in re-pulping. When these papers were used in older systems, they produced black-specked board, and "bleeder" spots were created. Considerable effort has gone into the development of improved systems.[1140-1142] It is possible to disperse the asphalt to adjoining fibers during the secondary fiber pulping so it is not noticeable in the finished product and will not produce bleeding in papers, such as dry-wall gypsum board. The most commonly used system in the United States is shown in Figure 4-129. After the thickener in a continuous wastepaper system, such as that illustrated in Figure 4-124, the stock is dewatered to 12 to 16% with an inclined-screw thickener, pressed to 35% consistency in a cone-type press, and discharged by means of a feed conveyor to a continuous digester. A plug of stock at about 50% consistency is formed in the throat of the feeder, which helps to contain the steam pressure in the digester tube. Steam at about 515 kPa (75 psi) is introduced into the continuous digester. This steam raises the temperature of the stock well above the melting point of the asphalt. As the stock is tumbled by the action of the internal flighted screw of the digester tube, the softened asphalt is deposited on and spread over the surface of the fibers. The digester tube has a variable speed drive that controls the exposure time of

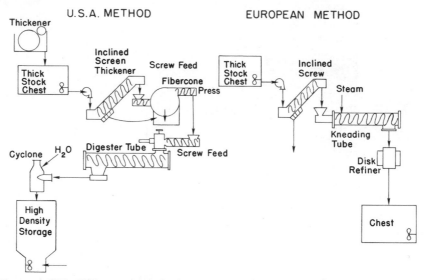

Figure 4-129. Bitumen asphalt dispersal systems, showing methods used in the United States and in Europe.

the stock to the steam and the tumbling action. At the discharge end of the digester tube, a special orifice releases the stock from the pressure vessel to atmospheric pressure. The accompanying steam propels the stock to the cyclone, where dilution and cooling water is added to return the stock to the desired temperature and consistency. The power required in this system is 59 kWh (3.3 hpd) per ton; the steam requirements are 0.4 kg (0.9 lb) of steam/0.45 kg (1 lb) of fiber.

In the European method a different approach is used as shown in Figure 4-129. The stock from the thickeners is dewatered and added to a kneading tube, where in the presence of steam, the stock is thoroughly kneaded to disperse the asphalt. A disk refiner at the discharge end of the kneading machine disperses the asphalt thoroughly among all of the fibers.

Deinking of Old Papers

In order to produce a white pulp from wastepaper that will be suitable for book and other similar papers, it is necessary to remove the ink from the wastepaper. The process of doing this is known as deinking. There are two basic steps in deinking: (1) dissolving or loosening the ink by chemical means and (2) removing the ink from the pulp by mechanical washing. All deinking systems have the following stages:[1143-1148]

1. Pulping or defibering in the presence of chemicals.
2. Cleaning and screening.

3. Washing.

4. Dewatering or thickening.

Ink removal is sometimes accomplished through flotation. Bleaching and bleach washing are included if required to produce white pulp.

Types of Papers Used in Deinking

The quality of deinked pulp is primarily determined by the kind of wastepaper used in the deinking plant. Thus it is desirable to obtain only the brightest grades of wastepaper, all of the same general kind. In some cases, wastepaper can be obtained from large paper-using plants that is so well segregated that it requires no sorting. The trend is away from hand-sorting (because of excessive labor costs) and toward the use of deinking methods based on the proper balance of chemical and mechanical operations to produce the desired end results from a variety of papers. The most desirable papers for deinking are fine shavings, cuttings, ledger stock, and magazine stock. The lower grades of mixed papers can be sorted to remove undesirable papers and other contaminants, such as carbon paper, waxed papers, impregnated papers, glassine, parchment, bits of cloth, typewriter ribbon, bits of wood and dirt, highly colored covers and posters, and wet-strength papers. But sorting is costly and disappearing from the industry.

Purchase specifications for wastepaper for deinking sometimes limit the groundwood content because groundwood is difficult to deink; it turns brown during the deinking, and it cannot be bleached with a straight hypochlorite bleach. Old groundwood papers are particularly difficult to defiber and tend to form small, hard clumps of fibers. Originally, groundwood papers were excluded by most deinking mills, but with the more widespread utilization of groundwood in printing papers and other grades, it has become increasingly difficult to obtain enough groundwood-free wastepaper to satisfy the demands of deinking plants. Therefore, special processes have been developed to handle groundwood papers. When the groundwood content exceeds 10%, the deinking method must be designed to handle groundwood. Under these circumstances, a special bleaching process is required to bring brightness levels to 70 or more. Papers containing groundwood can be identified by the color reaction obtained when the paper is stained with aniline sulphate (yellow), phloroglucinol (red), or a solution of sodium hydroxide (yellow). West-strength paper is objectionable because of the resistance of the paper to fiber separation, and high temperatures and low pH are required for the disintegration of these papers. Glassine and parchment papers are difficult, if not impossible, to defiber. Waxed, resin-impregnated, and resin-coated papers are resistant to water and cannot be defibered by ordinary deinking methods. Cellophane will not disperse, but unless it is excessively brittle, it will remain large enough to be removed by screening.

Papers treated with rubber-like or thermoplastic materials cause problems in deinking. As little as 1 kg of rubber-like material can ruin over 100 ton of pulp[1149,1150] if it is not properly dispersed. Pigment-coated book papers are readily deinked. Colored papers present somewhat of a problem in deinking, particularly if the dyes used in the paper are resistant, or fast, to chemicals. Most of the basic and acid dyes are destroyed by a caustic cook and can also be reduced with zinc hydrosulfite, although basic dyes tend to reoxidize on long standing. Most of the direct dyes can also be stripped with either caustic, chlorine, or hydrosulfite, but there are exceptions, such as the stilbene yellows, oranges, and turquoise blues, which are regarded as nondeinkable.[1151] The pigment types differ: (1) the chrome yellows and iron blues are destroyed by caustic, (2) the azoic types are regarded as nondeinkable, but can be discharged by direct chlorination, and (3) the phospho-tungstic-molybdic lakes present no problem since they are easily destroyed by hypochlorite bleach.

Chemicals Used in Deinking

Much deinking is done with plain alkali, but detergents and dispersing agents, such as soaps, sulfonated oils, bentonite, sodium metasilicate or silicate penthydrate, and other surface-active substances are sometimes used in combination with alkali. An ideal deinking formula would include an alkali to saponify the varnish or vehicle of the printing ink, a detergent to aid in the wetting of the pigment in the ink, a dispersing agent to prevent agglomeration of the pigment particles after release from the paper, and an absorption agent to bind the pigment and prevent redeposition on the fiber.

Alkali is used in the deinking formula for two purposes: (1) to remove rosin sizing from the paper and (2) to saponify the ink vehicle and release the pigment in the ink. There is generally about 0.5 to 2.0% ink on the weight of the paper; this must be completely removed if white pulp is to be produced. From the standpoint of ease of deinking, there are four principal types of inks:

1. Drying, oil-base inks.
2. Non-drying, oil-base inks.
3. Inks having a synthetic resin base.
4. Metallic inks with latex base.

Drying, oil-base inks that are slightly oxidized can be readily saponified by alkali. However, completely oxidized oil-base inks; nondrying, oil-base inks; and inks having a synthetic resin base cannot be completely saponified by alkali of ordinary concentration. Consequently special methods of deinking must be used for papers containing these inks. The various high-gloss and metallic inks in use today are extremely difficult to remove with alkali. Solvents (e.g., tri- or tetrachloroethylene, benzene, or carbon tetrachloride) or soaps and detergents can be used to aid in the deinking of these papers. Rosin is readily removed by saponification with alkali, and even the waxes used in sizing paper are melted and readily removed. Solvents are sometimes used to remove wax and polyethylene.[1152]

The type and amount of alkali required in deinking depends on the type of mechanical treatment and the temperature and time of cooking. Sodium carbonate and sodium hydroxide are widely used in deinking. Sodium carbonate is a milder agent than sodium hydroxide and results in less oxidation of the fiber and less fiber loss. Sodium hydroxide results in faster pulping. However, too strong caustic soda may seriously attack the cellulose, since it is in a particularly susceptible condition, having a high area of surface exposure. About 5% sodium hydroxide is the maximum used. A typical caustic soda deinking formula for white ledger paper is 3% sodium hydroxide solution at a temperature of 71°C.[1153] Concentrated sodium hydroxide should always be diluted if it is added directly to the fibers. A concentration greater than 17% should never be added to a mixture of fibers and water because caustic of this concentration will dissolve the cellulose and contaminate the batch with sticky scabs, similar to soft, hot melts and latices. If strong caustic must be used, it should be added to the pulper batch water before the paper is added. If used alone, 3 to 8% sodium carbonate is sufficient for the deinking of most papers. Some prefer a mixture of 2.5% sodium carbonate and 0.5% sodium hydroxide.[1154] Only a small part of the alkali is consumed in the cook. Laboratory results have shown that the consumption of alkali is in the range of 0.25 to 1.0%, based on the weight of the paper,[1155] so that most of the original alkali appears in the waste liquor. In deinking ordinary rosin-sized papers, the alkali reacts with the rosin to form a rosin soap, which acts as a detergent for the ink particles. Sequestering agents may be helpful in preventing the formation of calcium soaps. If unsized papers are used, no rosin soap is formed, and special detergents may be helpful. Plain soap appears to be helpful under some circumstances.[1156]

Sodium silicate is sometimes used for part of the alkali. Silicate is effective at a lower pH than sodium hydroxide. This is important in the deinking of groundwood papers to prevent yellowing. The more alkaline grades of silicate are preferred. A suitable grade is one containing 1 part Na_2O to 1.6 to 1.7 parts SiO_2, or sodium metasilicate ($Na_2 SiO_3 \cdot 5H_2O$). The amount of silicate varies from 2% to 9%. High-grade wastepaper can be deinked by using 3% of 42° Baumé (Bé) silicate and 1.5% sodium hydroxide at a temperature of 66°C for 45 min.[1157]

Once the ink pigment has been released, dispersing and absorptive agents are desirable to prevent the pigment from becoming redeposited on the surface of the fibers, which are strongly absorptive. One patented process calls for a mixture of high-silica sodium silicate and a fatty acid.[1158] Bentonite may be used for the same purpose[1159] and to prevent the agglomeration of carbon-black particles around curds of calcium or magnesium soaps formed during deinking. Other clays may be used, or the clay may be obtained from the filler in the paper. Waste book and magazine stock have been successfully deinked for years, and part of the ease of deinking these papers can be attributed to the presence of clay fillers. Bragg[1160] points out that 0.75% bentonite completely absorbs the ink and scum in a pulp containing unsaponified ink. Removal of ink by absorption on a hydrophobic particulate solid has been shown to be effective,[1161] but any soap or other surface-active agent must first be deactivated.

Peroxide has been proposed as an aid to deinking operations. Peroxides are particularly effective in combination with silicate. A suitable formula is 1 to 3% hydrogen peroxide, 3 to 6% silicate (58.5°Bé, 1.6 ratio), and 0.5 to 2.0% sodium hydroxide to be used at a temperature of 71°C for a period of 35 to 90 min.[1162,1163] If sodium peroxide is used in place of hydrogen peroxide, no sodium hydroxide need be added. Regular alkaline liquors have a detrimental effect on the brightness of groundwood containing pulps, but peroxide helps to prevent darkening and color reversion of the pulp. There is no advantage in using peroxide, unless the groundwood content of the wastepaper is over 15% or thereabout. Zinc hydrosulfite has also been used in deinking groundwood papers to improve the brightness. It is used to aid in the destruction of dyes when colored papers are present. About 0.5 to 1.5% will improve the brightness of groundwood papers.

Most of the formulas used in commercial deinking processes are relatively simple. For colored ledgers, tab card, computer printout, and other selected chemical fiber waste, some typical formulas are:

1. Sodium hydroxide, 4%.

2. Sodium hydroxide, 2.5%, plus 2.5% sodium silicate, plus 3% sodium carbonate.

3. Sodium hydroxide, 3%, plus 2% sodium silicate.

4. Sodium hypochlorite, 0.8%, expressed as chlorine, plus 4% sodium hydroxide.

For colored ledgers and for coated carbonless papers that are made with activated clay and encapulated ink, formula number 4 is best. The hypochlorite is used to strip the color. It is added to 60°C water and the batch defibered for 5 min. The sodium hydroxide diluted to not more than 15% concentration is then added and the defibering continued for an additional 15 min. Tab card and ledgers of low groundwood content can be processed in formulas 2 or 3. Newspapers can be deinked in solutions containing (1) 2% sodium peroxide, plus 5% sodium silicate, (2) 2% sodium peroxide, plus 3% sodium silicate, plus diatomaceus earth, or (3) 1% sodium hydroxide, plus 1.5% sodium silicate and 0.7% hydrogen peroxide. The temperature should be kept below 54°C when deinking groundwood. A formula for the deinking of groundwood papers that can be used at elevated temperatures is 1% sodium hydroxide, plus 8% sodium perborate, plus 2% sodium carbonate. After pulping, the stock is dewatered to 20% consistency with inclined-screw extractors, and rediluted for cleaning in high-pressure-drop centrifugal cleaners and screened through pressure screens having 0.014-inch slots. Care must be taken not to dewater the stock beyond 20% density because the ink, which is partially saponified and soft, will blacken the fibers by the rubbing pressure.

Pulping Conditions Used in Deinking

Pulping may be either batch or continuous. The batch method permits better control of water, chemicals, and wastepaper added to the pulper. The batch at a predetermined consistency, temperature, and chemical concentration is processed for a predetermined length of time, thus providing positive control of pulping conditions, regardless of the size of the perforations in the pulper through which the stock is to be withdrawn. Defibering is generally 98 to 100% completed and sufficient time is allowed for chemical reaction.[1164-1166] Samples can be taken before dumping the batch to insure that the pulping is complete. Overall, the batch process is considered the best for deinking, even though productivity will be 25% less than that obtainable with continuous operation.

Continuous pulping allows maximum productivity for a given size of pulping unit. Water, chemicals, and wastepaper are added at a controlled rate commensurate with the pulping capacity and the system demand. The wastepaper is exposed to the chemical solution and to the defibering action of the rotor until the particles are small enough to be withdrawn through the holes in the extraction plate. Retention time in the pulping unit varies and not all the wastepaper has the same exposure to the chemical action and defibering action. The degree of defibering is dependent on the rate of extraction and the demand. Sampling is possible but is not always representative of the entire batch.

Originally deinking was done by cooking in globe digesters at consistencies ranging from a low of 6% to a high of 35%, temperatures of 79 to 93°C and cooking time of 3 to 8 hr. Steam was generally introduced directly into the charge by means of a perforated pipe running throughout the width of the boiler. Heating with saturated steam rather than superheated steam was used, particularly if sodium hydroxide was used as the cooking agent.[1167] In extreme cases as long as 8 to 10 hr at 275 to 345 kPa (40 to 50 psi) steam pressure was used,[1168] but this resulted in a loss of yield due to degradation and solution of the carbohydrates. At the end of the cooking period, the stock was dumped, in the form of a pulpy mass, to a pit underneath the cooker, where the liquor was drained away. In most cases the liquor was saved for reuse. Most deinking is done in open pulpers at temperatures of 60 to 82°C and a retention time of 45 min to 1 hr 30 min. The power required is used for defibering; there is no actual "beating" of the stock in the sense of "hydrating" the fiber. Pulping consistencies vary from 7 to 25%. High-consistency pulping speeds up defibering, reduces the amount of chemical required, and results in pulp of higher freeness, compared with low-consistency pulping. Cooking liquor can be recovered by extraction immediately following the dump chest. A typical flow diagram is shown in Figure 4-130 and a shortcut system is shown in Figure 4-131.

The use of a rotating drum that is perforated along part of its length has been proposed for defibering and prescreening of wastepaper, generally using 1 to 2% sodium hydroxide.[1168 a]

After cooking, the quality of the deinked stock can be evaluated by making handsheets and examining by both reflected and transmitted light. Particles

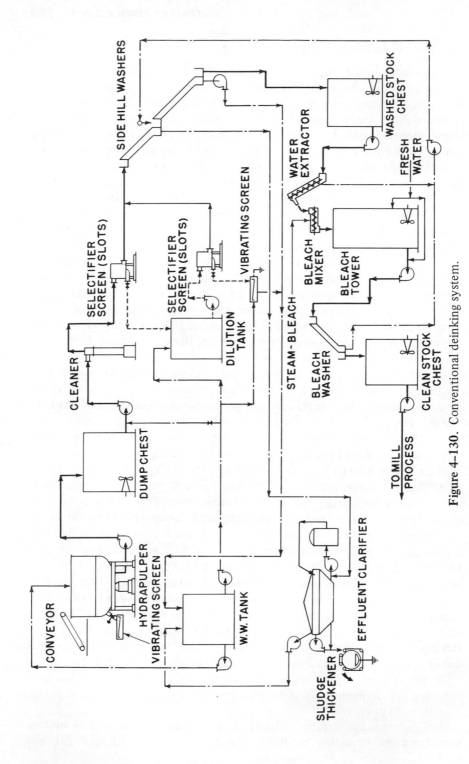

Figure 4-130. Conventional deinking system.

584

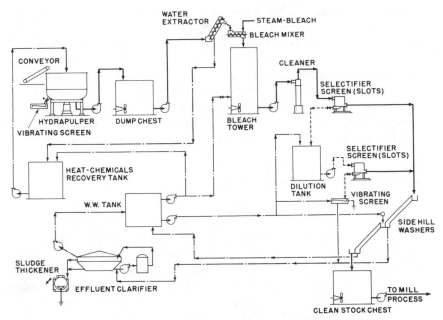

Figure 4-131. Shortcut deinking system.

in the sheet should be examined to determine their source. Some of the most undesirable materials are ink balls (ink from papers used by the printers for wiping), plastic particles (from adhesives and coatings), latex or oxidized rubber particles (from impregnants or binders), asphalt (from laminated papers), undispersed paper particles (from high groundwood content or wet-strength papers), and hot melt adhesives (from the bindings of cups and square-edge magazines).

Cleaning Stage of Deinking Process

The cleaning step follows the defibering of the wastepaper in the pulper. Centrifugal cleaners, which act on the principle of converting pressure to velocity thus providing the centrifugal force for separation, are the type generally used. The stock enters the top of the cleaner. While gravitational force remains constant, the velocity of the stock is slowed down because of friction with the sidewalls. The clean stock returns up through the center area of low or no pressure and is discharged at the top of the separator. The heavy contaminants separate at the bottom of the cleaner and are removed.

The two important variables in conical separators are stock consistency and pressure-drop in the separators. High-pressure-drop cleaners, having a pressure differential of 205 to 275 kPa (30 to 40 psi), which are used at consistencies ranging from 0.4 to 0.8%, will remove metal matter, sand, large particles of ink, and small contaminants. They are used in multistage arrangement after low-pressure-drop cleaners in the process. Low-pressure-drop cleaners are used at a pressure drop of 48 to 138 kPa (7 to 20 psi) and a consistency of 0.8 to 5.0%. They are usually in single stage arrangement followed by screening and then high-pressure drop cleaners. Low-pressure-drop cleaners remove contaminants that are 0.42 cm (1/16 in.) in diameter and larger.

Later in the process, core bleed cleaners, such as Celleco and Tri-Clean, can be used to remove materials of very low specific gravity, such as plastic particles, hot melts, and metallic inks. These operate at 0.7% consistency and below.

Screening Stage of Deinking Process

After centrifugal cleaning, the stock is screened before washing. Pressure screens having small holes or slots are used. When maximum cleaning is desired, pressure screens in series are used, the slots or holes being smaller in the secondary than in the primary screen.

Primary screening can be done at consistencies ranging from 0.3 to 4.5% depending on the size of the hole or slot. Most primary screening is done at 1.0 to 1.5% consistencies. There are many types of pressure screens. Most operate at inlet pressures of 138 to 275 kPa (20 to 40 psi) and a slot size of 0.036 cm (0.014 in.). The reject flow from the primary screens should be re-diluted and rescreened to increase the yield. Fine screening is done by means of rotary vibrator screens that vibrate at high frequency, while the cylinder is partially submerged in the stock. These screens have slots ranging from 0.010 to 0.020 cm (0.004 to 0.008 in.) in width. The consistency is generally 0.8 to 1.0%. Rejects are disposed of as refuse or passed to a tailing screen. The accept flow from both the primary and the secondary screens are combined and passed to the washers.

Washing Stage of Deinking Process

The stock after cleaning and screening must be washed to remove the dispersed inks, clay, and chemicals. The type of washing depends on individual preference, water availability, effluent-handling system, and initial-investment limitation. There are five different methods of washing. These are:

The Lancaster Washer.

Sidehill Washer.

Inclined-Screw Washer.

Vacuum Washer.

The American Disk Filter.

Washing processes are based on the simple principle of draining or pressing the water in the stock through a screen. Success depends on how finely divided and dispersed the ink is in the stock. Typical washers remove about 85% of the ink from the stock. Theoretically, it is possible by repeated washings to remove up to 99% of the ink, but in commercial practice this is not feasible. The ink particles are generally fairly large and tend to become entrapped by the fibers during the washing stage. Another effect is the redeposition of the ink particles on the fiber surfaces; this can occur in any washing operation. If deinking and washing have been done well, stock from old magazines should contain not more than 3% inorganic solids, be completely free of large dark-colored material, and have a brightness of around 50 GE.[1169]

The problems of waste water treatment from the washing stage in the deinking process have deterred many aspiring companies from using a washing process, despite its usefulness in removing fillers and unwanted fines from the stock. Removal of fines is important to the subsequent papermaking process because below a certain size, fines cease to contribute to web strength; they reduce drainage on the paper machine; they increase the drying cost; and they cause dust and lint during printing of the paper. To treat waste water from the deinking process satisfactorily, some means of treating the inky water is required, preferably by means of ink-sludge concentration. Large quantities of water are required to wash the pulp thoroughly. Excessive washing has the disadvantage of causing large fiber losses, but the pulp that is insufficiently washed will darken when alum is added because of the precipitation of suspended impurities on the fibers. Foaming should be held to a minimum, since the foam and froth formed during washing has a strong affinity for carbon-black particles. If the foam is allowed to break on the mat of stock, it may deposit particles of carbon, thereby resulting in a darkening of the stock.[1170] Antifoams are sometimes added to keep down foam.

Lancaster Washer. In this method a revolving cylinder covered with wire mesh (20/20, 40/40, 40/60 depending on wastepaper) is partially immersed in the dilute stock, and a mat is formed on the cylinder surface due to the controlled differential head. The water, ink, and clay flow from the outside into the center and discharge through the ends of the cylinder. The differential head is controlled by adjustable weirs in the discharge boxes. It is considered the best method of washing because the mat is thin and allows the water to flow freely into the center of the cylinder. A couch roll is used to aid in water removal; this pressed-out water combines with the naturally drained water. The pressing action dewaters the mat to 8 to 10% consistency. The stock is transferred from the cylinder face to the rubber-covered couch roll and a metal or micarta-lipped doctor removes the mat from the couch roll.

Lancaster washers are costly and space-consuming. Their capacity is low; about 0.020 to 0.034 l/cm² (5 to 9 gal/ft²) of cylinder area can be realized. Drainage rate is controlled by stock temperature, wire mesh, freeness of the stock, and differential head. A 594-cm (60-in.) diameter Lancaster will handle 0.14 ton/cm (0.36 ton/in.) of face; thus a 152- by 594-cm (60- by 234-in.) face will handle 85 ton of pulp. Two or more Lancasters arranged in series with repulper mixing chambers between them can be used for multiple washings. Losses are in the range of 8% for the first stage and 6% for the second stage.

Sidehill Washer. The sidehill, or slide-wire washer, considered by many as a crude type, does a very good job of washing. The stock at a consistency of 0.6 to 1.0% enters a headbox at the top of the washer, the velocity is dampened by a weir, and the stock then overflows onto the inclined, wire-mesh surface, which is inclined at an angle of 38°. The water drains through the wire, and the stock slides or tumbles down the inclined-wire sufrace. The water is collected in a compartment under the wire, which extends to the bottom of the washer. As the fibers tumble down the inclined wire, new areas are constantly exposed for water removal. The wire mesh is usually 58 × 80 or 58 × 100 and in some rare instances 60 × 60. The stock is collected at 3 to 7% consistency in a discharge box at the bottom of the washer.

The sidehill washer is characterized by low initial cost, low operational cost, and low maintenance cost. Freeness of the stock has little effect on the operation, compared to mat-type washers. Sidehill washers are commonly made in a length of 3.65 m (12 ft) and a width of 9.14 m (30 ft). Such a washer will handle 100 ton of stock, or 0.11 ton/cm (0.28 ton/in.) of face. Sidehill washer losses can be 10% in the first stage and 4% in the second stage, but this naturally depends on the type of stock being washed.

Inclined-Screw Washer. The inclined-screw washer consists of an inlet section that supplies stock to several inclined screws housed in a common casing or perforated cylinder having holes of 0.157 cm (0.062 in.) in diameter. As the screw rotates, the stock is pushed upward, the water is drained through the perforated cylinder, and the thickened stock is pushed out a separate discharge port, which is opposite the inlet section. The screw may have nylon brushes on the leading edges to keep the perforated holes clear or, in some models, this is achieved by maintaining a very close mechanical clearance. The screw carries the ever-thickening stock upward to a point where the flight stops. From this point to the discharge opening only the pressure of the upcoming stock pushes the thickened-stock plug toward the discharge opening. As the plug of stock leaves the perforated cylinder, a special breaker arm, which is secured to the shaft of the screw, breaks up the plug and causes it to fall out of the discharge opening.

Inclined-screw thickeners come in 22.8-cm (9-in.) and 40.6-cm (16-in.) diameters. The capacity of a 22.8-cm (9-in.) size is nominally 20 ton and the outgoing consistency ranges from 16 to 25%. This method of washing does not approach the Lancaster or Sidehill efficiency because ink particles are entrapped in the 7.6-cm (3-in.) thick mat of stock and cannot be expelled with the effluent.

Vacuum Washer. Vacuum washers are used for continuous operation and high productive capacity. A drum is immersed in the stock slurry and a vacuum applied to the submerged portion, causing the fibers to deposit on the outer surface as the drum rotates. The drum is divided into compartments that are connected to a rotating valve. Vacuum is applied through the valve to remove the effluent of ink, clay, and water. Several compartments may be connected to the vacuum port to form a progressively thicker and thicker mat. As the compartment leaves the vacuum area, air is permitted to enter to assist in the stock removal.

Water sprays may be used on the drum surface to assist in washing. Discharge consistency can be 12 to 17%. The drum washer does a reasonably good job, but due to the thick mat of fibers, some of the ink and clay particles are entrapped. Fiber loss is very low.

Disk-Type Washer. The disk washer operates on much the same principle as the drum vacuum cylinder. The washing areas are arranged in disks rather than a large drum. Each disk may be independently removed without disturbing the other disks and each disk can be operated as a separate washer. The disk has eight or more segments that are under vacuum as they pass through the slurry, thus forming a heavy mat on each side of the washer surface. As the segments pass the last vacuum port, air is introduced into the center of the segment and jets of water are applied to loosen the stock mat and roll it from the disk surface. The jet also cleans the wire or cloth surface.

The disk washer has a high dewatering capacity, which ranges from 0.008 to 0.020 l/cm^2 (2 to 5.0 gal/sq ft) of area. It has the same disadvantage as the drum vacuum washer, in that the heavy mat acts as a filter to retain ink and clay particles.

Flotation Deinking and Washing System

The flotation washing system consists of a cell or tank, a high-speed agitator, an overflow for froth removal, a mechanical paddle for removing the froth, and a discharge pipe to pass the stock to the next cell in the flotation line. The high-speed agitator induces a partial vacuum, which, in turn, causes air to enter the system and combine with the stock and flotation agents to form small air bubbles. Chemicals are used to create a suitable environment for the attraction of ink particles and pigments to the air bubbles, which are generated at the bottom of the cell and pass upward through the stock slurry to become froth at the surface. Because froth flotation depends on a difference in surface characteristics rather than density of the materials to be separated, it can be used for many materials regardless of their densities. As the ink-laden air bubbles reach the surface they are swept into a separate chamber by means of a two-armed rotating paddle that insures the removal of a uniform volume of froth from each revolution. The froth is withdrawn from the primary cells and pumped to secondary cells, where the ink is further concentrated. The fibers pass through the secondary units and are returned to the primary cells.

The froth removed by the secondary cells is centrifuged to a semidry mass and disposed of. The water from the centrifuge goes to the effluent system. Although ink is more easily floated than fibers, it is necessary to pass the stock through a number of cells in series to remove all the ink. The foam from these primary cells must be passed through a number of secondary cells to reclaim the fiber.

Flotation is considered by some as the best system for recovering newspapers because all the fibers and fillers pass through the cell and only the ink is removed. The yield is higher than that obtained by washing, but there are disadvantages. The retention time required in a flotation cell is in the range of 13 to 15 min. The consistency is usually 0.8%, which means that thickening is required after flotation. Cruea[1171] has pointed out that flotation and washing are both required for best results. In the case of fine paper, flotation helps to improve brightness, but washing is essential for pulp quality and clay removal. Flotation removes the specks that are too small to remove by screening and too large to be removed by washing. In the case of news, washing is essential to obtain high strength and to obtain an increase in freeness. Flotation-washing is most efficient when applied to news or high groundwood papers but can also be applied to ledger, chemical pulp papers with slightly less efficiency. A complete system for pulping and recovery, using flotation, is presented below.

The chemicals used for newsprint or high groundwood paper are 2% sodium peroxide, 4 to 5% sodium silicate, 0.1 to 0.3% foamer (Trilon BASF), 0.05 to 0.25% foamer (Decolort R), and 0.3 to 0.8% Decolort S, or Biancal Collector, all at a pH of 9.5. The chemicals used for chemical papers are 5% sodium peroxide and 0.8% sodium hydroxide at the pulping stage, plus the addition at the flotation cells of 3 to 5% fatty acid soap and 1% calcium chloride. Hot water and the chemicals for dispersing, flotation, and bleaching are added at the pulper before the wastepaper is added. This sequence is important because if the wastepaper is pulped first, before the chemicals are added, some printing inks could be beaten into the fiber, where they would not be floatable and thus impossible to remove. The formula needs to be exact because a definite alkalinity should remain for the following stages of the process. The pulped stock is dumped into a two-batch capacity chest, cleaned at high density and predeflaked. This tends to separate foil, magazine glued with latex, wet-strength papers, as well as any fibers adhering to metallic clips or staples. A second high-density cleaner removes the metal. A vibrating screen removes contaminants, such as foil, plastic, latex particles, and low-specific-gravity material. The stock is then fully deflaked and stored in a chest until any bleach added is exhausted. Storage longer than 2 hr will darken the stock because ink particles will redeposit on the fibers or will produce ink clots that are difficult to remove. Low-density cleaners may also be used, some with core bleed to remove all small contaminants. The stock is pumped to a distribution box, the consistency regulated to 0.8 to 1.2%, depending on the quality desired, and the stock is then metered into the flotation system. A constant flow and consistency is required for flotation.

The air bubbles generated in the flotation cell are stabilized by the foaming

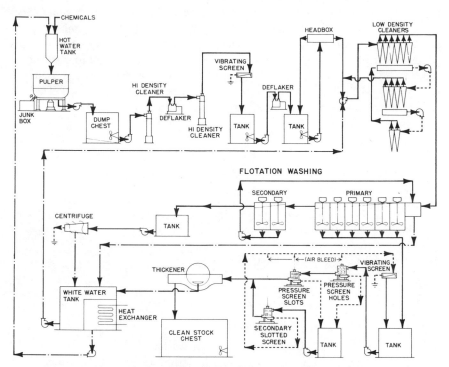

Figure 4-132. Flotation deinking and washing system.

chemicals through a decrease in the surface tension of the water. Although the pigments are heavier than water, they remain suspended because of their small size and hydrophobic properties. The collector chemical attaches the ink particles to the air bubbles so they can be floated to the surface to form a layer of foam that is directed toward the skimmer for removal. The amount and type of foamer and collector chemicals are selected to provide the desired buoyancy. If the exact ratio is not used, the performance can be greatly affected. The floated stock is then screened through holes of 0.16 to 0.20 cm (0.062 to 0.079 in.) in diameter, followed by screening through slots of 0.010 to 0.018 in. and then thickened. A complete flow diagram of a flotation deinking and washing system is shown in Figure 4-132. Horacek[1171a] has pointed out that dispersed ink particles in the range of 8 to 10 um can be removed by press washing at high consistency (up to 35%) which reduces the effluent that needs to be processed and/or treated.

Shrinkage of Wastepaper in Deinking

Shrinkage, which is the ratio of the total loss to the total furnish, is an important economic factor in deinking of wastepaper. Among the factors that affect shrinkage are the composition of the wastepaper, losses caused by mechanical treatment, and losses due to chemical treatment. Shrinkage on washing varies from about 8 to 10% for ledger stock to 20 to 35% for book and magazine stock. The total loss of stock in deinking usually ranges from 15 to 40%. Heavily coated book paper when used alone may result in as much as 40 to 60% shrinkage.[1172] Fiber loss is relatively low when long-fibered ledger stock is used, but is very high when short-fibered, pigment-coated book papers are used. The data needed to calculate shrinkage, all data being on a moisture-free basis, are as follows:

1. Weight of paper furnished, including rejects (A).
2. Weight of material in water furnished to the pulper (B).
3. Weight of chemicals added (C).
4. The total loss, including all sorting rejects, soluble, and insoluble matter in the plant effluent (D).

The percent shrinkage is then calculated by the formula:

$$\frac{D}{A + B + C} \times 100$$

The relative pulp yield for various grades of wastepaper are shown below.[1173] High ash content in the wastepaper is particularly important in its effect on the yield.

Type of Paper	Ash (%)	Yield (%)
Bond	2	90
Ledger	5	85
Offset	12	81
Book	20	66
Coated	25	58
Coated	30	50

Bleaching of Deinked Stock

Well-washed deinked stock ranges in color from a fairly bright, blue-white for ledger stock to a dull gray when mixed papers are used. The brightness may be as high as 60 GE. As mentioned above, peroxide may be used in the cooking operation to improve the brightness of the pulp, particularly when groundwood papers are present. Peroxide may also be used as a bleaching agent after cooking, using about (1) 3.0% sodium hydroxide and 1.25% hydrogen peroxide or (2) 1.5% sodium hydroxide and 1.5% sodium peroxide. If brightness in the 80 plus range is desired when deinking chemical fibers, a bleaching stage using hypochlorite can be used. There are three common systems, which are described below.

Single-stage Sodium Hypochlorite. Thickened stock from the washers is de-watered with an inclined-screw thickener to 20% consistency, mixed with 0.8% chlorine as sodium hypochlorite, heated to 50 to 60°C, and held for a retention time of 60 min. Sometimes a lesser amount of chlorine is used, especially if the bleached stock is to be used in low-grade napkin or toweling.

Chlorination Plus Hypochlorite. The thickened stock at 4 to 5% consistency is treated with chlorine gas and pumped through an uptower at a rate to provide a 15-min retention time, and then sodium hydroxide is added to make a hypo-chlorite stage. A retention time of 60 min is provided. The disadvantage of this system is the low consistency required, the relatively high losses in the chlorine stage, and the low degree of effectiveness in the hypochlorite stage. Normally a 7 point brightness increase is the best that can be attained.

Three-stage Bleach System. This consists of direct chlorination in a tower, caustic extraction in a tower, caustic washing, and finally a hypochlorite stage. The brightness increase using this method is usually 10 to 11 points.

A bleaching process in which the hypochlorite is added to thickened stock before washing is described below. After pulping with chemicals, the stock is dewatered to 25% consistency and simultaneously heated to 49 to 60°C. The pressed-out chemicals and hot water are stored for reuse. The chemicals, es-pecially sodium hydroxide, are only about 15% consumed in the pulping process, and, therefore, the cost of chemicals is reduced considerably by this action. The thickened stock is mixed with enough hypochlorite bleaching liquor to bring it to the desired brightness after washing. The degree of bright-ness obtained is directly related to the amount of chlorine added, as shown in Figure 4-133, where brightness is plotted against chlorine used in a deinked stock made from chemical papers. The amount of bleach and the cost of bleach-ing in this manner is not substantially different from that of a conventional

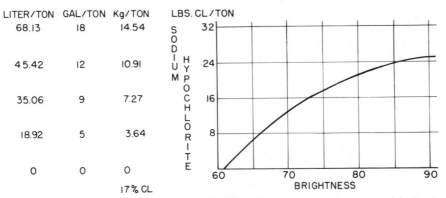

Figure 4-133. Brightness obtainable in bleaching process when hypochlorite is added to thickened stock at 25% consistency.

system. Although it is unorthodox to add bleach liquor immediately after dewatering, while the caustic, ink, and pigments are still present, the method has proven successful. This shortcut system is based on the fact that at pH 8 the chlorine absorption is small and the ink, being inert, will not absorb chlorine. The remaining color will be the same as in a conventional system. Using this system, newsprint can be deinked and processed into a good groundwood substitute. With this process the newsprint is pulped with selected chemicals at a temperature of 54°C, subjected to defibering for 45 min, dumped into a storage chest, dewatered to not more than 20% consistency, bleaching chemical added, the stock rediluted in a tower, and the stock then cleaned, screened, and washed.

Bleached deinked stock has a tendency to revert in color. Reversion depends on the conditions used in cooking (caustic soda being worse than other deinking agents), as well as the conditions of bleaching.[1174] High pH in hypochlorite bleaching reduces reversion. The use of peroxide for cooking or bleaching is also helpful. High temperatures up to 71°C in hypochlorite bleaching help to reduce reversion.

Properties of Deinked Stock

Most deinked stock has a slightly grayish cast because of the small amount of carbon retained by the fibers. The stock made from chemical papers is used for much the same purposes as soda pulp and is generally competitive with soda pulp in price. It is used in book papers and in coating rawstock, where it improves the bulk, opacity, softness, and formation of the paper. The pulp requires no beating. In fact, beating injures the stock, and, for this reason, deinked stock should always be added after the refining when used in mixed furnishes. Ordinarily, the final ash content of deinked stock is in the range of 2.5 to 4.0%. Even excessive washing will not reduce the ash content below about 1.0%.[1175] The pH is generally around 8.0 to 8.5.

PULPING OF RAGS AND COTTON LINTERS

H. P. DIXSON, Ph.D.

*Technical Advisor to President
Fox River Paper Company
Appleton, Wisconsin*

Rags were one of the earliest raw materials used in paper manufacture and prior to 1860 constituted the only significant source of papermaking fiber. Today, however, rags and cotton linters constitute only a relatively small part of the total raw material used for papermaking because of their limited supply and high cost compared with wood.

Use of Rags for Papermaking

High-grade cotton and, to some extent, linen rags are used to make the best grades of bond, writing, and technical papers, where permanence, high strength, and distinctive quality are of interest. Low-grade rags are used for the manufacture of felts used as a base for floor coverings and roof-base materials. In addition to rags, the paper industry utilizes other fiber wastes, such as rope, twine, tent material, and burlap for making strong wrapping papers. This material is generally sold as hard waste fiber, which may contain jute, sisal, manila, caroa, or hemp fibers. The availability of this material varies quite widely and depends on market condition.

Selection of Rags

Practically no old rags are used for high-grade paper that must have extreme cleanliness, permanence, and extra high-strength properties. The major source of rag fiber is waste from textile-mill weaving or from garment manufacturers.

The selection of rags suitable for high-grade papers has become exceedingly difficult due to contamination from synthetic fibers, such as nylon, cellulose acetate, and polyester. Rayon is also classed as a synthetic fiber by rag-pulp producers in relation to cotton. Fiber cloth clippings containing 100% synthetic fiber are easily detected by experienced rag sorters, but synthetic fibers blended with cotton in the weaving processes are sometimes difficult to detect and are unusable. The increased number of dyes that cannot be stripped without severe damage to the fiber further reduces the available fiber for the cotton-pulp producer. Wool or wool-cotton mixtures cannot be used to make pulp because the alkaline cooking condition used in pulping destroys the wool. Another situation making the selection of rags that are suitable for papermaking extremely difficult has been the increase in the use of synthetic finishes for permanent press and water repellency. These materials, such as vinyl resins, acrylates, urea-formaldehyde resins, and latices cannot easily be removed by the usual rag processing methods. These materials, when removed from the rag fiber, tend to agglomerate and escape the centrifugal cleaners and later plug the felts and wires and appear in the paper as shiny spots especially after machine calendering.

Raw cotton fiber from the gin could be used to make pulp. The only technical problem is the difficulty in mechanically shortening the fiber so that it can be processed in conventional rag-preparation equipment. However, the high cost of raw-cotton fiber, compared to traditional cotton-waste materials, prohibits its use. Although some cotton waste is procured directly from garment manufacturers or textile mills, it is usually procured from a dealer. Dealers frequently contract for the entire waste from a factory and then sort it into cotton, wool, blends, or straight synthetic. The cotton is further sorted into categories, such as unbleached muslin, white shirt, underwear, slashers, thrums, bleachable denim, unbleachable denim, mixed light prints, and other categories that are too numerous to list. In general, mills arrange for the degree of sorting that meets their needs, depending on how much of the final sorting they prefer to do. Old rags are sorted in much the same fashion, but separation is not as sharp. Typical categories are old white, mailbags, thirds, and blues. Even the best packed new rags are examined superficially as bales and are opened to check for tramp metal and hardwood bobbins. The clippings are then conveyed through several rag cutters in series to reduce the clippings to pieces about 2.5-cm (1-in.) square, before they are packed in the rag boilers. Hendrich[1176] recommends reducing the clippings even further to a flock for most uniform cooking.

Cooking Conditions in Rag Mills

Rags must be cooked to remove waxes, resins, finishing, and loading materials and to destroy or "start" the color. Rotary horizontal boilers or spherical boilers are used to cook the rags with direct steam. Cooking conditions vary widely, depending on the rags and equipment in any given mill. Moss[1177] described the cookers as steel vessels that hold 3 to 6 ton of rags. The cooker is mounted to rotate at 1 to 2 rpm on hollow trunions through which steam is introduced. The cooker is loaded through manholes, the liquor added, the manholes closed, and the vessel rotated throughout the cooking action.

Cooking is carried out in alkaline liquor at pressures from about 140 to 275 kPa (20 to 40 psi) although pressures up to 415 kPa (60 psi) are used. Liquor ratio is about 3:1. Cooking time varies from 2 to 15 hr. Yield of rag pulp varies greatly depending on the amount of loading material in the rags, severity of cook, quality of the fiber, and losses in the washing process. The loss from bale-weight to finished pulp will generally be in the range of 5 to 20%.

Rag-pulp preparation could undoubtedly be carried out in continuous cookers, washers, and bleaching equipment used for wood and for nonwoody materials. Even the drastic shortening of rag fibers to the length suitable for papermaking could probably be carried out in refiners modified to prevent plugging. Little effort has been made to do this because the use of rags is so small and rag mills are so small that the cost of new equipment and necessary modification of that equipment would be prohibitive for the tonnage required. In addition, rag clippings, even after extensive grading, are quite variable and each lot requires changes in chemicals used, cooking time, and bleaching requirements. It would be very difficult to adapt these conditions to a finely tuned continuous pulping system.

A typical rag-pulp preparation system is described by Felton[1178] in his review of nonwood pulping. The system consists of a conveyor belt on which the clippings are placed and roughly inspected for foreign material that would injure rag-cutter knives. The rags may be passed through dusting drums before the rag cutters. After cooking, the rags are sent to a spent-liquor draining pit and then to a washer-beater. This is followed by either another drainer or a wet-lap machine.

Chemicals Used in Rag Cooking

Sodium hydroxide, sodium carbonate, and lime are the usual chemicals used for cooking rags. Lime can be used as the sole cooking agent at about 15% on the weight of the fiber. However, lime produces insoluble reaction products, which are difficult to remove by washing. Some of these products remain in the pulp and fill up wires and felts on the paper machine and cause sizing problems in closed white-water systems. Because of these effects and pressure on mills to reduce effluent solids, lime has become a minor cooking agent.

With careful control of cooking temperatures and accurate measurement of chemicals and water, sodium hydroxide or sodium carbonate will produce satisfactory pulp without degradation of the cotton fiber. Rags for saturating felts are sometimes cooked with 6 to 8% sodium carbonate at about 300 kPa (30 psi) for 6 to 8 hr. Sodium carbonate is frequently used for unloaded white rags at 2 to 4% on weight of the fiber. Sodium hydroxide is seldom used above 5% on the weight of the fiber in order to avoid fiber degradation.

Other alkaline cooking mixtures have been proposed and some are probably used on specific rags. It has been claimed[1179] that hydrogen peroxide in the presence of sodium silicate and sodium hydroxide and cooking at a 20% consistency for 2.5 hr at 275 kPa (40 psi) will produce rag pulps of 86% brightness.

Vat dyes, which are very resistant to oxidizing bleach chemicals, must be stripped with a reducing agent which solubilizes the dye. Denim, a very popular rag clipping, is vat-dyed. Cooking is first done with caustic and the liquor is blown out with as little introduction of air as possible. A second cook is made with caustic, plus a small amount (1%) of sodium hydrosulfite and the liquor is blown down again. This may be repeated once more if color removal appears poor. When washed in a rag washer, instead of using a typical oxidizing bleaching agent, more sodium hydrosulfite is added.

Washing and Bleaching Rag Pulps

The washing of rag pulps is quite different from the washing of wood pulps. Rag washing is carried out in a special machine, called a washer-beater, which is similar to a Hollander beater, except that a revolving wire-covered drum (usually octagonal) turns in the stock and removes dirty water at the same rate that fresh water is added.

The washer functions not only as a washing machine, but also as a beater. When the water removed by the washer drum is fairly clear, the beater roll is lowered against the bedplate to loosen the remaining dirt and to brush out the rag pieces into individual fibers. Temperature, pH, and water characteristics are important in rag washing, just as in the washing of chemical wood pulps, but the degree of beating or mechanical action on the fibers is probably of the greatest importance. If the beater roll is lowered too soon, the dirt is beaten into the fibers, instead of being removed. The total washing period is about 4 to 10 hr. Water consumption is at the rate of 225 to 380 l/min (60 to 100 gal/min). The consistency during washing is about 3.5%.

Bleaching is generally done by adding 0.3 to 0.6% available chlorine as sodium or calcium hypochlorite to the partially beaten fibers in the washer. Many mills add 1 to 3 pints of sulfuric acid to the washer after the bleach liquor to reduce the pH to about 5.0. The beating should not be too hard during this period since the fibers are in a highly susceptible condition. When bleaching is considered complete (about 1 hr), the pulp is washed again for an hour with the roll down. The completed batch is then flushed to a drainer, which is a chest with a loose brick floor laid over a drainage grid. In 2 to 7 d the pulp has drained enough that it can be dug out in chunks. Most mills dump the batches in a large chest that holds several washer loads so that the pulp can be blended. The pulp may then be wet-lapped on a traditional wet-lap machine. Moss [1177] reports that sodium chlorite, actuated with either hypochlorite or acid, has been used to produce high brightness with minimum degradation.

Use of Cotton Linters for Papermaking

Cotton rags were the principal source of raw material for rag and rag-content papers for years. However, following World War II, the rapid development of faster dyes, textile finishes, and synthetic fibers greatly reduced the amount of rags suitable for papermaking. At that time the rag-content paper mills began to use cotton linters. Before this time some linter pulp had been used for specialty papers, such as blotting and saturating grades. This linter pulp was manufactured by the chemical pulp producers. This pulp was made with stringent viscosity and chemical specifications required for conversion to cellulose derivatives and rayon and was so highly purified it could not be beaten to produce strong paper. To overcome this, rag mills began to purchase raw linters and to process them into a more suitable product for papermaking than the "chemical linters." Now over 50% of the cotton pulp used in paper manufacture is from linters.

Properties of Cotton Linters

Cotton linters are the short, fuzzy fibers, plus lint-fiber stumps that are firmly attached to the cotton seed. They are not removed by ginning and hence a special-cutter operation is necessary to remove the fiber from the seed so the seed can be processed for oil. Linter removal is done at the oil mill. Removal of linters from the seed can be done in one pass through the linter "saws." This product is called mill run. However, for economic purposes it is usually desirable to take a first cut followed by a second cut to remove the rest of the material. The first cut is generally used for mattress material or upholstery and constitutes 30% of the linter production. The remaining 70% is used to make cellulose derivatives and for papermaking, although an increasing percentage of first cut and mill run are being processed into paper pulp. The average fiber length of cotton linters is 6 to 7 mm for first-cut linters, 3 mm for second-cut linters, and 3.5 to 5.0 mm for mill-run linters.

Cotton linters produce pulps of better color and higher alpha-cellulose content than cotton rag pulps. Pulp made from first-cut linters is stronger than that made from second-cut linters, primarily due to the fiber length. About 70 to 90% of cotton-linter fibers are mature fibers, of secondary growth, and the remaining percentages are stumps from the long cotton hairs, called lint. The 70 to 90% linter fuzz differs from the staple fiber stumps, in that they have greater wall thickness and cross section than lint fibers.

Second-cut linters average 80 to 85% cellulose, 1 to 1.5% ash, 3% lignin, 1% ether extractable, and 6% moisture. Second-cut linters contain hulls and seed particles nicked from the seed by the linter saws. First-cut linters contain field trash, which frequently is more difficult to remove to produce a clean pulp than seed particles in second-cut linters. Hulls and seed particles not removed by mechanical dry cleaning and cooking are usually removed by riffles or centrifugal cleaners normally used in paper mills for dirt removal.

Cooking, Washing, and Bleaching of Cotton Linters

Cooking is necessary to prepare cotton linters for papermaking in order to remove the oils, waxes, pectates, lignin, and other foreign matter. Cotton linters can be cooked in typical rag-cooking boilers; they can also be cooked in vertical stationary digesters or continuous cookers because of their much shorter fiber length compared to rag fiber. A typical second-cut cotton linter is cooked with 6 to 8% caustic with a liquor ration of 3:1. A small amount of wetting agent is used to improve removal of cotton oil. Cooking time and steam pressure are balanced to produce the minimum fiber damage and the maximum impurity removal. Each mill selects the optimum for its requirements. A typical condition is 310-kPa (45-psi) steam pressure for 4 to 8 hr.

Cooked linters can be washed and bleached in rag washers, but much less washwater and more effective bleaching can be carried out in bleaching equipment and bleaching sequences used for wood pulp. Rag mills that have not installed wood-pulp-type bleaching methods or modifications of such processes have, by and large, elected to purchase bleached, linter pulp from large linter-pulp producers who have developed pulps especially designed for papermaking.

Linters cooked in a rag mill under the conditions outlined above can be placed in a typical rag washer-beater and washed like rag stock until the washwater is essentially clear of cooking liquor. Very light roll-pressure is employed. After washing, the batch is heated to about 38°C and the stock is treated with approximately 1.0% sodium hypochlorite at pH 6.0 until the desired brightness is obtained. The stock is washed until free of residual chlorine and is then wet-lapped.

A much more satisfactory pulp is obtained by washing the cooked linter pulp on a screen and then removing as much remaining hull "pepper" and field residue as possible by means of a typical wood-pulp riffle or centrifugal cleaner. This stock is then thickened to 6 to 8% consistency and fed to a typical wood-pulp bleaching tower, where it is heated to 40 to 45°C and subjected to direct chlorination for 10 min. The amount of chlorine used is very low compared to that used with wood pulp because there is very little lignin present. The chief purpose is to treat the field trash residue so that subsequent hypochlorite treatment can readily attack this material. The chlorinated pulp can be washed at this stage, but because of the small amount of chlorinated product, the first-stage pulp can be treated with excess chlorine and, at the proper time, sodium hydroxide can be added to the unwashed pulp to convert the excess chlorine to sodium hypochlorite (90 to 120 min) and then washed to remove the soluble compounds and residual bleach. The total chlorine for all stages is about 1.0% in this procedure.

Large commercial bleachers of linters, like wood-pulp bleachers, use several stages of washing and bleaching for better control of conditions which is necessary for the production of a pulp to be used for chemical derivatives. Most linter pulp is manufactured for the preparation of cellulose derivatives, such as cellulose acetate, methyl cellulose, ethyl cellulose, and rayon. Some pulp made by the bleachers is especially processed to improve its papermaking properties. This includes a modified cooking procedure. Then in the paper mill wide bar refining is used to promote fibrillation without fiber shortening.

In an effort to improve linter pulp for papermaking the Writing Paper Association's Rag Content Group and some commercial bleachers supported an extensive study at the Institute of Paper Chemistry, Appleton, Wisconsin. Part of this work involved a chemical alteration of the linter fiber to improve fiber bonding. The effect of hydroxyethylation of linters on sheet strength is reported by Ward, Murray, and others.[1180] There was a dramatic increase in folding endurance as the degree of hydroxyethylation increased. The big problem was a large drop in sheet opacity. Blends of hydroxyethylated linters and untreated material showed properties intermediate between the two components, although folding endurance and breaking length were generally higher than expected.[1181] It was concluded that by blending it is possible to attain a more desirable balance between fold and tensile on one hand and opacity and tear on the other. However, hydroxyethylation adds appreciably to the cost of linter pulp and small rag mills cannot afford to carry out this operation. A large commercial producer offered a series of these pulps commercially. This development allows rag paper mills to blend first-, second-, and mill-run linter pulps with the chemically modified pulps as well as to blend with rag pulps.

Characteristics of Rag and Linter Pulps

Cotton (*Gossypium hirsutium*) consists of long, flat fibers, which appear twisted under low magnification. There are 150 to 300 spirals to the inch, twisting being less pronounced where the cell wall is thickest. The fiber is rounded at one end, but torn on the other end where the fiber was attached to the cotton seed. Longitudinal and spiral striations are common. Cotton staple fibers have an average length of about 18 mm and a diameter of about 20 μ. Native cotton fiber is too long to achieve the good formation required for bond and ledger papers. Much cutting must be done in the refining process to obtain the proper fiber length, which requires a great consumption of power. One of the reasons for the use of the shorter-fibered cotton-linter fiber is to produce a balanced cotton pulp where the amount of power can be reduced without a serious loss of strength.

Both rag and cotton-linter pulps are characterized by high alpha-cellulose content (often as high as 96 to 98%), low copper number, and high cuprammonium or cupraethylenediamine viscosity. Because of these chemical characteristics, both rag and linter pulps are used for the manufacture of high-grade papers for permanent records. Rag paper is very strong and durable and is the best grade of paper for use where the paper receives considerable handling. Rag pulps are well suited for making light-weight papers containing a high filler content. Pulps made from new rag cuttings may have as high as 300 cuprammonium viscosity. According to Dobo and Kobe[1182] linter pulps develop lower sheet strength than rag pulps due to different fiber properties, such as a larger fiber cross section and shorter fiber lengths. Many of these lower physical characteristics are due to the 70 to 90% short, thick fuzz fibers that are absent from textile lint cotton. Ward, and others,[1183] also concluded that the factors influencing sheet strength differences between rag and linters were physical rather than chemical. The lower sheet-strength properties of linters were associated with larger fiber cross section, shorter length, and lower intrinsic fiber strength. They did establish that lint and linters of similar background tend to develop similar hydrodynamic surface-area characteristics and drainage properties on beating. With the latitude of choice between rag and different grades of linter pulps, plus the option in choice of bar width and pattern in refiners, rag mills have many routes to secure papers of various specifications.

REFERENCES

1. Compiled in collaboration with W. W. Marteny.
1a. G. J. Kubes, *Trend* by P.P. Res. Inst. Can., Spring issue 29, 4 (1979).
2. R. J. Auchter, *Tappi,* **59** (4), 50 (1976).
3. W. R. Patton, *Pulp and Paper,* **51** (5), 70 (1977).
4. B. E. Helberg, E. D. Neuberger, and W. B. Simmons, *Tappi,* **60** (2), 70 (1977).
5. Rust Engineering Co., *Pulp and Paper,* **51** (3), 126 (1977).
6. R. W. Hisey, *Tappi,* **58** (8), 112 (1975).
7. S. R. Boyette, *Pulp and Paper,* **51** (4), 78 (1977).
8. W. H. Copeland, *Tappi,* **60** (8), 34 (1977).
9. G. F. Underhay, *Tappi,* **58** (7), 16A (1975).
10. T. Lindgren, *Paper Trade J.,* **139** (16), 33 (1953).

11. T. C. Smyth and E. G. Meckfessel, *Tappi*, **41** (1), 129A (1958).
12. C. D. McCulloch, *Pulp Paper Mag. Can.*, **60** (1), T-21 (1959).
13. W. H. de Montmorency and E. Koller, *Pulp Paper Mag. Can.*, **58** (3), 206 (1957). [Convention issue.]
14. W. H. de Montmorency, *Paper Trade J.*, **142** (7), 23 (1958).
15. W. H. de Montmorency, *Pulp Paper Mag. Can.*, **203** (1958). [Convention issue.]
16. F. S. Hohlik, *Can. Pulp Paper Ind.*, **12** (1), 31 (1959).
17. K. H. Klemm, *Paper Trade J.*, **141** (41), 62 (1957).
18. R. S. Magruder, *Tappi*, **32** (1), 29 (1949).
19. K. H. Klemm, *Paper Trade J.*, **141** (38), 56 (1957).
20. W. Brecht and W. Muller, *Papier-Fabr.*, **36** (23), 198; **36** (24), 257 (1938).
21. K. H. Klemm, *Paper Trade J.*, **141** (51), 36 (1957).
22. D. Atack and W. D. May, *Pulp Paper Mag. Can.*, 265 (1958). [Convention issue.]
23. I. H. Andrews, S. A. Collicutt, and R. C. Bedsoe, *Pulp Paper Mag. Can.*, **47**, 91 (1946). [Convention Issue.]
24. K. H. Klemm, *Paper Trade J.*, **141** (42), 35 (1957).
25. W. A. Barrett, *Pulp Paper Mag. Can.*, **59** (8), 131 (1958).
26. H. J. Perry, *Pulp Paper Mag. Can.*, **50** (5), 94 (1949).
27. O. S. Johansson and T. Pettersson, *Svensk Papperstidn.*, **55** (15), 497 (1952).
28. W. H. de Montmorency, *Pulp Paper Mag. Can.*, **58** (8), 141 (1057).
29. K. H. Klemm, *Paper Trade J.*, **141** (49), 52 (1957).
30. K. H. Klemm, *Paper Trade J.*, **141** (51), 36 (1957).
31. R. S. Magruder, *Tappi*, **32** (1), 29 (1949).
32. H. J. Perry, *Pulp Paper Mag. Can.*, **50** (5), 94 (1949).
33. J. H. Perry, *Tappi*, **32** (2), 67 (1949).
34. J. Edwards, G. D. O. Jones, G. J. C. Potter, and H. W. Johnston, *Tech. Assoc. Papers*, **20**, 201 (1937).
35. W. Brecht, *Papier-Fabr.*, **33** (13), 119; **33** (14), 121; **33** (15), 129 (1935).
36. See TAPPI Data Sheets, Technical Association of the Pulp and Paper Industry, Atlanta, Ga.
37. H. J. Perry, *Pulp Paper Mag. Can.*, **50** (5), 94 (1949).
37a. *Tappi*, **61** (7), 13 (1978).
38. J. Edwards, G. D. O. Jones, G. J. C. Potter, and H. W. Johnston, *Pulp Paper Mag. Can.*, **38** (2), 121 (1937).
39. R. S. Magruder, *Tappi*, **32** (1), 29 (1949).
40. R. S. Magruder, *Tappi*, **32** (1), 29 (1939).
41. J. H. Thickens, USDA Bulletin **127**, U.S. Government Printing Office, Washington, D.C., 1913.
42. J. Edwards, G. D. O. Jones, G. J. C. Potter, and H. W. Johnston, *Tech. Assoc. Papers*, **20**, 201 (1937).
43. K. H. Klemm, *Paper Trade J.*, **141** (50), 46 (1957).
44. G. von Alfthan, *Svensk Papperstidn.*, **58** (14), 503 (1955).
45. K. H. Klemm, *Paper Trade J.*, **141** (38), 56 (1957).
46. J. C. Jaeger and J. L. Somerville, *Paper Trade J.*, **129** (11), 99 (1944).
47. C. D. Demars, *Pulp Paper Mag. Can.*, **42**, 378 (1941).
48. E. R. Schafer and J. C. Pew, *Paper Trade J.*, **101** (13), 167 (1935).
49. K. H. Klemm, *Pulp Paper Mag. Can.*, **56** (12), 178 (1955).
50. P. T. Davis, *Tappi*, **40** (12), 203A (1957).
51. R. S. Magruder, *Tappi*, **23** (1), 29 (1949).
52. I. H. Andrews, *Pulp Paper Mag. Can.*, **41**, 89 (1940) [Convention issue.]
53. K. H. Klemm, *Paper Trade J.*, **141** (51), 36 (1957).

54. E. R. Schafer and J. C. Pew, *Paper Trade J.,* **105** (18), 266 (1937).
55. E. R. Schafer and J. C. Pew, *Paper Trade J.,* **101** (13), 167 (1935).
56. I. H. Andrews, *Pulp Paper Mag. Can.,* **42** (2), 117 (1941).
57. R. I. Wynne-Roberts, *Tech. Assoc. Papers,* **20,** 258 (1937).
58. H. A. Paterson, *Pulp Paper Mag. Can.,* **37,** 79 (1936).
59. R. I. Wynne-Roberts, *Tech. Assoc. Papers,* **20,** 258 (1937).
60. N. J. Lindberg and D. J. MacLaurin, *Tappi,* **39** (9), 663 (1956).
61. N. J. Lindberg and D. J. MacLaurin, *Paper and Timber (Finnish),* **37,** (8), 409 (1955).
62. K. D. Running, *Pulp Paper Mag. Can.,* **42,** 104 (1941).
63. E. R. Schafer and A. Hyttinen, *Tappi,* **33** (7), 335 (1950).
64. J. E. Atchison, *Tappi,* **33** (7), 335 (1950).
65. C. E. Libby and F. W. O'Neil, *Tappi,* **33** (4), 161 (1950).
66. A. J. Watson and W. E. Cohen, *Aust. Pulp Paper Ind. Tech. Assoc. Proc.,* **5,** 188 (1951).
67. K. H. Klemm, *Paper Trade J.,* **141** (32), 50 (1957).
68. R. I. Wynne-Roberts, *Tech. Assoc. Papers,* **20,** 258 (1937).
69. F. A. Whitman, *Tappi,* **40** (1), 20 (1957).
70. F. A. Whitman, *Pulp Paper Mag. Can.,* **56** (12), 152 (1955).
71. W. S. Morris, *Pulp Paper Mag. Can.,* **44,** 501 (1943).
72. H. P. Dahm, J. Brandal, and K. Helge, *Paper and Timber (Finnish),* **40,** (7), 355 (1958).
73. D. W. Glennie and H. Schwartz, *Paper Ind.,* **34** (6), 738 (1952).
74. L. G. Cottrall, *Paper-Maker,* **125** (1), 34 (1953).
75. U. Ullman, *Svensk Papperstidn.,* **61** (14), 445 (1958).
76. U. Ullman, *Paper Trade J.,* **143** (2) 50 (1959).
77. C. E. Curran, E. R. Schafer, and J. C. Pew, *Paper Trade J.,* **101** (8), 91 (1935).
78. B. Mannstrom, *Tappi,* **55** (4), 551 (1972).
79. L. G. Sears, *Pulp Paper Mag. Can.,* **46** (3), 185 (1945). [Sixth wartime issue.]
80. S. Coppick, *Paper Trade J.,* **117** (26), 269 (1943).
81. K. H. Klemm, *Paper Trade J.,* **142** (19), 36 (1958).
82. W. Brecht and H. Weiss, *Das Papier,* **11,** 82 (1957).
83. J. A. Cochrane, *Paper Trade J.,* **140** (17), 30 (1956).
84. J. D. Rue, *Paper Trade J.,* **81** (16), 157 (1925).
85. P. V. Derveer, *Paper Trade J.,* **141** (38), 36 (1957).
86. *Paper,* **17** (21), 18 (1916).
87. *Paper,* **18** (6), 19 (1916).
88. C. G. Schwalbe, *Paper Mill,* **28** (44), 32 (1915).
89. E. H. Lougheed, *Pulp Paper Mag. Can.,* **46** (3), 165 (1945). [Sixth wartime issue.]
90. A. Hyttinen and E. R. Schafer, *Tappi,* **32** (2), 79 (1949).
91. A. E. Dentremont, *Paper Trade J.,* **139** (8), 32 (1955).
92. *Paper Ind.,* **37** (3), 244 (1955).
93. D. F. Pollard, *Paper Trade J.,* **140** (44), (1956).
94. C. E. Libby and F. W. O'Neil, *Tappi,* **33** (4), 161 (1950).
95. F. Neumann and K. Herb, *Papier Darmstadt* **11,** (17, 18), 391 (1957).
96. W. Jensen, L. Nordman, and V. Eravuo, *Paper and Timber (Finnish),* **39** (4a), 165 (1957).
97. C. Mills and L. Beath, *Pulp and Paper Mag. Can.,* **76** (12), T–343 (1975).
98. R. J. Rankin, *Proc. Intl. Mech. Pulping Conf.,* Helsinki, Finland, (1977), p. 28.
99. W. R. Finn, *Pulp Paper Mag. Can.,* **76** (8), T–228 (1975).

100. J. L. Keays and D. A. Harper, *Proc. Intl. Mech. Pulping Conf.*, San Francisco Calif., (1975), p. 245.
101. W. D. May, *Pulp Paper Mag. Can.*, **74** (1), T-2 (1973).
102. D. Atack and P. N. Wood, *Proc. Intl. Mech. Pulping Conf.*, Stockholm, Sweden, (1973), p 12:1.
103. D. Wild and S. Steeves, *Pulp Paper Mag. Can.*, 73 (10), T-305 (1972).
104. W. G. Michellich, D. J. Wild, S. B. Beaulieu and L. Beath, *Pulp Paper Mag. Can.*, **71** (5), T-140 (1972).
105. A Asplund, *Svensk Papperstidn.* **56**, 550 (1953).
106. D. Atack, *Proc. Symp. Mech. Pulping*, Oslo, Norway, 1976.
107. D. A. Goring, *Proc. Tech. Sect., British Paper Board Maker, Assoc.*, London, England, 1966, p. 555.
108. D. A. Goring, *Pulp Paper Mag. Can.*, **64** (12), T-517 (1963).
109. J. A. Kurdin, *Appita,* **30** (4), 347 (1977).
110. H. Hoglund and G. Tistrtad, *Proc. Intl. Mech. Pulping Conf.*, Stockholm, Sweden, 1973, **3**, p 1.
111. O. L. Forgach, *Pulp Paper Mag. Can.*, **64** (2), T-89 (1963).
112. Ulla-Britt Mohlin, *Proc. CPPA Ann. Meeting*, Montreal, Canada, 1977, p A 23.
113. W. C. Frazier, *Proc. CPPA Ann. Meeting,* Montreal, Canada, 1977, p A 91.
114. D. H. Breck, W. D. May, and M. N. Tremblay, *Trans., Tech. Sect., CPPA,* 1975, p 89.
115. H. W. Giertz, *Proc. Can. Wood Chem. Symp.,* Mont Gabriel, Canada, 1976, p 25.
116. J. A. Kurdin, *Proc. Intl. Mech. Pulping Conf.*, San Francisco, Calif., 1975, p 123.
117. J. A. Kurdin, *Paper Age,* (1), 29 (1975).
118. M. T. Charters, *Tappi,* **58** (9), 122 (1975).
119. M. T. Charters, *Proc. Intl. Mech. Pulping Conf.*, Stockholm, Sweden, 1973, p 16:1.
120. A. Asplund, *Proc. Intl. Mech. Pulping Conf.*, Stockholm, Sweden, 1973, p 15:1.
121. V. Peterson, G. Dahlquist, and B. Engstrom, *Pulp Paper Intl.*, **15**, 43 (1973).
122. V. Peterson, G. Dahlquist, and B. Engstrom, *Proc. Intl. Mech. Pulping Conf.*, San Francisco, Calif., 1975, supl 11.
123. S. Beaulieu, A. Karnis, D. J. Wild, and J. R. Wood, *Pulp Paper Mag. Can.*, **78** (3), 61 (1977).
124. D. J. Wild and S. Beaulieu, *Proc. Intl. Mech. Pulping Conf.*, Helsinki, Finland, 1977, p 28.
125. V. Peterson and G. Dahlquist, *Proc. Intl. Mech. Pulping Conf.*, Stockholm, Sweden, 1973, p 14:1.
126. W. D. May, K. Miles, and R. C. Jeffreys, *Proc. Intl. Mech. Pulping Conf.*, Stockholm, Sweden, 1973, p 13:1.
127. D. Atack and M. I. Stationwala, *Proc. Intl. Mech. Pulping Conf.*, San Francisco, Calif., 1975, p. 135.
128. W. G. Mihelich, *Proc. Intl. Mech. Pulping Conf.*, Stockholm, Sweden, 1973, p 21:1.
129. H. G. Gavelin, *Paper Trade J.*, **150** (5), (1966).
130. H. Gustafsson and P. Vihmari, *Proc. Intl. Mech. Pulping Conf.*, Helsinki, Finlnd, 1977, p 41.
131. D. H. Breck, D. L. Young, and W. D. May, *Proc. Intl. Mech. Pulping Conf.*, Helsinki, Finland, 1977, p 10.

132. L. R. Beath, M. T. Neill, and F. A. Masse, *Pulp Paper Mag. Can.*, **67** (10), 423 (1966).
133. H. W. Jones, *Pulp Paper Mag. Can.*, **67** (6), 238 (1966).
134. D. H. Page, *Pulp Paper Mag. Can.*, **67** (1), T–2 (1966).
135. J. S. Cowan, R. S. Allan, and O. L. Forgach, *Proc. Tech. Sect. CPPA*, 1966, p 69.
136. C. W. Skeet and R. S. Allan, *Pulp Paper Mag. Can.*, **69** (7), T–222 (1968).
137. J. A. Kurdin, *Paper Trade J.*, **155** (3), 45 (1971).
138. G. Gavelin, *Paper Trade J.*, 150 (6), (1966).
139. W. H. Montmorency and E. Koller, *Pulp Paper Mag. Can.*, **58**, T–206 (1957).
140. D. J. Wild, *Proc. CPPA Ann. Meeting*, Montreal, Canada, 1971, p. D 96.
141. E. Stenberg and G. Norberg, *Proc. Intl. Mech. Pulping Conf.*, Helsinki, Finland, 1977, p 14.
142. A. Wong, *Pulp Paper Mag. Can.*, **78** (6), T–132 (1977).
143. J. M. Leach and A. N. Thakore, *Tappi* **59**, (2), 129 (1976).
144. T. E. Howard and C. C. Welden, *Proc. Intl. Mech. Pulping Conf.*, San Francisco, Calif., 1975, p 197.
145. J. A. Kurdin, *Pulp and Paper*, (3), 92 (1974).
146. J. A. Kurdin, *Proc. Intl. Mech. Pulping Conf.*, Stockholm, Sweden, (1973), p 22:1.
147. J. E. MacDonald, *Pulp Paper Mag. Can.*, **75**, T–105 (1974). [Convention issue, no. C, 95–99.]
148. H. W. Geirtz, *Eucepa Conf.*, Madrid, Spain, 1974, p 243.
149. R. S. Allan, C. W. Skeet, and O. L. Forgach, *Pulp Paper Mag. Can.*, **69**, T–351 (1968).
150. R. A. Leask, *Tappi*, **51** (12), 117A (1968).
151. I. Ito, *Proc. Intl. Mech. Pulping Conf.* San Francisco, Calif., 1975, p 185.
152. J. R. Wood, M. Pouw, and A. Karnis, *Pulp Paper Mag. Can.*, **78** (11), T–249 (1977).
153. B. Orgill, *Proc. Intl. Mech. Pulping Conf.*, San Francisco, Calif., 1975, p. 127.
154. B. C. Kloetzli, *Proc. CPPA Ann. Meeting*, Montreal, Canada, 1977. p A 67.
155. D. Atack, C. Heitner, and M. I. Stationwala, *Proc. Intl. Mech. Pulping Conf.*, San Francisco, Calif., 1975, p 135.
156. H. Hoglund and B. Frederikkson, *Proc. Intl. Mech. Pulping Conf.* Helsinki, Finland, 1977, p 17 B.
157. B. Frederikkson and H. Hoglund, *Proc. APPITA Gen. Conf.*, 1977, p. 17 B.
158. H. W. Giertz, *Proc. Intl. Mech. Pulping Conf.*, Stockholm, Sweden, 1973, p 6:1
159. J. R. Wood and A. Karnis, *Proc. Intl. Mech. Pulping Conf.*, Helsinki, Finland, 1977, p 20.
160. M. Jackson, S. A. Collicutt, and R. G. Meret, *Tappi*, **60** (12), 95 (1977).
161. R. G. Franzen, *Proc. Intl. Mech. Pulping Conf.*, Helsinki, Finland, 1977, p 30.
162. W. C. Frazier, M. Jackson, and S. A. Collicutt, *Pulp Paper Mag. Can.*, **77** (2), T–18 (1976).
163. C. T. Charters, *Proc. Intl. Mech. Pulping Conf.*, Helsinki, Finland, 1977, p 38.
164. A. Arjas, *Proc. Intl. Mech. Pulping Conf.*, Helsinki, Finland, 1977, p 19.
165. T. Fossum, A. Lindahl, L. Rudstrom, and L. Smedman, *Pulp Paper Mag. Can.*, **77** (2) T–24 (1976).
166. C. A. Richardson, *Proc. CPPA Annual Meeting*, Montreal, Canada, 1977, p B 49.

167. E. Bohmer, S. Havan, H. E. Hoydahl, and N. Soteland, *Pulp Paper Mag. Can.*, **77** (4), T–65 (1976).
168. E. Bohmer, *Proc. Intl. Mech. Pulping Conf.*, Stockholm, Sweden, 1973, p 7:1.
169. R. A. Leask, *Proc. CPPA Ann. Meeting*, Montreal, Canada, 1977, p B 171.
170. *Tappi*, **60** (7), 24 (1977).
171. R. W. Emery, *Pulp Paper Mag. Can.*, **61** (4), T–175 (1967).
172. H. E. Worster, *Paper Trade J.*, **157** (34), 31 (1973).
173. E. Hagglund, *Tappi*, **34** (2), 545 (1951).
174. K. L. Snyder and R. A. Premo, *Tappi*, **40** (11), 901 (1957).
175. P. Columbo, D. Corbetta, L. Gaetani, A. Pirotta, and A. Sartori, *Tappi*, **40** (9), 205A (1957).
176. B. M. Dillard, R. J. Gilmer, J. D. Kennedy, *U. S. Patent* 3954533 (1976).
177. (R. W. Snow, 1971, unpublished data.)
178. R. M. Husband, *Tappi*, **38** (10), 577 (1955); R. M. Husband, *Tappi*, **40** (6) 412 (1957).
179. K. E. Olsen, *U. S. Patent* 3003909 (1961).
180. R. S. Magruder, *Tappi*, **41** (12), 58A (1958).
181. N. C. S. Chari, *Ind. Eng. Chem., Process Des. Develop.* **7**, 372 (1968).
182. L. J. Schied, *Paper Trade J.*, **139** (45), 28 (1955).
183. H. Sadler and Trantino, *Papier*, **12** (21, 22) 553 (1959).
184. J. M. Loving, *Paper Industry*, **37** (10), 933 (1956).
185. E. Santiago, 1971, unpublished data.
186. Stone Container Corporation, Coshocton, Ohio (1977).
187. A. E. Rosenblad, *Chem. Eng. Progr.*, **72** (4) 53 (1976).
188. W. J. Nolan, *Tappi*, **37** (5), 152A (1954).
189. E. L. Keller and J. N. McGovern, *Tappi*, **34** (6), 262 (1951).
190. S. F. Galeano, *U. S. Patent* 3622443 (1971).
191. R. A. Nugent and R. A. Boyer, *Paper Trade J.*, **140** (18), 29 (1956); **140** (19), 30 (1956).
192. Data from: Bjorkvist, Gustafsson, and Jorgensen, *Pulp Paper Mag. Can.*, **55** (2), 68 (1954).
193. Data from: U. S. Forest Products Laboratory.
194. G. H. Chidester, *Paper Trade J.*, **129** (21), 451 (1949).
195. *Pulp and Paper*, **51**, 30 (1977). [Profile issue.]
196. *Tappi Standard T-1201 os-61*, "Definition of Terms Used in the Sulfite Process," *Tappi*, **54** (12), 2083 (1971); Data Sheet C-00, Tech. Sect., CPPA, June 1955 (Revised, September 1965); *Standard K 4P*, "Standard Terms Used in the Sulfite Pulping Process," Tech. Sect., CPPA, October 1962.
197. C. G. Palmrose, *Tech. Assn. Papers*, **18**, 309 (1935).
198. Standard J-1, "Analysis of Bisulfite Liquor," Tech. Sect. CPPA (September, 1974); Tappi Standard 604 pm-79 "Sulfur Dioxide in Sulfite Cooking Liquor", Tappi, **53** (7), 1373 (1970).
199. W. Boyd Campbell and O. Maass, *Can. J. Res.*, **2**, 42 (1930).
200. S. Rydholm, *Svensk Papperstidn.*, **57** (12), 427 (1954); **58** (8), 273 (1955).
201. O. V. Ingruber, *Pulp Paper Mag. Can.*, **55** (9), 124 (1954).
202. O. V. Ingruber, *Svensk Papperstidn.*, **65** (11), 448 (1962).
203. O. V. Ingruber, *Pulp Paper Mag. Can.*, **66** (4), T–215 (1965).
204. N. -H. Schöön and L. Wannholt, *Svensk Papperstidn.*, **72** (16), 489 (1969).
205. W. B. Beazley, W. B. Campbell, and O. Maass, *Dom. For. Serv. Bull.*, No. 93, Ottawa, Canada, 1938, p 38.

206. H. P. Markant, N. D. Phillips, and I. S. Shah, *Tappi,* **48** (11), 648 (1965).

207. R. B. Kesler and S. T. Han, *Tappi,* **45** (7), 534 (1962).

208. J. L. Date, F. LaFond, and I. M. Thompson, *Pulp and Paper,* **38** (23), 44 (1964).

209. R. A. Whitney and S. T. Han, *Tappi,* **35** (12), 569 (1952).

210. G. A. Crowe, *Tappi,* **37** (3), 154A (1954).

211. R. K. Strapp, W. D. Kerr, and K. E. Vroom, *Pulp Paper Mag. Can.,* **58,** 277 (1957). [Convention issue.]

212. G. H. Richter and L. H. Pancoast, Jr., *Tappi,* **37** (6), 263 (1954).

213. N. Hartler, P. Rönström, and L. Stockman, *Svensk Papperstidn.,* **64** (19), 699 (1961).

214. H. Mamers and N. C. Grove, *Appita,* **28** (3), 159 (1974).

215. G. R. Harris, *Pulp Paper Mag. Can.,* **58,** 284 (1957). [Convention Issue.]

216. N.-E. Virkola, A. A. Alm, M. Mäkinen, and R. Soila, *Paperi Puu,* **42** (12), 665 (1960).

217. F. M. Ernest and S. M. Harman, *Tappi,* **50** (12), 110A (1967).

218. P. K. Christensen and J. B. Hellum, *Norsk Skogind.,* **18** (10), 366 (1964).

219. R. N. Miller and W. H. Swanson, "Chemistry of the Sulfite Process," Lockwood Trade Journal Co., New York, N.Y., 1928.

220. G. H. Tomlinson and G. H. Tomlinson II, U.S. Patent 3046182 (1962).

221. E. Hägglund, *Paper Trade J.,* **85** (21), 203 (1927).

222. G. H. Chidester and J. N. McGovern, *Tech. Assoc. Papers,* **24,** 226 (1941).

223. R. M. Husband, *Tappi,* **38** (10), 577 (1955).

224. R. M. Dorland, R. A. Leask, and J. W. McKinney, *Pulp Paper Mag. Can.,* **59,** 226 (1957.) [Convention Issue.] U.S. Patent 2906659 (1959).

225. G. H. Tomlinson, G. H. Tomlinson II, J. R. G. Bryce, and N. G. M. Tuck, *Pulp Paper Mag. Can.,* **59,** 247 (1958). [Convention issue.]

226. L. Hall and L. Stockman, *Svensk Papperstidn.,* **61,** 871 (1958).

227. H. W. Giertz, *Norsk Skogind.,* **13** (5), 146 (1959).

228. L. T. Welch, T. G. Sheridan, and G. H. Tomlinson II, *Pulp and Paper,* **33** (5), 61 (1959).

229. H. Dyer, *Paper Trade J.,* **145** (41), 21 (1961).

230. D. Burrows, *Paper Trade J.,* **148** (21), 30 (1964).

231. H. P. Dahm, *Pulp and Paper Intl.,* **3** (1), 30 (1961).

232. G. W. F. Robinson, *Pulp Paper Mag. Can.,* **68** (4), T–179 (1967).

233. W. D. Kerr, *Proc. 1st Intl. Sulfite Pulping Conference* (CPPA/TAPPI) Chicago, Ill., June, 1961.

234. T. W. Maloney, V. Gibbs, and D. H. Andrews, *Pulp Paper Mag. Can.,* **74** (11), T–378 (1973).

235. O. V. Ingruber and G. A. Allard, *Pulp Paper Mag. Can.,* **74** (11), T–354 (1973); U. S. Patent 3630832 (1971).

236. J. V. A. Peltonen, Presented at 2nd Intl. Sulfite Pulping Conf., Boston, Mass., October 1972.

237. J. Peckham and V. Van Drunen, *Tappi,* **44** (5), 374 (1961).

238. N. Sanyer, T. Itoh, and E. L. Keller, *Tappi,* **47** (6), 323 (1964).

239. D. D. Hinrichs, *Svensk Papperstidn.,* **73** (5), 122 (1970); *Tappi,* **54** (7), 1105 (1971).

240. P. E. Shick, *Tappi,* **53** (8), 1451 (1970).

241. J. Bryce, F. Perry, N. Liebergott, and R. Mosher, "Bibliography of Multistage Pulping of Various Wood Species," Tech. Sect., CPPA (1969).

242. S. A. Rydholm, *Papier* (Darmstadt), **14** (10a), 535 (1960).

243. S. Lagergren, *Svensk Papperstidn.,* **67** (6), 238 (1964).

244. S. Lagergren and B. Lundén, *Pulp Paper Mag. Can.,* **60** (11), T–338 (1959).

245. G. E. Annergren and S. A. Rydholm, *Svensk Papperstidn.*, **62** (20), 737 (1959).
246. G. E. Annergren, I. Croon, B. F. Enström, and S. A. Rydholm, *Svensk Papperstidn.*, **64** (10), 386 (1961).
247. I. Croon, *Svensk Papperstidn.*, **66** (1), 1 (1963).
248. P. J. Nelson, *Appita*, **28** (4), 232 (1975).
249. A. Ogait, *Papier* (Darmstadt), **17** (11), 639 (1963).
250. A. Vethe, V. Lorås, and F. Løschbrandt, *Norsk Skogind.*, **14** (5), 167 (1960).
251. E. L. Bailey, *Tappi*, **45** (9), 689 (1962).
252. T. A. Pascoe, J. S. Buchanan, E. H. Kennedy, and G. Sivola, *Tappi*, **42** (4), 265 (1959).
253. J. R. G. Bryce and G. H. Tomlinson II, *Pulp Paper Mag. Can.*, **63** (7), T–355 (1962); G. H. Tomlinson II, U. S. Patent 3092535 (1963).
254. R. G. Lightfoot and O. Sepall, *Pulp Paper Mag. Can.*, **66** (5), T–279 (1965).
255. N. Sanyer, E. L. Keller, and G. H. Chidester, *Tappi*, **45** (2), 90 (1962).
256. T. H. Ulmanen, R. Räsänen, and M. J. Rantanen, *Tappi*, **46** (11), 151A (1963).
257. L. Ahlén, *Tappi*, **46** (11), 143A (1963).
258. I. Hassinen and R. Räsänen, *Pulp Paper Mag. Can.*, **75** (1), T–13 (1974).
259. N. M. Bikales and L. Segal, "Cellulose and Cellulose Derivatives," Part V, New York, N.Y.: Wiley-Interscience, 1971.
260. W. H. Bush, O. J. Walker, and K. C. Logan, *Pulp Paper Mag. Can.*, **53**, 286 (1952). [Convention Issue.]
261. J. W. Purdey, A. Marcoux, W. G. Nettleton, L. R. Beath, J. O. Kelley, and R. C. Bledsoe, *CPPA, Tech. Sect. Proc.*, 253 (1959).
262. J. E. Lambert, *Paper Trade J.*, **144** (43), 36 (1960).
263. L. Pamén, *Svensk Papperstidn.*, **63** (17), 550 (1960).
264. J. S. Hart and J. M. Woods, *Pulp Paper Mag. Can.*, **56** (8), 95 (1955).
265. J. S. Hart, *Tappi*, **41** (5), 218A (1958).
266. A. Bølviken and H. W. Giertz, *Norsk Skogind.*, **10** (10), 344 (1956).
267. G. Jayme, L. Broschinski, and W. Matzke, *Papier* (Darmstadt) **18** (7), 308 (1964).
268. J. R. G. Bryce, R. D. Mosher, H. J. Laberge, A. N. Fleming and G. A. Menier, CPPA Tech. Sec. Proc., 1972, D 11–22.
269. A. Hulteberg, *Paper Trade J.*, **150** (28), 58 (1966).
270. W. H. deMontmorency, K. E. Vroom, E. Koller, and G. A. Shaw, *Pulp Paper Mag. Can.*, **57** (3), 213 (1956).
271. F. P. Silver and L. R. Beath, *Pulp Paper Mag. Can.*, **54**, 216 (1953). [Convention Issue].
272. R. A. Leask, J. R. MacDonald, and A. C. Shaw, *Pulp Paper Mag. Can.*, **75** (10), T–361 (1974).
273. "The Burning of Sulfur," in "The Sulfur Manual," Texas Gulf Sulfur Co., New York, N.Y., 1963, chap. 7.
274. G. G. Vincent and A. Cook, *Pulp Paper Mag. Can.*, **47** 129 (1946), [Convention Issue.]
275. "Chemipulp-KC Jet-Type Sulfur Burner," Bulleton 100–A1, Chemipulp/Jensen Div., Watertown, N.Y.
276. J. W. DeVos, *Tappi*, **39** (12), 883 (1956).
277. A. R. Roche, *Can. Pulp Paper Ind.*, **16** (5), 33 (1963).
278. B. L. Browning and O. Kress, *Tech. Assoc. Papers*, **18**, 213 (1935).
279. *Tappi Standard 603 m 45*, "Analysis of Sulfur Burner Gas" Technical Association of the Pulp and Paper Industry, Atlanta, Ga. Tech. Sect., CPPA, *Standard J. 23P*, "Analysis of Sulphur Burner Gas", 1947.

280. W. T. Butler, *Pulp Paper Mag. Can.*, **52**, 243 (1951). [Convention Issue.]
281. "Spray Type SO_2 Gas Cooling System," Chemipulp/Jensen Div., The Stebbins Engineering & Mfg. Co., Watertown, N.Y.
282. T. Gerace, *Paper Trade J.*, **124** (14), 57 (1947).
283. R. J. Campney, *Pulp Paper Mag. Can.*, **64** (11), T–485 (1963).
284. *Paper Processing/Chemicals* **26**, 14 (May 1966).
285. H. R. Douglas, I. W. A. Snider, and G. H. Tomlinson II, *Pulp Paper Mag. Can.*, **66** (6), T–316 (1965); *Chem. Eng. Prog.*, **59** (12), 85 (1963).
286. W. J. M. Douglas, *Chem. Eng. Progr.*, **60** (7), 66 (1964).
287. J. M. Richards, *Paper Trade J.*, **149** (38), 69 (1965).
288. C. K. White, J. E. Vivian, and R. P. Whitney, *Tech. Assoc. Papers*, **31**, 141 (1948).
289. D. E. Marriner and R. P. Whitney, *Tech. Assoc. Papers*, **31**, 143 (1948).
290. C. W. Spalding and S. T. Han, *Tappi*, **45** (3), 192 (1962).
291. F. L. Wells and R. H. MacLaren, *Tappi*, **38** (11), 668 (1955).
292. K. Hagfeldt, T. Simmonds, and U. Söderlund, *Svensk Papperstidn.*, **76** (8), 292 (1973).
293. G. Tanny and O. V. Ingruber, *Pulp Paper Mag. Can.*, **67** (9), T–407 (1966).
294. R. A. McIlroy, N. Soltys, J. L. Clement, and A. J. Hornfeck, *Tappi*, **53** (10), 1950 (1970).
295. J. M. Jopp, *Pulp Paper Mag. Can.*, **58** (8), 148 (1957).
296. E. F. Tucker, *Tappi*, **33** (1), 29 (1950).
297. B. Thomas, *Pulp Paper Mag. Can.*, **58** (11), 139, (1957).
298. B. Thomas and R. C. Flynn, *Paper Industry*, **47** (4), 71 (1965).
299. M. A. Scheil, *Tappi*, **36** (6), 241 (1953).
300. J. E. Finsen, *Tappi*, **42** (2), 104 (1959).
301. A. Christensen, *Pulp Paper Mag. Can.*, **48**, 110 (1947). [Convention Issue.].
302. S. Svensson, Swedish Patent 68988 (1930), through CA. 2055 (1932).
303. J. M. McEwen, *Tappi*, **35** (8), 361 (1952).
304. V. P. Edwardes, *Pulp Paper Mag. Can.*, **45** (11), 891 (1944).
305. J. M. Currie, *Pulp and Paper*, **39** (37), 52 (1965).
306. D. E. Honer, *Paper Trade J.*, **151** (35), 55 (1967).
307. C. J. Myszak, *Pulp Paper Mag. Can.*, **69** (12), T–294 (1968).
308. D. M. Moody, *Tappi*, **52** (3), 448 (1969).
309. N. S. Lea and E. A. Christoferson, *Pulp and Paper*, **39** (42), 48 (1965).
310. W. D. Johnson and H. Gansler, *Pulp and Paper*, **45** (12), 54 (1971).
311. T. N. Kleinert and C. S. Joyce, *Tappi*, **40** (10), 813 (1957).
312. C. S. Joyce and T. N. Kleinert, *Pulp Paper Mag. Can.*, **58** (5), 131 (1957).
313. H. Augustin and D. Helmke, *Papier* (Darmstadt) **29** (9), 398 (1975).
314. A. Küng, *Svensk Papperstidn.*, **49** (7), 145 (1946).
315. H. K. van Eyken, *Pulp Paper Mag. Can.*, **65** (9), 103 (1964).
316. G. Rowlandson, *Tappi*, **55** (6), 959 (1972).
317. I. Hassinen, *Pulp and Paper*, **49** (11), 92 (1975).
318. G. Annergren and A. Backlund, *Pulp Paper Mag. Can.*, **67** (4), T–220 (1966).
319. J. E. Stone, *Pulp Paper Mag. Can.*, **57** (6), 139 (1956).
320. R. deMontigny and O. Maass, *Dom. For. Serv. Bull.*, No. 87, Ottawa, Canada, 1935.
321. W. B. Beazley, H. W. Johnson, and O. Maass, *Dom. For. Serv. Bull.*, No. 95, Ottawa, Canada, 1939.
322. A. J. Stamm, USDA Tech. Bull., No. 929, U.S. Government Printing Office, Washington, D.C., 1946.
323. H. K. Burr and A. J. Stamm, *J. Phys. Colloid Chem.*, **51** (1), 240 (1947).

324. A. J. Stamm, *Tappi,* **32** (5), 193 (1949).

325. A. J. Stamm, *Pulp Paper Mag. Can.,* **54** (2), 54 (1953).

326. N. I. Paranyi and W. Rabinovitch, *Pulp Paper Mag. Can.,* **56**, 163 (1955). [Convention Issue.]

327. J. H. Sutherland, H. W. Johnston, and O. Maass, *Can. J. Res.,* **10** (1), 36 (1934).

328. E. J. Howard, *Pulp Paper Mag. Can.,* **52** (7), 91 (1951).

329. J. H. Ross, *Pulp Paper Mag. Can.,* **54** (7), 103 (1953).

330. J. S. Hart, *Tappi,* **37** (8), 331 (1954).

331. N. I. Woods, *Pulp Paper Mag. Can.,* **57** (4), 142 (1956).

332. E. Vilamo, *Paperi Puu,* **36** (10), 401 (1954).

333. A. Alm and L. Stockman, *Svensk Papperstidn.,* **61** (1), 10 (1958).

334. R. Aurell, L. Stockman, and A. Teder, *Svensk Papperstidn.,* **61** (21), 937 (1958).

335. J. N. McGovern and G. H. Chidester, *Tech. Assn. Papers,* **17**, 292 (1934).

336. J. H. Feiner and W. Gallay, *Pulp Paper Mag. Can.,* **63** (9), T–435 (1962).

337. J. O. Murto et al., "Selluloosanpuun Lastatus" I–V. [Chipping of Pulpwood], report from the Central Laboratory, Helsinki, Finland, 1951, through Rydholm, "Pulping Processes," Interscience, New York, N.Y., 1965, p 270.

338. H. Green and F. H. Yorston, *Pulp Paper Mag. Can.,* **40** (3), 244 (1939).

339. H. Green and F. H. Yorston, *Pulp Paper Mag. Can.,* **41**, 123 (1940). [Convention Issue.]

340. J. E. Stone and L. F. Nickerson, *Pulp Paper Mag. Can.,* **59** (6), 165 (1958).

341. J. E. Stone and L. F. Nickerson, *Pulp Paper Mag. Can.,* **62** (6), T–317 (1961).

342. N. Hartler and O. Sundberg, *Svensk Papperstidn.,* **63** (8), 263 (1960).

343. H. Bausch and N. Hartler, *Svensk Papperstidn.,* **63** (9), 279 (1960).

344. N. Hartler, *Svensk Papperstidn.,* **66** (18), 696 (1963).

345. N. Hartler, *Svensk Papperstidn.,* **66** (11), 443 (1963).

346. K. C. Logan, O. J. K. Sepall, J. L. Chollet, and T. B. Little, *Pulp Paper Mag. Can.,* **61** (11), T–515 (1960).

347. "Soderhamn H-P Chipper," Bulletin No. 777A, Forano Limited, Plessisville, Quebec, Canada.

348. N. Hartler, *Svensk Papperstidn.,* **66** (17), 650 (1963); *Pulp and Paper,* **40** (1), 17 (1966).

349. W. J. Nolan, *Tappi,* **44** (7), 484 (1961).

350. H. Freedman, *Paper Trade J.,* **148** (28), 38 (1964).

351. A. Vethe, *Norsk Skogind.,* **21** (5), 180 (1967).

352. H. H. Saunderson and O. Maass, *Can. J. Res.,* **10** (1), 24 (1934).

353. G. A. Richter, *Tappi,* **32** (12), 553 (1949).

354. W. H. Swanson, *Lockwood Trade J.,* 113 (1928) Chap. 12 "Temperature Schedule in Sulfite Pulping".

355. R. N. Miller and W. H. Swanson, *Paper Trade J.,* **78** (15), 178 (1924).

356. J. S. Hart and J. M. Woods, *Pulp Paper Mag. Can.,* **57** (4), 158 (1959).

357. D. Rusten, *Norsk Skogind.,* **16** (8), 328 (1962).

358. C. Gustafson, *Paperi Puu,* **38** (9), 383 (1956).

359. F. H. Yorston, *Dom. For. Serv. Bull.,* No. 97, Ottawa, Canada, 1942, p 56.

360. W. O. Yean, J. H. Ross, and K. E. Vroom, *Pulp Paper Mag. Can.,* **58** (6), 197 (1957).

361. E. Ott, Ed., "Cellulose and Cellulose Derivatives," Interscience, New York, N.Y., 1943, p 501.

362. J. N. McGovern, *Paper Trade J.*, **103** (41), 297 (1936).
363. F. H. Yorston and N. Liebergott, *Pulp Paper Mag. Can.*, **66** (5), T-272 (1965).
364. W. F. Holzer, *Tech. Assoc. Papers*, **27**, 276 (1944).
365. O. V. Ingruber and G. A. Allard, *Tappi*, **50** (12), 597 (1967).
366. O. V. Ingruber, *Pulp Paper Mag. Can.*, **60** (11), T-346 (1959).
367. H. Makkonen, O. Järvela, and N. E. Virkola, *Paperi Puu*, **47** (2), 59 (1965).
368. O. V. Ingruber, *Pulp Paper Mag. Can.*, **58** (9), 161 (1957); *Tappi*, **41** (12), 764 (1958).
369. R. M. Husband, *Tappi*, **40** (6), 452 (1957).
370. E. Hägglund, *Svensk Kem. Tidskr.*, **37**, 116 (1925); **38**, 177 (1926); E. Hägglund, "Chemistry of Wood," Academic, New York, N.Y., 1951, pp 215-18.
371. J. M. Calhoun, F. H. Yorston, and O. Maass, *Can. J. Res.*, **15** (11), 457 (1937).
372. J. M. Calhoun, J. J. R. Cannon, F. H. Yorston, and O. Maass, *Can. J. Res.*, **16** (7), 242 (1938).
373. N. Liebergott and F. H. Yorston, *Pulp Paper Mag. Can.*, **77** (11), T-211 (1976).
374. J. H. Ross, J. S. Hart, R. K. Strapp, and W. Q. Yean, *Pulp Paper Mag. Can.*, **52** (9), 118 (1951).
375. N. Hartler, L. Stockman, and O. Sundberg, *Svensk Papperstidn.*, **64** (2), 33 (1961).
376. N. Hartler, L. Stockman, and O. Sundberg, *Svensk Papperstidn.*, **64** (3), 67 (1961).
377. R. M. Husband, *Tappi*, **40** (6), 412 (1957).
378. R. M. Dorland, R. A. Leask, and J. W. McKinney, *Pulp Paper Mag. Can.*, **57** (7), 122 (1956).
379. J. H. Feiner and W. Gallay, *Pulp Paper Mag. Can.*, **66** (2), T-77 (1965).
380. B. Hagberg and N. H. Schöön, *Svensk Papperstidn.*, **76** (15), 561 (1973); **77** (4), 127 (1974); **77** (15), 557 (1974); B. Hagberg, N. H. Schöön, L. Ljung, and K. Martinsson, *Svensk Papperstidn*, **78** (2), 57 (1975).
381. L. Ericksson and O. Lehtikoski, *Paperi Puu*, **52** (7), 435 (1970) **52** (12), 813 (1970).
382. F. H. Yorston, *Dom. For Serv. Bull.*, No. 97, Ottawa, Canada, 1942.
383. E. Hägglund, *Paper Trade J.*, **85**, 41 (1927).
384. E. Hägglund, "Chemistry of Wood," Academic, New York, N.Y., 1951, pp 443-48.
385. H. Erdtman, *Annulen*, **539**, (2), 116 (1939).
386. H. Erdtman, *Cellulosechem.*, **18** (4), 83 (1940).
387. H. Erdtman, *Svensk Papperstidn.*, **43** (13), 241 (1940).
388. J. C. Pew, *Tappi*, **32** (1), 39 (1949).
389. W. G. Hoge, *Tappi*, **37** (9), 369 (1954).
390. E. Adler, *Svensk Papperstidn.*, **54** (13), 445 (1951).
391. E. Adler and L. Stockman, *Svensk Papperstidn.*, **54** (14), 477 (1951).
392. H. Erdtman, *Tappi*, **32** (7), 303 (1949).
393. R. M. Dorland, R. A. Leask, and J. W. McKinney, *Pulp Paper Mag. Can.*, **58** (5), 135 (1957).
394. R. J. Stevens, *Pulp Paper Mag. Can.*, **59** (1), 96 (1958).
395. K. Bitschnau, R. Patt, and W. Sanderman, *Papier* (Darmstadt), **24** (3), 130 (1970).
396. R. M. Samuels, *For. Prod. J.*, **10** (3), 119 (1960).
397. H. P. Dahm, *Norsk Skogind.*, **15** (10), 421 (1961).

398. G. A. Richter, *Ind. Eng. Chem., 33* (4), 532 (1941).
399. G. Utaka, K. Oku, H. Matsuura, and A. Sakai, *Tappi,* **48** (5), 273 (1965).
400. R. M. Dorland, R. A. Leask, and J. W. McKinney, *Pulp Paper Mag. Can.,* **60** (2), T–37 (1959).
401. N. E. Virkola, *Paperi Puu,* **48** (9), 483 (1966).
402. H. W. Giertz and L. T. Lunde, *Norsk Skogind.,* **14** (10), 369 (1960).
403. J. Temler and J. R. G. Bryce, *Pulp Paper Can.,* **76** (2), T–48 (1975).
404. A. B. Gorham, *Pulp Paper Mag. Can.,* **56,** 160 (1955). [Convention Issue.]
405. E. Hägglund, "The Chemistry of Wood" Academic, New York, N.Y., 1951, pp 415–24.
406. K. Freudenburg, F. Sohns, and A. Janson, *Annelen,* **518** (1), 62 (1935).
407. H. Erdtman, G. Aulin-Erdtman, and B. O. Lindgren, *Svensk Papperstidn.,* **49** (9), 199 (1946).
408. H. Erdtman, *Svensk Papperstidn.,* **48** (4), 75 (1945).
409. H. Erdtman, *Tappi,* **32** (2), 75 (1949).
410. F. E. Brauns, *Paper Trade J.,* **111** (40), 177 (1940). [Tappi section.]
411. E. Hägglund, *Svensk Kem. Tidskr.,* **37,** 116 (1925).
412. E. Hägglund and T. Johnston, *Biochem. Z,* **202,** 439 (1928).
413. S. Häggroth, B. O. Lindgen, and U. Saéden, *Svensk Papperstidn.,* **56** (17), 660 (1953).
414. S. Rydholm and S. Lagergren, *Svensk Papperstidn.,* **62** (4), 103 (1959).
415. B. O. Lindgren, *Svensk Papperstidn.,* **55** (3), 78 (1952).
416. E. Adler and B. O. Lindgren, *Svensk Papperstidn.,* **55** (16½), 563 (1952).
417. K. Forss, K.-E. Fremer, and B. Stenlund, *Paperi Puu,* **48** (11), 669 (1966).
418. G. Gellerstadt and J. Gierer, *Svensk Papperstidn.,* **74** (5), 117, (1971).
419. W. Q. Yean and D. A. I. Goring, *Pulp Paper Mag. Can.,* **65,** T–127 (1964). [Convention Issue.]
420. W. Q. Yean and D. A. I. Goring, *Svensk Papperstidn.,* **71** (20), 739 (1968).
421. P. W. Lange, *Svensk Papperstidn.,* **50** (11B), 130 (1947); **57** (15), 525 (1954).
422. A. R. Procter, W. Q. Yean, and D. A. I. Goring, *Pulp Paper Mag. Can.,* **68** (9), T–445 (1967).
423. J. K. Hamilton, *Proc. Wood Chemistry Symp.,* (Applied Chem. Div., Intl. Union of Pure and Applied Chem.), London, England (1962), pp. 197–217.
424. J. Janson and E. Sjöström, *Svensk Papperstidn.,* **67** (19), 764 (1964).
425. K. Pfister and E. Sjöström, *Paperi Puu,* **59** (11), 711 (1977).
426. H. Meier, *Svensk Papperstidn.,* **65** (8), 299 (1962); **65** (16), 589 (1962).
427. J. Sundman, *Paperi Puu,* **32** (9), 267 (1950).
428. O. E. Öhrn and I. Croon, *Svensk Papperstidn.,* **63** (18), 601 (1960).
429. G. E. Annergren and I. Croon, *Svensk Papperstidn.,* **64** (17), 618 (1961).
430. J. K. Hamilton and H. W. Kircher, J. Am. Chem. Soc., **80** (17), 4703 (1958). See also reference 454.
431. H. Makkonen, *Paperi Puu,* **49** (6), 383 (1967).
432. G. Annergren and S. A. Rydholm, *Svensk Papperstidn.,* **63** (18), 591 (1960).
433. H. Makkonen and N.-E. Virkola, *Paperi Puu* 46 (11), 651 (1964).
434. I. Croon, B. F. Enström, and S. A. Rydholm, *Svensk Papperstidn.,* **67** (5), 196 (1964).
435. E. Heuser, *Tappi,* **33** (3), 118 (1950).
436. N. S. Thompson and O. A. Kaustinen, *Tappi,* **49** (12), 550 (1966).
437. O. Samuelson, N.-H. Schöön, and I. Edlund, *Svensk Papperstidn.,* **58** (22), 825 (1955).

438. N. A. Rozenberger and Z. S. Napkhanenko, *Bum. Prom.*, **1961** (47), 10 through CA **58**, 1631 (1963) Chem, Abst.
439. O. C. Bøckman, *Norsk Skogind.*, **16** (8), 320 (1962).
440. T. N. Kleinert, *Holzforschung*, **18** (5), 139 (1964).
441. E. Hägglund, "The Chemistry of Wood" Academic, New York, N.Y., 1951, pp 425–31.
442. L. Stockman, *Svensk Papperstidn.*, **54** (18), 621 (1951).
443. N.-H. Schöön, *Svensk Papperstidn.*, **65** (19), 729 (1962).
444. N. Hartler, L. Lind, and L. Stockman, *Svensk Papperstidn.*, **64** (5), 160 (1961).
445. J. Rexfelt, O. Samuelson, S. Hägglund, and T. Öhgren, *Svensk Papperstidn.*, **76** (12), 448 (1973).
446. L. Stockman and E. Hägglund, *Svensk Papperstidn.*, **54** (7), 243 (1951).
447. S. O. Regestad and O. Samuelson, *Svensk Papperstidn.*, **61** (18B), 735 (1958).
448. M. Goliath and B. O. Lindgren, *Svensk Papperstidn.*, **64** (12), 469 (1961).
449. N.-H. Schöön, *Svensk Papperstidn.*, **64** (17), 624 (1961).
450. O. L. Forgacs, *Tappi*, **44** (2), 112 (1961).
451. L. G. A. Kull, *Tappi*, **46** (11), 198A (1963).
452. I. Croon and B. Swan, *Svensk Papperstidn.*, **66** (20), 812 (1963).
453. J. Janson and E. Sjöström, *Svensk Papperstidn.*, **69** (4), 107 (1966).
454. J. K. Hamilton and N. S. Thompson, *Pulp Paper Mag. Can.*, **61** (4), T–263 (1960).
455. G. Annergren, S. Rydholm, and S. Vardheim, *Svensk Papperstidn.*, **66** (6), 196 (1963).
456. H. Makkonen, *Paperi Puu*, **49** (7), 437 (1967).
457. E. Back, *Svensk Papperstidn.*, **63** (22), 793 (1960).
458. L. H. Allen, *Pulp Paper Can.*, **76** (5), T–139 (1975).
459. S. A. Rydholm, "Pulping Processes," Chapter 16, I. "Deresinification," pp 1042–1036; Interscience Publishers, a division of John Wiley and Sons, Inc., New York–London–Sydney (1965).
460. D. L. Vincent, *Pulp Paper Mag. Can.*, **58** (11), 150 (1957).
461. D. B. Mutton, *Tappi*, **41** (11), 632 (1958).
462. S. K. Kahila, *Paperi Puu*, **46** (11), 615 (1964).
463. A. Assarsson and G. Åkerlund, *Svensk Papperstidn.*, **70** (6), 205 (1967).
464. *Tappi Standard T-204 os-76*, "Alcohol-Benzene and Dichloromethane Solubles in Wood and Pulp," (Revised, 1976).
465. L. H. Allen, *Trans. Tech. Sect.*, *CPPA*, **3** (2), TR–32 (1977).
466. P. Luner and D. W. Ferguson, *Tappi*, **47** (4), 205 (1964).
467. J. Fellegi, J. Polcin, and V. Farkašová, *Papier* (Darmstadt), **21** (10A), 659 (1967).
468. J. Janson, *Paperi Puu*, **58** (4), 315 (1976).
469. E. Adler and S. Häggroth, *Svensk Papperstidn.*, **53** (12), 321 (1950).
470. P. Monzie and D. Monzie-Guillemet, *Tappi*, **46** (7), 179A (1963).
471. W. Jensen, N.-E. Virkola, and A. A. Alm, *Paperi Puu*, **45** (8), 391 (1963).
472. N.-E. Virkola and A. A. Alm, *Papier* (Darmstadt), **15** (10a), 522 (1961).
473. H. W. Giertz and S. Wennerås, *Norsk Skogind.*, **14** (9), 322 (1960).
474. H. W. Giertz, *Papier* (Darmstadt), **17** (5), 191 (1963).
475. J. R. G. Bryce, S. Lamed, and G. H. Tomlinson II, *Pulp Paper Mag. Can.*, **65** (3), T–155 (1964).
476. W. H. Pitkin and H. L. Crosby, *Tappi*, **40** (11), 204A (1957).
477. J. Foster *Pulp Paper Mag. Can.*, **53**, 259, (1952). [Convention Issue.]
478. S. W. Hooper, *Pulp Paper Mag. Can.*, **59** (7), 115 (1958).
479. Ø. Ellefsen, *Norsk Skogind.*, **16** (5), 175 (1962).
480. E. L. Rastatter and A. H. Croup, *Tappi*, **35** (5), 223 (1952).

481. G. H. Tomlinson II and N. G. M. Tuck, *Pulp Paper Mag. Can.,* **53** (11), 109 (1952).
482. D. J. Whittle, *Tappi,* **54** (7), 1074 (1971).
483. R. P. Whitney, R. M. Elias, and M. N. May, *Tappi,* **34** (9), 396 (1951).
484. M. N. May, *Tappi,* **35** (11), 511 (1952).
485. L. G. Sillén and T. Andersson, *Svensk Papperstidn.,* **55** (16½), 622 (1952).
486. T. W. Bauer and R. M. Dorland, *Can. J. Technol.,* **32** (3), 91 (1954).
487. E. Rosen, *Trans. Roy. Inst. Techol.,* Stockholm, Sweden, No. 223, (1964), and earlier references by the same author cited in this publication.
488. C. Rosenblad, *Pulp Paper Mag. Can.,* **51** (5), 85 (1950).
489. B. Brunes, *Paper Trade J.,* **133** (25), 24 (1951); **133** (26), 18 (1951).
490. S.-E. Jonsson, "Evaporation of Sulfite Waste Liquor," in *Proc. IUPAC-Eucepa Symp. on Recovery of Pulping Chemicals,* Helsinki, Finland, 1968, Finnish Pulp and Paper Res. Inst., 1969, p 95.
491. G. F. Leitner, *Tappi,* **52** (7), 1296 (1969).
492. T. Simmons, *Svensk Papperstidn.,* **56** (4), 121 (1953).
493. B. Huldén, "On the combustion of Ca-base Sulfite Waste Liquor in Finland," in *Proc. IUPAC-Eucepa Symp. on Recovery of Pulping Chemicals,* Helsinki, Finland, 1968, Finnish Pulp and Paper Res. Inst., 1969, p. 387.
494. A. Scholander, *Svensk Papperstidn.,* **53** (2), 35 (1950).
495. G. H. Tomlinson and L. S. Wilcoxson, *Paper Trade J.,* **110** (15), 209 (1940).
496. R. S. Hatch, *Pulp Paper Mag. Can.,* **47** (8), 80 (1946).
497. R. J. Stanton, *Tappi,* **56** (7), 112 (1973).
498. J. L. Clement, *Tappi,* **49** (8), 127A (1966).
499. S. Ahman and H. Arne, *Pulp and Paper,* **49** (7), 80 (1975).
500. J. E. Hanway, E. B. Henby, and G. R. Smithson, *Tappi,* **50** (10), 64A (1967).
501. J. C. Laberge, *Tappi,* **46** (9) 538 (1963).
502. A. Ahonen, *Paperi Puu,* **56** (8), 619 (1974).
503. J. C. Kleinegger, *Tappi,* **52** (7), 1291 (1969).
504. J. E. Walther, H. Hamby III, and H. R. Amberg, *Tappi,* **56** (4), 144 (1973).
505. *Amer. Paper Ind.,* **56** (11), 30 (1974).
506. J. L. Clement and W. L. Sage, *Tappi,* **52** (8), 1449 (1969).
507. G. G. Copeland, *Pulp and Paper,* **42** (37), 29 (1968).
508. W. Q. Hull, B. C. Smith, J. H. Hull, and W. F. Holzer, *Tappi,* **37** (10), 14A (1954).
509. L. C. Jenness, *Tappi,* **37** (8), 137A (1954).
510. O. Samuelson and H.-H. Schöön, *Svensk Papperstidn.,* **60** (7), 259 (1957).
511. T. T. Collins, Jr., and P. E. Shick, *Paper Trade J.,* **155** (2), 41 (1971) and preceding articles listed therein.
512. R. P. Whitney, S. T. Han, and J. L. Davis, *Tappi,* **40** (7), 587 (1957).
513. R. P. Whitney, S. T. Han, R. B. Kesler, and J. F. Bakken, *Tappi,* **48** (1), 1 (1965).
514. J. W. Rice and N. Sanyer, *Tappi,* **51** (7), 321 (1968).
515. H. P. Markant, *Tappi,* **43** (8), 699 (1960).
516. K. N. Cederquist, N. K. G. Ahlborg, B. Lundén, and T. O. Wentworth. *Tappi,* **43** (8), 702 (1960).
517. M. W. Lüthgens, *Tappi,* **45** (11), 837 (1962).
518. J. Gullichsen, E. Saiha, and E. N. Westerberg, *Tappi,* **51** (9), 395 (1968).
519. L. R. Berry and A. D. Larsen, *Tappi,* **45** (11), 887 (1962).
520. R. E. Peterson and J. A. Krauth, *Pulp and Paper,* **43** (9), 133 (1969).
521. W. H. Gauvin and J. J. O. Gravel, *Tappi,* **43** (8), 678 (1960).

522. A. Bergholm, *Svensk Papperstidn.,* **66** (4), 125 (1963).

523. W. R. Cook, *Tappi,* **57** (9), 94 (1974); **60** (4), 106 (1977).

524. G. G. Copeland and J. E. Hanway, Jr., *Tappi,* **47** (6), 175A (1964).

525. A. Erdman, Jr., *Tappi,* **50** (6), 110A (1967).

526. B. Lundén, *Tappi,* **53** (9), 1726 (1970).

527. W. G. Farin, *Tappi,* **56** (9), 69 (1973).

528. J. D. Smith, *Tappi,* **59** (5), 72 (1976).

529. D. A. Holder, A. B. Mindler, and D. F. Manchester, *Pulp Paper Mag. Can.,* **66** (2), T–55 (1965).

530. W. R. Effer, E. W. Hopper, and H. B. Marshall, *Pulp Paper Mag. Can.,* **62** (10), T–447 (1961).

531. S. F. Ali, G. G. Wilson, and W. H. Whitney, *Tappi,* **51** (7), 69A (1968).

532. F. J. Zimmerman and D. G. Diddams, *Tappi,* **43** (8), 710 (1960).

533. G. A. Dubey, T. R. McElhinney, and A. J. Wiley, *Tappi,* **48** (2), 95 (1965).

534. K. Goel, R. Paquin, Y. Mehta, and Y. Lemay, *Trans. CPPA,* **3** (1), TR–6 (1977); **3** (1), TR–11 (1977).

535. J. J. Garceau, S. N. Lo, and L. Marchildon, *Pulp Paper Can.,* **77** (10), T–174 (1976).

536. T. G. Zitrides, *Pulp and Paper,* **49** (9), 126 (1975).

537. J. M. Pepper, *Pulp Paper Mag. Can.,* **46** (2), 83 (1945).

538. I. A. Pearl, *Tappi,* **52** (7), 1253 (1969).

539. H. M. McFarlane, *Pulp Paper Mag. Can.,* **62** (12), T–515 (1961).

540. V. F. Felicetta, M. Lung, and J. L. McCarthy, *Tappi,* **42** (6), 496 (1959).

541. J.-M. Perret, J. J. Garceau, G. Pineault, and S. N. Lo, *Pulp Paper Can.,* **77** (11), T–228 (1976).

542. J. W. Collins, L. A. Boggs, A. A. Webb, and A. J. Wiley, *Tappi,* **56** (6), 121 (1973).

543. Y. L. Bar-Sinai and M. Wayman, *Tappi,* **59** (3), 112 (1976).

544. F. A. Price, *Pulp Paper Mag. Can.,* **59** (3), 88 (1958); **61** (9), 116 (1960).

545. H. Schwartz, *Pulp Paper Mag. Can.,* **45** (8), 675 (1944).

546. K. C. Shen, *For. Prod. J.,* **24** (2), 38 (1974).

547. G. H. Tomlinson II and H. Hibbert, *J. Am. Chem. Soc.,* **58** (2), 345 (1936).

548. L. T. Sandborn, J. R. Salvesen, and G. C. Howard, U. S. Patent 2057117 (1938); L. T. Sandborn, U. S. Patent 2104701 (1937).

549. H. Hibbert and G. H. Tomlinson II, U. S. Patent 2069185 (1937).

550. J. R. Salveson, D. R. Brink, D. G. Diddams, and R. Orozorski, U.S. Patent 2434636 (1948).

551. D. Craig and C. D. Logan, *Pulp and Paper,* **45** (9), 137 (1971).

552. E. Avela, A. Hase, and R. Soila, *Paperi Puu,* **51** (7), 565 (1969).

553. K. Forss, *Paperi Puu,* **56** (3), 174 (1974).

554. A. B. Gorham and J. T. Charlton, *Pulp Paper Mag. Can.,* **74** (4), T–148 (1973).

555. A. J. Wiley, G. A. Dubey, B. F. Lueck, and L. P. Hughes, *Ind. Eng. Chem.,* **42** (9), 1830 (1950).

556. J. M. Holderby and W. A. Moggio, *J. Water Pollution Cont. Fed.,* **32** (2), 171 (1960).

557. J. C. Mueller, *Pulp Paper Mag. Can.,* **71** (22), T–488 (1970).

558. R. E. Simard and A. Cameron, *Pulp Paper Mag. Can.,* **75**, T–117 (1974). [Convention Issue.]

559. A. W. J. Dyck, *Paper Industry,* **39** (4), 26 (1957).

560. K. G. Forss, O. Gadd, R. O. Lundell, H. W. Williamson, M. Lampila, and J. Partanen, presented at International Sulfite Pulping and Recovery Con-

ference, Boston, Mass., October–November 1972, TAPPI/Tech. Sect., CPPA.

561. *Tappi Standard T-1203 os-61,* "Standard Terms Used in the Sulfate Process" (1961); *Standard K-2,* Tech. Sect., CPPA, "Standard Terms Used in the Alkaline Pulping Process," (1960) gives a corresponding series of definitions.

562. *Tappi Standard 1202 os-61,* "Standard Terms Used in the Soda Pulping Process" (1961).

563. G. E. Martin, *Tappi,* **33** (2), 84 (1950).

564. A. Teder and D. Tormund, *Svensk Papperstidn.,* **76** (16), 607 (1973).

565. L. Regnford and L. Stockman, *Svensk Papperstidn.,* **59** (14), 509 (1956).

566. *Standard J-12,* Tech. Sect., CPPA, "Analysis of Sulphate Green and White Liquors" (1961).

567. *Tappi Standard T-624 os-68,* "Analysis of Soda and Sulfate Green and White Liquors" (1968).

568. M. Wetherhorn and J. C. Scott, *South. Pulp Paper Mfr.,* **38** (3), 16 (1975).

569. G. W. Seymour, *Tappi,* **60** (7), 130 (1977).

570. G. Wallina and S. Noréus, *Svensk Papperstidn.,* **76** (9), 329 (1973).

571. K. Wilson, *Svensk Papperstidn.,* **71** (11), 446 (1968).

572. P. B. Borlew and T. A. Pascoe, *Paper Trade J.,* **122** (10), 99 (1946). [Tappi section.]

573. J. W. Rice and M. Zimmerman, *Tappi,* **50** (2), 72 (1967).

574. H. P. H. Cooper, *Tappi,* **58** (6), 59 (1975).

575. T. M. Grace, D. G. Sachs, and H. J. Grady, *Tappi,* **60** (4), 122 (1977).

576. J. Libert and G. Äkerlund, *Svensk Papperstidn.,* **74** (19), 619 (1971).

577. *Tappi Standard T-625, ts-64,* "Analysis of Soda and Sulfate Black Liquor" (1964).

578. O. Isämaki, *Paperi Puu,* **51** (11), 829 (1969).

579. U. Edberg, L. Engström, and N. Hartler, *Svensk Papperstidn.,* **76** (14), 534 (1973).

580. U. Edberg and S. Eskilsson, *Svensk Papperstidn.,* **75** (12), 467 (1972).

581. R. B. Reid, *Appita,* **31** (2), 128 (1977).

582. G. E. Annergren, S. O. Backlund, and K. Wiik, *Svensk Papperstidn.,* **76** (9), 324 (1973).

583. S. Backlund and I. Termén, *Pulp Paper Intl.,* **18** (10), 58 (1976).

584. L. Stockman and L. Ruus, *Svensk Papperstidn.,* **57** (22), 831 (1954).

585. C. B. Christiansen and J. B. Lathrop, *Pulp Paper Mag. Can.,* **55** (11), 113 (1954).

586. B. Haeglund and B. Roald, *Norsk Skogind.,* **9** (10), 351 (1955).

587. J. W. Hassler, *Tappi,* **38** (5), 265 (1955).

588. W. A. Mueller, *Can. J. Technol.,* **34** (3), 162 (1956).

589. J. Rennel and L. Stockman, *Svensk Papperstidn.,* **66** (21), 863 (1963).

590. L. A. Dela Grange, *Tappi,* **38** (6), 347 (1955).

591. W. A. Mueller, *Pulp Paper Mag. Can.,* **60** (1), T-3 (1959).

592. T. R. B. Watson, *Pulp Paper Mag. Can.,* **63** (4), T-247 (1962).

593. J. Richter, *Svensk Papperstidn.,* **61** (18B), 741 (1958).

594. S. Rydholm, "Continuous Pulping Processes," Special Tech. Assoc. Publ. No. 7, TAPPI, 1970, Lectures 7–9.

595. T. Knutsson and L. Stockman, *Tappi,* **41** (11), 704 (1958).

596. G. E. Annergren, O. E. Öhrn, and S. A. Rydholm, *Svensk Papperstidn.,* **66** (4), 110 (1963).

597. L. B. Jansson, *Tappi,* **46** (5), 296 (1963).

598. O. Fineman, *Svensk Papperstidn.,* **60** (11), 425 (1957).

599. R. P. Hamilton, *Tappi,* **44** (9), 647 (1961).

600. W. B. MacKay, *Pulp Paper Mag. Can.*, **65**, T–107 (1964). [Convention Issue.]
601. N. Hartler and J. Libert, *Svensk Papperstidn.*, **75** (2), 65 (1972).
602. N. Hartler and J. Libert, *Svensk Papperstidn.*, **76** (12), 454 (1973).
603. H. M. Canavan and Z. S. Blanchard, *Tappi*, **57** (3), 99 (1974).
604. G. Annergren, A. Backlund, J. Richter, and S. Rydholm, *Tappi*, **48** (7), 52A (1965).
605. P. J. Kleppe, *Tappi*, **58** (8), 172 (1975).
606. A. R. Sloman, *Pulp Paper Mag. Can.*, **65**, T–97 (1964) (Convention Issue).
607. E. McKenna and B. Prough, *Tappi*, **60** (5), 102 (1977).
608. J. W. C. Evans, *Paper Trade J.*, **152** (44), 68 (1968).
609. L. A. Carlsmith and R. H. Rasch, *Tappi*, **43** (12), 1013 (1960).
610. T. E. Winstead, *Paper Trade J.*, **156** (42), 52 (1972).
611. W. R. Patton, *Pulp and Paper*, **51** (5), 70 (1977).
612. J. E. Stone and H. V. Green, *Pulp Paper Mag. Can.*, **59** (10), 223 (1958).
613. K. H. Ekman and B. C. Fogelberg, *Paperi Puu*, **48** (4a), 175 (1966).
614. A. B. Wardrop and G. W. Davies, *Holzforschung*, **15** (5), 129 (1961).
615. J. E. Stone, *Tappi*, **40** (7), 539 (1957).
616. P. B. Borlew and R. L. Miller, *Tappi*, **53** (11), 2107 (1970).
617. N. Hartler and K. Östberg, *Svensk Papperstidn.*, **62** (15), 524 (1959).
618. O. Maass, *Pulp Paper Mag. Can.*, **54** (7), 98 (1953).
619. T. N. Kleinert and L. M. Marraccini, *Tappi*, **48** (3), 165 (1965).
620. E. J. Dostal, L. M. Marraccini, and T. N. Kleinert, *Pulp Paper Mag. Can.*, **61**, T–167 (1960). [Convention Issue.]
621. R. M. Screaton and S. G. Mason, *Svensk Papperstidn.*, **60** (10), 379 (1957).
622. J. V. Hatton, *Information Report VP-X-139*, Can. For. Serv., Western Forest Products Laboratory, Vancouver, Canada, 1975.
623. J. V. Hatton, *Pulp Paper Can.*, **76** (7), T–217 (1975).
624. J. V. Hatton, *Tappi*, **60** (4), 97 (1977).
625. K. J. Ljungkvist, *Svensk Papperstidn.*, **70** (12), 393 (1967).
625a. A. Lapointe, Paper Trade Journal **163** (18), 28 (1979).
626. J. V. Hatton and J. L. Keays, *Pulp Paper Mag. Can.*, **74** (1), T–11 (1973).
627. N. Hartler and W. Onisko, *Svensk Papperstidn.*, **65** (22), 905 (1962).
628. N. Hartler, *Paperi Puu*, **44** (7), 365 (1962).
629. P. Colombo, D. Corbetta, A. Pirotta, and G. Ruffini, *Svensk Papperstidn.*, **67** (12), 505 (1964).
630. J. V. Hatton, *Pulp Paper Can.*, **78** (3), T–57 (1977).
631. J. V. Hatton, *Tappi*, **59** (6), 151 (1976); **60** (8), 107 (1977).
632. J. V. Hatton, *Pulp Paper Can.*, **76** (8), T–232 (1975).
633. N. Hartler and Y. Stade, *Svensk Papperstidn.*, **80** (14), 447 (1977).
634. W. J. Nolan, *Tappi*, **51** (9), 378 (1968).
635. W. J. Bublitz and R. O. McMahon, *Tappi*, **59**, (12), 125 (1976).
636. R. L. Miller and C. W. Rothrock, Jr., *Tappi*, **46** (7), 147A (1963).
637. P. Colombo, D. Corbetta, A. Pirotta, and G. Ruffini, *Svensk Papperstidn.*, **63** (15), 457 (1960).
638. A. Wawer and M. N. Misra, *Pulp Paper Can.*, **78** (11), T–264 (1977).
639. U. Edberg, L. Engström, and N. Hartler, *Svensk Papperstidn.*, **76** (14), 529 (1973).
640. U. Edberg and S. Eskilsson, *Svensk Papperstidn.*, **75** (12), 467 (1972).
641. J. E. Stone and L. F. Nickerson, *Pulp Paper Mag. Can.*, **59** (6), 165 (1958).
642. G. F. Allo, *Pulp Paper Mag. Can.*, **47**, 116 (1946). [Convention Issue.]
643. V. Mattson, *Tappi*, **39** (2), 77 (1956).

644. J. C. Dean, C. M. Saul, and C. H. Turner, *Paperi Puu,* **38** (4a), 189 (1956).
645. M. N. May and J. R. Peckham, *Tappi,* **41** (2), 90 (1958).
646. W. J. Nolan and D. W. McCready, *Paper Trade J.,* **102** (4), 48 (1936). [Tappi section.]
647. H. F. Lewis and E. R. Laughlin, *Paper Trade J.,* **59** (51), 259 (1930). [Tappi section.]
648. M. W. Bray, *Tech. Assoc. Papers,* **12,** 268 (1929).
649. W. Surewicz, *Tappi,* **45** (7), 570 (1962).
650. D. D. Hinrichs, *Tappi,* **50** (4), 173 (1967).
651. R. Aurell and N. Hartler, *Svensk Papperstidn.,* **68** (4), 97 (1965).
652. R. Aurell, *Svensk Papperstidn.,* **67** (3), 89 (1964).
653. H. M. Banfill, J. G. Wheeler, and J. M. Ferguson, *Pulp Paper Mag. Can.,* **59,** 157 (1958). [Convention Issue.]
654. F. E. Brauns and W. S. Grimes, *Tech. Assoc. Papers,* **22,** 574 (1939).
655. T. Enkvist, *Svensk Papperstidn.,* **60** (17), 616 (1957).
656. M. W. Bray and C. E. Curran, *Paper Trade J.,* **97** (5), 52 (1933). [Tappi section.]
657. S. L. Schwartz and M. W. Bray, *Paper Trade J.,* **107** (12), 140 (1938). [Tappi section.]
658. C. R. Mitchell and J. H. Ross, *Pulp Paper Mag. Can.,* **33** (3), 35 (1932); U. S. Patent 1922262 (1933).
659. S. D. Wells, U. S. Patent 1949669 (1934).
660. S. D. Wells, G. E. Martin, and D. R. Moltzau, *Paper Trade J.,* **119** (19), 188 (1944). [Tappi section.]
661. S. D. Wells and K. A. Arnold, *Paper Trade J.,* **113** (9), 103 (1941). [Tappi section.]
662. C. B. Christiansen and G. W. Legg, *Pulp Paper Mag. Can.,* **59,** 149 (1958). [Convention Issue.]
663. J. E. Stone and D. W. Clayton, *Pulp Paper Mag. Can.,* **61** (6), T–307 (1960).
664. E. Hägglund, *Tappi,* **32** (6), 241 (1949).
665. J. S. Hart and R. K. Strapp, *Pulp Paper Mag. Can.,* **49,** 151 (1948). [Convention Issue.]
666. M. W. Bray, J. S. Martin, and S. L. Schwartz, *Tech. Assoc. Papers,* **22,** 382 (1939).
667. K. G. Chesley and P. L. Gilmont, *Tappi,* **38** (5), 279 (1955).
668. P. K. Paasonen and I. Lassenius, *Paperi Puu,* **53** (4a), 147 (1971).
669. G. W. Legg and J. S. Hart, *Tappi,* **43** (5), 471 (1960).
670. G. W. Legg and J. S. Hart, *Pulp Paper Mag. Can.,* **61** (5), T–299 (1960).
671. P. B. Borlew and T. A. Pascoe, *Tech. Assoc. Papers,* **30,** 570 (1947).
672. L. Munk, Z. Todorski, J. R. G. Bryce, and G. H. Tomlinson II, *Pulp Paper Mag. Can.,* **65** (10), T–411 (1964).
673. T. N. Kleinert, *Pulp Paper Mag. Can.,* **65** (7), T–275 (1964).
674. J. Morris, *Tappi,* **38** (12), 753 (1955).
675. M. W. Bray, *Paper Trade J.,* **87** (49), 220 (1928). [Tappi section.]
676. J. S. Martin, M. W. Bray, and C. E. Curran, *Paper Trade J.,* **97** (46), 242 (1933). [Tappi section.]
677. L. Stockman and E. Sundkvist, *Svensk Papperstidn.,* **61** (18B), 746 (1958).
678. E. J. Daleski, *Tappi,* **48** (6), 325 (1965).
679. G. L. Laroque and O. Maass, *Can. J. Res.,* **B19** (1), 1 (1941).
680. C. W. Carroll, *Tappi,* **43** (6), 573 (1960).
681. K. E. Vroom, *Pulp Paper Mag. Can.,* **58,** 228 (1957). [Convention Issue.]
682. L. J. Carver, *Paper Trade J.,* **158** (45), 38 (1974).

683. W. J. Nolan, *The Paper Industry*, **37** (1), 926 (1956).
684. W. J. Nolan, *Tappi*, **40** (3), 170 (1957).
685. T. N. Kleinert, *Tappi*, **48** (8), 447 (1965).
686. C. A. Wilson and A. J. Kerr, *Appita*, **30** (1), 55 (1976).
687. H. D. Wilder and E. J. Daleski, *Tappi*, **47** (5), 270 (1964); **48** (5), 293 (1965).
688. A. J. Kerr, *Appita*, **24** (3), 180 (1970).
689. L. Edwards and S. -E. Norberg, *Tappi*, **56** (11), 108 (1973).
690. S. LéMon and A. Teder, *Svensk Papperstidn.*, **76** (11), 407 (1973).
691. C. C. Smith and T. J. Williams, "Modelling and Control of Kraft Production Systems," Instrument Society of America, Pittsburgh, Pa., 1975, Chap. 3.
692. C. H. Wells, E. C. Johns, and F. L. Chapman, *Tappi*, **58** (8), 177 (1975).
693. A. J. Kerr and J. M. Uprichard, *Appita*, **30** (1), 48 (1976).
694. J. G. Lebrasseur, *South. Pulp Paper Mfr.*, **39** (9), 15 (1976).
695. J. V. Hatton, *Tappi*, **56** (8), 108 (1973).
696. F. P. Lodzinski and T. Karlsson, *Tappi*, **59** (9), 88 (1976).
697. A. J. Kerr, *Tappi*, **59** (5), 89 (1976).
698. H. O. Rogers, *Pulp and Paper*, **50** (9), 86 (1976).
699. J. J. Rettinger, *Tappi*, **58** (12), 94 (1975).
700. G. E. Coombes and P. C. Kohn, *Appita*, **30** (2), 148 (1976).
701. D. S. Brucker, K. Kostelic and E. W. Voightman, *Tappi*, **59** (9), 93 (1976).
702. G. Wallin and S. Noréus, *Svensk Papperstidn.*, **76** (9), 329 (1973).
703. J. W. T. Battershill, *Pulp Paper Mag. Can.*, **74** (5), T–174 (1973).
704. P. C. Bavadas and E. M. Cohen, *Pulp and Paper*, **49** (11), 64 (1975).
705. R. Aksela, S. Andtbacka, and M. Sanaksenaho, *Paperi Puu*, **59** (6, 7), 433 (1977).
706. S. A. Rydholm, "Pulping Processes," Interscience, New York, N.Y., 1965, pp 610–17.
707. J. V. Hatton and J. L. Keays, *Pulp Paper Mag. Can.*, **71** (11, 12), T–259 (1970).
708. J. L. Keays and J. V. Hatton, *Pulp Paper Mag. Can.*, **73** (10), T–300 (1972).
709. J. V. Hatton, *Tappi*, **56** (7), 97 (1973).
710. J. S. Martin and K. J. Brown, *Tappi*, **35** (1), 7 (1952).
711. R. M. Samuels and D. W. Glennie, *Tappi*, **41** (5), 250 (1958).
712. K. J. Brown, *Tappi*, **39** (6), 443 (1956).
713. R. A. Horn, *Tappi*, **58** (9), 142 (1975).
714. R. A. Horn and R. J. Auchter, *Paper Trade J.*, **156** (45), 55 (1972).
715. H. E. Young, *Pulp and Paper*, **48** (12), 46 (1974).
716. J. L. Keays, "Complete Tree Utilization of Mature Trees," in "Advances in processing and utilization of forest products," Ronald W. Rousseau, Editor, American Institute of Chemical Engineers, Symposium Series No. 139, **70**, 67 (1974).
717. R. E. Johnson, *Preprints, 61st Ann. Meet., Tech. Sect.*, CPPA, January 1975, p B–79.
718. P. Koch and S. J. Coughran, *Forest Products J.*, **25** (4), 23 (1975).
719. S. Eskilsson, *Svensk Papperstidn.*, **77** (5), 165 (1974).
720. J. L. Keays, *Pulp Paper Can.*, **75** (9), 43 (1974).
721. A. Wawer, *Pulp Paper Can.*, **76** (7), T–200 (1975).
722. J. W. Wilson, H. Worster, and D. O'Meara, *Pulp Paper Mag. Can.*, **61** (1), T–19 (1960).
723. K. Hunt and R. D. Whitney, *Information Report 0-X-217*, Can. For.

Serv., Dept. of Environment, Western Forest Products Laboratory, Vancouver, B.C. 1974.
724. W. C. Feist, E. L. Springer, and G. J. Hajny, *Tappi,* **56** (4), 148 (1973).
725. M. A. Hulme, *Tappi,* **58** (12), 118 (1975).
726. J. V. Hatton and K. Hunt, *Tappi,* **55** (1), 122 (1972).
727. G. J. Hajny, *Tappi,* **49** (10), 97A (1966).
728. R. A. Somsen, *Tappi,* **45** (8), 623 (1962).
729. R. J. Slinn, "Availability of Residuals in the U.S. Pulp and Paper Industry," in "Forest Product Residuals," W. K. Lautner, Editor, American Institute of Chemical Engineers, Symposium Series No. 146, **71**, 4 (1975).
730. H. L. Keaton and R. M. Rogan, Jr., *South. Pulp Paper Mfr.,* **36** (8), 18 (1973).
731. J. F. Oettinger, *Paper Trade J.,* **158** (23), 30 (1974).
732. T. G. Taylor, *Pulp Paper Can.,* **78** (1), T–14 (1977).
733. W. J. Bublitz, and T. Y. Yang, *Tappi,* **58** (3), 95 (1975).
734. I. Isotalo, L. Göttsching, N. -E. Virkola, and L. Nordman, *Paperi Puu,* **46** (3), 71 (1964); **46** (4a), 237 (1964).
735. T. Enkvist, *Tappi,* **37** (8), 350 (1954).
736. C. E. Ahlm, *Paper Trade J.,* **113** (13), 175 (1941). [Tappi section.]
737. T. Enkvist, B. Alfredsson, and E. Hägglund, *Svensk Papperstidn.,* **55** (16½), 588 (1952).
738. T. Enkvist, T. Ashorn, and K. Hästbacka, *Paperi Puu,* **44** (8), 395 (1962).
739. J. Gierer, *Svensk Papperstidn.,* **73** (18), 571 (1970).
740. J. Gierer and I. Norén, *Acta Chem. Scand.,* **16** (7), 1713 (1962).
741. J. Gierer and L. A. Smedman, *Acta Chem. Scand.,* **19** (5), 1103 (1965).
742. J. Gierer, S. Söderberg, and S. Thorén, *Svensk Papperstidn.,* **66** (23), 990 (1963).
743. E. Adler and B. Wesslen, *Acta Chem. Scand.,* **18** (5), 1314 (1964).
744. J. G. McNaughton, W. Q. Yean, and D. A. I. Goring, *Tappi,* **50** (11), 548 (1967).
745. J. E. Stone and A. M. Scallan, *Pulp Paper Mag. Can.,* **69** (10), T–288 (1968).
746. B. J. Fergus and D. A. I. Goring, *Holzforschung,* **24** (4), 118 (1970).
747. A. R. Procter, W. Q. Yean, and D. A. I. Goring, *Pulp Paper Mag. Can.,* **68** (9), T–445 (1967).
748. J. R. Wood, P. A. Ahlgren, and D. A. I. Goring, *Svensk Papperstidn.,* **71** (1), 15 (1972).
749. A. J. Kerr and D. A. I. Goring, *Can. J. Chem.,* **53** (7), 952 (1975).
750. A. J. Kerr and D. A. I. Goring, *Svensk Papperstidn.,* **79** (1), 20 (1976).
751. S. Axelsson, I. Croon, and B. Enström, *Svensk Papperstidn.,* **65** (18), 693 (1962).
752. R. Simonson, *Svensk Papperstidn.,* **74** (21), 691 (1971).
753. J. Saarnio and C. Gustafsson, *Paperi Puu,* **35** (3), 65 (1953).
754. S. Yllner, K. Östberg, and L. Stockman, *Svensk Papperstidn.,* **60** (21), 795 (1957).
755. R. L. Casebier and J. K. Hamilton, *Tappi,* **48** (11), 664 (1965).
756. S. Yllner and B. Enström, *Svensk Papperstidn.,* **59** (6), 229 (1956).
757. J. -Å. Hansson, *Svensk Papperstidn.,* **73** (3), 49 (1970).
758. R. Aurell, *Tappi,* **48** (2), 80 (1965).
759. R. Aurell, *Svensk Papperstidn.,* **66** (11), 437 (1963).
760. S. Dillén and S. Noréus, *Svensk Papperstidn.,* **71** (15), 509 (1968).
761. W. M. Corbett and J. Kenner, *J. Chem. Soc.,* **1955** (5), 1431.
762. J. W. Green, I. A. Pearl, K. W. Hardacker, B. D. Andrews, and F. C. Haigh, *Tappi,* **60** (10), 120 (1977).

763. D. B. Mutton, *Pulp Paper Mag. Can.*, **65** (2), T–41 (1964).
764. B. Alfredsson, L. Gedda, and O. Samuelson, *Svensk Papperstidn.*, **64** (19), 694 (1961).
765. H. Meier, *Svensk Papperstidn.*, **65** (16), 589 (1962).
766. D. W. Clayton, *Svensk Papperstidn.*, **66** (4), 115 (1963).
767. M. H. Johansson and O. Samuelson, *Svensk Papperstidn.*, **80** (16), 519 (1977).
768. M. H. Johansson and O. Samuelson, *Wood Sci. Technol.*, **11** (4), 251 (1977).
769. B. Lindberg, *Svensk Papperstidn.*, **59** (15), 531 (1956).
770. W. M. Corbett and G. N. Richards, *Svensk Papperstidn.*, **60** (21), 791 (1957).
771. N. Hartler, J. Libert, and A. Teder, *Ind. Eng. Chem., Process Des. Develop.*, **6** (4), 398 (1967).
772. E. Bilberg and P. Landmark, *Norsk Skogind.*, **15** (5), 221 (1961).
773. R. G. Barker, *Tappi*, **53** (6), 1087 (1970); U.S. Patent 3470061 (1969).
774. R. P. Green and Z. C. Prusas, *Pulp Paper Can.*, **76** (9), T–272 (1975).
775. P. J. Kleppe and K. Kringstad, *Norsk Skogind.*, **17** (11), 428 (1963).
776. P. Hilmo and K. Johnsen, *Pulp and Paper*, **43** (5), 73 (1969).
777. B. Alfredsson, O. Samuelson, and B. Sandstig, *Svensk Papperstidn.*, **66** (18), 703 (1963).
778. N. Sanyer and J. F. Laundrie, *Tappi*, **47** (10), 640 (1964).
779. D. W. Clayton and A. Sakai, *Pulp Paper Mag. Can.*, **68** (12), T–619 (1967).
780. M. G. Vinje and H. E. Worster, *Tappi*, **52** (7), 1341 (1969); U. S. Patent 3520773 (1970).
781. A. R. Procter and R. H. Wiekenkamp, *J. Polym. Sci.*, Part C, 1969 (28), 1.
782. A. R. Procter and K. A. Mohr, *Pulp Paper Can.*, **76** (10), 103 (1975).
783. A. R. Procter and G. E. Styan, *Tappi*, **57** (10), 123 (1974).
784. A. R. Procter, *Pulp Paper Canada* 77 (12), T–238 (1976); A. R. Procter, G. E. Styan, and M. G. Vinje, Can. Patent 971312 (1975).
785. N. Hartler, *Svensk Papperstidn.*, **62** (13), 467 (1959).
786. R. Aurell and N. Hartler, *Tappi*, **46** (4), 209 (1963).
787. H. H. Holton, *Pulp Paper Can.*, **78** (10), T–218 (1977); U.S. Patent 4012280 (1977).
788. B. Bach and G. Fiehn, *Zellst. Papier*, **21** (1), 3 (1972).
789. L. Löwendahl and O. Samuelson, *Svensk Papperstidn.*, **80** (17), 549 (1977).
789a. B. I. Fleming, G. J. Kubes, J. M. MacLeod and H. I. Bolker, Tappi, **61** (6), 43 (1978); **62** (7), 55 (1979).
790. P. Ahlgren, L. A. Olsson, and B. Vikström, *Svensk Papperstidn.*, **78** (3), 95 (1975).
791. M. G. Vinje and H. E. Worster, *Pulp Paper Mag. Can.*, **70** (21), T–431 (1969).
792. R. Aurell and N. Hartler, *Svensk Papperstidn.*, **68** (3), 59 (1965).
793. I. Croon and B. Enström, *Svensk Papperstidn.*, **65** (16), 595 (1962).
794. R. L. Casebier and J. K. Hamilton, *Tappi*, **50** (9), 441 (1967).
795. R. Aurell, *Svensk Papperstidn.*, **67** (2), 63 (1964).
796. H. Meier, *Svensk Papperstidn.*, **65** (8), 299 (1962).
797. O. Samuelson, G. Grangård, K. Jönsson, and K. Schramm, *Svensk Papperstidn.*, **56** (20), 779 (1953).
798. O. Franzon and O. Sameulson, *Svensk Papperstidn.*, **60** (23), 872 (1957).
799. R. R. Affleck and R. G. Ryan, *Pulp Paper Mag. Can.*, **70** (22), T–563 (1969).

800. I. Croon, *Svensk Papperstidn.*, **68** (10), 378 (1965).
801. J. W. Swanson and R. H. Cordingley, *Tappi*, **39** (10), 684 (1956).
802. B. E. Bergmann and S. Rying, *Tappi*, **58** (4), 147 (1975).
803. E. Abrams and J. R. Nelson, *Tappi*, **60** (10), 109 (1977).
804. N. Hartler and H. Norrström, *Tappi*, **52** (9), 1712 (1969).
805. J. Janson and I. Palenius, *Paperi Puu*, **54** (6), 343 (1972).
806. T. N. Kleinert, *Holzforschung*, **19** (6), 179 (1965).
807. J. Janson, I. Palenius, B. Stenlund, and P. -E. Sågfors, *Paperi Puu*, 57 (5), 387 (1975).
808. R. D. Cardwell and S. B. Cundall, *Appita*, **29** (5), 349 (1976).
809. C. B. Christiansen, J. S. Hart, and J. H. Ross, *Pulp Paper Mag. Can.*, **58** (1), 103 (1957).
810. W. F. Holzer and K. G. Booth, *Tappi*, **33** (2), 95 (1950).
811. R. M. Kellogg and E. Thykeson, *Tappi*, **58** (4), 131 (1975).
812. T. Ekström, *Pulp Paper Intl.*, **18** (6), 37 (1976).
813. O. Sosanwo, *Paperi Puu*, **58** (11), 791 (1976).
814. H. E. Worster and M. Bartels, *South. Pulp Paper Mfr.*, **39** (2), 32 (1976).
815. K. Hunt and J. V. Hatton, *Pulp Paper Can.*, **77** (12), T–271 (1976).
816. W. D. Kerr and J. S. Hart, *Tappi*, **42** (3), 254 (1959).
817. K. Kringstad and B. Vikström, *Svensk Papperstidn.*, **79** (2), 52 (1976).
818. J. McK. Limerick, *Tappi*, **35** (7), 297 (1952).
819. D. F. Morton, J. K. Perkins and G. E. Sanford in R. G. MacDonald, Ed., "The Pulping of Wood," Vol. I, 2nd Ed., "Pulp and Paper Manufacture," McGraw-Hill, New York, N.Y., 1969, pp. 478–504.
820. R. E. Millar, Jr., J. M. Piette, and J. J. Owen, Jr., *Tappi*, **38** (10), 604 (1955).
821. A. L. Wiley, N. R. Phillips, and E. A. Henry, Jr., *Tappi*, **38** (10), 600 (1955).
822. D. B. Rogers, *Tappi*, **59** (5), 81 (1976).
823. N. Hartler, O. Danielsson, and G. Ryrberg, *Tappi*, **59** (9), 105 (1976).
824. J. A. Kurdin, *Pulp and Paper*, **49** (9), 92 (1975).
825. R. E. Carvill, *South. Pulp Paper Mfr.*, **41** (2), 18 (1978).
826. W. G. Dedert and H. K. Waters, *Paper Trade J.*, **141** (2), 34 (1957).
827. R. Baldus, B. Warnqvist, and L. L. Edwards, *Trans., Tech. Sect.*, CPPA **3** (2), TR–56 (1977).
828. J. K. Perkins, H. S. Welsh, and J. H Mappus, *Tappi*, **37** (3), 83 (1954).
829. H. K. Waters and R. E. Bergstrom, *Tappi*, **38** (3), 169 (1955).
830. H. V. Nordén, V. J. Pohjola, and R. Seppänen, *Pulp Paper Mag. Can.*, **74** (10), T–329 (1973).
831. J. R. Phillips and J. Nelson, *Pulp Paper Can.*, **78** (6), T–123 (1977).
832. A. Rosen, *Tappi*, **58** (9), 156 (1975).
833. N. Hartler and S. Rydin, *Svensk Papperstidn.*, **78** (10), 367 (1975).
834. L. Edwards and S. Rydin, *Svensk Papperstidn.*, **78** (16), 577 (1975).
835. P. M. K. Choi, W. Q. Yean, and D. A. I. Goring, *Preprints, 62nd Ann. Meet., Tech. Sect.*, CPPA, January 1976, p 473.
836. J. K. Perkins, *Paper Trade J.*, **153** (8), 30 (1969).
837. R. V. Gossage and J. M. McSweeney, *Tappi*, **60** (4), 110 (1977).
838. S.-E. Jönsson, *Svensk Papperstidn.*, **64** (15), 565 (1961).
839. F. Kommonen, *Paperi Puu*, **50** (6), 347 (1968).
840. M. Perron and B. Lebeau, *Trans., Tech. Sec.*, CPPA, **3** (1), TR–1 (1977).
841. A. Tomiak, *Pulp Paper Mag. Can.*, **75** (9), T–331 (1974).
842. S. Rydholm, "Continuous Pulping Processes," Special Technical Association Publication No. 7, TAPPI, Chaps. 10, 11.
843. S. W. McKibbins, *Tappi*, **43** (10), 801 (1960).

844. D. A. Williams. S. W. McKibbin, and J. W. Riese, *Tappi*, **48** (9), 481 (1965).
845. J. Richter, *Tappi*, **49** (6), 48A (1966).
846. T. E. Jenkin, *Tappi*, **59** (6), 138 (1976).
847. J. P. Morgan, *Pulp Paper Can.*, **76** (2), 57 (1975).
848. J. Gullichsen, *Pulp Paper Mag. Can.*, **74** (8), T–266 (1973).
849. J. Gullichsen and H. Östman, *Tappi*, **59** (6), 140 (1976).
850. L. Edwards and S. Rydin, *Svensk Papperstidn.*, **79** (11), 354 (1976).
851. R. E. Carvill, *Tappi*, **51** (9), 104A (1968).
852. A. C. Martin, *Pulp and Paper*, **49** (8), 78 (1975).
853. B. Steenberg, *Svensk Papperstidn.*, **56** (20), 771 (1953).
854. E. Cowan in "The Pulping of Wood," Vol. I, 2nd Ed., "Pulp and Paper Manufacture," R. G. MacDonald, Ed., McGraw-Hill, New York, N.Y., 1969, Chap. 12.
855. G. H. Tomlinson II and N. G. M. Tuck, *Pulp Paper Mag. Can.*, **53** (11), 109 (1952).
856. *Tappi Standard T-213 os-77*, "Dirt in Pulp."
857. J. Hill, H. Hoglund, and E. Johnsson, *Tappi*, **58** (10), 120 (1975).
858. B. Arhippainen and B. Jungerstam, *Tappi*, **52** (6), 1095 (1969).
859. C. Gudmundson, H. Alsholm, and B. Hedström, *Svensk Papperstidn.*, **75** (19), 773 (1972).
860. H. L. Smith, Jr., *Tappi*, **46** (12), 182A (1963).
861. L. R. Berry, *Tappi*, **49** (4), 68A (1966).
862. T. M. Grace, *South. Pulp Paper Mfr.*, **40** (8), 16 (1977).
863. E. Venemark, *Svensk Papperstidn.*, **63** (12), 387 (1960).
864. J. H. Stacie and L. A. Wilhelmsen, *Tappi*, **52** (7), 1278 (1969).
865. R. E. Bergstrom and J. R. Lientz, *Tech. Assn. Papers*, **29**, 515 (1946).
866. W. R. Wills and B. Hedström, *Svensk Papperstidn.*, **66** (17), 667 (1963).
867. P. H. West, H. P. Markant, and J. H. Coulter, *Tappi*, **44** (10), 710 (1961).
868. E. H. Backteman and R. Y. Marr, *Pulp Paper Mag. Can.*, **72** (12), T–381 (1971).
869. H. G. Lankenau and A. R. Flores, *Pulp Paper Mag. Can.*, **70** (2), T–20 (1969).
870. R. Källenfors, *Pulp and Paper*, **49** (11), 144 (1975).
871. G. H. Tomlinson, *Pulp Paper Mag. Can.*, **49** (6), 65 (1948).
872. R. E. Chamberlain and C. E. Cairns, *Pulp Paper Mag. Can.*, **73** (1), T–13 (1972).
873. A. V. Craig, *Pulp Paper Mag. Can.*, **66** (7), T–293 (1965).
874. A. Vegeby, *Tappi*, **49** (7), 103A (1966).
875. W. J. Darmstadt, D. D. Wangerin, and P. H. West in "Chemical Recovery in Alkaline Pulping Processes," R. P. Whitney, Ed., Tappi Monograph No. 32, 1968, Chap. 3, p 59.
876. J. L. Clement, J. H Coulter, and S. Suda, *Tappi*, **46** (2), 153A (1963).
877. G. H. Tomlinson, G. H. Tomlinson II, J. N. Swartz, H. D. Orloff, and J. H. Robertson, *Pulp Paper Mag. Can.*, **47** (8), 71 (1946).
878. F. W. Hochmuth and W. J. Darmstadt, *Tappi*, **43** (7), 128A (1960).
879. A. Borg, A. Teder, and B. Warnqvist, *Tappi*, **57** (1), 126 (1974).
880. P. O. Karjalainen, J. E. Lofkrantz, and R. D. Christie, *Pulp Paper Mag. Can.*, **73** (12), T–393 (1972).
881. W. A. Moy and G. E. Styan, Can. Patent 958158 (1974).
882. L. M. Roberts, C. E. Beaver, and W. H. Blessing, *Tech. Assoc. Papers*, **31**, 651 (1948).
883. E. J. Malarkey and C. Rudosky, *Pulp Paper Mag. Can.*, **69** (22), T–434 (1969).

884. N. L. Heberer, *Tappi,* **59** (2), 105 (1976).
885. "New Recovery Processes for Black Liquor" in *Proceedings, Gunnar Sundblad Seminar,* Billingehus, Skövde, Sweden, May 1976, Svenska Träforskningsinstitutet, Stockholm, Sweden, September, 1976.
886. I. J. W. Johnson, H. V. Miles, and S. L. McClesky, Jr., in "Chemical Recovery in Alkaline Pulping Processes," R. P. Whitney, Ed., Tappi Monograph No. 32, 1968, Chap. 4, p 100.
887. E. G. Kominek and J. M. Kahn, *Tech. Assoc. Papers,* **29,** 365 (1946).
888. N. E. Crouse and C. E. Stapley, Pulp and Paper Canada, **80,** T–93 (1979). (Convention Issue).
889. J. F. Kuehl in *Proceedings of the Symp. on Recovery of Pulping Chemicals,"* Helsinki, Finland, 1968, p 523.
890. S. Rydin, P. Hagglund, and E. Mattson, *Svensk Papperstidn.,* **80** (2), 54 (1977).
891. W. Christiani, D. C. Gillespie, and R. P. Logan, *Tappi,* **43** (6), 211A (1960).
892. C. H. Knight and C. F. Cornell, *Pulp Paper Mag. Can.,* **64,** T–133 (1963). [Convention Issue.]
893. A. E. Reed and W. F. Gillespie, *Tappi,* **32** (12), 529 (1949).
894. A. Boniface and R. J. Mattison, *Paper Trade J.,* **150** (22), 28 (1966).
895. K. Kinzer in *Proceedings of the Symp. on Recovery of Pulping Chemicals,* Helsinki, Finland, 1968, p 279.
896. L. M. Bingham and P. A. Angevine, *Tappi,* **60** (11), 111 (1977).
897. H. H. Stuart and R. E. Bailey, *Tappi,* **48** (5), 104A (1965).
898. L. M. Bingham, *Tappi,* **52** (1), 59 (1969).
899. J. W. T. Battershill, *Pulp Paper Mag. Can.,* **67,** T–149 (1966). [Convention Issue.]
900. A. F. O'Donnell and P. W. King, *Appita,* **23** (6), 434 (1970).
901. G. B. Hughey, L. K. Herndon, and J. R. Withrow, *Paper Trade J.,* **114** (9), 105 (1942). [Tappi section.]
902. J. A. Murray, H. C. Fisher, and D. W. Sabean, *Paper Trade J.,* **132** (35), 22 (1951).
903. C. W. Rothrock, Jr., *Tappi,* **41** (6), 241A (1958).
904. K. V. Sarkanen, B. F. Hrutfiord, L. N. Johanson, and H. S. Gardner, *Tappi,* **53** (5), 766 (1970).
905. *Proc. Int. Conf. on Atmospheric Emissions from Sulfate Pulping,* E. R. Hendrickson, Ed., Sanibel Island, Fla., U.S. Public Health Service, Natl. Coun. for Stream Improvement, and U. of Florida, 1966.
906. J. Weiner and L. Roth, Institute of Paper Chemistry Bibliography No. 237 "Air Pollution in the Pulp and Paper Industry," Appleton, Wis., 1969; Supplement No. I, 1970.
907. L. Louden and J. Weiner, "Odors and Odor Control," Institute of Paper Chemistry *Bibliography* No. 267, Appleton, Wis., 1976.
908. G. Leonardos, D. Kendall, and N. Barnard, *J. Air Pollution Control Assoc.,* **19** (2), 91 (1969).
909. R. Cederlöf, M. L. Edfors, L. Friberg, and T. Lindvall, *Tappi,* **48** (7), 405 (1965).
910. T. T. C. Shih, B. F. Hrutfiord, K. V. Sarkanen, and L. N. Johanson, *Tappi,* **50** (12), 630 (1967).
911. K. Kringstad, F. deSousa, and U. Westermark, *Svensk Papperstidn.,* **79** (18), 604 (1976).
912. W. T. McKean, B. F. Hrutfiord, K. V. Sarkanen, L. Price, and I. B. Douglass, *Tappi,* **50** (8), 400 (1967).
913. I. B. Douglass, *Proc. Int. Conf. on Atmospheric Emissions from Sulfate Pulping,* in reference 905, p 41.

914. G. G. DeHaas and G. A. Hansen, *Tappi,* **38** (12), 732 (1955).

915. J. L. Morrison, *Tappi,* **52** (12), 2300 (1969).

916. G. Collin, *Appita,* **31** (2), 131 (1977).

917. D. N. Carter and L. Tench, *Pulp Paper Mag. Can.,* **75** (8), T–297 (1974).

918. I. B. Douglass, M. -L. Kee, R. L. Weichman, and L. Price, *Tappi,* **52** (9), 1738 (1969).

919. J. P. Morgan and F. E. Murray, *Pulp Paper Mag. Can.,* **73** (5), T–124 (1972).

920. R. S. Rowbottom and J. G. Wheeler, *Pulp Paper Can.,* **76** (2), T–44 (1975).

921. M. J. Matteson, L. N. Johanson, and J. L. McCarthy, *Tappi,* **50** (2), 86 (1967).

922. I. B. Douglass, *J. Air Pollution Control Assoc.,* **18** (8), 541 (1968).

923. A. Tirado and V. Gonzales, *Tappi,* **52** (5), 853 (1969).

924. E. Thomas, S. Broaddus, and E. W. Ramsdell, *Tappi,* **50** (8), 81A (1967).

925. W. R. C. Barrett and K. L. Kemeys, *Appita,* **30** (4), 334 (1977).

926. A. A. Coleman, *Tappi,* **41** (10), 166A (1958).

927. S. Lindberg, *Svensk Papperstidn.,* **69** (15), 484 (1966).

928. W. F. Beckwith, *Tappi,* **57** (12), 147 (1974).

929. F. E. Murray, *Pulp Paper Mag. Can.,* **69** (1), T–26 (1968).

930. G. N. Thoen, G. G. DeHaas, R. G. Tallent, and A. S. Davis, *Tappi,* **51** (8), 329 (1968).

931. R. D. Christie, *Pulp Paper Mag. Can.,* **73** (10), T–274 (1972).

932. F. E. Murray, *Tappi,* **42** (9), 761 (1959).

933. F. Martin, *Pulp and Paper,* **43** (6), 126 (1969).

934. R. D. Christie and P. R. Stuber, *Pulp Paper Mag. Can.,* **73** (5), T–118 (1972).

935. G. Hawkins, *Paper Trade J.,* **146** (9), 38 (1962).

936. D. H. Padfield, *Tappi,* **56** (1), 83 (1973).

937. R. C. Tobias, G. C. Robertson, D. E. Schabauer, and B. Dickey, *Amer. Paper Ind.,* **53** (10), 36 (1971).

938. J. P. Morgan, D. F. Sheraton, and F. E. Murray, *Pulp Paper Mag. Can.,* **71** (6), T–120 (1970).

939. H. B. H. Cooper, *Chem. Eng. Prog.,* **71** (3), 74 (1975).

940. S. F. Galeano and C. D. Amsden, *Tappi,* **53** (11), 2142 (1970).

941. R. G. Winklepleck and S. Suda, *Tappi,* **56** (6), 73 (1973).

942. T. M. Grace, *Tappi,* **60** (11), 132 (1977).

943. F. E. Murray and H. B. Rayner, *Pulp Paper Mag. Can.,* **69** (5), T–137 (1968).

944. R. E. Chamberlain, J. E. Lofkrantz, R. G. Norris, A. R. Smith, and A. R. Wostrodowski, *Pulp Paper Can.,* **79** (2), T–44 (1978).

945. I. B. Douglass and L. Price,*Tappi,* **51** (10), 465 (1968).

946. D. L. Brink, J. F. Thomas, and K. H. Jones, *Tappi,* **53** (5), 837 (1970).

947. C. J. Lang, G. G. DeHaas, J. V. Gommi, and W. Nelson, *Tappi,* **56** (6), 115 (1973).

948. F. E. Murray and H. B. Rayner, *Tappi,* **48** (10), 588 (1965).

949. J. E. Walther and H. R. Amberg, *Tappi,* **55** (8), 1185 (1972).

950. L. L. Canovali and S. Suda, *Tappi,* **53** (8), 1488 (1970).

951. J. L. Clement and J. S. Eliott, *Pulp Paper Mag. Can.,* **70** (3), T–24 (1969).

952. J. V. Gommi, *Tappi,* **54** (9), 1523 (1971).

953. C. Oloman, F. E. Murray, and J. B. Risk, *Pulp Paper Mag. Can.,* **70** (23), T–499 (1969).

954. S. P. Bhatia and S. Prahacs, *Pulp Paper Mag. Can.,* **73** (10), T–266 (1972).

955. S. P. Bhatia, T. L. C. deSouza, and S. Prahacs, *Tappi,* **56** (12), 164 (1973).

956. R. W. Brown, J. W. Diluzio, D. E. Kuhn, and C. G. Rapp, *Tappi,* **58** (4), 94 (1975).
957. I. Palenius and P. Hiisverta, *Pulp Paper Mag. Can.,* **71** (21), T-465 (1970).
958. H. E. Worster, M. F. Pudek, and R. E. Harrison, *Pulp Paper Mag. Can.,* **72** (12), T-371 (1971).
959. J. -E. Carles, R. Angelier, E. Muratore, and P. Monzie, *ATIP,* **27** (2), 147 (1973).
960. I. Palenius, *Paperi Puu,* **58** (9), 551 (1976).
961. J. J. Renard, D. M. Mackie, H. G. Jones, H. I. Bolker, and D. W. Clayton, *Trans., Tech. Sect., CPPA,* **1** (1), (1975).
962. A Brodén and O. Samuelson, *Paperi Puu,* **57** (9), 607 (1975).
963. J. L. Minor and N. Sanyer, *Tappi,* **58** (3), 116 (1975).
963a. J. S. Fujii and M. A. Hannah, *Tappi,* **61** (8), 37 (1978).
964. H. H. Holton and F. L. Chapman, *Tappi,* **60** (11), 121, (1977).
964a. K. L. Ghosh, V. Venkatesh, and J. S. Gratzl, *Tappi,* **61** (8), 57 (1978).
965. G. H. Tomlinson and G. H. Tomlinson II, Can. Patent 448476 (1948).
966. I. A. Pearl, *Tappi,* **52** (7), 1253 (1969).
967. C. H. Hoyt and D. W. Goheen in "Lignins: Occurrence, Formation, Structure, and Reactions," K. V. Sarkanen and C. H. Ludwig, Eds., Wiley-Interscience, 1971, Chap. 20.
968. J. J. Keilen and A. Pollak, *Ind. Eng. Chem.,* **39** (4), 480 (1947).
969. J. H. Keilen, W. K. Dougherty, and W. R. Cook, *Ind. Eng. Chem.,* **44** (1), 163 (1952).
970. R. A. V. Raff and G. H. Tomlinson II, *Can. J. Res. F.,* **27** (11), 399 (1949).
971. "Sulfate Turpentine Recovery," J. Drew, J. Russell, and H. Bajah, Eds., Pulp Chemicals Assoc., New York, N.Y., 1971.
972. J. M. Conley, E. P. Crowell, D. H. McMahon, and R. G. Barker, *Tappi,* **60** (12), 114 (1977).
973. I. A. Stine and J. B. Doughty, *Forest Prod. J.,* **11** (11), 530 (1961).
974. J. Drew and G. D. Pylent, *Tappi,* **49** (10), 430 (1966).
975. W. L. Thornburg, *Tappi,* **46** (8), 453 (1963).
976. E. L. Springer, M. Benjamin, W. C. Feist, L. L. Zoch, Jr., and G. J. Hajny, *Tappi,* **60** (2), 88 (1977).
977. D. C. Tate, *Tappi,* **50** (4), 110A (1967).
978. H. H. White, *Tappi,* **57** (6), 123 (1974).
979. L. B. Jansson, *Tappi,* **50** (4), 114A (1967).
980. A. A. R. Rönnholm, *Pulp Paper Intl.,* **16** (2), 45 (1974).
981. W. J. Roberts and A. R. Day, *J. Am. Chem. Soc.,* **72** (3), 1226 (1950).
982. A. M. Patureau, *Tappi,* **48** (9), 68A (1965).
983. J. M. Derfer, *Tappi,* **46** (9), 513 (1963).
984. J. M. Derfer, *Tappi,* **49** (10), 117A (1966).
985. W. S. Shelow and A. L. Pickens, Jr., *Tappi,* **41** (5), 148A (1958).
986. M. T. Still, *Tappi,* **46** (2), 127A (1963).
987. K. F. Sylla, *Svensk Papperstidn.,* **67** (3), 75 (1964).
988. S. J. Ryczak, *Tappi,* **46** (2), 129A (1963).
989. J. P. Krumbein, *Tappi,* **47** (5), 142A (1964).
990. L. A. Agnello and E. O. Barnes, *Ind. Eng. Chem.,* **52** (9), 727 (1960).
991. P. Knoer, *South. Pulp Paper Mfr.,* **38** (9), 18 (1975).
992. *Tappi Standard T-689pm-76,* "Analysis of Crude Tall Oil."
993. E. Abrams, *Tappi,* **46** (2), 136A (1963).
994. T. L. Chang, *Anal. Chem.,* **40** (6), 989 (1968).
995. E. P. Crowell and B. B. Burnett, *Tappi,* **49** (7), 327 (1966).
996. R. W. Ellerbe, *Paper Trade J.,* **157** (26), 40 (1973).

997. S. R. Boyette, *Pulp and Paper,* **51** (4), 78 (1977).

998. J. Drew, E. R. Joyce, J. R. Rhyne, and J. B. Cox, *Chem. Eng. Prog.,* **69** (9), 80 (1973).

999. W. M. Hearon, *Chem. Eng. Prog.,* **60** (9), 91 (1964).

1000. H. B. Marshall and C. A. Sankey, U. S. Patent 2686120 (1954).

1001. K. Takubo, *Forest Prod. J.,* **10** (7), 373 (1960).

1002. T. Enkvist, J. Turunen, and T. Ashorn, *Tappi,* **45** (2), 128 (1962).

1003. R. E. Mark, *Tappi,* **60** (8), 31 (1977).

1004. S. Rydholm, *"Pulping Processes,"* Interscience, New York, N. Y., 1965.

1005. R. L. Jones, "Pulp and Paper Manufacture," Vol. 1, McGraw-Hill, New York, N.Y., 1969.

1006. F. W. Herrick and H. L. Hergert, "Recent Advances in Phytochemistry," Vol. II, Plenum, New York, N.Y., 1977.

1007. J. Rauch, "The Kline Guide to the Pulp and Paper Industry," 3rd ed., Charles H. Kline, Fairfield, N.J., 1976.

1007a. L. C. Bratt, *Pulp and Paper* **53** (6), 102 (1979).

1008. "Feasibility Study of Production of Chemical Feedstock from Wood Waste," Pulp and Paper Research Institute of Canada, Pointe Claire, Quebec, Canada, 1975.

1009. I. S. Goldstein, *Science,* **189** (9), 847 (1975).

1010. Katzen, "Chemicals from Wood Waste," AN 441, USDA Forest Products Laboratory, Madison, Wis., 1975.

1011. E. V. Anderson, *Chem. Eng. News,* **56** (31), 8 (1978).

1012. D. E. Campbell, *Pulp and Paper* 51 (14), 138–9 (1977).

1013. K. G. Forss, K. Jokinen, and M. Savolainen, "Pekilo Fermentation of Wood Hydrolyzates," TAPPI Forest Biology/Wood Chemistry Conference, Atlanta, Georgia.

1014. G. M. Barton and B. F. MacDonald, "A New Look at Foliage Chemicals," TAPPI Forest Biology/Wood Conference, Atlanta, Georgia, 1977.

1015. R. E. Dickson and P. R. Larson, "Muka from Populus Leaves: a High-Energy Feed," TAPPI Forest Biology/Wood Conference, Atlanta, Georgia, 1977.

1016. Proceedings Advisory Committee on Pulp and Paper, FAO Rome, Italy, May 1972.

1017. J. E. Atchison, "Fiber Conservation and Utilization" in *Proceedings, Pulp and Paper Seminar,* Chicago Ill., May 1974, p 206.

1018. B. Saiha, 9th Tech. Conf., Budapest, Hungary, October 1975.

1019. J. E. Atchison, *IP&P,* **17** (12), 681 (1963); **18** (2), 159 (1963).

1020. E. C. Lathrop and S. I. Aronovsky, *Tappi,* **32** (4), 145 (1949).

1021. E. C. Lathrop and S. I. Aronovsky, "Pulp and Paper Prospects in Latin America," United Nations, N. Y., 1955, p 298.

1022. E. R. Gremler and J. N. McGovern, *Tappi,* **43** (8), 200 (1960).

1023. "Raw Materials for More Paper," FAO, Rome, Italy, 1952.

1024. D. K. Misra, *Tappi CA Report No. 43,* 119 (1972).

1025. M. Idrees and H. Veeramani, *IP&P,* **31** (6), 17 (1977).

1026. A. Panda, *Pulp Paper Intl.,* p 65 (1964).

1027. J. Salaber and F. Maza, *Tappi CA Report No. 43,* 55 (1972).

1028. H. Veeramani, Ph.D. Thesis, U. of Washington, Seattle, Wash., 1970.

1029. K. Kalyanasundaram, R. Ganesh, N. S. Jaspal, and R. L. Bhargava, *Tappi CA Report No. 43,* 79 (1972)..

1030. H. D. MacMurray and W. E. Bryden, *Proc. Tech. Sect., Brit. Paper and Board Makers Assoc.,* 32 (2), 525 (1951).

1031. P. Lengyel, *Zellst. Papier,* 9 (3), 89 (1960).

1032. H. Frey, *Celuloza Hirtie,* 6 (11), 391 (1957).

1033. V. A. Grabovsky and Ven-Ci Ljan, *Trudy Leningr. Technol. Inst. Celljulozno-Bumazn,* Prom. **12** (164).

1034. L. J. Jacobs, Combustion Engineering Boiler Operations. Analysis of Greek straw pulp liquor (August 22, 1972).

1035. H. Reiche, *Das Papier.* **11** (21), 834 (1967).

1036. R. Madej/*Przeglad Papirniczy,* **15** (6), 168 (1959).

1037. L. Puhakka, "*Höyryvoimalaitostekniikka*", Osa I (1965), 41, s. Insinöövijärj, Koul. Keskous, Julkaisu 22, Helsinki, Finland.

1038. Copeland Process Limited, Montreal, Canada.

1039. L. A. Pradt, *Pulp Paper Intl.,* (Oct. 1969).

1040. J. E. Morgan, *Tappi CA Report No. 52,* 55 (1973).

1041. D. K. Misra, *Tappi CA Report No. 58,* 7 (1975).

1042. D. K. Misra, *Tappi CA Report No. 53,* 19 (1974).

1043. D. K. Misra, *Tappi CA Report No. 40,* 279 (1971).

1044. E. L. Ritman, F. M. Muller, and J. G. Dijkhuls, *Paper Industry,* **29** (12), 1751 (1948).

1045. T. Jeyasingam, *Tappi CA Report No. 53,* 7 (1974).

1046. S. A. Rydholm, "Pulping Processes," Interscience, New York, N.Y., (1965).

1047. S. I. Aronovsky, "Annual Crop Fibers" in "Pulp and Paper Manufacture," Vol. II, McGraw-Hill, New York, N.Y.

1048. D. K. Misra, *Tappi CA Report No. 52,* 7 (1973).

1049. O. Paez, E. Trauwitz, and J. Varela, *Tappi CA Report No. 52,* 91 (1973).

1050. Y. Fouad, *Tappi CA Report No. 52,* 37 (1973).

1051. S. I. Aronovsky, A. J. Ernst, H. M. Sutcliffe, and G. H. Nelson. *Tech. Assoc. Papers,* **31,** 299 (1948).

1052. O. Constantinescu and C. Savineanu, *Celuloza Hirtie,* **14** (6), 219 (1965).

1053. D. K. Misra, *Tappi CA Report No. 40,* 51 (1971).

1054. J. Rutkowski and W. Surewicz, *Przeglad Papirniezy* 28 (6), 193 (1972).

1055. G. Viola, *Paper Trade J.,* 74 (1966).

1056. R. Pouchelle, "A New Process for the Production of Corrugating Medium, Termed the H. F. Method (Højbygaard Papir Fabrikker Process)," *ATIP Bull.,* **18** (4), 184 (1964).

1057. M. N. Batheja, *Ippta,* 307 (1964). [Souvenir no.]

1058. H. Ibrahim, *Tappi CA Report No. 58,* 101 (1975).

1059. T. Macdonald, *Pulp Paper Intl.,* (Sept. 1963).

1060. J. Salaber and F. Maza, *Tappi CA Report No. 40,* 89 (1971).

1061. E. C. Lathrop and S. I. Aronovsky, *Tappi,* **37** (23), 24A (1954).

1062. S. B. Knapp, R. A. Watt, and J. D. Wethern, *Tappi,* **40** (8), 595 (1957).

1063. J. E. Atchison, *Tappi CA Report No. 40,* 1 (1971).

1064. J. E. Atchison, *Tappi CA Report No. 34,* 21 (1970).

1065. H. A. Captein, S. B. Knapp, R. A. Watt, J. D. Wethern, *Tappi,* **40** (8), 620 (1957).

1066. G. Viola, *Paper Trade J.,* **149** (39), 40 (1965).

1067. E. J. Villavicencio, *Pulp Paper Mag. Can.,* **70** (2), 46 (1969).

1068. D. S. Cusi, *Pulp and Paper Intl.,* **1** (3), 42 (1959).

1069. E. C. Lathrop, *Tappi,* **40** (10), 788 (1957).

1070. D. F. J. Lynch and M. J. Goss, *Ind. Eng. Chem.,* **24** (11), 1249 (1932).

1071. D. D. Hinrichs, S. B. Knapp, J. H. Milliken, and J. D. Wethern, *Tappi,* **40** (8), 626 (1957).

1072. J. Salaber, *Tappi CA Report No. 53,* 41 (1974).

1073. J. W. Emerson, *Pulp and Paper,* 72 (Feb. 1972).

1074. Y. C. Tsou, *Conference on Pulp and Paper Development in Asia and the Far East,* Tokyo, Japan, October 1960.

1075. E. Reichmann, O. Constantinescu, and P. S. Fischgold, Rumanian Patent 51779 (1969).
1076. Y. A. Fouad, presented at UNIDO Expert Group Meeting on Pulp and Paper, Vienna, Austria, September, 1971.
1077. S. R. Krishnaswamy, presented at UNIDO Expert Group Meeting on Pulp and Paper, Vienna, Austria, September, 1971.
1078. P. W. R. Jolley and D. S. Cusi, presented at UNIDO Expert Group Meeting on Pulp and Paper, Vienna, Austria, September, 1971.
1079. E. J. Villavicencio, presented at UNIDO Export Group Meeting on Pulp and Paper, Vienna, Austria, September, 1971.
1080. "Man-made Forests and Role of Industry," West Coast Paper Mills, Dandeli (Karnataka State), India, 1977.
1081. H. J. Nieschlag, *Tappi,* **43** (12), 993 (1960).
1082. F. N. Tamoland, E. O. Mabesa, M. A. Eusebio, M. J. Sagrado, and B. A. Lomibao, *Tappi,* **40** (8), 671 (1957).
1083. T. F. Clark, "Annual Crop Fibers and the Bamboos" in "Pulp and Paper Manufacture," Vol. II, McGraw-Hill, New York, N.Y., 1969. p. 33.
1084. M. P. Bhargava and C. Singh, *Utilization (New Series), Indian For. Bull.,* No. 127, 1942, Dehra Dun For. Res. Inst. (1947).
1085. W. Raitt, *"The Digestion of Grasses and Bamboos for Papermaking,"* Crosby, Lockwood and Son, London, England, 1931.
1086. R. L. Bhargava and M. B. Jauhari, *Tappi CA Report No. 40,* 209 (1971).
1087. S. C. Chen and S. J. Lin, *Co-operative Bull. Taiwan Forestry Res. Inst.,* No. 21, **18** (1973).
1088. J. A. Semana, J. O. Escolano, and M. R. Mosalud, *Tappi,* **50** (8), 416 (1967).
1089. J. N. McGovern, *Tappi,* **50** (11), 63A (1967).
1090. S. -C. Chao and T. T. Pan, *Co-operative Bull. Taiwan Forestry Res. Inst.,* No. 19, **10** (1972).
1091. H. L. Crosby, Dorr Oliver, Inc., "Report on Bamboo Pulp Bleaching at Seshasayee Paper and Boards Mill in India," (1963).
1092. M. H. Jauhari, N. S. Jaspal, R. L. Bhargava, *Tappi CA Report No. 53,* 63 (1974).
1093. V. P. Gohel and L. Thoria, *IP&P,* **11** (1), 33 (1956).
1094. D. K. Misra, private communications from mill executives engaged in producing dissolving grade pulps, (1961).
1095. M. K. Gupta, *Tappi CA Report No. 53,* 55 (1974).
1096. A. Wiedermann, *Tappi CA Report No. 40,* 309 (1971).
1097. N. Lenyuk and Yu. N. Nepenin, *Mater. Nauch. -Tekh. Konf. Khim. - Tekhnol. Fak., Leningrad Lesotekh. Akad.,* 3–9 (1967).
1098. E. Reichmann, O. Constantinescu, I. Anton, and T. Burova, *Celuloza Hirtie,* **10** (7, 8) 277 (1961).
1099. I. A. Chivu, *Proc. Regional Symposium, Pulp and Paper Research Middle East and N. Africa,* Beirout 1963, p 227.
1100. J. N. McGovern, *Paper Trade J.* **155** (35), 27 (1971).
1101. V. S. Smith, *Paper Ind.,* **17** (5), 315 (1935).
1102. J. N. McGovern and J. Grant, *Tappi,* **45** (5) 343 (1962).
1103. D. C. Tapadar, *J. Proc. Inst. Chemists, India.,* **35**, Pt. 2, 63 (1963).
1104. F. M. Gomez, Republic of the Philippines Patent 2699 (1966); Patent 2812 (1966); Patent 3661 (1967), and Patent 3921 (1967).
1105. R. J. Kulik, *Paper Maker* (London), **161** (2) 42 (1971).
1106. William T. Heyse, *Tappi CA Report No. 52,* 33 (1973).
1107. G. L. Bidwell, U. S. Patent 1531728 (1925).
1108. W. H. Rapson, Ed., *The Bleaching of Pulp,* Tappi Monograph No. 27, 1963, p 236.

1109. *Paper,* **187** (10), 528 (1977).
1110. B. Biswas, *IP&P,* **14** (2) 117 (1959).
1111. J. B. Calkin and G. S. Witham, *"Modern Pulp and Paper Making."* Reinhold, New York, N.Y., 1957, p 40.
1112. H. J. Nieschlag, G. H. Nelson, and I. A. Wolff, *Tappi,* **44** (7) 515 (1961).
1113. T. Jeyasingham, *Paper Trade J.,* **158** (46), 28 (1974).
1114. G. A. White, D. G. Cummins, E. L. Whitely, W. T. Fike, J. K. Greig, J. A. Martin, G. B. Killinger, J. J. Higgins, and T. F. Clark, "Cultural and Harvesting Methods for Kenaf," Production Research Service, USDA, U.S. Government Printing Office, April 1970.
1115. G. A. White, *Crops Soil Mag.,* **21,** 13 (1968).
1116. G. A. White, *Tappi,* **52** (4), 656 (1969).
1117. M. O. Bagby, T. F. Clark, W. C. Adamson, and G. A. White, *Tappi CA Report No. 58,* 69 (1975).
1118. T. F. Clark, M. O. Bagby, R. L. Cunningham, G. F. Touzinsky, and W. H. Tallent, *Tappi CA Report No. 40,* 267 (1971).
1119. G. H. Nelson and J. A. Leming, *Tappi,* **40** (10), 846 (1957).
1120. M. O. Bagby, G. H. Nelson, E. G. Helman, and T. F. Clark, *Tappi,* **54** (11), 1876 (1971).
1121. M. O. Bagby, R. L. Cunningham, and T. F. Clark, *Tappi,* **58,** 121 (1975).
1122. T. F. Clark, R. L. Cunningham, and I. A. Wolff, *Tappi,* **54,** 63 (1971).
1123. M. A. Abou-State, T. Zimaity, and A. M. El-Masry, *J. Appl Chem. Biotechnol.,* **23** (7) 511 (1973).
1124. D. C. Tapadar, *Ippta,* 137 (1964). [Souvenir no.]
1125. W. J. Bublitz, *Tappi CA Report No. 40,* 163 (1971).
1126. S. R. D. Guha, R. Pant, and B. G. Karira. *Indian Forester,* **99,** (12), 717 (1973).
1127. R. V. Bhat, *Tappi,* **35** (4) 183 (1952).
1128. S. R. D. Guha, M. M. Singh, K. Kumar, and A. G. Jadhav, *Indian Forester,* **100** (10), 627 (1974).
1129. R. V. Bhat and S. C. Asthana, *Indian Forester,* **80** (7), 409 (1954).
1130. S. B. Sutfin and H. J. Bettendorf, *Fiber Container,* **43** (8), 99 (1958).
1131. J. J. Forsythe, *Tappi,* **59** (11), 27 (1976).
1132. *PSFF#77,* Paper Stock Institute of America, New York, N. Y.,
1133. *Paper Trade J.,* **143** (12), 30 (1959).
1134. J. Vernage, *Tappi,* **41** (12), 211A (1958).
1135. P. Seifert, R. Secor, and P. Cruea, *Tappi,* **59** (6), 111 (1976).
1136. Black Clawson Co., Middletown, Oh.
1137. Voith Morden Co., Portland, Oreg.
1138. Beloit Jones Corp., Pittsfield, Mass.
1139. Escher Wyss Company, Nashua, N.H.
1140. J. J. Higgens, *Tappi,* **42** (1), 132A (1959).
1141. Fibre Conservation Corp., Chicago, Ill.
1142. *Fibre Container,* **43** (12), 112 (1958).
1143. W. Sooy, *Tappi,* **35** (8) 28A (1952).
1144. J. P. Weidner, *Tappi,* **35** (8), 20A (1952).
1145. C. A. Shubert, *Paper Trade J.,* **93** (13), 137 (1931).
1146. H. A. Morrison, *Paper Trade J.,* **94** (18), 239 (1932).
1147. H. O. Ware and R. T. Washburn, *Paper Trade J.,* **122** (26), 283 (1946).
1148. J. J. O'Conner, *Tech. Assoc. Papers,* **10,** 121 (1927).
1149. W. A. Kirkpatrick and J. A. Wise, *Paper Trade J.,* **134** (9), 26 (1952).
1150. M. R. Kinne, *Tappi,* **39** (8), 168A (1956).
1151. R. Kumler, *Paper Ind.,* **39** (11), 916 (1958).
1152. S. W. McMyler, *Tappi,* **59** (11), 25 1976.

1153. R. W. Hynes, *Tappi*, **35** (8), 102A (1952).
1154. F. C. Clark, *Paper Trade J.*, **116** (13), 147 (1943).
1155. H. P. Bailey and A. H. Nadelman, *Paper Trade J.*, **125** (2), 13 (1947).
1156. P. Kajanne and V. Ostring, *Paper & Timber* (Finnish) **40** (4A), 201 (1958).
1157. R. C. Merrill, *Tappi*, **32** (11), 520 (1949).
1158. F. H. Snyder and S. F. M. Maclaren, *Paper Trade J.*, **98** (17), 216 (1934).
1159. S. D. Wells, *Paper Mill News*, **67** (20), 12 (1944).
1160. A. O. Bragg, *Paper Trade J.*, **112** (17), 207 (1941).
1161. B. D. Sparks and I. E. Puddington, *Tappi*, **59** (11), 117 (1976).
1162. R. C. Merrill, *Tappi*, **32** (11), 520 (1949).
1163. E. W. Turley, *Paper Ind.*, **31** (3), 340 (1949).
1164. H. P. Bailey, A. H. Nadelman, and E. F. Andrews, *Tech. Assoc. Papers*, **31**, 265 (1948).
1165. J. W. Morrow, *Tappi*, **38** (1), 132A (1955).
1166. J. D. Lohnas, *Tappi*, **38** (1), 134A (1955).
1167. A. O. Bragg, *Paper Trade J.*, **112** (17), 207 (1941).
1168. F. C. Clark, *Paper Trade J.*, **116** (13), 147 (1943).
1168a. F. Sundman and N. Venho, *Tappi*, **61** (8), 73 (1978).
1169. H. A. Morrison, *Paper Ind.*, **33** (10), 1180 (1952).
1170. S. D. Wells, *Paper Mill News*, **67** (20), 12 (1944).
1171. R. P. Cruea, *Tappi*, **61** (6), 27 (1978).
1171a. R. G. Horacek, *Tappi* **62**(7), 39 (1979).
1172. R. O'Donoghue, *Paper Ind.*, **22** (9), 911 (1940).
1173. J. J. Forsythe, *Tappi*, **39** (10), 719 (1956).
1174. H. P. Bailey, J. J. Forsythe, and E. F. Andrews, *Tappi*, **37** (1), 14 (1954).
1175. H. O. Ware and R. T. Washburn, *Paper Trade J.*, **122** (26), 283 (1946).
1176. W. G. Hendrich, *Paper Trade J.*, **120** (17), 163 (1945).
1177. L. A. Moss, "The Bleaching of Pulp," Tappi Monograph No. 10, 1953, Chap. 10.
1178. A. Felton, *Tappi*, **59** (1), 112 (1976).
1179. F. J. T. Harris, *Proc. Tech. Sect.*, British Paper Board Makers Assoc. **36** (3), 639 (1955).
1180. K. Ward, Jr., M. L. Murray, A. J. Morak, and M. H. Voelker, *Tappi*, **50** (9), 462 (1967).
1181. K. Ward, Jr., A. J. Morak, and M. H. Voelker, *Tappi*, **52** (3), 479 (1969).
1182. E. J. Dobo and K. A. Kobe, *Tappi*, **40** (7), 573 (1957).
1183. K. Ward, M. H. Voelker, and D. J. MacLaurin, *Tappi*, **48** (11), 657 (1965).

5 BLEACHING

V. LORÅS

Research Director
The Norwegian Pulp and Paper Research Institute
Oslo, Norway

History of Bleaching

The first bleaching agent was probably sunlight. The yellow color of textiles was removed by fading in the sun. The word bleach comes from the Anglo-Saxon word blaecan. The meaning of this word is "to fade." The first known commercial bleaching was carried out in The Netherlands in the seventeenth century. The Dutch received linen from surrounding countries and bleached it on the grassy meadows in the summer. The climate in The Netherlands was considered to be especially suited for this bleaching. Until World War II bleaching of linen on the snow in the spring was common practice in Scandinavia. This bleaching method would probably not have been very useful for the bleaching of wood pulp because sunlight contains considerable ultraviolet radiation.

At the end of the eighteenth century chlorine and hypochlorite became available and the commercial sun bleaching of textiles was soon replaced by bleaching with hypochlorite. Calcium hypochlorite, called bleaching powder, became the most useful bleaching compound because it could be prepared and transported as a powder. For a long time calcium hypochlorite was the only bleaching agent for rag and other fibers used for papermaking. The next change in bleaching came about 1930 when elementary chlorine was commercially applied as a bleaching agent for pulp. Multistage bleaching of sulfite pulp, using chlorination, alkaline extraction, and hypochlorite expanded rapidly. The bleaching of kraft pulp by this method came after it was discovered that hot alkali extraction dissolved the products formed by chlorination and that several bleaching stages would bleach kraft pulp to a high brightness. The effectiveness of chlorine as an agent for lignin-removal in pulp had been known for a long time, but there had been no suitable equipment for carrying out the chlorination process on an industrial scale. The breakthrough for chlorination came with the development of stainless steel, which could resist the attack of chlorine. The next major advance in pulp bleaching, the introduction of chlorine dioxide, was also closely connected with the development of more corrosion-resistant construction materials. Chlorine dioxide had been known as a bleaching agent for pulp from about 1920, but the first commercial application was not until 1946. Besides the problems with equipment materials there also were problems arising from the

633

fact that chlorine dioxide is highly toxic and explosive when the concentration in the gas phase exceeds 15%.

Bisulfite was used in the nineteenth century for the brightening of mechanical pulp. Dithionite (hydrosulfite) was introduced later for the brightening of mechanical pulp and also to give a slight brightening to kraft pulp. Peroxide, as sodium peroxide or hydrogen peroxide, was introduced as a bleaching agent for pulp about 1940.

Oxygen has been well known for many years as a strong oxidizing agent. Separation of oxygen from air is also relatively cheap in large-scale plants. It was logical, therefore, that oxygen should be tried as a pulping and bleaching agent, but the first really promising results with oxygen bleaching were not obtained until about 1955. The first commercial installation for oxygen bleaching was made in 1970, and there are a significant number of other plants in operation or under construction.

Bleaching of pulp is in a changing state. The effluents from bleaching plants are among the most troublesome ones in the pulp and paper industry and nobody can at the present time predict what the final solution will be. There is general agreement that we shall see great changes in the next decade. New bleaching agents will probably be introduced. Ozone appears to be on the threshold of industrial application and chlorine monoxide is a potential bleaching agent. New methods for the application of the conventional bleaching agents are being introduced and even newer methods can be expected in the future.

Reasons for Bleaching

The principal reason for bleaching is to obtain a brighter pulp. This is achieved in the case of chemical pulps primarily through the removal of lignin. Removal of other undesirable compounds can be an additional aim of bleaching. In the case of sulfite pulp the removal of resins, fatty acids, fatty acid esters, and other extractives may be desirable. For some grades of papermaking pulps and for pulps to be used in the manufacture of absorbent products, removal of a portion of the hemicellulose is beneficial. For dissolving pulps, removal of the hemicellulose is very important in order to obtain a pulp with the required properties. One could believe that the easiest method of bleaching would be to brighten the colored compounds in the pulp rather than removing them. The fact is that bleaching of pulp to a high brightness without removal of other substances is very difficult.

The removal of lignin in the bleaching process can be considered as a continuation of the cooking process. Usually it is not possible to remove the lignin in the cooking process to the extent that the brightness is satisfactory. The yield as well as the strength properties of chemical pulps would be reduced excessively if all the lignin were removed in the cooking process and it therefore becomes important to remove the remaining lignin during the bleaching process. This is especially true for kraft pulps where the lignin has become so dark during the cooking process that the last traces of lignin must be removed during bleaching in order to obtain a high-brightness pulp.

Optical Changes in the Bleaching Process

The bleaching of pulp can be considered as a removal of light-absorbing sub-stances from the pulp. In terms of the Kubelka-Munk theory[1,2,3,4] the light-absorption coefficient is reduced, whereas the light-scattering coefficient is generally changed very little. The relationship among the reflectance factor, the light-absorption coefficient, and the light-scattering coefficient, according to Kubelka and Munk , is as follows:

$$\frac{k}{s} = \frac{(1 - R_\infty)^2}{2R_\infty}$$

where R_∞ = reflectance factor with sheets of the same pulp as backing
k = light-absorption coefficient
s = light-scattering coefficient

In this formula R_∞ is used as a decimal fraction instead of the customary usage as percent. The dimension of k and s in the SI system is m^2/kg; before the SI system was adopted the dimension of these coefficients was cm^2/g so that the SI system values are one-tenth of those calculated by the older system.

The color of unbleached pulp is yellow or brown. This means that the absorp-tion of light is greatest in that part of the visible spectrum where yellow or brown have their complementary color, which is the blue part of the spectrum. Therefore, the reflectance of blue light will be lower than that of light in the other parts of the visible spectrum as shown in Figure 5-1. The purpose of bleaching is to remove color and it, therefore, is logical to measure the effect of the bleaching in that part of the spectrum where the changes are most pro-nounced. Brightness, or more correctly papermaker's brightness, is based on this concept. The reflectance of light in a specified part of the spectrum is measured and given as percentage of the reflectance of a standard under the same illumi-nating and viewing conditions. Figure 5-2 shows the section of the visible spectrum used for the determination of brightness and the curve for spectral sensitivity to light for a normal observer according to the CIE system [Com-mission International de l'Eclairage (International Commission on Illumination)]. It should be pointed out that neither the brightness nor the reflectance of light, weighted according to the visible sensitivity, give an unambiguous measure for the impression of whiteness. Visual whiteness is subjective and so far there is no general agreement on whiteness formulas based on physical measurements.

The original brightness scale was the TAPPI or GE brightness.[5] The optical geometry for the TAPPI method is 45° for the illuminating light and normal viewing. The SCAN-brightness is based on diffuse illumination and normal obser-vation[6] and the ISO brightness, which was introduced in 1977, is also based on this optical geometry.[7] TAPPI and SCAN brightness use magnesium oxide as the reference standard; a standard sample prepared from the smoke of burning mag-nesium is given a reflectance value of 100% for all parts of the visible spectrum.

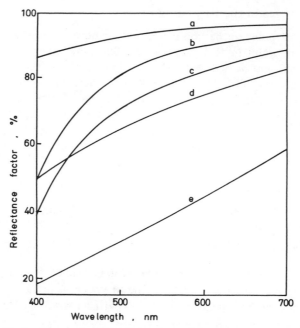

Figure 5-1. Reflectance of visible light for different pulps. (a) Bleached chemical pulp. (b) Peroxide-bleached groundwood pulp. (c) Unbleached groundwood pulp. (d) Unbleached sulfite pulp. (e) Unbleached sulfate pulp.

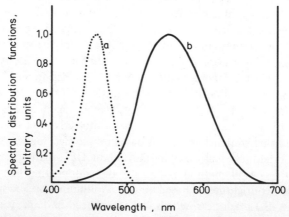

Figure 5-2. Relative spectral distribution for evaluating brightness and opacity. (a) Brightness. (b) Opacity.

Because of the difference in optical geometry the SCAN method gives slightly higher values than the TAPPI method for chemical pulps. Handsheets used for the determination of TAPPI brightness are dried at room temperature, whereas handsheets for SCAN brightness are usually dried at 60°C. The ISO brightness uses an absolute reflecting diffuser as 100% reflectance value. This means that values obtained by the ISO method are lower than values obtained by the TAPPI or the SCAN method. The relationship between ISO brightness and SCAN brightness is as follows:[8]

$$\text{ISO brightness} = 0.974 \cdot \text{SCAN brightness} + 0.1$$

In determining ISO brightness, handsheets are dried at room temperature.

The result of bleaching is usually measured by the brightness obtained. Sometimes when a measure for the efficiency of the bleaching is desired, the difference in brightness between the bleached and the unbleached pulp—the brightness gain—is used as a measure. Another and more scientific method is to calculate the decrease in light absorption caused by the bleaching. In order to calculate the light-absorption and the light-scattering coefficients the following must be done: translucent sheets with good formation must be prepared, the basis weight of the sheet must be measured, and two independent reflectance measurements or one reflectance and one transmittance measurement must be carried out. Usually two reflectance measurements are used, one with a black backing and the other with sheets of the same pulp as backing or with a white backing having a known reflectance factor. Another method for characterizing bleaching efficiency is to calculate the decoloration number (DC)[9] as follows:

$$\text{DC} = (k/s) \text{ unbleached} - (k/s) \text{ bleached}$$

The calculated figures will be proportional to the decrease in the light-absorption coefficient if the light-scattering coefficient remains unchanged in the bleaching process. This method should not be used for comparison of pulps that differ considerably in their light-scattering properties. The brightening obtained from bleaching can also be expressed as the *relative* decoloration number (DC) using the following formula:[9]

$$[\text{DC}] = \frac{(k/s) \text{ unbleached} - (k/s) \text{ bleached}}{(k/s) \text{ unbleached}}$$

If s is assumed to be constant this formula can be written:

$$[\text{DC}] = \frac{k \text{ unbleached} - k \text{ bleached}}{k \text{ unbleached}}$$

This means that the relative decoloration number is independent of the light-scattering coefficient and inversely proportional to the light-absorption coefficient of the unbleached pulp. The relative DC obtained in this way is more closely related to the brightness-gain than is the decoloration number itself. Figure 5-3 shows examples of the relationship between the initial brightness, brightness gain, and decrease in light-absorption coefficient at two different levels of light-scattering coefficient.

Another change in optical properties caused by bleaching is a lowering of the opacity of paper made from bleached pulp. The opacity of paper is determined by the light-absorption coefficient, the light-scattering coefficient, and the basis weight of the paper. The opacity increases with increasing values of all three of these properties. Because the light-absorption coefficient decreases with bleaching and the light-scattering coefficient remains practically unchanged, bleached pulp will have a lower opacity than unbleached pulp at the same basis weight. Sometimes figures for opacity of pulp are given in the literature without specifying the basis weight of the sheets used for the measurement. Such figures have no value because opaque sheets can be obtained from all pulps if the sheets are thick enough.

Light-scattering coefficients and opacity are usually measured at a relative spectral distribution comparable to the spectral sensitivity of a normal human eye. The maximum for these curves is in the green to yellow part of the spectrum. The light-scattering coefficient is almost independent of wavelength in the visible spectrum, whereas the light-absorption coefficient may change very much. In evaluating the effect of bleaching the light-absorption coefficient should be measured in the same wavelength range used for measuring brightness.

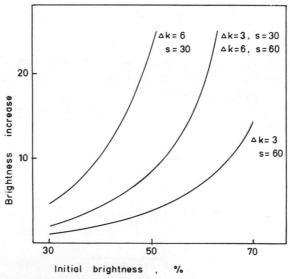

Figure 5-3. Relationship between brightness of unbleached pulp, light-scattering coefficient, decrease in light-absorption coefficient, and brightness gain.

The light-absorption coefficient in the opacity range is also reduced by bleaching, but the decrease is not as large as in the brightness range. The Kubelka-Munk theory is not valid if the light-absorption coefficient is very high or very low.[9a]

BLEACHING OF MECHANICAL PULPS

The bleaching of mechanical pulps is greatly affected by the properties of the wood from which the pulp is produced. Some of the most important properties of the wood are discussed in the following sections. These properties apply for the most part to the bleaching of stone groundwood pulps and the bleaching of disk refiner pulps. The bleaching of both types of pulps are discussed.

Brightness of Wood

The color of wood species differs considerably.[10] Ebony is a very dark wood and aspen is an example of a very light-colored wood. Spruce, which is the main raw material for the production of stone groundwood, is also a bright species, but the brightness of the wood can vary considerably within the same tree.[11] Similar variations are also found for Loblolly pine (*Pinus taeda*).[12] The difference in color between springwood and summerwood of spruce is easily noticeable. The light-scattering coefficient is lower in summerwood than in springwood[11] and the light absorption is higher. Thus the color difference between summerwood and springwood is the result of a higher light absorption and a lower light scattering in the summerwood.

A significant negative correlation exists between the color and brightness of spruce and its lignin content.[11,12] In the case of Douglas fir a good correlation was found between the absorbance at 400 nm and at 280 nm.[13,14] If the absorbance at 280 nm is taken as a measure of lignin content, then the yellow color corresponding to the absorbance at 400 nm must be closely correlated to the lignin content. Heartwood had a higher absorbance at 400 nm than the sapwood for the same absorbance at 280 nm. The main components of the lignin structure are fairly well known.[15,16] Most of these structures do not have a significant absorption of visible light. One group contributing to the color of lignin is the coniferylaldehyde end group.[17,18] Other possible chromophores are the o- and p-quinones.[19,20] The o-quinone structure, which in Björkman lignin from spruce is about 0.7%, has been estimated to be responsible for 35 to 60% of the color in lignin.[21]

The carbohydrate components of wood—cellulose and hemicellulose—are almost colorless and their contribution to the color of wood is negligible.[22] Many of the compounds in the extractives have a yellow color.[23,24] The light-absorption coefficient was found to be about 0.1 m^2/kg for a 1 percent solution of extractives in dichloromethane.[22] Measurements made on shavings of spruce before and after extraction indicated a light-absorption coefficient of 0.2 m^2/kg from the extractives in spruce.[25] In the case of Loblolly pine, extraction of the sapwood with ethanol-benzene had no influence on the light-absorption coefficient.[12]

The main contributer to color in the brighter wood species seems to be lignin. Table 5-1 shows values for light-absorption and light-scattering coefficients determined for wood from spruce, fir and pine.[11,12,22,26,27]

Changes in Wood Color during Storage

The color of wood tends to change during storage. The usual change is to a deeper yellow or to a brown color, but if the wood is attacked by fungi the color may change to a blue or red. The method of log storage influences the brightness of the wood. Storage of unbarked spruce logs in air will result in a lower brightness in both the unbleached and bleached pulps than will storage in the debarked condition[28,29]. The effects of storage on the brightness of wood are shown in Figure 5-4. Some of the color developed during storage can be removed by bleaching, but the final brightness will always suffer. Storage of unbarked logs in water has been considered to be the best way of preserving the brightness of the wood. For unbleached groundwood pulp this method gives the best brightness, but if the pulp is to be bleached with peroxide, water storage of unbarked logs may not be the best method. If unbarked spruce logs are stored in water through the summer, the outer part of the wood will be stained with compounds from the bark. Some of these compounds have a structure similar to the pinosylvine in pine heartwood and cause difficulties in sulfite pulping.[30] Moreover, wood that has been bark-stained will turn green-to-black in alkaline medium. Because peroxide bleaching is carried out under alkaline conditions, the dark color formed will influence the bleaching.[29] Storage of the logs in water after debarking would be a better method when maximum brightness gain by peroxide bleaching is the

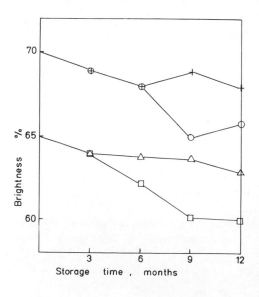

Figure 5-4. Change in pulp brightness by storage of wood. (□ unbarked, unbleached; △ debarked, unbleached; ○ unbarked, bleached; + debarked, bleached.)[28]

TABLE 5-1 OPTICAL PROPERTIES OF WOOD

Wood Species	Brightness (%)	Light-Scattering Coefficient (m²/kg)	Light-Absorption Coefficient (m²/kg)	Reference
Spruce, *Picea abies*, earlywood	45.3–51.7	9.4–14.0	2.8–4.1	11
Spruce, *Picea abies*, latewood	29.1–41.4	6.6–11.5	3.1–9.5	11
Spruce, *Picea abies*, radial section	–	20.7	4.6	22
Spruce, *Picea abies*, tangential section	–	13.7	3.1	22
Black spruce, *Picea mariana*	67.5	25.0	3.2	26
Black spruce, *Picea mariana*	46.2	18.8	6.0	27
Balsam fir, *Abies balsamea*	49.8	25.0	6.2	27
Jack pine, *Pinus banksiana*	46.3	21.6	6.7	27
Loblolly pine, *Pinus taeda*	49.5	20.2	5.3	12

aim. The brightness of Loblolly-pine samples decreases by about 5 units after 9-mo storage, mainly as a result of an increase in light-absorption.[12]

The reactions that cause the discoloration of wood during storage are not completely understood. It is quite certain that microorganisms play an important role; oxidative reactions from the oxygen in the atmosphere are also involved. Brown-rot fungi can demethylate lignin and form catechol structures,[31,32] which in turn can be oxidized to quinones, thereby producing an increase in color. The p-hydroxybenzyl aryl ethers in lignin can rearrange to o-hydroxydiphenyl-methanes, which can be dehydrogenated by enzymes or oxidizing agents to give stable o-quinonemethides that may be responsible for part of the color in pulps.[33] Metal ions in the wood can also influence the color. Iron has an especially strong influence both on the color of wood and the color of pulp.

The raw material used for refiner mechanical pulping and for thermomechanical pulping is usually inferior to the logs that are used in the production of stone groundwood. A large part of the chips used for refiner and thermomechanical pulping comes from sawmill slabs, and this raw material contains more bark than drum-debarked logs. Handling and transporting of chips may also introduce impurities like dust and sand. A considerable part of this can be removed in chip washers, but some of it will remain in the chips and be mixed into the pulp. Even if the bleachability of the pulp is not affected by these impurities, the brightness of both the unbleached and the bleached pulp will suffer.

It is well known that chips discolor rapidly when stored for long periods and that the dark color will influence the brightness of the pulp. How much of this discoloration remains after bleaching is not known, but it can be assumed that the discoloration caused by chip storage will influence the brightness of the bleached pulp in an unfavorable way.

Relation of Wood Brightness to Pulp Brightness

Green spruce springwood appears very bright to the eye, but if a brightness reading is made, it will be lower than that obtained on stone groundwood pulp made from the wood.[11,26] The reason is that the light scattering in wood is low compared to the light scattering of stone groundwood.[11,22,26] The increase in light scattering obtained by grinding is a result of the formation of small fiber particles or fines that increase the total surface area in the sheet made from the pulp.

The light-absorption coefficient of stone groundwood pulp is usually higher than that of the wood from which it is made.[11,12,26,27] One cause is the high temperature in the grinding zone, which produces a discoloration and even a charring of the wood under unfavorable grinding conditions. Another factor is that knots, which are considerably darker than the rest of the wood, are ground to a fine powder, and thereby have a maximum effect on the light absorption. The logs also contain bark and other impurities that have a major effect on the light absorption. Determination of optical properties of wood has usually been carried out on green wood, but the wood used for the production of groundwood pulp will normally have been stored for some time, resulting in an increase in the light-absorption coefficient. It is important to remember that the Kubelka-Munk theory may give quite large errors when it is applied to wood without corrections.[27]

Brecht and Meltzer[34] found a good correlation between pulp brightness and the brightness of the wood used. Schweizer[35] reported that spruce wood attacked by fungi gave a pulp with low brightness compared with pulp from fresh wood.

The operating conditions used in the disk refining process can affect the brightness to an appreciable degree. If the temperature in the fiberizing zone is high, as it is likely to be in pressurized refining in thermomechanical pulping, the brightness of the wood used. Schweizer[35] reported that spruce wood attacked initial brightness as well as the bleachability of the pulp by charring of the wood material and by adding to the iron content of the pulp. Pressurized refining may result in a loss of about 2 units of brightness compared to open-discharge refining.[36] By increasing the presteaming temperature in thermomechanical pulping from 115 to 135°C, the brightness of the bleached pulp can be reduced by about 2 units.[37]

Processes Used in Bleaching of Mechanical Pulp

There are two principal methods used for the bleaching of mechanical pulps: reductive and oxidative bleaching. These two methods can be used singly or combined in a two-stage bleaching with oxidative bleaching in the first stage and reductive in the second stage or vice versa. The reductive bleaching agents are bisulfite, dithionite, and borohydride. The oxidative bleaching agents are peroxide, hypochlorite, peracetic acid, and ozone.

The first bleaching chemical used for the brightening of stone groundwood pulp was sulphur dioxide or bisulfite. It was used as far back as the nineteenth century. The effectiveness of bisulfite on stone groundwood is slight, but bisulfite brightening gained new interest after refiner pulping was introduced because these pulps can be increased in brightness somewhat by adding bisulfite in the pulping process. The next reductive agent used was dithionite, which is a stronger reducing chemical and has a greater bleaching action than bisulfite. Borohydride is a very strong reducing agent, but its brightening effect is no better than dithionite. Trivalent uranium is a strong reducing agent that has a significant bleaching effect.

Peroxide is the best known of the oxidative bleaching agents and almost the only oxidative agent used today. Hypochlorite has been used to some extent, but chlorine compounds usually discolor mechanical pulp unless large amounts of active chlorine are used. Peracetic acid and ozone have so far been used only in the laboratory or in the pilot plant.

Reductive Bleaching of Mechanical Pulps

Mill-scale reductive bleaching is usually carried out with dithionite, but bisulfite or a mixture of sulfite and bisulfite are sometimes used in order to obtain a slight brightening of the pulp.

Bisulfite Bleaching

Bleaching of stone groundwood with bisulfite after grinding has been used.[38] The addition of bisulfite during grinding will improve the brightness about 2 units, but this procedure is seldom used. Far more interesting is the use of bisulfite for the brightening of disk refiner pulps. The brightness of these pulps is usually lower than that of stone groundwood due partly to their lower light-scattering coefficient, especially for thermomechanical pulps. The reduction of light absorption brought about by the addition of bisulfite is sufficient to increase the brightness of disk refiner pulps by 2 to 4 units.[39] The bisulfite can be applied in two ways in the disk refiner and thermomechanical processes. The simplest method is to add the chemical in solution at the refiner either with the chips or with the water used for control of the consistency. The other method involves pretreatment of the chips prior to the refining and/or presteaming, but this procedure is more a part of the pulping process than the bleaching process.

The temperature in refining can be 100°C or slightly above while the pulp is between the refiner plates and even higher with presteaming and pressurized refining. The consistency is normally 20 to 25% and sometimes as high as 40%. Because the retention time in the refiner is very short, leaving only a short period for the bleaching chemical to perform its function, there may be an additional effect from the sulfite after the pulp has left the refiner.

Dithionite Bleaching

Dithionite bleaching is the reductive bleaching method that is most used. Sodium dithionite ($Na_2S_2O_4$) is the preferred form. Zinc dithionite, which has been (and still is) used to some extent, is more stable than the sodium form and can be produced at the mill site, but it has the drawback of being toxic to fish. Calcium dithionite has given good bleaching results[40] but has not been used commercially. Aluminium dithionite has a fairly good bleaching effect but its production is inefficient.[41]

Bleaching Conditions with Dithionite. Dithionite bleaching is usually carried out at low consistency. Maximum bleach response is obtained at about 4% consistency.[42] Dithionite is easily oxidized by air, and at high consistencies the pulp will usually contain enough air so that a significant amount of the bleaching agent will be lost through air oxidation. At very low pulp concentrations, the water will contain sufficient oxygen to oxidize partly the reducing agent. Figure 5-5 shows the effect of air on dithionite bleaching.[43] Tyminski[44] obtained 3

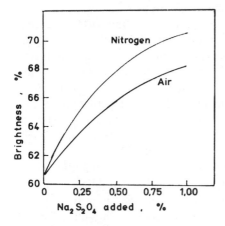

Figure 5-5. Effect of air on dithionite bleaching of groundwood pulp.[43]

units increase in brightness when nitrogen gas was used to purge the pulp suspension of air. The bleaching response is considerably influenced by temperature. Low temperature is unfavorable and the probable reason is that the pulp slurry contains too much oxygen. Maximum brightness is usually obtained at 70°C. Figure 5-6 shows the influence of temperature on bleaching of stone groundwood pulp with sodium dithionite.[44] Good bleaching results can be obtained at high temperature if no oxygen is present. Such conditions will exist when mechanical pulps are produced in disk refiners because the atmosphere between the disks is almost saturated with steam at 100°C.

Dithionic acid is a rather weak acid and the sodium salt produces an alkaline

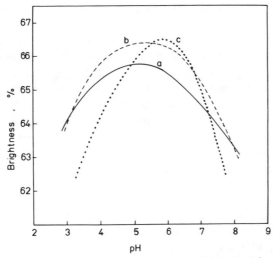

Figure 5-6. Influence of pH and temperature on dithionite bleaching of groundwood pulp at (a) 50°C, (b) 60°C, and (c) 80°C.[44]

solution. The pH decreases rapidly when the dithionite is mixed with pulp due to the acidic groups formed as a result of the bleaching reaction and the reaction between the dithionite and the oxygen in the pulp.[45] Without buffering or addition of alkali, the pH will be 5 or lower at the end of the bleaching. Figure 5-6 also shows the effect of pH on brightness in buffered systems of sodium dithionite. Maximum brightness is obtained between pH 5 and 6. The optimum pH for zinc dithionite bleaching is about one pH unit lower than that for sodium dithionite bleaching.[46] The reaction rate of dithionite bleaching is high and most of the brightening is finished after 10 to 15 minutes.[42] After this initial bleaching period there is a slow increase in brightness for at least 2 hr. The bleaching in this stage is probably caused by decomposition products of dithionite. Treatment of the pulp with SO_2 after dithionite bleaching gives an additional brightening and this brightening is almost independent of the bleaching time.[42] The slow postbleaching effect is therefore probably not caused by the sulfite formed during the dithionite bleaching.

A fairly good bleaching response can be obtained with small amounts of dithionite, but the bleaching seems to be somewhat unstable with low dosages. Normally from 0.5 to 1.0% of dithionite is used, and dosages above 1% give only a very slight additional brightening.[42,47]

Influence of Metals in Dithionite Bleaching. Wood contains many elements; in pine, for example, as many as 27 trace elements have been detected.[42,48-50] The amount of metal components varies with the wood species and some of the metals seem to be firmly bound in the wood. In spruce, manganese is the most abundant transition metal. The content of iron is somewhat lower, but the iron content in pulp can be high because of contamination from the process water and equipment.[43] In groundwood, the concentration of metal ions, such as iron and copper, may differ considerably from that in the wood due to contamination during processing.[42] Ferric iron is yellow and has an obviously bad effect on the brightness of the pulp. However, both ferrous and ferric iron have a detrimental effect on dithionite bleaching.[51,52] Manganese has almost no effect on dithionite bleaching. Copper, nickel, zinc, lead, and chromium are usually present in such low concentrations that they have no effect. Aluminum has an effect similar to that of iron, but not as great.[51]

The effect of metal ions on dithionite bleaching can be reduced by addition of sequestering agents. The most effective sequestering agents are the salts of aminocarboxylic acids, such as di- or tetra-sodium ethylenediamine tetracetate (EDTA) and pentasodium diethylenetriamine pentacetate (DTPA). Other sequestering agents are nitrilotriacetic acid (NTA), sodium tripolyphosphate (STP), and citrates. Phosphates and citrates have a buffering as well as a sequestering effect and it been shown[52] at least in the case of the acetate that an improved bleaching response can be obtained from the buffering action equal to that obtained with sequestering agents. EDTA and DTPA are stable compounds and their influence on the ecology has been questioned. Sodium tripolyphosphate is a nutrient for botanical life and may cause eutrophication. Addition of sequestering agents to the dithionite bleach will normally increase the brightness by 2 to 4 units.[43,53] The effect is less for zinc dithionite bleaching than for sodium dithionite bleaching, especially if the aminocarboxylic acids are used since these form complexes with zinc that reduce their effectiveness.[52] Aluminium ions act similarly to zinc and if large amounts of white water from the paper mill are used for dilution, the aluminium can have a deleterious effect in dithionite bleaching. The sequestering agent must be present during bleaching in order to obtain maximum brightness gain. Removal of metal ions from the pulp before bleaching does not always improve the bleachability.[52]

Effect of Wood Species in Dithionite Bleaching. Rapson and coworkers[42] investigated the response of dithionite bleaching on groundwood from nine Canadian wood species and found that the brightness level as well as brightness gain varied considerably with wood species. Groundwood made from the sapwood of Jack pine (*Pinus banksina*) and Western hemlock (*Tsuga heterophylla*) and bleached with zinc dithionite was found to have a slightly higher brightness than groundwood made from the heartwood.[54] Groundwood from pine is very difficult to bleach to a high brightness and special precautions must be taken in grinding and bleaching of such pulp.[55]

Groundwood from eucalyptus is also especially difficult to bleach. Groundwood from old eucalyptus trees having a high concentration of tannins must be extracted with alkali or hot water before bleaching in order to obtain a reasonable increase in brightness. The use of sequestering agents in normal dosages will have no effect on the bleaching of this wood because the tannin has a strong tendency to react with the metal ions.[56] If the trees are cut at the right time of the year and not stored for a long time, fairly good bleaching results can be obtained.[57] Storage of logs usually decreases the brightness of the unbleached as well as the bleached pulp,[28,29] but the brightness gain obtained with most species by dithionite bleaching is usually not reduced by storage. However, with aspen, storage and defects of the wood may influence the bleaching response considerably.[58]

Influence of Defiberizing Process in Dithionite Bleaching. Refiner and thermomechanical pulps have a lower light-scattering coefficient than stone groundwood. From a theoretical point of view the brightness gain with these pulps should be greater than with groundwood because pulp of low light-scattering coefficient should increase more in brightness for the same reduction in light-absorption coefficient. In practice, refiner pulps bleached with dithionite show a brightness-gain that is as high or higher than that for stone groundwood.[59] Refiner pulp from thinnings of *Pinus radiata* shows a good response to dithionite bleaching.[60] However, thermomechanical pulp generally shows a lower brightness-gain than normal stone groundwood,[37] probably because of changes in the character and amount of colored substances produced during the pretreatment of the chips with heat.

Methods of Dithionite Bleaching. Dithionite bleaching of stone groundwood can be carried out in a stock chest or in a bleaching tower. When the bleaching is carried out in a chest, bleached and unbleached pulp will be mixed to some extent and this will result in a lower brightness compared to bleaching in a tower. Also air oxidation of the dithionite is more likely to take place in a stock chest than in a tower.[61] The consistency in tower bleaching with dithionite is from 2 to 5%; upflow towers are used. Thorough and rapid mixing of pulp and bleaching chemicals is very important in order to obtain maximum bleaching response. The pulp should be as free from air as possible before addition of the chemicals and the mixing should be carried out without introducing air into the pulp. The usual method is to add chemicals on the suction side of the stock pump of the bleaching tower, using a level control to avoid suction of air into the system.[43] The bleached pulp is removed at the top of the tower.

Dithionite bleaching solution can be prepared by metering the dry powder into water immediately before it is mixed with the pulp. Sequestering agents can be added to the pulp or to the bleaching solution. The dithionite solution can also be prepared in batches, but the solution should not be kept more than one day and alkali should be added for stabilizing the dithionite. If zinc dithionite is produced on site from zinc dust and sulfur dioxide or if sodium dithionite is made on site from sodium borohydride, sodium hydroxide, and sulfur dioxide, the chemicals will be in solution and can be metered directly. A good control system is necessary in order to keep the production of bleaching chemicals at the appropriate rate for the bleach requirement without using large storage tanks for the dithionite solution. If the bleaching is carried out in a stock chest, the dithionite is sometimes added as a dry powder into the chest. This method usually gives a poor bleach response because the mixing of pulp and chemicals will be too slow and air entrainment cannot be avoided.

If white water is recycled for dilution of the pulp before the bleaching, the pH will be lowered. In such cases sodium hydroxide should be added to keep the pH at the optimum level for bleaching. Dithionite solutions are corrosive, and the white water from bleached pulps can attack paper-machine wires and other equipment in the mill. All pumps and piping should be constructed of acid-proof material, preferably stainless steel. Wooden towers have proved to be satisfactory for dithionite bleaching; stainless steel and brick-lined concrete towers are also well suited.

In the case of refiner and thermomechanical pulps, dithionite bleaching can be carried out in the refiner. The bleaching chemicals should be added as close as possible to the refiner disks in order to avoid decomposition before the dithionite can act on the fiberized wood. The atmosphere between the refiner disks will have a low-oxygen content because of the high steam pressure in the refining zone. At the high consistency and the high temperature generally used in refining, the bleaching is very rapid and almost finished when the pulp leaves the refiner. One drawback with this method is that the bleached pulp must pass through the screening system, where it may lose some of the brightness gained in the bleaching. This loss in brightness may be caused by air oxidation of some of the decolorized compounds back to their original chromophoric structure. Oxidation of ferrous iron to ferric is also a possible cause for the color reversion.

Use of Other Reductive Bleaching Agents

Sodium borohydride is a strong reducing agent, but the bleaching effect is surprisingly low compared with dithionite.[62] The borohydride salt forms an alkaline solution, but this alkalinity is not sufficient to keep the pH above the critical range for self-decomposition of borohydride during bleaching. Somewhat better results are obtained with higher alkalinity and high consistency.[63] The presence of a sequestering agent (DTPA) and sodium silicate also seems to improve the bleaching. Sodium borohydride is too expensive to be used, as such, for bleaching of pulp but can be used for on-site production of sodium dithionite.

Thioglycolic acid has a bleaching effect comparable to dithionite.[64] However, the chemical cost is higher than for dithionite and the bleached pulp has an unpleasant smell. Trivalent uranium is a powerful reducing chemical, and in research work it has produced very bright pulps.[65] In addition to its high price, uranium salt is toxic and the bleaching must be carried out in rather strong acid medium. Therefore, this reagent has significance only for scientific work. Hydrogen has been proposed as a bleaching agent for mechanical pulp. The reaction involved is probably the addition of hydrogen to double bonds. The brightness increase is small but some stabilization of brightness can be obtained.

Reactions in Reductive Bleaching

In reductive bleaching, dithionite is oxidized to sulfite according to the following reaction:

$$S_2O_4{}^{2-} + H_2O = 2HSO_3^- + 2H^+ + 2e$$

This equation indicates that the reduction should be more efficient with increasing pH and the maximum effect should be obtained at a pH of about 9.[66] However, as mentioned before, the optimum pH for bleaching is between 5 and 6.[44] The reason for this discrepency may be that the chromophoric structures react more easily under acid conditions[44] or that new chromphoric groups are formed more rapidly at high pH than they are removed by dithionite. According to the above equation, the reducing reaction should proceed more easily at low than at high concentrations of sulfite. However, addition of sulfite seems to improve the bleaching result to the extent that some of the dithionite can be replaced with sulfite.[67] Several other reactions unquestionably take place, in addition to the formation of sulfite, as a result of the bleaching reaction. Reaction with the oxygen in the air probably produces bisulfite, sulfate, and hydrogen ions. Thiosulfate, sulfide, and polythionates are formed as a result of decomposition and this is accentuated by the increase in acidity of the solution.[46] Thiosulfate in acid medium is very corrosive and contributes to the corrosive effect of spent bleach liquor from dithionite bleaching.

Very little is known about the chromophoric groups that are reduced in dithionite bleaching. One effect may be addition of bisulfite to carbonyl groups. Reduction of orthoquinones and coniferylaldehyde groups is also a possibility. However, the light absorption of coniferyladehyde at about 350 nm seems to be only slightly influenced by dithionite, whereas sodium borohydride causes a considerable reduction of the light absorption in this part of the ultraviolet spectrum.[68,69] Model quinoid structures are easily reduced by dithionite, but condensed quinones react substantially more slowly.[70] Colored compounds of the flavon type will be reduced by dithionite,[71] but such compounds have not been detected in spruce wood. Refiner and thermomechanical pulps produced from sawmill by-products usually contain some bark and the colored compounds in spruce bark are only bleached slightly by dithionite.[37]

Fleury and Rapson[65,72] studied the absorption of visible and ultraviolet light by stone groundwood and model compounds by treatment with dithionite, borohydride, divalent chromium, and trivalent uranium. The results were interpreted as follows: dithionite and borohydride may reduce quinones and quinone methides, but the main part of the alpha-carbonyls remain; trivalent uranium reduces both quinone compounds and alpha-carbonyl structures and, therefore, gives a better brightness-gain than dithionite and borohydride. A study of the reaction of trivalent uranium on lignin-model compounds showed that the reaction with colored compounds, such as phenol-alpha-carbonyl, was very rapid, whereas colorless beta-carbonyl was not attacked.[73] The changes in ultraviolet, visible, and infrared spectra of stone groundwood by treatment with reducing agents were studied by Polcin and Rapson.[74] The results indicate that the wood chromophores differ with wood species and that different reducing agents act in different ways. Dithionite mainly attacks quinoid and alpha-beta-unsaturated carbonyls, while borohydride and trivalent uranium can also reduce ring-conjugated carbonyls.

If trivalent iron is present in the pulp, some of the brightening obtained with reducing agents can be caused by reduction of the iron to the ferrous stage.[70]

Oxidative Bleaching of Mechanical Pulps

Oxidative bleaching of mechanical pulp was introduced later than reductive bleaching, but oxidative bleaching has certain advantages over dithionite bleaching. The brightness gain obtained with oxidative bleaching can be much higher than with dithionite bleaching and the corrosion problems are less for peroxide than for dithionite bleached pulp. Several oxidizing agents have been tried for bleaching of mechanical pulp, but peroxide is the only one used in industrial bleaching.

Peroxide Bleaching of Mechanical Pulps

There is no fundamental difference in bleaching with sodium peroxide or hydrogen peroxide. With sodium peroxide the pH will be too high for good bleaching and sulfuric acid is, therefore, added to lower the pH. If hydrogen peroxide is used, the pH will be lower than required for a good reaction rate so an alkali has to be added. Another way of adjusting the pH is to mix sodium peroxide and hydrogen peroxide to obtain optimum alkalinity, which should be a starting pH between 10 and 11.

Additional Chemicals Used with Peroxide. A simple mixture of hydrogen peroxide and alkali will not give the best bleaching results. One or more additional chemicals usually are added in order to improve the result. The most common chemical used is sodium silicate; others include magnesium sulfate and sequestering agents, such as DTPA and phosphates.

Sodium silicate has a buffering action in the pH range where peroxide is most active as a bleaching agent. Another effect of sodium silicate is to inactivate metal ions. Magnesium sulfate has a similar inactivation effect and is especially useful when the hardness of the water is low. Sodium silicate and magnesium seem to act synergistically, but the mechanism for this is not known. Calcium has an effect similar to that of magnesium. Rather large quantities of sodium silicate are used in peroxide bleaching of mechanical pulp. Usually a 40% solution of sodium silicate is used and about 5% of this is added based on pulp weight (2.0% silicate on pulp weight). Magnesium sulfate (Epsom salt) is used in small quantities, generally 0.05 to 0.2% based on the pulp.

Influence of Metals on Peroxide Bleaching. Metal ions, such as copper, iron, manganese, and nickel are catalysts that cause decomposition of hydrogen peroxide.[75] The origin of these metals can be the original wood, [42,48,49] the water,[49,76] the bleaching compounds, contaminations of the wood during transport, or the corrosion of equipment used for producing the pulp.[49,76] Manganese has the most serious effect on peroxide stability. Copper and iron show some catalytic effect, but zinc reduces the rate of decomposition.[49] Aluminium ions in the pulp can also influence the peroxide bleaching in an unfavorable way if the concentration is high.[77] If white water from the paper mill is used in the production of mechanical pulp, the content of aluminium in the pulp can be high enough to affect peroxide bleaching.

Sequestering or removal of metal ions is very important in obtaining maximum bleach response. The usual method is to pretreat the pulp before bleaching with a chelating agent. Pentasodium diethylenetriamine pentacetate (DTPA) and trisodium 2-hydroxyethyl-ethylenediamine triacetate (HEDTA) have proved to be the best agents for improving the bleach response.[78] However EDTA, sodium tripolyphosphate, chlorine, hypochlorite, and calcium chloride[76,79] also have a beneficial effect. Manganese is easy to remove by pretreatment with chelating agents, but some of the iron is firmly held in the pulp and cannot be removed by a single pretreatment with chelating agents.[76] The effectiveness of pretreatment before bleaching is dependent on removal of metal ions together with white water by thickening of the pulp.[80] As environmental restrictions make it necessary to reuse more white water, the effect of pretreatment with chelating agents will diminish. For DTPA, which is comparatively stable under the conditions of peroxide bleaching, some improvement can be obtained by addition during bleaching, but it has been shown that addition of DTPA without dewatering can have a deleterious effect on the bleaching.[77]

Bleaching Conditions with Peroxide. Peroxide bleaching of mechanical pulp is carried out in alkaline medium. Some of the alkali will be supplied by the sodium silicate, but additional alkali must be used if hydrogen peroxide is the bleaching agent. Optimum bleaching results, for a given dosage of peroxide, are dependent upon alkalinity, consistency, temperature, and retention time. These factors have been studied for the peroxide bleaching of aspen groundwood.[81] Taking all factors into consideration a considerable amount of work would be required to find the optimum conditions for each separate pulp. However, for practical work the conditions have become relatively well fixed.

Consistency is an important factor and the brightness gain will increase with increasing consistency,[82] as shown in Figure 5-7. Some of this effect can be the result of the removal of harmful metal ions in the dewatering process.[77] Peroxide bleaching is usually carried out at consistencies between 10 and 20%. Temperature, retention time, and alkalinity are important variables after the consistency is fixed. There is a general rule that some peroxide should be left when the bleaching is terminated. The optimum residual peroxide is a function of the dosage of peroxide, consistency, total alkali, temperature and bleaching time. Usually one-tenth of the added peroxide should be left when the bleaching is stopped. Temperatures for peroxide bleaching are normally between 40 and 60°C. Optimum retention time depends on the dosage of peroxide but is usually between 1 to 3 hr.

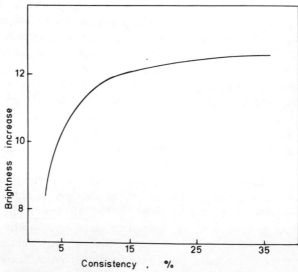

Figure 5-7. Brightness gain versus consistency in peroxide bleaching.[82]

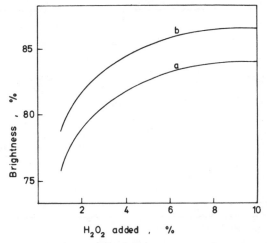

Figure 5–8. Brightness gain versus charge of hydrogen peroxide in bleaching of groundwood. (a) Single stage H_2O_2. (b) Two stages; H_2O_2 plus $Na_2S_2O_4$.[63]

The brightness gain with peroxide is much more dependent on the dosage of the bleaching agent than it is in dithionite bleaching. The increase in brightness obtained with increasing amounts of peroxide for spruce groundwood is shown in Figure 5–8. The brightness gain above a dosage of 3% is marginal, but there is a slight increase up to 10%.[63] The maximum brightness obtainable is about 85% SCAN. Peroxide bleaching of poplar groundwood has given brightness values of about 86% GE, but the dosage of peroxide is very high.[42]

Effect of Wood Quality in Peroxide Bleaching. Spruce, aspen, and poplar pulps are comparatively easy to bleach, but pulps from pine and some fir species can be very difficult to bleach to high-brightness values.[42] One reason is that the heartwood is much darker in pine than in spruce, and the brightness for the unbleached pine pulp will, therefore, be rather low, especially in the case of stone groundwood, where the whole log must be used. In addition to this, the heartwood of pine contains compounds that affect the peroxide bleaching. Jack pine groundwood that was bleached with peroxide—with and without pretreatment with chelating agents and with and without extraction with acetone—showed that the extractives influence the bleaching and that the combination of extractives and metal ions is especially unfavorable.[83]

Groundwood that was prepared separately from heartwood and sapwood of jack pine and Western hemlock showed that the bleachability was much poorer for the heartwood pulp than for the sapwood pulp.[70] The difference in bleachability between sapwood and heartwood of Western hemlock was less noticeable than for jack pine. Pulp containing heartwood of Scotch pine also has poor bleachability[84] as does the heartwood of Douglas fir.[85] The difficulty in bleach-

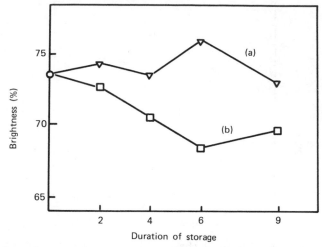

Figure 5-9. Influence of log storage on brightness of peroxide bleached ground-wood pulp. (a) Stored debarked in sea water. (b) Stored unbarked in sea water.

ing heartwood increases with the age of the tree; trees below 40 yr of age give pulps that are easier to bleach than pulp from older trees.[85]

The effect of wood storage on the bleachability of mechanical pulp has been discussed. Storage will usually reduce the brightness of the wood and thereby the brightness of the unbleached pulp. Even if some of the color formed during storage is removed by bleaching, the final brightness will be lower than for the corresponding pulp from green wood.[28] Figure 5-9 shows the influence of bark-staining on the bleachability of spruce (*Picea abies*) groundwood.[29]

Influence of Defiberizing Process in Peroxide Bleaching. As mentioned in the section on dithionite bleaching, the lower light-scattering coefficient of refiner and thermomechanical pulps compared to stone groundwood should tend to give a higher increase in brightness for the same reduction in light-absorption coefficient. For spruce thermomechanical pulp such a tendency has been found,[37] but this may not be the general rule. In some cases, thermomechanical pulp has shown a poor bleachability with peroxide.

The difference in bleachability of thermomechanical pulps may be caused partly by the wood species used. It has been shown that high pretreatment temperature reduces the bleachability,[37] and this effect may be more or less pronounced, depending on the wood used. Chips that contain bark produce a low-brightness unbleached pulp[37,86] that is difficult to bring to a high brightness by bleaching. Iron contamination from the disks in refiner and thermomechanical pulping can also influence the bleachability, and this effect can be serious if the refining is carried out in such a way that metal-to-metal contact occurs between the disks. If the wood material becomes charred as a result of adverse refining conditions, the pulp will be very difficult to bleach.

Methods of Peroxide Bleaching.   Bleaching of mechanical pulp with peroxide is usually carried out in a tower that may be upflow, downflow, or a combination of upflow and downflow. The most common type is downflow[43,61] because the pulp consistency must be high in order to achieve a good utilization of the bleaching agent. In a downflow tower, the holding time can be controlled by adjusting the pulp level in the tower. A press or a filtering device is used to thicken the pulp before the bleaching. If chelating agent is used, it is added before the thickening, and the main part of the complexed metals will be removed by the thickening. After thickening, the pulp can be heated with steam to the required temperature, which usually is between 40 and 60°C.

The chemicals are metered by pump as individual solutions and are mixed together just before addition to the pulp.[61] In high-consistency bleaching, special mixers are normally used for mixing the bleaching agent with the pulp, but the mixing can also be carried out in a high-consistency transport pump. The consistency in bleaching is dependent on the dewatering equipment and the equipment used for mixing the pulp and bleach. Consistencies between 10 and 20% are used and 15% is about average. As mentioned before, the brightness gain increases with consistency, but the additional increase in the brightness obtained must be balanced against the additional cost of equipment required for the higher consistency. The retention time is usually from 1.5 to 3 hr. At the bottom of the tower the pulp is diluted with water to a consistency between 3 to 4% before it is discharged from the tower. As the pulp is alkaline, the pH of the slurry must be reduced in order to avoid color reversion. Continuous measurements of pH and automatic control to a preset value is common practice in modern mills. Usually the pH is lowered to about 5.5, which is the same as for the unbleached pulp. If sulfur dioxide or bisulfite is used for lowering the pH, the residual peroxide will also be consumed. For acidification without reduction of peroxide, sulfuric acid and alum can be used. When large dosages of peroxide are used, a considerable amount of peroxide remains in the pulp after bleaching. Part of this peroxide can be recovered and used for prebleaching at the grinder.[87] In order to use the residual peroxide for bleaching, sodium silicate and sodium hydroxide must be added to the recirculated liquor.[88] Another method of peroxide bleaching is to add peroxide in excess and bleach at a consistency of 4 to 5%. After a short retention time the consistency of the pulp is increased by pressing and the white water returned to the process.[89] A reduction in peroxide consumption can be obtained in this way.

A disk refiner is a very efficient mixer, and bleaching in the refiner was proposed in the early stages of development of refiner pulping.[90] If the bleaching is carried out in the pulping stage, the temperature will be high and the bleaching time short. So far, this method has not been used to any extent, but results obtained in pilot plants are promising.[80,86] Bleaching in a postrefining stage should also be possible for refiner pulp as well as for stone groundwood. The temperature would be somewhat lower than in the pulping stage and a longer time would be needed before the bleaching had reached its optimum.[91]

Other Oxidative Bleaching Agents for Mechanical Pulps

Persulfates and perborates are well-known bleaching agents, but for bleaching of pulp they do not have any advantage over hydrogen and sodium peroxide, and they would be more costly to use. Organic peroxide compounds act as bleaching agents, but they are more expensive than hydrogen peroxide. Peracetic acid is a fairly good bleaching agent. Its use in laboratory investigations has been reported[92-95] and brightness values above 80 have been obtained.[92] Peracetic acid can also act as a delignifying agent[96-98] and as a strength-improving agent.[95] Bleaching with organic peroxides prepared by oxidation with air has been proposed.[99] Acetone peroxides have also been tried.[100] The bleaching effect with organic peroxides is comparable to that with hydrogen peroxide bleaching, but the cost is higher.

Ozone is a very strong oxidizing agent. However, its bleaching effect is very slight on softwood pulps. The brightness may actually be decreased by ozone treatment due to a reduction in the light-scattering coefficient resulting from better bonding. Hardwood mechanical pulps can be bleached with ozone.[101-102] If peroxide and sodium silicate are added before the ozone treatment, a brightening effect may be obtained.[103] Hypochlorite has a bleaching effect on hardwood pulps,[104] but softwood mechanical pulps are discolored by hypochlorite. Pretreatment with hypochlorite has a beneficial effect on peroxide bleaching if the pulp has been attacked by fungi.[105]

Reactions in Peroxide Bleaching of Mechanical Pulps

The active component in alkaline bleaching with peroxide is the hydrogen peroxyanion, HO_2^-, which is formed in the first stage in the dissociation of the acid H_2O_2 as follows:

$$H_2O_2 \rightleftharpoons HO_2^- + H^+$$

The dissociation constant, K_A, for this acid is $3.55 \cdot 10^{-12}$ at $35°C$.

$$K_A = \frac{(H^+)(HO_2^-)}{(H_2O_2)}$$

At the pH where peroxide bleaching is carried out, only a few percent of the hydrogen peroxide is present as the hydrogen peroxyanion. Peroxide is unstable at the pH used for pulp bleaching and some of the peroxide decomposes in the bleaching process. Catalytic decomposition of hydrogen peroxide produces hydroxyl and hydroperoxyl radicals, but these are intermediate compounds of short duration.

The peroxide oxidation is not selective, and some of the oxidizing agent will therefore be used for reactions with nonchromophoric compounds in the pulp. Jones[106] investigated the consumption of peroxide in the bleaching of spruce groundwood and concluded that the extractives consumed 1 to 4%, the carbohydrates nearly 60%, and the lignin about 40%. It is generally believed that only a small part of the peroxide consumed is used for decoloration of chromophoric structures. The bleaching of groundwood with peroxide is very probably a decoloration of p- and o-quinones and coniferylaldehyde structures, although other unknown reactions are undoubtedly involved. Reeves and Pearl[107] studied the reaction on lignin-model substances. For coniferylaldehyde a possible reaction mechanism is shown in Figure 5–10. The conjugated system is destroyed in the reaction and the color thereby reduced. The aldehyde groups formed are subsequently oxidized to carboxyl groups in the alkaline medium. Oxidation of quinone with peroxide has been studied by Corbett.[108] The initial reaction was found to be an epoxidation of a double bond similar to the reaction with coniferylaldehyde. This intermediate product was then further oxidized to low-molecular mono- and di-basic acids.

Dence and coworkers[109] made an extensive investigation of the reactions of alkaline peroxide with softwood and hardwood lignin-model compounds, with spruce groundwood, and with groundwood lignin. An intermediate product, 4-methyl-o-quinone, was formed. Oxidation of alpha-methylvanillyl alcohol probably produces a corresponding o-quinone and this alcohol is also converted to methoxy-p-benzoquinone by displacement of the side chain and oxidation of the corresponding hydroquinone. Further degradation results in a mixture of maleic, malonic, oxalic, oxalacetic, and methoxy- or hydroxy-succinic acids. Demethylation also takes place. Spruce groundwood and spruce milled-groundwood lignin also gave the same acids under similar treatment with alkaline

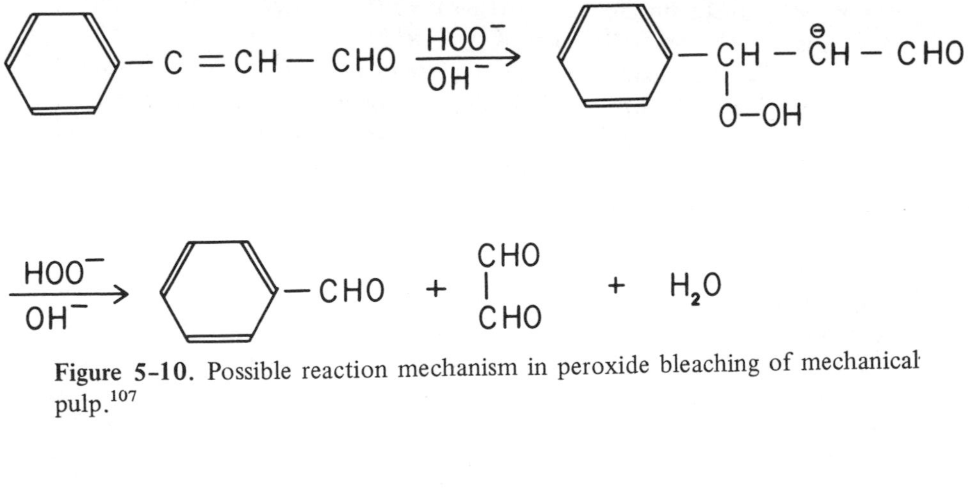

Figure 5–10. Possible reaction mechanism in peroxide bleaching of mechanical pulp.[107]

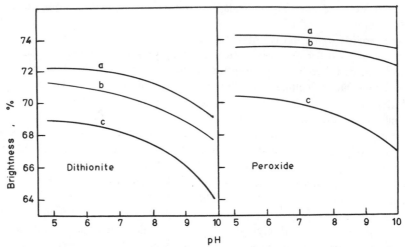

Figure 5-11. Influence of pH and temperature on discoloration of groundwood · pulp at (a) 20°C, (b) 50°C, and (c) 85°C.

hydrogen peroxide. The results indicate that a considerable amount of the peroxide is consumed in fragmentation of unetherified phenolic units.

The brightness gain obtained in the peroxide bleaching of groundwood can partly be ascribed to elimination of chromophoric groups on the lignin side chain. Removal of *o*- and *p*-quinoid structures in the lignin may play a considerable role in the removal of color,[21,109] but new *o*- and *p*-quinones can be formed under peroxide bleaching conditions.[109] Spittler and Dence[110] treated milled-wood lignin obtained from a mixture of spruce and balsam groundwood, using methylation, catalytic hydrogenation, and reduction with borohydride or diborane. All of these treatments resulted in a reduction of structures absorbing at 350 nm. Peroxide treatment reduced the absorptivity still further, except for the lignin treated with diborane. The results substantiated the role of carbonyl groups as contributors to the color of high-yield pulps.

The model compounds that are sometimes used for hardwood lignin are 4-methylsyringol and alpha-methylsyringyl alcohol.[111] It has been shown that methylsyringol is degraded to acetone, carbon dioxide, and low-molecular-weight dicarboxylic acids by alkaline peroxide treatment and that *o*-quinone is produced as an intermediate product. Alpha-methylsyringyl alcohol yielded 2, 6-dimethoxy-*p*-benzo-quinone, acetic acid, aldehydes, carbon dioxide, and low-molecular-weight dicarboxylic acids.

In the attack on the phenolic structure, the free phenol probably plays an important role. Improved bleaching is obtained if the free phenol is tied up before the bleaching.[83] Studies[112] of changes in ultraviolet and infrared spectra indicate that attack on the benzene ring has a limtied effect on color removal and that the decoloration is caused mainly by reactions with the aldehydes and with the conjugated double bonds in side chains.

During peroxide bleaching new colored compounds are formed as a result of side reactions. Discoloration of mechanical pulp in alkali is well known and this discoloration increases with temperature and pH, as Figure 5–11 shows. The competing reactions that occur in the peroxide bleaching process—decoloration on the one hand and the formation of new chromophoric groups on the other hand—are the reason why the bleaching process must be stopped before all of the peroxide is consumed. From experience it is known that the necessary residual peroxide increases with the size of the dosage.

Bleaching of Mechanical Pulp in Two or More Stages

Chemical pulps are usually bleached in more than one stage and a multistage process seems logical for mechanical pulps if high-brightness values are required. Results from laboratory and mill bleaching in two stages are reported in the literature.[42,43,54,61,63,83,113–117]

Oxidative bleaching with peroxide followed by reductive bleaching with dithionite is the usual method for multistage bleaching of mechanical pulps. Because peroxide bleaching is terminated before all the peroxide is consumed, the residual peroxide should be destroyed before the dithionite is added. An adjustment of the pH of the pulp must also be made before the dithionite is added. Sulfur dioxide and sulfite liquor are suitable for both purposes. As already mentioned, new colored structures are formed in peroxide bleaching. Some of these chromophores can be reduced with dithionite and this will give an additional brightness gain of 2 to 4 units. The opposite bleaching sequence, using dithionite in the first stage and peroxide in the second stage, has also been tried. However, the results are inferior to those obtained with peroxide in the first stage.[43,63,113,118] The reason is that some of the chromophoric groups reduced by the dithionite are subsequently oxidized with peroxide, thereby offsetting part of the benefits of the reductive bleaching.

Two-stage bleaching with borohydride and peroxide will produce high-brightness values.[63] Borohydride and peroxide can also be used together in a single-stage bleaching; it is rather surprising but these two chemicals do not react with each other under normal bleaching conditions. The effect is not as good as that obtained in two-stage bleaching with the same chemicals, but it is better than with peroxide alone.[63,119] By bleaching in three stages with borohydride-peroxide-dithionite, SCAN brightness values of 87 to 88% were obtained. This seems to be about the maximum obtainable brightness because bleaching in four stages did not give higher values.[63] In general, two-stage bleaching of mechanical pulp is not attractive because the increase in brightness is so small in the second stage that it usually will pay to add more peroxide in a single-stage process to obtain the required brightness.

BLEACHING OF SEMICHEMICAL AND CHEMIMECHANICAL PULPS

In the range between mechanical and chemical pulps there are several types of pulps that are produced by a combination of chemical and mechanical pulping. These processes are discussed in the pulping chapter under semichemical (NSSC) pulping and refiner and thermomechanical pulping. As explained there, the chemical treatment used in pulping may be minimal and the mechanical treatment very high, or the pulping process may be similar to a full chemical pulping, except that conditions are used to produce an extremely high yield.

The conventional neutral sulfite semichemical pulp (NSSC) has a dark-brown color and normally this pulp is used where high brightness is not a requirement. If this pulp could be bleached without removal of the lignin, a much broader field of application would be possible. Semichemical pulps can be bleached by removal of the lignin, but this is not an economical way of producing bleached pulps. The consumption of bleaching chemicals is very high and the loss in yield by bleaching is so large that the final yield based on wood is almost as low as for normal bleached chemical pulps. The reader is referred to the section on bleached NSSC pulp in the pulping chapter. The combination of chemical pretreatment of chips and thermomechanical pulping, the process known as chemithermomechanical pulping, produces pulps with interesting properties. Bleaching of such pulps has already started, but very little has been published.

Bleaching of Semichemical and Chemimechanical Pulps Prepared by Sulfite Treatment

The bleachability of pulps prepared by treatment with sulfite or bisulfite is dependent on the pulping conditions, the degree of pulping and the wood species. Giertz[120] found that a high-yield bisulfite pulp made from spruce was impossible to bleach with peroxide. On the other hand birch pulp with a yield above 80% could be bleached with peroxide to a fairly high brightness. Luner and Ferguson[121] studied the effects of pH and yield in sulfite pulping on the bleachability, using spruce and birch. Starting pH values were 4.5, 9.8, and 13, and maximum pulping temperatures were 145, 180, and 170°C. A minimum in brightness was obtained at a pulping yield between 80 and 60%, both before and after bleaching with peroxide. For spruce bisulfite pulp and birch alkaline pulp, the minima became more pronounced after the bleaching. The bisulfite pulp made from spruce with a yield below 85% could not be bleached at all with peroxide or borohydride. Palenius[122] prepared birch pulps at an initial pH of 7 in the temperature range of 80 to 160°C and with unbleached yields between 70 and 95%. The increase in brightness by dithionite bleaching was about the same at all yields. The effect of peroxide bleaching was dependent on the yield and best results were obtained when the yield was above 80%. Brightness values above 80 were obtained. Christensen[123] studied the influence of sulfite pulping conditions for spruce on bleachability and final brightness. Pulps were prepared in the yield range 70 to 95% at three initial pH values, 4.5, 9, and 12. The brightness gain with dithionite as well as with peroxide bleaching increased with increasing pH in the pulp preparation. The final brightness after bleaching was highest when pulping was carried out at pH 4.5 and with high yield. With 2% hydrogen peroxide a brightness of 75 was obtained and bleaching in two stages with peroxide and dithionite gave a brightness of 80. Norrström[124,125] investigated the influence of bisulfite concentration on brightness, absorption of light, and bleachability of high-yield pulp from spruce. These pulps have a comparatively low absorption of light, but the color is unstable. The specific light absorption of lignin changes during sulfite delignification, first showing a rapid initial drop and then increasing with decreasing yield. As lignin is removed from the pulp during the cooking, the specific absorption coefficient of the pulp goes through a minimum and then through a maximum at about 75% yield. The light absorption of the pulp increases with temperature and decreases with increasing bisulfite concentration. The bleach response with dithionite as well as peroxide shows a minimum between 70 and 80% yield. Increased temperature and amount of bisulfute during the cooking causes a reduction in the bleach response. Treatment of the pulps at conditions simulating the bleaching conditions, but without bleaching agent, shows that a strong discoloration occurs for this type of pulp, especially in the alkaline medium used for peroxide bleaching. The bleach response is, therefore, a result of competing reactions, involving the formation of new chromophoric groups caused by temperature and pH during the bleaching and the removal of chromophoric groups by the bleaching agent.

The conclusion from the different investigations carried out with sulfite or bisulfite high-yield pulp is that for softwoods the yield must be above 85% and preferably above 90% in order to obtain a good bleach response. For hardwoods, the best bleaching results are also obtained when the pulp yield is above 85%, but hardwood pulps can be bleached to some degree even in the most difficult range of pulp yield between 70 and 80%.

Chemimechanical pulps, where the chemical treatment is slight, are more interesting for production of bright pulps than the normal semichemical pulps. Kindron[59] found that refiner pulp from chips impregnated with sulfite or bisulfite responded well to peroxide bleaching. Similar results have been reported for peroxide bleaching of chemimechanical hardwood sulfite pulp.[126] Pretreatment of chips with chemicals before refining has, therefore, become an interesting field of research. Chemical pretreatment, followed by preheating before the mechanical treatment, can produce pulps with fairly good brightness and strength properties, but the brightness gain by bleaching is lower than for refiner pulp.[127] Richardson[128] has reported that high-yield refiner pulps from aspen give high brightness when bleached with peroxide. Case[116] studied the bleachability of spruce refiner pulps prepared from chips pretreated with sulfite. The brightness gain by peroxide bleaching was lower than that for stone groundwood and for normal refiner pulps, but the unbleached brightness was higher and the final brightness was at least as good as that for the two other types of mechanical pulps.

Pretreatment of chips with sulfite will have a small reductive bleaching effect.

One would, therefore, expect that subsequent bleaching with a reducing agent would give a lower brightness gain than that from the bleaching of pulp from untreated chips. The final brightness should, however, be equal to or even somewhat better because of less formation of chromophoric compounds when sulfite is present during the refining. Peroxide bleaching of pulp from sulfite or bisulfite pretreated chips will probably give highly variable bleach response, depending on the conditions of pretreatment and the amount of reducing chemical remaining in the unbleached pulp. If a large amount of sulfite is present, some of the peroxide will be used to oxidize it, and a comparatively high dosage of peroxide will be necessary to obtain the required brightness. In such a case, washing of the pulp before bleaching may improve the bleachability. The presence of reduced chromophoric groups may also require peroxide to oxidize them to their original chromophoric state before bleaching can take place.

Bleaching of Chemimechanical Pulps Prepared by Alkaline Treatment

Alkaline sulfite treatment has, as mentioned above, been investigated as a method for producing chemimechanical pulps of good brightness and bleachability. Sodium sulfite has advantages over other alkalis because the sulfite can prevent excessive discoloration in the pulping process. However, pretreatment of chips with sodium hydroxide before refining, employing the cold soda process, is used with hardwood on an industrial scale to produce bleachable pulps. In stone grinding the use of an alkaline medium was found to improve strength and brightness for eucalyptus pulp.[129] When the pulp was washed before bleaching with zinc dithionite, the brightness was increased 12 units to a final brightness level above 70.[129,130] Eucalyptus pulp made by the cold soda process was bleached with hypochlorite.[131] An additional treatment with sulfur dioxide or dithionite further increased the brightness of the hypochlorite bleached pulp, but washing between stages was necessary to obtain maximum brightness-gain. If the reducing agent was used in the first stage, the final brightness was less.

Cold soda pulp prepared from birch responds very well to brightening with sodium borohydride.[132] Brightness gains above 20 units can be obtained with less than 1% consumption of sodium borohydride.[133] Peroxide bleaching of cold soda pulp in a disk refiner increased the brightness by about 20 units with an addition of 1.25% hydrogen peroxide,[134] using a mixture of beech, birch, and maple.

Very good results have been obtained by pretreatment of chips with alkali and peroxide before mechanical defibration. Poplar pulp with high brightness and good strength properties was obtained in this way.[135,136] Birch and beech give good results with peroxide pretreatment, but hornbeam did not respond favorably to this method.[135]

BLEACHING OF CHEMICAL PULPS

Unbleached chemical pulps are too dark in color for use in the manufacture of high-grade white papers. The dark color is derived mainly from the lignin in the pulp.

The brightness of pulp cooked by the sulfate process is much lower than that of the wood from which it is produced, even though the light scattering is higher in the unbeaten pulp than in the wood. As most of the lignin is removed in cooking, the lower pulp brightness is caused by the large increase in the light absorption of the remaining lignin, which increases the overall light-absorption coefficient of the pulp. Sulfite pulping also increases the light-absorption coefficient of the lignin remaining in the pulp, but the brightness of the pulp is normally higher than that of the wood because most of the lignin is removed. The light-absorbing configurations of pulp are not completely known and the methods used for removing the color are based mainly on experience and not on theoretical considerations.

Bleaching of chemical pulps to a high brightness without removal of lignin has not been successful so far. The lignin that remains after the cooking has been completed must be removed in the bleaching process if a pulp of high brightness is desired. Bleaching can therefore be considered as a continuation of the pulping process. Single-stage bleaching of chemical pulp is seldom used today and almost all bleaching is multistage. The bleaching sequence can be varied to some extent. In the early development of bleaching, chlorination was always used in the first stage. Multistage bleaching normally includes an alkali-extraction stage. A general principle in the bleaching of chemical pulp is to alternate between acid and alkaline stages. In the acid stage, an oxidizing agent is applied; in the alkaline stage, the products formed in the oxidative stage are extracted from the pulp. Exceptions from this rule are hypochlorite, oxygen, and peroxide bleaching, where the oxidation is carried out in alkaline medium. When these agents are used, a combination of oxidative bleaching and alkali extraction can be carried out in the same stage.

Bleachability Measurement

The bleaching of chemical pulps is mainly a removal of residual lignin and, therefore, a relationship should exist between lignin content and chemical requirement in bleaching. This has proved to be fairly accurate and a measurement of the lignin content of pulp is a very good indication of the amount of bleaching chemical required. However, direct methods of lignin determination are too time-consuming for routine use and generally indirect methods are used for determination of bleachability.

If the first bleaching stage is chlorination, a measurement of the chlorine consumption of the pulp should be a logical method for characterizing the bleachability. The Roe chlorine number is based on this reasoning. The original Roe method[137] has been modified somewhat and the definition given by TAPPI is as follows: the chlorine number is defined as the number of grams of chlorine consumed by 100 g of moisture-free pulp in 15 min at 25°C.[138] In the test procedure, 0.5 to 2.0 g of shredded pulp having a moisture content between 40 to 60% is chlorinated with chlorine gas in a special apparatus. A similar test is also used for the standard Canadian method.[139] TAPPI also has a shorter version in which the reaction time is reduced from 15 to 5 min.[140] For kraft pulps a straightline relationship exists between the Roe chlorine number and the lignin content, up to a lignin content of 14%. The equation for this relationship is:[141]

$$\text{Roe number} \times 0.811 = \text{lignin content, \%.}$$

Addition of chlorine as a gas, as in the Roe test, is not a very practical procedure, and several simpler tests based on the reaction between chlorine or chlorine compounds and lignin have been proposed. One method, which has become a standard, is based on the generation of chlorine by acidification of sodium hypochlorite. In this method, a test-sample of pulp is treated for 15 min at a temperature of $25 \pm 1°C$, with chlorine generated *in situ* by acidification of a sodium hypochlorite solution. The residual chlorine, which should be more than 50% of the amount added, is determined by iodometric titration. A correction factor is used to calculate the consumption at 50% residual chlorine.[142] This method has been adopted in many countries, in some cases in a slightly modified version, as a standard method for determining the bleachability of pulp or as a measure of the degree of cooking.[143-145] The chlorine number determined in this way corresponds fairly well with the Roe number, although a different temperature was originally used (20°C) for determination of the Roe number.[146] A good correlation exists between this test and the lignin content as shown in the following equation:[147]

$$\text{lignin content, \%} = 0.90 \times \text{chlorine number}$$

Hypochlorite, as such, is sometimes used as a bleachability test. The best known method is the Sieber chlorine number.[148] Hypochlorite is sometimes used as a rapid test for control of cooking in the sulfite process. The hypochlorite is added to a pad of pulp and the pulping degree judged from the color obtained—the color being darker, the higher the lignin content. This method obviously gives only an approximate indication of the bleachability.

In the pulp mill it is very important to have a rapid and reliable test for measuring bleachability. Because permanganate reacts rapidly with lignin, but slowly with carbohydrates in acid solutions, several bleachability tests have been developed based on the consumption of permanganate by the pulp. For a while a multiplicity of different procedures were used, which made a comparison of bleachability figures from one mill to another very difficult, but about 1955 work on a standardized permanganate method was started.[149,150] This resulted in the kappa number. The kappa number is defined as the number of ml of 0.1 N potassium permanganate solution consumed by 1 g of pulp in 10 min at 25°C. The results are corrected to 50% consumption of the permanganate added.[151-154] For kraft pulp the relationship between kappa number and lignin content is:[155]

$$\text{lignin}, \% = 0.147 \times \text{kappa number}$$

The relationship between lignin content and kappa number is very good for bleachable pulps. The main reason for this good relationship is that a specified excess of permanganate is used in the kappa method, whereas some of the other permanganate test methods operate with a fixed permanganate dosage. Some of the other methods show a fairly good relation to the kappa number.[156] Figure 5-12 shows the relationship between several different bleachability test methods for sulfite and sulfate pulps.[157] Modifications in the method for determination of kappa number have been proposed for pulps with high lignin content. However, these are not bleachability tests, but rather are tests for measuring the degree of cooking.

Tests for measuring bleachability in single-stage bleaching with hypochlorite also exist.[158,159] These methods can also be used for determining the requirement of hypochlorite in a multistage bleaching process if the test is carried out on pulp from the stage preceding the hypochlorite stage.

Composition and Behavior of Chlorine-Water Systems

Chlorine and compounds of chlorine are the best established and most widely used bleaching agents for chemical pulps. It seems advisable to discuss the general chemistry of chlorine-water systems before taking up the specific use of chlorine bleaching agents.

When chlorine is dissolved in water an equilibrium is reached almost instantaneously between chlorine and hypochlorous acid as follows:

$$Cl_2 + H_2O \rightleftharpoons HOCl + H^+ + Cl^-$$

All the oxidizing power of the chlorine molecule is transferred to the hypochlorous acid, while the other chlorine atom is reduced to chloride. Hypochlorous acid and hypochlorite ion, therefore, contain two equivalents of available chlorine. The equilibrium constant for the reaction at 25°C is:[160]

$$K_1 = \frac{(HOCl) \cdot (H^+) \cdot (Cl)}{(Cl_2)} = 3.9 \cdot 10^{-4}$$

The hydrolysis of chlorine is dependent on the concentration, as shown in

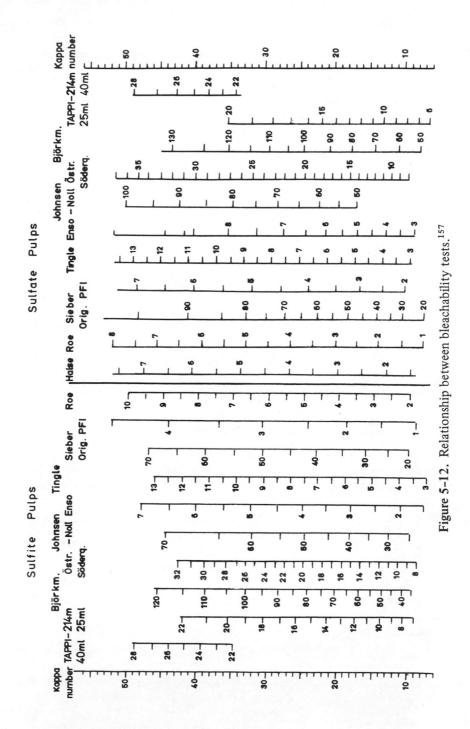

Figure 5-12. Relationship between bleachability tests.[157]

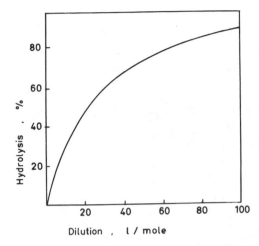

Figure 5-13. Hydrolysis of chlorine versus dilution.[161]

Figure 5-13, and dilution of the system causes an increase in hypochlorous acid and a relative decrease in chlorine.[161] The effect of temperature on the equilibrium constant is shown in Table 5-2, where it can be seen that hydrolysis increases when the temperature is raised.[160] The final equilibrium is also dependent on the dissociation of hypochlorous acid, which is a weak acid. The equilibrium constant for this acid at 25°C is:[162]

$$K_2 = \frac{(H^+) \cdot (OCl^-)}{(HOCl)} = 5.6 \cdot 10^{-8}$$

The pH of the solution has an obvious effect on the equilibrium. At a pH about 7.3, hypochlorous acid and hypochlorite ion are present in nearly the same concentration. Figure 5-14 shows the equilibrium concentrations of chlorine, hypochlorous acid, and hypochlorite ion for varying pH up to pH 9.[163] At pH 1,

TABLE 5-2 THE EFFECT OF TEMPERATURE ON THE HYDROLYSIS CONSTANT OF AQUEOUS CHLORINE SOLUTIONS[a]

Temperature °C	$K_1{}^b$	pK_1
0	$1.46 \cdot 10^{-4}$	3.84
15	$2.81 \cdot 10^{-4}$	3.55
25	$3.94 \cdot 10^{-4}$	3.40
35	$5.10 \cdot 10^{-4}$	3.29
45	$6.05 \cdot 10^{-4}$	3.22

[a] See reference 160.

$$^b K_1 = \frac{[HOCl] \cdot [H^+] \cdot [Cl^-]}{[Cl_2]}$$

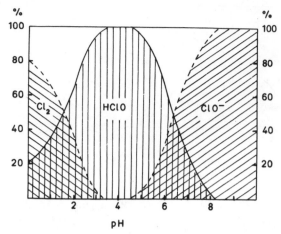

Figure 5-14. Influence of pH on the ratio of chlorine compounds.[163]

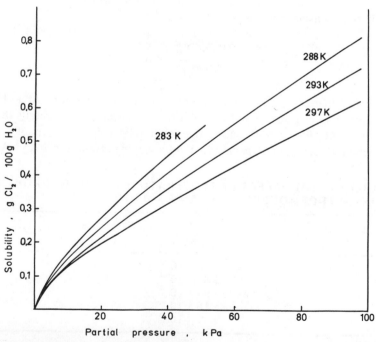

Figure 5-15. Solubility of chlorine versus partial pressure at different temperatures (°Kelvin).[166]

668

chlorine is the dominant component; at pH 4, the solution contains almost all hypochlorous acid; for pH values above 9, the oxidizing agent is practically 100% hypochlorite ion.

Hypochlorous acid can also react in another way and form dichlorine monoxide (Cl_2O):[164,165]

$$2\ HOCl \rightleftharpoons Cl_2O + H_2O$$

Dichlorine monoxide is very reactive, but it is not known how it reacts with pulp.

Hypochlorous acid and hypochlorite ion react at pH about 7 and form chlorate. This reaction is used in the industrial production of chlorate. Chlorate does not react with pulp except in very strong acid solution.

The solubility of chlorine in water under conditions of different partial pressure and temperature is shown in Figure 5-15.[166]

Multistage Bleaching of Chemical Pulps

Multistage bleaching of chemical pulps can be carried out with many different combinations of stages. As many as eight stages are sometimes used. To write the name of each stage in a long sequence is time-consuming and impractical. Consequently, it has become common practice to use a single letter as a symbol for a particular stage. The symbols in most common use and the stage which they represent are:

C   = chlorination

E   = alkaline extraction

H   = hypochlorite bleaching

D   = chlorine dioxide bleaching

P   = peroxide bleaching

O   = oxygen bleaching

C_D = chlorination, with addition of a small amount of chlorine dioxide

D/C   = sequential bleaching, with chlorine dioxide and subsequent chlorination without washing between the addition of chemicals

C + D = chlorination, with a mixture of chlorine and chlorine dioxide, with chlorine as the larger part

A   = acid treatment

Washing between the stages is indicated by a hyphen. No washing between application of two chemicals is indicated by an inclined line as shown for the sequential treatment with chlorine dioxide followed by chlorine.

Chlorination

The principle of using chlorine as a delignifying agent is an old idea. Because of the lack of suitable equipment for handling this corrosive chemical, the use of chlorine for industrial bleaching was not practiced until 1930. Chlorination of pulp, as it is usually carried out, does not show any decoloring effect, and, in fact, the color of the pulp may increase with chlorination. The important function of chlorine in bleaching is to convert the lignin in the pulp to compounds that are soluble in water or alkali. Chlorine is never used as the only bleaching agent but is always used in conjunction with alkali extraction, and it is generally followed by the use of bleaching agents, such as hypochlorite, chlorine dioxide, or peroxide.

Chlorination is usually carried out at low consistency, generally at 3 to 4%, and at the temperature of the production water which tends to give a different reaction rate depending on the time of the year and the climate. The influence of temperature and reaction time on chlorine consumption is shown in Figure 5-16 for sulfate pulp[167] and in Figure 5-17 for sulfite pulp.[168] Chlorine reacts very rapidly with the pulp and most of the chlorine is consumed in a few minutes. However, usually between 30 min- to 1 hr-retention-time is used for chlorination in order to allow the chlorinated compounds to decompose and to allow the degradation products to diffuse out of the fiber.

The correct dosage of chlorine is very important. If too little chlorine is added, the pulp will be very difficult to bleach to a high brightness; if too much chlorine is used, the chlorine will attack the carbohydrates and a bleached pulp with low physical strength will be the result. The dosage of chlorine is based on chlorine or permanganate number. For sulfite pulp, a 0.95 × Roe number was found to be the optimum dosage of chlorine.[169,170] The attack on the carbohydrates is especially noticeable at high temperature. To avoid or reduce this attack, chlorine dioxide, sulfamic acid, or ammonium compounds can be added.[171-176]

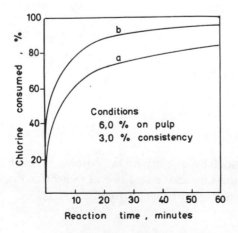

Figure 5-16. Rate of reaction in chlorination of sulfate pulp at (a) 3°C and (b) 20°C.[167]

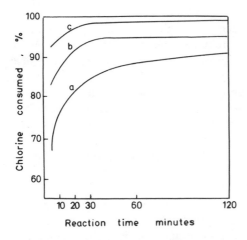

Figure 5-17. Rate of reaction in chlorination of sulfite pulp at (a) 2°C, (b) 20°C, and (c) 50°C.[168]

The low consistency generally used for chlorination has two obvious disadvantages. One, as mentioned, is the difficulty of controlling the temperature to maintain a constant reaction rate; the other is the large volume of effluent from the chlorination process. Higher consistencies would, therefore, be desirable and high-consistency chlorination has been studied in the laboratory and pilot plant. Gas-phase chlorination takes place rapidly at consistencies up to 60%,[177-179] but the heat generated in the process imposes a limit on the consistency of about 30 to 35% when concentrated chlorine gas is used. Medium-consistency chlorination at 9 to 11% seems feasible when efficient mixing equipment is used, and when displacement bleaching is applied in the subsequent stages.[180]

Chlorination at temperatures up to 50°C seems feasible without loss in viscosity of the pulp if reaction time is adjusted to reaction rate at the applied temperature. Addition of small amounts of chlorine dioxide improves the viscosity as well as the yield in high-temperature chlorination.[181] Gas-phase chlorination can be carried out at high temperature if the chlorine is diluted with an inert gas, and temperatures as high as 65°C have been proposed for this type of chlorination.[182] Laboratory trials indicate that chlorine can be used in gas-phase bleaching in the third or fifth stage.[183]

Alkaline Extraction

Extraction of pulp with alkali is not a bleaching in the sense of whitening the pulp. In fact, alkaline extraction darkens the pulp and this is especially noticeable when the extraction follows a chlorination stage. However, alkaline extraction serves several useful purposes in the multistage bleaching of chemical pulps.

Usually the alkaline extraction is carried out after the chlorination or another acid stage. The first effect of the alkaline treatment is to neutralize the acids or acidic groups formed in the preceding stage. For some types of sulfite pulp, such as those used for greaseproof papers, the alkali is used merely to neutralize and extract the degradation products from lignin, and as much as possible of the hemicellulose is left in the pulp since this is beneficial in the subsequent beating of the pulp. In other grades of pulp, a greater amount of alkali is used to extract some or most of the hemicellulose. Removal of hemicellulose is desirable for pulp that is to be used in the manufacture of paper for absorbent qualities. Most of the hemicellulose should be removed from pulp to be used for rayon manufacture and for some other speciality uses.

Many of the compounds formed by chlorination are not soluble in water but are soluble in alkali. Chlorinated lignin contains a high amount of acidic groups and these are soluble in their anionic form. Some of these chlorinated compounds are broken down by alkali to lower-molecular-weight compounds, which are more soluble in the extraction process. Sulfate pulps are more difficult to extract than sulfite pulps, and good extraction with alkali was found to be the answer to the full bleaching of these pulps.[184] During the alkali treatment, soaps are formed from the fatty acids and these aid in the extraction of other compounds from the pulp. Surface-active agents are sometimes added with alkali for the same purpose.

Chlorination of the resins in the pulp makes them difficult to remove, and, therefore, an alkaline extraction before chlorination has been proposed for sulfite pulps[185] and for sulfate pulps.[186] The use of alkali in the first stage can have an advantage over chlorination from an environmental point of view[187] if sodium is used as the base in the pulping liquor. Under these conditions, the compounds extracted with sodium hydroxide can be evaporated and burned with the pulping liquor without special problems.

Sodium hydroxide is practically the only alkali used for industrial extraction of pulp. Calcium hydroxide has been used for neutralization after chlorination, and it has been shown to be effective in hot alkaline extraction.[188] Kraft white liquor has been proposed for hot as well as cold alkaline purification. Hydroxides of other alkali metals are also good extraction agents, but they are not competitive for economic reasons.[189] In gas-phase bleaching, ammonia would be the best agent to use as alkali.[190] Among the substances tried for hot alkaline purification of pulp are sodium sulfide, acetate, carbonate, phosphate, sulfite, and borax.[186]

Alkali extraction is usually carried out at comparatively low alkali concentrations. However, the purification of the pulp with alkali in a cold alkali treatment can be considered as a special case of alkali extraction. Cold alkali purification requires a very high alkali concentration. This causes swelling of the fibers and the excess alkali is hard to remove from the fibers. The high consumption of chemical and the equipment necessary for this process makes cold alkali extraction very expensive and consequently it is used only for special grades of dissolving pulps that are produced mainly for nitration and acetylation purposes. Most of the hemicelluloses are removed by cold alkali treatment, and pulp with a high alpha-cellulose content can be prepared.[191]

In the conventional alkaline extraction, which is carried out subsequent to the chlorination stage, the temperature of the pulp must usually be increased considerably to obtain the desired effect. The temperature range is from $25°C$ to more than $100°C$; for most paper pulps the temperature is above $50°C$ and for dissolving pulps above $70°C$. To reduce steam consumption the pulp consistency should be high during alkaline extraction, especially when extracting at high temperatures. Usually a consistency between 10 to 18% is used, but for hot alkali purification of dissolving pulps, consistencies up to 35% have been used. The reaction time for paper pulps is usually from 30 min to 1 hr and for dissolving grades from 1 to 5 hr.

The most important factor in alkaline extraction is the amount of alkali used. For paper grades, the amount of alkali is usually between 1 to 2% on pulp weight after a first stage chlorination and about 0.5% after a chlorine dioxide stage. For dissolving grades, up to 5% sodium hydroxide based on pulp weight is applied. The loss of yield is highly dependent on the alkali charge.[192] The effect of the alkaline treatment can be improved by addition of oxidizing agents. Sodium hypochlorite[193] and peroxide have been used for this purpose.[193,194]

Hypochlorite Bleaching

Calcium hypochlorite was the first agent used for the bleaching of chemical pulp. It has been widely used and a substantial part of the bleached pulp is still being treated in one or two stages with hypochlorite. The importance of hypochlorite as a bleaching agent was lessened after chlorine dioxide was introduced as a bleaching agent. The reason is that hypochlorite attacks the cellulose much more than chlorine dioxide.

Hypochlorite can be used for single-stage bleaching of either sulfite or sulfate pulp. However, for sulfate pulp a severe attack on the cellulose will occur if the brightness is raised above 50%. Brightness values of 80% can be obtained with sulfite pulp, but the lignin content before bleaching must be low. Sulfite pulp of high lignin content is discolored by small dosages of hypochlorite so that large dosages are necessary to obtain a real bleaching. High dosages of hypochlorite usually attack the cellulose to a considerable degree. In multistage bleaching, hypochlorite is normally used after chlorination and alkaline extraction. Bleaching with hypochlorite in the first stage of multistage bleaching can also be used.[195] The addition of hypochlorite in the alkaline extraction stage will improve the brightness. Hypochlorite as a final stage after chlorine dioxide bleaching can be used for two reasons: (1) to obtain a more color-stable pulp when the bleaching is carried out at high pH and (2) to reduce the viscosity of the cellulose in the case of dissolving pulp. Hypochlorite bleaching is carried out in the consistency range from 4 to 18%. The influence of consistency on the reaction is shown in Figure 5-18.[196] The main effect of increased consistency is that the concentration of hypochlorite is increased and the reaction rate thereby increased.

The rate of reaction between hypochlorite and pulp is doubled for each $7°C$

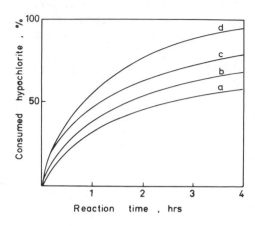

Figure 5-18. Rate of reaction in hypochlorite bleaching at 40°C and (a) 6% consistency, (b) 10% consistency, (c) 15% consistency, and (d) 25% consistency.[196]

rise in temperature.[197] For paper grades, the bleaching temperature is usually between 30 and 45°C; for dissolving pulps, temperatures up to 60°C are used. Because the reaction rate is affected by temperature the required retention time is dependent on temperature. However, retention time is also dependent on pH and on the stage in the bleaching process at which the hypochlorite is applied. The pH plays a very important role in hypochlorite bleaching because it affects the quality of the pulp produced. Hypochlorite solutions should always be alkaline when they are prepared. If sodium hypochlorite is being used, the alkalinity of the bleach solution can be adjusted to the range required for the bleaching reaction. If calcium hypochlorite is being used, the alkalinity of the bleaching agent will be limited by the solubility of calcium hydroxide and the alkalinity will be decreased with increasing available chlorine in the solution, as shown in Figure 5-19.[197a]

The reactions between pulp and hypochlorite produce acidic groups, and the pH will gradually drop during the bleaching. Clibbens and Ridge[198] showed in 1927 that cellulose was severely attacked by hypochlorite at a pH around 7. Both sulfite and sulfate pulps are attacked[199] and the heaviest attack on both pulps is in the range of 6 to 7 pH. In order to keep the pH above this critical region, additional alkali sometimes must be used especially if the hypochlorite stage follows an acid stage. In continuous bleaching, all alkali must be added at the start of the bleaching and the amount used should be sufficient to keep the final pH above 8. Sulfamic acid can be used in hypochlorite bleaching[200-203] as a protective agent to reduce the effect of low pH; the pH can be reduced to 7 without too severe attack on the pulp when this chemical is present. The reaction rate in hypochlorite bleaching can be increased by addition of Cu, Ni, and Br compounds,[204-207] but these have not been used to any extent industrially, and there is some doubt about their usefulness.

The combined effect of the many variables in hypochlorite bleaching is rather difficult to predict and statistical methods have been used to determine the effect of each variable.[208,209]

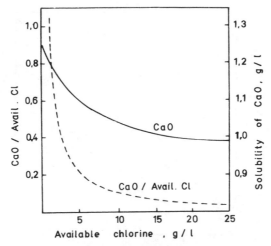

Figure 5-19. Influence of calcium hypochlorite concentration on solubility of lime.[197a*]

Chlorine Dioxide Bleaching

Commercial bleaching of pulp with chlorine dioxide started in 1946 in Canada and Sweden. The use of chlorine dioxide developed slowly in the first years after its introduction but picked up considerably soon thereafter. In the last two decades there probably has not been a plant built for fully bleached sulfate pulp that does not include one or more stages for chlorine dioxide treatment. The advantage of chlorine dioxide over hypochlorite is that it is more specific in its reactions and mainly attacks the lignin and extractive compounds.

The bleaching effect of chlorine and chlorine dioxide mixtures was discovered almost 150 years ago,[210] but the composition of the mixture was not understood at the time. The bleaching effect of chlorine dioxide on wood fibers was reported by Schmidt, in 1921, who used it for the preparation of holocellulose by stepwise treating wood with chlorine dioxide and alkali. References to his early work on chlorine dioxide bleaching is given in a more recent publication.[211] A combined pulping and bleaching process using chlorine dioxide and alkaline extraction in several stages has been proposed.[212] The physical properties of chlorine dioxide deterred its early use. It is a rather unstable compound. It has an unpaired electron, which makes it very reactive and it is toxic. In the gas phase, at concentrations above 12 to 15%, it is explosive; its explosiveness is intensified by heat or light. The solubility of chlorine dioxide in water[213] is shown in Figure 5-20, which points up its volatility and the fact that low temperature is necessary to keep it in solution. At high temperature there is a risk of getting too high a concentration of chlorine dioxide in the gas phase. Chlorine dioxide is also very corrosive and this was one of the main deterrents to its industrial use. In spite of all these drawbacks, progress in bleaching technology has made chlorine dioxide bleaching a fairly safe operation.

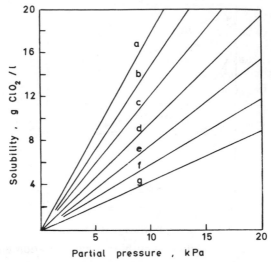

Figure 5-20. Solubility of chlorine dioxide versus partial pressure at (a) 0°C, (b) 5°C, (c) 10°C, (d) 15°C, (e) 20°C, (f) 25°C, and (g) 30°C.[213]

When first introduced as a bleaching agent, chlorine dioxide was used in the final stages of multistage bleaching, but, as mentioned in the section on chlorination, addition of chlorine dioxide in the chlorination stage is now common.[171,172,176,214-219] Sequential application of chlorine dioxide and chlorine is sometimes used.[173,220-223] Chlorine dioxide as a first stage in a multistage bleaching without application of chlorine has also been proposed.[185,224] Often when chlorine dioxide is used, it is applied in more than one stage. In order to obtain a high brightness with sulfate pulp two stages of chlorine dioxide with an alkaline extraction between the stages is the usual procedure.[225,226] Chlorine dioxide after a hypochlorite stage is often used for sulfate as well as sulfite pulps.

Consistency has very little effect on the bleaching with chlorine dioxide. The reaction rate is almost the same over the consistency range of 4 to 15% if the other parameters are kept constant. The consistency used in mill operations is usually between 10 and 12%. When chlorine dioxide is used in the first stage, the temperature usually is low. Chlorine dioxide bleaching in the final stages is carried out between 60 and 80°C, with 70°C being considered the optimum temperature. The rate of reaction is rapid at high temperature, but there is a slow increase in brightness after the initial reaction, and retention times from 3 to 5 hr are usually allowed at 70°C. The reaction is influenced by pH. Poor bleaching results are obtained under alkaline conditions, and although bleaching can be carried out around the neutral point,[227] a pH between 3 and 5 is normally used in mill practice.

All methods for the manufacture of chlorine dioxide produce some chlorine. A small amount of chlorine mixed with chlorine dioxide has no harmful effect, but a large amount may attack the cellulose especially if used in the final stage.[228,229]

The effect of chlorine dioxide in different sequences of bleaching of sulfite and sulfate pulps has been reported in several publications.[193,215,218,219,225,226,230-232]

Chlorite Bleaching

In alkaline or neutral solution, sodium chlorite has no bleaching effect and it must be activated by acidification to a pH below 4 or by reaction with chlorine or hypochlorous acid. Under these conditions chlorine dioxide is formed and the bleaching reactions are about the same as when chlorine dioxide is applied directly. Activation with chlorine produces chlorine dioxide according to the reaction:

$$2ClO_2{}^- + Cl_2 \rightarrow 2ClO_2 + 2Cl^-$$

Sodium chlorite can be transported and stored at the mill, unlike chlorine dioxide, and pulp mills with a limited requirement of chlorine dioxide have used chlorite. However, because chlorite is manufactured from chlorine dioxide, bleaching with chlorite will generally be more expensive than chlorine dioxide bleaching.

Peroxide Bleaching

The industrial use of peroxide as a bleaching agent was first applied on mechanical pulps, but peroxide can also be used for the bleaching of chemical pulps. Sulfite pulps with low lignin content can be bleached to some extent in a single-stage bleaching, but for sulfate pulps the peroxide is used in one of the stages of multistage bleaching.

The consistency should be high in peroxide bleaching, about 10 to 20%. The temperature for bleaching sulfate pulp with peroxide in a multistage process is usually between 60 and 80°C and the reaction time is 2 to 4 hr. The alkalinity for optimum bleaching effect is higher for chemical pulps than for mechanical pulps.[233] Sodium silicate is often used with peroxide for the bleaching of chemical pulps, but the presence of this chemical is not as important as for the bleaching of mechanical pulp. If peroxide is used in the final stage, sulfur dioxide is used to reduce the remaining peroxide and to acidify the pulp.

In the case of sulfate pulps, peroxide is usually applied in the final bleaching stage, but it can also be used in the alkaline extraction stage.[194] The purpose of using peroxide in the extraction stage is to improve the brightness without adding more bleaching stages. Semibleached sulfate pulp can be prepared in two stages, using chlorination and peroxide,[234] or in a four-stage process, using the sequence C-E-H-P.[235,236] For fully bleached sulfate pulps the addition of peroxide in the final stage has shown an advantage in brightness stability compared with chlorine dioxide.[233] Peroxide as a final stage, following treatment with chlorine dioxide, gives pulps with high and very stable brightness.[233,237,238] Peroxide can also be applied in combination with chlorine or hypochlorite to obtain fully bleached pulps using fewer stages.[239] Optimum conditions for the peroxide stage in the bleaching of sulfate pulps using the sequence C-E-H-D-P has been determined by a computer program.[240] Optimization of chemicals used in the C, H, and D stages of the C-E-D-D-P bleaching sequence for sulfate pulp was studied,[241] and the final brightness was found to be highly dependent on the amount of hypochlorite and chlorine dioxide applied.

Reactions in Chlorination and Alkaline Extraction

Chlorine is a very reactive chemical and it reacts with most of the compounds present in pulp. However, the rate of reaction differs greatly with the different compounds. Hardell and Lindgren[242] compared the rates of reaction given in the literature[165,243-245] between chlorine and four different organic compounds. The values differ by a ratio of 1 to 10^8, as shown in Table 5-3.

Chlorine can react with organic compounds in two fundamentally different ways:[246] (1) by formation of radicals as intermediate products or (2) by formation of polarized chlorine or hypochlorous acid. Radicals can be formed under the influence of light or by reaction with organic compounds as follows:

$$Cl_2 + RH \rightarrow Cl^{\cdot} + R^{\cdot} + HCl$$

Initially the reaction rate is slow, but because each chlorine atom starts a chain reaction, the total reaction can be rapid. The principal reactions in the chain reactions are:

$$RH + Cl^{\cdot} \rightarrow R^{\circ} + HCl$$

$$R^{\cdot} + Cl_2 \rightarrow RCl + Cl^{\cdot}$$

TABLE 5-3 RATE OF REACTION BETWEEN CHLORINE AND ORGANIC COMPOUNDS[a]

Compound	Reaction Rate $M^{-1} s^{-1}$	pH
Phenol[b]	10^4	2.0
$CH_2 = CH \cdot CH_2 OH$[c]	10^{-1}	4.7
$CH_3 = CH = CH \cdot COOH$[d]	10^{-2}	4.7
Metyl betaglucoside[e]	10^{-4}	2.0

[a]See reference 242.
[b]See reference 243.
[c]See reference 244.
[d]See reference 165.
[e]See reference 245.

These ractions can be slowed or stopped by radical scavengers, such as chlorine dioxide[247] and oxygen.[246] Chlorine can react simultaneously in either radical or nonradical reactions. However, no evidence has been obtained to show that chlorine in radical form reacts with phenols or phenol ethers so the reactions between chlorine and lignin are, therefore, probably nonradical reactions.

The known reactions between chlorine and lignin or between chlorine and lignin-model compounds are substitution, oxidation, and addition. When one chlorine atom is substituted in the benzene ring, one chloride ion is formed simultaneously. Oxidation produces two chloride ions per molecule of chlorine and no chlorine in the lignin molecule. Addition of chlorine to a double bond should produce no chloride and two chlorine atoms in the lignin molecule, according to older literature on this subject. But the addition of chlorine to double bonds in aqueous medium produces a chlorine atom and a hydroxyl group in the lignin and a chloride ion in the solution[242] so that determination of chlorine in the pulp and chloride ion in the water phase cannot distinguish between substitution and addition. Therefore, determination of chlorine in the pulp and chloride in the aqueous solution after chlorination does not give exact information on the importance of the different forms of reaction between chlorine and lignin. Using Li as the symbol for the nonreacting part of the lignin molecule, the above reactions can be written as follows:[242]

Substitution $\quad$ $LiH + Cl_2 \rightarrow LiCl + HCl$

Oxidation $\quad$ $LiH + Cl_2 + H_2O \rightarrow LiOH + 2HCl$

Addition $\quad$ $Li-CH=CH-Li + Cl_2 + H_2O \rightarrow LiCHCl-CH(OH)-Li + HCl$

Chlorination of pulp results in a loss of methoxyl groups from the lignin. Demethylation has been shown to be a chlorine-catalyzed hydrolysis that produces methanol and a free-phenolic-type hydroxyl group in the lignin. The free phenolic group makes the lignin more hydrophilic and more readily soluble in water and alkali.[248,249] Substitution of chlorine into the aromatic portion of the guaiacylpropane unit is a rapid reaction and is very probably the dominating reaction in the initial phase of pulp chlorination. The reaction is so rapid that diffusion of chlorine into the pulp probably determines the overall rate of reaction. From chlorination trials in the laboratory and from experience with chlorination in mills, it is well known that a rapid and thorough mixing is necessary in order to obtain uniform chlorination of the pulp.

The rate of the chlorination reaction decreases with increasing pH and since hypochlorous acid should cause more oxidation and less substitution than chlorine, oxidation must be a slower reaction than substitution. However, oxidation is an important part of the chlorination reaction because acidic groups are formed and these make the lignin more readily soluble in the subsequent alkaline extraction. The formation of dicarboxylic acids by chlorination probably goes by way of orthoquinones. Quinones are easily polymerized to dark-colored compounds and these could be responsible for the color of chlorinated pulp. The importance of addition reactions during chlorination of pulp is rather uncertain, but chlorine can be added to unsaturated side chains in the lignin molecule, and in sulfite pulps addition of chlorine to unsaturated extractives is very probable. Addition of hydrogen chloride to quinones formed during the chlorination is also a possible reaction.

Investigations on the spent liquor from chlorination have shown that the aromatic structure of the lignin is destroyed to a considerable degree.[250,251] Among the low-molecular-weight compounds formed during chlorination are methanol, acetaldehyde, oxalic, glucolic, malonic, maleic, fumaric, succinic, chloromaleic, and chlorofumaric acids, all of which have been detected in the chlorination waste liquor.[252] Lindström and Nordin[253] isolated a number of phenolic compounds from chlorinated sulfate pulp, as shown in Table 5–4. Nonpolar chlorinated compounds have also been detected in spent liquor from bleaching of sulfite and sulfate pulp.[253a] Erickson and Dence[254] studied the occurrence of phenolic oligomers in spent liquor from chlorination, and 11 oligomers were detected in the case of sulfate pulp. Mono-, di- and tri-chloro-lignin oligomers were detected. The presence of 5- and 6-linked condensed phenolic units containing from 0 to 2 chloro substituents per phenolic ring was also shown.

Less than 10% of the chlorine compounds are organically bound in the spent liquor from chlorination of sulfate pulp.[255,256] Chlorination of isolated lignin

TABLE 5–4 SUBSTANCES FOUND IN SPENT LIQUOR FROM CHLORINATION AND ALKALINE EXTRACTION.[a]

Number of Isomers Found	Compound	Concentration			
		Chlorination Stage		Extraction Stage	
		(mg/m^3)	(g/t pulp)	(mg/m^3)	(g/t pulp)
–	Dichlorophenol	+	–	+[b]	–
–	2, 4, 6-Trichlorophenol	25	0.9	115	1.8
–	2, 6-Dibromophenol	+	–	+	–
3	Dichloroguiacol	+	–	+	–
3	Trichloroguiacol	+	–	35	0.5
–	Trichloroguiacol	+	–	168	2.7
–	Trichloroguiacol	+	–	–	–
–	Dichlorocatechol	+	–	+	–
2	Trichlorocatechol	18	0.6	+	–
–	Trichlorocatechol	83	3.0	+	–
–	Tetrachloroguiacol	+	–	137	2.2
–	Monochloropropiovanillone	+	–	+	–
–	Tetrachlorocatechol	32	1.6	42	0.9

[a]See reference 252.
[b]+ indicates compounds detected but not determined quantitatively.
Pulp: pine kraft pulp, kappa number 31.6.
Chlorination: 3.5% consistency, 20°C, 60 min, 6.64% Cl$_2$.
Extraction: 8% consistency, 50°C, 120 min, 2.8% NaOH, final pH 10.6.

showed that the dissolved lignin had a higher degree of chlorine substitution and was more oxidized than the fraction that was not soluble in water.[250] The composition of the lignin remaining in the pulp after chlorination is not known, but some of the configurations found in dissolved lignin are probably also present in the chlorinated pulp. If chlorine dioxide is applied prior to chlorination, oxidation will be a more important part of the reaction. The lignin will be more soluble in water and will contain fewer chlorinated compounds. The use of a mixture of chlorine and chlorine dioxide may also give a lower content of chlorinated substances. Chlorination of pulp is a heterogenous process, and the rate of reaction is dependent on physical conditions, such as fiber morphology and the rate of mixing of pulp and chlorine. In the first stage of chlorination, diffusion into the fibers determines the rate of reaction.[257]

In the alkaline extraction of chlorinated sulfate pulp, the principal reaction is one in which the lignin compounds are dissolved. This is brought about partly by ionization of the acidic groups formed in the lignin during chlorination and partly by degradation of these compounds. Kempf and Dence[258] found that 60 to 90% of the chlorine bound to the lignin is split off in the alkaline extraction. Chlorine attached to an aromatic or alkene carbon atom is usually stable in alkali, whereas alkane chlorine bonds can be broken. If a double bond is conjugated with a carbonyl group, the chlorine can be split off by alkaline hydrolysis. Such alkali-labile chlorine atoms have been shown to be present in chlorinated quinones[259,260] and in chlorinated chatecols.[259] When chlorine is split off from such compounds, it is partly replaced by hydroxyl groups, which increase the solubility in alkali.

The alkaline extract from chlorinated sulfate pulp contains the same monomeric compounds as the spent liquor from the chlorination stage and some of them in considerably higher concentrations.[253] Thirteen phenolic and chlorophenolic oligomers were detected in the alkaline extract from chlorinated sulfate pulp[254] and some of these have been isolated and tested for toxicity.[261] The compounds found in the alkaline extract have a higher average molecule weight than the compounds in the chlorination spent liquor. The probable reason for this is that high-molecular-weight compounds are dissolved more easily in alkaline than in acid solution. An increase in molecular weight by condensation during the extraction may also be another explanation. Alkaline extraction, especially in the case of sulfate pulp, results in an increase in the light-absorption coefficient and a reduction in brightness. The structure of the colored compounds in the extracted pulp and in the extract is not known. Hardell and deSousa[255] found that the color of the dissolved substances was mainly attributable to high-molecular-weight compounds. The color of the extracted pulp is improved if oxidative compounds, such as hypochlorite and peroxide are used in the alkaline extraction.[194,262]

Leach and Thakore[261] isolated chlorinated resin compounds in the alkaline extract of sulfate pulp. In a well-washed sulfate pulp, the resins and fats should be removed as soaps formed in the cooking process,[263] but if washing is inadequate, some of these compounds can be left in the unbleached pulp and end up as chlorinated compounds. Sulfite pulps will always contain some extractives and it is a well known fact that when these compounds are chlorinated they become very sticky and may cause pitch trouble. The chlorinated extractives from hardwood pulps are especially difficult to remove.[264] In the alkaline extraction the alkali will form soaps with the fatty acids and these will solubilize some of the nonsaponifiable compounds. A similar effect will be obtained if surface-active agents are used as additives in the alkaline extraction.

The reaction between chlorine and the carbohydrates in the pulp is much slower than with lignin. However, chlorine will attack the carbohydrates even under normal conditions of chlorination and the effect can be severe at high temperature. The extent of attack is usually determined by viscosity measurement, which tends to exaggerate the effect because the viscosity determination is usually carried out in an alkaline medium. The carbohydrates can be attacked by chlorine radicals or by other radicals formed during chlorination, as shown in a series of investigations in Sweden.[245,247,265,266] The formation of radicals is influenced by light and the attack on the model carbohydrate, beta-D-glucoside, is considerably faster when the samples are illuminated than when they are kept in darkness.[247] Formation of radicals can also occur without illumination. The total effect on the carbohydrates by chlorine cannot be explained by radical reactions because some degradation takes place even in the presence of chlorine dioxide or other radical scavengers. The nonradical reactions during chlorination are probably oxidation and hydrolysis,[267] and mechanisms for such reactions have been proposed.[268] The reaction between the carbohydrates and chlorine is not specific and radical reactions and oxidation can attack at more than one site in the chain unit.[269,270] Carbonyl and carboxyl groups are formed in carbohydrates by chlorination. The carbonyl groups are alkali labile. If the carbonyl group is in the beta position to the carbon atom, which connects the glucose unit to the next unit, the carbohydrate chain will be easily split. It has been shown that colored compounds can be formed by alkali treatment of oxidized carbohydrate models.[271,272] The peeling reaction, where the alkaline degradation starts from an aldehyde end group and proceeds stepwise, is an important reaction in the cooking of sulfate pulp. Under normal condition of alkaline extraction this reaction probably does not take place because it requires a temperature above 90°C.[242]

Reactions with Hypochlorite

Hypochlorite is a rather unstable compound especially in the pH region 6 to 7, where it decomposes into chloride and chlorate as follows:

$$3HOCl \rightarrow ClO_3{}^- + 2Cl^- + 3H^+$$

This is the reaction used in the manufacture of chlorate on an industrial scale.

Bromine will increase the rate of reaction by forming hypobromite, which acts as a catalyst in the reaction.[273,274] Hypochlorite can also decompose into oxygen and chloride as follows:

$$2OCl^- \rightarrow 2Cl^- + O_2$$

This reaction is accelerated by light.

The reactions in hypochlorite bleaching were studied extensively in the early days of pulp bleaching when hypochlorite was the principal bleaching agent used. In the last decade, chlorination, alkaline extraction, and chlorine dioxide bleaching have received far more research attention. Hypochlorite bleaching has not been investigated with the more sophisticated tools and techniques that have been applied to these newer bleaching processes. The lack of knowledge is especially noticeable for reactions between lignin and hypochlorite.

Changes in brightness, permanganate number, and lignin content during hypochlorite bleaching show that the lignin is attacked and that the reactions involve a decoloration and also a lignin removal. At very high hypochlorite consumption, the absorption peak at 280 nm disappears, indicating that the aromatic structure is destroyed.[275] Chlorine analysis of lignin preparations treated with hypochlorite at pH 11 indicates that chlorination is a possible reaction under alkaline conditions. This would be expected inasmuch as phenols are chlorinated under similar conditions.[275] Oxidation of guaiacol with hypochlorite gives methanol, oxalic acid, and maleic acid. Four carboxyl equivalents were formed from the consumption of 6 moles of HOCl.[276] Reactions with model substances indicate that hypochlorite attacks lignin through the free phenolic hydroxyl group or through phenol ethers attached to the alpha or beta position in the phenylpropane structure.[275,277,278] Carbon dioxide and organic acids are formed when hypochlorite reacts with lignin.

The reactions with carbohydrates are very important in hypochlorite bleaching. The degrading action of hypochlorite on cotton cellulose was investigated between 1920 and 1930 and it was found that the maximum rate of degradation occurs between pH 6 and 7[198] The number of carbonyl groups in the cellulose shows a maximum in the same pH range. The carbonyl groups are oxidized to carboxyl groups and the rate of oxidation is dependent on the pH. At high pH, carbonyl groups are oxidized more rapidly than they are formed, whereas between pH 6 and 7 the formation of carbonyl groups is faster than their oxidation to carboxyl groups.[279,280] Theander[281,282] found a maximum rate of hypochlorite consumption at approximately the neutral point, using methyl beta-D-glucopyranoside as a model, but the maximum yield of neutral compounds was obtained at pH 4, where hypochlorous acid is almost the only active chlorine compound present. Carbonyl groups were found to exist at carbon atoms 2, 3, and 6 and, as can be seen from Figure 5–21, all three of these are alkali-labile groups and can result in chain splitting.

Oxidation of cellulose with hypochlorite has been shown to produce D-arabinose; D-glucose; gluconic, glyoxylic, and erythronic acids;[281,283] oxalic

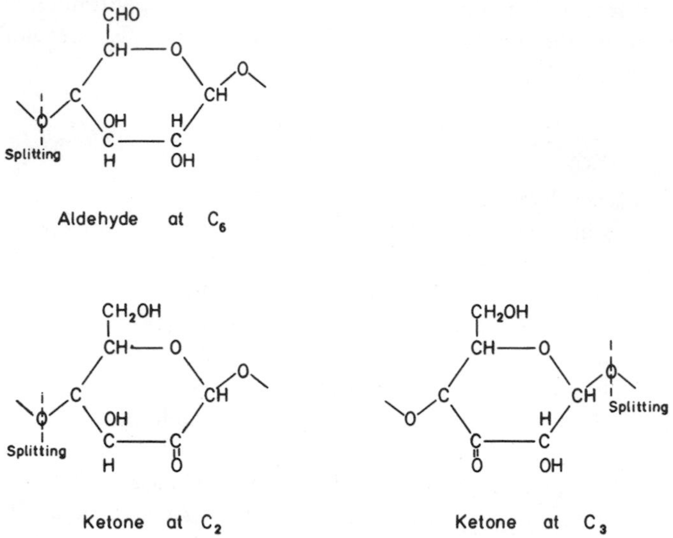

Figure 5-21. Beta-elimination in oxidized anhydroglucose units.

acid; carbon dioxide;[283] and, as already pointed out, aldehyde, ketone, and carboxyl groups are formed.[284] Hydrolysis of cellulose oxidized with hypochlorous acid gives erythrose, arabinose, and xylose, the formation of xylose probably being caused by decarboxylation of glucuronic acid.[285] Oxidation of cotton with hypochlorite at pH 8.2, followed by partial hydrolysis of the oxidized product showed that carboxyl groups were produced at the carbon atoms in the 1, 2, and 6 positions.[286] Andersson and Samuelson[287] found that oligomers having 2 to 7 glucose units were produced in hypochlorite oxidation of cellulose. Small amounts of oligomers with arabinose or erythrose end-units were also produced. The most abundant monomer was glucose, but arabinose and erythrose were also detected. Considerable amounts of oligomers having 2 to 8 units that were acidic in nature were found. Glyceric, glycolic, and formic acids were the most abundant of the monobasic acids, and oxalic acid was the most common dibasic acid. Gluconic, arabinonic, erythronic, and several other acids were isolated and identified.

A free radical mechanism has been proposed for one of the reactions taking place in the oxidation of the model compound, methyl beta-cellooligoside, using hypochlorite at pH 9.[288] Compounds of nickel, cobalt, and copper accelerate the oxidation of carbohydrates with hypochlorite. Bromide and hypobromide also accelerate the reaction between pulp and hypochlorite.[205] Catalysis of hypochlorite bleaching, using metallic compounds, is somewhat uncertain and is difficult to predict.[205,206] Sulfamic acid acts as an inhibitor for hypochlorite oxidation. Urea has a dual effect: it increases the reaction rate in the first phase of bleaching and inhibits the rate of reaction at the end of the bleaching.[289]

Reactions with Chlorine Dioxide and Chlorite

If chlorine dioxide is reduced in acid medium, the final chlorine compound should be chloride. This indicates that chlorine dioxide has five oxidation equivalents, and if chlorine dioxide is calculated as available chlorine, the content of chlorine dioxide as ClO_2 must be multiplied by 2.63.

The chemistry of chlorine dioxide bleaching is, however, rather complicated and chlorite, chlorate, and hypochlorite are formed in the reactions. In alkaline medium, chlorine dioxide is mainly reduced to chlorite, which is inactive as a bleaching agent in this pH range; or it can be oxidized to chlorate, which also is inactive in this pH range. In acid medium, chlorous acid can be formed. Chlorous and hypochlorous acids are only intermediates because they react to produce chlorine dioxide as follows:[290]

$$HOCl + 2HClO_2 \rightarrow 2ClO_2 + H^+ + Cl^- + H_2O$$

A similar reaction takes place between chlorine and chlorite:

$$2ClO_2^- + Cl_2 \rightarrow 2ClO_2 + 2Cl^-$$

The reaction between chlorite and hypochlorous acid or chlorite and chlorine produces chlorate as follows:

$$ClO_2^- + HOCl \rightarrow ClO_3^- + Cl^- + H^+$$

$$ClO_2^- + Cl_2 + H_2O \rightarrow ClO_3^- + 2Cl^- + 2H^+$$

Formation of chlorate in chlorine dioxide bleaching represents a loss of bleaching capacity.[291] Spent liquor from chlorine dioxide bleaching of sulfate pulp usually contains chlorate corresponding to 20 to 36% of the chlorine dioxide applied.[292] Chlorine dioxide reacts with lignin and chlorite is formed.[293] At low pH, the chlorite will be present as its corresponding acid, and the chlorous acid reacts rapidly with lignin and forms chlorate as one of the reaction products. As the pH rises the concentration of chlorite increases. Optimum brightness is achieved at the pH where the sum of chlorate and chlorite is at a minimum. The presence of chloride ions inhibits the formation of chlorate from chlorite.

Chlorine dioxide is applied in either the first or the final stages of bleaching and its reaction with the components of the pulp differs depending on the stage in which it is used. The role of chlorine dioxide as radical scavenger during chlorination has already been mentioned. In this role chlorine dioxide is oxidized to chlorate and the chlorine radical reduced to chloride.[242]

Under normal bleaching conditions the carbohydrates in the pulp are not attacked by chlorine dioxide to any noticeable degree and this is the main advantage of chlorine dioxide bleaching. Because chlorine dioxide does not normally degrade the carbohydrates in the pulp, most of the studies have been directed toward the reactions with lignin. Lignin-model compounds have been used in most of the work carried out in this field.[294-300] Phenolic-model substances react rapidly with chlorine dioxide, but the reaction between non-phenolic-model compounds and chlorine dioxide is comparatively slow. The results indicate that chlorine dioxide attacks phenolic groups in the lignin during the first part of the bleaching process. Model substances of the guaiacyl-type reduce chlorine dioxide to chlorite and chlorine. The chlorine leads to the formation of chlorine substituted compounds in the reaction products. Because chlorite reacts with chlorine in acid medium the chlorine should exist only temporarily if the bleaching is started with pure chlorine dioxide, but substitution of chlorine in lignin or in lignin-model compounds is so rapid that this substitution reaction is possible in the presence of chlorite. Commercial chlorine dioxide always contains some chlorine, and in the initial phase of bleaching the concentration of chlorite may be insufficient to react with all the chlorine present.

Quinones are probably formed by chlorine dioxide treatment of lignin-model compounds. Derivatives of labile quinones have been prepared by reaction with benzensulfinic acid.[299] Dark-colored products are formed when guaiacyl compounds are oxidized with chlorine dioxide. These products are probably formed by polymerization of o-quinones, and the dark color indicates that quinone groups are present in the polymerized compounds. Lindgren[299] has proposed a mechanism for the reaction of guaiacyl compounds with chlorine dioxide, whereby the compounds are first oxidized to phenoxy radicals and the chlorine dioxide is reduced to chlorite. These radicals then react with chlorine dioxide, hypochlorous acid is produced, and the lignin-model substances are oxidized to muconic acid structures. Oxidation of stilbene compounds with chlorine dioxide does not show any general reaction scheme, but indications are that conjugated double bonds are split by chlorine dioxide and CO groups are formed.[301] The reactions of chlorite with lignosulfonic acid and the lignin in sulfate pulp are retarded by compounds that react with chlorine and are accelerated by chlorine and formaldehyde. Formaldehyde and chlorite react and form chlorine.[302] Because chlorite and chlorine react to form chlorine dioxide, the main reactions probably are the same as those with chlorine dioxide. The formation of chlorate as a result of the reaction of chlorine dioxide with lignin model compounds is higher in the presence of aliphatic double bonds than with phenolic groups.[303] The mechanism of chlorate formation is probably through reduction of chlorine dioxide by the lignin to produce chlorine or the ClO radical, which in turn oxidize the remaining chlorine dioxide to chlorate.[304]

Reactions between lignin and chlorine dioxide have been studied for the delignification of wood,[305] but the reaction mechanism for chlorine dioxide bleaching in the final stages has apparently not been studied to any extent. The rate of reaction in the last stage bleaching of sulfite pulp with chlorine dioxide has been determined, and a maximum rate was found at pH 4.5 and a minimum at pH 7.[306] The reactions between chlorine dioxide and residual lignin in the final bleaching stages are difficult to study because the lignin content is low and the lignin at that stage is modified so much that the common model compounds for lignin do not represent the true situation. The main effect is a decoloration. Some of the compounds formed are apparently only slightly soluble in acid because a two-stage chlorine dioxide treatment with an intermediate alkaline extraction is required in order to obtain a high-brightness pulp.

Reactions in Peroxide Bleaching of Chemical Pulps

The reactions between lignin and peroxide have mainly been studied in connection with the bleaching of high-yield pulps. The mechanism of color removal when peroxide is used for multistage bleaching of chemical pulps is not known. Investigation of these reactions in the final bleaching stages is difficult because the amount of colored substances is low, and the chemical structure of these compounds is unknown. Oxidation of carbonyl to carboxyl groups will take place and this may impart a stabilizing effect. Because peroxide bleaching is carried out in alkaline medium, a degradation of the carbohydrates will result if carbonyl groups are introduced in a preceding stage.

Oxygen Bleaching of Chemical Pulps

Bleaching of chemical pulps has always employed one or more oxidizing stages. Oxygen is the most abundant oxidizing agent, and delignification and bleaching with oxygen were tried at an early stage of pulp bleaching. Nikitin and Akim[307] reported successful bleaching with oxygen and alkali in 1956. However, the carbohydrates were degraded too much, and the strength properties of the pulp were not comparable to those of conventionally bleached pulp. The next step forward was the discovery that certain inorganic compounds acted as protectors for the carbohydrates during oxygen bleaching, magnesium compounds having a specially favorable effect in this connection.[308,309] Since then, research on oxygen bleaching has been carried out in almost every laboratory where pulp bleaching is studied. The industrial application of oxygen bleaching was brought about primarily by environmental problems resulting from the conventional chlorine bleaching processes. The first full-scale, commercial oxygen bleaching was done in 1970[310] and a large-scale pilot plant was started the same year.[311] Today a large number of mills are using oxygen bleaching or have decided to use this method. Most of these mills use oxygen in the first stage, followed by a conventional bleaching sequence using chlorination subsequent to the oxygen stage.

Oxygen bleaching must be carried out in alkaline medium in order to obtain an acceptable rate of reaction without using a very high oxygen pressure. For ecological reasons the washwater from the oxygen stage should be used for washing the unbleached pulp, and the chemicals should be recovered together with those in the spent liquor from the cooking process. The alkali used in the oxygen stage should, therefore, have the same cation as that used in the cooking liquor. For the bleaching of sulfate pulp, sodium hydroxide, sodium carbonate, and white liquor can be used as the alkali, but if white liquor is used, it should be oxidized before the bleaching.[312] For the bleaching of sulfite pulp, sodium hydroxide, ammonia, magnesium and calcium hydroxides have been used as the alkali.[313-318] At one time, a pH of 11 was considered to be the lower limit for removal of lignin in oxygen bleaching,[319] but this limit is dependent on the temperature and to some extent on the oxygen pressure used in bleaching.

For sulfate pulps, oxygen bleaching is carried out between 90 and 120°C.[320] Bleaching of sulfite pulp, using sodium hydroxide as the alkali, requires about the same temperature as sulfate pulp. If magnesium hydroxide is used as the alkali, the pH and the rate of reaction will be lower and a reaction temperature between 130 and 140°C must be used if reasonable reaction times are to be obtained.[317] Reaction times in oxygen bleaching, ranging from 20 min to 4 hr have been used in laboratory work, but in mill-scale bleaching the retention time is 30 to 60 min.

Pulp consistency is an important factor in oxygen bleaching. Low as well as high consistency can be used, but oxygen bleaching is difficult to carry out between 15 and 20% consistency. All full-scale, oxygen bleaching plants have so far used the high-consistency range between 22 and 30%. Because the bleaching agent is a gas, the pulp must be fluffed to obtain a sufficiently large contact surface between the pulp and the gas. The gas space in the pulp is dependent on the consistency and on the pressure. The consistency of the pulp must, therefore, be adjusted to the depth of the pulp bed.[321] For bleaching at low to medium consistency, another technique can be used to obtain an interaction between the pulp and the oxygen,[322] whereby the oxygen is mixed with the pulp in a disk refiner or other type of intensive mixing equipment. This disperses the gas in small bubbles and mixes it intimately with the pulp, which are important factors for obtaining a sufficiently high rate of reaction. Oxygen bleaching can be carried out at low consistency, and laboratory and pilot scale trials at 3% consistency indicate that such a process can be used on a mill scale.[323]

The removal of lignin during oxygen bleaching, expressed as a reduction in the kappa number, is mainly a function of alkali charge, retention time, and the temperature used in the bleaching operation.[324] If the alkali charge is high, delignification will take place at lower pressures,[325] but the normal range of oxygen pressure is between 390 and 780 kPa. The consumption of oxygen is seldom determined in laboratory work, and the values reported in the literature are mainly from full-scale production.[322,326,327] The amount of oxygen required is related to the reduction desired in kappa number. There is also a relationship between the amount of oxygen required and the amount of alkali used in the bleaching, but this relationship is not linear[328] due partly to the bleeding off of gas in order to avoid explosions caused by too high concentrations of turpentines and carbon monoxide in the gas phase above the pulp bed. The carbon monoxide content should be kept below 1%.[329] Loss of oxygen through oxidation of residual black liquor in the pulp and from the discharging of the pulp from the reactor makes the consumption of oxygen appreciably higher than required for the delignification. For commercial scale operation, between 24 to 30 kg of oxygen per ton of pulp are used.[326,327]

The brightness of oxygen-bleached sulfate pulp is related to the kappa number[324] but, as one might expect, this relationship is not linear. Oxygen is not a specific delignification agent, and the carbohydrates are attacked simultaneously with the lignin. As a result, there is a relationship between viscosity and kappa number in oxygen-bleached pulps. The reduction in viscosity with reduction in kappa number increases with the bleaching temperature.[324] The decrease in viscosity is also dependent on the amount of alkali used when the bleaching is carried so far that the residual alkali is low.[324] A kappa number between 15 and 20 is generally considered to be the lower limit in the oxygen bleaching of softwood sulfate pulp without noticeable reduction of the physical strength of the pulp. Hardwood sulfate pulp can be bleached to a kappa number of 10 without a decrease in physical strength in the pulp.[327] For sulfite pulp, the kappa number can be reduced below 10 and even below 5 without appreciable effect on the strength properties.[313,315,316]

In the case of sulfate pulp, the yield is about the same with oxygen bleaching as with conventional bleaching. A gain in total yield can be obtained if the cooking is stopped at higher kappa number because the removal of lignin is more selective with oxygen bleaching than in the last phase of sulfate pulping.[330-333] Cooking to a high yield and completing the delignification with oxygen and alkali has been studied for sulfate[334] and soda pulps[335,336] from southern pine. For sulfate pulp, oxygen-alkali delignification gives pulp of similar quality as sulfate digestion to the same kappa number, but pulps prepared by soda cooking and oxygen-alkali delignification are inferior to sulfate pulp. Birch pulps prepared by the same sequence gave promising results.[337]

For sulfite pulp, bleaching with oxygen and sodium hydroxide gives a lower yield than conventional bleaching, but if magnesium hydroxide is used as the alkali, the yield will be the same as with conventional bleaching.[317] Delignification of high-yield monosulfite and neutral sulfite pulps, using oxygen and alkali, produces pulps with a yield up to 13% higher than that obtained with sulfate pulping at the same lignin content.[338] Compared with linerboard sulfate pulp the gain in yield was 10 to 17%.[339,340] The properties of birch sulfite pulp can be changed in the direction of sulfate pulp by oxygen treatment.[341]

Reactions with Lignin in Oxygen Bleaching

Oxidation of lignin in alkaline medium has been studied with several different objectives in mind: (1) to determine the structure of lignin, (2) to understand the reactions during oxygen bleaching, and (3) to produce new chemicals from lignin. Production of vanillin from sulfite spent liquor is an example of the latter case. In this case, air is used for oxidation at 180 to 200°C. For oxygen bleaching of pulp, the temperature is usually between 100 and 120°C.

Kratzl, and others,[342,343] studied the oxidation of lignin-model compounds in alkali at 70°C. Their results indicated that the primary reaction is formation of resonance-stabilized phenoxylradicals and oxidative coupling to *o-o*-diphenyls. Oxidation of 2-methoxy-4-methylphenol (creosol) yielded about 60% of a dicreosol plus small amounts of dark-colored condensation products. For guaiacyl derivatives with alpha-keto- or alpha-hydroxy groups in the side chain the oxidative attack may start at the alpha atom in the side chain. Eckert, and others,[344] treated lignin-model compounds with oxygen in alkali at 50°C and concluded that the reaction between oxygen and lignin is exceedingly complex with several reactions apparently taking place simultaneously. In addition to the reactions proposed above,[342] formation of epoxidated structures and ring cleavage with carboxylic acids as final products were found.

Gierer, and others[345] studied reactions between oxygen and dihydroxystilbene structures in alkali. Dihydroxystilbene structures were used because these are formed in lignin during alkaline delignification. Quinonemethide radicals were common intermediates in these reactions, arising from a one-electron oxidation step. Radicals from 2, 4-dihydroxystilbene underwent ring closure to give 2-arylcoumaran-3-yl radicals, and this intermediate either dimerized or was oxidized to 2-arylcoumarones. Quinonemethide radicals from 4, 4-dihydroxystilbene were oxidized to 4, 4-stilbenequinones, which were subsequently oxidized to benzils.

Gierer, and others,[346] also studied the degradation of phenolic lignin units of the beta-arylether type with oxygen and alkali. Formation of the degradation products could be explained by three different pathways. Oxidative elimination of the side chain by oxygen was the dominant process, but side-chain elimination by hydrogen peroxide is also a possible reaction mechanism under more drastic conditions. Hydrogen peroxide may be formed by autoxidation of certain lignin structures and intermediates. Oxidative cleavage by oxygen of the bond between the alpha and beta carbon atoms is another possible pathway. The primary oxidation products were aliphatic and aromatic aldehydes and *p*-quinones, which by further oxidation were degraded to aliphatic acids.

Oxidation of isolated sulfate lignin with oxygen in alkaline medium showed that the consumption of alkali was proportional to the amount of degraded lignin.[347] Potassium hydroxide was more effective in this process than sodium hydroxide. Spectrophotometric studies of oxidized lignin indicated that the aromatic structure was at least partly preserved in the reaction products and that conjugated double bonds and alpha-carbonyl groups were present. Oxidation in the presence of possible catalysts did not increase the amount of degraded lignin.[348]

Reactions with Carbohydrates in Oxygen Bleaching

The reactions between carbohydrates and oxygen in alkaline medium have been studied using model compounds, cotton cellulose, and hemicellulose. In studies[349,350] in which hydrocellulose was treated with oxygen, only a small amount of metasaccharinic acid was formed and the end groups consisted mainly of aldonic acids, such as arabinonic, mannonic, and erythronic acids. Cellobiose used as a model compound for the study of the degradation of carbohydrates by oxygen and alkali was found to produce more than 20 nonvolatile acids, formic acid, and carbon dioxide.[351] Comparative experiments were carried out with D-glucosone and monosaccharides, and the results seem to confirm the suggestion of Samuelson and Stolpe[349] that D-glucosone is an intermediate product in the degradation with oxygen. Ericsson, and others,[352] also studied the oxidation of D-glucosone and found that arabinonic acid and carbon dioxide were the main products from the oxidation. These results support the assumption that oxidation of glucose with oxygen partly goes by way of glucosone.

The stability of aldonic acid end groups in alkali was studied and compared with the stability of cellobiitol. In the absence of oxygen, glycosyl-arabonic acid was degraded almost completely in 2 hr in 1% NaOH at 135°C. Cellobionic and glucosyl-erythronic acid were only slightly degraded under the same conditions. Oxygen did not significantly influence the degradation rate for glucosyl-arabonic acid, but cellobionic and glucosyl-erythronic acids and especially cellobiitol were degraded much faster in the presence of oxygen than when a nitrogen atmosphere was used.[353]

Rowell[354] studied the stopping reactions when cellobiose was degraded in oxidative and nonoxidative alkaline solutions. In nonoxidative solutions, cellobiose was completely degraded to monomeric units. Under oxidative conditions as much as 27% of aldonic acids was formed, the principal acids being D-arabonic, D-mannonic, and D-erythronic, which were present in the ratios 6:2:1. Samuelson and Stolpe[355] used celobiitol as a model compound and found that it was oxidized under conditions used for oxygen bleaching. When cellobiitol was used alone, the initial reaction was slow, but the presence of glucose and sulfate lignin increased the rate of reaction. Peroxide is formed in the initial reaction and is consumed in a later stage in oxidation of carbohydrates. The content of peroxide increases in the presence of magnesium carbonate. Methyl glucosides have been used as model substances for studying the attack on the glucose unit within the cellulose chain.[356-359] Some of the reaction products indicate that nonterminal carboxyl groups are introduced into the cellulose chain by treatment with oxygen and alkali.[358] Using glucose containing ^{14}C, Kratzl and Schwarz[360] concluded that the oxidation of degradation products formed by reaction with alkali is a more dominant reaction than the direct oxidative attack on the glucose molecule.

Kolmodin and Samuelson[361] impregnated cotton cellulose with alkali and sulfate lignin and oxidized it under conditions similar to those used in oxygen bleaching. The results indicated that hydroxy groups in the cellulose chain were oxidized and that cleavage of the chain by beta-alkoxyelimination, followed by further oxidation, fragmentation, and rearrangement, gave arabonic, mannonic, and erythronic acids. Akim[362] found that oxygen-alkali degradation of cellulose requires a source for the formation of hydroperoxides and conditions for the decomposition of peroxides with formation of free radicals. Model compounds of hemicellulose have also been studied under conditions simulating oxygen bleaching. The reactions taking place are similar to those found for cellulose model substances.[363-366]

An important and active field of research since the discovery of the protective action of magnesium salt and other inorganic salts[308,309] has been the study of the influence of catalytic and protecting compounds in oxygen bleaching. Ericsson, and others,[367] showed that salts of cobalt, iron, and copper had a catalytic effect on the oxygen degradation of cotton linters. Phenolic-lignin models did not themselves act as catalysts, but they increased the catalytic effect of certain heavy metals. Lignin, xylan, and glucose also acted as catalysts, but magnesium salts had an inhibiting or protective effect under all conditions. The sequestering agents ethylenediamine tetracetic acid (EDTA) and nitrilotriacetic acid (NTA) accelerated the degradation of cotton linters but retarded the degradation of sulfate pulp. Samuelson and Stolpe[368] found that magnesium complexes were better protectors than magnesium salts against oxidative degradation of cellobiitol. They also found that complexes of intermediate stability were better protectors than extremely stable complexes, especially when iron was present.

Other protective agents against the degradation of carbohydrates by oxygen and alkali are cerium and silver salts and uranium dioxide. Triethanolamine is almost as effective as magnesium salts.[369] The effect of protecting substances is apparently associated with their ability to inhibit the formation of reactive intermediates by the action of peroxides. Addition of formaldehyde and glucitol results in a more selective delignification during oxygen bleaching, and the effect is additional to the protective action of magnesium salts. The results indicate that degradation depends on a combination of effects produced by catalytic transition metal ions and peroxides.[370] Iodide has a protecting effect that is additive to the effect of magnesium compounds.[371]

Samuelson, and others,[368,372-375] studied the degradation mechanism of carbohydrates with oxygen and alkali, using cellobiitol as a model compund[372-374] and using cotton and hydrocellulose.[375,376] The degradation of cellobiitol in the presence of iron was effectively retarded by magnesium and lanthanum salts, due apparently to a scavenging of transition metals. If magnesium hydroxide was added in excess, an additional protection effect was obtained. The coprecipitation of iron compounds with magnesium hydroxide is an important factor.[375] As noted by others, the attack increased with an increase in the concentration of copper and iron salts added. On the other hand, the catalytic effect of cobalt was highest at low concentrations, and this is explained by the fact that precipitated cobalt hydroxide produced at higher concentrations acts as an inhibitor. In the presence of iron salts the initial consumption of oxygen was slow, but if glucose was added the reaction increased rapidly.

In oxygen bleaching of cellulose, iron and cobalt compounds increase the formation of carboxyl groups along the cellulose chain. Hydrocellulose and cotton contain similar carboxylic end groups when oxidized with oxygen in alkaline medium.[376] Oxygen treatment of cellobiose showed that iron and cobalt salts, pH and temperature had significant influence on the formation of dicarboxylic acid in the reaction product.[377]

The protective effect of magnesium salts against oxygen attack on cellulose has been proposed as an interaction with a hexulopyranoside unit.[378] Sinkey and Thompson[379] showed in an investigation using methyl beta-glucoside as a model compound that oxygen-alkali degradation first produces hydrogen peroxide as an intermediate product, and then more stable organic peroxides are formed. Magnesium compounds stabilize the intermediate hydrogen peroxide by deactivating any catalytic ions. A mechanism by which the hydrogen peroxide produces free radicals that initiate a chain degradation of the carbohydrates has been proposed. The formation of hydrogen peroxide and organic peroxides in oxygen bleaching is discussed by McClosky, and others.[380] Experiments with kraft pulp showed that the degradation of the pulp is related to the consumption of oxygen. In the presence of transition metal ions, the oxygen consumption is increased,[381] but manganese (II) salts have a positive effect on the carbohydrates in oxygen-alkali bleaching.[381a]

The formation of carbon monoxide in oxygen-alkali reaction has been studied by means of model compounds. Carbon monoxide is produced from aldehyde sugars, and ferric ions accelerate this process. A parallelism exists between the formation of carbon monoxide and carbon dioxide. The reaction mechanism is not known except that it is a radical reaction.[382]

If bicarbonate is used as the alkali in oxygen bleaching instead of sodium hydroxide, the rate of reaction is lower, and higher temperatures must be used in order to obtain a rapid delignification. Copper salts have a strong catalytic effect both on delignification and degradation of carbohydrates in sulfate pulp. Cobalt salts act as a catalyst for the degradation of the cellulose, as has been shown to be the case with oxygen-sodium hydroxide in the presence of lignin.[383] The catalytic effect of transition metal ions in oxygen-alkali degradation of cotton linters showed striking similarity to the processes in the aging of alkali cellulose.[384]

A survey of the literature on the reaction mechanism taking place in oxygen bleaching has been published.[385] The reactions have obvious similarities to those that occur in oxygen pulping. Of the papers that have been published on alkali-oxygen pulping, only a few references have been given in this chapter.[386,387]

Reactions with Extractives in Oxygen Bleaching

Oxygen bleaching of sulfite pulp reduces the content of extractives to low values.[314] Pusa and Virkola[388-390] have in a series of investigations studied the removal of extractives in several bleaching sequences for sulfite pulp. The content of extractives in the pulp decreased to 0.2 to 0.3% when the bleaching sequences O-D-P and O-D/C-E-D-P were used. If the fines were removed by fractionation before the bleaching, the content of extractives was in the order of 0.1 to 0.2%. Oxygen bleaching by itself will not reduce the content of extractives to the desired level for all types of pulp, but in multistage bleaching if the sequences are started with oxygen, the pulp will have the minimum of extractives.

Other Bleaching Agents for Chemical Pulps

Sulfur dioxide is sometimes used as a final stage in pulp bleaching after an oxidative bleaching stage. The purpose of using sulfur dioxide is: to reduce any remaining oxidants, to acidify the pulp and thereby increase the brightness or to acidify to a low enough pH for removal of metallic ions. Sulfur dioxide treatment after bleaching achieves all three of these purposes. In chlorine dioxide bleaching, the pulp is acid and sulfur dioxide is used to prevent corrosion by the remaining chlorine dioxide. The sulfur dioxide is usually applied on the last shower or in the pulp storage tank. Sulfur dioxide or sodium sulfite can also be used between two bleaching stages, for example between chlorination and alkaline extraction.

Dithionite has been used to obtain a slight brightening of sulfate pulp. A few units increase in brightness can also be obtained with sulfite pulp using dithionite. Sodium borohydride has been proposed for a final treatment of bleached pulp to reduce the carbonyl groups and to stabilize the pulp against color reversion,[391] but this is not used commercially.

Several oxidative bleaching agents have been tried in the laboratory. Bromine can be used instead of chlorine or in combination with chlorine. Fluorine has also been proposed. Chlorine monoxide reacts very rapidly with pulp[392] and it possibly plays a role in chlorination and hypochlorite bleaching. Several peroxycompounds can be used for bleaching, but hydrogen and sodium peroxide are the only peroxide compounds used commercially.

Ozone has been tried for bleaching as well as for pulping.[393] The reaction is rapid but nonselective. Ozone has been tried in place of chlorine in the first stage of multistage bleaching of sulfate hardwood pulp. The degradation of carbohydrates resulted in low viscosity, but the strength properties were on the same level as for conventionally bleached pulp.[394] Softwood pulp is somewhat more attacked than hardwood pulp in bleaching sequences starting with ozone, but satisfactory pulp could be obtained from softwood as well as hardwood by oxygen pulping and ozone bleaching.[395] Ozone can also be used after an oxygen bleaching stage for either sulfate or sulfite pulp. Two-stage bleaching with oxygen and ozone will bleach sulfite pulp to a brightness above 90 and sulfate pulp to a brightness of 70 with as good strength properties as bleaching in the conventional way to the same brightness.[396,397]

Sequences Used in Multistage Bleaching of Chemical Pulps

The number of bleaching stages required in multistage bleaching is dependent on the wood species used, the type of pulping process, the degree of pulping, the quality of the process water and the level of brightness required in the finished product.

For pulps cooked by the sulfite and bisulfite processes, three to five stages are used to obtain fully bleached pulp. Common sequences for softwood sulfite pulp are C-E-H, C-E-H-H, and C-E-H-D. Other sequences that have been proposed or used commercially are C-C-E-H, C-C-E-H-H, C-E-H-D-H, C-E-D-D/H, and H-C-E-H. For environmental reasons one mill changed from the sequence C-E-H-D to E-C-H-D.[187] (See the section that discusses multistage bleaching of chemical pulps for explanation of bleaching symbols.) Sulfite and bisulfite pulp from hardwood is commonly bleached with the same sequences as sulfite pulp from softwood. However, because the resin in hardwood can be very difficult to remove after chlorination, it is desirable to use an oxidative stage before chlorination. Hypochlorite can be used for this purpose, but chlorine dioxide is more efficient.[185] Oxygen bleaching has so far not been used on an industrial scale for sulfite pulp, but laboratory work has shown sequences consisting of O-P, O-P-D, and O-D-P to be feasible.[313,315]

Sulfate pulp from softwood used in the production of newsprint, magazine paper, and other types of paper with a high content of mechanical pulp is usually bleached to the same brightness level as the mechanical pulp. The brightness level should, therefore, be in the region 60 to 70, but lower brightness values for the chemical pulp component can be used if the mechanical pulp is especially high in brightness. At these brightness levels the sulfate pulp is termed semibleached. A common bleaching sequence for semibleached sulfate pulp used in mixtures with mechanical pulp is C-E-H. Other sequences sometimes used are C-E-H-H and C-H-E-H. If peroxide is used as one of the stages, sequences of C-P, C-P-H-P, or C-P-H-D can be used to obtain the desired brightness level.[234−236,238] If oxygen is used, semibleached pulp can be obtained with sequences O-D, C-O, O-H, and O-C-E-H.[320]

Bleaching of softwood sulfate pulp to high brightness requires the use of many bleaching stages. Before chlorine dioxide became a common bleaching agent the sequences generally used were C-E-H-E-H, C-E-H-E-H-H, C-E-C-E-H-H, and C-C/H-E-H-H—these are still used to some extent. After chlorine dioxide and hydrogen peroxide were introduced as bleaching agents, the choice became more varied and among the sequences now used are C-H-E-D-E-D, C-E-H-D-E-D, C-E-H-E-H-D, C-H-D-E-D, C-E-D-E-D, C-E-H-D-P, C-E-H-P-D, C-E-D-P-D, C-E-H-E-D-P, and C-E-H-D-E-D-P.[236,238,400] Sequences with a mixture of chlorine and chlorine dioxide in the first stage are also being used and there is at least one company using chlorine dioxide as the main component in the first bleaching stage of softwood sulfate pulp.[401] With oxygen as one of the stages in full bleaching of softwood sulfate pulp, the bleaching sequences O-C/D-E-D-E-D and O-C-E-D-E-D are used. Oxygen bleaching combined with gas-phase bleaching will produce fully bleached pulp with the sequences $O-C_g-E-D_g$ or $O-D_g-E-D_g$,[402] where the subscript $_g$ represents a gas-phase bleaching with chlorine or chlorine dioxide.

Bleaching of hardwood sulfate pulp is somewhat easier than bleaching of softwood sulfate pulp. The same sequences as for softwood pulp can be used, but usually fewer stages are required. Extractives remaining in the pulp can be more troublesome in the case of hardwoods and chlorination is, therefore, avoided in some mills. In such cases, bleaching is usually started with a chlorine dioxide stage instead of a chlorination stage. Some of the sequences used for bleaching hardwood sulfate pulps are C-E-D-E-D, C-E-H-D-E-D, and D-E-D-E-D. If oxygen bleaching is used, the sequences can be D/C-O-D[403] and O-C-E-D-E-D.[327] The sequences E-D-P-D and P-D-P-D have also been proposed for bleaching of hardwood sulfate pulp.[404] For eucalyptus soda pulp the sequences C-E-H-H and C-E-D-H can be used.[405]

Process Procedures Used in Mill-Scale Bleaching of Chemical Pulps

Bleaching of chemical pulps is carried out to a much larger extent than bleaching of mechanical pulps. The technical developments in bleaching have also been mainly in connection with the bleaching of chemical pulps.

All bleaching methods have certain requirements in common. Among these are mixing of the pulp and chemicals and washing of residual chemical and by-products of the bleaching reaction from the pulp. If the required bleaching temperature is higher than the temperature of the pulp, hot water or steam must be applied to raise the temperature. Most bleaching reactions are comparatively slow and the pulp must be kept at the reaction conditions for the required time. Gas-phase chlorination and bleaching of refiner or thermomechanical pulp during pulping are exceptions, and in these cases the retention time can be very short. Bleaching of chemical pulp is a lignin-removal and to a lesser extent a carbohydrate-removal process, and the dissolved substances must be removed by washing after bleaching.

Bleaching of pulp was started as a batch process. Even chlorination began as a batch process. Continuous processes were introduced between 1930 and 1940, and the advantages over the batch methods soon resulted in continuous processes being used in all new bleach plants. The batch processes—using the Hollander, the Bellmer bleacher, or a vertical bleacher—had one advantage in that the pulp and bleach solution were in movement and there was some relative movement between pulp and liquid. In contrast, in the early continuous processes, the pulp moved in a tower in plug flow and there was almost no relative movement between pulp and bleaching solution. This condition still exists in the conventional bleaching tower.

A different concept of bleaching was introduced with the introduction of the idea of dynamic bleaching.[406] The idea was that a relative movement between pulp and bleach solution would increase the rate of bleaching. This proved to be true, but it was difficult to carry out on a mill scale with existing equipment. However, at about the same time, the continuous diffuser, which operates on the principle described in the section on washing processes, was introduced,[407] and this equipment was used for dynamic or displacement bleaching with success.[408,409] The first bleach plant based on this principle was started in 1975.[410]

The equipment used in the different bleaching stages depends on the bleaching chemical used, the time and temperature required, and the objectives to be achieved. Mixing of pulp and chemicals is a very important part of bleaching. Thorough and, usually, rapid mixing are necessary to obtain maximum bleaching efficiency. The most desirable mixer will depend on the pulp consistency and to some extent on the reaction rate of the process. Mixing can be carried out in special mixers or in pumps. When pumps are used, they serve two functions: mixing and transportation of the pulp. The latter is an important part of the continuous bleaching process. Mixing is also necessary when the temperature has to be increased with water or steam. Steam and chemicals are sometimes added to the pulp in the same mixer.

Washing of the pulp to remove the reaction products from the bleaching is an operation that is common to both batch and continuous bleaching. Only in a few cases is bleaching carried out without washing between the stages.

The process procedures used in the different stages of multistage bleaching of chemical pulps are discussed in the following sections.

Chlorination Process. Chlorination is usually carried out at low consistencies in the range from 2 to 4%. Mixing is comparatively easy in this concentration range if an ordinary chemical solution is used. However, chlorine is usually applied as a gas, which causes special mixing problems. Even aqueous solutions of chlorine can be difficult to mix with the pulp because of the low solubility and the high volatility of chorine. The rapidity of the reaction between chlorine and pulp requires a very fast and thorough mixing in order to obtain uniform chlorination of the pulp.

Upflow towers or a combination of upflow and downflow towers are used for chlorination.[411,412] In the early days of commercial chlorination, the towers were equipped with single mixers in the bottom of the tower and these did not provide sufficient mixing to obtain uniform chlorination.[413-415] Today, external mixers of several different designs are used to provide adequate mixing.[180,416-419a] Dispersion of the chlorine gas with water jets is also used for improving the uniformity of chlorination.[413,420] Poor mixing of chlorine and pulp can cause channeling in the tower.[414,421]

Chlorination towers are usually made of steel or concrete lined with tile, brick, rubber, or polyester resin.[420] Some of the newer towers are made of polyester resin reinforced with fiberglass.[410,422,423] The mixers are lined with rubber, tile, polyvinyl chloride, or polyester resin. The rotors of the mixers can be rubber-coated steel, 316 stainless steel, or Hastelloy C. Piping, valves, and other equipment in contact with wet chlorine must be constructed of corrosion-resistant materials.[424] If a mixture of chlorine and chlorine dioxide is used, rubber-lined equipment is not satisfactory. Titanium has proved to be resistant to mixtures of chlorine and chlorine dioxide[425,426] and it is being increasingly used in mixers and piping for wet chlorine and chlorine dioxide. However, titanium should not be used in contact with dry chlorine.[427,428]

Low consistency chlorination is usually carried out at the process water temperature because of the large amount of hot water or steam required to obtain a higher temperature. Medium consistency chlorination at a temperature of about 50°C has been proposed as a practical procedure. If a mixture of chlorine and chlorine dioxide is used in this process, the results seem to be promising.[180-182]

Gas-phase chlorination at a consistency of 30% has proved to be feasible in pilot-plant trials.[179] Gas-phase chlorination and medium-consistency chlorination at elevated temperature require only short reaction time and the reactors can, therefore, be much smaller in size than the towers used in conventional low-consistency chlorination.[429,430]

Alkaline Extraction Process. Hot alkaline extraction after a conventional chlorination requires a considerable energy input in the form of hot water or steam. High consistency in the extraction stage is therefore necessary in order to keep the energy consumption as low as possible. The alkali can be added to the pulp at: (1) the last shower on the preceding washer, (2) in a shredder-conveyor between the filter and the extraction tower, or (3) in the steam mixer or mixer of special design.

Because of the high consistency used, the extraction towers are usually the downflow type. The pulp is diluted at the bottom of the tower before it is pumped to the next washer, using washer water for dilution, as is commonly done in all washing processes. Upflow towers may be used for alkaline extraction. This is generally done when the bleaching towers are upflow and it is desired to have all towers in the bleach plant of the same type. A rapid extraction process consisting of a mix tank and a press has been proposed.[431] Extraction towers are usually steel towers lined with tile or rubber. The corrosion problems are not as severe in the extraction stages as in the other bleaching stages.

Hypochlorite Bleaching Process. Hypochlorite bleaching is carried out in either upflow or downflow towers. High-density upflow towers are used in some mills and such towers can also be used for alkaline extraction. A scraper removes the pulp from the top of the tower, and the pulp is diluted with white water before the washer. Hypochlorite towers are usually constructed of concrete and lined with tile or acid-proof brick. Reinforced polyester may also be used for hypochlorite towers.[432] Pumps are made of stainless steel, rubber-lined steel, PVC, or PVC-lined steel.

Chlorine Dioxide Bleaching Process. When chlorine dioxide bleaching is carried out in a downflow tower, there is a risk of accumulation of the gas in the top of the tower. This can cause puffs or explosions and, therefore, chlorine dioxide bleaching is usually carried out in upflow towers, although combinations of upflow and downflow towers are sometimes used. There are two types of combined upflow-downflow towers. One type has a small diameter upflow section outside the main downflow tower. The other type has a central upflow section inside the main tower; the stock enters the central section at the bottom of the tower and is scraped into the outer section at the top.

Because chlorine dioxide is very corrosive, high-grade stainless steel or titanium must be used in mixers and washers. The towers are constructed of tile-lined steel or concrete. Piping is made of high-grade stainless steel, polyvinyl chloride, or fiberglass-reinforced polyester.[433]

Peroxide Bleaching Process. Peroxide bleaching of chemical pulp is carried out in much the same way as peroxide bleaching of mechanical pulp. (See the discussion in the section bleaching of mechanical pulps.)

Washing Processes in Multistage Bleaching. Washing of the pulp is an important part of the bleaching operation and washing procedures must be closely scrutinized. In past years there were few, if any, environmental restrictions on the spent liquor from the bleach plant. This situation has changed. It is expected that dissolved organic compounds must be destroyed and the remaining chemicals recovered or removed from the effluent water. These environmental changes have a considerable influence on the methods of washing in bleach plants.

There are two principles used for the removal of soluble substances by washing: (1) dilution followed by thickening and (2) displacement. Theoretically, thickening from 1 to 10% should remove 90% of the impurities, but in practice this is far from achieved. Repeated dilution and thickening can and is being used for washing, but a large consumption of water is necessary to obtain good washing by this procedure. If the bleaching process is carried out at high consistency, the attendant thickening, which is done in presses, can be considered as part of the washing process.

Filters are the most common piece of equipment used for washing in the bleach plant. The washing effect on a filter may be achieved by thickening and by displacement. If the water used to dilute the pulp before it enters the filter is less contaminated than the water in the pulp, a washing effect by thickening will be achieved. Usually, however, the filtrate from the filters is used to dilute the pulp coming from the bleaching tower, and because the amount of water circulated in this way is usually very large, the washing effect by thickening is slight. A displacement washing effect is obtained on the filter by using showers. The amount of water used in the showers is usually far less than that used for dilution of the incoming pulp. Vacuum and pressure filters are used for washing. The most common type is the valveless vacuum filter, where a vacuum is obtained by means of channels in the filter drum. Other filter types have vacuum pumps or utilize drop-leg suction. Some newer types of presses operate on a combination of thickening, displacement, and rethickening. In this way, efficient washing is obtained without a large consumption of water.

Water consumption in the bleach plant can be reduced considerably by countercurrent washing. This procedure has been used for years in many bleach plants, but not to the extent proposed for plants with completely or almost completely closed white-water systems. Countercurrent washing will be covered in more detail in the discussion on the environmental aspects of bleaching.

Displacement washing can be carried out in a diffuser. This is an old technique for washing of unbleached kraft pulp, but it is rather new for the washing of bleached pulp. It was introduced with the continuous diffuser,[407] in which the washing is carried out at about 10% consistency at the top of the bleaching tower. The washing unit consists of concentric rings with perforated plates on both sides of each ring. Suction is applied between these plates, and the rings move upward at the same speed as the pulp for about 15 cm. The suction is then stopped and the rings are rapidly brought back to their starting position. In this way, clogging of the perforated plates is avoided. The washing water is transported in rotating arms and distributed through slots between the suction rings. There are several advantages to this washing method. First, no dilution of the pulp is required if the bleaching is carried out at about 10% consistency. The washing period is several times longer than for filter washing. Another advantage is that the washing can be carried out at the same temperature as the bleaching so that in a bleaching sequence with consecutive stages at the same temperature the heat requirement will be low. Some saving in electrical energy will also be obtained because pumping of large quantities of water is avoided.[434]

Displacement Bleaching Process. Displacement bleaching is closely related to displacement washing. The same type of equipment is used in both operations and the first experiments with large scale displacement bleaching were a continuation of displacement washing experiments.[408]

In its most advanced state, displacement bleaching uses no washing between bleaching stages, the reaction products from one stage being displaced by the chemical solution for the next stage.[409] In one example of displacement bleaching, chlorinated pulp is mixed with alkali before the displacement tower. In the displacement tower the bleaching sequence is E-D-E-D-W. The extract from the first alkali stage is sewered. Most of the chemical solution used in each of the following stages is recirculated and only that portion corresponding to the volume of makeup chemical solution is sewered. The consumption of chemicals is about the same in displacement and conventional bleaching, but the power and steam consumption are considerably lower in displacement bleaching and water consumption is only one-fifth of that in conventional bleaching. The space requirement is also much smaller than that for conventional bleaching based on towers and filters.[410]

Displacement bleaching with hypochlorite and especially with chlorine dioxide requires corrosion-resistant construction. The tower can be lined with acid-resistant brick. The diffuser must be made of titanium and other parts of the equipment should be made of such material as Hastelloy C or fiberglass.[410]

Another type of "dynamic" bleaching is pulse bleaching. The bleaching is carried out in alternating short periods from one to three minutes. A very rapid bleaching is obtained in this way, but the method has not been developed to an industrial stage so far.[435]

Gas-Phase Bleaching. Gas-phase bleaching with chlorine, ammonia, and chlorine dioxide has been carried out on a pilot scale, using the sequence C_g-NH_3-D_g-NH_3-D_g. Brightness and physical properties of the pulp are comparable to those obtained with conventional bleaching, and the costs for chemicals and steam are considerably lower for gas-phase bleaching.[179]

Mill-scale Oxygen Bleaching. High-consistency bleaching with oxygen in the range of 20 to 30% requires special dewatering of the pulp before bleaching. An acid pretreatment of the pulp is also necessary if it contains a high level of the metallic ions—manganese, iron, or copper.[436] The acid treatment can be carried out by addition of acid before thickening and washing on a filter.

The alkali required for oxygen bleaching should be mixed evenly into the pulp in order to obtain optimum bleaching response. Mixing of pulp and alkali can be carried out by: (1) adding in a mixer after thickening of the pulp, (2) adding before the thickening and recirculating the liquid from the thickening of the pulp, or (3) spraying on the fluffed pulp. The thickening can be carried out with a vacuum filter or a press. Fluffing of the thickened and shredded pulp is usually carried out after feeding into the top of the pressurized reactor. The reason for this is that the pulp should form a plug to avoid gas leakage during the feeding.

The high-consistency reactors used in oxygen bleaching are downflow towers. The Kamyr reactor has a series of decks and segments on which the pulp is cascaded down the tower.[327,329] The pressure on the pulp bed is thus kept low, and a consistency as low as 22% can be used.[327] The Sunds reactor is without special arrangements for reducing the pressure on the pulp bed, and the pressure on the pulp in the lower part of the reactor is so high that a consistency above 25% is required in order to obtain a sufficiently high gas volume in the pulp.[321] In the bottom of the reactor, the pulp is diluted to about 3% before discharging. Oxygen is charged at the top of the tower and the pressure is kept at a nearly constant value. Oxygen pressure in the reactors has been reported to be 750 to 950 kPa. Oxygen at this pressure reacts rapidly with pulp of high consistency. High-consistency oxygen bleaching is carried out at 115 to 125°C.

Reaction gases are carbon monoxide, carbon dioxide, and turpentines. Carbon dioxide causes no problem because it is absorbed in the alkaline liquor. Carbon monoxide and turpentines can give explosive mixtures with oxygen. Usually the reactor must be vented to avoid too high a concentration of these gases, and some oxygen will be lost in this operation. More oxygen will, therefore, be required than what is consumed in the bleaching process.[327] In one plant, the concentration of carbon monoxide has been kept below one percent without purging of the reactor.[329] In the vented gas, carbon monoxide and turpentines can be oxidized catalytically and the gases returned to the reactor.[436 a]

Fires and explosions are a potential source of trouble in oxygen bleaching. In the consistency range where oxygen bleaching is carried out, the pulp will not ignite even with an electrical heated filament. However, in the consistency range of 38 to 40%, the pulp can be ignited by an electrical heating filament in an oxygen atmosphere simulating the bleaching process. Therefore, the accumulation of dried pulp clinging to the reactor wall is a possible source of ignition, but contamination of the pulp with oil[437] or gas explosions are probably the greater cause of fires and puffs in the reactors. To prevent trouble, the reactors are equipped with multiple safety devices. If the pressure or temperature should rise above a safe level, the feed of oxygen is stopped, a valve is opened for reduction of pressure, and water showers are released. In addition, conventional mechanical safety valves and rupture disks are provided. Each of the systems is activated at different pressure.[437] Special lubricants must be used on equipment in contact with the oxygen in the reactor.

Oxygen in alkaline medium is not very corrosive to carbon steel. However, in the top zone of the reactor, where the oxygen is in the gas phase, pulp sticking to the wall of the reactor will gradually become acid and can cause corrosion. This was observed on the first mill-scale reactor, and a new reactor was lined with stainless steel 304 L in the exposed area.[329] Stainless steel 304 is also used for full lining of the reactor.[327] Because metallic ions affect the bleaching, the presses and fluffers should be made of corrosion-resistant materials.

Oxygen bleaching at medium consistency around 8% has been tried on a large scale.[322] Pulp and oxygen are mixed in a disk refiner and the mixture enters a central inverted cone of an upflow-downflow tower. The reaction is almost finished when the pulp reaches the edge of the cone and is scraped into the outer part of the tower. The pulp is diluted to 3% consistency before it is discharged from the reactor. The temperature ranges from 105 to 115°C and a reaction time from 30 to 40 min has been used. The oxygen pressure used is 500 kPa, and this pressure level is maintained by means of an air cushion. The turpentine content is held below 0.25% by degassing. The loss of oxygen caused by degassing is low, and the formation of carbon monoxide is appreciably lower than in high-consistency oxygen bleaching. The oxygen charge can be used to control the reduction in kappa number.

CONTROL PROCEDURES IN THE BLEACHING PROCESS

Process control in bleaching has been important from the very beginning, but the degree of control and the methods used have changed considerably over the years. Improved control has been one of the main advances in bleaching technology. A considerable part of the progress has been in connection with new bleaching processes and development of new process equipment.

General Process Control Procedures

Fundamental to all bleaching process control is the flow of pulp and chemicals.[438] The flow of pulp is usually measured by a magnetic flowmeter equipped with indicating instruments and recorders. The flow measured is the sum of pulp and water, and in order to know the pulp flow a consistency measuring device is needed. The consistency meter is usually a consistency controller that keeps the consistency at a preset value by addition of water. The pulp coming to the dilution point must have a consistency as high or higher than the preset value in order to obtain the desired result. The consistency controller should be equipped with an indicating or recording instrument as well as an alarm instrument for warning if the incoming consistency is too low. When the consistency is kept constant the flowmeter will give a signal proportional to the pulp flow. Flowmeters for chemicals have been used for many years. The common type of flowmeter is a slightly conical cylinder with a float. The same principle is used for gases as well as liquids. The addition of chemicals is often linked to the flow of pulp in such a way that the flow of chemicals is automatically changed to the appropriate value when the flow of pulp is changed.

Temperature control has always been an important part of pulp bleaching. The instruments used today may be somewhat more refined than earlier instruments, but the principles are the same. Control of pH is required in many bleach stages. Clogging of the electrodes with pulp or other substances is often a problem and several devices for cleaning the electrodes have been developed. At high consistency, pH measurement cannot be used in the bleaching stage proper, and the measurement has to be made after dilution of the pulp. This may result in a considerable time lag, and only changes taking place over a considerable time span can be detected if the dilution in the tower is carried out with water from the subsequent thickening stage.

Level control is necessary in many parts of the bleach plant including storage tanks for pulp and bleaching chemicals. In downflow bleaching towers, the level can be controlled by several different methods. One method is to measure the distance from the top of the tower to the pulp surface by measuring the time for the reflection of sound waves. Another method is to use radiation sources and receivers at two or three levels in the tower to keep the pulp-air interface within certain limits.[439] Strain gauges in the supporting construction of the tower can also be used to measure the weight of the pulp, thereby indicating the pulp level. Differential pressure cells are probably the most common device for level control in downflow towers. The level of the dilution zone at the bottom of the tower must also be controlled because otherwise all the pulp in the tower may be diluted. As the temperature of the liquid used for dilution is usually lower than the temperature in the bleaching stage, temperature control of the dilution zone will indicate if too much or too little liquid is being used for dilution. Automatic correction based on this information is possible. Level control is also used in washer seal tanks and other water tanks in the bleaching system. Such control becomes more important when reduction of water consumption and countercurrent washing have to be applied.

Channeling can take place in both upflow and downflow towers and it can result in a variation in retention time for different portions of the pulp. If the total volume of the pulp in the tower, the consistency, and the flow of pulp is known, the theoretical retention time can be calculated. The real retention time can be determined by making a sudden large change in the dosage of one of the chemicals and measuring the change at the end of the bleaching stage. Salts, such as potassium or lithium chloride, are often used as indicators because they can be easily measured. Sodium and calcium salts can also be used if they are not being used as part of the bleaching stage to be investigated. Radioactive tracer elements are also used, but special equipment, trained personnel, and safety precautions against nuclear radiation are necessary. The retention time can also be determined by introducing pulp impregnated with fluorescent dye and observing when the dyed pulp leaves the bleaching tower.[440]

Process Control in Chlorination

The control methods mentioned so far are more or less common to all aspects of bleaching and to other operations in the pulp and paper industry. More specific problems are encountered in the chlorination step, particularly in regulating the correct dosage of chlorine. Several factors cause difficulties in this control. The first is the pulp itself, which can vary greatly in its bleachability. Variations in the washing efficiency of the brown stock will influence the requirement of bleaching chemicals, and temperature changes in the process water will influence the reaction rate. It is very important to add the correct charge of chlorine to obtain good results in bleaching. If the dosage is too low, it is almost impossible to attain the required brightness, whereas an overdose of chlorine may adversely affect the strength of the pulp.

The most difficult problem to handle is the variation in lignin content of the pulp, which directly affects the requirement of chlorine. The direct measurement of bleach requirement, continuous or intermittent, before the actual chlorination is not practical because of the length of time required to make the test. Indirect methods are, therefore, used to control the chlorination process. There are two principal methods used for control: (1) measurement of the color of the chlorinated pulp and (2) measurement of residual chlorine.

Measurement of the color of the pulp as a control test was first used in the so-called "Black Widow,"[441] where the reflectance of light in one region of the spectrum was measured. In order to obtain an unambiguous answer from this type of measurement the lignin content of the unbleached pulp must be constant. Color measurement of the pulp can also be made by special instruments, such as the "Opticlor" and "Polybrite" systems, which are used in North America.[442–448] The earlier type of instrument was designed to measure the reflectance of light on the washer after chlorination is completed, but this procedure results in a considerable delay before the information can be transmitted to the control valve for chlorine dosage. The newer type of "Opticlor" measures the reflectance from the pulp shortly after the injection of chlorine.[444,448] The reflectance is measured in two different regions of the visible spectrum, one in blue light and the other in red light.[444] Optical sensors initially caused problems arising from pulp, dirt, and condensation collecting on the optical system. Automatic cleaning and heating of the equipment to a temperature higher than the temperature of the surroundings have reduced these problems. Instruments with dual wavelength sensors are less affected by contamination of the optics. Reflectance measurements on washer drums are influenced by moisture content of the pulp, and the relationship between the reflectance factor of wet and dry pulp will differ with the light-scattering coefficient of the dry pulp.[449]

Determination of residual chlorine in connection with chlorination control is carried out in three ways: (1) measurement of oxidation-reduction potential (ORP)[450-458], (2) polarographic measurements[459-463], and (3) calorimetric measurements.[464] Measurement of chemiluminescence has also been proposed but has not been used commercially.[465] ORP measurements have the disadvantage that the change in potential increases rapidly at a low content of chlorine and then changes relatively slowly with increasing chlorine concentration. The main information obtained from this type of sensor is whether or not there is any residual chlorine present. Mixtures of chlorine and chlorine dioxide may cause problems. Polarographic methods give a different response for chlorine and chlorine dioxide. In the Polarox system a pulp flow of about 2 m/sec is required in order to remove reaction products from the electrodes. In the Polarodor system, vibration keeps the electrodes in operating condition. The wear on the electrodes has been a problem, but this seems to have been solved in a satisfactory way.[463] The calorimetric method is a newer method and no long-time experience with the equipment is reported. A drawback is that it requires additional chemical to obtain a response, and the sample and the reagent must be metered very accurately. A few minutes is required before a signal is obtained, but this delay is probably of minor importance. The signal from the sensors for residual chlorine is usually applied for feedback control of the charge of chlorine. However, it can also be used for forward control of the alkaline extraction stage and subsequent stages.

All sensors used for chlorination are dependent on good mixing of pulp and chlorine. The reaction time will influence the signal from the sensor, and, therefore, the time from addition of chlorine to the sensing element should be constant. If the flow-rate changes, the reaction time will change. To control the reaction time a separate sample can be taken and pumped to the sensing device at a constant flow-rate. Temperature is important for the reaction rate. The influence of temperature variations can be corrected if a computer system is used, but normally changes in temperature are slow and are not corrected.

Process Control in Alkaline Extraction and Final Bleaching Stages

Temperature, level, consistency, and flow control of these stages are described in the first part of this section. If only upflow towers are used, the pulp flow will be determined by the charging of the first tower in the bleaching sequence. After alkaline extraction, the kappa number of the pulp is often measured but this is done manually. Optical measurements of the alkaline extracted pulp have been used for feed control of the hypochlorite stage.[466] Optical measurements on washers[467-472] or on the pulp slurry,[448,473,474] can be used in the hypochlorite stage and in the chlorine dioxide stage.[475] Optical sensors have also been used for automatic control of dithionite bleaching of mechanical pulp.[476]

Process Control in Oxygen Bleaching

Pressure and temperature control is obviously of importance in oxygen bleaching. Kappa number before and after the oxygen stage is of importance for control of the process. The charge of alkali should be varied with the kappa number. Monitoring of carbon monoxide and turpentine in the reactor is also important for the control of the process. The safety aspects of oxygen bleaching are discussed in the section on mill-scale bleaching.

Laboratory Control Procedures

Some of the process control is carried out manually in the laboratory or by the operators in the bleach plant. The concentration of the bleach solution and the amount of residual bleaching chemical at the end of a bleach stage are common tests. Other standard and useful tests are described in publications as follows:

Kappa number	TAPPI	T–236 os–76
	SCAN-test	SCAN–C1:77
	ISO	R 302–1963
Permanganate number	TAPPI	T–214 su–71
	CPPA	G. 2U
Chlorine consumption or chlorine number	TAPPI	UM 209
	TAPPI	T–219 m–54
	SCAN-test	SCAN–C29:72
	ISO	3260–2975 (E)
	CPPA	G. 16, November, 1964
	CPPA	G. 1U
Hypo number	TAPPI	T–253 pm–75
Available chlorine residual	TAPPI	T–678 su–71
	CPPA	J.22 May 1968
Chlorine dioxide	CPPA	J.14 April 1963
Process water	TAPPI	T–620 m–55
	CPPA	H. 4P (a–n), March 1967
Hypochlorite	CPPA	J. 2H, November 1964
Peroxide bleach liquor and spent liquor	CPPA	J. 16P, November 1962

Computer Control Procedures

A complete control system requires that all available information be put together and evaluated in order to obtain optimum results from bleaching. Analog computers have been used for bleaching control, but they are of limited value where long and varying time delays are encountered in the system. Digital computers are, therefore, the usual type used. Small microcomputers have been introduced.[477] Such computers have a sufficient capacity for a bleach plant control system. The control strategy will differ from one plant to another, but for all systems a good knowledge of the basic data for the bleach plant is necessary.[478] Usually some simplifications have to be used in models for computer control,[479] and manual control data and general background knowledge of the bleaching process are usually included in the computer program.

When chlorination is used as the first bleaching stage, the control of this stage is of the utmost importance. The information obtained in this stage can be used for feedback control of the chlorine added and feedforward control of the other stages. Oxidation-reduction potential has been used for indicating the residual chlorine and computing from this the correct dosage of chlorine to be added.[480] If the residual chlorine is determined after a well-defined reaction time, the chlorination process can be used as a chlorine-number analyzer.[463] The residual chlorine can be kept constant and the dosage of chlorine applied through the automatic control procedure will then be nearly proportional to the lignin content of the pulp. A correction can be applied in such a way that the residual chlorine value is increased when the ratio of chlorine to the pulp is increased because this indicates that the pulp has a higher lignin content, and the residual chlorine should then be higher. Any changes in consistency and flow-rate must be corrected in the system. The temperature of the chlorination will influence the rate of reaction and changes in the temperature should be used to correct the model. However, when conventional low-temperature chlorination is used, the changes will be slow and good results are reported when one program is used for the summer period and another program for the winter period.[459] The application of computer control in bleach plants has been reported in several papers and some of the problems encountered in the use of these systems are discussed.[448,471,481–491] Successful computer control of mechanical pulp bleaching with peroxide has been reported.[492]

Product Control

Brightness measurements on-line can be used as a process control and a product control. On-line measurements of moisture content and basis weight are sometimes used for market pulp, but these have very little connection with the bleaching of the pulp.

The most important product test on bleached pulp is brightness. The brightness value must, however, be interpreted in relation to the other properties of the pulp in order to obtain a correct picture of the quality. The additional quality tests depend on the type of pulp. Visible impurities, such as bark, pitch specks, and poorly bleached fiber bundles, should be low in all types of pulp. The usual way of determining specks is by visual counting and classification according to size. Automatic counting of specks in pulp sheets has been used and such methods have certain advantages compared with visual counting.[493,494] Reduction of shives and bark in bleached pulp can be achieved by optimization of the bleaching process and choice of bleach sequence. In order to reduce specks to an acceptable level in pulps that are not well defibrated, considerably more bleaching chemicals must be used than is necessary for the bleaching of fully defibrated pulp.[494,495]

Depositable extractives that can result in pitch problems are undesirable in any type of pulp. There is no standard method for determining the amount of troublesome extractives. Usually the total content of extractives is determined and used as an indication of pitch problems. Determination of pitch deposited by agitation in a copper vessel[496] or by beating in the PFI beater[497] has been used for the evaluation of the tendency of the pulp to cause pitch trouble. For dissolving grades of pulp the content of extractives is a quality criterion. Alpha-, beta-, and gamma-cellulose as well as alkali solubility are also quality tests for dissolving pulps. Ash, metallic ions, and chloride content are important for dissolving pulps and for other types of pulp.

Pulp to be used for the production of printing and writing papers should have good light-scattering properties. Light scattering is determined on laboratory handsheets. Physical strength of the pulp and energy consumption required for beating of the pulp to the desired strength are standard tests for paper pulps. Bleaching will seldom improve the strength and may severely reduce the strength of chemical pulps if bleaching is excessive. Viscosity is often used as a measure of the degradation of the cellulose during bleaching. Viscosity is related to the strength of the pulp if the viscosity is reduced below a certain level;[498,499] the critical viscosity depends on the pulp and the bleaching sequence. For dissolving pulps, a certain amount of degradation of the cellulose in the bleaching process is a requirement, but the final viscosity should be kept at a constant level. A list of tests for product control on bleached pulps is given below together with their publication source.

Brightness	TAPPI	T–217 m–48
	TAPPI	T–218 os–75
	CPPA	E.I, April 72
	SCAN-test	SCAN–C11:75
	ISO	ISO 3688–1977 (E)
Light scattering	SCAN-test	SCAN–C27:76
Impurities	TAPPI	T–213 ts–65
	TAPPI	T–246 su–70
Extractives	TAPPI	T–204 os–76
	CPPA	G. 13, November 1964
	SCAN-test	SCAN–C7:62
	SCAN-test	SCAN–C8:62
	ISO	ISO 624–1974
Ash	TAPPI	T–211 m–58
	TAPPI	T–244 su–70
	CPPA	G. 10, September 1960
	SCAN-test	SCAN–C6:62
	ISO	ISO 776–1974
	ISO	ISO 1762–1974
Iron	TAPPI	T–242 su–69
	SCAN-test	SCAN–C13:62
	ISO	R 779–1968

Copper	TAPPI	T–243 su–69
	SCAN-test	SCAN–C12:62
	ISO	R 778–1968
Manganese	TAPPI	T–241 su–69
	SCAN-test	SCAN–C14:62
	ISO	R 1830–1970
Calcium	TAPPI	T–247 su–71
	SCAN-test	SCAN–C10:62
	ISO	R 777–1968
Silicates	TAPPI	T–245 os–70
	SCAN-test	SCAN–C9:62
Chloride	TAPPI	T–256 pm–76
Sulfur	CPPA	G. 28, October 1970
Alpha-, beta-, gamma- cellulose	TAPPI	T–203 os–74
	CPPA	G. 29P, April 1972
Alkali resistance	CPPA	G. 27P, September 1965
	ISO	ISO 699–1974
Alkali solubility	TAPPI	T–235 os–76
	CPPA	G. 26, May 1966
	SCAN-test	SCAN–C2:61
	ISO	ISO 692:1974
Pentosans	TAPPI	T–223 os–71
	CPPA	G. 12, June 1962
	SCAN-test	SCAN–C4:61
Viscosity	TAPPI	T–206 os–63
	TAPPI	T–230 os–76
	TAPPI	T–254 pm–76
	CPPA	G. 24P, March 1968
	CPPA	G. 4U
	SCAN-test	SCAN–C15:62
Copper number	CPPA	G.22, June 1956
	SCAN-test	SCAN–C22:66
Disintegration for testing	CPPA	C. 6, June 73
	SCAN-test	SCAN–C18:65
Drainage properties	TAPPI	T–221 os–63
	CPPA	C. 1, April 72
	SCAN-test	SCAN–C19:65
	SCAN-test	SCAN–21:65
Beating	TAPPI	T–200 ts–66
	TAPPI	T–224 su–68
	TAPPI	T–225 os–75
	CPPA	C.2, November 1962
		C. 3, January 1964
		C. 7, April 1972

	SCAN-test	SCAN-C23:67
	SCAN-test	SCAN-C24:67
	SCAN-test	SCAN-C25:76
Forming of handsheets for physical tests	TAPPI	T–205 os–71
	CPPA	C.4, May 1950
		C5, June 1973
	SCAN-test	SCAN-C26:76
Strength properties	TAPPI	T–220 os–71
	CPPA	D. 12, September 1960
	SCAN-test	SCAN-C28:76

COLOR REVERSION OF BLEACHED PULPS

Brightness stability during storage of the pulp, during processing, and finally during the life of the end product is an important quality criterion. The required stability depends very much on the use and the lifetime of the products prepared from the pulp. Paper for documents and other long-lived products must have good stability during storage. Pulp used as filler in plastic products must be stable to the heat encountered during the processing of these products, and usually such pulps should show good stability to irradiation with light. Pulp used for the manufacture of paper or paperboard that is to be used for display or is exposed to light should not be affected too much by irradiation. Dissolving pulps should have good color stability in alkali in order to avoid too strong a yellow color in the mercerized pulp. On the other hand, color stability is of very little importance for newsprint and ordinary magazine paper, and demands for high-stability pulp for such products can be a waste of resources.

All pulp is unstable and from a thermodynamic point of view will ultimately end as water and carbon dioxide. Fortunately, this process is very slow under normal conditions of storage and use. However, after a period of storage, changes can easily be detected in appearance and other properties. The easiest tests for detecting changes in pulp are viscosity and brightness measurements and any decrease in these values shows that a change has taken place. A reduction of the brightness indicates that yellow compounds are formed. This does not mean that the color formed is the same as the color removed by bleaching, and color reversion is, therefore, not quite the correct expression for the yellowing of bleached pulps.

The stability of bleached chemical pulps is usually much better than the stability of mechanical pulps, especially when the pulps are exposed to light. The two types of pulp will, therefore, be treated separately.

Determination of Color Reversion

The color changes in pulp and paper products are caused by storage, heat during processing, or exposure to light. Paper products are seldom exposed to outdoor sunlight but are subjected to irradiation from indoor daylight and artificial light.

Long-time storage is not a practical way of testing and heat treatment is used to simulate the effect of storage. Results obtained by heating at high temperatures will obviously give an indication of the heat resistance of the pulp. Heating in a humid atmosphere will indicate the stability during drying in the case of market pulp or to drying during the paper manufacturing process. Exposure to daylight cannot be used as a standard testing method because of the frequent changes in intensity of the light and variations with the season of the year. Artificial light from mercury lamps and from carbon arcs has been used but light from xenon lamps is the common source for testing of light-induced color reversion. With suitable filters, these lamps give a spectral distribution of emitted light, which is similar to indoor or outdoor daylight. For most pulp and paper products the indoor spectral distribution is preferred. A good correlation was found for the fading of colored paper when the samples were irradiated with xenon light and light from a carbon arc.[500]

Because the color formed by reversion is yellow, the change in reflectance of visible light is most pronounced in the blue part of the spectrum. Measurement of the blue reflectance, or brightness, is, therefore, well suited for determination of the change taking place. The decrease in brightness is the most commonly used measure for color reversion. However, the decrease in brightness is not only a function of the colored compounds formed but also is affected by the starting brightness and the light-scattering coefficient of the pulp. The increase in light-absorption coefficient resulting from the treatment should be determined in order to obtain a more accurate measure of the color formed. This is somewhat complicated because translucent sheets with a good formation have to be prepared, and the basis weight and reflectance factor with black backing and with white or paper backing must be determined. The light-scattering coefficient usually remains almost unchanged during color reversion and this makes it possible to calculate figures that are proportional to the change in light-absorption coefficient. Based on the Kubelka-Munk theory, the following equation is used where k is the light-absorption coefficient, and s is the light-scattering coefficient, and where R_∞ is the reflectance factor obtained with paper backing made with sheets from the same pulp:

$$\frac{k}{s} = \frac{(1 - R_\infty)^2}{2 R_\infty}$$

R_∞ must be inserted as a decimal fraction, not as percent, in this equation. The relationship between k/s and R_∞ can be found in tables.[501] If k/s before the color reversion is subtracted from k/s after the reversion, a figure is obtained that is proportional to the change in light absorption. This figure multiplied by 100 is called the postcolor number or pc-number.[502]

$$\text{pc-number} = 100 \left[\left(\frac{k}{s} \right) \text{after color reversion} - \left(\frac{k}{s} \right) \text{before color reversion} \right]$$

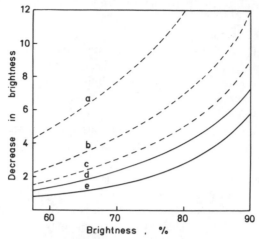

Figure 5-22. Relationship between initial brightness, light-scattering coefficient (m^2/kg), increase in light-absorption coefficient (m^2/kg), and drop in brightness: (a) $s = 20$, $\Delta k = 1.0$; (b) $s = 40$, $\Delta k = 1.0$ or $s = 20$, $\Delta k = 0.5$; (c) $s = 60$, $\Delta k = 1.0$; (d) $s = 40$, $\Delta k = 0.5$; and (e) $s = 60$, $\Delta k = 0.5$.

However, there should be a warning against use of this number for comparison of color reversion of pulps that differ noticeably in light-scattering coefficients. The pc-number is inversely proportional to the s-value, and if two pulps with s-values in the ratio of one to two are compared, the pulp with the higher s-value will give a pc-number one-half of the value of the other pulp for the same increase in light-absorption coefficient. Figure 5-22 shows examples of the relationship between brightness, drop in brightness, light-scattering co-efficient, and increase in light-absorption coefficient.

The conditions used for the testing of brightness stability often vary from one laboratory to another, and the figures obtained can show great differences. Within any one laboratory, however, the brightness test can be very useful for determining the influence of pulping and bleaching methods, wood species, and contamination of the pulp. The trends found in a series of investigations are often more important than the actual figures themselves. There is no stan-dard method for determination of color reversion, but useful methods are proposed.[503,504]

Increasing the temperature and increasing the humidity will accelerate the color reversion.[505] Temperatures from 80 to 160°C are used in testing stability, and the relative humidity used may vary from almost 0 to 100%. Components other than water in the atmosphere can influence the stability; chlorine has a negative effect on the stability and sulfur dioxide may be oxidized and fixed as sulfate or sulfuric acid. The stability is reduced in acid medium. Metallic ions can act as catalysts for the oxidation of sulfur dioxide.[506] For pulp stored in bales, only the outer part of the bales will be exposed to the atmosphere to a noticeable degree.

Color Reversion of Mechanical Pulp

The stability of mechanical pulp is poor in comparison to that of fully bleached chemical pulp. This difference is most noticeable when the pulps are exposed to daylight or other sources containing shortwave light. When stored in darkness, mechanical pulp has a fairly good stability. Bleached mechanical pulp sheets stored in a book at ordinary room conditions for three years showed a brightness loss of only 1.5 to 2 units.[507] A comparison of the decrease in brightness obtained by irradiation is shown in Figure 5-23[508] for unbleached, for dithionite bleached, and for peroxide bleached groundwood pulp. The change in brightness is least for the unbleached pulp and highest for the dithionite bleached pulp.[508,509]

The color stability of groundwood pulp is usually measured by heating in the temperature range of 80 to 120°C in dry or humid atmosphere. The decrease in brightness is not as rapid as that obtained by irradiation, but prolonged heating results in considerable reduction in brightness. Table 5-5 shows the relationship between rapid color reversion tests and the results of storage in light or in darkness.[507]

The color stability of mechanical pulp is influenced by the raw material. In the case of unbleached Western hemlock, the color reversion is approximately the same for the sapwood and the heartwood, whereas in the case of jack pine, the heartwood pulp is discolored much more than the sapwood pulp.[510] Removal of metallic ions with EDTA treatment improves the thermal stability to some extent, and a subsequent extraction with ethanol-benzene improves

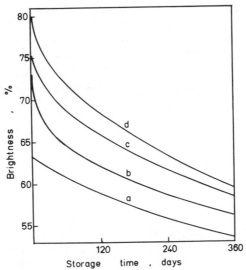

Figure 5-23. Decrease in brightness during storage of groundwood pulp in daylight: (a) unbleached; (b) dithionite bleached; (c) peroxide bleached; and (d) peroxide bleached.[507]

TABLE 5-5 COLOR REVERSION IN STORAGE OF GROUNDWOOD
PULP COMPARED WITH ACCELERATED TESTS

Accelerated Test	Storage	Storage Time (d)		
		Unbleached Pulp	Bleached with Dithionite	Bleached with Peroxide
3 hr in Xenotest	Daylight, south	7	6	7
	Daylight, north	9	9	10
	Indirect daylight	140	145	165
	In darkness	280	280	x[a]
18 hr, 85°C, 95% RH	Daylight, south	3	3	4
	Daylight, north	7	6	7
	Indirect daylight	95	60	90
	In darkness	170	150	230
18 hr, 105°C, dry atmosphere	Daylight, south	10	8	8
	Daylight, north	13	12	11
	Indirect daylight	285	335	250
	In darkness	x[a]	x[a]	x[a]

[a]x-more than 360 days.

the stability further, especially for pulp from jack-pine heartwood. Similar results were obtained for the stability of bleached pulps from these wood species.[54] Rapson, and others,[42] investigated the color stability of groundwood from nine wood species. Using thermal treatment as the test condition, alder pulp—unbleached or bleached with dithionite as well as peroxide bleached—yellowed considerably more than the other pulps. Pine pulp also showed poor stability; Western hemlock pulp, bleached with dithionite, yellowed almost as much as pine pulp. Pulp from cottonwood showed the best stability. Bark-staining of spruce reduces the stability to light in the case of unbleached as well as dithionite bleached pulp. The yellowing effect of bark-staining compounds is removed by peroxide bleaching.[29]

Reactions in Light-Induced Color Reversion of Mechanical Pulp

The changes in mechanical pulp caused by irradiation have been studied since 1940.[511] It has been observed that only light of wavelength shorter than 350 nm causes noticeable color reversion. Irradiation at wavelengths longer than 480 nm actually increases the reflectance slightly. It was found that the effect of wavelength on the yellowing of groundwood is almost identical to the ultraviolet absorption curve of lignin. This indicates that the absorption of light by lignin is responsible for an appreciable part of the color reversion.

Leary[512] observed that yellowing by irradiation does not occur when air is completely removed from the pulp. In an atmosphere of nitrogen or carbon dioxide, no yellowing was observed after 500 hr of irradiation. This indicates that oxygen is necessary for the reactions taking place in yellowing by irradiation, or, in other words, that photochemical oxidation is principally responsible for the color reversion. Irradiation of mechanical pulp and wood by ultraviolet light causes chemical changes. Forman[513] found that the content of acid-insoluble lignin is decreased by irradiation. The methoxyl content is decreased when wood meal is irradiated. Leary[514] found a linear relation between the rate of loss of methoxyl groups and the rate of yellowing during the early stage of irradiation in air. The loss in methoxyl content was less when newsprint was irradiated in vacuum than when it was irradiated in air.[512] Carbon monoxide, carbon dioxide, and methanol are formed when wood is irradiated; the formation is considerably lower in nitrogen atmosphere than in air.[515]

Studies with electron-spin resonance spectroscopy have shown that ultraviolet light produces free radicals in wood.[516] These free radicals are stable in vacuum and in helium but decay in the presence of oxygen. The suggested mechanism of the reactions is as follows:

$$(1) \quad R \xrightarrow{h\nu} R^{\cdot}$$

$$(2) \quad R^{\cdot} + O_2 \rightarrow ROO^{\cdot}$$

$$(3) \quad ROO^{\cdot} + HR \rightarrow ROO + R^{\cdot}$$

$$(4) \quad \left. \begin{array}{l} 2R^{\cdot} \longrightarrow \\ ROO^{\cdot} + R^{\cdot} \longrightarrow \\ 2ROO \longrightarrow \end{array} \right\} \text{nonradical products}$$

Lin and Kringstad[68] found that only a few structural units in lignin are able to absorb ultraviolet light of the frequencies present in daylight. These are the aryl-alpha-carbonyl structures and the structures containing double bonds in conjugation with aromatic rings. Such compounds act as photosensitizers, giving rise to phenoxy radicals by a direct hydrogen abstraction mechanism[517] or by generation of singlet oxygen,[518] which can create phenoxy radicals. The formation of colored compounds in mechanical pulps by irradiation is probably mainly a result of reactions between intermediate phenoxy radicals and oxygen.[514,517] Gellerstedt and Pettersson[519] studied the influence of ultraviolet light on lignin model compounds. The main reaction observed was an oxidative elimination of side chains and a simultaneous formation of quinones. The reaction rate of chain elimination was dependent on pH and a minimum was found at pH about 6. Studies on Björkman lignin from spruce dissolved in methyl-cellosolve showed that the absorbance of near ultraviolet light decreases when the solution is irradiated with ultraviolet light of similar spectral composition as daylight.[520,521] In the presence of oxygen the absorbance of blue light in the visible spectrum is increased by this type of irradiation, whereas in the absence of oxygen no increase in absorbance of blue light was observed.

Irradiation of mechanical pulp tends to increase the copper number, increase the alkali solubility, and decrease the cellulose content, indicating that the irradiation is attacking the carbohydrates in the pulp.[522,523] Some observations indicate that there may be an interaction between the carbohydrates and the lignin by irradiation.[524] However, from investigations of light-induced color reversion of mechanical pulp and wood, it can safely be concluded that the lignin is the main contributor to the yellowing.

Some wood species contain hydroxystilbenes and Morgan and Orsler[525] have shown that such compounds are unstable in light and form dark compounds by irradiation. Heartwood of pine and bark-stained spruce contain compounds of the hydroxystilbene type, and it has been shown that bark-staining reduces the brightness stability of mechanical pulp.[29]

Stabilization of Mechanical Pulp against Light

Many attempts have been made to improve the stability of mechanical pulp. Methylation, acetylation, and benzoylation of the free phenolic hydroxyl group in the lignin improve the stability appreciably, but all these methods are expensive and adversely affect other properties of the pulp.[83,512,526-528] Lin and Kringstad[529] showed that the reduction of Björkman lignin with sodium borohydride improves the stability appreciably and that subsequent catalytic hydrogenation converts the lignin to a state where it is completely stable against irradiation. However, reduction of groundwood with sodium borohydride does not improve the stability; a possible explanation for this phenomenon is that irradiation of pulp results in an interaction between the lignin and the carbohydrates, leading to new radicals by irradiation. Another possibility for stabilizing the brightness is to use ultraviolet absorbants or antioxidants. Some effect can be obtained in this way, but the effective compounds are too expensive to be used in pulp or paper production.[524] Singlet oxygen quenchers, such as beta-carotene, seem to have a stabilizing effect against irradiation on lignin-model compounds.[519]

The color reversion by irradiation of stone groundwood papers takes place in the outer layer of the sheet. If a pad of thin sheets is irradiated, the pulp is bleached to some extent inside the pad, as shown in Figure 5-24.[508] The explanation of this phenomenon is that the ultraviolet light is absorbed strongly in the outer layer and that only the visible light penetrates to the inner layers. As is known, visible light can bleach stone groundwood.[511]

Titanum dioxide absorbs ultraviolet light very strongly and the reflectance of this pigment falls abruptly just below the lower wavelength of visible light. If titanium dioxide is used for the coating of paper containing mechanical pulp, an appreciable improvement in brightness stability is obtained.[508] Zinc compounds have a similar stabilizing effect.[530]

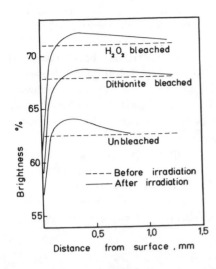

Figure 5-24. Yellowing and bleaching by irradiation of groundwood pulp.[508]

Reactions in Thermal-Induced Color Reversion of Mechanical Pulp

The reactions taking place in thermal-induced color reversion of mechanical pulp have not been studied as much as light-induced color reversion. For some wood species there apparently is a combination of several effects that discolor the pulp. Investigations of Western hemlock and jack pine showed that metallic ions and compounds in the extractives contributed very much to thermal-induced color reversion. This was especially noticeable for the heartwood of jack pine.[54,510,531] However, removal of the extractives had the same effect on both unbleached and bleached pulps, indicating that bleaching does not affect the color reversion caused by the extractives.

Color Reversion of Chemical Pulp

Unbleached chemical pulps, semichemical pulps, and chemimechanical pulps all have poor stability to irradiation with daylight or ultraviolet light. Such pulps often show a larger brightness-drop than stone groundwood. One reason for this is that the light-scattering is lower for pulps prepared by chemical treatment, or a combination of chemical and mechanical treatment, than it is for groundwood. For the same increase in light-absorption coefficient, the pulp with the lower light-scattering coefficient will show the larger drop in brightness. Another reason is that in the first part of color reversion of mechanical pulp by exposure to daylight, or light of similar spectral composition, two reactions apparently take place simultaneously: (1) yellowing caused by short wavelength light and (2) bleaching caused by longer wavelength light. With sulfite treatment, the pulp is bleached to some extent. This bleaching seems to be brought about by the same chromophoric groups that are bleached by daylight.[69] Therefore, the reactions that cause yellowing will not be partly offset by photochemical bleaching in the case of pulp prepared with sulfite.[69,532] The color reversion of unbleached chemical pulps has not been studied as extensively as for mechanical pulps and bleached chemical pulps. However, it is known that unbleached mechanical pulp and bisulfite pulps, yellowed by irradiation with light in the near ultraviolet (365 nm), can be bleached by irradiation with visible light. In the case of bisulfite pulps, the light-induced bleaching is reversed by storage in darkness.[532]

Bleaching of chemical pulp will usually improve the brightness stability considerably, but the improvement is dependent on the digestion process, the bleaching process, and the wood species. In the period of 1950 to 1970, color reversion of chemical pulps was one of the main fields of study in bleaching and these investigations resulted in many publications. Extensive reviews of the literature are available.[533-535]

Light-Induced Color Reversion of Chemical Pulp

Bleached chemical pulps have a low absorption of light in the visible range. However, the absorption coefficient increases with increasing frequency of visible light and even the brightest pulps have a higher light-absorption coefficient in the blue than in the red part of the spectrum. Absorption of light is the first condition for photochemical change in the pulp and, therefore, visible light can be expected to have only a slight influence on the color reversion of bleached chemical pulps.

In the near ultraviolet spectrum pure cellulose exhibits a noticeable absorption. However, the amount of ultraviolet light to which paper is subjected is limited. The ultraviolet light from solar radiation that reaches the earth's surface is restricted to a wavelength above 290 nm.[536] Paper products are normally used indoors and radiation below 320 nm is almost completely absorbed by window glass. Light from incandescent lamps usually has very little radiation in the ultraviolet region. Some types of fluorescent lamps emit light in the near ultraviolet, but only a small fraction is below 320 nm. Therefore, the irradiation that is of most importance in the near ultraviolet range has a wavelength from 320 to 400 nm. The effect of this irradiation, if it is absorbed, increases with decreasing wavelength or increasing frequency because the energy in the irradiation increases with the frequency.

Much of the earlier work on light-induced changes in pulp or cellulose is of minor interest because the effects of light and heat were not separated. Launer and Wilson[537] demonstrated the effect of temperature by irradiating cotton and wood fibers of low lignin content. When the samples were kept at 30°C, a bleaching occurred by irradiation, but without temperature control the samples yellowed. When the samples were heat treated after irradiation with light of 330 to 400 nm, the samples yellowed. The effect of heat treatment and storage on color after irradiation has also been reported in other articles.[538,539] A similar effect was found for cellulose oxidized with periodate, but further oxidation with chlorous acid or reduction with borohydride reduced the postirradiation effect considerably. As periodate oxidation opens the glucose ring and forms alde-hyde groups, the indications are that aldehyde groups formed by irradiation play an important part in the postirradiation effect.[540]

Light of short wavelength has a degrading effect on cellulose, as shown by reduction in viscosity and increase in copper number. Both these effects can be explained by the formation of aldehyde groups. Chemical pulp bleached without severe attack on the carbohydrates is comparatively stable to daylight.

From studies on mechanical pulp it is well known that lignin is unstable when irradiated with light of short wavelength. The lignin content in bleached chemical pulps is low and this lignin is certainly quite different from the lignin in mechanical pulp. There is no conclusive proof that the residual lignin in bleached chemical pulp causes yellowing, but the possibility cannot be ruled out. An indication that lignin is a factor is that sulfate pulp yellows more than sulfite pulp in simulated indoor daylight. Degradation of dissolving pulp by irradiation with ultraviolet light is lessened when the pulp is impregnated with lignin.[541] The strong absorption of ultraviolet light by lignin is a plausible explanation for this effect.

Heat-Induced Color Reversion of Chemical Pulp

Whereas light is a more important factor than heat in the yellowing of mechan-ical pulp, the opposite is true for chemical pulps. The effect of heat on bright-ness loss has been studied much more thoroughly than the effect of light in the case of chemical pulps. The results from several investigations show that the formation of colored compounds by heat treatment is influenced by many factors.

Influence of Pulping Process on Color Reversion in Chemical Pulp

The pulping method undoubtedly has some influence on the yellowing of bleached pulp, but the reactions that take place in multistage bleaching tend to disguise such effects. For example, sulfite pulps that contain a high amount of hemicellulose tend to yellow more than pulps with a low content of hemi-cellulose and the reason for this is that the hemicellulose is probably attacked more than the cellulose in the bleaching process. Histed[542] found a correlation between pc-number and pentosan content for sulfite pulps bleached with the sequence C-E-H.

When sulfite and sulfate pulps are compared, the sulfate pulps will usually show a stronger formation of yellow compounds than sulfite pulps.[543,544] There is no obvious reason for this. It is well known that the hemicelluloses are af-fected differently in sulfate cooking than in sulfite cooking. In the sulfate cook the hemicellulose is cross-linked to the extent that it will not dissolve in strong alkali. One should think that such crosslinking would stabilize the pulp, but very little is known about these reactions.

Effect of Lignin on Color Reversion of Chemical Pulp

In fully bleached pulp the lignin content is very low, but correlations between lignin content and color reversion have been found for sulfite pulp from spruce[543] as well as sulfate pulp from birch.[545] It is also possible that bleached pulps contain degradation products from lignin and that these products cannot be characterized as lignin. This could be an explanation for the difference in color reversion between sulfate and sulfite pulp. The lignin content in semi-bleached pulp is so high that it can be a significant factor in color reversion.

Effect of Resin on Color Reversion of Chemical Pulp

Resins in sulfite pulp have a considerable effect on color reversion by heat treatment. Extracting bleached sulfite pulps with organic solvents has been shown to improve the color stability.[391,502,546−548] Resins, as such, are probably of minor importance in color reversion, but the resins take up chlorine during the chlorination stage in bleaching and when the pulp is heated, hydrochloric acid is split off and this acid can attack the cellulose to produce yellow compounds.[549,550]

Effect of Oxidized Carbohydrates on Color Reversion of Chemical Pulp

The formation of carbonyl and carboxyl groups as a result of oxidation has a significant effect on the color reversion of pulp.[551,552] Virkola and Lehti-koski[553] showed that oxidation of cotton linters with chlorine/hypochlorous acid increases the color reversion and the increase is greater at pH 2.5 and 6 than it is at pH 4, where the oxidizing agent is almost 100% hypochlorous acid. Copper number and content of carboxyl groups are also higher at pH 2.5 and 6 than at pH 4. Color reversion from chlorine/hypochlorous acid oxidation of cellulose is mainly caused by the formation of carbonyl groups. Chlorination of pulp under normal conditions where lignin is present in relatively large amount probably has very little effect on the color reversion that is caused by the oxidation of carbohydrates.

Hypochlorite bleaching can have a strong influence on color stability of pulp. The strength decrease by bleaching at neutral pH is well known and the color stability is also reduced by bleaching in this pH region.[553] However, hypochlorite bleaching at high pH improves the color stability, and at pH 10 to 11 the pulp is stabilized by hypochlorite bleaching, even if it has been attacked before this treatment. The effect of hypochlorite is most pronounced in the final bleaching stage when the lignin content of the pulp is low. The reaction between hypochlorite and lignin is more rapid than the reaction with carbohydrates, and hence the attack on the carbohydrates will be limited if an appreciable amount of lignin is present in the pulp, unless large excess of the bleaching agent is used.

Rapson and Corbi[554] found that compounds with carbonyl, carboxyl, and hydroxyl groups in the same molecule readily form yellow substances. Carboxyl and carbonyl groups have a synergistic effect and pulps with high contents of these two groups yellow rapidly.[555]

Chlorine dioxide has a more specific attack on lignin than hypochlorite and only small amounts of oxidized groups are introduced by chlorine dioxide bleaching.[279] Chlorine dioxide bleaching has, therefore, a favorable effect on brightness stability. Peroxide bleaching in the final stage results in good color stability. The improvement in stability increases with pH and the amount of peroxide applied.[233,556,557] By peroxide bleaching, a decrease in yellowing was found in spite of an increase in carbonyl content.[233] This is rather difficult to explain from the general knowledge of the effect of oxidized carbohydrates on color reversion. Oxygen bleaching has a favorable effect on brightness stability. Semibleached pulp with comparatively stable brightness can be prepared by this bleaching method. The alkaline medium in which oxygen bleaching is carried out is of importance in this connection because carbonyl groups are eliminated in alkaline medium.

Alkaline extraction has a beneficial effect on brightness stability. By extraction with alkali, the resins and reaction products from the previous bleaching stage are removed, and these are compounds that very likely would cause color reversion in the pulp. Another effect of alkaline extraction is the destruction of alkali-labile carbonyl groups that are known to form yellow compounds during heat treatment.

Reduction of bleached pulp with sodium borohydride reduces the color reversion markedly.[391] The relationship between color reversion and carboxyl content is apparently almost linear at the same copper number.[534]

Effect of Cations on Color Reversion of Chemical Pup

Czepiel[558] has shown that iron and copper ions affect the pulp in two ways: (1) inorganic colored compounds can be formed, for example by oxidation of ferrous to ferric salts, and (2) the metallic ions can act as catalysts that increase the rate of color reversion. Manganese seems to improve the stability under some conditions, although Sihtola and Sumiala[559] found in the case of cotton linters oxidized with dinitrogen tetraoxide that manganese increased the color reversion when calcium or aluminium ions were replaced, but decreased the color reversion when sodium ions were replaced.

Brightness reversion in the presence of iron and copper ions was studied for cellulose that contained aldehyde and carboxyl groups.[560] In the case of dicarboxyl cellulose, copper retarded the formation of yellow color, whereas iron increased the yellowing. For dialdehyde cellulose, copper and iron in low concentration had a stabilizing effect, but iron in high concentration accelerated the color reversion.

Effect of Process Variables on Color Reversion

A reversion of brightness can occur during the bleaching process if the bleaching agent is consumed before the pulp is cooled and neutralized. The best known example of this effect is in the bleaching of mechanical pulps with peroxide, where the alkaline medium causes yellowing. High temperature applied after the bleaching chemical is exhausted can also cause color reversion in the case of chemical pulps, and such an effect has been observed in both hypochlorite and chlorine dioxide bleaching.

Some brightness-loss in the drying of pulp is to be expected, especially in driers using steam-heated cylinders. In such a case the outer layer of the pulp sheet may be noticeably more yellow than the inner portion. One reason for this is that the temperature is higher in the outer layer. Another cause may be the migration of soluble colored compounds to the surface of the sheet, especially if the pulp has not been washed sufficiently well. Brightness reversion can also take place in bales of pulp during storage. This effect is most noticeable if the pulp is baled at high temperature. The inner part of the bale will cool very slowly in such a case; some mills have installed a sheet cooler between the drying machine and the baling equipment to avoid this problem.

Reactions Involved in Color Reversion of Chemical Pulp

Investigations of color reversion have mainly been concerned with the effect on such properties as brightness, increase in acidity, formation of reducing groups and degradation of the cellulose. The formation and structure of the compounds that cause the yellow color have only been studied to a limited extent.

Photodegradation of cellulose has been studied in the textile field, but much of the work has been carried out with ultraviolet radiation of higher frequencies than those normally encountered with pulp and paper products. The direct photolysis of carbon-carbon or carbon-oxygen bonds requires an energy of 335 to 377 kJ/mole and the removal of hydrogen requires an energy of 420 to 460 kJ/mole.[561] Ultraviolet light with a wavelength shorter than 340 nm has the required energy for cleavage of these bonds. The relationship between energy, E, and frequency, ν, for one photon is: $E = h \cdot \nu$, where h is the Planck's constant. The effect of light can be intensified by sensitizing agents in the pulp or paper.

Some of the degradation products produced by irradiation of cellulose are carbon monoxide, carbon dioxide, and hydrogen. Other compounds produced include acetaldehyde, propionaldehyde, acetone, and methanol.[561] Mono-, di- and tri-carbohydrates have also been detected in irradiated cellulose. It has been suggested that the photolytic reaction takes place at the primary alcohol groups on the C_6 atom of the anhydroglucose units or at the secondary alcohol groups on the carbon atoms 2 and 3, and this may be the source of formation of carbon monoxide, carbon dioxide, and aldehydes.[562] A mechanism has been proposed in which alcohol groups are photolyzed to carbonyl groups with the liberation of hydrogen.[563] A combination of hydrolysis and oxidation is also a possible reaction mechanism,[564] but the formation of yellow compounds is not explained by this theory.[565]

Heat-induced color reversion has been extensively studied, but very little is known about the yellow compounds formed. Hydrolysis of carbohydrates to oligomeric and monomeric compounds is part of the process.[566] Hydrolysis in acid medium is known to produce furan compounds, furfural from pentosans, and hydroxymethyl furfural from cellulose.[558] These compounds have a pronounced tendency to condense and form dark substances, such reactions are a possibility in the color reversion of pulp.

When free aldoses are exposed to prolonged heating in cellulose sheets, epimerization generally occurs. The epimerization is dependent on the formation of enediols and this is probably the first step in the reactions leading to yellow compounds. From the intermediate 1, 2-enediol, a 2-furaldehyde is formed[567] and furaldehydes readily form colored compounds. Modified aldoses undergo epimerization when the reducing power is retained. If the reducing power is eliminated and epimerization is stopped, there will be no structural modification and no color reversion. A carboxyl group together with a reducing group and the ability to epimerize will cause color reversion to take place. Glucuronolacetone impregnated in a pulp sheet and heated to 105°C for 24 hr was completely consumed and produced a very strong yellowing.[568] By the formation of furan compounds in this manner, free radicals are not involved in the reaction, but other reactions leading to colored compounds from carbohydrates can possibly be initiated by free radicals.

It has been shown that heat treatment of a 6-aldehydocellulose in dry as well as humid atmosphere causes very rapid yellowing. A 6-carboxyl cellulose is appreciably more stable than the carbonyl and a reduced 6-aldehydocellulose is more stable than the carboxyl compound.[569] Theander[570] studied the mechanism of the reactions and found that beta-elimination takes place at 80°C down to pH 3.5. Esterification of carboxyl groups with propyleneoxide stabilizes pulp to heat treatment in dry as well as humid atmosphere.[571] Heat-aged pulp has been shown[572] to contain arabinose, xylose, glucose, and lactic acid. Compounds detected in heat-treated oxidized cellulose are acetaldehyde, propionaldehyde, acetone, and methanol.[555] Heating of bleached sulfate pulp at 130°C brings about a rapid consumption of oxygen during the first 10 to 15 min and then consumption ceases.[573]

The amount of extractable material in bleached pulp is changed by heat treatment. When ethanol was used for extraction, the amount of extracted material increased and a considerable part of the dissolved substance was inorganic salts. When dichloromethane was used for extraction, the amount of extracted material decreased.[566]

ENVIRONMENTAL ASPECTS OF BLEACHING

Bleaching in the conventional way requires large quantities of water. In a multistage bleach plant the consumption of water can amount to several hundred cubic meters per ton of pulp. Most pulp mills have reduced their water consumption considerably by reuse of water from one stage to another and figures as low as 26.5 m^3 of water/ton of pulp are reported.[574]

The content of organic matter in spent bleach liquor is low even for highly closed water systems. A loss in yield of 10% is a high figure in bleaching. The concentration of organic substance from this loss will be below 4 kg/m^3 for a water consumption of 26.5 m^3/ton of pulp. Because of the low content of organic matter, a separate recovery system for spent bleaching liquor will require much thermal energy for evaporation; very little energy can be produced by burning the concentrated liquor.

As long as there were no environmental restrictions on discharges from the bleach plant, the disposal of spent liquor did not cause any problems to the mills, but this situation has changed drastically. Today, the effluents from bleach plants are the most serious problem in the pulp and paper industry. Investigation of the composition of spent bleach liquor and study of methods to abate this source of environmental contamination have the highest priority in pulp research in many countries (see the discussion on environmental control).

Discharges to the Atmosphere from Bleaching

Air pollution from bleach plants is a minor problem compared with water pollution. However, some bleaching agents including chlorine, chlorine dioxide, and sulfur dioxide are quite toxic gases. Under normal operating conditions the discharges to the atmosphere from bleach plants are so low that they cause no external environmental problem. Accidents or incorrect operation can, however, cause severe and dangerous conditions in the area surrounding bleach plants. The atmosphere inside the bleach plant often contains gases from the bleaching chemicals. These gases all have a characteristic smell that is readily noticeable at levels below the dangerous limits and leakages are, therefore, easily detected. Concentrations of chlorine dioxide as high as 0.07 ppm have been detected at the sheet forming and drying machine, and carbon monoxide is apparently formed in the chlorine dioxide stage.[574a] In new bleach plants most equipment is located outdoors. Units, such as filters, where residual gases can escape, are equipped with hoods.

All chemicals used in a bleach plant should be handled with care according to prescribed safety instructions. Chlorine and chlorine dioxide are corrosive in addition to being toxic. Chlorine dioxide is explosive in gas concentrations above 12 to 15%. Oxygen at high pressure reacts rapidly with organic compounds and may cause fires and explosions. The liquid and solid chemicals used in bleaching should also be handled with care. Chlorate, hydrogen peroxide, and sodium peroxide are very active chemicals that react readily with organic substances and may cause fires or an explosion. Caustic soda is an active chemical and any contact with the eyes or skin must be prevented. Concentrated sulfuric acid is well known for its destructive effect in contact with organic substances.

Discharges to Water from Bleaching

As already indicated, the main environmental problem in bleaching arises from the liquid waste. Untreated spent liquor from the cooking process contains a much larger amount of organic matter than bleach liquor waste, but only a minor part of the organic substance from sulfate cooking leaves the mill as effluent, and for sulfite mills restrictions on spent-liquor discharge have or will considerably reduce the load from these sources. Bleach plant effluents can, therefore, be the largest part of the total organic waste load from the pulp mill. Table 5-6 shows the amount of organic substances in conventional kraft bleach plant waste.[256] The source of organic waste is mainly in the chlorination and the first alkaline extraction stages. The color of these effluents, especially that of the alkaline extract of chlorinated kraft pulp, is very dark and may be an environmental problem.

The type of waste in the bleach plant liquid effluent is more important than the amount of waste. The effect on BOD of the water to which the waste is sent is not the main problem. The most troublesome waste products are the chlorinated aromatic compounds dissolved in the chlorination and the subsequent alkaline extraction stages. Many substances in this group are toxic and also can have carcinogenic or mutagenic effects. The toxic effect of bleach plant effluents is well recognized, but the long-term effect on fish and other aquatic life by bioaccumulation is not fully known. A number of the organic compounds from chlorination and alkaline extraction have been identified, and the quantities produced have also been determined for some of the compounds. The chemical composition of some of these substances is given in the section on chlorination and alkaline extraction.[252,253–256,261]

TABLE 5-6 DISSOLVED ORGANIC COMPOUNDS BY BLEACHING OF SULFATE PULP[a]

Wood Species	Bleaching Stage	COD (kg O_2/t pulp)	BOD_7 (kg O_2/t pulp)	Color (kg Pt/t pulp)
Pine	C	22–24	6.6–7.7	12–24
Pine	E	44–54	4.2–5.9	144–148
Pine	D/C	24	8.6	30
Pine	E	40	4.8	98
Pine	D	22	6.5	23
Pine	E	53	4.8	68
Birch	C	16	5.5	12
Birch	E	16	5.3	31
Birch	C + D	17	4.4	15
Birch	E	16	2.9	27
Birch	D	13	4.9	11
Birch	E	32	4.3	24

[a]See reference 256.

Toxicity of Bleach Plant Effluents

The toxicity of bleach plant effluents has been studied extensively. Canada has been the leading country in this research, and toxicity to fish has been the main subject. Lethal dosage levels have been determined and sublethal effects have been studied.[575-578] Howard and Walden[579] in a comprehensive study determined the BOD, TOC, and the toxicity for more than 1000 samples taken from the process streams and the combined effluents from 7 Canadian mills producing bleached kraft pulp. The most toxic process stream was that from the first alkaline extraction. Unbleached white water and the effluent from the first chlorination stage were next in toxicity. Combined effluents were less toxic than the individual bleaching process streams. Regression analysis has been used to correlate toxicity, analytical data, and process parameters.[580]

Free chlorine in the effluent is toxic to fish, even in low concentration. Seppovaara[581] found that rainbow trout were killed in a dilution of 1:150 when 0.5 mg/1 active chlorine was present in the effluent. When this effluent was mixed with effluents from other bleaching stages, the toxicity was reduced substantially. The compounds from gaseous chlorination showed less toxicity than the effluents from aqueous chlorination.[582] Hypochlorite, chlorine dioxide, and chlorate are all toxic compounds, but normally the concentration of these compounds will be low in bleach plant effluents. Chlorate is not as toxic to fish as it is to plants.

Leach and Thakore[261] determined the toxicity of compounds isolated from the alkaline extract of kraft pulp. The 96-hr median-lethal-concentration (LC_{50}) for these compounds was measured in static bioassays on juvenile rainbow trout. The most toxic compounds were: 3,4,5-trichloroguaiacol (0.75 mg/l); 3,4,5,6-tetrachloroguaiacol (0.32 mg/l); monochlorodehydroabietic acid (0.6 mg/l); dichlorodehydroabietic acid (0.6 mg/l); and 9,10-epoxystearic acid (1, 5 mg/l). All of these compounds were found in effluents from seven western Canadian kraft pulp mills. Bioassays using synthetic compounds gave results that agreed very well with the results obtained with mill effluents. The toxicity is very dependent on the concentration of chlorinated phenolic compounds in the effluents. A small increase in concentration of chlorinated phenolic compounds can change the effect from zero toxicity to fish to 100% lethal effect. Because chlorinated phenolic compounds are the most toxic substance in the bleach plant effluent, the toxicity of the effluent will depend on the lignin content of the pulp or, in other words, on the kappa number.

Resin acids are toxic compounds that are present in kraft pulping liquor from softwood.[583] The amount of these resin acids in natural and chlorinated form that end up in the bleach plant effluent depends on the wood species and on the degree of washing of the unbleached pulp. Large differences in the amount of resin acids are found in the effluents from the alkaline extraction in different mills. Epoxystearic acid may possibly be formed by chlorination of oleic acid and subsequent conversion to the epoxy form during the alkaline extraction.[261] The effluents from chlorination and first alkaline extraction stages are the main contributors to the toxicity of bleach plant effluents, and the effluent from chlorination as well as from the alkaline extraction has been reported as the more toxic of these two process streams.[579,584]

The sublethal effects of bleach plant effluents are probably of more importance for the environment than the lethal effect because they show the long-term effect from the accumulation of toxic substances in the organism. Lethal concentrations seldom occur when there is adequate dilution of the effluent. However, accidents in the operation of the bleach plant can cause the concentration of toxic substances in the diluted effluent to be too high. Howard and Walden[585] compared the effects of short-term exposure (12 hr) and long-term exposure (25 d) of bleach plant effluent on Coho salmon. Results from these two test conditions differed significantly. Red blood-cell counts were not influenced by short-term exposure but decreased significantly by long-term exposure. Juvenile salmon exposed to bleach plant effluents from kraft mills showed changes in respiration pattern and swimming ability above a certain concentration.[586] Webb and Brett[587] found that growth rate and gross conversion efficiency for under-yearling salmon were unaffected at or below a concentration of 2.5% of effluent from a bleached sulfate pulp mill but were significantly affected at a concentration of 25%. The threshold concentration for sublethal effects has been shown to be about 10 to 20% of the 96-hr LC_{50}-concentration for effluents from bleached kraft pulp mills.[588-590] Time-to-death of starved Coho salmon fingerlings decreased progressively when exposed to bleached kraft pulp mill effluent concentrations higher than 40% of the 96-hr LC_{50}-concentration.[591]

Seppovaara[581] studied the effects of long-time exposure of rainbow trout and crusian carp to diluted effluents from the bleaching of sulfate pulp. Deterioration in blood values and injuries to internal organs were found by long-term exposure to sublethal concentrations. The accumulation of chlorinated constituents from bleaching has been studied by ^{36}C-labelled compounds.[592] A summary of the knowledge about sublethal threshold concentations to effluents from kraft pulp bleaching and to the total effluent from bleached kraft mills is given by Davis.[593] The average threshold concentration was 0.16 of the 96-hr LC_{50}-concentration and the range was from 0.06 to 0.33. A review of toxicity of pulp mill effluents has been published by Walden and Howard.[594] Bleach plant effluents may contain mutagenic and carcinogenic compounds, and mutagenic effects have been detected in chlorination waste liquor.[594a]

The color of bleach plant effluents may cause disturbance in the ecology. The absorption of light by the colored effluents reduces the energy available for the growth of algae. Other substances in the effluents may also influence the growth of algae;[595,596] research work in this field is continuing.[594]

Abatement of Bleach Plant Pollution

The problem of water pollution from bleach plants can be attacked in several ways. Changing the method of bleaching is one way and this has been tried especially by the introduction of oxygen bleaching. To be effective, however, oxygen bleaching must be combined with evaporation and burning of the dissolved organic substances removed in the oxygen stage in order to obtain a reduction of the organic load in the effluent. A more drastic approach is to use the conventional bleaching chemicals, evaporate the waste liquor, and burn all the organic substances dissolved in the bleaching process. Purification of the effluents from the bleach plant has also been tried.

Modification of the conventional bleaching methods can be partially effective in reducing the problem with bleach plant effluents. When chlorine dioxide, or a mixture of chlorine and chlorine dioxide, is used in the first bleaching stage, fewer chlorinated compounds are formed and the oxidation reactions are favored.[299,300] Spent liquor from chlorine dioxide treatment is less toxic to fish than spent chlorination liquor.[575] Chlorination with a mixture of chlorine and chlorine dioxide also reduces the toxicity compared with conventional chlorination,[575] and sequential application of chlorine dioxide and chlorine also has a favorable effect.[597] Sodium hypochlorite in the first stage followed by a mild alkaline extraction and a second hypochlorite bleaching result in an effluent with low color and no toxic effect when used to produce semibleached sulfate pulp.[598] Gas-phase chlorination produces soluble compounds of lower toxicity than aqueous chlorination.[582] Cooking to a low lignin content reduces the amount of organic and toxic substances removed in the bleaching process.[599]

If sodium is used as the base in sulfite pulping, the first stage in the bleaching process can be an alkaline extraction with sodium hydroxide.[187] If this extract is used for washing the unbleached pulp and the extract is concentrated and burned together with the spent liquor from the pulping process, a substantial reduction in BOD and color from the bleach plant is obtained. This procedure is most effective when used for dissolving pulps in which the lignin content is low. Alkaline extraction in the first stage should also be possible when ammonia is used as the base in the sulfite process, but it is probably not feasible when calcium- or magnesium-base cooking liquor is used.

As mentioned in the discussion on oxygen bleaching, the industrial use of this process was mainly brought about by environmental restrictions. When the spent liquor from the oxygen stage is used for washing the brown stock, a considerable reduction in BOD, COD, and color is obtained.[312,600-603] Oxygen bleaching of sulfite pulp should give a better result than that obtained with sulfate pulp. Results obtained in laboratory work indicate that bleaching of sulfite pulps with oxygen without water pollution should be possible.

Purification of Bleach Plant Effluents

The methods used for the purification of bleach plant effluents can be divided into three categories: physical, chemical, and biological. Many of the processes used commercially include two, or even all three, of these methods. It is especially difficult to distinguish between chemical and physical effects; thus these two methods of effluent purification will be treated together.

The most noticeable characteristic of bleach plant effluents from mills using the sulfate cooking process is the dark color. Several chemical and physical methods have been tried for reduction of this color. One of the methods is based on precipitation with lime to remove the acidic compounds dissolved by alkaline extraction. Luner, and others,[604] studied the mechanism of this color removal and found that the phenolic and enolic groups in the colored substances react with calcium under alkaline conditions and form insoluble calcium salts. The precipitation is dependent on the molecular weight of the colored compounds. Two methods are used for the removal of color with lime: one based on the use of lime in excess;[605] the other on addition of lime in approximately stoichiometric proportion.[606] Cellulose fibers can be added to improve the settling of the calcium compounds.[607] The colored compounds can be burned in the lime kiln and the calcium recovered.[608] Addition of soluble calcium salts improves the reduction in color and COD by lime treatment.[608a] Aluminium and ferric salts used in conjunction with sodium hydroxide or lime may aid in the coagulation of bleach plant effluents.[609-612]

Removal of colored compounds by adsorption or by ion exchange has been proposed.[613-616] The ion exchange method developed by Lindberg, and others, is used commercially on the alkaline extract from sulfate pulp.[617-619] The colored compounds removed in the ion-exchange tower are subsequently eluted with sodium hydroxide and burned in the recovery furnace. Removal of 95% of the colored compounds from the bleaching of sulfate pulp is reported.[617] A nonpolluting bleach plant is possible by use of this method.[619a] Purification of bleach plant effluents by ultrafiltration and reverse osmosis has been studied and fairly good results have been obtained, but the costs are high.[620-622] Laboratory and mill-scale trials have shown a color reduction of 60%.[622]

Activated carbon has also been tried for adsorption of colored compounds and other constituents in bleach plant effluents. The carbon preferentially adsorbs compounds of low molecular weight and comparatively low color.[623] When used in combination with aeration and the addition of settling aids, such as alum and polyelectrolytes, 70 to 85% of the color and 20 to 40% of the BOD were removed.[624] Activated charcoal from bark has been reported to remove about 95% of the color and 75% of the COD from the alkaline extract of sulfate pulp bleaching.[625] Other methods that have been tried for removal of color and organic compounds from bleach plant effluents are: electrolytic coagulation,[626] addition of fly ash[627] and amines,[628] and mixing of alkaline extraction effluent with recycled chlorination filtrate.[629] Detoxification by foam separation has also been investigated. Surface-active toxic compounds are adsorbed at the gas-water interface, concentrated in the foam, and removed together with fiber in the effluent.[630-632] Reviews of color-removal technology are given in two papers.[606,633]

Biological purification of bleach plant effluents will reduce the BOD and the toxicity. Reduction in BOD seems to be rather unpredictable, the probable reason being that most of the tests have been on combined effluents from integrated mills, the composition of which has varied considerably. BOD reductions from 20 to 90% are reported, but the usual figure is 60%.[634-637] There seems to be no close correlation between BOD and toxicity.[638] Spills of toxic compounds from other process streams may explain some of the lack of correlation.[639]

Jank, and others,[640] investigated activated sludge treatment. They found that a two-stage process reduced toxicity and BOD better than a single-stage process. Liao and Dawson[641] carried out microbiological studies on a two-stage activated sludge system and showed that the major microbial groups found in the natural activated sludge were maintained when bleach plant effluent was treated. Berwart[642] reported good results for the activated sludge treatment of bleach plant effluents and condensates in a French mill producing bleached sulfate hardwood pulp, although the system used may not have been optimum. Experiments with the aeration of water consisting of condensate, washwater, and bleach plant effluent from an ammonium-base sulfite mill showed that a reduction of the BOD up to 90% could be obtained, but the treated water had a strong color.[643]

Chen, and others,[644] compared treatment processes involving activated sludge, rotating biological surface, plastic-medium trickling-filter, and aerated basin in pilot-plant trials. A mixture of bleach plant effluent and the discharge from a primary clarifier for the other effluents in an integrated mill was used. The activated-sludge and the rotating-biological-surface treatments removed 90% of the BOD. The trickling-filter system was very sensitive to variation in BOD loading, but up to 72% removal was achieved. A BOD removal of 83% was attained by use of the aerated basin. The recovery time for all systems was 2 to 3 ds if the system was upset by abrupt changes in the pH or by organic spills. Comparison of treatment of effluents from the bleaching of kraft pulp by activated sludge and in aerated lagoons showed that aerated lagoons were more effective in reducing toxicity than activated sludge systems.[645,646] The costs of operating biological-treatment systems have been compared in two Canadian reports.[647,648]

Biological treatment alone may not be sufficient if a low level of COD is required in the effluent. Chemical pretreatment with lime or other coagulant is needed in order to reduce the COD load.[649] The long-term stability of lignin compounds was studied by Bouveng and Solyom[650] and they found that the chlorinated lignin from sulfate pulp was degraded more rapidly in seawater than in fresh water.

Countercurrent Washing in Bleach Plants

As mentioned in the discussion on mill-scale bleaching methods, countercurrent washing has been used to some extent in bleach plants. Historically, there has been one general rule for reuse of water in bleach plants—it is that acid and alkaline streams should not be mixed. Acccording to this rule, spent liquor from a chlorine dioxide stage can be used in a preceding chlorine dioxide stage or for washing of chlorinated pulp. Spent liquor from alkaline extraction can be used for dilution or washing in a preceding alkaline stage. Rapson[651] introduced new ideas about the recovery of spent bleach liquor in 1967. In his proposed washing system, the pulp from each bleaching stage is washed with spent liquor from the succeeding stage. Fresh water should only be used in the last-stage washer. The spent liquor from the first bleaching stage should be used for washing of the unbleached pulp, and the dissolved organic substances in this liquor are then burned in the recovery furnace together with the black liquor.

Histed and coworkers[652-655] in an extensive study have investigated the possibilities of water reuse in the bleach plant and the use of spent liquor from the bleach plant for washing of brown stock. Their conclusion is that it should be possible to use mixed bleach plant effluents for dilution in the screening operation and for countercurrent washing of unbleached pulp. One problem is that any calcium in the liquor will precipitate and will cause trouble if a closed water system is used. The consumption of bleaching chemicals will increase somewhat because of the large amount of dissolved organic compounds that are carried along with the pulp. Special precautions must be taken to prevent heavy carry-over of chemical and dissolved organic compounds from one stage to the next and to keep any chlorine-consuming substances dissolved in the alkaline extraction from entering the bleach plant with the unbleached pulp.[654]

A schematic flowsheet of an optimum bleach plant countercurrent washing system is shown in Figure 5-25.[656] (A similar system, as well as alternative systems, are shown in the discussion on environmental control.) The volume of effluent water from the bleach plant is estimated at 18 m^3/ton AD pulp. About two-thirds of this water can be used for washing of the unbleached pulp and thereby introduced into the recovery system for the cooking chemicals. The filtrate from the first bleaching stage is primarily used for the washing of the unbleached pulp. In order to avoid evolution of sulfides and to prevent corrosion, the liquor used for washing of the brown stock is kept at a pH between 8 and 10 by addition of sodium hydroxide. The remaining water from the bleach plant can be used for green-liquor preparation, for slaking of lime or for dilution of concentrated white liquor. The filtrate from the first-stage alkaline extraction is primarily used for dilution of white liquor. A fraction of the chlorination filtrate is used in the lime-kiln scrubber thus purging calcium from the recovery system.

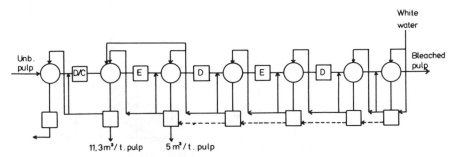

Figure 5-25. Countercurrent washing.

Removal and Recovery of Chloride from Bleach Plant Effluents

The high content of chlorides in bleach plant effluents causes problems in the recovery process. This problem is especially acute in coastal mills, in the western part of North America, that use pulpwood stored in seawater. Content of sodium chloride as high as 100 g/l in the white liquor has been reported for one mill.[657] Figure 5-26 shows the theoretical amounts of sodium chloride to be expected in bleach plant effluents from the different bleaching sequences of kraft pulp bleaching.[658] In a conventional C-E-H-D-E-D bleaching, the amount of chlorides in the bleach plant effluent will be about 80 kg/ton of AD pulp when the unbleached pulp has a kappa number of 35[659]. Cooking to lower kappa number followed by an oxygen bleaching stage or by the use of a high percentage of chlorine dioxide in the first bleaching stage may reduce the amount of chlorides in the bleach plant effluent to 15 to 20 kg/ton of pulp.[659]

The chloride content will rapidly increase in the recovery system unless some method is used to remove it. Several methods have been proposed. One method is based on removal of gas dust from the flue by washing with water.[660] This method has been modified by leaching the dust with sulfuric acid to remove sodium chloride and recover sodium sulfate.[661] Another method is based on the formation of hydrochlorine in the flue gas by reaction between sodium chloride, sulfur dioxide, and oxygen in the temperature range of 700 to 900°C followed by removal with water.[662] Suitable reaction conditions for the reaction

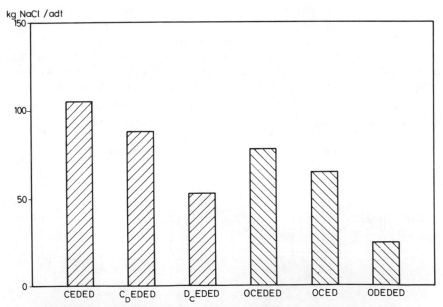

Figure 5-26. Sodium chloride produced in different bleaching sequences.

exist near the furnace outlet and in the flue gas ducts. The other product is sodium sulfate.

Henricson, and others,[658] propose the removal of chloride from the green liquor. The liquor is cooled, and sodium carbonate then crystallizes and most of it is removed from the liquor. The remaining solution containing the chlorides is causticized and all the lime required for the total causticizing is slaked in this solution. The crystallized sodium carbonate is redissolved by heating and causticized separately with the required amount of lime from the first causticizing stage. The excess solution from the first causticizing is evaporated, whereby sodium chloride is crystallized and removed. The remaining solution from evaporation/crystallization, containing sodium sulfide, is mixed with the causticized sodium carbonate liquor.

Rapson and coworkers[663,664] have proposed several methods for removal of chlorides from the recovery system. In one system the white liquor is concentrated in two stages. In the first stage the white liquor is concentrated to 26 to 30% NaOH and Na_2S. Sodium carbonate and sodium sulfate crystallize and are separated from the white liquor and fed into the black liquor. The remaining white liquor is fed into the second stage and concentrated to 36 to 42% NaOH plus Na_2S at a temperature of 74 to 93°C to crystallize sodium chloride, sodium carbonate, and a double salt of sodium carbonate and sulfate. The concentrated white liquor remaining from the second stage has a relatively low concentration of sodium chloride, sodium carbonate, and sodium sulfate. This liquor is diluted with bleach plant effluent or other spent liquors and then used for cooking. The crystallized sodium chloride is purified in three stages, first by classification of the crystals, then by leaching the sodium chloride at 60 to 90°C, and finally filtering the leached slurry and washing with water. The leach liquor is returned to the first stage of evaporation.

Other proposed methods of salt recovery are smelt leaching combined with white liquor evaporation/crystallization, and green-liquor and white-liquor evaporation/crystallization.[665] A full-scale mill using the evaporation/crystallization of white liquor has been constructed in Canada.[401]

Effluents from Bleaching of Mechanical Pulp

In the reductive bleaching of mechanical pulp the load of dissolved organic substances in the effluent is low. The biochemical oxygen demand is about 5 kg O_2 per ton of pulp, and the chemical oxygen demand about 13 kg O_2 per ton of pulp. If zinc dithionite is used for bleaching, it may cause an environmental problem because of the toxicity of this chemical to fish and other organisms in the water. The organic reaction products from the bleaching are present in low concentrations and it is unlikely that they will cause environmental problems. The sequestering agents used in reductive bleaching are not toxic, but they may cause secondary effects when discharged from the mill. For example, the strong complexing effect of amino-carboxylic acids (EDTA and DTPA) may influence the natural level of free metallic ions in the water

TABLE 5-7 CHEMICAL AND BIOCHEMICAL OXYGEN DEMAND IN WASHWATER FROM PEROXIDE BLEACHING OF THERMOMECHANICAL PULP

H_2O_2 Applied	COD ($kg\ O_2/t$ pulp)	BOD_7 ($kg\ O_2/t$ pulp)
1.0	23	15
2.0	26	20
3.0	38	26

to which the discharge is made. These compounds are degraded very slowly. Phosphates used as sequestering agents have the disadvantage of being a nutrient for vegetative organisms in the water and may lead to a heavy production of biomass and a secondary pollution problem.

In the oxidative bleaching of mechanical pulp, more organic substance is dissolved than in reductive bleaching. The oxygen demand in the effluent increases with the amount of bleaching agent used and to some extent with the bleaching conditions. Table 5-7 shows values for the biochemical and the chemical oxygen demand of the washwater from the peroxide bleaching of thermomechanical pulp from spruce. These figures are for the bleaching process only and do not include the load from the pulping process but they are maximum values because some of the dissolved organic compounds will follow the pulp in the mill. Sequestering agents may be used in peroxide bleaching, but they will have the same effects in discharged water as sequestering agents from reductive (dithionite) bleaching.

PRODUCTION OF BLEACHING CHEMICALS

Although most of the chemicals used in bleaching may be purchased rather than produced at the pulp mill, it is helpful to the pulp mill chemist to know how they are manufactured. The methods for the production of the bleaching agents used in bleached pulp manufacture are described below.

Sodium Sulfite. Sodium sulfite may be produced at the mill by reacting sodium hydroxide or sodium carbonate with sulfur dioxide. Sulfur dioxide may be obtained by burning of sulfur, by roasting of sulfide ores, or by purchase as liquid sulfur dioxide. The pH of the solution used for bleaching is usually below 7. The solution is often referred to as a bisulfite solution, but the pH of bisulfite is about 4.5, and most bleaching solutions used for the brightening of mechanical pulp are mixtures of bisulfite and sulfite. Sodium sulfite and bisulfite can also be purchased as a dry powder. Sodium metabisulfite, $Na_2S_2O_5$, is a by-product from other industries and a commercial type of bisulfite.

Zinc Dithionite. The easiest way to produce dithionite is by the reduction of sulphur dioxide with zinc dust:

$$Zn + 2SO_2 \longrightarrow ZnS_2O_4$$

The production was often carried out at the pulp mill at a relatively low price. However, as zinc is somewhat toxic to fish, the use of zinc dithionite for pulp bleaching has been restricted in some countries.

Sodium Dithionite. One of the oldest methods of sodium dithionite production is to convert zinc dithionite to sodium dithionite with sodium hydroxide:

$$ZnS_2O_4 + 2NaOH \longrightarrow Na_2S_2O_4 + Zn(OH)_2$$

Another old method is based on the reduction of sulfur dioxide in aqueous solution with a sodium amalgam produced by electrolysis:

$$4Na + 2SO_2 + 2H_2O \longrightarrow Na_2S_2O_4 + 2NaOH + H_2$$

A method that was introduced in the 1960s is based on the reduction of sulfur dioxide with sodium borohydride:

$$NaBH_4 + 8NaOH + 8SO_2 \longrightarrow 4Na_2S_2O_4 + 6H_2O + NaBO_2$$

This method is used only for production of the bleaching agent at the mill site.[115] Sodium borohydride is stable in alkaline medium, but it decomposes very rapidly in acid medium. The yield of sodium dithionite is best about pH 7 and very good pH control is necessary in order to obtain maximum efficiency.[115,666] Sodium borohydride can be prepared from sodium, hydrogen, and boronoxide. Pure dry sodium borohydride is very expensive, but as an alkaline solution (Borol) it is considerably cheaper, and this alkaline solution is used for production of sodium dithionite.

Production of sodium dithionite as a dry powder is expensive. Dithionite reacts very rapidly with oxygen and the concentration, crystallization, and drying of the salt must be done in the absence of oxidizing agents. The content of $Na_2S_2O_4$ in commercial products is about 90%. Sometimes sequestering and stabilizing compounds are added to the commercial product. The powder is packaged in plastic bags inside steel drums. If the sodium dithionite becomes moist during transport or storage, it may ignite spontaneously when the drums are opened. The reactions leading to this behavior is not well understood.

Sodium Peroxide. Sodium peroxide is produced in two stages. In the first stage, sodium is burned to sodium oxide in a stream of dry air in a rotary steel burner. In the second stage, the sodium oxide is converted to peroxide in a rotating steel finisher where oxygen is fed countercurrently to the sodium oxide at a temperature of 240 to 390°C. Sodium peroxide is transported as dry granules in drums or special containers. The content of Na_2O_2 is approximately 90%.

Hydrogen Peroxide. There are several methods of producing hydrogen peroxide. The two main methods are the electrolytic and the autoxidation processes.

The electrolytic method involves the electrolysis of sulfuric acid or a mixture of ammonium sulfate and sulfuric acid. Persulfuric acid and ammonium persulfate are formed, and when these are concentrated, they are hydrolyzed to hydrogen peroxide and sulfuric acid or ammonium sulfate. The hydrogen peroxide is vaporized together with water and rectified to give a product of commercial strength. The electrolysis process is still used in older plants, but all new plants use the autoxidation process.[667]

The common autoxidation process is based on the reaction between anthrahydroquinone and oxygen to give anthraquinone and hydrogen peroxide. The oxidation is carried out with air. The anthraquinone compounds, usually 2-alkylanthraquinone, and the corresponding anthrahydroquinone are dissolved in organic solvents. The hydrogen peroxide formed is extracted from the organic solvent with water and the hydrogen peroxide concentrated by distillation. The remaining anthraquinone solution is hydrogenated after the removal of water. The hydrogenation is carried out with hydrogen gas, using nickel or palladium as the catalyst. The anthrahydroquinone formed by the hydrogenation is used in a new cycle of peroxide formation. The raw materials for production of hydrogen peroxide in this way are hydrogen and oxygen. Alcohols and aldehydes can also be used in the production of hydrogen peroxide by autoxidation processes.

Commercial hydrogen peroxide is usually concentrated to 50% by weight, but sometimes it is shipped in a concentration of 35%. The solution is stabilized by addition of complexing agents, such as 8-hydroxyquinoline, pyrophosphate, or colloidal compounds, such as hydrated stannic oxide. The stabilization is a result of inactivation of metal catalysts that decompose peroxide. Acidification to a low pH also has a stabilizing effect. Large quantities of hydrogen peroxide are transported and stored in containers of aluminum. Smaller quantities are transported in polyethene drums or bottles. Hydrogen and sodium peroxide react with organic compounds and any contact between concentrated solutions of peroxide or between dry sodium peroxide and organic compounds must be avoided.

Chlorine and Alkali. Two of the principal heavy chemicals used in bleaching are chlorine and sodium hydroxide. The production of these two chemicals is carried out in the same process. All industrial production of chlorine and sodium hydroxide is based on electrolysis of sodium chloride. Two methods have been used for many years: the mercury cathode method[668] and the diaphragm method.[669] The diaphragm method has been most used in North

America, while the mercury-cell method has been widely used in Europe. A more recent process is based on a membrane method.

In the diaphragm method, the anode and cathode are separated by a diaphragm of asbestos. The anode is graphite and the cathode is steel screen or wire. The reaction taking place at the anode is:

$$2Cl^- - 2e \longrightarrow Cl_2$$

and at the cathode is:

$$2H^+ + 2e \longrightarrow H_2$$

Chlorine escapes from the anode, and sodium hydroxide and hydrogen are formed at the cathode. The sodium chloride is fed to the anode compartment, where the level is controlled. The cathode compartment will contain as much sodium chloride as sodium hydroxide and the solution must be evaporated in order to purify the alkali and recover the salt. The sodium hydroxide obtained by this method usually contains about one percent sodium chloride.

In the mercury-cell method, sodium ions are discharged at the mercury cathode and form an amalgam. The amalgam is used as the anode in a second cell, where the sodium hydroxide produced becomes the electrolyte and the cathode is graphite. The reaction in this cell is:

$$2\,Na + 2H_2O \longrightarrow 2NaOH + H_2$$

The sodium hydroxide is produced at a concentration of 50% without evaporation and it is not contaminated with sodium chloride. After the danger of mercury in the food chain was discovered, steps were taken to reduce the pollution. Methods have been found to eliminate almost all of the mercury in the caustic and water effluents, but loss of mercury with the vented air has been difficult to avoid. For this reason the mercury-cell method will probably not be used in any new plant.

In the membrane method, the anode and cathode are separated by a selective ion-exchange membrane. This membrane is a fluorocarbon plastic that allows the positive ions H^+ and Na^+ to permeate the membrane, but rejects the negative ions Cl^- and OH^-. A sodium hydroxide almost free of chloride can be prepared in these cells. In principle this method is similar to the diaphragm method, but the alkali as well as the chlorine are purer in the membrane process.[670,671]

The hydrogen gas from chlorine-alkali production is either used in chemical plants or burned to utilize the energy. In some small plants the hydrogen is vented to the air. Chlorine and hydrogen produce an explosive mixture and these two gases from the electrolysis must be kept apart. For each ton of chlorine produced, 1.1 ton of sodium hydroxide is obtained. This requires 1.8 tons of sodium chloride and about 11.5 GJ (3.200 kWh)/ton of chlorine. The cell voltage is between 3.5 and 4.5 V.

Sodium hydroxide is usually shipped at a concentration of 45 to 50%. At this concentration it solidifies at 12°C. A 20% solution of sodium hydroxide will freeze at about –20°C. Sodium hydroxide is often stored at this concentration in cold climate.

The chlorine gas from the anode can be used as it comes from the cell if the pulp mill has its own chlorine-alkali plant. If the gas must be transported, water and impurities are removed before liquefaction of the gas. Part of the water is removed by cooling, but the gas should not be cooled below 10°C because a hydrate, $Cl_2 \cdot 8H_2O$, can be formed below this temperature. Scrubbing with concentrated sulfuric acid is used to remove the remaining water. Liquefaction is carried out by compression or refrigeration or by a combination of these methods. Dry chlorine can be handled in low carbon steel equipment.

Chlorine is shipped in cylinders, tank cars, or tank barges. The pulp industry, which is a large consumer of chlorine, generally purchases chlorine in tank cars. Liquid chlorine has a high-temperature expansion coefficient and the tanks should not be filled to more than 80% of their volume in order to avoid liquid pressure in the tank. Tank cars can be used for storage, but usually the chlorine is unloaded to a special storage tank. Liquid chlorine can be unloaded by its own vapor pressure, but at low temperature the unloading rate will be low and compressed air is used to increase the pressure. This air must be dry and clean. Chlorine tanks are equipped with a pipe for unloading as a gas and a pipe that goes to the bottom of the tank for unloading as a liquid. When the storage tank is being filled, the gas pipe must be open and the gas leaving the tank applied in the process.

For the chlorination of pulp and for the preparation of hypochlorite, the chlorine is generally used as a gas. The liquid is gasified in special chlorine evaporators. These evaporators are heated with hot water to a temperature of 60°C. Liquid chlorine may contain sulfuric acid and other contaminants. These are removed with sedimentation traps and filters as the chlorine is gasified. The pipe for chlorine gas should have an atmospheric loop with the highest point at least 10.5 m above the point where the chlorine is applied in the aqueous phase in order to prevent suction of water back to the liquid chlorine by vacuum in the pipe. In all handling of chlorine a potential risk is involved, and the operations for unloading and use of chlorine should be done according to the specifications for such work.

Hypochlorite. Sodium and calcium hypochlorite are used for the bleaching of pulp. These hypochlorites are prepared from chlorine and sodium or calcium hydroxide. The reactions are:

$$2NaOH + Cl_2 \longrightarrow NaOCl + NaCl + H_2O$$
$$2Ca(OH)_2 + 2Cl_2 \longrightarrow Ca(OCl)_2 + CaCl_2 + 2H_2O$$

Calcium hypochlorite is more difficult to prepare than sodium hypochlorite.

Burnt lime is the usual source of alkali for calcium hypochlorite. It is first converted into calcium hydroxide in a slaker followed by a grit settler. A settling tank for the hypochlorite solution is needed in order to remove suspended particles. The pH of the solution must be kept on the alkaline side during the preparation. If the pH drops to the neutral point, the formation of chlorate is favored; the rate of this reaction is highest at a certain proportion of HClO to ClO$^-$ in the solution. The addition of chlorine should, therefore, be stopped before all hydroxide is consumed. The amount of excess alkali when calcium is used is limited by the solubility of calcium hydroxide. As the lime reacts with the chlorine and forms hypochlorite and chloride ions, the concentration of calcium ions increases in the solution because much of the slaked lime is not a true solution. The consequence of this is that the concentration of hydroxyl ions will decrease because the concentration of hydroxyl ions is inversely proportional to the square root of the concentration of calcium ions. Figure 5–19 shows the relationship between free lime and available chlorine in calcium hypochlorite solutions. Sodium hypochlorite has the advantage that the amount of excess alkali can be high.

Hypochlorite can be prepared in a batch or in a continuous process. The continuous process is the more common method. In the batch process, the end point is determined by a pH indicator, determination of available chlorine, or by the temperature rise in combination with the testing of the original alkali concentration. In the continuous process, the reaction can be controlled either by the temperature rise or by an oxidation-reduction measurement. In order to obtain stable production, the concentration and flow of alkali must be controlled. The concentration of alkali can be controlled by the density of the solution, and in the case of sodium hydroxide, conductivity measurement can also be used. The flow of chlorine is controlled by the oxidation-reduction potential. The chlorine flow can also be controlled by the alkali flow in such a way that a failure in the alkali supply will stop the flow of chlorine. If the lime contains manganese, an overchlorination of the bleach liquor will be noticed by a pink color caused by formation of permanganate.

Chlorate. Chlorate, as such, is not used as a bleaching agent, but in all industrial processes for chlorine dioxide generation, sodium chlorate is the raw material. Like the chlorine-alkali process, the production of sodium chlorate is by the electrolysis of sodium chloride. The difference is that in the chlorine-alkali process the chlorine and sodium hydroxide formed at the electrodes are separated. In the chlorate process, the chlorine and alkali are allowed to react to form hypochlorite as follows:

$$Cl_2 + 2OH^- \longrightarrow ClO^- + Cl^- + H_2O$$

In a second step the hypochlorite is transformed into chlorate.

The rate of the reaction by which hypochlorite is converted into chlorate is very dependent on the pH. About one part hypochlorite to two parts hypochlorous acid produce the optimum condition:

$$2HClO + ClO^- \longrightarrow ClO_3^- + 2Cl^- + 2H^+$$

The dissociation constant for hypochlorous acid is about $5 \cdot 10^{-8}$; this means that the optimum pH for chlorate formation is about pH 7. In order to maintain this pH some hydrochloric acid must be added. Chlorate can also be formed by anodic discharge of hypochlorite ion and reaction with water:

$$ClO^- \longrightarrow ClO + e$$

$$6ClO + 3H_2O \longrightarrow 2ClO_3 + 4Cl^- + 30 + 6H^+$$

This reaction consumes electrical energy and evolves oxygen which should be prevented. The formation of chlorate is also dependent on temperature. A high temperature favors the electrolysis and the formation of chlorate, but the consumption of the graphite electrodes increases rapidly with temperature.[672] Titanum with special coating can be used as the anode; the cathodes are made of carbon steel. Optimum temperature for chlorate formation is about 60°C. Sodium dichromate is used as an additive in the chlorate process to reduce corrosion. For production of one ton of sodium chlorate about 550 kg of sodium chloride and 21.6 GJ (6000 kWh) are required.

Hydrogen gas from the electrolysis is usually diluted with air below its explosive concentration and expelled immediately. The hydrogen gas can be collected and used for fuel, but this involves a safety risk.[672]

The chlorate solution from on-site production at the pulp mill can be used for chlorine dioxide preparation without crystallization of the salt. When the chlorate must be transported from a chemical plant to the pulp mill, the sodium chlorate is usually produced in dried crystallized form. Sodium chlorate is sometimes shipped as a concentrated solution or as a slurry in tank cars or trucks. Most pulp mills purchase chlorate, but some large mills make their own chlorate.[672,673] Dry sodium chlorate reacts explosively with organic material or finely divided metals, but handling of chlorate is relatively safe as long as it is in solution.

Chlorine Dioxide. Chlorine dioxide is a yellow-green to orange gas at room temperature. The gas has a characteristic smell, is toxic, and is spontaneously explosive at high concentrations. Chlorine dioxide can be liquefied at 11°C and solidified at –59°C. Pure chlorine dioxide is very unstable and even when it is diluted with air to 15% it may cause explosions. Chlorine dioxide is, therefore, always prepared at the mill site.

All commercial methods for chlorine dioxide preparation are based on the reduction of sodium chlorate. Several agents are used for the reduction of the chlorate. Sulfur dioxide or chloride are used in the most common processes, but methanol and chromic sulfate are also used. The chemistry of chlorine dioxide preparation is rather complicated. The overall reaction with sulfur dioxide can be written:

$$2NaClO_3 + SO_2 \longrightarrow 2ClO_2 + Na_2SO_4$$

However, the formation of chlorine dioxide requires a strongly acidic medium and more reaction steps are involved. Moreover, several side reactions can take place. The most important reactions probably are:[674-676]

$$HClO_3 + HCl \longrightarrow HClO_2 + HClO$$
$$HClO + HCl \longrightarrow Cl_2 + H_2O$$
$$HClO_3 + HClO_2 \longrightarrow 2ClO_2 + H_2O$$
$$2HClO_2 + Cl_2 \longrightarrow 2ClO_2 + 2HCl$$

In the strongly acidic solution the chlorate is converted into chloric acid, which is reduced to chlorine dioxide. It is believed that the reaction between chloric acid and chloride is the primary one for all processes where sodium chlorate is used as the starting material. The effect of adding sulfur dioxide is to reduce the chlorine and hypochlorous acid to chloride and thereby obtain a chlorine dioxide with a low concentration of chlorine.

Some of the commercial processes for the preparation of chlorine dioxide are:

1. Kesting or Day process
2. Persson or Columbia Southern process
3. Holst process
4. Mathieson process
5. Solvay process
6. Rapson R 2 process
7. Rapson R 3 process

The Holst[677,678] and Mathieson[676,679,680] processes use sulfur dioxide as a reductant. The Kesting-Day,[680-683] Rapson R 2[684,685] and R 3[687-689] processes use chloride, the Solvay process uses methanol,[680,690,691] and the Persson-Columbia Southern process uses chromic sulfate, which is regained by reduction with sulfur dioxide.[680,692] The Holst process is operated batchwise. The other processes are continuous.

The spent liquors from the chlorine dioxide preparation are strongly acidic, contain chloride, and chlorate—most of them also contain sulfates. In the Kesting-Day and Persson-Columbia Southern processes, the spent liquors are recycled and there is no discharge of effluents to water. In the Holst, Mathieson, and R 2 processes, the spent liquor can to some extent be used in kraft mills for acidification of tall oil soaps or as makeup chemical in the pulping recovery process. However, for kraft mills using hardwoods there will be no need for acidification of soaps, and as the loss of sodium sulfate in the washing and recovery processes has to be reduced owing to environmental restrictions, all the sulfuric acid and sodium sulfate from the chlorine dioxide plant cannot be utilized in the mill. The contaminated sulfuric acid is not a commercial product and most mills have no market for sodium sulfate. Therefore, environmental problems will play an important role in the choice of method for production of chlorine dioxide. The amount of sodium sulfate and the excess of sulfuric acid from the different methods of chlorine dioxide preparation are given in Table 5-8 in tons per ton of ClO_2 produced. The Kesting process is, as mentioned, a closed process without discharge to water. With the R 3 process, a chemical balance in a mill can be obtained with low chemical losses in the pulping and recovery processes.

In the Kesting process, the preparation of chlorine dioxide is combined with chlorate production. The chlorate-chloride solution from the chlorate plant is fed to the chlorine dioxide plant, where hydrochloric acid is added to obtain the acidic medium required. The liquor from the chlorine dioxide process has a high concentration of chlorate but this is returned to the chlorate plant. This solution is acidic and must be neutralized with sodium hydroxide to produce sodium chloride, which serves as the raw material for the chlorate electrolysis. The hydrochloric acid required for the chlorine dioxide process is produced by reacting hydrogen from the chlorate production with purchased chlorine or with chlorine from the chlorine-alkali electolysis at the mill. The investment in the Kesting process is high and, therefore, this method is suitable only for large mills with a high consumption of chlorine dioxide. The economy of the process also depends very much on the price of electrical energy.

There are two processes based on the Rapson R 3 principle: Hooker's single vessel process (SVP) in the United States and the ERCO process in Canada. Figure 5-27 shows the main steps of the ERCO process. A solution of sodium chlorate and sodium chloride is fed to the reactor together with sulfuric acid. The reactor is operated under reduced pressure and at fairly high temperature.

TABLE 5-8 SODIUM SULFATE PRODUCED AND EXCESS OF SULFURIC ACID IN CHLORINE DIOXIDE PRODUCTION.[a]

Reducing Agent	Na_2SO_4 Produced[b]	Excess of H_2SO_4 (93%)	Total Equivalent Na_2SO_4 [b]
Methanol	1.20	1.65	3.42
SO_2	1.20	1.65	3.42
Chloride (R 2)	2.30	3.40	6.88
Chloride (R 3)	2.30	0.00	2.30

[a] See reference 693.
[b] Tons per ton of ClO_2.

TABLE 5-9 RAW MATERIALS AND BY-PRODUCTS IN THE R 3 AND KESTING PROCESSES

	Process	
	R 3	Kesting
Raw materials		
ton/ton ClO_2		
$NaClO_3$	1.66	—
NaCl	0.95	—
H_2SO_4	1.70	—
HCl (calculated as 100%)	—	1.50–1.56
NaOH	—	1.1
By-products		
ton/ton ClO_2		
Cl_2	0.58	0.78
Na_2SO_4	2.30	0.00

The concentration of sulfuric acid is higher in this process than in the processes using sulfur dioxide. However, the sulfuric acid is recycled after removal of the sodium sulfate which crystallizes in the bottom of the reactor. If one wishes to reduce the sulfate production to a lower level, sulfuric acid can partly be replaced by hydrochloric acid.[693,694]

Table 5-9 shows the amounts of raw materials required and the by-products produced in the R 3 and the Kesting processes. Both of these processes produce rather high quantities of chlorine as by-product. Part of this chlorine can be used

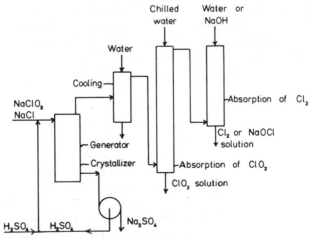

Figure 5-27. Chlorine dioxide production in the R3 process.

TABLE 5-10 SOLUBILITY DATA FOR SODIUM CHLORATE IN WATER

Temperature ($^\circ$C)	$NaClO_3$ (%)	Density (kg/m^3)
-15	43.1	1380
0	44.0	1390
20	50.4	1430
40	56.0	1472
60	61.6	1515
80	66.8	1558
100	71.5	1600

"Pulp Bleaching," Bulletin 200, Hooker Electrochemical Co., Niagara Falls, N.Y.

in bleaching: in combination with chlorine dioxide, as chlorine water for chlorination, or for preparation of hypochlorite. Sodium chlorate can also be regenerated from the sodium hypochlorite.[695]

When sodium chlorate is purchased in the solid form, it must be dissolved. If it is shipped in tank cars, it can be unloaded in the form of a slurry by recirculating hot water between the car and a dissolver-storage tank. The solubility of sodium chlorate is shown in Table 5-10. The dissolving of chlorate is an endothermic process and the water used for making up the solution should therefore be heated to 60°C or more. The storage tanks are usually made of stainless steel. The alkaline chlorate solution is not a particularly strong oxidizing agent, but chloric acid decomposes spontaneously at concentrations over about 30% in the presence of organic matter.

The chlorine dioxide gas produced is always diluted and is then transported by air, by steam, or by a mixture of sulfur dioxide and the remaining gases from the air after burning of sulfur. As mentioned, the concentration of ClO_2 should be kept below 15 volume percent in order to avoid explosions. Even if the chlorine dioxide is diluted, explosions may occur but they are rather mild and are known as puffs. These puffs are caused by high temperature; contaminants in the raw material, such as oil, paper, wood, string and rust; or by leakages from the sulfuric acid pipe being spattered into the gas phase. In order to avoid damage by these puffs, the chlorine dioxide reactors are equipped with pressure-relief lids. The decomposition of gaseous chlorine dioxide is inhibited by water vapor up to a temperature of 90°C. A large surface in proportion to the volume is favorable to the stability.[696] The rapid decomposition of chlorine dioxide probably occurs by way of flame propagation up to a chlorine dioxide concentration of 40%. The decomposition front can be stopped by a pulp bed.[697]

The gas from the chlorine dioxide reactor is passed into the bottom of a packed absorption tower through which cold water is circulated. A solution containing 6 to 8 g ClO_2/liter is usually obtained. The solubility of ClO_2 in water is very dependent on the temperature, as can be seen from Figure 5-20. In tropical areas or in the summertime in subtropical areas, the water for absorption of chlorine dioxide must be chilled.[698] Chlorine dioxide in water decomposes rapidly by irradiation especially by irradiation with ultraviolet light. The final products from this decomposition are chloride and chlorate.[699]

Control of the chlorine dioxide process is very important. The metering of reagents and the flow of diluting gas have to be accurate. The reaction temperature must be kept within certain limits, and the supply of chemicals must be shut off if the temperature goes out of control. The concentration of chlorine dioxide gas can be determined continuously by measuring the absorbance of ultraviolet light in the gas at 436 nm.[700] Chlorate and chlorine dioxide in solution can be determined by calorimetric methods[463,701] using the reaction with ferrous ions for chlorate determination and with sulfite for chlorine dioxide determination. The acidity of the chlorine dioxide solution can be determined calorimetrically by neutralization.

The production of chlorine dioxide involves some hazards,[702-704] but improvement in equipment and control of the process have reduced the problems. Today chlorine dioxide generation can be considered as a fairly safe operation. The high toxicity of chlorine dioxide makes it necessary to control the concentration in the internal and external environment. Instruments for determining concentrations in the air inside and outside the plant are on the market, but so far no instrument seems to be satisfactory for determining the concentration of chlorine dioxide in the tail gases leaving the process equipment.

Corrosion has been a severe problem in the production and use of chlorine dioxide.[705-707] Stainless steel can be attacked heavily by chlorine dioxide when chloride is present,[708] although chlorine dioxide alone does not corrode acid-proof steel. Titanium is very resistant to chlorine dioxide and the use of this metal has increased substantially.[707,709,710] Other construction materials used in chlorine dioxide generators are steel shells lined with PVC, acid-proof brick, tile, or lead.[705,706] Unplasticized PVC has also proved to be satisfactory in the chlorine dioxide generator.[677] Piping and valves can be of Hastelloy C, PVC, or porcelain. For piping, glass-lined pipe, polyester fiberglass, PVC, and lead are used. Absorption towers are made of PVC, stoneware, porcelain, or steel that is lined with glass or brick. The same type of materials are also used for storage of chlorine dioxide solutions.[677,705,706] Anodic protection of stainless steel has improved the resistance towards chlorine dioxide in pulp washers.[711]

Sodium Chlorite. Sodium chlorite can be used instead of chlorine dioxide for bleaching of pulp. Chlorite is produced in crystalline form and can be transported. Sodium chlorite is made from chlorate, but the first product in the production is chlorine dioxide. Production of chlorite is, therefore, more expensive than the production of chlorine dioxide. The oxidation capacity is also reduced by one to five in transforming chlorine dioxide into chlorite. The reaction is:

$$4NaOH + Ca(OH)_2 + C + 4ClO_2 \longrightarrow 4NaClO_2 + CaCO_3 + 3H_2O$$

Calcium carbonate is removed by filtration and the sodium chlorite in crystalline form at a purity of 90 to 98% is obtained by evaporation.[712] Precautions in transport and handling should be the same as for chlorate.

Oxygen. There are many potential sources of oxygen, but the easiest way of obtaining oxygen is from air, which contains 21% oxygen.

The usual method of separating oxygen from the other components of air is by the Linde or cryogenic process.[713] In this process, the air is compressed and cooled with water, the carbon dioxide and impurities are removed, the air is cooled to about –193°C by expansion, and the oxygen is separated from the nitrogen and argon in a rectification column. Oxygen produced in this way usually has a purity about 99.5%.

Another method of separating oxygen from air is by the pressure swing absorption process. The principle of this process is that zeolite has a significantly higher adsorption capacity for nitrogen than for oxygen and argon. Nitrogen is adsorbed at high pressure, and when the zeolite is almost saturated, the pressure is reduced to desorb nitrogen from the zeolite and regenerate the adsorption capacity. Two or more adsorption units are used: one for adsorption, while the other is being regenerated. A gas containing 70 to 95% oxygen can be produced by this method. The investment is lower than for the cryogenic process and this process may be competitive for relatively small units below 100 ton/day.[714,715]

Oxygen is transported either as compressed air or as liquid. For large-scale transport, liquid oxygen is used. Storage of liquid oxygen is in a Dewar tank and the temperature is kept sufficiently low by evaporation of a small amount of oxygen. When oxygen is used for bleaching, it is passed through an evaporator and transported to the oxygen reactor as a gas. Concentrated oxygen has a strong oxidizing power. Oils and fat are easily ignited in pure oxygen at high pressure. Special compounds must be used as lubricants in bearings and valves that are in contact with oxygen or where there is a risk of such contact.

Ozone. Ozone is prepared by silent electrical discharges in dry air or in oxygen[716] at a voltage of about 15,000 V. Pyrex glass is used as a dielectric, and water is used for removal of the energy generated. The gas leaving the generator will contain 1 to 4% ozone and because of the rapid decomposition of ozone it must be used without storage. If oxygen is used to produce ozone, the remaining oxygen must be purified, dried, and returned to the process. Ozone is a toxic gas that can be smelled in very low concentrations. It is comparatively easy to detect ozone in the air, and the production can be shut off automatically if the concentration in the ambient air reaches a preset value.

REFERENCES

1. P. Kubelka and F. Munk, *Zh.* (tech. Phys.,) **12** (11a), 593 (1931).
2. P. Kubelka, *J. Opt. Soc. Amer.,* **38** (5), 448 (1948).
3. J. A. Van den Akker, *Tappi,* **32** (11), 498 (1949).

4. D. B. Judd and G. Wyszecki, "Color in Business, Science and Industry," 3rd ed., Wiley, New York, N.Y., 1975, p 420.
5. *TAPPI Standard T-217 m-48.*
6. *SCAN-test SCAN-C11:62.*
7. International Organization for Standardization, *ISO 3688-1977.*
8. Å. Stenius, J. Rydberg, and L. Söderhjelm, *Svensk Papperstidn.,* **78** (11), 403 (1975).
9. J. Polcin and W. H. Rapson, *Tappi,* **52** (10), 1960 (1969).
9a. A. Teder and D. Tormund, *CPPA Trans. Tech. Sect.,* **3** (2), 41 (1977).
10. W. Sanderman and F. Schlumbom, *Holz als Roh-und Werkstoff,* **20** (7), 245 (1962).
11. V. Lorås and G. Wilhelmsen, *Norsk Skogind.,* **27** (12), 346 (1973).
12. M. D. Wilcox, *Svensk Papperstidn.,* **78** (1), 22 (1975).
13. M. Douek, C. Heitner, L. Lamandé, and D. A. I. Goring, *CPPA Trans. Tech. Sect.,* **2** (3) 78 (1976).
14. M. Douek and D. A. I. Goring, *CPPA Trans. Tech. Sect.,* **2** (3), 83 (1976).
15. K. Freudenberg and A. C. Neish, "Constitution and Biosynthesis of Lignin," Springer Verlag, Berlin, West Germany, 1968.
16. I. A. Pearl, "The Chemistry of Lignin," Marcel Dekker, New York, N.Y., 1967.
17. E. Adler, *Paperi Puu,* **43** (11), 634 (1961).
18. J. C. Pew and W. J. Connors, *Tappi,* **54** (2), 245 (1971).
19. J. C. Pew and W. J. Connors, *J. Org. Chem.,* **34** (3), 580 (1969).
20. W. J. Connors, J. S. Ayers, K. V. Sarkanen, and J. S. Gratzl, *Tappi,* **54** (8), 1284 (1971).
21. F. Imsgard, S. I. Falkehag, and K. Kringstad, *Tappi,* **54** (10), 1680 (1971).
22. H. Norrström, *Svensk Papperstidn.,* **72** (2), 25 (1969).
23. W. E. Hillis, "Wood Extractives and Their Significance to the Pulp and Paper Industries," Academic, New York, N.Y., 1962, p 405.
24. G. M. Barton, *Tappi,* **56** (5), 115 (1973).
25. V. Lorås, H. Nickelsen, and J. Brandal, The Norwegian Pulp and Paper Research Institute Report 1965.
26. V. N. Gupta and D. B. Mutton, *Pulp Paper Mag. Can.,* **68** (C), T-107 (1967).
27. H. G. Jones and C. Heitner, *Pulp Paper Mag. Can.,* **74** (5), T-182 (1973).
28. F. A. Whitman, *Tappi,* **40** (1), 20 (1957).
29. V. Lorås and G. Wilhelmsen, *Norsk Skogind.,* **28** (5), 121 (1974).
30. H. P. Dahm, *Norsk Skogind.,* **17** (5), 178 (1963).
31. T. K. Kirk and E. Adler, *Acta Chem. Scand.,* **23** (2), 705 (1969).
32. T. K. Kirk and E. Adler, *Acta Chem. Scand.,* **24** (9), 3379 (1970).
33. J. M. Harkin, "Lignin Structure and Reactions," Advances in Chemistry Series No. 59, American Chemical Society, Washington, D.C., 1966, p 65.
34. W. Brecht and K. -P. Meltzer, *Wochbl. Papierfabr.,* **93** (18), 777 (1965).
35. G. Schweizer, *Wochbl. Papierfabr.,* **100** (11, 12), 433 (1972).
36. M. T. Charters and R. O. Ward, *Proc. Intl. Mechanical Pulping Conf.,* Stockholm, Sweden, 1973, p 16: 1.
37. V. Lorås, *Tappi,* **59** (11), 99 (1976).
38. K. H. Klemm, *Paper Trade J.,* **142** (12), 52 (1958).
39. L. R. Beath, W. G. Mihelich, S. B. Beaulieu, and D. J. Wild, *Proc. Intl. Mechanical Pulping Conf.,* Stockholm, Sweden, 1973, 21: 1.
40. F. E. Murray, S. S. Rawalay, and J. V. Hatton, *Pulp Paper Mag. Can.,* **68** (1), T-28 (1967).
41. J. W. Smith, D. F. Cooper, L. E. Seeley, and C. B. Anderson, *CPPA, Tech. Sect. Paper and Discussion,* **3,** D-79 (1972).

42. W. H. Rapson, M. Wayman and C. B. Anderson, *Pulp Paper Mag. Can.,* **66** (5), T-255 (1965).
43. D. H. Andrews, *Pulp Paper Mag. Can.,* **69** (11), T-273 (1968).
44. A. Tyminski, *Pulp Paper Mag. Can.,* **68** (6), T-268 (1967).
45. S. Back, *Tappi,* **39** (4), 170 A (1956).
46. O. V. Ingruber and J. Kopanidis, *Pulp Paper Mag. Can.,* **68** (6), T-258 (1967).
47. D. B. Sparrow, *Tappi,* **39** (7), 486 (1956).
48. S. A. Rydholm, "Pulping Processes," Interscience, New York, N.Y., 1965, p 232.
49. D. W. Read, B. D. Eade, and N. R. Slingsby, *Pulp Paper Mag. Can.,* **69** (13), T-296 (1968).
50. T. Fossum, N. Hartler, and J. Libert, *Svensk Papperstidn.,* **75** (8), 305 (1972).
51. V. N. Gupta, *Pulp Paper Mag. Can.,* **71** (18), T-391 (1970).
52. F. A. Abadie-Maumert, V. Lorås, and N. Soteland, *Norsk Skogind.,* **23** (12), 361 (1969).
53. V. N. Gupta and D. B. Mutton, *Pulp Paper Mag. Can.,* **70** (11), T-174 (1969).
54. M. Wayman, C. B. Anderson, and W. H. Rapson, *Pulp Paper Mag. Can.,* **69** (9), T-225 (1968).
55. G. G. Bouzane and M. Wayman, *Pulp Paper Mag. Can.,* **68** (12), T-653 (1967).
56. J. W. Tardif, *Appita,* **13** (2), 58 (1959).
57. A. Bosia, *Cell Carta,* **12** (4), 5 (1961).
58. E. Masak, Jr., *Tappi,* **43** (1), 166 A (1960).
59. R. R. Kindron, *Tappi,* **46** (12), 152 A (1963).
60. I. A. Crawford, *Appita,* **23** (3), 177 (1969).
61. R. Jensen, *Wochbl. Papierfabr.,* **96** (11, 12), 399 (1968).
62. W. C. Mayer and P. Donofrio, *Pulp Paper Mag. Can.,* **59** (10), 157 (1958).
63. V. Lorås and N. Soteland, *Norsk Skogind.,* **26** (10), 255 (1972).
64. K. Kringstad and F. A. Abadie-Maumert, *Norsk Skogind.,* **22** (7), 230 (1968).
65. R. A. Fleury and W. H. Rapson, *Pulp Paper Mag. Can.,* **69** (6), T-154 (1968).
66. A. Teder, *Nordic Symp. in Lignin Preserving Bleaching,* Oslo, Norway, 1969.
67. T. Ploetz and H. Meyer, *Papier* (Darmstadt), **19** (10 a), 655 (1965).
68. S. Y. Lin and K. P. Kringstad, *Tappi,* **53** (4), 658 (1970).
69. V. Lorås and M. J. Rengård, *Norsk Skogind.,* **23** (9), 239 (1969).
70. J. Polcin and W. H. Rapson, *Pulp Paper Mag. Can.,* **73** (1), T-2 (1972).
71. P. C. Trotter, *Tappi,* **45** (6), 449 (1962).
72. R. A. Fleury and W. H. Rapson, *Pulp Paper Mag. Can.,* **70** (23), T-517 (1969).
73. S. H. Hawthorne and W. H. Rapson, *Pulp Paper Mag. Can.,* **75** (6), T-234 (1974).
74. J. Polcin and W. H. Rapson, *Pulp Paper Mag. Can.,* **72** (3), T-103 (1971).
75. W. D. Nicoll and A. F. Smith, *Ind. Eng. Chem.,* **47** (12) 2548 (1955).
76. R. H. Dick and D. H. Andrews, *Pulp Paper Mag. Can.,* **66** (3), T-201 (1965).
77. P. K. Paasonen, *Paperi Puu,* **52** (9), 555 (1970).
78. R. D. Spitz, *Tappi,* **44** (10), 731 (1961).
79. H. Schröter, *Wochbl. Papierfabr.,* **96** (20), 725 (1968).
80. M. Solinas, *Pulp Paper Can.,* **77** (3), T-50 (1976).

81. M. L. Slove, *Tappi*, **48** (9), 535 (1965).
82. P. C. Holladay and R. J. Solari, TAPPI Monograph No. 27, 1963, p 186.
83. D. H. Andrews and P. Des Rosiers, *Pulp Paper Mag. Can.*, **67** (C), T–119 (1966).
84. S. Hauan, H. E. Höydahl, V. Lorås, and E Böhmer, *Papier* (Darmstadt), **29** (7), 269 (1975).
85. M. K. Gupta, *Tappi*, **59** (11), 114 (1976).
86. V. Lorås, J. Carles, and G. Papageorges, *ATIP*, **30** (9), 322 (1976).
87. A. Lindahl, Swedish Patent Application 15973 (1971).
88. F. A. Abadie-Maumert and V. Lorås, *Norsk Skogind.*, **24** (12), 412 (1970).
89. R. Jensen, *Wochbl. Papierfabr.*, **100** (19), 708 (1972).
90. C. K. Textor, U. S. Patent 3023140, (1962).
91. V. Lorås and N. Soteland, *Norsk Skogind.*, **28** (4), 90 (1974).
92. M. Wayman, C. B. Anderson, and W. H. Rapson, *Tappi*, **48** (2), 113 (1965).
93. T. J. McDonough, *Pulp Paper Mag. Can.*, **73** (C), T–79 (1972).
94. C. B. Christiansen, G. L. Burroway, and W. F. Parker, *Tappi*, **49** (2), 49 (1966).
95. W. P. Stevens and R. Marton, *Tappi*, **49** (10), 452 (1966).
96. K. V. Sarkanen and J. Suzuki, *Tappi*, **48** (8), 459 (1965).
97. Y. -Z. Lai and K. V. Sarkanen, *Tappi*, **51** (10), 449 (1968).
98. K. Sakai, K. -I. Kuroda, and S. Kishimoto, *Tappi*, **55** (12), 1702 (1972).
99. J. Gierer and T. Norin, Swedish Patent 324285 (1970).
100. K. Kratzl, J. S. Gratzl, E. Schäfer-Risnyovszky, and A. Hruschka, *Papier*, (Darmstadt) 22 (10 A), 686 (1968).
101. N. Soteland and K. Kringstad, *Norsk Skogind.*, **22** (2), 46 (1968).
102. N. Soteland, *Norsk Skogind.*, **25** (3), 61 (5), 135 (1971).
103. N. Liebergott, *Pulp Paper Mag. Can.*, **73** (9), T–214 (1972).
104. R. M. Kingsbury, E. S. Lewis, and F. A. Simmonds, *Paper Trade J.*, **126** (24), T–272 (1948).
105. R. T. Mills, W. S. Hinegardner, and W. O. Stauffer, *Paper Trade J.*, **122** (20), T–216 (1946).
106. G. W. Jones, *Tappi*, **33** (3), 149 (1950).
107. R. H. Reeves and I. A. Pearl, *Tappi*, **48** (2), 121 (1965).
108. J. F. Corbett, *J. Chem. Soc.*, (C), 2308 (1966).
109. C. W. Bailey and C. W. Dence, *Tappi*, **52** (3), 491 (1969).
110. T. D. Spittler and C. W. Dence, *Svensk Papperstidn.*, **80** (9), 275 (1977).
111. A. W. Kempf and C. W. Dence, *Tappi*, **58** (6), 104 (1975).
112. J. Polcin and W. H. Rapson, *Pulp Paper Mag. Can.*, **72** (3), T–114 (1971).
113. R. W. Barton, *Tappi*, **41** (3), 161 A (1958).
114. H. Jacques and R. W. Barton, *Pulp Paper Mag. Can.*, **59** (10), 154 (1958).
115. C. A. Richardson, D. C. Johnson, and T. G. Goodart, *Tappi*, **53** (12), 2275 (1970).
116. R. L. Case, *Tappi*, **50** (3), 57 A (1967).
117. B. A. Marshall, *Can. Pulp Paper Ind.*, **28** (9), 28 (1975).
118. H. Schröter, *Wochbl. Papierfabr.*, **97** (23, 24), 1023 (1969).
119. G. E. Smedberg, U. S. Patent 3100732 (1963).
120. H. W. Giertz, *Tappi*, **44** (1), 1 (1961).
121. P. Luner and D. W. Ferguson, *Tappi*, **47** (4), 205 (1964).
122. I. Palenius, *Pulp Paper Mag. Can.*, **69** (2), T–44 (1968).
123. P. K. Christensen, *Pulp Paper Mag. Can.*, **69** (2), T–56 (1968).
124. H. Norrström, *Svensk Papperstidn.*, **73** (15), 455 (1970).
125. H. Norrström, *Svensk Papperstidn.*, **73** (23), 761 (1970).
126. G. L. K. Hoh, N. J. Stalter, and F. L. Fennel, *Paper Trade J.*, **150** (47), 57 (1966).

127. D. Atack, C. Heitner, and M. I. Stationwala, *Preprint, Intl. Mechanical Pulping Conf.*, Helsinki, Finland, 1977, p 7: 1.
128. C. A. Richardson, *Tappi*, **45** (12), 139 A (1962).
129. A. J. Pearson, J. L. Somerville, and A. E. Elder, *Appita*, **18** (1), 16 (1964).
130. A. J. Pearson and I. H. Coopes, *Appita*, **18** (1), 26 (1964).
131. L. B. Johnson and J. L. Somerville, *Appita*, **16** (1), 4 (1962).
132. P. Luner, *Tappi*, **43** (10), 819 (1960).
133. P. Luner and R. Supka, *Tappi*, **44** (9), 620 (1961).
134. J. F. Jahne and C. E. Price, *Tappi*, **43** (7), 226 A (1960).
135. A. Meinecke, *Pulp Paper Mag. Can.*, **73** (7), T–180 (1972).
136. W. Ginzel, S. M. Saad, and G. Stegmann, *Holzforschung*, **23** (5), 152 (1969).
137. R. B. Roe, *Paper Trade J.*, **79** (15), 43 (1924).
138. *TAPPI Standard T–202 os–69*.
139. *CPPA Standard G. 16*, 1964.
140. *TAPPI Useful Method 209; TAPPI*, **54** (11), 1924 (1971).
141. W. B. Beazley, J. H. E. Herbst, and J. A. Histed, *Pulp Paper Mag. Can.*, **58** (9), 133 (1957).
142. International Organization for Standardization *ISO 3260–1975 (E)*.
143. "SCAN-test," *Svensk Papperstidn.*, **75** (12), 485 (1972); *Paperi Puu*, **54** (5), 313 (1972); *Norsk Skogind.*, **26** (5), 128 (1972).
144. *Zellst. Papier*, Merkblatt IV/53/71.
145. *TAPPI Standard T–253 pm-75*.
146. B. Kyrklund and G. Strandell, *Paperi Puu*, **49** (3), 99 (1967).
147. B. Kyrklund and G. Strandell, *Paperi Puu*, **51** (4a), 299 (1969).
148. R. Sieber, "Die Chemisch-Technischen Untersuchungs-Methoden der Zellstoff-und Papier-Industri," Springer-Verlag, Berlin, West Germany, 1951, p 426.
149. V. Berzins and J. E. Tasman, *Pulp Paper Mag. Can.*, **58** (10), 145 (1957).
150. C. Valeur and I. Törngren, *Svensk Papperstidn.*, **60** (22), 829 (1957).
151. International Organization for Standardization *ISO R 302–1963 (E)*.
152. *TAPPI Standard T–236 os-76*.
153. *CPPA Standard G. 18*, 1967.
154. *SCAN-test SCAN-C1:77*.
155. *Data Sheet D–6*, Tech. Sect., CPPA, 1968.
156. J. V. Hatton, *Tappi*, **58** (10), 150 (1975).
157. P. Ålander, I. Palenius, and B. Kyrklund, *Paperi Puu*, **45** (8), 403 (1963).
158. *TAPPI Useful Method 206; Tappi*, **54** (11), 1919 (1971)
159. *TAPPI Useful Method 207; Tappi*, **54** (11), 1921 (1971)
160. R. E. Connick and Y.-T. Chia, *J. Am. Chem. Soc.*, **81** (6), 1280 (1959).
161. C. W. Dence, TAPPI Monograph No. 27, 1963, p 41.
162. J. C. Morris, *J. Phys. Chem.*, **70** (12), 3798 (1966).
163. H. W. Giertz in "Chemistry of Wood," E. Hägglund, Academic, New York, N.Y., 1951, p 508.
164. C. G. Swain and D. R. Christ, *J. Am. Chem. Soc.*, **94** (9), 3195 (1972).
165. D. A. Craw and G. C. Israel, *J. Chem. Soc.*, 550 (1952).
166. R. P. Whitney and J. E. Vivian, *Ind. Eng. Chem.*, **33** (6), 741 (1941).
167. E. P. Duncan and W. H. Rapson, *Pulp and Paper*, **34** (11), 100 (1960).
168. N. E. Virkola, Y. Hentola, M. Mäkinen, and R. Soila, *Svensk Papperstidn.*, **62** (14), 477 (1959).
169. J. A. Histed, *Tappi*, **47** (11), 669 (1964).
170. J. A. Histed, *Tappi*, **48** (10), 562 (1965).
171. W. H. Rapson and C. B. Anderson, *Pulp Paper Mag. Can.*, **67** (1), T–47 (1966).

172. J. V. Hatton, *Pulp Paper Mag. Can.*, **68** (4), T–181 (1967).
173. J. V. Hatton, *Pulp Paper Mag. Can.*, **69** (1), T–11 (1968).
174. W. Q. Jack and L. D. Feller, *Pulp Paper Mag. Can.*, **68** (9), T–461 (1967).
175. L. C. Aldrich, *Tappi*, **51** (3), 71 A (1968).
176. N. Hartler, H. Norrström and P. -O. Östling, *Svensk Papperstidn.*, **72** (9), 289 (1969).
177. D. D. Hinrichs, *Tappi*, **45** (10), 765 (1962).
178. N. Liebergott and F. H. Yorston, *Tappi*, **48** (1), 20 (1965).
179. N. Liebergott, H. G. Barclay, J. Merka, D. P. W. Pounds, H. I. Bolker, and D. W. Clayton, *Trans., Tech. Sect., CPPA*, **1** (2), 49 (1975).
180. J. Gullichsen, *Tappi*, **59** (11), 106 (1976).
181. H. Norrström and G. Åkerlund, *Pulp Paper Mag. Can.*, **75** (6), T–219 (1974).
182. G. E. Annergren, P. O. Lindblad, and B. Pettersson, *Pulp Paper Can.*, **78** (7), T–149 (1977).
183. C. Christiansen and R. D. Friedt, *Paper Trade J.*, **159** (24), 44 (1975).
184. F. Löschbrandt, *Kamyr Nachrichten*, No. 2–4 (1939).
185. P. O. Lidby and L. Stockman, *Svensk Papperstidn.*, **60** (23), 878 (1957).
186. G. A. Richter, *Tappi*, **38** (3), 129 (1955).
187. R. Brännland, R. Gustafsson, and B. Hultman, *Svensk Papperstidn.*, **79** (18), 591 (1976).
188. A. Assarsson, L. Stockman, and O. Theander, *Svensk Papperstidn.*, **62** (23), 865 (1959).
189. M. Wayman, *TAPPI Monograph No. 27*, 1963, p 100.
190. N. Liebergott and F. H. Yorston, *Pulp Paper Mag. Can.*, **68** (11), T–563 (1967).
191. M. Wayman and D. L. Sherk, *Tappi*, **39** (11), 786 (1956).
192. T. Bergek, S. Gustavsson, and E. Lindwall, *Svensk Papperstidn.*, **50** (11 B), 22 (1947).
193. H. Sihtola, Y. Hentola, N. Hakkarainen, C. Backlund, A. Saarinen, G. Wigren, T. Ulmanen, and E. Saxén, *Paperi Puu*, **39** (10), 477 (1957).
194. M. G. Delattre, *Appita*, **28** (2), 89 (1974).
195. W. A. Moy, K. Sharpe, and G. Betz, *Pulp Paper Mag. Can.*, **76** (5), T–166 (1975).
196. S. A. Rydholm, "Pulping Processes," Interscience, New York, N.Y., 1965, p 959.
197. K.-E. Lekander and L. Stockman, *Svensk Papperstidn.*, **58** (21), 775 (1955).
197a. E. Opfermann and E. Hochberger, "Die Bleiche des Zellstoffs," Vol. 3 1st part Otto Elsner Verlagsgesellschaft, Berlin, Germany, 1935. p 96 [Present address: Darmstadt, West Germany.]
198. D. A. Clibbens and B. P. Ridge, *J. Textile Inst.*, **18**, T–135 (1927).
199. W. H. Rapson, *Tappi*, **39** (5), 284 (1956).
200. R. G. Guide and R. R. Holmes, *Tappi*, **47** (8), 515 (1964).
201. T. Hosoi, G. Greeno, and A. Hatano, *Tappi*, **48** (5), 281 (1965).
202. R. U. Tobar, *Paper Trade J.*, **150** (26), 36 (1966).
203. G. Fiehn, *Zellst. Papier*, **17** (3), 67 (1968).
204. J. H. E. Herbst, *Pulp Paper Mag. Can.*, **63** (10), T–501 (1962).
205. J. H. E. Herbst, *Tappi*, **45** (11), 833 (1962).
206. H. F. J. Wenzl, *Tappi*, **46** (3), 95 A (1963).
207. V. Valtchev and E. Valtcheva, *Cellulose Chem. Technol.* **1** (5), 551 (1967).
208. J. W. Smith and W. L. Thornburg, *Tappi*, **43** (6), 596 (1960).
209. D. O'Meara and C. E. J. Mair, *Tappi*, **45** (7), 578 (1962).
210. W. H. Rapson, *TAPPI Monograph No. 27*, 1963, p 130.

211. E. Schmidt, *Papier* (Darmstadt), **19** (10a), 728 (1965).
212. L. A. Cox and H. E. Worster, *Tappi,* **54** (11), 1890 (1971).
213. J. F. Haller and W. W. Northgraves, *Tappi,* **38** (4), 199 (1955).
214. W. H. Rapson, *Pulp and Paper,* **32** (1), 46 (1958).
215. R. P. Singh and D. H. Andrews, *Pulp Paper Mag. Can.,* **66** (12), T–628 (1965).
216. J. V. Hatton, F. E. Murray, and T. P. Clark, *Pulp Paper Mag. Can.,* **67** (4), T–241 (1966).
217. I. Croon and S. Dillén, *Tappi,* **51** (5), 97 A (1968).
218. J. A. Histed and G. C. Kaufmann, *Pulp Paper Mag. Can.,* **71** (20), T–443 (1970).
219. B. J. Fergus, *Tappi,* **56** (1), 114 (1973).
220. J. V. Hatton, *Pulp Paper Mag. Can.,* **68** (9), T–411 (1967).
221. J. V. Hatton and B. Eade, *Pulp Paper Mag. Can.,* **68** (9), T–420 (1967).
222. W. Q. Jack and L. D. Feller, *Pulp Paper Mag. Can.,* **68** (9), T–461 (1967).
223. B. A. Edblad, E. S. Becker, R. Nicholson, and P. R. Thomas, *Pulp Paper Mag. Can.,* **74** (3), T–66 (1973).
224. P. I. Alsefelt and R. Brännland, *Svensk Papperstidn.,* **74** (13, 14), 402 (1971).
225. J. L. Reneaut, *Pulp Paper Mag. Can.,* **69** (6), T–147 (1968).
226. B. J. Fergus and R. N. Jones, *Appita,* **27** (5), 341 (1974).
227. O. Sepall, P. Berg-Johannessen, and M. G. Duffy, *Tappi,* **48** (1), 25 (1965).
228. W. H. Rapson and C. B. Anderson, *Tappi,* **40** (5), 307 (1957).
229. H. C. Scribner, C. B. Anderson, and W. H. Rapson, *Pulp Paper Mag. Can.,* **69** (19), T–367 (1968).
230. N. -E. Virkola, *Paperi Puu,* **41** (11), 526 (1959).
231. W. H. Rapson and C. B. Anderson, *Pulp Paper Mag. Can.,* **61** (10), T–495 (1960).
232. D. D. Hinrichs, *Pulp Paper Mag. Can.,* **62** (9), T–437 (1961).
233. N. Hartler, E. Lindahl, C.-G. Moberg, and L. Stockman, *Tappi,* **43** (10), 806 (1960).
234. K. Bjerketveit and P. K. Christensen, *Norsk Skogind.,* **23** (4), 104 (1969).
235. C. Mlakar and J. Peltonen, *Paperi Puu,* **50** (11), 629 (1968).
236. P. K. Paasonen, *Paperi Puu,* **56** (9), 696 (1974).
237. F. L. Fennel and N. J. Stalter, *Tappi,* **51** (1), 62 A (1968).
238. M. G. Delattre, *Paperi Puu,* **53** (3), 117 (1971).
239. P. K. Christensen, *Pulp Paper Mag. Can.,* **73** (2), T–44 (1972).
240. M. L. Slove, *Tappi,* **47** (12), 170 A (1964).
241. B. J. Fergus, *Tappi,* **56** (1), 118 (1973).
242. H.-L. Hardell and B. O. Lindgren, *SCAN-Forsk, Report No. 96,* 1975.
243. E. Grimley and G. Gordon, *J. Phys. Chem.,* **77** (8), 973 (1973).
244. G. C. Israel, *J. Chem. Soc.,* 1289, 1950.
245. P. S. Fredricks, B. O. Lindgren, and O. Theander, *Svensk Papperstidn.,* **74** (19), 597 (1971).
246. M. L. Poutsma, *J. Am. Chem. Soc.,* **87** (19), 4293 (1965).
247. P. S. Fredricks, B. O. Lindgren, and O. Theander, *Acta Chem. Scand.,* **21** (10), 2895 (1967).
248. C. W. Dence and K. V. Sarkanen, *Tappi,* **43** (1), 87 (1960).
249. K. V. Sarkanen and R. W. Strauss, *Tappi,* **44** (7), 459 (1961).
250. J. B. Van Buren and C. W. Dence, *Tappi,* **53** (12), 2246 (1970).
251. D. J. Bennett, C. W. Dence, F. L. Kung, P. Luner, and M. Ota, *Tappi,* **54** (12), 2019 (1971).
252. M. Ota, W. B. Durst, and C. W. Dence, *Tappi,* **56** (6), 139 (1973).
253. K. Lindström and J. Nordin, *J. Chromatogr.,* **128** (1), 13 (1976).

253a. A. Björseth, G. Lunde, and N. Gjös, *Acta Chem. Scand.*, **B 31** (9), 797 (1977).

254. M. Erickson and C. W. Dence, *Svensk Papperstidn.*, **79** (10), 316 (1976).

255. H.-L. Hardell and F. de Sousa, *Svensk Papperstidn.*, **80** (4), 110 (1977).

256. H.-L. Hardell and F. de Sousa, *Svensk Papperstidn.*, **80** (7), 201 (1977).

257. E. M. Karter and E. G. Bobalek, *Tappi*, **54** (11), 1882 (1971).

258. A. W. Kempf and C. W. Dence, *Tappi*, **53** (5), 864 (1970).

259. S. A. Braddon and C. W. Dence, *Tappi*, **51** (6), 249 (1968).

260. A. Ivancic and S. A. Rydholm, *Svensk Papperstidn.*, **62** (16), 554 (1959).

261. J. M. Leach and A. N. Thakore, *J. Fisheries Res. Board Can.*, **32** (8), 1249 (1975).

262. "The SSVL Environmental Care Project, Technical Summary," Stiftelsen Skogsindustriernas Vatten-och Luftvårdsforskning, Stockholm, Sweden, 1974, p 67.

263. B. Lindgren and T. Norin, *Svensk Papperstidn.*, **73** (5), 143 (1969).

264. B. Leopold and D. B. Mutton, *Tappi*, **42** (3), 218 (1959).

265. P. S. Fredrikcs, B. O. Lindgren, and O. Theander, *Cellulose Chem. Technol.*, **4** (5), 533 (1970).

266. P. S. Fredricks, B. O. Lindgren, and O. Theander, *Tappi*, **54** (1), 87 (1971).

267. N. N. Lichtin and M. H. Saxe, *J. Am. Chem. Soc.*, **77** (7), 1875 (1955).

268. K. V. Sarkanen and C. W. Dence, *J. Org. Chem.*, **25** (5), 715 (1960).

269. B. Alfredsson and O. Samuelson, *Svensk Papperstidn.*, **77** (12), 449 (1974).

270. R. L. Whistler, T. W. Mittag, T. R. Ingle, and G. Ruffini, *Tappi*, **49** (7), 310 (1966).

271. O. Theander, *Tappi*, **48** (2), 105 (1965).

272. A. E. Luetzow and O. Theander, *Svensk Papperstidn.*, **77** (9), 312 (1974).

273. L. Farkas, M. Lewin and R. Bloch, *J. Am. Chem. Soc.*, **71** (6), 1988 (1949).

274. M. Lewin and M. J. Avrahami, *J. Am. Chem. Soc.*, **77** (17), 4491 (1955).

275. S. A. Rydholm, "Pulping Processes," Interscience, New York, N. Y., 1965, p 948.

276. H. Richtzenhain and B. Alfredsson, *Acta Chem. Scand.*, **7** (8), 1177 (1953).

277. H. Richtzenhain and B. Alfredsson, *Acta Chem. Scand.*, **8** (9), 1519 (1954).

278. H. Richtzenhain and B. Alfredsson, *Acta Chem. Scand.*, **10** (5), 719 (1956).

279. O. Samuelson and C. Ramsel, *Svensk Papperstidn.*, **53** (6), 155 (1950).

280. H. Sihtola and G. R. Gustafsson, *Paperi Puu*, **32** (10), 295 (1950).

281. O. Theander, *Svensk Papperstidn.*, **61** (18), 581 (1958).

282. O. Theander, *Svensk Kem. Tidskr.*, **71** (1), 1 (1959).

283. J. T. Henderson, *J. Am. Chem. Soc.*, **79** (19), 5304 (1957).

284. J. H. E. Herbst, *TAPPI Monograph No. 27*, 1963, p 119.

285. J. W. Daniel, *Tappi*, **42** (7), 534 (1959).

286. I. Norstedt and O. Samuelson, *Svensk Papperstidn.*, **68** (17), 565 (1965).

287. S.-I. Andersson and O. Samuelson, *Cellulose Chem. Technol.*, **10**, (2), 209 (1976).

288. M. L. Wolfram and W. E. Lucke, *Tappi*, **47** (4), 189 (1964).

289. S. Hernadi, *Zellst. Papier*, **24** (5), 147 (1975).

290. F. Emmenegger and G. Gordon, *Inorg. Chem.*, **6** (3), 633 (1967).

291. R. Soila, O. Lehtikoski, and N.-E. Virkola, *Svensk Papperstidn.*, **65** (17), 632 (1962).

292. T. Nilsson and L. Sjöström, *Svensk Papperstidn.*, **77** (17), 643 (1974).

293. W. H. Rapson and C. B. Anderson, *CPPA Trans. Tech. Sect.*, **3** (2), 52 (1977).

294. R. M. Husband, C. D. Logan, and C. B. Purves, *Can. J. Chem.*, **33** (1), 68 (1955).
295. C. D. Logan, R. M. Husband, and C. B. Purves, *Can. J. Chem.*, **33** (1), 82 (1955).
296. K. V. Sarkanen, K. Kakehi, R. A. Murphy, and H. White, *Tappi*, **45** (1), 24 (1962).
297. C. W. Dence, M. K. Gupta, and K. V. Sarkanen, *Tappi*, **45** (1), 29 (1962).
298. T. Ishikawa, M. Sumimoto, and T. Kondo, *Jap. Tappi*, **23** (3), 117 (1969).
299. B. O. Lindgren, *Svensk Papperstidn.*, **74** (3), 57 (1971).
300. B. O. Lindgren and B. Ericsson, *Acta Chem. Scand.*, **23** (10), 3451 (1969).
301. B. O. Lindgren and T. Nilsson, *Acta Chem. Scand.*, B-28 (8), 847 (1974).
302. B. O. Lindgren and T. Nilsson, *Svensk Papperstidn.*, **75** (5) 161 (1972).
303. B. O. Lindgren and T. Nilsson, *Svensk Papperstidn.*, **78** (2), 66 (1975).
304. B. O. Lindgren, *Papier*, (Darmstadt) 28 (10 A), V-29 (1974).
305. N. G. V. Linden and G. A. Nicholls, *Tappi*, **59** (11), 110 (1976).
306. K. Baczynska, *Zellst. Papier*, **20** (6), 170 (1971).
307. V. M. Nikitin and G. L. Akim, *Trudy Leningr. Lesotekhn. Akad.*, **75**, 145 (1956); *BIPC*, **28** (12), 1662, 1958.
308. A. Robert, P. Rerolle, A. Viallet, and C. Martin Borret, *ATIP*, **18** (4), 151 (1964).
309. A. Robert, A. Viallet, P. Rerolle, and J.-P. Andreolety, *ATIP*, **20** (5), 207 (1966).
310. G. Rowlandson, *Tappi*, **54** (6), 962 (1971).
311. A. Jamieson, S. Noreus, and B. Pettersson, *Tappi*, **54** (11), 1903 (1971).
312. A. G. Jamieson and L. Å. Smedman, *Svensk Papperstidn.*, **76** (5), 187 (1973).
313. H. Makkonen, M. Pitkänen, and M. Nikki, *Paperi Puu*, **55** (12), 947 (1973).
314. H. Makkonen and M. Ranua, *Papier*, (Darmstadt) 29 (10 A), V-25 (1975).
315. P. K. Christensen, *Norsk Skogind.*, **28** (2), 39 (1974).
316. P. K. Christensen, *Norsk Skogind.*, **28** (8), 220 (1974).
317. P. K. Christensen, *Norsk Skogind.*, **29** (4), 84 (1975).
318. P. K. Christensen, *CPPA Trans. Tech. Sect.*, **3** (2), TR-47 (1977); *Pulp Paper Can.*, **78** (6) (1977).
319. A. Robert and A. Viallet, *ATIP*, **25** (3), 237 (1971).
320. I. Croon and D. H. Andrews, *Tappi*, **54** (11), 1893 (1971).
321. H.-E. Engström, L. Ivnäs, and B. Pettersson, *Tappi*, **54** (11), 1899 (1971).
322. P. J. Kleppe, Å. Backlund, and Y. Schildt, *Tappi*, **59** (11), 77 (1976).
323. I. A. Yrjälä, L. J. Suoninen, and P. K. Paasonen, *Preprint, Intl Pulp Bleaching Conf.*, Chicago, 1976, p 147.
324. N. Hartler, H. Norrström, and S. Rydin, *Svensk Papperstidn.*, **73** (21), 696 (1970).
325. H.-M. Chang, J. S. Gratzl, and W. T. McKean, *Tappi*, **57** (5), 123 (1974).
326. J. Kalish, *Pulp Paper Intl.*, **17** (3), 53 (1975).
327. E. David, A. Cannone, J. Carles, and M. Durand, *Pulp Paper Intl.*, **18** (6), 47 (1976).
328. G. Collin, presented at Norwegian Pulp and Paper Technical Association, Oslo, Norway, 1970.
329. B. Coetzee, *Pulp Paper Mag. Can.*, **75** (6), T-223 (1974).
330. A. G. Jamieson and L. Å. Smedman, *Tappi*, **57** (5), 134 (1974).
331. H. Makkonen, M. Pitkänen, and T. Laxén, *Tappi*, **57** (2), 113 (1974).
332. P. J. Kleppe, H.-M. Chang, and R. C. Eckert, *Pulp Paper Mag. Can.*, **73** (12), T-400 (1972).
333. A. G. Jamieson and G. Fossum, *Appita*, **29** (4), 253 (1976).
334. H.-M. Chang and P. J. Kleppe, *Tappi*, **56** (1), 97 (1973).

335. H.-M. Chang, W. T. McKean, and S. Seay, *Tappi,* **56** (5) 105 (1973).
336. H.-M. Chang, W. T. McKean, J. S. Gratzl, and C.K. Lin, *Tappi,* **56** (9), 116 (1973).
337. I. Palenius and L. Hiisvirta, *Pulp Paper Mag. Can.,* **71** (21), T–465 (1970).
338. H. E. Worster and M. F. Pudek, *Pulp Paper Mag. Can.,* **74** (9), T–318 (1973).
339. H. E. Worster and M. F. Pudek, *Tappi,* **57** (3), 138 (1974).
340. H. E. Worster and M. F. Pudek, *Pulp Paper Mag. Can.,* **75** (1), T–2 (1974).
341. B. Lönnberg, I. Palenius, and M. Ranua, *Tappi,* **59** (11), 102 (1976).
342. K. Kratzl, J. Gratzl, and P. Claus, "Lignin Structure and Reactions," Advances in Chemistry Series 59, American Chemical Society, Washington, D.C., 1966, p 157.
343. K. Kratzl, W. Schäfer, P. Claus, J. Gratzl, and P. Schilling, *Monatsh. Chem.,* **98** (3), 891 (1967).
344. R. C. Eckert, H.-M. Chang and W. P. Tucker, *Tappi,* **56** (6), 134 (1973).
345. J. Gierer, I. Pettersson, and I. Szabo-Lin, *Acta Chem. Scand.,* B–28 (10), 1129 (1974).
346. J. Gierer, F. Imsgard, and I. Norén, *Acta Chem. Scand.,* B–31 (7), 561 (1977).
347. J. Farkas and J. Janci, *Papir Celul.,* **30** (2), V–3 (1975).
348. J. Farkas and J. Janci, *Papir Celul.,* **30** (11), V–75 (1975).
349. O. Samuelson and L. Stolpe, *Tappi,* **52** (9), 1709 (1969).
350. H. Kolmodin and O. Samuelson, *Svensk Papperstidn.,* **75** (9), 369 (1972).
351. R. Malinen and E. Sjöström, *Paperi Puu,* **54** (8), 451 (1972).
352. B. Ericsson, B. O. Lindgren, and O. Theander, *Cellulose Chem. Technol.,* **7** (5), 581 (1973).
353. R. Malinen, E. Sjöström, and J. Ylijoki, *Paperi Puu,* **55** (1), 5 (1973).
354. R. M. Rowell, *Pulp Paper Mag. Can.,* **72** (7), T–236 (1971).
355. O. Samuelson and L. Stolpe, *Svensk Papperstidn.,* **72** (20), 662 (1969).
356. B. Ericsson, B. O. Lindgren, and O. Theander, *Cellulose Chem. Technol.,* **8** (4), 363 (1974).
357. B. Ericsson and R. Malinen, *Cellulose Chem. Technol.,* **8** (4), 327 (1974).
358. R. Malinen and E. Sjöström, *Cellulose Chem. Technol.,* **9** (3), 231 (1975).
359. J. T. McCloskey, L. R. Schroeder, J. D. Sinkey, and N. S. Thompson, *Paperi Puu,* **57** (3), 131 (1975).
360. K. Kratzl and H. A. Schwarz, *Holzforschung,* **29** (1), 29 (1975).
361. H. Kolmodin and O. Samuelson, *Svensk Papperstidn.,* **73** (4), 93 (1970).
362. G. L. Akim, *Paperi Puu,* **55** (5), 389 (1973).
363. H. Kolmodin and O. Samuelson, *Svensk Papperstidn.,* **76** (2), 71 (1973).
364. R. Malinen and E. Sjöström, *Paperi Puu,* **56** (11), 895 (1974).
365. R. Malinen and E. Sjöström, *Paperi Puu,* **57** (3), 101 (1975).
366. R. Malinen, *Paperi Puu,* **57** (4a), 193 (1975).
367. B. Ericsson, B. O. Lindgren, and O. Theander, *Svensk Papperstidn.,* **74** (22), 757 (1971).
368. O. Samuelson and L. Stolpe, *Svensk Papperstidn.,* **74** (18), 545 (1971).
369. E. Sjöström and O. Välttilä, *Paperi Puu,* **54** (11), 695 (1972).
370. R. Gustavsson and B. Swan, *Tappi,* **58** (3), 120 (1975).
371. J. Minor and N. Sanyer, *Tappi,* **57** (2), 109 (1974).
372. O. Samuelson and L. Stolpe, *Acta Chem. Scand.,* **27** (8), 3061 (1973).
373. O. Samuelson and L. Stolpe, *Svensk Papperstidn.,* **77** (1), 16 (1974).
374. O. Samuelson and L. Stolpe, *Svensk Papperstidn.,* **77** (14), 513 (1974).
375. M. Manouchehri and O. Samuelson, *Svensk Papperstidn.,* **76** (13) 486 (1973).
376. L. Löwendahl and O. Samuelson, *Svensk Papperstidn.,* **77** (16), 593 (1974).

377. L. Löwendahl, G. Petersson, and O. Samuelson, *Acta Chem. Scand.*, B–29 (9), 975 (1975).
378. J. Defaye and A. Gadelle, *Pulp Paper Mag. Can.*, **75** (11), T–394 (1974).
379. J. D. Sinkey and N. S. Thompson, *Paperi Puu*, **56** (5), 473 (1974).
380. J. T. McClosky, J. D. Sinkey, and N. S. Thompson, *Tappi*, **58** (2), 56 (1975).
381. N. S. Thompson and H. M. Corbett, *Tappi*, **59** (3), 78 (1976).
381a. M. Manouchehri and O. Samuelson, *Svensk Papperstidn.*, **80** (12), 381 (1977).
382. K. Kratzl, H. Silbernagel, and F. W. Vierhapper, *Holzforschung*, **28** (3), 77 (1974).
383. M. Manouchehri and O. Samuelson, *Svensk Papperstidn.*, **78** (6), 211 (1975).
384. A. F. Gilbert, E. Pavlovova, and W. H. Rapson, *Tappi*, **56** (6), 95 (1973).
385. D. Lachenal, *ATIP*, **30** (6), 203 (1976).
386. T. N. Kleinert, *Tappi*, **59** (9), 122 (1976).
387. K. Abrahamsson and O. Samuelson, *Svensk Papperstidn.*, **79** (9), 281 (1976).
388. R. Pusa and N.-E. Virkola, *Paperi Puu*, **56** (8), 599 (1974).
389. R. Pusa and N.-E. Virkola, *Paperi Puu*, **56** (9), 687 (1974).
390. R. Pusa and N.-E. Virkola, *Paperi Puu*, **57** (2), 43 (1975).
391. I. Jullander and K. Brune, *Svensk Papperstidn.*, **62** (20), 728 (1959).
392. H. I. Bolker and N. Liebergott, *Pulp Paper Mag. Can.*, **73** (11), T–332 (1972).
393. Z. Osawa and C. Schuerch, *Tappi*, **46** (2), 79 (1963).
394. R. B. Secrist and R. P. Singh, *Tappi*, **54** (4), 581 (1971).
395. S. Rothenberg, D. H. Robinson, and D. K. Johnsonbaugh, *Tappi*, **58** (8), 182 (1975).
396. N. Soteland, *Pulp Paper Mag. Can.*, **75** (4), T–153 (1974).
397. N. Soteland, *Paper 184* (56) (1975). [World Res. and Dev. No.]
398. H. Lagergren, H. Norrstrøm, and U. Randahl, *Svensk Papperstidn.*, **72** (22), 745 (1969).
399. T. Krause, *Cellulose Chem. Technol.*, **5** (1), 83 (1971).
400. W. H. Rapson, TAPPI Monograph No. 27, 1963, p 150.
401. *Pulp and Paper*, **51** (3), 94 (1977).
402. N. Liebergott, H. I. Bolker, and D. W. Clayton, *Preprint, Intl. Pulp Bleaching Conf.*, Chicago, Ill., 1976 p 167.
403. J. C. W. Evans, *Paper Trade J.*, **157** (42), 32 (1973).
404. D. Lachenal, C. de Choudens, and P. Monzie, *ATIP*, **31** (1), 16 (1977).
405. I. A. Crawford, *Appita*, **23** (2), 115 (1969).
406. W. H. Rapson and C. B. Anderson, *Tappi*, **49** (8), 329 (1966).
407. J. Richter, *Tappi*, **49** (6), 48A (1966).
408. J. Gullichsen, *Tappi*, **56** (11), 78 (1973).
409. J. Gullichsen, *Tappi*, **57** (10), 111 (1974).
410. T. E. Jenkin and B. R. Mitchell, *Prep. nt, Intl. Pulp Bleaching Conf.*, Chicago, Ill., 1976, p 9.
411. J. K. Perkins, *Tappi*, **58** (1), 75 (1975).
412. R. G. Elliott, *South. Pulp Paper Mfr.*, **40** (4), 24 (1977).
413. B. L. Shera, *Paper Trade J.*, **147** (43), 28 (1963).
414. B. L. Shera, *Paper Trade J.*, **149** (15), 42 (1965).
415. E. S. Atkinson and H. de V. Partridge, *Tappi*, **49** (2), 66A (1966).
416. H. Freedman, *Paper Trade J.*, **152** (50), 43 (1968).
417. R. G. Elliott and T. D. Farr, *Tappi*, **56** (11) 68 (1973).
418. *Paper Trade J.*, **147** (30), 35 (1963).

419. *Svensk Papperstidn.*, **71** (7), 314 (1968).
419a. P. M. Swenson, *Pulp and Paper*, **49** (11), 125 (1975).
420. B. G. Klowak and E. Sanal, *Pulp Paper Mag. Can.*, **64** (C), T-137 (1963).
421. B. E. Helberg, *Tappi*, **48** (9), 94 A (1965).
422. *South. Pulp Paper Mfr.*, **40** (4), 15 (1977).
423. *Paper Trade J.*, **154** (20), 61 (1970).
424. J. L. Aitkens, *Tappi*, **48** (11), 104A (1965).
425. E. H. Morter and M. J. Kent, *Pulp Paper Mag. Can.*, **68** (12), T-659 (1967).
426. J. A. McMaster and R. L. Kane, *Tappi*, **54** (7), 1156 (1971).
427. F. C. Hanks, Jr., and H. G. Taylor, *Tappi*, **43** (8), 229A (1960).
428. *Tappi*, **44** (2), 183A (1961).
429. J. K. Perkins, *Tappi*, **56** (11), 87 (1973).
430. L. A. Carlsmith, R. W. Schleinkofer, and R. W. Hoag, *Tappi*, **60** (5), 106 (1977).
431. J. K. Perkins, *Tappi*, **59** (8), 63 (1976).
432. M. Vernerey, *ATIP*, **21** (2), 93 (1967).
433. R. Simonette and W. A. Szymanski, *Tappi*, **51** (2), 79A (1968).
434. P. Heitto, *Preprint, Intl. Pulp Bleaching Conf.*, Chicago, Ill., 1976, p 27.
435. T. Laxén, *Zellst. Papier*, **23** (6), 169 (1974).
436. C. J. Myburgh, *Tappi*, **57** (5), 131 (1974).
436a. R. Hillström, B. Lindqvist, L. Smedman, and A. Jamieson, *Preprint, Intl. Environmental Improvement Conf.* Montreal, Canada, 1976, p 61.
437. J. Serafin, and A. Jamieson, *Pulp Paper Mag. Can.*, **73** (10), T-281 (1972).
438. C. Owensby, *Paper Trade J.*, **151** (3), 38-41; (4), 38 (1967).
439. D. Carlson and G. Lewis, *Pulp and Paper*, **51** (3), 100 (1977).
440. J. W. Snyder, *Tappi*, **49** (12), 105A (1966).
441. D. N. Obenshain, *Tappi*, **41** (1), 1 (1958).
442. J. Strom and H. Weyrick, *Pulp and Paper*, **44** (2), 68 (1970).
443. J. Strom and H. Weyrick, *Pulp Paper Intl.*, **14** (2), 54 (1972).
444. J. Strom, *Tappi*, **56** (11), 65 (1973).
445. J. A. Isbister, *Pulp Paper Can.*, **76** (C), T-94 (1975).
446. *Pulp and Paper*, **47** (9), 111 (1973).
447. T. Battershill and J. MacTaggart, *Pulp Paper Can.*, **76** (7), T-213 (1975).
448. G. G. Wire, *Pulp and Paper*, **51** (2), 96 (1977).
449. Å S:son Stenius, *Tappi*, **52** (5), 942 (1969).
450. B. L. Shera, *Tappi*, **46** (10), 172A (1963).
451. L. J. Champion, *South. Pulp Paper Mfr.*, **27** (11), 54 (1964).
452. O. Töppel, *Papier* (Darmstadt) **18** (10a), 568 (1964).
453. B. L. Shera and J. B. Heitman, *Tappi*, **48** (2), 89 (1965).
454. B. L. Shera, *Tappi*, **50** (2), 99A (1967).
455. B. Kyrklund, *Paperi Puu*, **49** (2), 65 (1967).
456. K. A. Kuiken and F. P. Smith, *Tappi*, **51** (1), 40 (1968).
457. G. Fiehn, *Zellst. Papier*, **17** (7, 8), 211; (9), 284 (1968).
458. D. G. Allen, *Tappi*, **55** (7), 1118 (1972).
459. P. Hyväri, *Paperi Puu*, **52** (8), 469 (1970).
460. O. Härkönen, *Zellst. Papier*, **20** (7), 219 (1971).
461. I. Yrjala and T. Forsten, *Pulp Paper Mag. Can.*, **75** (8), T-301 (1974).
462. I. Wartiovaara and H. Makkonen, *Paperi Puu*, **57** (4a), 232 (1975).
463. P. Landmark and K. Johnsen, *Preprint, Intl. Symp. on Process Control*, CPPA, Vancouver, Canada, 1977, p 126.
464. B. Hultman, A. Dalborg, and R. Bergstrøm, *CPPA Trans. Tech. Sect.*, **2** (3), 92 (1976).
465. U. Isacsson and G. Wettermark, *Svensk Papperstidn.*, **79** (18), 595 (1976).
466. B. D. Ritchey, *Paper Trade J.*, **151** (22), 49 (1967).

467. F. A. Christiansen, *Amer. Paper Ind.,* **49** (8), 74 (1967).
468. *Pulp Paper Intl.,* **10** (5), 68 (1968).
469. T. M. Chan, R. G. Lightfoot, and A. K. Chatterjee, *Pulp Paper Mag. Can.,* **71** (10), T–219 (1970).
470. P. Karjalainen and M. Otala, *Pulp Paper Intl.,* **14** (12), 52 (1972).
471. R. Soila and P. Hyväri, *Zellst. Papier,* **24** (2), 36 (1975).
472. H. W. Danforth, *Pulp and Paper,* **50** (10), 134 (1976).
473. S. Kanneko, *Jap. Tappi,* **19** (12), 629 (1965).
474. J. Strom, *Can. Pulp Paper Ind.,* **29** (6), 19 (1976).
475. R. N. Jones, R. D. Cardwell, and B. J. Fergus, *Appita,* **30** (5), 412 (1977).
476. *Pulp and Paper,* **43** (1), 147 (1969).
477. E. Narduzzi, *Pulp and Paper,* **51** (2), 102 (1977).
478. J. Gullichsen, H. Männistö and E. N. Westerberg, *Pulp Paper Mag. Can.,* **68** (11), T–555 (1967).
479. L. Edwards, G. Hovsenius, and H. Norrström, *Svensk Papperstidn.,* **76** (3), 123 (1973).
480. T. C. Burnett, *Pulp Paper Mag. Can.,* **71** (14), T–311 (1970).
481. G. L. Kott, *Tappi,* **50** (7), 41A (1967).
482. T. M. Stout, *Paper Trade J.,* **153** (15), 54 (1969).
483. B. O. Jender, *Paper Trade J.,* **153** (21), 48 (1969).
484. R. Burkhardt and M. Asgari, *Paper Trade J.,* **154** (38), 74 (1970).
485. P. Hyväri, *Paperi Puu,* **55** (4a), 199 (1973).
486. W. I. Robinson, C. Y. Chai, R. S. Park, and J. J. Howarth, *Tappi,* **57** (2), 117 (1974).
487. E. C. Johns, *Pulp and Paper,* **48** (2), 76 (1974).
488. L. Moore and T. M. Haner, *Papeterie,* **96** (3), 138 (1974).
489. W. I. Robinson, D. B. Brewster, and C. Y. Chai, *Pulp Paper Mag. Can.,* **75** (4), T–135 (1974).
490. B. G. Freedman, *Tappi,* **57** (10), 92 (1974).
491. H. W. Danforth, D. T. Powell, and G. W. Seymour, *Tappi,* **58** (3), 91 (1975).
492. P. Kilpinen, *Paperi Puu,* **55** (4a), 168 (1973).
493. M. M. MacLeod and A. W. Kempf, *Tappi,* **60** (5), 118 (1977).
494. P. Axegård and A. Teder, *Pulp Paper Can.,* **78** (9), T–196 (1977).
495. G. E. Annergren and P. O. Lindblad, *Tappi,* **59** (11), 95 (1976).
496. C. Gustaffson, V. Tammela, and S. Kahila, *Paperi Puu,* **34** (4a), 121 (1952).
497. F. A. Abadie-Maumert and N. Soteland, *ATIP,* **20** (5), 191 (1966).
498. H. P. Dahm, *Norsk Skogind.,* **13** (6), 188 (1959).
499. H. P. Dahm, *Norsk Skogind.,* **13** (9), 294 (1959).
500. J. M. Penney, *Tappi,* **54** (8), 1274 (1971).
501. D. B. Judd and G. Wyszecki, "Color in Business, Science and Industry," 3rd ed., John Wiley, New York, N.Y., 1975, p 491.
502. H. W. Giertz, *Svensk Papperstidn.,* **48** (13), 317 (1945).
503. *TAPPI um 200.*
504. *CPPA E. 1 U.*
505. J. Gullichsen, *Paperi Puu,* **47** (4a), 215 (1965).
506. W. H. Langwell, *Tech. Bull.,* **29** (1), 21; **29** (2), 52 (1952).
507. F.-A. Abadie-Maumert and V. Loräs, *ATIP,* **31** (9), 334 (1977).
508. F.-A. Abadie-Maumert, E. Böhmer, and V. Loräs, *ATIP,* **28** (3), 117 (1974).
509. H. Schröter, *Papier* (Darmstadt) **21** (10A), 760 (1967).
510. J. Polcin, M. Wayman, C. B. Anderson, and W. H. Rapson, *Pulp Paper Mag. Can.,* **70** (21), T–410 (1969).

511. P. Nolan, J. A. Van den Akker, and W. A. Wink, *Paper Trade J.*, **121** (11), T-101 (1945).
512. G. J. Leary, *Tappi*, **51** (6), 257 (1968).
513. L. V. Forman, *Paper Trade J.*, **111** (21), T-266 (1940).
514. G. J. Leary, *Tappi*, **50** (1), 17 (1967).
515. M. A. Kalnins, *USDA Forest Service Research Paper*, FPL-57, 1966, p 23.
516. M. A. Kalnins, C. Steelink, and H. Tarkow, *USDA Forest Service Research Paper*, FPL-58, 1966 p 1.
517. K. P. Kringstad and S. Y. Lin, *Tappi*, **53** (12), 2296 (1970).
518. G. Gellerstedt and E.-L. Pettersson, *Acta Chem. Scand.*, **B-29** (10), 1005 (1975).
519. G. Gellerstedt and E.-L. Pettersson, *Svensk Papperstidn.*, **80** (1), 15 (1977).
520. S. Y. Lin and K. P. Kringstad, *Norsk Skogind*, **25** (9), 252 (1971).
521. K. P. Kringstad, *Papier* (Darmstadt), **27** (10A), 462 (1973).
522. H. F. Lewis and D. Fronmuller, *Paper Trade J.*, **121** (14), T-133 (1945).
523. H. F. Lewis, E. A. Reineck, and D. Fronmuller, *Paper Trade J.*, **121** (8), T-76 (1945).
524. K. Kringstad, *Tappi*, **52** (6), 1070 (1969).
525. J. W. W. Morgan and R. J. Orsler, *Holzforschung*, **22** (1), 11 (1968).
526. D. F. Manchester, J. McKinney, and A. A. Pataky, *Svensk Papperstidn.*, **63** (20), 699 (1960).
527. R. P. Singh, *Tappi*, **49** (7), 281 (1966).
528. V. Lorås, *Pulp Paper Mag. Can.*, **69** (2), T-49 (1968).
529. S. Y. Lin and K. P. Kringstad, *Tappi*, **53** (9), 1675 (1970).
530. G. Gellerstedt, presented at The Swedish Technical Association of Pulp and Paper Industry, Stockholm, Sweden, 1976.
531. J. Polcin and W. H. Rapson, *Pulp Paper Mag. Can.*, **73** (1), T-2 (1972).
532. S. Claesson, E. Olson, and A. Wennerblom, *Svensk Papperstidn.*, **71** (8), 335 (1968).
533. W. H. Rapson and I. H. Spinner, TAPPI Monograph No. 27, 1963, p 273.
534. I. H. Spinner, *Tappi*, **45** (6), 495 (1962).
535. J.-F. Le Nest, *Papeterie*, **93** (5), 422; (6), 540 (7), 644 (1971).
536. D. M. Gates, *Science*, **151** (3710), 523 (1966).
537. H. F. Launer and W. K. Wilson, *J. Res. Natl. Bur. Std.*, **30** (1), 55 (1943).
538. R. A. Stillings and R. J. Van Nostrand, *J. Am. Chem. Soc.*, **66** (5), 753 (1944).
539. H. F. Launer and W. K. Wilson, *J. Am. Chem. Soc.*, **71** (3), 958 (1949).
540. W. R. MacMillan, TAPPI Monograph No. 27, 1973, p 280. [M.S. Thesis,]
541. T. N. Kleinert, *Papier* (Darmstadt) **24** (4), 207 (1970).
542. J. A. Histed, *Pulp Paper Mag. Can.*, **68** (11), T-546 (1967).
543. E. J. Howard and J. A. Histed, *Tappi*, **47** (11), 653 (1964).
544. V. Lorås, *Norsk Skogind.*, **21** (10), 368 (1967).
545. I. Croon, S. Dillén, and J.-E. Olsson, *Svensk Papperstidn.*, **69** (5), 139 (1966).
546. E. O. Aalto, *Paperi Puu*, **36** (3), 71 (1954).
547. W. H. Rapson and C. B. Anderson, *Pulp Paper Mag. Can.*, **61** (10), T-495 (1960).
548. J. Bergman, N. Hartler, and L. Stockman, *Svensk Papperstidn.*, **66** (15), 547 (1963).
549. I. Croon and J.-E. Olsson, *Svensk Papperstidn.*, **67** (21), 850 (1964).
550. J. Bergman, *Svensk Papperstidn.*, **68** (9), 339 (1965).
551. N.-E. Virkola and H. Sihtola, *Norsk Skogind.*, **12** (3), 87 (1958).
552. W. H. Rapson, C. B. Anderson, and G. F. King, *Tappi*, **41** (8), 442 (1958).

553. N.-E. Virkola and O. Lehtikoski, *Paperi Puu*, **42** (11), 559 (1960).
554. W. H. Rapson and J.-C. Corbi, *Pulp Paper Mag. Can.*, **65** (11), T–459 (1964).
555. J.-C. Corbi and W. H. Rapson, *Pulp Paper Mag. Can.*, **65** (11), T–467 (1964).
556. C. Raaka, M. N. May, R. J. Solari, and F. A. Foderaro, *Tappi*, **40** (3), 201 (1957).
557. F. L. Fennell, G. E. Smedburg, and N. J. Stalter, *Tappi*, **43** (11), 903 (1960).
558. T. P. Czepiel, *Tappi*, **43** (4), 289 (1960).
559. H. Sihtola and R. Sumiala, *Paperi Puu*, **45** (2), 43 (1963).
560. I. H. Spinner, M. H. MacKinnon, and J. W. Lilley, *Pulp Paper Mag. Can.*, **67** (C), T–114 (1966).
561. R. L. Desai, *Pulp Paper Mag. Can.*, **69** (16), T–322 (1968).
562. J. H. Flynn and W. L. Morrow, *J. Polym. Sci.*, **A–2** (1), 91 (1964).
563. J. H. Flynn, W. K. Wilson, and W. L. Morrow, *J. Res. Natl. Bur. Std.*, **60** (3), 229 (1958).
564. B. C. Bera, S. D. Kapila and D. A. L. Rao, *J. Sci. Ind. Res.*, **21D** (12), 444 (1962).
565. C. Simonescu and G. Rozmarin, *Chem. Ind.* (London), (15), 627 (1966).
566. S. Dillén, T. Popoff, and O. Theander, *Svensk Papperstidn.*, **71** (9), 377 (1968).
567. E. F. L. J. Anet, *Advan. Carbohydrate Chem.*, **19**, 181 (1964).
568. A. Beélik, *Pulp Paper Mag. Can.*, **68** (3), T–135 (1967).
569. A. E. Luetzow and O. Theander, *Svensk Papperstidn.*, **77** (9), 312 (1974).
570. O. Theander, *Tappi*, **48** (2), 105 (1965).
571. E. Sjöström and E. Eriksson, *Tappi*, **51** (1), 16 (1968).
572. J. G. Treffers, *Papierwereld*, **15** (3), 69 (1960).
573. S. Hernadi, *Svensk Papperstidn.*, **79** (13), 418 (1976).
574. G. G. Nelson, Jr., A. A. Yankowski, G. H. Trostel, and M. P. Clark, *Tappi*, **55** (6), 933 (1972).
574a. B. Ahling, *Svensk Papperstidn.*, **80** (13), 415 (1977).
575. T. E. Howard and C. C. Walden, *Tappi*, **48** (3), 136 (1965).
576. J. B. Sprague and D. W. McLeese, *Pulp Paper Mag. Can.*, **69** (24), T–431 (1968).
577. B. S. Das, S. G. Reid, J. L. Betts, and K. Patrick, *J. Fisheries Res. Board Can.*, **26** (11), 3055 (1969).
578. T. E. Howard and C. C. Walden, *Tappi*, **48** (3), 136 (1965).
579. T. E. Howard and C. C. Walden, *Pulp Paper Mag. Can.*, **72** (1), T–3 (1971).
580. C. C. Walden, T. E. Howard, and W. J. Sheriff, *Pulp Paper Mag. Can.*, **72** (2), T–81 (1971).
581. O. Seppovaara, *Paperi Puu*, **55** (10), 713 (1973).
582. A. Wong, M. LeBourhis, R. A. Wostradowski, and S. Prahacs, *Preprint, Intl. Pulp Bleaching Conf.*, Chicago, Ill., 1976.
583. J. M. Leach and A. N. Thakore, *J. Fisheries Res. Board Can.*, **30** (4), 479 (1973).
584. J. L. Betts and G. G. Wilson, *Pulp Paper Mag. Can.*, **68** (2), T–53 (1967).
585. T. E. Howard and C. C. Walden, *Pulp Paper Pollution Abatement*, Project Rept. 9–1 (1971), Can. Dept. Environment; *ABIPC*, **43** (1), 62 (1972).
586. T. E. Howard, D. J. McLeay, and C. C. Walden, *Pulp Paper Pollution Abatement*, Project Rept. 9–2 (1971), Can. Dept. Environment; *ABIPC*, **43** (1), 61 (1972).
587. P. W. Webb and J. R. Brett, *J. Fisheries Res. Board Can.*, **29** (11), 1555 (1972).

588. J. C. Davis, *J. Fisheries Res. Board Can.*, **30** (3), 369 (1973).

589. J. C. Davis, *CPPA Tech. Sect. Papers, Discussions*, No. 1, **D-13** (1974).

590. T. E. Howard, *J. Fisheries Res. Board Can.*, **32** (6), 789 (1975).

591. D. A. Brown and D. J. McLeay, *J. Fisheries Res. Board Can.*, **32** (12), 2528 (1975).

592. O. Seppovaara and T. Hattula, *Paperi Puu*, **59** (8), 489 (1977).

593. J. C. Davis, *J. Fisheries Res. Board Can.*, **33** (9), 2031 (1976).

594. C. C. Walden and T. E. Howard, *Tappi*, **60** (1), 122 (1977).

594a. P. Ander, K.-E. Eriksson, M.-C. Kolár, K. Kringstad, U. Rannung, and C. Ramel, *Svensk. Papperstidn.*, **80** (14), 454 (1977).

595. R. N. Soniassy, J. C. Mueller, and C. C. Walden, *Pulp Paper Can.*, **78** (8), T-179 (1977).

596. V. Eloranta, *Vatten*, **32** (1), 20 (1976).

596a. L. Landner and T. Larsson, *Svensk Papperstidn.*, **80** (12), 371 (1977).

597. R. J. Gall and F. H. Thompson, *Tappi*, **56** (11), 72 (1973).

598. W. A. Moy, K. Sharpe, and R. G. Betz, *Pulp Paper Can.*, **76** (5), T-166 (1975).

599. C.-J. Alfthan, H. Norrström, and G. Åkerlund, *Svensk Papperstidn.*, **79** (6), 180 (1976).

600. J. Höök and H. Norrström, *Papier* (Darmstadt), **26** (10A), 646 (1972).

601. A. Jamieson and L. Smedman, *Tappi*, **56** (6), 107 (1973).

602. B. Michaels, *Can. Pulp Paper Ind.*, **29** (6), 17 (1976).

603. A. Jamieson, S. Noreus, and B. Pettersson, *Tappi*, **54** (11), 1903 (1971).

604. P. Luner, C. Dence, D. Bennett, and F.-L. Kung, *NCASI Tech. Bull.*, No. 239, 1970; *ABIPC*, **41** (11) 1029 (1971).

605. R. S. Wright, J. L. Oswalt and J. G. Land, Jr., *Tappi*, **57** (3), 126 (1974).

606. I. Gellman and H. F. Berger, *Tappi*, **57** (9), 69 (1974).

607. M. Gould, *Chem. Eng.*, **78** (2), 55 (1971); Canadian Patent 878980 (1971).

608. E. L. Spruill, *Tappi*, **56** (4), 98 (1973).

608a. F. M. A. Nicolle, R. Shamash, K. V. Nayak, and J. A. Histed, *Pulp Paper Can.*, **78** (9), T-210 (1977).

609. J. Clarke and M. W. Davis, Jr., *Pulp and Paper*, **43** (2), 135 (1969).

610. N. E. Tejera and M. W. Davis, Jr., *Tappi*, **53** (10), 1931 (1970).

611. T. Ploetz, *Papier* (Darmstadt), 28 (10A), V39 (1974).

612. H. S. Dugal, J. O. Church, R. M. Leekley, and J. W. Swanson, *Tappi*, **59** (9), 71 (1976).

613. G. D. Button, *Paper Trade J.*, **155** (46), 84 (1971).

614. D. C. Kennedy and L. R. McConnell, *South Pulp Paper Mfr.*, **37** (1), 26 (1974).

615. S. L. Rock, A. Bruner, and D. C. Kennedy, *Tappi*, **57** (9), 87 (1974).

616. R. L. Sanks, *J. WPCF*, **47** (7), 1924 (1975).

617. S. Lindberg, *Paper Trade J.*, **157** (51), 36 (1973).

618. L. G. Anderson, B. Broddevall, S. Lindberg, and J. Phillips, *Tappi*, **57** (4), 102 (1974).

619. L. G. Anderson and S. Lindberg, *Pulp Paper Intl.*, **16** (6), 46 (1974).

619a. K. A. Andersson, *Tappi*, **60** (3), 95 (1977).

620. J. S. Johnson, Jr., R. E. Minturn, and G. E. Moore, *Tappi*, **57** (1), 134 (1974).

621. M. Fels, D. Smith, C. Miller, and P. Miller, *Can. Pulp Paper Ind.*, **27** (9), 50 (1974).

622. E. Muratore, M. Pichon, and P. Monzie, *Svensk Papperstidn.*, **78** (16), 573 (1975).

623. C. W. Dence, P. Luner, J. Chang, W. Durst, J. Hsieh, and M. Klinkhammer, *NCASI Stream Imp. Tech. Bull*, No. 273 (1974); *ABIPC*, **45** (5), 504 (1974).

624. A. Wong, T. Tenn, J. Dorica, and S. Prahacs, *Preprint, CPPA Environment Improvement Conf.,* Vancouver, Canada, 1975, p 69.
625. D. G. MacDonald and T. G. Nguyen, *Pulp Paper Mag. Can.,* **75** (5), T–199 (1974).
626. D. O. Herer and F. E. Woodard, *Tappi,* **59** (1), 134 (1976).
627. M. S. Nasr, R. G. Gillies, N. N. Bakhiski, and D. G. MacDonald, *Can. Pulp Paper Ind.,* **28** (9), 30 (1975).
628. S. Prahacs, *Pulp Paper Pollution Abatement,* Project Rept. 1-1 (1971), Can. Dept. Environment; *ABIPC,* 43 (1), 66 (1972).
629. K. V. Nayak, F. M. A. Nicolle, and J. A. Histed, *Pulp Paper Can.,* **76** (4), T–111 (1975).
630. K. S. Ng, J. C. Mueller, and C. C. Walden, *Pulp Paper Mag. Can.,* **74** (5), T–187 (1973).
631. K. S. Ng, J. C. Mueller, and C. C. Walden, *Pulp Paper Mag. Can.,* **75** (7), T–262 (1974).
632. K. S. Ng, J. C. Mueller, and C. C. Walden, *J. WPCF,* **48** (3), 458 (1976).
633. M. A. Tyler and A. D. Fitzgerald, *CPPA Tech. Sect. Papers and Discussions,* No. 4, D–116 (1973).
634. J. A. Servizi, E. T. Stone, and R. W. Gordon, *Intl. Pacific Salmon Fisheries Comm.,* Rept. No. 13 (1966); *ABIPC,* **40** (10), 874 (1970).
635. G. E. Charles and G. Decker, *J. WPCF,* **42** (19), 1725 (1970).
636. A. Wong and S. Prahacs, *CPAR Project Rept.,* 228 (1974); *ABIPC,* **46** (5), 522 (1975).
637. J. Rennerfelt and S. von Schantz, *Svensk Papperstidn.,* **68** (5), 129 (1965).
638. J. C. Mueller and C. C. Walden, *Pulp Paper Mag. Can.,* **75** (8), T–274 (1974).
639. C. C. Walden and T. E. Howard, *Pulp Paper Mag. Can.,* **75** (11), T–370 (1974).
640. B. E. Jank, D. W. Bissett, V. W. Cairns, and S. Metikosh, *CPPA Air Stream Improvement Conf.,* Montreal, Canada, 1974, p 125.
641. C. F.-H. Liao and R. N. Dawson, *J. WPCF,* **47** (10), 2384 (1975).
642. G. Berwart, *Technique Eau, Assainissement,* **335,** 51 (1974); *ABICP,* **46** (9), 959 (1976).
643. A. Chantefort, M. Marchand, and J. Sechet, *Tribune CEBEDEAU,* **367/368,** 294 (1974); *ABIPC,* **46** (9), 960 (1976).
644. H. T. Chen, E. E. Fredrickson, J. F. Cormack, and S. R. Young, *Tappi,* **57** (5), 111 (1974).
645. J. C. Mueller and C. C. Walden, *J. WPCF,* **48** (3), 502 (1976).
646. *Can. Pulp Paper Ind.,* **29** (10), 23 (1976); Report by P. H. M. Guo, W. K. Bedford and B. E. Yank, Wastewater Technology Centre, Burlington, Canada.
647. J. E. G. Sikes and G. A. Nieminen, *Preprint, CPPA Environment Improvement Conf.,* Vancouver, Canada, 1975, p 9.
648. S. C. Mosher, *Preprint, CPPA Environment Improvement Conf.,* Vancouver, Canada, 1975, p 93.
649. D. Ruzickova, *Papir Celul.,* **31** (10), 220 (1976).
650. H. O. Bouveng and P. Solyom, *Svensk Papperstidn.,* **79** (7), 224 (1976).
651. W. H. Rapson, *Pulp Paper Mag. Can.,* **68** (12), T–635 (1967).
652. J. A. Histed and F. M. A. Nicolle, *Pulp Paper Mag. Can.,* **74** (12), T–386 (1973).
653. J. A. Histed and F. M. A. Nicolle, *Pulp Paper Mag. Can.,* **75** (5), T–171 (1974).
654. F. M. A. Nicolle and J. A. Histed, *Pulp Paper Can.,* **76** (9), T–253 (1975).

655. J. A. Histed and G. André, *Preprint, Intl. Pulp Bleaching Conf.*, Chicago, Ill., 1976 p 103.
656. D. W. Reeve, G. Rowlandson and W. H. Rapson, *Preprint Intl. Pulp Bleaching Conf.*, Chicago, Ill., 1976, p 117.
657. P. O. Karjalainen, J. E. Lofkrantz, and R. D. Christie, *Pulp Paper Mag. Can.*, **73** (12), T–393 (1972).
658. K. Henricson, E. Kiiskilä and N.-E. Virkola, *Paperi Puu*, **57** (10), 643 (1975).
659. P. E. Ahlers, H. Norrström, and B. Warnqvist, *Preprint Intl. Pulp Bleaching Conf.*, Chicago, Ill., 1976, p 19.
660. R. H. Wright, *Pulp Paper Mag. Can.*, **57** (12), 171 (1956).
661. W. A. Moy, P. Joyce and G. E. Styan, *Pulp Paper Mag. Can.*, **75** (4), T–150 (1974).
662. B. Warnqvist and H. Norrström, *Tappi*, **59** (11), 89 (1976).
663. D. W. Reeve and W. H. Rapson, *Pulp Paper Mag. Can.*, **74** (1), T–19 (1973).
664. D. W. Reeve, J. A. Lukes, K. A. French, and W. H. Rapson, *Pulp Paper Mag. Can.*, **75** (8), T–293 (1974).
665. D. W. Reeve and W. H. Rapson, *Pulp Paper Mag. Can.*, **71** (13), T–274 (1970).
666. F. G. Sellers, *Pulp and Paper*, **47** (12), 80 (1973).
667. R. E. Kirk and D. E. Othmer, "Encyclopedia of Chemical Technology," Vol. 11, 2nd ed., Wiley-Interscience, New York, N.Y., 1972, p 396.
668. R. B. MacMullin in J. S. Sconce, ed., "Chlorine: Its Manufacture, Properties and Uses," Reinhold, New York, N.Y., 1962, p 127.
669. M. S. Kircher in J. S. Sconce, ed., "Chlorine: Its Manufacture, Properties and Uses," Reinhold, New York, N.Y., 1962, p 81.
670. S. A. Dahl, *Chem. Eng.*, **82** (17), 60 (1975).
671. N. R. Iammartino, *Chem. Eng.*, **83** (13), 86 (1976).
672. H. V. Casson, G. J. Crane, and G. E. Styan, *Pulp Paper Mag. Can.*, **69** (1), T–31 (1968).
673. D. G. Elliott, *Tappi*, **51** (4), 88A (1968).
674. F. Lenzi and W. H. Rapson, *Pulp Paper Mag. Can.*, **63** (9), T–442 (1962).
675. E. Uusitalo, *Paperi Puu*, **44** (4a), 235 (1962).
676. M. J. Jalkanen, *Norsk Skogind.*, **14** (12), 520 (1960).
677. R. Hallan, *Norsk Skogind.*, 16 (8), 310 (1962).
678. J. H. A. Kleyn, *Appita*, **23** (2), 124 (1969).
679. *Paper Trade J.*, **145** (35), 32 (1961).
680. W. H. Rapson, *Can. J. Chem. Eng.*, **36** (6), 262 (1958).
681. E. Kesting, *Tappi*, **36** (4), 166 (1953).
682. *Allg. Papier Rdsch.*, **1957** (21), 1074 (1957).
683. J. C. W. Evans, *Paper Trade J.*, **141** (7), 44 (1957).
684. W. H. Rapson, *Tappi*, **41** (4), 181 (1958).
685. H. de V. Partridge and W. H. Rapson, *Tappi*, **44** (10), 698 (1961).
686. J. W. Robinson, *Tappi*, **46** (2), 120 (1963).
687. H. de V. Partridge, E. S. Atkinson, and A. C. Schulz, *Tappi*, **54** (9), 1484 (1971).
688. H. de V. Partridge and E. S. Atkinson, *Pulp Paper Mag. Can.*, **73** (8), T–203 (1972).
689. R. J. Read, *Pulp Paper Mag. Can.*, **68** (10), T–506 (1967).
690. F. M. Ernest, *Paper Trade J.*, **143** (34), 46 (1959).
691. J. Schuber and W. A. Kraske, *Pulp and Paper*, **27** (10), 104 (1953).
692. P. Conrad and G. Carlson, *Pulp Paper Mag. Can.*, **58** (11), 160 (1957).

693. R. K. Bradley and G. N. Werezak, *Tappi,* **57** (4), 98 (1974).
694. W. H. Rapson, *Pulp Paper Mag. Can.,* **75** (10), T–351 (1974).
695. E. L. Harrington, *Tappi,* **48** (8), 110A (1965).
696. R. A. Crawford and B. J. DeWitt, *Tappi,* **51** (5), 226 (1968).
697. L. Torregrossa, D. Justice, R. Schleinkofer, K. Vogel, and O. Luthi, *Tappi,* **59** (11), 92 (1976).
698. J. Knoble, *Paper Mill News,* **85** (28), 8 (1962).
699. S. Dardelet and A. Robert, *ATIP,* **22** (2), 123 (1968).
700. C. Knutson, *Tappi,* **58** (2), 115 (1975).
701. L. Uhlin, A. Berglund, A. Dalborg, and B. Hultman, *Svensk Papperstidn.,* **78** (15), 544 (1975).
702. C. Matzenger, Jr., *Pulp and Paper,* **31** (13), 73 (1957).
703. R. Hirschberg, *Papier* (Darmstadt), **16** (9), 419 (1962).
704. P. L. Gilmont, *Tappi,* **51** (4), 62A (1968).
705. W. H. Rapson and R. J. Neale, *Pulp Paper Mag. Can.,* **58** (12), 135 (1957).
706. R. E. L. Wheless, *Tappi,* **46** (7), 141A (1963).
707. H. B. Meier, *South. Pulp Paper Mfr.,* **23** (6), 88 (1960).
708. K. Passinen and P.-E. Ahlers, *Paperi Puu,* **48** (9), 549 (1966).
709. G. W. Bauer, *Tappi,* **43** (7), 240A (1960).
710. C. W. Field, *Materials Protect.,* **5** (10), 47 (1966).
711. J. M. Robinson, C. O. Stiles, and M. Hutchison, *Tappi,* **50** (5), 91A (1967).
712. S. A. Rydholm, "Pulping Processes," Interscience, New York, N.Y., 1965, p 1092.
713. R. E. Kirk and D. E. Othmer, "Encyclopedia of Chemical Technology," Vol. 14, 2nd ed., Wiley-Interscience, New York, N.Y., 1972 p 394.
714. *Chem. Eng.,* **77** (21), 54 (1970).
715. J. C. Davis, *Chem. Eng.,* **79** (23), 88 (1972).
716. R. E. Kirk and D. E. Othmer, "Encyclopedia of Chemical Technology," Vol 14, 2nd ed., Wiley-Interscience, New York, N.Y., 1972, p 421.

APPENDIX

APPENDIX *UNITS OF MEASUREMENT AND CONVERSION FACTORS. UNITS OF MEASUREMENT IN THIS BOOK ARE GIVEN IN THREE SYSTEMS: THE NORTH AMERICAN (ENGLISH) UNITS, OLDER METRIC UNITS, AND THE NEW INTERNATIONAL SYSTEM OF UNITS (SI). THIS HAS BEEN NECESSARY BECAUSE THE USE OF SI UNITS HAS NOT BECOME ESTABLISHED AS THE SINGLE SYSTEM. THE FOLLOWING TABLE COMPARES THE NORTH AMERICAN (ENGLISH) UNITS WITH THE SI UNITS AND SHOWS THE CONVERSION FACTORS USED IN CONVERTING TO SI UNITS.*

Quantity or Test	Common North American (English) Unit	SI Unit	Conversion
Woodyard Measures	Cord = 85 ft^3 solid wood Stacked = 128 ft^3 total Straw-pile = 190 to 200 ft^3 total Cunit = 100 ft^3 solid wood	Cubic meter	$M^3 = 0.0283 \times ft^3$ (Specify solid, stacked, or straw-piled)
Density	Pounds per cubic foot	Kilograms per cubic meter	$kg/m^3 = 16.02 \times lb/ft^3$
Capacity	Cubic feet Gallon (U.S.)	Cubic meter Liter	$m^3 = 0.0283 \times ft^3$ liter = 3.785 × gal
Flow	Gallons per minute	Cubic meters per minute	$m^3/mm = 3.785 \times 10^{-3} \times gal/min$
Pressure	Pounds per square inch Inches of mercury	Kilopascal	kPa = 6.89 × psi kPa = 3.377 × in Hg
Nip pressure	Pounds per inch of width (lineal inch)	Kilonewtons per meter	kN/m = 0.175 × pli
Sheet tension	Pounds per inch of width (lineal inch)	Newtons per meter	N · m = 1.356 × ft-lb
Power	Horsepower (electric and mechanical)	Kilowatt	kW = 0.746 × hp

Energy	Kilowatt-hour	Megajoule	$MJ = 3.6 \times kw\text{-}hr$
	Horsepower-hour	Joule	$MJ = 2.686 \times hp\text{-}hr$
	British Thermal Units (BTU)		$J = 1055 \times BTU$
Specific energy	Horsepower-days per short ton	Megajoules per kilogram	$MJ/kg = 7.0 \times 10^{-2}\ hp\text{-}days/ton$
Speed	Feet per minute	Meters per minute	$m/min = 0.3048 \times ft/min$
Torque	Foot-pounds	Newton-meter	$N \cdot m = 1.356 \times ft\text{-}lb$
Mass	Ton (2000 lb)	Metric ton	$t = 0.907 \times t$
	Pound	Kilogram	$kg = 0.454 \times lb$
	Ounce	Gram	$g = 28.3 \times oz$
Force	Kilogram	Newton	$N = 9.81 \times kg$
Volume	Ounce	Cubic centimeter	$cm^3 = 29.6 \times oz$
	Gallon	Cubic meter	$m^3 = 0.00379 \times gal$
	Liter	Cubic centimeter	$cm^3 = 1000 \times l$
	Cubic inch	Cubic centimeter	$cm^3 = 16.4 \times in.^3$
	Cubic foot	Cubic meter	$m^3 = 0.0283 \times ft^3$
	Cubic yard	Cubic meter	$m^3 = 0.765 \times yd^3$
Area	Square inch	Square centimeter	$cm^2 = 6.45 \times in.^2$
	Square foot	Square meter	$m^2 = 0.0929 \times ft^2$
Length	Angstrom	Newton	$0.1 \times A^\circ$
	Micron	Micrometer	$\mu m = 1 \times \mu$
	Mil	Millimeter	$mm = 0.0254 \times mil$
	Foot	Meter	$m = 0.305 \times ft$
Mass per unit volume	Ounces per gallon	Kilograms per cubic meter	$kg/m^3 = 7.49 \times oz/gal$
	Pounds per cubic foot	Kilograms per cubic meter	$kg/m^3 = 16.0 \times lb/ft^3$
Viscosity	Poises	Pascal · seconds	$Pa \cdot sec = 10^{-1}\ poise$

APPENDIX TABLE (CONTINUED)

Quantity or Test	Common North American (English) Unit	SI Unit	Conversion
Basis weight	Pounds per 3000 square feet	Grams per square meter	$g/m^2 = 1.63 \times lb/3000\ ft^2$
	Pounds per 1000 square feet	Grams per square meter	$g/m^2 = 4.88 \times lb/1000\ ft^2$
	Pounds per ream		
	$17 \times 22 \times 500$	Grams per square meter	$g/m^2 = 3.760 \times lb/ream$
	$22.5 \times 28.5 \times 500$	Grams per square meter	$g/m^2 = 2.193 \times lb/ream$
	$22 \times 34 \times 500$	Grams per square meter	$g/m^2 = 1.880 \times lb/ream$
	$24 \times 36 \times 500$	Grams per square meter	$g/m^2 = 1.627 \times lb/ream$
	$24 \times 36 \times 480$	Grams per square meter	$g/m^2 = 1.695 \times lb/ream$
	$24 \times 37 \times 500$	Grams per square meter	$g/m^2 = 1.583 \times lb/ream$
	$25 \times 38 \times 500$	Grams per square meter	$g/m^2 = 1.480 \times lb/ream$
	$25 \times 40 \times 500$	Grams per square meter	$g/m^2 = 1.406 \times lb/ream$
	$30 \times 40 \times 500$	Grams per square meter	$g/m^2 = 1.172 \times lb/ream$
	$25 \times 40 \times 1000$	Grams per square meter	$g/m^2 = 0.703 \times lb/ream$
Caliper	Mil	Millimeter	$mm = 0.0254 \times mil$

Property			
Bulk (apparent density)	grams per cubic meter	grams per cubic meter	1.000
Bursting strength	Pounds per square inch	Kilopascal	$kPa = 6.89 \times psi$
Burst index	Gram force per square centimeter $\times$ square meter per gram $(g/cm^2 \cdot m^2/g)$	Kilopascal $\times$ square meter per gram	$kPa \cdot m^2/g = 9.81 \times 10^{-2} \times g/cm^2 \cdot m^2/g$
Tear strength	Grams force	Millinewtons	$mN = 9.81 \times g$
Tear index	100 gram force $\times$ square meter per gram $(100\ g \cdot m^2/g)$	Millinewtons $\times$ square meter per gram	$mN \cdot m^2/g = 9.81 \times 10^{-2} \times 100\ g\ force \times m^2/g$
Tensile breaking load	Pounds per inch width	Kilonewtons per meter	$kN/m = 0.175 \times p$
	Kilograms per 15 mm	Kilonewtons per meter	$kN/m = 0.654 \times kg/15\ mm$
Concora crush	Pounds force	Newton	$N = 4.448 \times lb_f$
Edge crush	Pounds force per inch	Kilonewton per meter	$kN/m = 0.175 \times pi$
Flat crush	Pounds force per square inch	Kilopascal	$kPa = 6.89 \times psi$
Internal bond (Scott)	Foot-pound force per square inch	Kilojoule per square meter	$kJ/m^2 = 2.102 \times ft\text{-}lb/in.^2$
Stiffness (Gurley)	Milligrams force	Millinewton	$mN = 9.807 \times 10^{-3} \times mg_f$
Stiffness (Taber)	Taber units	Millinewtons	$mN = 2.03 \times units$

INDEX

771